Bozenna Pasik-Duncan (Editor)

Stochastic Theory and Control

Proceedings of a Workshop held in Lawrence, Kansas

With 54 Figures

Springer

Editor

Prof. Bozenna Pasik-Duncan
University of Kansas
Dep. of Mathematics
405 Snow Hall
Lawrence, KS 66045
USA

Cataloging-in-Publication Data applied for
Die Deutsche Bibliothek – CIP-Einheitsaufnahme
Stochastic theory and control : proceedings of a workshop held in Lawrence, Kansas / Bozenna Pasik-Duncan (ed.). - Berlin ; Heidelberg ; New York ; Barcelona ; Hong Kong ; London ; Milan ; Paris ; Tokyo : Springer, 2002
(Lecture notes in control and information sciences ; 280)
(Engineering online library)
ISBN 3-540-43777-0

ISBN 3-540-43777-0 Springer-Verlag Berlin Heidelberg New York

Springer-Verlag Berlin Heidelberg New York
a member of BertelsmannSpringer Science + Business Media GmbH

http://www.springer.de

Typesetting: Digital data supplied by author. Data-conversion by PTP-Berlin, Stefan Sossna e.K.
Cover-Design: design & production GmbH, Heidelberg
Printed on acid-free paper SPIN 10881686 62/3020Rw - 5 4 3 2 1 0

Lecture Notes
in Control and Information Sciences 280

Editors: M. Thoma · M. Morari

Springer

Berlin
Heidelberg
New York
Barcelona
Hong Kong
London
Milan
Paris
Tokyo

Dedicated to Tyrone Duncan,
an important contributor
to stochastic control and filtering theory,
in honor of his sixtieth birthday.

Preface

This volume contains almost all of the papers that were presented at the Workshop on Stochastic Theory and Control that was held at the University of Kansas, 18–20 October 2001. This three-day event gathered a group of leading scholars in the field of stochastic theory and control to discuss leading-edge topics of stochastic control, which include risk sensitive control, adaptive control, mathematics of finance, estimation, identification, optimal control, nonlinear filtering, stochastic differential equations, stochastic partial differential equations, and stochastic theory and its applications. The workshop provided an opportunity for many stochastic control researchers to network and discuss cutting-edge technologies and applications, teaching and future directions of stochastic control. Furthermore, the workshop focused on promoting control theory, in particular stochastic control, and it promoted collaborative initiatives in stochastic theory and control and stochastic control education.

The lecture on "Adaptation of Real-Time Seizure Detection Algorithm" was videotaped by the PBS. Participants of the workshop have been involved in contributing to the documentary being filmed by PBS which highlights the extraordinary work on "Math, Medicine and the Mind: Discovering Treatments for Epilepsy" that examines the efforts of the multidisciplinary team on which several of the participants of the workshop have been working for many years to solve one of the world's most dramatic neurological conditions.

Invited high school teachers of Math and Science were among the participants of this professional meeting. They were motivated and inspired by the First NSF Workshop for High School Teachers of Math and Science that took place in June of 2000 in Chicago. These teachers joined control researchers in their love and fascination for stochastic theory and control. The teachers at the meeting seemed to be really excited to be invited to such a specialized technical meeting. Furthermore, a number of graduate students were invited to broaden their exposure in stochastic control education.

On October 19, the workshop honored the sixtieth birthday of Tyrone Duncan, an important contributor to stochastic control theory. 110 people attended the reception to honor Tyrone Duncan. The reception was held at the University of Kansas Spencer Museum of Art. The Master of Ceremony was John Baillieul from Boston University, the Keynote Speakers were Sanjoy Mitter from Massachusetts Institute of Technology and Charles Himmelberg from University of Kansas. Music was provided by Cellist, Ed Laut, Professor of Music and Dance, from University of Kansas.

The workshop had three important aspects:

1. the outreach program—it was videotaped by PBS
2. the control education program—it brought several high school teachers together with control researchers and students

3. the interdisciplinary research programs—it brought together mathematicians, engineers, economists, and biomedical scientists together to discuss control systems applications.

The workshop was supported by the National Science Foundation and the University of Kansas. We especially thank Dr. R. S. Baheti, NSF Program Director and Jack R. Porter, Chairman of the Department of Mathematics, University of Kansas. We also thank Diana Carlin and Robert Weaver from the College of Liberal Arts and Sciences, George Bittlingmayer and Prakash Shenoy from the School of Business, Victor Frost from the Information and Telecommunication Technology Center, Robert Barnhill and Jim Roberts from the KU Center for Research for providing additional important funds and Cheryl Schrader, 2001 Vice President for Conferences, IEEE Control Systems Society, for providing important co-technical sponsorship.

We thank the members of the Program Committee: John Baillieul, Sergio Bittanti, Wendell Fleming, P. R. Kumar, Steven Marcus, William McEneaney, Sean Meyn, Lukasz Stettner, Pravin Varaiya, George Yin, Qing Zhang, and Xun Yu Zhou, and the members of the Organizing Committee and Local Advisory Board: Robert Barnhill, Mark Frei, Victor Frost, Ivan Osorio, Jack Porter, Prakash Shenoy, John Westman and Fred Van Vleck. Kerrie Brecheisen, Kathleen Brewer, Monica McKinney, Gloria Prothe, Sandra Reed, Yi Yan, and Yiannis Zacharious were the local assistants whom we thank for their generous help with the administration of the workshop. We also thank all reviewers for their important contributions to the quality of this volume. Larisa Martin did an outstanding job typesetting this manuscript. Last, but not least, we would like to thank all the participants for making this workshop successful and memorable. Very special thanks go to two participants, Lukas Stettner and Tyrone Duncan, for their outstanding assistance in preparing this volume during the last two months.

It has been my pleasure, honor, and joy to work with all the participants and the outstanding authors of this volume. I hope that the readers of this volume will enjoy reading it as much as I did. I would like to thank my husband and colleague, Tyrone, and my daughter, Dominique, for their help and support during my work on this project.

Bozenna Pasik-Duncan

List of Participants

Eyad Abed, University of Maryland; abed@isr.umd.edu
Grazyna Badowski, University of Maryland; grazyna@math.wayne.edu
John Baillieul, Boston University; johnb@bu.edu
John Baras, University of Maryland; baras@isr.umd.edu
Bernard Bercu, University of Paris-Sud; Bernard.Bercu@math.u-psud.fr
Tomasz Bielecki, Northeastern Illinois University; t-bielecki@neiu.edu
Robert Buche, Brown University; buche@cfm.brown.edu
Andrew Bucki, Oklahoma School of Science and Mathematics; abucki@ossm.edu
Kathleen Brewer, University of Kansas; brewer@math.ukans.edu
Peter Cains, McGill University; peterc@cim.mcgill.edu
Charalambos Charalambous, University of Ottawa; chadcha@site.uottawa.ca
Zhisheng Chen, Sprint; Zhishen.Chen@mail.sprint.com
Ronald Diersing, University of Notre Dame; rdiersin@nd.edu
Jiangxia Dong, University of Kansas; jxdong@math.ukans.edu
Tyrone Duncan, University of Kansas; duncan@math.ukans.edu
David Elliott, University of Maryland; delliott@isr.umd.edu
Robert Elliott, University of Calgary; relliott@ucalgary.ca
Wendell Fleming, Brown University; whf@cfm.brown.edu
Evelyn Forbes, Parsons High School, Parsons, Kansas; eforbes@vikingnet.net
Mark Frei, Flint Hills Scientific, L.L.C.; frei@sound.net
Laszlo Gerencser, MTA SZTAKI; gerencser@szatki.hu
Shane Hass, Massachusetts Institute of Technology; shaas@mit.edu
Floyd Hanson, University of Illinois; hanson@uic.edu
Kurt Helmes, Humboldt University of Berlin; helmes@wiwi.hu-berlin.de
Daniel Hernandez-Hernandez, Centro de Investigacion en Matematica; dher@cimat.mx
Yaozhong Hu, University of Kansas, hu@math.ukans.edu
Jianyi Huang, University of Illinois; jhuang@control.csl.uiuc.edu
Yasong Jin, University of Kansas; jinyasong@hotmail.com
Biff Johnson, University of Kansas; biff@math.ukans.edu
Connie Johnson, Summer Academy, Kansas City, Kansas; srj12201@aol.com
Rafail Khasminskii, Wayne State University; rafail@math.wayne.edu
Faina Kirillova, National Academy of Science of Belarus; kirill@nsys.minsk.edu
P. R. Kumar, University of Illinois; prkumar@uiuc.edu
Tze Leung Lai, Stanford University; lait@stat.stanford.edu
E. Bruce Lee, University of Minnesota; eblee@ece.umn.edu
Francois LeGland, IRISA; Francois.Le_Gland@irisa.fr
David Levanony, Ben Gurion University; levanony@ee.bgu.ac.il
Andrew Lim, Columbia University; lim@ieor.columbia.edu

Xiaobo Liu, University of Kansas; xliu@math.ukans.edu
Steven Marcus, University of Maryland; marcus@glue.umd.edu
Mihaela Matache, University of Nebraska; dmatache@unomail.unomaha.edu
William McEneaney, University of California at San Diego; wmcenean@math.ucsd.edu
Sanjoy Mitter, Massachusetts Institute of Technology; mitter@lids.mit.edu
Daniel Ocone, Rutgers University; ocone@math.rutgers.edu
Ivan Osorio, M.D., Kansas University Medical Center; iosorio@kumc.edu
Tao Pang, Brown University; pang@cfm.brown.edu
Bozenna Pasik-Duncan, University of Kansas; bozenna@math.ukans.edu
Khanh Pham, University of Notre Dame; kpham@nd.edu
Vahid Ramezani, University of Maryland; rvahid@isr.umd.edu
Michael Sain, University of Notre Dame; sain.1@nd.edu
Lonnie Sauter, Sprint; Lonnie.1.sauter@mail.sprint.com
Lukasz Stettner, Polish Academy of Sciences; stettner@impan.gov.pl
Richard Stockbridge, University of Wisconsin at Milwaukee; stockbri@uwm.edu
Jordan Stoyanov, University of Newcastle; jordan.stoyanov@ncl.ac.uk
Allanus Tsoi, University of Missouri; tsoi@math.missouri.edu
Fred Van Vleck, University of Kansas; vanvleck@math.ukans.edu
Zhennong Wang, Boeing; Zhennong.wang@pss.boeing.com
Ananda Weerasinghe, Iowa State University; weera@math.iastate.edu
Kathy Welch-Martin, Wainwright Middle School, Lafayette, Indiana; k3dm@aol.com
John Westman, University of California at Los Angeles; jwestman@math.ucla.edu
Xi Wu, University of Illinois; wu@math.uic.edu
Yi Yan, Sprint; Yan.x.yi@mail.sprint.com
Stephen S.-T. Yau, University of Illinois; yau@uic.edu
George Yin, Wayne State University; gyin@math.wayne.edu
Austin Yuen, Sprint; Austin.yuen@mail.sprint.edu
Yiannis Zachariou, University of Kansas; yiannis@math.ukans.edu
Yong Zeng, University of Missouri at Kansas City; zeng@mendota.umkc.edu
Qing Zhang, University of Georgia; qingz@math.uga.edu
Peter Zimmer, West Chester University; pzimmer@wcupa.edu

Contents

Nonlinear and Stochastic Stability Problems in Gated Radar Range Trackers

E. H. Abed[1], R. E. Gover[2], A. J. Goldberg[2], and S. I. Wolk[2]

[1] Department of Electrical Engineering and the Institute for Systems Research, University of Maryland, College Park, MD 20742, USA
[2] Advanced Techniques Branch, Tactical Electronic Warfare Division, Code 5753, Naval Research Laboratory, Washington, D.C. 20375, USA

Abstract. The stability of a model of the dynamics of a class of split-gate radar range trackers is considered, under both deterministic and stochastic target models. The emphasis is on stability as a means toward studying the dynamics of the tracker in the presence of both an actual target and a decoy target. The model employed reflects Automatic Gain Control action for noise attenuation as well as nonlinear detector laws for target resolution. The deterministic and stochastic stability of track points are studied using deterministic and stochastic Liapunov functions, and stochastic bifurcation issues are discussed.

1 Introduction

Electronic Warfare (EW) encounters involving a tracking radar and multiple closely spaced targets have long been a rich source of interesting problems in identification, estimation and control [4]. Depending on the driving physical problem, a variety of theoretical questions can arise. For example [6], in multi-target tracking one is concerned with the development of algorithms for tracking the motion of several targets simultaneously. There are also situations in which a tracking radar is expected to resolve a single target from among targets plus decoys. Interference of the reflected signals from the targets and decoys, and significant levels of noise, are among the factors making analysis of these EW problems a formidable task.

To be more specific about the motivating application for this work, consider the following scenario. A tracking radar is attempting to track a particular target, such as a ship. Suppose that the ship, to confuse the radar, modifies the returned electromagnetic signal to the radar. For example, the ship may emit a chaff cloud which would also be subtended by the radar and would return a reflected signal along with the echo signal due to the ship. The radar dynamics are such that it will track a single target. These dynamics also are in force even if two targets (ship and chaff) are actually present. Due to the wind and to the ship's motion, the separation of the ship and the chaff cloud changes with time. At some point, the radar must commit to one of these targets. A long-term goal of this work is to gain insights on

B. Pasik-Duncan: Stochastic Theory and Control, LNCIS 280, pp. 1–17, 2002.

how this commitment occurs, and to find fast algorithms for estimating the probability of committing to either target.

Determining the probabilities of committing to either target is unwieldy. This is equivalent to solving for the evolving probability density function for a complicated nonlinear stochastic system.

In this paper, we focus on track point stability in split-gate tracking systems. These systems involve the generation of two gates, the early and the late gates. These gates are positioned in time so that a portion of the echo pulse passes through each gate. The tracking system adjusts the gates so as to drive an associated error voltage to zero.

The nonlinear model studied in this paper was introduced in the authors' previous work [1] and Monte Carlo studies of the model were pursued in [9]. The general model of [1] allows for deterministic as well as stochastic target fluctuations. Along with the model derivation, [1] includes a linearized deterministic stability analysis of track points. In particular, the dependence of the resolving capability of a radar range tracking system on the separation of the targets was considered. This resolving capability was studied by considering the linearized stability of equilibria of the model corresponding to the targets and to an artificial centroid target.

Although linearized stability analysis was useful in the previous study, the information it can yield is rather limited. Besides the necessity to rely on deterministic models, another limitation is that no global information on transient behavior can be deduced from an inherently local linearized stability analysis.

Here, we continue the work in [1] and [9] in the following two directions. First, we extend the linearized deterministic stability analysis of [1] to achieve Liapunov functions useful for estimating domains of attraction of stable track points. Second, we consider the stochastic stability analysis of the system using stochastic Liapunov functions. In addition, some ideas for future work are given that make contact with stochastic bifurcation studies in other fields.

The remainder of the paper is organized as follows. In Section 2, a model of gated range tracker dynamics presented in [1] is recalled, and then specialized to the case of a constant target and a stochastic target. In Section 3, deterministic stability of the tracker model (without noise) is considered using quadratic-type Liapunov functions. Not surprisingly, stronger statements can be made for deterministic stability than for stochastic stability. Stochastic stability of the tracker model is studied in Section 4. In Section 5, issues for further research related to stochastic stability and bifurcation for split-gate radar trackers are briefly discussed.

This paper is dedicated to Professor Tyrone Duncan on the occasion of his sixtieth birthday.

2 Nonlinear Radar Range Tracker Models

In [1], a simple null-tracking system is studied. In this section, we recall from [1] a nonlinear model of this gated radar range tracking system. The tracking system is driven by the echo signal E. This is the received sum channel voltage signal analytic envelope after passing through the receiver intermediate frequency (IF) filter but before AGC normalization. The tracking system is composed of a first-order tracker coupled with an automatic gain control (AGC). The main output of the system is the estimated relative slant range ρ.

The basic model obtained in [1] is a nonlinear, discrete-time model which describes the evolution of two essential tracker variables with time: the slant range ρ and V, the AGC voltage. Although results on stability of discrete-time stochastic systems are available [2], the analysis tends to be smoother in the continuous-time setting. Thus we devote the rest of the paper to the continuous-time setting.

In the derivation of the basic model in [1], time is decomposed into a discrete component k which identifies the received pulse being processed, and a continuous component σ which is associated with detailed processing of a given pulse. By taking the limit of the general model as the pulse repetition interval becomes vanishingly small [1], a continuous-time version of the model is obtained. The variable σ does not appear in the continuous-time model, while k is replaced by a continuous time variable t.

In this section, we recall how the general discrete-time model yields an approximate continuous-time model, and how the continuous-time model specializes for the case of a single constant or stochastically fluctuating point target.

2.1 General discrete-time model

The basic discrete-time model of the gated range tracking system studied in [1] is given by the following system of four first-order difference equations:

$$\rho[k+1] = \rho[k] + K\beta^2 T_{\text{PRI}} e^{-2V_d[k]} \int_{-T_1}^{T_1} |E(\sigma, k-1)|^2 w_D(\sigma - \rho_d[k]) d\sigma \tag{1}$$

$$\rho_d[k+1] = \rho[k] \tag{2}$$

$$V[k+1] = V[k] + \frac{T_{\text{PRI}}}{T_{\text{AGC}}} \times \left(\beta^2 e^{-2V_d[k]} \int_{-T_1}^{T_1} |E(\sigma, k-1)|^2 w_S(\sigma - \rho_d[k]) d\sigma - 1 \right) \tag{3}$$

$$V_d[k+1] = V[k] \tag{4}$$

Here, $\rho[k]$ is the relative slant range estimate at time k and $V[k]$ is the AGC voltage at time k. The variables ρ_d and V_d are one-step delayed versions of ρ and V, respectively. The integrals appearing in this model are the difference and sum detector laws. The function $w_D(\sigma)$ is the difference weighting pattern and $w_S(\sigma)$ is the sum weighting pattern. The dynamical system (1)-(4) is driven by the echo signal $E(\sigma, k)$, the received sum channel voltage signal analytic envelope after passing through the receiver IF filter but before AGC normalization. Other parameters appearing in (1)-(4) include the width of the weighting patterns T_1, the pulse repetition interval length T_{PRI}, the AGC time constant T_{AGC}, the tracker gain K, and the effective AGC gain β.

The model (1)-(4) can be stochastic or deterministic, depending on the received voltage $E(\sigma, k)$. If this signal depends randomly on k, then the system is stochastic. For further results on the discrete-time model, see [1]. In this paper we focus on a continuous-time model which results from (1)-(4) in the limit that the pulse repetition interval T_{PRI} becomes vanishingly small.

2.2 Continuous-time model

In [1], a continuous-time model was obtained from (1)-(4) by considering the limiting case of small T_{PRI}. The continuous-time model thus obtained is a system of two first-order ordinary differential equations, which takes the form

$$\frac{d\rho(t)}{dt} = K\beta^2 e^{-2V(t)} \int_{-T_1}^{T_1} |E(\sigma,t)|^2 w_D(\sigma - \rho(t)) d\sigma \tag{5}$$

$$T_{\text{AGC}} \frac{dV(t)}{dt} = \beta^2 e^{-2V(t)} \int_{-T_1}^{T_1} |E(\sigma,t)|^2 w_S(\sigma - \rho(t)) d\sigma - 1 \tag{6}$$

There is a slight abuse of notation in (5)-(6). The quantity $E(\sigma, t)$ is not obtained by replacing k with t in the signal $E(\sigma, k)$. To be precise, we should have written $E(\sigma - T_{\text{PRI}} + [t/T_{\text{PRI}}]T_{\text{PRI}})$ instead (here, $[x]$ for a real number x denotes the greatest integer less than or equal to x). However, the form used above is useful in facilitating approximations to be made later in the paper. In (5)-(6), t is real time and replaces the discrete time k, and the state variables ρ and V now depend on t. Also, the received signal analytic envelope E now is a function of the delay variable σ and real time t, as we have just discussed.

The model (5),(6) is stochastic if the received signal E is a random process. However, if $E(\sigma, t)$ is periodic in t with period T_{PRI}, then this model reduces to a deterministic, autonomous model:

$$\frac{d\rho(t)}{dt} = K\beta^2 e^{-2V(t)} \int_{-T_1}^{T_1} |E_0(\sigma)|^2 w_D(\sigma - \rho(t)) d\sigma \tag{7}$$

$$T_{\mathrm{AGC}}\frac{dV(t)}{dt} = \beta^2 e^{-2V(t)} \int_{-T_1}^{T_1} |E_0(\sigma)|^2 w_S(\sigma - \rho(t))d\sigma - 1 \tag{8}$$

2.3 Detector laws for a periodic echo signal

It is convenient to introduce notation for the difference and sum detector laws appearing in the deterministic continuous-time model (7)-(8). For any real ρ, denote

$$\mathcal{D}(\rho) := \int_{-T_1}^{T_1} |E_0(\sigma)|^2 w_D(\sigma - \rho)d\sigma \tag{9}$$

$$\mathcal{S}(\rho) = \int_{-T_1}^{T_1} |E_0(\sigma)|^2 w_S(\sigma - \rho)d\sigma \tag{10}$$

Then $\mathcal{D}(\rho)$ is the difference detector law and $\mathcal{S}(\rho)$ is the sum detector law. Typical difference and sum weighting patterns for split-gate tracking systems are shown in detail in [1]. Note that these weighting patterns have support on a subset of the interval $-T_1 \leq \sigma \leq T_1$. Thus, $w_S(\sigma)$ is a symmetric function, positive on its support, and $w_D(\sigma)$ is an antisymmetric function, and is positive (respectively, negative) for all positive (respectively, negative) values of σ on its support.

To specify the shape of the signal $E_0(\sigma)$, assume the transmitted pulse is rectangular and symmetric. Then, for a point target, the reflected (received) pulse will also be rectangular and symmetric. Assuming an IF filter which is matched to the transmitted pulse, we find that $E_0(\sigma)$ is a symmetric triangle pulse. Note that the pulse width of $E_0(\sigma)$, denoted by T_p, in general differs from T_1 [1].

Under these assumptions on the weighting patterns and the received signal, it is straightforward to see that the difference detector law $\mathcal{D}(\rho)$ and the sum detector law $\mathcal{S}(\rho)$ take the following forms: $\mathcal{D}(\rho)$ is an antisymmetric function, and positive values of $\mathcal{D}(\rho)$ occur for negative values of ρ. Also, $\mathcal{S}(\rho)$ is a nonnegative symmetric function.

3 Deterministic Stability of the Radar Range Tracking System

In this section we study stability of the dynamic models of Section 2 in the case of a single, constant point target. Liapunov functions are obtained in two steps. First, a simplified model obtained under the assumption of an instantaneous AGC is addressed. A singular perturbation result of Saberi and Khalil [16] is then employed to extend the result to the case of a fast AGC.

3.1 Continuous-time deterministic Liapunov functions

In this section we construct Liapunov functions for the continuous-time deterministic model (7)-(8). The construction is carried out under the assumption of a fast-acting AGC. With this assumption, the system may be viewed as consisting of slow and fast subsystems, and Liapunov functions may be obtained using results of Saberi and Khalil [16]. Before proceeding with the construction, we recall the main result of [16]. We actually state a slightly modified result to facilitate application to the radar range tracking model (7)-(8).

Liapunov functions for singularly perturbed systems We now recall a result of Saberi and Khalil [16] on the construction of Liapunov functions for singularly perturbed systems of ordinary differential equations. Two aspects of the statement below have been modified from [16] to be more readily applicable to the system under consideration. These modifications do not represent a major modification of the result of [16]. They will be noted when they arise.

Consider a nonlinear system

$$\dot{x} = f(x, y) \tag{11}$$

$$\epsilon \dot{y} = g(x, y) \tag{12}$$

where $x \in \mathbb{R}^n$ and $y \in \mathbb{R}^m$, f and g are nonlinear functions (at least Lipschitz continuous), and $\epsilon > 0$. Suppose that this system is defined for $x \in B_x \subset \mathbb{R}^n$ and $y \in B_y \subset \mathbb{R}^m$, and that the origin is an equilibrium point of (11),(12), i.e., that $f(0,0) = g(0,0) = 0$. Let $(0,0) \in B_x \times B_y$.

System (11),(12) is said to be singularly perturbed if ϵ is small. Associated with (11),(12) are two other systems: the reduced system and the boundary layer system. The reduced system is defined as

$$\dot{x} = f(x, y) \tag{13}$$

$$0 = g(x, y) \tag{14}$$

Suppose that in $B_x \times B_y$ Eq. (14) has a unique solution $y = h(x)$. Then the reduced system can be written in the more compact form

$$\dot{x} = f(x, h(x)) =: f_r(x) \tag{15}$$

The boundary layer system is defined as

$$\frac{dy}{d\tau} = g(x, y(\tau)) \tag{16}$$

where $\tau = t/\epsilon$ is the stretched time scale. The variable x plays the role of a fixed parameter vector in the boundary layer system.

To recall the main result of Saberi and Khalil [16] as it applies in our setting, we state the following three assumptions. These are equivalent (though not identical) to corresponding assumptions in [16], for systems (11), (12). Had the assumed form of the system equations (11), (12) allowed dependence of the function $g(x, y)$ on ϵ, assumption (A3) below would have included an additional inequality, as in [16]. This generality is not required in addressing the radar range tracker models of interest in this work.

(A1) The reduced system (15) has a Liapunov function $\mathcal{V}(x)$ such that for all $x \in B_x$

$$(\text{grad}_x\ \mathcal{V}(x))^T f_r(x) \le -\alpha_1 \psi^2(x) \tag{17}$$

for some $\alpha_1 > 0$, and some scalar function $\psi(x)$ that vanishes at $x = 0$ but differs from 0 for all other $x \in B_x$.

(A2) The boundary layer system (16) has a Liapunov function $\mathcal{W}(x, y)$ such that for all $x \in B_x$ and $y \in B_y$

$$(\text{grad}_y\ \mathcal{W}(x, y))^T g(x, y) \le -\alpha_2 \phi^2(g(x, y)) \tag{18}$$

for some $\alpha_2 > 0$, and some scalar function $\phi(g(x, y))$ that vanishes for $g(x, y) = 0$ but differs from 0 for all other $x \in B_x$, $y \in B_y$.

(A3) The following two inequalities hold for all $x \in B_x$, $y \in B_y$:

(a)

$$(\text{grad}_x\ \mathcal{W}(x, y))^T f(x, y) \le c_1 \phi^2(g(x, y)) + c_2 \psi(x) \phi(g(x, y)); \tag{19}$$

(b)

$$(\text{grad}_x\ \mathcal{V}(x))^T [f(x, y) - f_r(x)] \le \beta_1 \psi(x) \phi(g(x, y)) \tag{20}$$

where c_1, c_2 and β_1 are nonnegative constants .

We can now state the following result, a special case of Theorem 1 of [16].

Theorem 1. *Suppose assumptions (A1)-(A3) hold, and let d be a number in the interval* $(0, 1)$. *Introduce the quantity*

$$\epsilon^*(d) := \frac{\alpha_1 \alpha_2}{\alpha_1 c_1 + \beta_1^2 / 4d(1 - d)} \tag{21}$$

Then, or all $\epsilon < \epsilon^*(d)$, *the origin* $(x = 0, y = 0)$ *is an asymptotically stable equilibrium point of (11), (12), and*

$$v(x, y) := (1 - d)\mathcal{V}(x) + d\mathcal{W}(x, y) \tag{22}$$

is a Liapunov function for (11), (12).

Stability of the reduced and boundary layer models System (7)-(8) is a special case of the general singularly perturbed model (11)-(12), where $x = \rho$, $y = V$, and $\epsilon = T_{\text{AGC}}$. The reduced model of the system (7)-(8) is obtained by setting $T_{\text{AGC}} = 0$ in (8), solving the resulting algebraic equation for V in terms of the other state variable ρ, and substituting the result in

(7). The resulting reduced model may be written, using the notation $\mathcal{D}(\rho)$ and $\mathcal{S}(\rho)$ for the difference and sum detector laws (see (9), (10)), as follows:

$$\begin{aligned}\frac{d\rho(t)}{dt} &= K\mathcal{D}(\rho)/\mathcal{S}(\rho)\\ &=: f_r(\rho)\end{aligned}\tag{23}$$

A Liapunov function for this first-order model is given by

$$\mathcal{V}(\rho) := -\int_0^\rho \frac{\mathcal{D}(u)}{\mathcal{S}(u)}du \tag{24}$$

Then it is straightforward to verify that under the stated assumptions, $\mathcal{V}(\rho)$ is locally positive definite near $\rho = 0$. Moreover, the time-derivative of $\mathcal{V}$ along trajectories of the reduced model (23) is given by

$$\dot{\mathcal{V}}(\rho) = -K\frac{\mathcal{D}^2(\rho)}{\mathcal{S}^2(\rho)} \tag{25}$$

and is thus locally negative definite.

Consider next the boundary layer system of (7)-(8). This system involves the single variable V, and is given by

$$\frac{dV(\tau)}{d\tau} = \beta^2 e^{-2V(\tau)}\mathcal{S}(\rho) - 1 \tag{26}$$

By a simple calculation we find that the following qualifies as a Liapunov function for the boundary layer system:

$$\mathcal{W}(\rho, V) := (\beta^2 e^{-2V}\mathcal{S}(\rho) - 1)^2 \tag{27}$$

Stability of the full continuous-time model In this subsection we combine the Liapunov functions for the reduced and boundary layer systems obtained above to yield a Liapunov function for the full system (7)-(8). This is achieved through an application of Theorem 1. Thus, we verify the applicability of Assumptions (A1)-(A3) to (7)-(8).

Define sets B_ρ and B_V (B_x and B_y of Theorem 1, respectively) by

$$B_\rho := \mathbb{R} \tag{28}$$

and

$$B_V := \{V : \quad V_1 \le V \le V_2\} \tag{29}$$

where V_1 and V_2 are constants defining a reasonable range for the voltage magnitude. Also, introduce constants

$$\bar{\mathcal{S}} := \max_\rho \mathcal{S}(\rho) \tag{30}$$

and

$$\bar{\mathcal{S}'} := \max_\rho |\mathcal{S}'(\rho)| \tag{31}$$

From Eq. (25), it is easy to see that Assumption (A1) holds with

$$\psi(\rho) := |\mathcal{D}(\rho)/\mathcal{S}(\rho)| \tag{32}$$

and

$$\alpha_1 := K \tag{33}$$

To check the validity of Assumption (A2), we proceed as follows:

$$\begin{aligned}(\text{grad}_V\ \mathcal{W}(\rho,V))^T g(\rho,V) &= -4\beta^2 e^{-2V}\mathcal{S}(\rho)(\beta^2 e^{-2V}\mathcal{S}(\rho)-1)^2\\ &\leq -4\beta^2 e^{-2V_2}\bar{\mathcal{S}}(\beta^2 e^{-2V}\mathcal{S}(\rho)-1)^2\\ &=: -\alpha_2\phi^2(g(\rho,V))\end{aligned} \tag{34}$$

where

$$\alpha_2 := 4\beta^2 e^{-2V_2}\bar{\mathcal{S}} \tag{35}$$

and

$$\begin{aligned}\phi(g(\rho,V)) &:= |g(\rho,V)|\\ &= |\beta^2 e^{-2V}\mathcal{S}(\rho)-1|\end{aligned} \tag{36}$$

Assumption (A3) consists of two inequalities that must be satisfied. To verify inequality (19) we derive an upper bound on $(\text{grad}_\rho\ \mathcal{W}(\rho,V))^T f(\rho,V)$:

$$\begin{aligned}(\text{grad}_\rho\ \mathcal{W}(\rho,V))^T f(\rho,V) &= 2K\mathcal{S}'(\rho)(\beta^2 e^{-2V})^2(\beta^2 e^{-2V}\mathcal{S}(\rho)-1)\mathcal{D}(\rho)\\ &\leq 2K|\mathcal{S}(\rho)||\mathcal{S}'(\rho)|(\beta^2 e^{-2V})^2|\beta^2 e^{-2V}\mathcal{S}(\rho)-1|\left|\frac{\mathcal{D}(\rho)}{\mathcal{S}(\rho)}\right|\\ &\leq 2K\bar{\mathcal{S}}\bar{\mathcal{S}}'(\beta^2 e^{-2V_1})^2|\beta^2 e^{-2V}\mathcal{S}(\rho)-1|\left|\frac{\mathcal{D}(\rho)}{\mathcal{S}(\rho)}\right|\end{aligned} \tag{37}$$

Thus the first inequality of Assumption (A3) holds with $c_1 := 0$ and $c_2 := 2K\bar{\mathcal{S}}\bar{\mathcal{S}}'(\beta^2 e^{-2V_1})^2$.

Next the second inequality of Assumption (A3) is verified by deriving an upper bound on $(\text{grad}_\rho\ \mathcal{V}(\rho))^T[f(\rho,V)-f_r(\rho)]$:

$$\begin{aligned}(\text{grad}_\rho\ \mathcal{V}(\rho))^T[f(\rho,V)-f_r(\rho)] &= -K\frac{(\mathcal{D}(\rho))^2}{(\mathcal{S}(\rho))^2}(\beta^2 e^{-2V}\mathcal{S}(\rho)-1)\\ &\leq K\left(\max_\rho\left|\frac{\mathcal{D}(\rho)}{\mathcal{S}(\rho)}\right|\right)\left|\frac{\mathcal{D}(\rho)}{\mathcal{S}(\rho)}\right|\ |\beta^2 e^{-2V}\mathcal{S}(\rho)-1|\end{aligned} \tag{38}$$

Thus the second inequality of (A3) holds with $\beta_1 := K\ (\max_\rho |\mathcal{D}(\rho)/\mathcal{S}(\rho)|)$.

Theorem 1 therefore applies, and we have that the origin of the system (7), (8) is locally asymptotically stable for all $\epsilon < \epsilon^*$, where

$$\epsilon^* := \frac{8\beta^2 e^{-2V_2}\bar{\mathcal{S}}}{K\max_\rho^2 |\mathcal{D}(\rho)/\mathcal{S}(\rho)|} \tag{39}$$

This result is obtained from Theorem 1 by maximizing the expression (21) for the upper bound $\epsilon^*(d)$ over d in the interval $(0,1)$, which amounts in this case to setting $d = 0.5$.

4 Stochastic Stability of the Radar Range Tracking System

In this section we study stochastic stability of the range tracking system in the case of a single fluctuating target.

4.1 Stochastic target models

As noted above, the continuous-time model (5)-(6) can be stochastic or deterministic, depending on the nature of the received signal $E(\sigma, t)$. Denote the nominal received signal, i.e., the (periodic) received signal $E(\sigma, t)$ in the absence of noise, as $E_0(t)$.

A reasonable approximation for the case of a single randomly fluctuating target is to take

$$E(\sigma, t) = n(t)E_0(\sigma) \tag{40}$$

where $n(t)$ is a random signal whose statistics are discussed below. The tracker system (7)-(8) then takes the form

$$\frac{d\rho}{dt} = K\beta^2 e^{-2V(t)}|n(t)|^2 \int_{-T_1}^{T_1} |E_0(\sigma)|^2 w_D(\sigma - \rho)d\sigma \tag{41}$$

$$T_{\mathrm{AGC}}\frac{dV}{dt} = \beta^2 e^{-2V(t)}|n(t)|^2 \int_{-T_1}^{T_1} |E_0(\sigma)|^2 w_S(\sigma - \rho)d\sigma - 1. \tag{42}$$

In our study of stochastic stability of the system (41)-(42) we consider two alternatives for the noise process $n(t)$. The first alternative, which we state roughly at this point, is to view $|n(t)|^2 - 1$ as a white Gaussian noise process. Strictly speaking this would violate the obvious requirement $|n(t)|^2 - 1 > -1$, but we are willing to pay this small price to obtain a tractable stochastic model. The second alternative entails that all realizations of $n(t)$ are continuous and bounded away from zero and infinity. In this case we assume a fast AGC, i.e., we let T_{AGC} be small. This permits application of singular perturbation results and the problem can be treated using a sample path approach.

It is interesting to observe that with the presence of noise this model admits *no equilibrium points*. Thus we would expect that if the system is stable in some sense then we would have a stronger type of convergence for ρ than for V. For example, it may happen that ρ converges asymptotically with probability 1 while V converges only in distribution.

4.2 Stochastic stability

The purpose of this section is to briefly summarize some pertinent results on stochastic differential equations and stochastic stability. Consider a general Ito stochastic differential equation

$$dx(t) = f(x(t))dt + g(x(t))dw(t) \tag{43}$$

where the state vector $x(t) \in \mathbb{R}^n$, and $w(t)$ is a scalar-valued Wiener process. Suppose the initial condition $x(0) = c$ is a given constant. Let f and g be defined and measurable on $\mathbb{R}^n$.

To study stability of Eq. (43), let us make the following additional assumption:

$$f(0) = 0, \quad g(0) = 0 \tag{44}$$

That is, the origin $x = 0$ is the unique solution of (43) satisfying the initial condition $c = 0$. It is possible, and of significant interest in the application problem considered here, to state stochastic stability results that do not require this assumption. Instead, one sometimes wishes to assume the milder condition that $f(0) = 0$ but $g(0) \neq 0$. This will be discussed at the end of this section, in the context of partial stochastic stability problems.

A general theory of stochastic stability of systems (43) exists which parallels the deterministic Liapunov stability theory. This theory uses the idea of *stochastic Liapunov function.* A stochastic Liapunov function is a scalar function $v(x)$ for which certain sign definiteness conditions are satisfied. As in the deterministic Liapunov stability theory, the time-derivative of the stochastic Liapunov function along trajectories of Eq. (43) is of crucial importance. In the stochastic setting, it is the expectation of this time-derivative that is of relevance. In this paper, we will take a direct approach to the analysis of stochastic stability of the radar range tracker model, so that we will not need to recall detailed theorems on stochastic Liapunov functions. The definitions below will suffice.

Denote by $x(t; c)$ the solution $x(t)$ to Eq. (43) satisfying the initial condition $x(0) = c$.

Definition 1. [3] The equilibrium 0 of (43) is stochastically stable if, for every $\epsilon > 0$,

$$\lim_{c \to 0} \mathrm{P}\ [\sup_{0 \le t < \infty} |x(t; c)| \ge \epsilon] = 0$$

Otherwise, it is stochastically unstable. The origin is stochastically asymptotically stable if it is stochastically stable and

$$\lim_{c \to 0} \mathrm{P}\ [\lim_{t \to \infty} x(t; c)] = 1$$

The origin is stochastically asymptotically stable in the large if it is stochastically stable and

$$\mathrm{P}\ [\lim_{t \to \infty} x(t; c) = 0] = 1$$

for all $c \in \mathbb{R}^n$.

There are interesting examples of systems that exhibit a form of stochastic stability that applies to a part of the state variables rather than to the state vector as a whole. Consider a system (43) for which 0 is not necessarily an equilibrium point, and for which the state vector $x(t)$ consists of two subvectors, $y(t)$ and $z(t)$. Suppose that $y(t)$ behaves in a stable fashion in some neighborhood of a value y_0, but that the behavior of $z(t)$ is of much less importance or is allowed to be erratic. Proving that the behavior of y is stable in some sense is a (stochastic) *partial stability problem.* Results on partial stability of deterministic systems have been reported in the literature (see, e.g., Vorotnikov [18] and Rouche, Habets and Laloy [15]). The stochastic stability result developed next for the radar tracking system studied here is an example of a partial stability result.

4.3 Tracker stability in the presence of white noise

The situation in which the received power signal is a nominal signal modulated by a white noise process is considered next. In this case, we view the system (41)-(42) as representing the random differential equation:

$$\dot{\rho} = K\beta^2 e^{-2V}\mathcal{D}(\rho) + \epsilon K\beta^2 e^{-2V}\mathcal{D}(\rho)\Psi(t) \tag{45}$$

$$T_{\mathrm{AGC}}\dot{V} = \beta^2 e^{-2V}\mathcal{S}(\rho) - 1 + \epsilon\beta^2 e^{-2V}\mathcal{S}(\rho)\Psi(t) \tag{46}$$

where Ψ is a standard scalar-valued white noise process.

Eqs. (45)-(46) include a white noise input. Note that we have not assumed that ϵ is small, so the noise is not necessarily small. As discussed in Section 3.1.2, to analyze such a system using the Ito calculus one needs to include the Wong-Zakai correction term in an Ito version of the equations. The stochastic Ito differential equations equivalent to the random differential equations (45)-(46) are given by

$$d\rho = \left\{ K\beta^2 e^{-2V}\mathcal{D}(\rho) + \frac{1}{2}\epsilon^2 K^2\beta^4 e^{-4V}\mathcal{D}(\rho)\mathcal{D}'(\rho) - \frac{\epsilon^2 K\beta^4}{T_{\mathrm{AGC}}} e^{-4V}\mathcal{D}(\rho)\mathcal{S}(\rho) \right\} dt + \epsilon K\beta^2 e^{-2V}\mathcal{D}(\rho)\, dw \tag{47}$$

$$dV = \left\{ \frac{1}{T_{\mathrm{AGC}}}(\beta^2 e^{-2V}\mathcal{S}(\rho) - 1) + \frac{1}{2T_{\mathrm{AGC}}}\epsilon^2 K\beta^4 e^{-4V}\mathcal{D}(\rho)\mathcal{S}'(\rho) - \frac{1}{T_{\mathrm{AGC}}^2}\epsilon^2\beta^4 e^{-4V}\mathcal{S}^2(\rho) \right\} dt + \frac{\epsilon\beta^2}{T_{\mathrm{AGC}}} e^{-2V}\mathcal{S}(\rho)\, dw \tag{48}$$

where w is a scalar-valued Wiener process.

Now we prove a stability result for the system (47), (48) above, under the assumptions that V is bounded and ϵ is small. Note that $\mathcal{D}(\rho)/\rho$ is globally bounded. Define the nonnegative function

$$v(\rho) := -\int_0^\rho \mathcal{D}(u)du \tag{49}$$

Letting $\alpha(t) := K\beta^2 e^{-2V(t)}$, the solution to (47) may be written as

$$\begin{aligned}\rho(t) &= c \exp\left\{\epsilon \int_0^t \alpha(s)\frac{\mathcal{D}(\rho(s))}{\rho(s)}dw_s - \frac{1}{2}\int_0^t \epsilon^2\alpha^2(s)\frac{\mathcal{D}^2(\rho(s))}{\rho^2(s)}dw_s \right.\\ &\qquad + \int_0^t \alpha(s)\frac{\mathcal{D}(\rho(s))}{\rho(s)}ds \\ &\qquad \left. + \frac{1}{2}\int_0^t \epsilon^2\alpha^2(s)\left[\frac{\mathcal{D}(\rho(s))\mathcal{D}'(\rho(s))}{\rho(s)} - \frac{\mathcal{D}(\rho(s))\mathcal{S}(\rho(s))}{KT_{\mathrm{AGC}}\rho(s)}\right]ds\right\} \\ &=: c\, e^{a(t)}\end{aligned} \tag{50}$$

Now, denoting by γ a constant whose numerical value changes from line to line,

$$\begin{aligned}\mathrm{P}\left[\sup_{0\le t} v(\rho(t)) \ge \mu\right] &\le \mathrm{P}\left[\sup_{0\le t}\rho(t) \ge \gamma\right] \\ &\le \mathrm{P}\left[\sup_{0\le t} a(t) \ge \gamma\right]\end{aligned} \tag{51}$$

However, since V is assumed bounded, $\alpha(s) > \gamma$, and, moreover, $-\mathcal{D}(x)/x \ge \gamma$ is a region $[-\delta, \delta]$. Thus,

$$\begin{aligned}\mathrm{P}\left[\sup_{0\le t} a(t) \ge \gamma\right] &\le \mathrm{P}\left[\sup_{0\le t}\ \epsilon \int_0^t \alpha(s)\frac{\mathcal{D}(\rho(s))}{\rho(s)}dw_s \ge \gamma t + \gamma\right] \\ &\le \mathrm{P}\left[\sup_{0\le t<1}\epsilon\ w(t) \ge \gamma\right] + \mathrm{P}\left[\sup_{1\le t}\ \epsilon\frac{w(t)}{t} > \gamma\right]\end{aligned} \tag{52}$$

It is now clear that, for sufficiently small ϵ, the right side can be made arbitrarily small. The stability result is established.

4.4 Tracker stability in the presence of bounded noise

If the noise process $n(t)$ appearing in the model (7)-(8) is continuous and bounded in magnitude away from zero and infinity, and if the AGC is fast, an alternate stability analysis is possible. This is a sample path stability analysis in the single-target case, with minimal probabilistic content. Singular perturbation theory is used in the analysis. Since the approach is rather straightforward, only a brief outline is given here.

If the AGC time constant T_{AGC} is very small, for any given sample path of (7)-(8), singular perturbation results may be applied. This involves writing reduced and boundary-layer models for the system. The reduced model is obtained by formally setting $T_{\text{AGC}} = 0$ in (8), solving the resulting equation for e^{-2V} in terms of ρ and $n(t)$, and substituting the result in (7). Since the AGC is included to minimize the effect of noise on tracker performance, it is not surprising that the reduced model does not depend on the noise process $n(t)$. The reduced model is found to be given explicitly by

$$\dot{\rho}(t) = K\frac{\mathcal{D}(\rho)}{\mathcal{S}(\rho)} \tag{53}$$

The boundary-layer model is obtained by changing the time scale from t to $\tau := t/T_{\text{AGC}}$ and setting $T_{\text{AGC}} = 0$ in the resulting model. Thus, the boundary-layer model is

$$\frac{dV}{d\tau} = \beta^2 e^{-2V}|(n(t_0)|^2\mathcal{S}(\rho_0) - 1 \tag{54}$$

where t_0 and ρ_0 are viewed as constant parameters.

Hoppensteadt [7] has given easily verifiable conditions for the solution of the original model (7)-(8) to be given approximately by that of the reduced model (53). These conditions are satisfied for the model under consideration, except for a boundedness condition. This condition is also satisfied if one appropriately limits the values of the AGC exponential nonlinearity. Stability analysis of the reduced model (53) is straightforward, since $v(\rho)$ of (49) qualifies as a Liapunov function for the reduced model. Thus, stability analysis of the range tracker in the case of bounded noise is rather straightforward.

5 Directions for Further Research

The analysis in the previous sections addresses simplified models and represents a modest attempt to begin the nonlinear and stochastic stability analysis of gated range trackers. In this section, we mention additional research problems involving stochastic stability and bifurcation.

The deterministic and stochastic stability analysis in the foregoing sections should be extended to models explicitly containing the effects of two or more targets. In this way, estimates of domains of attraction (target "relative strengths") can be obtained.

To present the next issue for further work, consider an EW scenario in which a tracking radar is approaching a target and a decoy. If the decoy is a chaff cloud, as mentioned in Section I, then the position of the decoy changes continuously with time. This, along with the motion of the missile carrying the radar itself, implies that a more realistic model would involve swept paramaters, rather than the constant parameters used in the preceding sections.

In bifurcation theory, typically one considers changes in steady-state behavior as a parameter changes quasistatically. However, simulations have shown considerable impact of time-varying separation distance and noise on target commitment.

To indicate the type of analysis that may be possible to help in understanding the stochastic bifurcation issues that occur in this application, we briefly recall a simple example from [14]. Consider the deterministic laser model

$$\dot{x}(t) = x\left[-1 + \frac{A}{(1+x^2)}\right] \tag{55}$$

Here, x is the normalized amplitude of the electric field and A is the pump parameter. For constant A, the system is autonomous, and the equilibria are $x = 0$ and $x^2 = A - 1$. (Note that the nonzero equilibria occur only for $A > 1$.)

For $A = 1$, two stable solutions bifurcate from the trivial solution. In the laboratory, this transition to lasing is commonly studied by sweeping A through the critical value $A = 1$:

$$A = A(t) = A_0 + vt, \qquad A_0 < 1, v > 0. \tag{56}$$

Linearizing the system at the origin using this $A(t)$, we get

$$\dot{E}(t) = E(1 - A) \tag{57}$$

for the dynamics of the error. This implies

$$E(t) = E(0)e^{[-t(A_0+vt/2-1)]} \tag{58}$$

Let t_{cr} be the equivalent critical time for loss of stability in the autonomous case. Thus, t_{cr} solves

$$A_0 + vt_{cr} = 1 \tag{59}$$

For the swept case, the error propogation law is given by (58) above. Let t^*_{cr} be the critical time for loss of stability in the swept case. This, t^*_{cr} solves

$$A_0 + vt^*_{cr}/2 = 1 \tag{60}$$

We find that $t^*_{cr} = 2t_{cr}$ *regardless of the sweep rate* v.

The special result above for sweeping in the noise-free case no longer holds in the presence of noise. It holds in the limit of very low noise, or high sweep rate. The delay becomes smoothed for the noise driven system. See [14] for details.

An interesting research problem is to obtain a similar understanding of the combined effects of parameter slewing and noise amplitude on target resolution in gated radar trackers.

The effects of noise correlation time on target resolution has been considered using numerical simulation in our previous work [9]. It was found that, for noise correlation times much smaller than the tracker time constant, a unimodal probabality density splits at a critical separation (stochastic bifurcation). For noise correlation time on the order of the tracker time constant, the probabilty density of the range estimate does not show distinct splitting. (Density flattens out for a relatively small target seperations.) In this case, the emergence of the two distinct modes that indicates target resolution does not occur until a separation distance that is larger than the separation distance at which the fast noise bifurcation occurs.

Acknowledgment

This research has been supported by the Naval Research Laboratory.

References

1. Abed, E. H., Goldberg, A. J., and Gover, R. E. (1991) Nonlinear modeling of gated radar range tracker dynamics with application to radar range resolution. IEEE Trans. Aerospace Electron. Syst. **(27)**, 68–82.
2. Agarwal, R. P. (1992) Difference Equations and Inequalities. Marcel Dekker, New York, NY.
3. Arnold, L. (1992) Stochastic Differential Equations: Theory and Applications. Krieger, Malabar, FL. (Reprint of 1974 John Wiley edition.)
4. Barton, D. K., ed. (1975) Radars, Volume IV: Radar Resolution and Multipath Effects. Artech House, Dedham, MA.
5. Blankenship, G. L. and Papanicolaou, G. C. (1978) Stability and control of stochastic systems with wide-band noise disturbances, I. SIAM J. Appl. Math. **(34)**, 437–476.
6. Bar-Shalom, Y. and Fortmann, T. E. (1988) Tracking and Data Association. Academic Press, San Diego, CA.
7. Hoppensteadt, F. C. (1966) Singular perturbations on the infinite interval. Trans. Amer. Math. Soc. **(123)**, 521–535.
8. Gihman, I. I. and Skorohod (1972) Stochastic Differential Equations. Springer-Verlag, Berlin. (English translation of 1968 Russian edition.)
9. Gover, R. E., Abed, E. H., and Goldberg, A. J. (1992) Stochastic stability analysis of nonlinear gated radar range trackers. In: Stochastic Theory and Adaptive Control, Duncan, T.E. and Pasik-Duncan, B., eds., Lecture Notes in Control and Information Sciences 184, Springer-Verlag, Berlin, 225-239.
10. Has'minskii, R. Z. (1980) Stochastic Stability of Differential Equations, Sijthoff and Noordhoff, Alphen aan den Rijn, The Netherlands. (English translation of 1969 Russian edition.)
11. Khalil, H. K. (1992) Nonlinear Systems. Macmillan, New York.
12. Kushner, H. J. (1967) Stochastic Stability and Control. Academic, New York, NY.
13. Kushner, H. J. (1990) Weak Convergence Methods and Singularly Perturbed Stochastic Control and Filtering Problems. Birkhäuser, Boston, MA.

14. Lugiato, L. A., Broggi, G., Merri, M., and Pernigo, M. A. (1989) Control of noise by noise and applications to optical systems. In: Noise in Nonlinear Dynamical Systems: Vol. 2, Theory of Noise Induced Processes and Special Applications. Moss, F. and McClintock, P.V.E., eds., Cambridge University Press, Cambridge, 293–346.
15. Rouche, N., Habets, P., and Laloy, M. (1977) Stability Theory by Liapunov's Direct Method. Springer-Verlag, New York.
16. Saberi, A. and Khalil, H. K. (1984) Quadratic-type Lyapunov functions for singularly perturbed systems. IEEE Trans. Automat. Contr. **(29)**, 542–550.
17. Vidyasagar, M. (1993) Nonlinear Systems Analysis, Second Edition, Prentice Hall, Englewood Cliffs, NJ.
18. Vorotnikov, V. I. (1988) The partial stability of motion. PMM U.S.S.R.**(52)**, 289-300 (English translation).
19. Wong, E. and Hajek, B. (1985) Stochastic Processes in Engineering Systems. Springer-Verlag, New York.

Asymptotic Properties and Associated Control Problems of Discrete-Time Singularly Perturbed Markov Chains

Dedicated to Professor Tyrone Duncan on the occasion of his 60th birthday

G. Badowski[1], G. Yin[2], and Q. Zhang[3]

[1] University of Maryland, College Park, MD 20742, USA
[2] Wayne State University, Detroit, MI 48202, USA
[3] University of Georgia, Athens, GA 30602, USA

Abstract. This work is concerned with asymptotic properties of singularly perturbed Markov chains in discrete time with finite state spaces. We study asymptotic expansions of the probability distribution vectors and derive a mean square estimate on a sequence of occupation measures. Assuming that the state space of the underlying Markov chain can be decomposed into several groups of recurrent states and a group of transient states, by treating the states within each recurrent class as a single state, we define an aggregated process, and show that its continuous-time interpolation converges to a continuous-time Markov chain. In addition, we prove that a sequence of suitably scaled occupation measures converges to a switching diffusion process weakly. Next, control problems of large-scale nonlinear dynamic systems driven by singularly perturbed Markov chains are studied. It is demonstrated that a reduced limit system can be derived, and that by applying nearly optimal controls of the limit system to the original one, nearly optimal controls of the original system can be obtained.

1 Introduction

Markov chains are used frequently to model uncertainty in networks, control and optimization problems, manufacturing and production planning, and communication systems. Very often, the underlying systems have large state spaces with many elements. As a result, to obtain the desired optimal control policies or to carry out optimization tasks, the computation required is quite intensive. Therefore, a straightforward implementation of numerical schemes could render the computation infeasible. It is imperative to find systematic approximation procedures to reduce the complexity of the underlying systems. One of the possible resolutions is the use of hierarchical decomposition.

In their work, Ando and Fisher proposed the so-called nearly completely decomposable matrix models; see [5] and the references therein. Such a notion has subsequently been applied to queueing networks for organizing the

B. Pasik-Duncan: Stochastic Theory and Control, LNCIS 280, pp. 19–34, 2002.

resources in a hierarchical fashion, to computer systems for aggregating memory levels, to economics for reduction of complexity of complex systems (see [5,23]), and to manufacturing systems for hierarchical production of planning (see [22]). The main idea is based on the following observation: Within a large network, different states often vary at different rates; some of them change rapidly and others vary slowly. To capture such a feature, we introduce a small parameter into the system to highlight the contrasts of different rates of changes. Then the system is naturally decomposed into a number of subsystems with smaller dimensions. We can aggregate the states in each of the subsystem into one state. Using this aggregation scheme, we can further obtain a limit system that is an average with respect to certain invariant measures, which leads to a substantial reduction of complexity for the underlying system. This is our main motivation of the study and exploration of singularly perturbed Markovian systems.

As was mentioned in the previous paragraph, hierarchical modeling and analysis yield practically feasible solutions for many large-scale systems. Nevertheless, to utilize the hierarchical approach for stochastic networks, it is foremost that we have a thorough understanding of the structural properties of the Markov chains. Continuous-time Markovian models were treated in [11–13,19,20,22,25,27,28,30], and discrete-time systems were considered in [1,4,9,18,24]. Different from the existing literature on discrete-time singularly perturbed systems, we aim to understanding the structural properties of the underlying singularly perturbed Markov chains; see [2,3,26,29].

In this paper, we review some of the recent development and use the asymptotic properties to treat large-scale optimal control problems. For simplicity, we focus on stationary problems. In addition, we work with such chains that contain only recurrent states. However, use the ideas presented in [26], the formulation and results can also be extended to certain nonstationary problems and for Markov chains including transient states.

The rest of the paper is arranged as follows. Section 2 gives the formulation of the problem. In Section 3, we proceed with the examination of unscaled and scaled occupation measures. Using asymptotic expansions of the probability distribution vectors, a mean square estimate is obtained for a sequence of occupation measures. Next, an aggregated process is defined. Although the aggregated process may be non-Markovian, it is proved that its continuous-time interpolation converges weakly to a continuous-time Markov chain whose generator is an appropriate average with respect to the associated stationary distributions. Then, we proceed to examine further a sequence of scaled occupation measures, and derive a functional central limit result. It is shown that the limit process is a switching diffusion. Section 4 is devoted to nearly optimal controls of large-scale discrete-time nonlinear dynamic systems driven by singularly perturbed Markov chains with a large state space. Our main goal is to reduce the complexity of the underlying systems. We obtain a reduced or limit system, such that the original system is averaged out with respect to the invariant measures. We show that by using the optimal control of the limit problem, we can construct nearly optimal controls of the original problem. Finally, Section 5 makes further remarks.

2 Formulation

Suppose T is a positive real number, $\varepsilon > 0$ is a small parameter, and α_k^ε, for $0 \le k \le \lfloor T/\varepsilon \rfloor$, is a discrete-time Markov chain with finite state space $\mathcal{M} = \{1, \ldots, d\}$, where $\lfloor z \rfloor$ denotes the integer part of a real number z. For notational simplicity, in what follows, we often simply write T/ε in lieu of $\lfloor T/\varepsilon \rfloor$. Let

$$f(\cdot,\cdot,\cdot) : \mathbb{R}^n \times \mathbb{R}^{n_1} \times \mathcal{M} \mapsto \mathbb{R}^n,$$

$$c(\cdot,\cdot,\cdot) : \mathbb{R}^n \times \mathbb{R}^{n_1} \times \mathcal{M} \mapsto \mathbb{R}.$$

Consider a stochastic dynamic system with the state $x_k^\varepsilon \in \mathbb{R}^n$ and control $u_k \in \Gamma \subset \mathbb{R}^{n_1}$. The problem of interest is:

$$\begin{aligned} &\text{minimize: } J^\varepsilon(x,\alpha,u) = \varepsilon E \sum_{k=0}^{T/\varepsilon} c(x_k^\varepsilon, u_k, \alpha_k^\varepsilon), \\ &\text{subject to: } x_{k+1}^\varepsilon = x_k^\varepsilon + \varepsilon f(x_k^\varepsilon, u_k, \alpha_k^\varepsilon),\ x_0^\varepsilon = x. \end{aligned} \tag{1}$$

where $E = E_{x,\alpha}$ is the expectation given $\alpha_0^\varepsilon = \alpha$ and $x_0^\varepsilon = x$.

The control problem (1) may be obtained from a continuous-time problem via discretization. Discrete-time control problems also come from many applications directly since observations and data can often only be obtained in discrete time. Unlike the usual stochastic control problems for controlled diffusions, there is no diffusion involved here. The sample paths of the disturbance is pure jump type. As a result, instead of one real-valued cost function, we have d ($d = |\mathcal{M}|$, the cardinality of $\mathcal{M}$) real-valued cost functions. As alluded to previously, in many stochastic networks, d is inevitably large, as a consequence, one must deal with a large number of value functions. Therefore, any attempt on reduction of complexity will be welcomed.

To study the proposed control problem, it is necessary to use certain properties of discrete-time Markov chains. Thus, we will consider a Markov chain α_k^ε with the transition probability matrix P^ε given by

$$P^\varepsilon(\varepsilon k) = P(\varepsilon k) + \varepsilon Q(\varepsilon k), \tag{2}$$

where for each $\le 0 \le k \le T/\varepsilon$, $P(\varepsilon k)$ itself is a transition probability matrix and $Q(\varepsilon k) = (q_{ij}(\varepsilon k))$ is a generator, i.e., for each $i \ne j$, $q_{ij}(\varepsilon k) \ge 0$ and for each i, $\sum_j q_{ij}(\varepsilon k) = 0$ or $Q(\varepsilon k)\mathbb{1}_d = 0$. Here and hereafter, $\mathbb{1}_\iota \in \mathbb{R}^{\iota \times 1}$ denotes a column vector with all components being 1. The motivation for using such models stems from the discretization of systems in continuous time. Consider a singularly perturbed, nonstationary, continuous-time Markov chain, whose forward equation is

$$\dot{p}^{\varepsilon}(t) = p^{\varepsilon}(t)\left(\frac{\widetilde{Q}(t)}{\varepsilon} + \widehat{Q}(t)\right),$$

where $p^{\varepsilon}(t) = (P(\alpha^{\varepsilon}(t) = 1), \dots, P(\alpha^{\varepsilon}(t) = d))$. Making change of variable $\zeta = t/\varepsilon$ and $p(\zeta) = p^{\varepsilon}(t)$ leads to

$$\frac{dp(\zeta)}{d\zeta} = p(\zeta)(\widetilde{Q}(\varepsilon\zeta) + \varepsilon\widehat{Q}(\varepsilon\zeta)).$$

Discretizing this equation with a step size $h > 0$ yields

$$p(k+1) = p(k)[I + h\widetilde{Q}(\varepsilon k)] + \varepsilon p(k)h\widehat{Q}(\varepsilon k).$$

Note that $h > 0$ can be properly chosen so that

$$I + h\widetilde{Q}(\varepsilon k) \stackrel{\text{def}}{=} P(\varepsilon k)$$

becomes a transition probability matrix. It is also clear that $h\widehat{Q}(\varepsilon k) \stackrel{\text{def}}{=} Q(\varepsilon k)$ is a generator. As a result, we obtain a transition probability matrix of the form (2). The above discussion reveals the connection of discrete-time and continuous-time Markov chains and illustrates that the discrete-time systems may arise from discretization. Discrete-time singularly perturbed Markov chains also arise in many discrete-time Markov decision processes and control and optimization problems directly.

It is easily seen that in (2), $P(\varepsilon k)$ is the dominating force for ε sufficiently small, so its structure is important. In accordance with the well-known Kolmogorov classification of states, in a finite-state Markov chain, there is at least one recurrent state, and either all states are recurrent or there is also a collection of transient states in addition to the recurrent states. Furthermore, (see for example, [10, p. 94]) a transition probability matrix of a finite-state Markov chain can be put into the form of either

$$P(\varepsilon k) = \text{diag}(P_1(\varepsilon k), \dots, P_l(\varepsilon k)), \tag{3}$$

or

$$P(\varepsilon k) = \begin{pmatrix} P_1(\varepsilon k) & & & & \\ & P_2(\varepsilon k) & & & \\ & & \ddots & & \\ & & & P_l(\varepsilon k) & \\ P_{*,1}(\varepsilon k) & P_{*,2}(\varepsilon k) & \cdots & P_{*,l}(\varepsilon k) & P_*(\varepsilon k) \end{pmatrix}, \tag{4}$$

where for each $0 \le k \le T/\varepsilon$ and each $i \le l$, $P_i(\varepsilon k)$ is a transition matrix within the ith recurrent class, and the last row $(P_{*,1}(\varepsilon k), \dots, P_{*,l}(\varepsilon k), P_*(\varepsilon k))$ in (4) corresponds to the transient states. [Hereafter, $\text{diag}(A_1, \dots, A_j)$ denotes a block diagonal matrix with matrix entries $A_1, \dots, A_j$ having appropriate dimensions.] A Markov chain with transition matrix given by (3) consists of l recurrent classes, whereas a Markov chain with transition matrix (4) has l recurrent classes and a number of transient states. Note that $P_{*,i}(\varepsilon k)$, $i = 1, \dots, l$, are the transition probabilities from the transient states to the recurrent states, and $P_*(\varepsilon k)$ is the transition probabilities within the transient states. The term $\varepsilon Q(\varepsilon k)$ facilitates the transitions among different recurrent classes. Due to the presence of the small parameter $\varepsilon > 0$, the transitions attributed to εQ represent "weak" interactions. For ease of presentation, in the rest of this paper, we concentrate on stationary Markov chains α_k^ε with time-dependent transition matrices, i.e. $P^\varepsilon(\varepsilon k) = P^\varepsilon$, that have transition matrix given by (2) with P specified by (3). More discussions on the nonstationary case can be found in [26] and [29]. In this paper, we mainly consider a Markov chain with transition matrix (2) and with $P^\varepsilon(\varepsilon k) = P = \text{diag}(P_1, \dots, P_l)$ for simplicity.

3 Asymptotic Properties

The probability vector

$$\xi_k^\varepsilon = (P(\alpha_k^\varepsilon = 1), \dots, P(\alpha_k^\varepsilon = m)) \in \mathbb{R}^{1\times m}$$

satisfies the vector-valued difference equation for each $k = 0, 1, \dots, T/\varepsilon$

$$\xi_{k+1}^\varepsilon = \xi_k^\varepsilon P_k^\varepsilon, \quad \xi_0^\varepsilon = \xi_0 \tag{5}$$

such that each component $\xi_{0,i} \ge 0$ and $\xi_0 \mathbb{1}_m = 1$. Note that x_0 is independent of ε. We assume:

(**A**) Both P^ε, and P are transition probability matrices such that for each $i \le l$, P_i is aperiodic and irreducible. In the case of inclusion of transient states, P_* has all eigenvalues inside the unit circle.

Making use of Fredholm alternative via verification of orthogonality conditions, asymptotic expansions of the probability vector and the corresponding transition probabilities were obtained in [26]. The following lemma summaries these results.

Lemma 1. *Under condition* (A),

(1) *for the probability distribution vector, we have*

$$\xi_k^\varepsilon = \theta(\varepsilon k)\mathrm{diag}(\nu^1, \dots, \nu^l) + O(\varepsilon + \lambda^k) \tag{6}$$

for some λ, $0 < \lambda < 1$, *where* ν^i *is the stationary distribution corresponding to the transition matrix* P_i, *and* $\theta(\varepsilon k) \in \mathbb{R}^{1\times l}$ (*with* $\theta(\varepsilon k) = (\theta_1(\varepsilon k), \dots, \theta_l(\varepsilon k))$ *satisfies*

$$\frac{d(\theta(\tau)}{dt} = \theta(\tau)\overline{Q}, \quad \theta_i(0) = \xi_0^i \mathbb{1}_{m_i}$$

where

$$\overline{Q} = \mathrm{diag}(\nu^1, \dots, \nu^l) Q \mathbb{1}, \tag{7}$$

and $\mathbb{1} = \mathrm{diag}(\mathbb{1}_{m_1}, \dots, \mathbb{1}_{m_l})$.

(2) *for* $k \le T/\varepsilon$, *the* k*-step transition probability matrix* $(P^\varepsilon)^k$ *satisfies*

$$(P^\varepsilon)^k = \Phi(t) + \varepsilon\widehat{\Phi}(t) + \Psi(k) + \varepsilon\widehat{\Psi}(k) + O\left(\varepsilon^2\right), \tag{8}$$

where

$$\begin{aligned} \Phi(t) &= \mathbb{1}\Theta(t)\widetilde{\nu} \\ \frac{d\Theta(t)}{dt} &= \Theta(t)\overline{Q}, \quad \Theta(0) = I. \end{aligned} \tag{9}$$

The proof of the first part of the lemma is in [26]. To prove the second part, we use a singular perturbation approach for the matrix-valued functions. We seek asymptotic expansion of the form

$$(P^\varepsilon)^k = \Phi(t) + \varepsilon\widehat{\Phi}(t) + \Psi(k) + \varepsilon\widehat{\Psi}(k) + e_k,$$

where $t = \varepsilon k$ and e_k is a matrix representing the approximation errors. Substituting the above expansion into the equation $(P^\varepsilon)^{k+1} = (P^\varepsilon)^k P^\varepsilon$ and comparing coefficients of like powers of ε, we obtain

$$\begin{aligned} &\Phi(t)(P - I) = 0, \\ &\widehat{\Phi}(t)(P - I) = \frac{d\Phi(t)}{dt} - \Phi(t)\overline{Q}, \\ &\Psi(k+1) = \Psi(k)P, \\ &\widehat{\Psi}(k+1) = \widehat{\Psi}(k)P + \Psi(k)Q. \end{aligned} \tag{10}$$

Using a similar approach as that of [26], it can be shown that $\Phi(t)$ is given by (9); for a treatment of continuous-time problems, see [25, Chapter 5 and Chapter 6].

It follows from the third equation of (10), $\Psi(k) = \Psi(0)P^k$. Choose the initial condition $\Psi(0) = I - \Phi(0)$. Moreover, similar to [26] (see also [25, Chapter 6]), we can show that there is a $0 < \lambda < 1$ such that

$$|\Psi(k)| \leq K\lambda^k.$$

Likewise, choosing $\widehat{\Phi}(0) + \widehat{\Psi}(0) = 0$, we obtain

$$\widehat{\Psi}(k+1) = \widehat{\Psi}(0)P^{k+1} + \sum_{j=0}^{k} \Psi(j)QP^{k-j},$$

and we can further show that $|\widehat{\Psi}(k)| \leq K\lambda^k$, where $K > 0$ denotes a positive constant. Finally, we can show that $e_k = O(\varepsilon^2)$. This is similar to the estimates in [26]. We thus omit the details.

To proceed, let us decompose the state space $\mathcal{M}$ into l subspaces according to the structure of P matrix. That is,

$$\begin{aligned} \mathcal{M} &= \{s_{11}, \ldots, s_{1d_1}\} \cup \{s_{21}, \ldots, s_{2d_2}\} \cup \ldots \cup \{s_{l1}, \ldots, s_{ld_l}\} \\ &= \mathcal{M}_1 \cup \mathcal{M}_2 \cup \cdots \cup \mathcal{M}_l, \end{aligned} \tag{11}$$

with $d = d_1 + d_2 + \cdots + d_l$.
The subspace $\mathcal{M}_i$, for each i with $i = 1, \ldots, l$, consists of recurrent states belonging to the ith recurrent class. In view of the decomposition, define an aggregated process $\overline{\alpha}^\varepsilon_k$ of α^ε_k by

$$\overline{\alpha}^\varepsilon_k = i \text{ if } \alpha^\varepsilon_k \in \mathcal{M}_i \text{ for } i = 1, \ldots, l. \tag{12}$$

Define continuous-time interpolations (for $t \in [0, T]$) by

$$\alpha^\varepsilon(t) = \alpha^\varepsilon_k, \quad \text{and} \quad \overline{\alpha}^\varepsilon(t) = \overline{\alpha}^\varepsilon_k \quad \text{if } t \in [\varepsilon k, \varepsilon(k+1)), \tag{13}$$

It is interesting to note that starting with a discrete-time system, owing to the scaling, the interpolated process (generally is non-Markovian) will converge weakly to a Markov chain with an appropriate generator.

Theorem 2. *Assume* (A). *Then as* $\varepsilon \to 0$, $\overline{\alpha}^\varepsilon(\cdot)$ *converges weakly to* $\overline{\alpha}(\cdot)$, *a Markov chain generated by* $\overline{Q}$ *given in* (7) *in* $D[0, T]$, *the space of functions that are right continuous and have left limits endowed with the Skorohod topology.*

To give some flavor of the argument of proof, we present the sketch of the proof below. First we show that $\overline{\alpha}^\varepsilon(\cdot)$ is tight. In fact,

$$E[(\overline{\alpha}^\varepsilon(t+s) - \overline{\alpha}^\varepsilon(s))^2 | \mathcal{F}^\varepsilon_s]$$

$$
\begin{aligned}
&= E[(\overline{\alpha}^{\varepsilon}(t+s) - \overline{\alpha}^{\varepsilon}(s))^2 | \alpha^{\varepsilon}(r), r \leq s] \\
&= E[(\overline{\alpha}^{\varepsilon}_{(t+s)/\varepsilon} - \overline{\alpha}^{\varepsilon}_{s/\varepsilon})^2 | \alpha^{\varepsilon}_1, \dots, \alpha^{\varepsilon}_{s/\varepsilon}] \\
&= E[(\overline{\alpha}^{\varepsilon}_{(t+s)/\varepsilon} - i)^2 | \alpha^{\varepsilon}_{s/\varepsilon} = s_{ij}] \text{ (Markov property)} \\
&= \sum_{p=1}^{l} (p-i)^2 P(\overline{\alpha}^{\varepsilon}_{(t+s)/\varepsilon} = p | \alpha^{\varepsilon}_{s/\varepsilon} = s_{ij}) \\
&\leq l^2 \sum_{p \neq i} \sum_{p_1=1}^{d_p} P(\alpha^{\varepsilon}_{(t+s)/\varepsilon} = s_{pp_1} | \alpha^{\varepsilon}_{s/\varepsilon} = s_{ij}) \\
&= l^2 \sum_{p \neq i} \sum_{p_1=1}^{d_p} \nu^p_{p_1} \theta_{ip}(t+s, s) + O(\varepsilon + \lambda^{t/\varepsilon}) \\
&= l^2 \sum_{p \neq i} \theta_{ip}(t+s, s) + O(\varepsilon + \lambda^{t/\varepsilon}).
\end{aligned}
$$

Since

$$
\lim_{t \to 0} \theta_{ik}(t+s, s) = 0 \text{ for } i \neq k,
$$

$$
\lim_{t \to 0} \lim_{\varepsilon \to 0} E[(\overline{\alpha}^{\varepsilon}(t+s) - \overline{\alpha}^{\varepsilon}(s))^2 | \mathcal{F}^{\varepsilon}_s] = 0.
$$

Thus $\overline{\alpha}^{\varepsilon}(\cdot)$ is tight by the tightness criterion [7,12].

Next, we establish the convergence of finite dimensional distributions of $\overline{\alpha}^{\varepsilon}(\cdot)$ to $\overline{\alpha}(\cdot)$. For any $0 \leq t_1 < t_2 < \dots < t_d \leq T$, and $i_1, \dots, i_d \in \overline{\mathcal{M}} = \{1, \dots, l\}$, we have

$$
\begin{aligned}
&P\left(\overline{\alpha}^{\varepsilon}(t_\iota) = i_\iota, \dots, \overline{\alpha}^{\varepsilon}(t_1) = i_1\right) \\
&\quad = P\left(\overline{\alpha}^{\varepsilon}_{t_\iota/\varepsilon} = i_\iota, \dots, \overline{\alpha}^{\varepsilon}_{t_1/\varepsilon} = i_1\right) \\
&\quad = P\left(\alpha^{\varepsilon}_{t_\iota/\varepsilon} \in \mathcal{M}_{i_\iota}, \dots, \alpha^{\varepsilon}_{t_1/\varepsilon} \in \mathcal{M}_{i_1}\right) \\
&\quad = \sum_{j_1=1}^{d_{i_1}} \cdots \sum_{j_\iota=1}^{d_{i_\iota}} P\left(\alpha^{\varepsilon}_{t_\iota/\varepsilon} = s_{i_\iota j_\iota}, \dots, \alpha^{\varepsilon}_{t_1/\varepsilon} = s_{i_1 j_1}\right) \\
&\quad = \sum_{j_1=1}^{d_{i_1}} \cdots \sum_{j_\iota=1}^{d_{i_\iota}} P(\alpha^{\varepsilon}_{t_1/\varepsilon} = s_{i_1 j_1}) \prod_{\kappa=2}^{\iota} P\left[\alpha^{\varepsilon}_{t_\kappa/\varepsilon} = s_{i_\kappa j_\kappa} | \alpha^{\varepsilon}_{t_{\kappa-1}/\varepsilon} = s_{i_{\kappa-1} j_{\kappa-1}}\right] \\
&\quad \longrightarrow \sum_{j_1=1}^{d_{i_1}} \cdots \sum_{j_\iota=1}^{d_{i_\iota}} \nu^{i_1}_{j_1} \theta_{i_1}(t_1) \prod_{\kappa=2}^{\iota} \nu^{i_\kappa}_{j_\kappa} \theta_{i_{\kappa-1} i_\kappa}(t_\kappa - t_{\kappa-1}) \quad \text{as } \varepsilon \to 0 \\
&\quad = \theta_{i_1}(t_1) \prod_{\kappa=2}^{\iota} \theta_{i_{\kappa-1} i_\kappa}(t_\kappa - t_{\kappa-1}) \\
&\quad = P(\overline{\alpha}(t_\iota) = i_\iota, \dots, \overline{\alpha}(t_1) = i_1).
\end{aligned}
$$

Thus, we have proved that $\overline{\alpha}^{\varepsilon}(\cdot)$ converges weakly to $\overline{\alpha}(\cdot)$, a Markov chain whose generator is $\overline{Q}$.

Many applications require the understanding of the amount of time the Markov chain spends in a given state. This leads to the study of occupation measures. For $k = 0, \ldots, T/\varepsilon$, define a sequence of occupation measures by

$$O^{\varepsilon}_{ij,k} = \varepsilon \sum_{r=0}^{k-1} \left(I_{\{\alpha^{\varepsilon}_r = s_{ij}\}} - \nu^i_j I_{\{\overline{\alpha}^{\varepsilon}_r = i\}} \right) \tag{14}$$

where ν^i_j denotes the jth component of the stationary distribution ν^i. Define a continuous-time interpolation:

$$O^{\varepsilon}_{ij}(t) = O^{\varepsilon}_{ij,k} \text{ if } t \in [k\varepsilon, (k+1)\varepsilon) \tag{15}$$

The use of $k-1$ is more or less for convenience. It allows us to write the difference equation

$$O^{\varepsilon}_{ij,k+1} = O^{\varepsilon}_{ij,k} + \varepsilon [I_{\{\alpha^{\varepsilon}_k = s_{ij}\}} - \nu^i_j I_{\{\overline{\alpha}^{\varepsilon}_k = i\}}]. \tag{16}$$

We can then obtain:

Theorem 3. *Assume* (A). *Then for* $i = 1, \ldots, l$, $j = 1, \ldots, d_i$,

$$\sup_{t \leq [0,T]} E|O^{\varepsilon}_{ij}(t)|^2 = O(\varepsilon).$$

The result above indicates that the mean squares error of the sequence of occupation measures is of the order $O(\varepsilon)$. One of the consequences is that this sequence converges in mean squares to 0 and another is that it suggests a $\sqrt{\varepsilon}$ scaling for a sequence of scaled occupation measures since

$$\sup_{t \in [0,T]} E|O^{\varepsilon}_{ij}(t)/\sqrt{\varepsilon}|^2 = O(1).$$

We provide a sketch of the proof. Note that

$$\begin{aligned}
E|O^{\varepsilon}_{ij}(t)|^2 &= E|O^{\varepsilon}_{ij,t/\varepsilon}|^2 \\
&= E\left(\varepsilon \sum_{r=0}^{t/\varepsilon - 1} [I_{\{\alpha^{\varepsilon}_r = s_{ij}\}} - \nu^i_j I_{\{\overline{\alpha}^{\varepsilon}_r = i\}}] \right)^2 \\
&= \varepsilon^2 E \sum_{r=0}^{t/\varepsilon - 1} \sum_{p=0}^{t/\varepsilon - 1} \{ I_{\{\alpha^{\varepsilon}_r = s_{ij}, \alpha^{\varepsilon}_p = s_{ij}\}} - \nu^i_j I_{\{\overline{\alpha}^{\varepsilon}_r = i, \alpha^{\varepsilon}_p = s_{ij}\}} \\
&\quad - \nu^i_j I_{\{\alpha^{\varepsilon}_r = s_{ij}, \overline{\alpha}^{\varepsilon}_p = i\}} + \nu^i_j \nu^i_j I_{\{\overline{\alpha}^{\varepsilon}_r = i, \overline{\alpha}^{\varepsilon}_p = i\}} \}
\end{aligned}$$

Let

$$\begin{aligned}\Phi^{\varepsilon}_{r,p} = & I_{\{\alpha^{\varepsilon}_r = s_{ij}, \alpha^{\varepsilon}_p = s_{ij}\}} - \nu^i_j I_{\{\overline{\alpha}^{\varepsilon}_r = i, \alpha^{\varepsilon}_p = s_{ij}\}} \\ & - \nu^i_j I_{\{\alpha^{\varepsilon}_r = s_{ij}, \overline{\alpha}^{\varepsilon}_p = i\}} + \nu^i_j \nu^i_j I_{\{\overline{\alpha}^{\varepsilon}_r = i, \overline{\alpha}^{\varepsilon}_p = i\}}\end{aligned} \tag{17}$$

Then

$$E|O^{\varepsilon}_{ij,t/\varepsilon}|^2 = \varepsilon^2 \left(\sum_{p \le r} \sum E\Phi^{\varepsilon}_{r,p} + \sum_{r<p} \sum E\Phi^{\varepsilon}_{r,p} \right).$$

Some detailed estimates allow us to obtain

$$E\Phi_{rp} = \begin{cases} O(\varepsilon + \lambda^{r-p}) & \text{for } p < r, \\ O(\varepsilon + \lambda^{p-r}) & \text{for } r < p. \end{cases}$$

The desired estimate then follows. To proceed, we further define a sequence of scaled occupation measures as follows. For each $i = 1, \dots, l, j = 1, \dots, d_i$, and $k \ge 0$ define the normalized occupation measure:

$$n^{\varepsilon}_k = \left(n^{\varepsilon}_{11,k}, \dots, n^{\varepsilon}_{1d_1,k}, \dots, n^{\varepsilon}_{l1,k}, \dots, n^{\varepsilon}_{ld_l,k}\right), \tag{18}$$

where

$$n^{\varepsilon}_{ij,k} = \sqrt{\varepsilon} \sum_{r=0}^{k-1} (I_{\{\alpha^{\varepsilon}_r = s_{ij}\}} - \nu^i_j I_{\{\overline{\alpha}^{\varepsilon}_r = i\}}) = \frac{1}{\sqrt{\varepsilon}} O^{\varepsilon}_{ij,k}. \tag{19}$$

Also define a continuous-time interpolation: $n^{\varepsilon}(t) = n^{\varepsilon}_k$ if $t \in [\varepsilon k, \varepsilon(k+1))$. By using the perturbed test function methods, we can prove the weak convergence of $n^{\varepsilon}(\cdot)$. To do so, working with the pair $(n^{\varepsilon}(\cdot), \overline{\alpha}^{\varepsilon}(\cdot))$, we verify that $(n^{\varepsilon}(\cdot), \bar{\alpha}^{\varepsilon}(\cdot))$ is tight, characterize the limit of any convergent subsequence of $(n^{\varepsilon}(\cdot), \bar{\alpha}^{\varepsilon}(\cdot))$ as a unique solution to a martingale problem, and show that the limit is a switching diffusion. The detailed proof can be found in [29].

Theorem 4. *Under* (A), $(n^{\varepsilon}(\cdot), \overline{\alpha}^{\varepsilon}(\cdot))$ *converges weakly to* $(n(\cdot), \overline{\alpha}(\cdot))$ *such that the limit is the solution of the martingale problem with operator* $\mathcal{L}$ *given by*

$$\mathcal{L} f(x,i) = \frac{1}{2} \sum_{j_1=1}^{d_i} \sum_{j_2=1}^{d_i} a_{j_1 j_2}(i) \frac{\partial^2 f(x,i)}{\partial x_{ij_1} \partial x_{ij_2}} + \overline{Q} f(x,\cdot)(i), \tag{20}$$

and the covariance of the limit process is given by $S(j) = \Sigma(j)\Sigma'(j)$, *where*

$$\Sigma(i) = \mathrm{diag}(0_{d_1 \times d_1}, \dots, 0_{d_{i-1} \times d_{i-1}}, \sigma(i), 0_{d_{i+1} \times d_{i+1}}, \dots, 0_{d_l \times d_l}), \tag{21}$$

$\sigma(i) \in \mathbb{R}^{d_i \times d_i}$ *satisfying* $\sigma(i)\sigma'(i) = A(i) = (a_{j_1 j_2}(i))$, *and*

$$A(i) = \nu^i_{\text{diag}} \Psi(0,i) + \nu^i_{\text{diag}} \sum_{k=1}^{\infty} \Psi(k,i) + \sum_{k=1}^{\infty} \Psi'(k,i) \nu^i_{\text{diag}}, \tag{22}$$

with $\nu^i_{\text{diag}} = \text{diag}(\nu^1_1, \dots, \nu^i_{d_i}) \in \mathbb{R}^{d_i \times d_i}$.

4 Optimal Control

In this section, we apply the results obtained in the previous section to derive a limit for the proposed control problem and obtain near optimal controls. To begin, we make the following assumptions concerning the control and the dynamics.

(C) (a) The control space Γ is a compact subset in $\mathbb{R}^{n_1}$.
(b) For each γ and each α, $f(\cdot, \gamma, \alpha)$ is continuously differentiable; for each α, $f(\cdot,\cdot,\alpha)$ is a continuous function; $|f(x,\gamma,\alpha)| \le K(1+|x|)$ for each (x, γ, α).
(c) for each α, $c(\cdot,\cdot,\alpha)$ is continuous, for each (γ,α), $c(\cdot,\gamma,\alpha)$ is Lipschitz continuous, and $|c(x,\gamma,\alpha)| \le K(1+|x|^{k_0})$ for some positive integer k_0 and for all triples (x,γ,α).

This assumption deals with the dynamics of the system and conditions on the cost function. By using (C), for each γ and each α, on bounded x-set, $f(\cdot,\gamma,\alpha)$ is Lipschitz continuous.

To solve the problem (1), we use the relaxed control formulation; see [8] and [16, pp. 47-54]. The problem in terms of the relaxed control is given by

$$\mathcal{P}_\varepsilon : \begin{cases} \text{minimize: } J^\varepsilon(x,\alpha,m^\varepsilon) = \varepsilon E \displaystyle\sum_{k=0}^{T/\varepsilon} \int c(x^\varepsilon_k, \gamma, \alpha^\varepsilon_k) m^\varepsilon_k(d\gamma), \\ \text{subject to: } x^\varepsilon_{k+1} = x^\varepsilon_k + \varepsilon \displaystyle\int f(x^\varepsilon_k,\gamma,\alpha^\varepsilon_k) m^\varepsilon_k(d\gamma), \\ x^\varepsilon_0 = x, \\ \alpha^\varepsilon_0 = \alpha, \end{cases} \tag{23}$$

where $\{m^\varepsilon_k(\cdot)\}$ is a sequence of discrete-parameter relaxed controls such that $m^\varepsilon_k(\cdot)$ is $\mathcal{F}_k$-measurable. We will use $\mathcal{R}_\varepsilon$ to denote the collection of admissible relaxed controls for $\mathcal{P}_\varepsilon$ on $\Gamma \times [0,T]$. Notice that under the relaxed control formulation, the system is linear in the control variable, and the compactness is easily obtainable.

To proceed, define a continuous-time interpolation of x^ε_k by

$$x^\varepsilon(t) = x^\varepsilon_k, \text{ if } t \in [\varepsilon k, \varepsilon k + \varepsilon). \tag{24}$$

To take care of the dependence of the control on the Markovian states, use the notation

$$U_0 = \{\Gamma = (\gamma^1, \ldots, \gamma^l) : \gamma^i = (\gamma^{i1}, \ldots, \gamma^{id_i}),\ \gamma^{ij} \in U,\ i = 1, \ldots, l\}.$$

It can be shown that although (23) is a discrete time problem, its associated limit is a continuous time problem given by

$$\mathcal{P}_0 : \begin{cases} \text{minimize: } J(x, \alpha, m) = E\displaystyle\int_0^T \int \overline{c}(x(s), \Gamma, \overline{\alpha}(s)) m_s(d\Gamma) ds, \\ \text{subject to: } x(t) = x + \displaystyle\int_0^t \int \overline{f}(x(s), \Gamma, \overline{\alpha}(s)) m_s(d\Gamma) ds, \\ x(0) = x, \\ \overline{\alpha}(0) = \alpha \mathbb{1}, \end{cases} \tag{25}$$

where $\overline{c}(\cdot)$, and $\overline{f}(\cdot)$ are defined as

$$\begin{aligned} \int \overline{c}(x, \Gamma, i) m_s(d\Gamma) &= \sum_{j=1}^{d_i} \int \nu_j^i(t) c(x, \gamma^{ij}, s_{ij}) m_s(d\gamma^{ij}), \\ \int \overline{f}(x, \Gamma, i) m_s(d\Gamma) &= \sum_{j=1}^{d_i} \int \nu_j^i(t) f(x, \gamma^{ij}, s_{ij}) m_s(d\gamma^{ij}), \end{aligned} \tag{26}$$

respectively. Denote by $\mathcal{R}_0$ the set of admissible controls for the limit problem, i.e., it is a collections of $m(\cdot)$, measures on $\mathcal{B}(U \times [0,T])$ such that $m(U,t) = t$ for all $t \geq 0$ and $m(\cdot)$ is $\mathcal{F}_t$-adapted, where $\{\mathcal{F}_t\}$ is a collection of increasing σ-algebras such that $\mathcal{F}_t$ measures at least $\{\overline{\alpha}(s); s \leq t\}$. The following theorem states the convergence result.

Theorem 5 *Suppose that* (A) *and* (C) *hold, and that $m^\varepsilon(\cdot)$ is a δ_ε-optimal admissible relaxed control for $\mathcal{P}_\varepsilon$. Then*

(a) *$\{x^\varepsilon(\cdot), m^\varepsilon(\cdot)\}$ is tight.*
(b) *Suppose that $(x(\cdot), m(\cdot))$ is the limit of a weakly convergent subsequence of $(x^\varepsilon(\cdot), m^\varepsilon(\cdot))$ (also indexed by ε for simplicity). Then $m(\cdot) \in \mathcal{R}_0$ and $(x(\cdot), m(\cdot))$ is a solution to $\mathcal{P}_0$.*
(c) *$J^\varepsilon(m^\varepsilon) \to J(m)$ as $\varepsilon \to 0$.*

Although we are dealing with relaxed controls, one can always use a nearly optimal feedback control to approximate a relaxed control. In addition, an optimal (or nearly optimal) control of the limit problem can be used in the original system, which preserves the near optimality. Since within the class of relaxed control, there is a minimizer, for each $\iota \in \mathcal{M}$ and each initial point x, let $v^\varepsilon(x, \iota)$ and $v^0(x, \iota)$ be the minimal costs for $\mathcal{P}_\varepsilon$ and $\mathcal{P}_0$, respectively, i.e., $v^\varepsilon(x, \iota) = \inf_{m^\varepsilon \in \mathcal{R}_\varepsilon} J^\varepsilon(x, \iota, m^\varepsilon)$, and $v^0(x, \iota) = \inf_{m \in \mathcal{R}_0} J(x, \iota, m)$. Using a version of the chattering theorem similar to that of [16, p. 59], for each $\delta > 0$, there is a finite set $U_\delta \subset U$, a $\Delta > 0$, and an admissible $u^\delta(t) = (u^{\delta,1}(t), \ldots, u^{\delta,l}(t)) \in U_0$, where $u^{\delta,i}(t) = (u^{\delta,i1}(t), \ldots, u^{\delta,id_i}(t))$, for the limit problem that is a δ-optimal for $\mathcal{P}_0$, i.e., $J(u^\delta, x, \iota) \le v^0(x, \iota) + \delta$. Define a feedback control by

$$\overline{u}(x, \alpha, t) = \sum_{i=1}^{l} \sum_{j=1}^{d_i} I_{\{\alpha = s_{ij}\}} u^{\delta,ij}(t),$$

and let $u^{\varepsilon,\delta}(t) = \overline{u}(x^\varepsilon(t), \alpha^\varepsilon(t), t)$.

Theorem 6. *Suppose that the conditions of Theorem* 5 *are fulfilled and that the differential equation in the limit problem* $\mathcal{P}_0$ *has a unique solution for each initial condition. Then*

$$\limsup_{\varepsilon \to 0} |J^\varepsilon(x, \iota, u^{\varepsilon,\delta}) - v^\varepsilon(x, \iota)| \le \delta \tag{27}$$

for each $\iota \in \mathcal{M}$.

Note that in lieu of one value function, we have a collection of $d = |\mathcal{M}|$ value functions,

$$v^\varepsilon(x, s_{11}), \ldots, v^\varepsilon(x, s_{1d_1}), \ldots, v^\varepsilon(x, s_{l1}), \ldots, v^\varepsilon(x, s_{ld_l}).$$

Using a dynamic programing approach, it may be shown that the limit v^0 of v^ε has the feature that for all initial s_{ij} with $i = 1, \ldots, l$, $j = 1, \ldots, d_i$, $v^\varepsilon(x, s_{ij}) \to v^0(x, i)$. That is, the limit depends only on which recurrent class $\mathcal{M}_i$ the initial point belongs to. Thus, the number of value functions corresponding to each recurrent class reduces from d_i to 1.

If transient states are also included, the value functions become

$$v^\varepsilon(x, s_{11}), \ldots, v^\varepsilon(x, s_{l1}), \ldots, v^\varepsilon(x, s_{ld_l}), v^\varepsilon(x, s_{*1}), \ldots, v^\varepsilon(x, s_{*m_*}).$$

For the transient states, have denote the limit of $v^\varepsilon(x, s_{*j})$ by $v(x, *j)$ and $v(x, *) = (v(x, *1), \ldots, v_d(x, *m_*))'$. Then it may be shown that

$$v(x, *) = a_1 v(x, 1) + \cdots + a_l v(x, l),$$

where $a_i = -(P_* - I)^{-1} P_{*,i} \mathbb{1}_i$. For a treatment of related problem of Markov decision processes, see [17]. Note that the a_i used above has an interpretation of probability vector. In fact, $a_{i,j}$, the jth component of a_i is the probability of entering the ith recurrent class starting from transient state s_{*j}.

Note that the reduced or limit system is an average of the original system with respect to the invariant measures. It turns out that in the limit problem, there are only l states (where l is the total number of irreducible recurrent classes in the original problem). By using the optimal control of the limit problem, we can then construct nearly optimal controls of the original problem leading to asymptotic optimality. Associated with the original problem, we have m value functions. In the limit system, the total number of value functions reduces to l. It is easily seen that if $l \ll d$, the complexity will be much reduced. The weak convergence approach in conjunction with the relaxed control representation allows us to use weaker conditions and handle more general nonlinearity.

5 Further Remarks

We have developed a number of asymptotic properties of singularly perturbed Markov chains in discrete time, and then applied those results to control problems. The results obtained can be extended to nonstationary models and Markov chains having transient states.

Acknowledgement

The research of G. Badowski was supported in part by Wayne State University; the research of G. Yin was supported in part by the NSF; the research of Q. Zhang was supported in part by the USAF and ONR.

References

1. Abbad, M., Filar, J. A., and Bielecki, T. R. (1992) Algorithms for singularly perturbed limiting average Markov control problems, IEEE Trans. Automat. Control **AC-37**, 1421–1425.
2. Badowski, G., Yin, G., and Zhang, Q. (2000) Discrete-time singularly perturbed Markov chains and applications to stochastic networks, in Proc. the 38th Annual Allerton Conference on Communication, Control, and Computing, 528–536.
3. Badowski, G., Yin, G., and Zhang, Q. Near-optimal controls of discrete-time dynamic systems driven by singularly perturbed Markov chains, J. Optim. Theory Appl., to appear.
4. Blankenship, G. (1981) Singularly perturbed difference equations in optimal control problems, IEEE Trans. Automat. Control **T-AC 26**, 911–917.
5. Courtois, P. J. (1977) Decomposability: Queueing and Computer System Applications, Academic Press, New York.
6. Delebecque, F. and Quadrat, J. (1981) Optimal control for Markov chains admitting strong and weak interactions, Automatica **17**, 281–296.
7. Ethier, S. N. and Kurtz, T. G. (1986) Markov Processes: Characterization and Convergence, J. Wiley, New York.
8. Fleming, W. H. (1982) Generalized solution in optimal stochastic control, Proc. URI Conf. on Control, 147–165.
9. Hoppensteadt, F. C. and Miranker, W. L. (1977) Multitime methods for systems of difference equations, Studies Appl. Math. **56**, 273–289.

10. Iosifescu, M. (1980) Finite Markov Processes and Their Applications, Wiley, Chichester.
11. Il'in, A., Khasminskii, R., and Yin, G. (1999) Asymptotic expansions of solutions of integro-differential equations for transition densities of singularly perturbed switching diffusions: Rapid switchings, J. Math. Anal. Appl. **238**, 516–539.
12. Khasminskii, R. Z., Yin, G., and Zhang, Q. (1996) Asymptotic expansions of singularly perturbed systems involving rapidly fluctuating Markov chains, SIAM J. Appl. Math. **56**, 277–293.
13. Khasminskii, R. Z., Yin, G., and Zhang, Q., (1997) Constructing asymptotic series for probability distribution of Markov chains with weak and strong interactions, Quart. Appl. Math. **LV**, 177-200.
14. Kumar, P. R. and Varaiya, P. (1984) Stochastic Systems: Estimation, Identification and Adaptive Control, Prentice-Hall, Englewood Cliffs, N.J.
15. Kushner, H. J. (1984) Approximation and Weak Convergence Methods for Random Processes, with Applications to Stochastic Systems Theory, MIT Press, Cambridge, MA.
16. Kushner, H. J. (1990) Weak Convergence Methods and Singularly Perturbed Stochastic Control and Filtering Problems, Birkhäuser, Boston.
17. Liu, R. H., Zhang, Q., and Yin, G. (2001) Nearly optimal control of singularly perturbed Markov decision processes in discrete time, Appl. Math. Optim. **44**, 105–129.
18. Naidu, D. S. (1988) Singular Perturbation Methodology in Control Systems, Peter Peregrinus Ltd., Stevenage Herts, UK.
19. Pan, Z. G., and Başar, T. (1995) H^{∞}-control of Markovian jump linear systems and solutions to associated piecewise-deterministic differential games, in New Trends in Dynamic Games and Applications, G. J. Olsder (Ed.), 61–94, Birkhäuser, Boston.
20. Phillips, R. G. and Kokotovic, P. V. (1981) A singular perturbation approach to modeling and control of Markov chains, IEEE Trans. Automat. Control **26**, 1087–1094.
21. Pervozvanskii, A. A. and Gaitsgori, V. G. (1988) Theory of Suboptimal Decisions: Decomposition and Aggregation, Kluwer, Dordrecht.
22. Sethi, S. P. and Zhang, Q. (1994) *Hierarchical Decision Making in Stochastic Manufacturing Systems*, Birkhäuser, Boston.
23. Simon, H. A. and Ando, A. (1961) Aggregation of variables in dynamic systems, Econometrica, **29**, 111–138.
24. Tse, D. N. C., Gallager, R. G., and Tsitsiklis, J. N. (1995) Statistical multiplexing of multiple time-scale Markov streams, IEEE J. Selected Areas Comm. **13**, 1028–1038.
25. Yin, G. and Zhang, Q. (1998) Continuous-time Markov Chains and Applications: A Singular Perturbation Approach, Springer-Verlag, New York.
26. Yin, G. and Zhang, Q. (2000) Singularly perturbed discrete-time Markov chains, SIAM J. Appl. Math., **61**, 834–854.
27. Yin, G., Zhang, Q., and Badowski, G. (2000) Asymptotic properties of a singularly perturbed Markov chain with inclusion of transient states, Ann. Appl. Probab., **10**, 549–572.
28. Yin, G., Zhang, Q. and Badowski, G. (2000) Singularly perturbed Markov chains: Convergence and aggregation, J. Multivariate Anal. **72**, 208–229.

29. Yin, G., Zhang, Q., and Badowski, G. (2000) Asymptotic properties of singularly perturbed Markov chains in discrete time, preprint.
30. Zhang, Q. and Yin, G. (1999) On nearly optimal controls of hybrid LQG problems, IEEE Trans. Automat. Control, **44**, 2271–2282.

Feedback Designs in Information-Based Control*

J. Baillieul

Dept. of Aerospace and Mechanical Engineering, Boston University, Boston, MA 02215, USA. johnb@bu.edu

Abstract. This paper reports a tight bound on the data capacity a feedback channel must provide in order to stabilize a right half-plane pole of a linear, time-invariant control system. The proof is constructive, and involves considering a general class of quantized control realizations of classical feedback designs. Even for the coarsest quantizations—with two-element control input sets, which we refer to as a *binary realization*—the bound is achievable in the scalar case. The open question of whether bounded trajectories in higher order systems could be produced by a binary realization is answered in the affirmative—again via an explicit construction for a system with two-dimensional state space. It is also shown how binary realizations of classical feedback designs organize the way in which the controller pays *attention* to different open-loop modes in the plant.

1 Introduction

This paper expands a part of the lecture I gave at the Kansas Workshop on Stochastic Theory and control. The topic was chosen in light of its interest to several participants. It is dedicated in friendship and respect to Tyrone Duncan on the occasion of his 60-th birthday and also to Bozenna Pasik-Duncan who made the workshop possible.

In a number of new technological settings we find feedback control designs are constrained by data rate limitations in the communication channels between sensors, controller, and actuators. Such constraints are a consideration, for instance, in current and future generations of MEMS arrays in which there can be as many as 10^4 to 10^6 actuators or sensors on a single chip, and on a larger physical scale they are of central interest in closing feedback loops through wireless links such as Bluetooth^{TM} or IEEE 802.11(b) and using network protocols such as CAN, [6]. Work is only just now beginning on the switching and encoding strategies that will be needed to produce stable closed-loop dynamics across a broad spectrum of applications. We are finding that many of the bandwidth assignment issues that are predominant

* Support from the Army Research Office under the ODDR&E MURI97 Program Grant No. DAAG55-97-1-0114 to the Center for Dynamics and Control of Smart Structures (through Harvard University) and from ODDR&E MURI01 Program Grant Number DAAD19-01-1-0465 to Boston University and the Center for Networked Communicating Control Systems is gratefully acknowledged.

B. Pasik-Duncan: Stochastic Theory and Control, LNCIS 280, pp. 35–57, 2002.

in managing the traffic in modern communications networks are also present in some form in networked actuator arrays.

This paper presents recent results on the ways in which control designs are constrained by communications bandwidth constraints on the feedback channels. One of the main results provides a formula for the minimum feedback channel data rate required to stabilize a liner time invariant (LTI) system with a prescribed open-loop pole. It is assumed that the state of our system is a continuous (i.e. not quantized) variable which is sampled at evenly spaced instants of time. (The assumption that sampling is at a constant rate is not essential.) The control variables, on the other hand, are assumed to take on only finitely many values (as would be the case with control values being generated by a D/A converter). Our data-rate result does not depend on the specific values of the control inputs, but it does depend on the cardinality of the control set and the amount of time required to process and transmit data over the feedback channels. It is thus a result about quantization and flow of information and how this is related to controlling the action of a physical system.

Issues of the type discussed here have been raised in a number of places in the recent literature. The data-rate bounds of Sections 2 and 3 below were announced in the workshop notes of the 1999 ARO Workshop on Smart Structures ([2]). A result similar to that of Section 2 has appeared in work by Nair and Evans ([7]), but in this case the bound was established in an entirely different way using methods of stochastic stability and asymptotic quantization theory. Extension of this result to the case of multidimensional LTI systems has also been recently reported by Nair and Evans ([8]). A similar bound appears in more recent work of Tatikonda and Mitter ([9]), again applying to the case of multidimensional LTI systems. It seems fair to say that all of this work has been inspired by results of Wong and Brockett ([10]) which treated such data-rate bounds in terms of some explicit encoding schemes.

An interesting concept which has emerged in this research is that of *attention*—a term which is intended to serve as both a descriptive characteristic of controlled dynamical systems in general and a figure of merit for closed-loop feedback designs. While *attention* has had a long history of study in the cognitive sciences, it is only following Brockett's 1997 paper ([4]) that the term has appeared in the control literature. Brockett's paper defined an attention functional which characterized the complexity of a feedback control law in terms of its rates of change with respect to both time and space. In the present note, we shall present a slightly different perspective in which the notion of *attention* involves the "cost of information processing and data transmission" required to implement a stable feedback law for an open loop unstable linear time-invariant system. More specifically, we are interested in the computational resources and channel capacity (measured in bits/second) that are required to produce a bounded response when there are right half-plane poles in the open loop system. While our discussion is carried out in language reminiscent of classical information theory, unlike information theory, our results are not probabilistic in nature. Indeed, our results are ground in geometry, and our goal is to find efficient ways of encoding the spatial variations that must be present in stabilizing control laws. One intuitively appealing result is that by our measure, the attention required to implement a stable feedback law increases as a function of how far open-loop poles extend into the right half-plane.

The paper is organized as follows. In the next section, we consider the problem of stabilizing an LTI system as depicted in Figure 1. Here a bound is established on the capacity required on the channel between G and H in order to stabilize a right half-plane pole in G. The bound is tight and the result has been proved for the case of first order systems. Section 3 discusses extensions to higher order systems. Section 4 treats the geometry of control coding and describes the way in which a binary realizations of constant gain feedback designs organize the attention paid by the controller to each of the system's open-loop modes. In terms of an explicit example, it is shown for the first time that a two-element control set suffices to produce bounded trajectories in a scalar-input control system whose state space has dimension > 1 and which is open loop unstable. It is also shown how the response of a control system to a finitely quantized set of admissible control levels may be expected to differ (in significant ways) from the response produced by classical (analog) designs.

2 Digital Control of LTI Systems With Uniform Sampling Rate—the Data Rate Bound in the Scalar Case

Classical feedback control theory is aimed at understanding design principles for integrating sensors and actuators for controlling a physical system in a way that it will perform its prescribed tasks efficiently and reliably. The basic principles are very simple as illustrated in Figure 1. Information about the state of the system of interest (called the plant G) is provided by sensor data. The control system (represented by the box labeled H together with the dotted interconnection) processes signals from sensor outputs, compares these values with specified operating goals, and "closes the loop" by sending appropriate actuator signals to keep the system operating as near as possible to its operating goals. There is today a vast array of mathematically sophisticated tools for design and analysis of the controller H. The tools apply to both continuous and discrete-time systems, and for many applications, they have become an essential aid for implementations of feedback control mediated by digital microprocessors. As we shall indicate below, however, many feedback control laws cannot be implemented in a satisfactory way if there

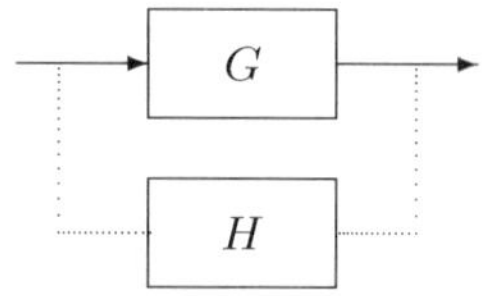

Fig. 1. A feedback system. G is a rational function plant, and H is the controller—possibly having digital components.

are constraints on the rate at which information is processed and transmitted through the feedback loop.

In this paper, we consider an explicit realization of the feedback system in Figure 1 of the form

$$\dot{x} = Ax + bu,$$

where (A, b) is a controllable pair. The case in which the control u is a scalar is the most interesting, and considering this case avoids extraneous complexity. It will also be useful to assume we have adopted the representation

$$A = \begin{pmatrix} a_1 & 0 & \dots & 0 \\ 0 & a_2 & \dots & 0 \\ \vdots & \vdots & \ddots & \vdots \\ 0 & 0 & \dots & a_n \end{pmatrix}, \quad b = \begin{pmatrix} 1 \\ 1 \\ \vdots \\ 1 \end{pmatrix},$$

wherein the controllability condition is equivalent to assuming the diagonal entries of A are distinct.

The case we shall treat involves digital control with a uniform sampling interval h. We assume control actuation is of the sample-and-hold type, so that control inputs to the system are constant over each sampling interval. The state of the plant evolves in discrete time according to

$$x(k+1) = F\,x(k) + \Gamma\,u(k), \tag{1}$$

where

$$F = e^{Ah}, \quad \text{and} \quad \Gamma = A^{-1}(e^{Ah} - I)b.$$

If a digital computer is used to implement feedback control, a sensor reading of the (typically analog) value $x(k)$ is digitized and stored in a computer register. Using whatever control algorithm has been implemented, the computer determines a value of the control input $u(k)$ (from a finite set U of admissible control values), and this in turn is converted to an analog value (e.g. a voltage or a current) which is then fed back to the plant.

For many applications, eight or fewer bits suffice to specify the values of the control function $u(\cdot)$, and hence the cardinality of the set U of admissible control values is typically ≤ 256. Whatever the cardinality, $|U|$, of the set of admissible controls, the number N of bits needed to uniquely encode each control value is such that $|U| \le 2^N$. The inequality may be strict since some encoding schemes will use a certain number of bits for purposes other than identifying control values.

We are interested in the case in which there is a limited rate at which data can be transmitted over the link connecting G to H, then used in a real-time calculation, with the resulting control signal transmitted back to G. There may be any number of reasons for the data rate to be limited—including the feedback loops being closed through a data network with congested communication channels, communication channels subject large amounts of noise, etc. While any of the data handling components around the feedback loop may limit the closed loop data rate in some way, it will be useful to assume in our analysis that the entire constraint on the data rate—to say R bits per second—occurs on the link between the controller H and the plant G.

Remarks on the Quantization of the Control Set

- There is a trade-off between the cardinality of the set of admissible control values (or levels) $u(k)$ and amount of time required on average to transmit digitally encoded data from the control computer to the plant. This problem is discussed in Wong and Brockett, [10], where both controller and sensor data are assumed to be encoded using *prefix codes*. Using known bounds on the average length of codewords in prefix codes, bounds are given on the minimum sampling interval, h, required to permit the transmission of control data between the sensors and actuators and the control computer.
- The present paper is concerned with a more primal relationship between the coarseness of the quantization of the admissible control actions and the coarseness of our quantization of time. For right half-plane poles of A, the corresponding modes $e^{a_i h}$ of F are monotonically increasing functions of h. Below we shall show that for stable operation of any digital implementation, the data-rate capacity R of the feedback channel must always exceed $a_i \log_2 e$. It will also be shown that the number of possible distinct control actions that can be taken in the time interval h cannot be less than $e^{a_i h}$. There is thus an interesting theoretical trade-off between spatial and temporal quantization. For smaller values of the sampling interval h (= faster sampling), fewer distinct possible control actions are needed. The minimum required data rate is the same in all cases, however.
- Stating our main result in somewhat imprecise terms, the degree of instability of a system may be quantified in terms of the amount of *attention* that must be paid to it by the controller. The greater the degree of in-

stability of the open-loop system, the finer the quantization of control values and the more information there is which must be communicated between the controller and plant in each time interval.

Classical notions of asymptotic stability must be modified in order to discuss control designs with quantized control values. This was noted in Wong and Brockett ([10]) where we were introduced to the notion of *containability* as a digital control concept corresponding to *asymptotic stability.* For the purposes of the present paper, we adopt a more primitive notion of stability:

Definition 1. A linear control system (1) with admissible control set U is *boundable* if there exists a bounded set S and an open set $M \subseteq S$ such that for each $x_0 \in M$ there is a control sequence $\mathcal{U} = \{u(k) \in U : k = 1, 2, \ldots\}$ such that the trajectory defined by (1) with $x(0) = x_0$ and control input sequence $\mathcal{U}$ remains in S for all time.

Boundability is a very weak notion of stability, and any system which is not boundable cannot be containable in the sense of Wong and Brockett ([10]).

Example 1. The case of binary control. Suppose the set U of admissible control values has two elements. These correspond to what we assume are precisely two different possible actuator commands. While these values might be different voltages or currents or other physical quantities of constant magnitude, there is no loss of generality in the model in assuming that $U = \{-1, 1\}$, since this may always be brought about for a system of the form (1) by an affine change of coordinates. To make the discussion of boundability interesting, we assume that (1) has at least one open-loop unstable mode. I.e. there is at least one element α_i on the diagonal of the (diagonal) matrix F such that $\alpha_i > 1$. In its most general form, a feedback law is defined by a *selection function* $f : \mathbb{R}^n \to U$ which for each $x \in \mathbb{R}^n$ chooses a control value so that the closed-loop evolution of (1) is prescribed by the formula

$$x(k+1) = Fx(k) + \Gamma f(x(k)). \tag{2}$$

Selection functions related to classical constant gain feedback designs will be treated in Section 4. An important feature in the present setting is that using binary control to implement a stabilizing feedback law illustrates the data-rate bound in a very clear way. Specializing the system (1) to the open-loop unstable scalar case ($n = 1$) with binary control gives closed-loop dynamics given explicitly by

$$x(k+1) = \begin{cases} \alpha\, x(k) + \beta \;\; x(k) \le 0 \\ \\ \alpha\, x(k) - \beta \;\; x(k) > 0 \end{cases}. \tag{3}$$

This transition function is depicted in Figure 2.

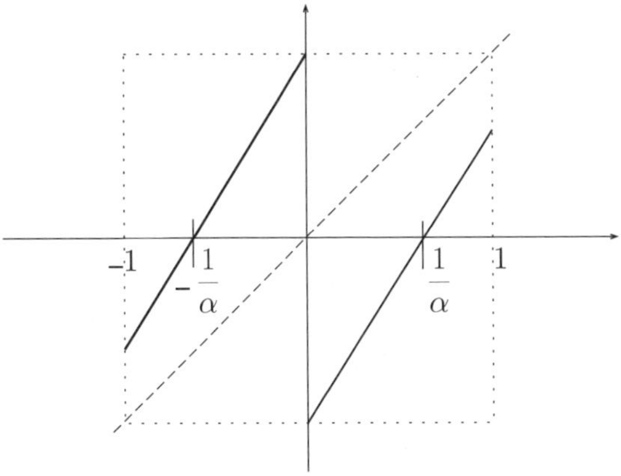

Fig. 2. Closed loop dynamics (3) of the scalar system with binary feedback in the case $\beta = 1$.

Proposition 1. *The closed loop system (3) admits an invariant interval if and only if $\alpha < 2$.*

Proof As seen in Figure 2, here are two fixed points of the mapping (3): $-\beta/(\alpha-1), \beta/(\alpha-1)$. In studying this closed loop transition function, we note that $\alpha < 2$ if and only if $\beta < \beta/(\alpha-1)$. If $\alpha < 2$, an easy calculation shows that the interval $[-\beta/(\alpha-1), \beta/(\alpha-1)]$ is invariant. (There are other invariant intervals too in this case.) Suppose on the other hand that there is an invariant interval. This can only happen if $\beta \leq \beta/(\alpha-1)$, because otherwise there is a subinterval of points containing the origin (specifically

$$\left(\frac{-\beta}{\alpha-1}(1-\frac{2}{\alpha}), \frac{\beta}{\alpha-1}(1-\frac{2}{\alpha})\right)\ )$$

such that no bounded trajectory of (3) can enter the interval. But the set of points in $(-\beta/(\alpha-1), \beta/(\alpha-1))$ which are initial points of trajectories entering this neighborhood can be shown to be dense. On the other hand, any trajectory started outside $[-\beta/(\alpha-1), \beta/(\alpha-1)]$ is clearly unbounded. This proves our statement. □

This result that (3) is boundable using a two-element control set $U = \{-1, 1\}$ if and only if $\alpha < 2$ is a special case of our more general result. Specifically, since $\alpha = e^{ah}$, the inequalities $\alpha < 2$ and $ah \log_2 e < 1$ are equivalent. The second more clearly expresses the bound which the product of the (right half-plane) pole magnitude and the sampling interval must satisfy in order for the system to be boundable using binary control. The main result of the section is the following:

Theorem 1. *Consider a first order system $G(s) = b/(s-a)$ whose sampled realization (1) specializes to*

$$x(k+1) = \alpha\, x(k) + \beta\, u(k), \tag{4}$$

where

$$\alpha = e^{ah}, \quad \text{and} \quad \beta = \frac{b}{a}(e^{ah} - 1).$$

If the channel capacity R is greater than $a \log_2 e$, then we may chose a sampling interval h and a set U of admissible control values such that $|U| > e^{ah}$ and such that the resulting feedback control implementation is boundable.

Remark 1. Control information versus control authority. An interesting aspect of the example and the main theorem is that the conditions for boundability and the existence of invariant intervals do not depend on β. They depend only on the amount of information that can be communicated between the controller and plant in one unit of time.

Remark 2. Remarks on the channel capacity conditions. *(i)* Theorem 1 was announced in [2] and is proved in [3]. The proof is constructive and involves the explicit construction of the set U of control values together with a feedback law which renders a compact set invariant under motions of the system (4). As indicated in the above example, the condition $a \log_2 e < R$ is also essentially necessary because the cardinality of the set U must be larger than e^{ah} in order to guarantee bounded motions of (4). The channel capacity R must in turn be large enough to transmit enough bits of data to uniquely identify control values in each time interval h. *(ii)* The theorem actually represents a crude lower bound on the amount of channel capacity one would like to have to implement feedback control of a physical system. It provides only a condition for the existence of a bounded response. To address the myriad standard control design issues such as rise time, gain and phase margins, etc., we shall need to have the capacity to transmit a great deal more data through the channel. Questions of this type involve coding strategies and geometric aspects of rate distortion theory. These matters are discussed in [3].

Remark 3. Design of the control value set. In our general formulation of the problem of quantized feedback control, we have assumed the set of admissible control values is symmetric about the origin: $U = \{-d_n, -d_{n-1}, \dots, -d_1, d_1, \dots, d_n\}$ (where the elements are listed in order of increasing magnitude). In real-time digital control implementations, an interesting design question is whether the control levels d_i should be evenly spaced. (This question comes up, for instance, in choosing whether to use integer [evenly spaced] or floating point arithmetic [logarithms of floating point numbers are evenly spaced].) Suppose we focus the discussion on the case in which a standard constant gain feedback design, $u = u(x) = -kx$ is being implemented in terms of our quantized finite set of control values. If we choose a control set U corresponding to the bounding case in which $|U| \sim e^{ah}$ in Theorem 1, one is then forced to take the control values to be more or less evenly spaced. Stable control with logarithmic spacing of the control values will thus require higher channel capacity in the feedback loops.

3 The Case of Higher Order and Multivariable Systems

We begin this section by noting that LTI systems with quantized feedback can typically be locally but not globally stabilized. This is not surprising since for state values of large magnitude, the feedback signal will be saturated at a control level (in U) of maximal magnitude. We note that for the system representation presented in the previous section, the domain to which bounded feedback trajectories are confined may be computed explicitly. This is made precise in the following.

Proposition 2. Local stabilization by quantized feedback. *For any feedback law implemented using quantized control with a finite set U of control values, the closed loop system is only locally stable in the sense that for the i-th open-loop unstable mode (associated with state component x_i), there is a compact interval containing the origin such that any trajectory of (2) whose i-th component leaves this interval must be unbounded.*

Proof Let $f : \mathbb{R}^n \to U$ be any selection function. The i-th component of the system (2) evolves according to the state transition equation

$$x_i(k+1) = \alpha_i\, x_i(k) + \beta_i\, f(x(k)),$$

where α_i is the i-th diagonal entry in the diagonal matrix F and β_i is the i-th entry in Γ. Assume (without loss of generality) that as x ranges over all of $\mathbb{R}^n$, $f(x)$ takes on all possible values in U. If $N = |U|$ (the cardinality of U), then the transition $x_i(k) \mapsto x_i(k+1)$ is described by a mapping of the form

$$x_i(k+1) = \alpha_i\, x_i(k) + \beta_i\, u,$$

where $u \in U$ is such that $u = f(x(k))$. Let $u_{\max}$ be the largest element in U and $u_{\min}$ be the least element in U. There is no loss of generality in assuming $u_{\min} < 0 < u_{\max}$. A simple calculation shows that the mapping $x \mapsto \alpha_i x + \beta_i u_{\min}$ has a fixed point given by $x_{fp} = -\beta_i u_{\min}/(\alpha_i - 1)$.

Note that after m transitions,

$$\begin{aligned} x_i(k+m) &= \alpha_i^m\, x_i(k) + \alpha_i^{m-1}\beta_i u(1) + \alpha_i^{m-2}\beta_i u(2) + \cdots + \beta_i u_{\min} \\ &\geq \alpha_i^m\, x_i(k) + \alpha_i^{m-1}\beta_i u_{\min} + \cdots + \beta_i u_{\min}. \end{aligned}$$

But the right hand side of this inequality is the result of iterating the mapping

$$x \mapsto \alpha_i\, x + \beta_i u_{\min} \tag{5}$$

n times with initial condition $x_i(k)$. Suppose $x_i(k) > x_{fp}$. Then under the mapping (5),

$$\begin{aligned} x_i(k+1) - x_{fp} &= \alpha_i\, x_i(k) + \beta_i u_{\min} - x_{fp} \\ &= \alpha_i\, x_i(k) + \beta_i u_{\min} - (\alpha_i\, x_{fp} + \beta_i u_{\min}) \\ &= \alpha_i (x_i(k) - x_{fp}), \end{aligned}$$

and a simple inductive argument shows that

$$x_i(k+m) - x_{fp} = \alpha_i^m (x_i(k) - x_{fp}).$$

This tends to ∞ as $m \to \infty$, and in turn implies any trajectory of (2) with $x_i(k) > x_{fp}$ is unbounded in the limit as $k \to \infty$. □

Remark 4. The above proof provides a bit more information than explicitly stated in Proposition 2. for each unstable mode, α_i, there are two fixed points, associated with $u_{\min}$ and $u_{\max}$ as indicated. These define explicit and sharp bounds on the domain in which bounded motions of (2) with feedback f are possible.

We have remarked that the channel capacity bound of Theorem 1 provides a crude lower bound guaranteeing the existence of bounded motions in the idealized case of first order systems. For higher order systems, the requirements on channel capacity are more severe if there is more than one right half-plane pole in the open loop (uncontrolled) system. The following result presents conditions under which bounded motions are assured for a system evolving in $\mathbb{R}^n$.

Theorem 2. *Consider the constant coefficient linear control system*

$$\dot{x}(t) = Ax(t) + bu(t) \tag{6}$$

where A is an $n \times n$ matrix and b is $n \times 1$. (Thus u is a scalar input.) Assume A has distinct real eigenvalues and (A, b) is a controllable pair. Let

$$\tau = e^{\lambda(A)_{\max} h},$$

where $\lambda(A)_{\max}$ is the largest eigenvalue of A. Then if

$$\log_2 \tau < 1/n,$$

there is a finite set U of control values such that the sampled version of (3) is boundable.

Proof The sampled system corresponding to (3) is

$$x(k+1) = Fx(k) + \Gamma u(k),$$

where $F = e^{Ah}$ and $\Gamma = A^{-1}(e^{Ah} - I)b$. Let U be a (finite) set of control values. A set K is invariant if for any $x \in K$, there is a control sequence $u_1, \ldots, u_m \in U$ such that

$$F^m x + F^{m-1}\Gamma u_1 + \cdots + \Gamma u_m \in K.$$

This is equivalent to

$$x \in F^{-m}\big(K - F^{m-1}\Gamma u_1 - \cdots - \Gamma u_m\big),$$

for some m; $u_1, \ldots, u_m \in U$, or

$$K \subseteq \bigcup_{m\geq 1} \bigcup_{u_1,\ldots,u_m\in U} [F^{-m}K - F^{-1}\Gamma u_1 - \cdots - F^{-(m-1)}\Gamma u_m]. \tag{7}$$

Under the hypothesis of the theorem, it is not difficult to show that the unit hypercube, $K = [0,1]^n$, can be made to be invariant by proper choice of control set U. Since $\lambda_{\max}(F) < 2^{1/n}$, $\lambda_{\min}(F^{-1}) > 2^{-1/n}$. From this it follows that

$$\mathrm{vol}(F^{-n}(K) > \frac{1}{2^n}\mathrm{vol}(K).$$

(Since both K and $F^{-n}(K)$ are rectangular, and the length of a side of $F^{-n}(K)$ $> 1/2$ the length of a side of K, the result is straightforward.) From this, it follows that we may chose 2^n control values such that equation (7) is satisfied. This proves the theorem. □

Remark 5. Remarks on quantized control of higher order systems. This result remains fairly constructive but appears to be somewhat conservative vis-a-vis other recently published bounds. Indeed, if

$$\log_2[e^{\lambda(A)_{\max}h}] < \frac{1}{n},$$

then it is straightforward to show

$$\sum_{i=1}^{n} \max\{0, \log_2 |e^{\lambda_i(A)}|\} < R.$$

(Cf. [8],[9].) In the next section we shall pursue our constructive analysis to examine an interesting case in which

$$\frac{1}{n} < \log_2[e^{\lambda(A)_{\max}h}] < 1.$$

Example 2. The inverted micropendulum. For control systems with only a single unstable pole, the results of the previous section apply. An interesting example is provided by a slight modification of the classical inverted pendulum example. Recently, using fundamentally nonlinear methods, we have stably balanced some very small ($\frac{1}{8}$-inch) pendulums. It is interesting to think about trying to balance pendulums of this size using digital implementations of classical feedback methods. Recall that for the inverted pendulum there is one right half-plane pole whose magnitude is roughly $\sqrt{g/\ell}$. Taking $g = 10$ (meters/sec.^{2}) and $\ell = 0.001$ (meters), we can use the bound of Theorem 1 to estimate the minimum channel capacity needed to implement this control. In this case $a = \sqrt{g/\ell} = 100$, and thus the minimum channel capacity required is $100 \log_2 e = 144.2695041...$ bits per second. As remarked in the previous section, this is a tight bound, but it is one which gives only a conservative estimate of the data capacity that is needed to support a moderately sophisticated control law (as prescribed by standard LQR or H^∞ techniques). Moreover, if sophisticated data encoding is used in the communications link between the plant and controller, the required channel capacity will be even higher. While a data rate of even 1000 bits/second is well within the bounds of even the slowest telephone modems, it suggests that there could be limits in trying to multiplex a large number of devices using a noisy channel such as might be encountered in a MEMS array.

4 The Concept of *Attention* and Binary Control of Multivariable Systems

We shall continue to assume, as in Section 2, that the communications bandwidth constraint affecting our system is localized on the link between the controller H and the plant G (depicted in Figure 1). There are several component parts to solving the joint feedback control and communication problem:

1. Choose an appropriate finite set U of admissible control values;
2. Implement a selection function $f : \mathbb{R}^n \to U$ which chooses an element of U for each state of the system;
3. Encode these choices of control elements for transmission from the controller H to the plant G.

All choices must be consistent with the data-rate bounds discussed in the preceding sections, and the tradeoffs that may be required to stay within these bounds. In general, it is desirable to choose the finite set U so that it provides a finely quantized partition of a real interval of control values. A fine quantization will require the cardinality of U to be large, however, and as a consequence, any encoding will require longer codewords or larger data packets. We refer to [10] for further details.

These three steps constitute the general procedure in solving the *coding problem* for information-based digital control. We may think of 1. and 2. together as constituting the *source coding* step while 3. involves *channel coding*.

Optimal encoding for information-based control is not yet well understood. In the present section, we shall treat the source coding problem using binary control (i.e. control using a 2-element control set) of (1). This represents the coarsest possible quantization of the control values, and it remains an open question as to whether being within the data rate bounds that have appeared in the literature is sufficient for stabilizing an open-loop unstable system using binary control. The general case will be treated elsewhere, but some of the issues involved may be framed in terms of 2-d systems of the form

$$\begin{pmatrix} x(k+1) \\ y(k+1) \end{pmatrix} = \begin{pmatrix} \alpha_1 & 0 \\ 0 & \alpha_2 \end{pmatrix} \begin{pmatrix} x(k) \\ y(k) \end{pmatrix} + \begin{pmatrix} 1 \\ 1 \end{pmatrix} u(k). \tag{8}$$

We assume $\alpha_1,\ \alpha_2 > 0$, and $u(k) \in \{-1, 1\}$ for $k = 1, 2, \ldots$. Our purpose is to consider *selection functions* $f : \mathbb{R}^2 \to U$ which implement stabilizing (bounding) control laws according to

$$u(k) = f(x(k)). \tag{9}$$

Remark 6. It is useful to keep in mind the closed-loop dynamics function depicted in Figure 2. This represents the qualitative form of the switching law which we would always use to choose control values if the states x and y could be controlled independently rather than being coupled through a common control channel. The challenge in control design in this setting is to choose binary sequences which will effectively resolve the conflicts that arise in trying to bound the two states simultaneously. The following result shows that there is no chance of resolving this conflict if the system (8) is not controllable.

Proposition 3. *If the system (8) is open-loop unstable (i.e. has open-loop pole outside the unit circle) and is not controllable, then it is not boundable.*

Proof Let x_0, y_0 be any two initial conditions with $x_0 \neq y_0$. The system (8) is controllable if and only if $\alpha_1 \neq \alpha_2$. Hence, under the hypothesis, $\alpha_1 = \alpha_2$. Both x and y must evolve according to the input sequence $u(1), u(2), \ldots$. Then writing $e(k) = x(k) - y(k)$, we have

$$e(k+1) = \alpha_1\, e(k).$$

The assumption that (8) is open-loop unstable implies $\alpha_1 > 1$, and hence $|e(k)| \to \infty$ as $k \to \infty$. □

How a binary feedback control law divides its *attention* among system components. In [4], Brockett introduced the notion of *attention* in control. This is a quantitative measure of the magnitude and rate of change of the control law operating in accordance with design specifications. In the context of the discrete-time control systems with quantized (finite sets of) control values studied in the preceding sections, we shall define the *attention* of a control implementation to be $(\log_2 |U|)/h$—the logarithm of the cardinality of the set of possible control values divided by the unit time interval. Defined in this way, attention has units bits-per-second. It will also be useful to be able to discuss in a quantitative way how much attention is being paid by a control law to a certain subregion of the state space. Let $S \subset \mathbb{R}^n$. We shall say that a control law specified by a selection function $f : \mathbb{R}^n \to U$ gives attention $(\log_2 |f(S)|)/h$ to the set S, where the notation $|f(S)|$ indicates the cardinality of the subset $f(S) \subset U$. We consider the way in which one choice of selection function pays attention to different parts of the state space for the system (8).

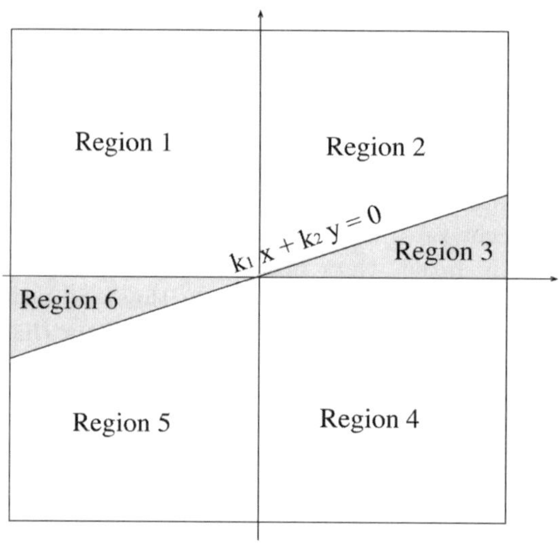

Fig. 3. This figure illustrates how the quantized (binary) realization of the feedback law $u(x) = \boldsymbol{k} \cdot x$ divides its attention between the two modes of the system. The selection function as defined in the text takes the value -1 above the line $k_1x+k_2y = 0$ and $+1$ on or below the line.

If the set of admissible control values U were an open subset of $\mathbb{R}$ containing the point 0, classical techniques would would specify choices of constant gains $\boldsymbol{k} \in \mathbb{R}^2$ such that with

$$u = u(x, y) = k_1x + k_2y,$$

the closed-loop system has $(x, y) = (0, 0)$ as an asymptotically stable rest point. The binary control which approximates the constant gain feedback is

$$u(x, y) = \begin{cases} -1 \text{ if } k_1 x + k_2 y < 0 \\ \\ 1 \text{ if } k_1 x + k_2 y \geq 0. \end{cases} \tag{10}$$

Suppose $\alpha_2 > \alpha_1 \geq 1$. Letting $u = \boldsymbol{k} \cdot \boldsymbol{x} = k_1 x + k_2 y$ be a constant-gain feedback law placing the closed-loop poles of (8) inside the unit circle, we must have $-k_2 > k_1 > 0$. The above binary realization of this law partitions the 2-dimensional state space into the six regions depicted in Figure 3. The switching rule given by u may be described in terms of these regions as follows: If the state $(x(k), y(k))$ lies in either Region 2 or Region 5, the state transition $(x(k), y(k)) \mapsto (x(k{+}1), y(k{+}1))$ prescribed by u applied to (8) coincides with what it would be if the two states x and y were independently controlled using the canonical scalar binary control law (3). When $(x(k), y(k)) \in$ Region 1 or Region 4, the control law u favors the state y in the sense that the transition $y(k) \mapsto y(k+1)$ is prescribed by the canonical binary law

$$y(k+1) = \begin{cases} \alpha_2\, y(k) + 1 \text{ if } y(k) < 0 \\ \\ \alpha_2\, y(k) - 1 \text{ if } y(k) \geq 0, \end{cases} \tag{11}$$

while the state x moves in the destabilizing direction. (I.e. $x(k) \mapsto \alpha_1 x(k) - 1$ if $x(k) < 0$ and $x(k) \mapsto \alpha_1 + 1$ if $x(k) \geq 0$.) Since the open-loop pole corresponding to the state y is farther from the unit circle than the pole for x, it is not surprising that the control law will favor a transition tending to stabilize y in Regions 1 and 4 where the objectives of stabilizing x and y are in conflict. It is somewhat counterintuitive that the feedback law never pays attention to the x-component in regions 1 and 4, because (as shown below) in order for trajectories of (8) to remain bounded, the control law cannot always favor the y-component. It is precisely in Regions 3 and 6, where the absolute value of the x-component is considerably larger than the absolute value of the y-component, that the two open-loop modes are treated equitably, and the control law relaxes its "preference" for keeping y bounded at the expense of x. It is somewhat surprising, however, that the way the control law pays attention to x in these regions is in making the transition (prescribed by (10)) move both components in the destabilizing direction. (I.e., for both x and y, the binary selection of the closed loop transition law (3) is reversed.)

To describe the operation of the prescribed selection function in terms of *attention*, consider the four canonical quadrants of $\mathbb{R}^2$ labeled in counter-clockwise S_1 through S_4, starting with the quadrant in which both x and y are positive. In terms of choosing a control value from the two-element set $U = \{-1, 1\}$, when both α_1 and $\alpha_2 > 1$, the objectives of keeping the x-

and y-components bounded are in conflict in S_2 and S_4, while the objectives can be simultaneously addressed in regions S_1 and S_2. With h normalized to be unity, the selection function associated with the feedback law depicted in Figure 3 gives attention 1 to region S_2 and region S_4 while each of S_1 and S_3 receives attention 2.

4.1 An Interesting Critical Case

We conclude by analyzing an interesting special case which illustrates what may be expected when one open-loop mode is unstable with the other being marginally stable. The choice of open loop modes $\alpha_1 = 1$ and $\alpha_2 = (1+\sqrt{5})/2 = 1.61803\ldots$ allows us to carry out some surprisingly explicit calculations regarding the qualitative dynamics and invariant sets of (8) under the control law (10). This level of explicitness appears in the proof of the following proposition, which makes contact with our older work on the ergodic behavior of "“chaotic” feedback designs, [1].

Proposition 4. *If* $\alpha_2 = (1+\sqrt{5})/2$*, the dynamical system* (11) *has the following two properties:*

(i) *Every initial condition* $y_0 \in (-1,1)$ *initiates a trajectory which takes on successive positive values at most twice and successive negative values at most twice;*
(ii) *With respect to Lebesgue measure, almost every initial condition* $y_0 \in (-1,1)$ *gives rise to a trajectory whose itinerary in the invariant interval* $(-1,1)$ *is characterized statistically by a density* $\rho(\cdot)$ *such that for every* $-1 < a < b < 1$*, the fraction of time the trajectory has spent in the subinterval* $[a,b]$ *is given by* $\int_a^b \rho(s)\,ds$.

Proof (i) The closed loop dynamics are given by successive iterations of the function whose graph is depicted in Figure 2. The subinterval $[0,1]$ is the union of subintervals $[0,1/\alpha]$ and $[1/\alpha,1]$. The subinterval $[0,1/\alpha]$ is mapped in a one-one fashion onto the interval $[-1,0]$, and the subinterval $[1/\alpha,1]$ is mapped in a one-to-one fashion onto $[0,1/\alpha]$ if and only if $1/\alpha = \alpha - 1$. The solution of this equation with $\alpha > 0$ is $\alpha = (1+\sqrt{5})/2$.

(ii) Let $F(\cdot)$ be the mapping $y(k) \mapsto y(k+1)$ defined by ( ). We say that a density $\rho : [-1,1] \to [0,1]$ is *invariant* under F if

$$\int_E \rho(s)\,ds = \int_{F^{-1}(E)} \rho(s)\,ds$$

for every measurable set $E \subset [-1,1]$. Let $g(\cdot)$ be any real valued function which is measurable on $[-1,1]$. By the Birkhoff ergodic theorem, if there exists a unique density ρ which is invariant under F, then for almost all $x \in [-1,1]$

$$\lim_{k\to\infty} \frac{1}{k}\sum_{j=0}^{k-1} g(F^j(x)) = \int_{-1}^{1} g(s)\rho(s)\,ds.$$

It is not difficult to see that an invariant density for our F satisfies the functional equation

$$\rho(y) = \frac{1}{\alpha}\rho\left(\frac{y-1}{\alpha}\right) + \frac{1}{\alpha}\rho\left(\frac{y+1}{\alpha}\right), \tag{12}$$

where we have dropped the subscript on α. We look for a piecewise constant solution of the form

$$\rho(y) = \begin{cases} a_0 & -1 \le x < -\frac{1}{\alpha} \\ a_1 & -\frac{1}{\alpha} \le x < 0 \\ a_2 & 0 \le x < \frac{1}{\alpha} \\ a_3 & \frac{1}{\alpha} \le x \le 1 \\ 0 & \text{elsewhere.} \end{cases}$$

We may investigate solutions of this form by observing that under our hypothesis that $1/\alpha = \alpha - 1$ the equation (12) may be rewritten as a system of homogeneous linear equations:

$$a_0 = \tfrac{1}{\alpha}\, a_2$$

$$a_1 = \tfrac{1}{\alpha}\, a_0 + \tfrac{1}{\alpha}\, a_2$$

$$a_2 = \tfrac{1}{\alpha}\, a_3 + \tfrac{1}{\alpha}\, a_1$$

$$a_3 = \tfrac{1}{\alpha}\, a_1$$

In general, we would expect such a homogeneous system to admit only the zero solution, but for the special value of $\alpha = (1+\sqrt{5})/2$, this admits the solution $a_0 = a_3 = t$ and $a_1 = a_2 = \alpha t$ for a free parameter t. Normalizing $\rho(\cdot)$ so that $\int_{-\infty}^{\infty} \rho\, ds = \int_{-1}^{1} \rho\, ds = 1$, we determine that $t = (1+\sqrt{5})/(4\sqrt{5}) \approx 0.361803398875$. This proves statement *(ii)* of the proposition and leaves us with an explicit formula for ρ:

$$\rho(y) = \begin{cases} \frac{1+\sqrt{5}}{4\sqrt{5}} & -1 \le y < -\frac{2}{1+\sqrt{5}} \\ \frac{3+\sqrt{5}}{4\sqrt{5}} & -\frac{2}{1+\sqrt{5}} \le y < \frac{2}{1+\sqrt{5}} \\ \frac{1+\sqrt{5}}{4\sqrt{5}} & \frac{2}{1+\sqrt{5}} \le y \le 1 \\ 0 & \text{elsewhere.} \end{cases} \tag{13}$$

□

Binary Implementation of a Neutrally Stable Design Results in Unbounded Trajectories. Utilizing the high level of explicitness permitted by the special case considered in Proposition 4, it is possible to provide a detailed analysis of the way in which the feedback law (10) divides its attention between the modes of the system (8). Consider the closed-loop matrix

$$\begin{pmatrix} \alpha_1 + k_1 & k_2 \\ k_1 & \alpha_2 + k_2 \end{pmatrix} \tag{14}$$

As simple root-locus calculation shows that with with $k_1 = 0$ and $-(3+\sqrt{5})/2 < k_2 < (1-\sqrt{5})/2$, the eigenvalues of this system are 1 and something strictly less than 1 in magnitude. For the case of a non-quantized control law in which the set of admissible values of the control is all real numbers, we find that any initial value x_0 is neutrally stable, although the x-trajectory is influenced by the asymptotically stable y-trajectory. In this case, setting $k_1 = 0$ results in the control law paying no attention to x save for an essentially transient coupling to the motion of y.

There is a qualitatively different way in which the y trajectory influences the motion of x for our digital (binary) control law (10). Because the y trajectory is not asymptotically stable in the case of quantized control, it has a more persistent influence on the motion of x. Using the fact that the evolution of $y(k)$ is, in light of Proposition 4, a stationary pseudo-random process governed by the probability density function ρ defined by equation (13), we shall show that the state transitions $x(k) \mapsto x(k+1)$ are those of a denumerable Markov chain. Since the dynamics are essentially a one-dimensional random walk, the x trajectory will, with non-zero probability, have excursions which exceed any given bound.

The analysis of the coupled x, y-dynamics proceeds as follows. As remarked in the proof of Proposition 1, $[-1/(\alpha_2-1), 1/(\alpha_2-1)]$ is an invariant interval for the process (11). It is not difficult to show, using an argument similar to the proof of Proposition 2, that this is a maximal invariant interval. (I.e., it is contained in no other and contains all others as subintervals.) It is also straightforward to show that $\big(-1/(\alpha_2-1), 1/(\alpha_2-1)\big)$ is the domain of attraction for the invariant interval $[-1,1]$ (the support of the density ρ), and that any trajectory of (11) which starts in $\big(-1/(\alpha_2-1), 1/(\alpha_2-1)\big)$ enters $[-1,1]$ after finitely many steps. We describe the motion of $x(k)$ from the time that $y(k)$ enters $[-1,1]$. We thus relabel the time index so that $k=0$ corresponds to the first time that the y trajectory has entered $[-1,1]$. Also we assume, without loss of generality, that $x_0 = 0$. We shall be interested in the record of $x(k)$ for even values of k, as this is somewhat simpler to describe than the entire record. Note that with $\alpha_2 = (1+\sqrt{5})/2$, we have $1/\alpha_2 = \alpha_2 - 1$, and $y(k)$ is such that

I. If $-1 \le y(k) < -\frac{1}{\alpha_2}$, the trajectory takes two steps to the right to arrive at y(k+2);

II. If $-\frac{1}{\alpha_2} \le y(k) < \frac{1}{\alpha_2}$, the trajectory takes one step to the right and one step to the left (not necessarily in that order) to arrive at $y(k+2)$;
III. If $\frac{1}{\alpha_2} \le y(k) \le 1$, the trajectory takes two steps to the left to arrive at $y(k+2)$.

By Proposition 4, $y(\cdot)$ is an ergodic system, and according to the statistical characterization given in terms of ρ:

$$\begin{aligned}
\text{Prob}\{y(k) \in [-1, -1/\alpha_2)\} &= \textstyle\int_{-1}^{-1/\alpha_2} \rho(s)\,ds = q\ (\approx 0.138197)\\
\text{Prob}\{y(k) \in [-1/\alpha_2, 1/\alpha_2)\} &= \textstyle\int_{-1/\alpha_2}^{1/\alpha_2} \rho(s)\,ds = p\ (\approx 0.723607)\\
\text{Prob}\{y(k) \in [1/\alpha_2, 1]\} &= \textstyle\int_{1/\alpha_2}^{1} \rho(s)\,ds = q\ (\approx 0.138197).
\end{aligned}$$

Since the transitions $x(k) \mapsto x(k+2)$ are slaved to the motion of $y(k) \mapsto y(k+2)$, but take integer steps, we have the following 2-step transition probabilities:

$$\begin{aligned}
\text{Prob}\{x(k+2) = x(k) - 2\} &= q\\
\text{Prob}\{x(k+2) = x(k)\} &= p\\
\text{Prob}\{x(k+2) = x(k) + 2\} &= q
\end{aligned}$$

Adopting standard Markov chain notation, we let p_{ij} denote the transition probability that $x(k+2) = j$ given that $x(k) = i$. Then from the above it is clear that

$$p_{ij} = \begin{cases} p & \text{if } j = i,\\ q & \text{if } j = i-2 \text{ or } i+2,\\ 0 & \text{otherwise.} \end{cases}$$

Standard Markov chain methods may now be used to show that with respect to the above probabilities defined in terms of ρ, for almost any initial condition $y_0 \in [-1, 1]$ and $x_0 = 0$, there is a nonzero probability that the value $x(2k)$ will be $\ge 2j$ for any $j \le k$. Figure 4 shows a typical trajectory of the x-component of (8) with $\alpha_1 = 1$, $\alpha_2 = (1+\sqrt{5})/2$, and the binary feedback law (10) with $k_1 = 0$ and $k_2 = -1$.

A Binary Feedback Law With Compact Invariant Set. By making k_1 nonzero, and thus allowing the feedback control to devote some attention to the x component of the system, we can establish explicit bounds on the magnitude of excursions of both components of the trajectory. Indeed, for the α_i's under consideration and placing the closed loop eigenvalues of (14)

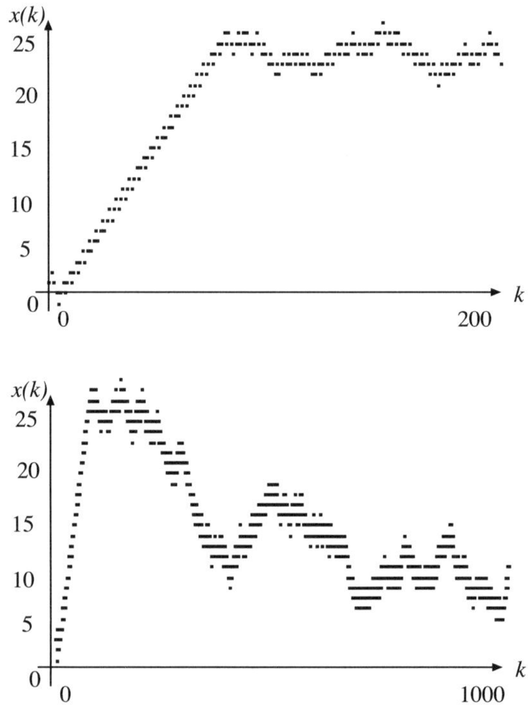

Fig. 4. The x-component trajectory of (8) (respective record lengths 200 and 1000) with $\alpha_1 = 1$, $\alpha_2 = (1+\sqrt{5})/2$, and the binary feedback law (10) with $k_1 = 0$ and $k_2 = -1$.

at 0.5 and 0.5 respectively, it is possible to explicitly construct a (compact) invariant set for (8). The precise statement is as follows.

Proposition 5. *Let* $\alpha_1 = 1$ *and* $\alpha_2 = (1+\sqrt{5})/2$, *and let* $k_1 = 1$ *and* $k_2 = -5$. *(This is the ratio of gains placing the eigenvalues of (14) at* $(0.5, 0.5)$*.) Then as depicted in Figure 5, five vertical line segments constitute an invariant set for the motion of (8) with feedback law (10) if the endpoints of these segments are specified as follows:*

$$A = \left(2, \frac{2}{5}\right) \quad B = (1, b) \quad C = \left(1, \frac{1}{5}\right) \quad D = (2, d) \quad E = (1, e) \quad F = (0, f),$$

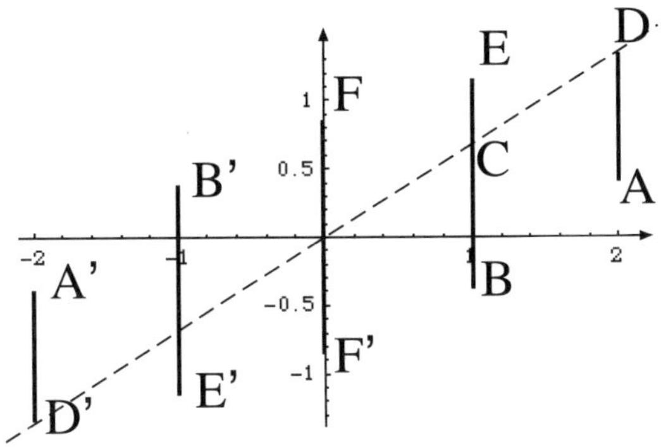

Fig. 5. This figure illustrates the invariant set of (8) under the (binary) realization of the feedback law $u(x, y) = k_1x+k_2y$, where $k_1 = 1$ and $k_2 = -5$. The coordinates of the points are given in the text. The dashed line is the locus of $u(x, y) = 0$.

where

$$\begin{aligned}
b &= \left\langle \alpha_2 \left(\alpha_2 \left\{ \alpha_2 \left[\alpha_2 \tfrac{1}{5} + 1 \right] - 1 \right\} - 1 \right) + 1 \right\rangle && \approx -0.37082 \\
d &= \alpha_2 \tfrac{1}{5} + 1 && \approx 1.32361 \\
e &= \alpha_2 \left[\alpha_2 \tfrac{1}{5} + 1 \right] - 1 && \approx 1.14164 \\
f &= \alpha_2 \left\{ \alpha_2 \left[\alpha_2 \tfrac{1}{5} + 1 \right] - 1 \right\} - 1 && \approx 0.847214,
\end{aligned}$$

with the primed symbols $A', \ldots, F'$ designating points which are the mirror images of $A, \ldots, F$ reflected in the origin.

Proof Let $\mathcal{F}$ denote the transition function of the closed loop system (8) with feedback law (10). First, note that under $\mathcal{F}$ the segment $\overline{DA}$ is mapped into the segment $\overline{EB}$. Similarly, the segment $\overline{A'D'}$ is mapped into $\overline{E'B'}$. To follow the motions of the points on the other line segments, we separately consider the subsegments above and below the (dashed) line $k_1x + k_2y = 0$. First note that $\mathcal{F}$ maps $\overline{FO}$ into the segment $\overline{E'B'}$, and by symmetry, $\overline{F'O}$ is mapped into $\overline{EB}$. $\mathcal{F}$ maps $\overline{EC}$ into the segment $\overline{FF'}$, and a similar statement could be made about the mirror images of these points. Finally, and this is where the special values of α_1 and α_2 come in, $\overline{BC}$ is mapped one-to-one onto $\overline{DA}$. This proves that the set is invariant. □

Remark 7. The special values $\alpha_1 = 1$, $\alpha_2 = (1 + \sqrt{5})/2$ allow the above construction of an invariant set to be carried out very explicitly. There are several important observations regarding the construction. First, the set is one dimensional, and while it is an attracting set for a larger set within which it is contained, this larger domain of attraction is also one dimensional (with the x-component taking only discrete integer values). For trajectories with x-components which do not have integer initial conditions, it is less straightforward to find invariant sets and corresponding domains of attraction. Numerical experiments indicate that trajectories starting within any rectangular region containing the origin will typically be unbounded if the initial value of the x-component is not close to being integer valued.

The main conclusion from these observations is that the binary encoded feedback law produces motions whose boundedness depends fragilely on initial conditions. An important open question is: How will control system performance degrade as achievable data-rates approach the minimum possible values that have been published in the literature? Our examples speak to this question only in the case of binary encoding of the feedback control laws we have discussed. The general problem of optimal coding of control laws for data-rate limited feedback channels remains a completely open question calling for future research.

5 Conclusion

Motivated by several current applications where feedback control designs need to operate using data links with limited channel capacity, we have described recent results on the minimum data rates required for the implementation of stabilizing feedback control laws. For systems with a single real right half-plane pole a, the requirement on the channel capacity R is that

$$R > a \log_2 e.$$

This is a fundamental minimum associated with the system, and the data rate must satisfy this inequality in order to implement any control law which produces a bounded response in the system. To study the rate required for information processing in a particular real-time control implementation, we have revisited a figure of merit for control design called the control law's *attention*, which was proposed and originally studied by Brockett ([4]). The *attention* of a control law is a measure of the amount of information that must be processed per unit time to produce desired system performance while operating in a given region of the state space. While the minimum data rate is prescribed in terms of the open-loop dynamics of the control system (and is thus independent of the choice of control design), *attention* is calculated in terms of a specific encoding of the feedback design.

One of the main contributions of the paper has been the examination of strategies for encoding control signals. Special attention has been given to binary realizations, wherein feedback laws are implemented using only two control levels (as would be the case using a 1-bit D/A converter). It has been shown that irrespective of the data rate, controllability of the underlying system is a necessary condition for a binary feedback realization to produce bounded trajectories in a system which is open loop unstable.

Whether controllability (together with the data rate inequality being satisfied) is also sufficient has been shown to be a delicate question. It depends both on how near the system is to operating at its minimum data rate as well as on the encoding of the feedback law. This has been illustrated for a system with a two dimensional state space and a binary realization of a classical feedback design. The design provides a neutrally stable response in its classical analogue implementation, but for almost all (Lebesgue measure) initial conditions leads to unbounded trajectories in response to the corresponding binary realization. For the same system, there is an alternative feedback design which provides the first example of which we are aware of a system with state space dimension greater than one for which a binary control law produces bounded trajectories confined to a compact invariant set.

References

1. Baillieul, J. (1985) Chaotic dynamics and nonlinear feedback control, in Proceedings of the XVI Banach Center Semester on Optimal Control. Banach Center Publications, **14**, 17–34.
2. Baillieul, J. (1999) Feedback designs for controlling device arrays with communication channel bandwidth constraints, ARO Workshop on Smart Structures, Penn. State University, August 16–18.
3. Baillieul, J. (2002) Feedback coding in information-based control," Boston University Preprint.
4. Brockett, R. W. (1997) Minimum attention control, in Proceedings of the 36th IEEE Conference on Decision and Control, Sand Diego, CA, December 10–12, 2628–2632.
5. Friedland, B. (1986) Control System Design: An Introduction to State-Space Methods, McGraw-Hill, New York.
6. Moyne, J. R., Lian, F.-L., and Tilbury, D. M. (2001) Performance evaluation of control networks: Ethernet, controlnet, and cevicenet, IEEE Control Systems Magazine, February, 66–83.
7. Nair, G. N. and Evans, R. J. (2000) Stabilization with data-rate-limited feedback: Tightest attainable bounds, Systems and Control Letters, **41**(1), 49–56.
8. Nair, G. N. and Evans, R. J. (2001) Exponential stabilizability of multidimensional linear systems with limited data rates, in Proceedings of the IFAC World Congress, July, 2002, to appear.
9. Tatikonda, S. and Mitter, S. K. (2001) Control under communication constraints, preprint.
10. Wong, W. S. and Brockett, R. W. (1999) Systems with finite communications bandwidth constraints II: Stabilization with limited information feedback, IEEE Trans. Automatic Control **44**, 1049–1053.

Ergodic Control Bellman Equation with Neumann Boundary Conditions

Dedicated to Professor Tyrone Duncan for his 60th anniversary

Alain Bensoussan[1] and Jens Frehse[2]

[1] University Paris-Dauphine and CNES, France
[2] Institut für Angewandte Mathematik der Universität Bonn, Germany

1 Abstract

Let O be an open bounded smooth domain of $\mathbb{R}^n$, and let $\Gamma = \partial O$ be its boundary. We denote by n the normal vector at the boundary Γ, oriented towards the outside of O. Let us consider the canonical process

$$\begin{aligned} \Omega &= C^0([0,\infty);\bar{O}), \\ y(t,\omega) &\equiv \omega(t), \qquad \text{if} \qquad \omega \in \Omega, \\ \mathfrak{F}^t &= \sigma(y(s),\ 0 \le s \le t). \end{aligned}$$

For any $x \in \bar{O}$, there exists a probability measure P^x on Ω, a standard Wiener process in $\mathbb{R}^n$, $w(t)$, such that $w(t)$ is an $\mathfrak{F}^t$ martingale, an increasing adapted process $\xi(t)$ such that

$$\begin{aligned} y(t) &= x + w(t) - \int_0^t \mathbf{1}_\Gamma(y(s)) n(y(s)) d\xi(s) \\ &\mathbf{1}_O(y(t)) d\xi(t) = 0, \qquad \forall t, \text{ a.s. } . \end{aligned} \tag{1}$$

Such a measure is uniquely defined. The processes $\xi(t)$ and $w(t)$ also uniquely defined.

Let next be functions

$$g(x,v),\ f(x,v) : \bar{O} \times V \to \mathbb{R}^n,\ R \tag{2}$$

measurable where $V \subset R^k$.

An admissible control is a process $v(t)$, which is adapted with values in V, and such that, if we consider the stochastic process

$$g(t) = g(y(t),\ v(t)), \tag{3}$$

and

$$\beta(t) = \exp\left\{ \int_0^t g(s) \cdot dw(s) - \frac{1}{2}|g(s)|^2 ds \right\}, \tag{4}$$

B. Pasik-Duncan: Stochastic Theory and Control, LNCIS 280, pp. 59–71, 2002.

then $\beta(t)$ is an $\mathfrak{F}^t$, P^x martingale. We can then define the change of probability measure

$$\left.\frac{dP^{x,v}}{dP}\right|_{\mathfrak{F}^t} = \beta(t), \tag{5}$$

and the process

$$w^{x,v}(t) = w(t) - \int_0^t g(s)ds. \tag{6}$$

On Ω, $P^{x,v}$, $\mathfrak{F}^t$, the process $w^{x,v}(t)$ is a standard Wiener process and $y(t)$ appears as the solution of

$$y(t) = x + \int_0^t g(s)ds + w^{x,v}(t) - \int_0^t \mathbf{1}_\Gamma(y(s))n(y(s))d\xi(s). \tag{7}$$

Let us now consider the process

$$f(t) = f(y(t), v(t)), \tag{8}$$

and we assume that

$$E\int_0^T |f(t)|dt < \infty, \qquad \forall T. \tag{9}$$

We define for any $x, v(.)$ admissible control the cost function

$$J^{x,v(.)} = \lim_{T\to\infty} \frac{1}{T} E^{x,v(.)} \int_0^T f(t)dt, \tag{10}$$

and the ergodic control consists in minimizing $J^{x,v(.)}$. The problem will be solved by the Bellman equation approach. Namely we look for a pair (z,ρ) where

$$z \in W^{2,s}(O), \quad s \gtrsim \frac{n}{2}, \quad \rho \text{ scalar}, \tag{11}$$

$$\begin{aligned} -\frac{1}{2}\Delta z + \rho = H(x, Dz) \quad &\text{in } O, \\ \left.\frac{\partial z}{\partial n}\right|_\Gamma = 0, \end{aligned} \tag{12}$$

where $H(x,p)$, called the Hamiltonian, is defined by

$$H(x,p) = \inf_{v\in V}\{f(x,v) + p\cdot g(x,v)\}. \tag{13}$$

Suppose that $\hat{v}(x)$ is a measurable solution which satisfies

$$H(x, Dz(x)) = f(x, \hat{v}(x)) + Dz(x) \cdot g(x, \hat{v}(x)), \tag{14}$$

a.e. and let

$$\hat{v}(t) = \hat{v}(y(t)). \tag{15}$$

Suppose $\hat{v}$ is admissible, then it is easy to check that

$$\rho = J^{x,\hat{v}(.)} = \inf_{v(.)} J^{x,v(.)}. \tag{16}$$

So the problem is to solve (12). We shall make the following assumption on the Hamiltonian

$$H(x,p) \text{ is a Caratheodory function }, \tag{17}$$

$$H(x,p) \geq -\lambda + \lambda_0 |p|^2, \tag{18}$$

$$H(x,p) \leq \bar{\lambda} + \bar{\lambda}_0 |p|^2 \tag{19}$$

with $\lambda,\ \bar{\lambda} \geq 0,\ \lambda_0,\ \bar{\lambda}_0 > 0$.

Note, that the function z is defined up to an additive constant. An other possibility of assumptions is the following:

$$-\bar{\lambda} - \bar{\lambda}_0 |p|^2 \leq H(x,p) \leq \lambda - \lambda_0 |p|^2. \tag{20}$$

2 Preliminaries

2.1 Green Function for the Neumann Problem

Consider the following equation

$$\begin{aligned} -\frac{1}{2}\Delta G + G &= \delta(x-\xi), \\ \left.\frac{\partial G}{\partial n}\right|_{\Gamma} &= 0, \end{aligned} \tag{21}$$

where ξ is a point in O. The solution of (2) is denoted by G^{ξ} and is the Green function of the operator $-(1/2)\Delta + I$, with Neumann boundary conditions. The existence of a solution of (2) (necessarily unique) can be shown by smoothing the Dirac measure on the right hand side. We want to show here that some of the basic properties of Green functions for Dirichlet problems carry over to this case. We have the

Proposition 1. *The solution of (2) satisfies*

$$\|G^\xi\|_{W^{1,\mu}(O)} \leq C, \qquad \forall 1 \leq \mu < \frac{n}{n-1}, \quad \forall \xi, \tag{22}$$

$$\|G^\xi\|_{L^\nu(O)} \leq C, \qquad \forall 1 \leq \mu < \frac{n}{n-2}, \quad \forall \xi. \tag{23}$$

Proof. We prove the a priori estimate, so the proof is formal. The rigorous proof goes through the approximation of (2) in which the Dirac measure is replaced by

$$\delta^h = \frac{\mathbf{1}_{B_h}}{|B_h|},$$

where B_h is a ball of radius h, centered in the singularity ξ. We test (2) with

$$\frac{G}{(1+G^s)^{\frac{1}{s}}}, \qquad 0 < s < 1,$$

which is smaller than 1, so we get

$$\int_O \frac{1}{2} DGD\Big[\frac{G}{(1+G^s)^{\frac{1}{s}}}\Big] dx + \int_O \frac{G^2}{(1+G^s)^{\frac{1}{s}}} dx \leq 1,$$

hence

$$\frac{1}{2}\int_O \frac{|DG|^2}{(1+G^s)^{\frac{1+s}{s}}} dx + \int_O \frac{G^2}{(1+G^s)^{\frac{1}{s}}} dx \leq 1. \tag{24}$$

Using the fact that for $0 < s < 1$,

$$\frac{1}{1+G} \leq \frac{2^{\frac{1-s}{s}}}{(1+G^s)^{\frac{1}{s}}},$$

it follows that

$$\frac{1}{2}\int_O \frac{|DG|^2}{(1+G)^{1+s}} dx + \int_O \frac{G^2}{(1+G)} dx \leq 2^{\frac{1-s}{s}}. \tag{25}$$

Therefore, as easily seen,

$$\|(1+G)^{\frac{1-s}{2}}\|_{H^1(O)} \leq (2 + 2^{\frac{1-s}{s}})^{\frac{1}{2}}.$$

Hence from Sobolev imbedding it follows

$$\|(1+G)^{\frac{1-s}{2}}\|_{L^{\frac{2n}{n-2}}(O)} \leq C_s, \tag{26}$$

which yields (4). Next, set

$$\beta = \frac{n(1-s)}{2(n-1-s)},$$

then writing

$$\int_O |DG|^{2\beta} dx = \int_O \left| \frac{DG}{(1+G)^{\frac{1+s}{2}}} \right|^{2\beta} (1+G)^{(1+s)\beta} dx,$$

and using Hölder's inequality as well as (25) and (26), we obtain

$$\int_O |DG|^{\frac{n(1-s)}{n-1-s}} dx < C_s, \tag{27}$$

which implies (2).

2.2 Approximation

We consider as an approximation the problem

$$-\frac{1}{2}\Delta u_\varepsilon + \varepsilon u_\varepsilon = H(x, Du_\varepsilon), \tag{28}$$

$$\left.\frac{\partial u_\varepsilon}{\partial n}\right|_\Gamma = 0, \tag{29}$$

which has a smooth solution $u_\varepsilon \in W^{2,s}(O)$, $\forall 2 \leq s < \infty$, see [3]. By classical maximum principle arguments, we get

$$-\lambda \leq \varepsilon u_\varepsilon \leq \bar{\lambda}. \tag{30}$$

Moreover testing with 1, and using the assumption (6) yields

$$\int_O |Du_\varepsilon|^2 dx \leq \frac{\lambda + \bar{\lambda}}{\lambda_0}. \tag{31}$$

Let θ_ε be a number such that

$$\begin{aligned} &\text{Meas}\{x | u_\varepsilon - \theta_\varepsilon \geq 0\} \geq \frac{1}{2} \text{Meas}\, O, \\ &\text{Meas}\{x | u_\varepsilon - \theta_\varepsilon \leq 0\} \geq \frac{1}{2} \text{Meas}\, O. \end{aligned} \tag{32}$$

From Poincare's inequality and (31) it follows that the function

$$z_\varepsilon = u_\varepsilon - \theta_\varepsilon \tag{33}$$

satisfies

$$\|z_\varepsilon\|_{H^1(O)} \leq C, \tag{34}$$

and z_ε satisfies, from (29),

$$-\frac{1}{2}\Delta z_\varepsilon + \varepsilon z_\varepsilon + \rho_\varepsilon = H(x, Dz_\varepsilon),$$
$$\left.\frac{\partial z_\varepsilon}{\partial n}\right|_\Gamma = 0 \tag{35}$$

where $\rho_\varepsilon = \varepsilon\theta_\varepsilon$.

3 Main Result

3.1 Statement of the Result

Our objective is to prove the following

Theorem 1. *We make the assumption (17), (18), (19). Then considering a subsequence z_ε, ρ_ε such that*

$$z_\varepsilon \rightharpoonup z \text{ in } H^1 \text{ weakly}, \tag{36}$$

$$\rho_\varepsilon \to \rho. \tag{37}$$

The pair (z, ρ) is necessarily a solution of (12). Moreover $z \in W^{2,s}(O)$, $\forall 2 \leq s < \infty$.

Periodic boundary conditions were considered in [1].
We begin by proving some additional estimates.

Lemma 1. *We have the estimate*

$$\|z_\varepsilon\|_{L^q(O)} \leq C_q, \qquad \forall\, 2 \leq q < \infty. \tag{38}$$

Proof. We proceed by induction. The result (12) is of course true with $q = 2$. We test (35) with $|z_\varepsilon|^q$, and obtain

$$\frac{q}{2}\int_O |Dz_\varepsilon|^2 |z_\varepsilon|^{q-2} z_\varepsilon dx + \int_O \varepsilon u_\varepsilon |z_\varepsilon|^q dx \geq$$
$$\geq \int_O (-\lambda + \lambda_0 |Dz_\varepsilon|^2)|z_\varepsilon|^q dx.$$

Hence

$$\frac{q}{2}\int_O |Dz_\varepsilon|^2 |z_\varepsilon|^{q-2} z_\varepsilon dx + (\bar{\lambda} + \lambda)\int_O |z_\varepsilon|^q dx \geq \lambda_0 |Dz_\varepsilon|^2 |z_\varepsilon|^q dx.$$

Using Young's inequality, we easily deduce the following estimate:

$$\frac{\lambda_0}{2}\int_O |Dz_\varepsilon|^2|z_\varepsilon|^q dx \le (\bar{\lambda}+\lambda)\int_O |z_\varepsilon|^q dx + \frac{1}{2}\Big(\frac{q-1}{\lambda_0}\Big)^{q-1}. \tag{39}$$

From Poincaré's inequality, it follows, since z_ε^+ and z_ε^- vanish on a set of measure larger than $\frac{1}{2}|O|$,

$$\begin{aligned}\Big(\int_O (z_\varepsilon^+)^{\frac{(q+2)n}{n-2}} dx\Big)^{\frac{n-2}{n}} &\le c\int_O |D(z_\varepsilon^+)^{\frac{q+2}{2}}|^2 dx \le \\ &\le \int_O |z_\varepsilon|^q\,|Dz_\varepsilon|^2 dx,\end{aligned}$$

and also

$$\Big(\int_O (z_\varepsilon^-)^{\frac{(q+2)n}{n-2}} dx\Big)^{\frac{n-2}{n}} \le c\int_O |z_\varepsilon|^q\,|Dz_\varepsilon|^2 dx,$$

therefore also

$$\Big(\int_O |z_\varepsilon|^{\frac{(q+2)n}{n-2}} dx\Big)^{\frac{n-2}{n}} \le c\int_O |z_\varepsilon|^q\,|Dz_\varepsilon|^2 dx, \tag{40}$$

and from (13) we deduce

$$\Big(\int_O |z_\varepsilon|^{\frac{(q+2)n}{n-2}} dx\Big)^{\frac{n-2}{n}} \le c_q\int_O |z_\varepsilon|^q dx + c_q. \tag{41}$$

This proves that if z_ε remains bounded in L^q, then it remains bounded in $L^{\frac{(q+2)n}{n-2}}$. Starting witn $q=2$, we deduce that z_ε remains bounded in L^q, for any $q\ge 2$.

3.2 L^∞ Estimate

We want to show the crucial estimate

Lemma 2. *We have*

$$\|z_\varepsilon\|_{L^\infty} \le c. \tag{42}$$

Proof. Using (3), we deduce from (35)

$$-\frac{1}{2}\Delta z_\varepsilon \geq -\lambda - \bar{\lambda}. \tag{43}$$

Consider the Green function G^ξ defined in (2), where $\xi \in O$. We test (43) with G^ξ and obtain

$$\frac{1}{2}\int_O Dz_\varepsilon \cdot DG^\xi dx \geq -(\lambda+\bar{\lambda})\int_O G^\xi dx.$$

On the other hand, if we test (2) with z_ε, we also obtain

$$\frac{1}{2}\int_O DG^\xi Dz_\varepsilon dx + \int_O z_\varepsilon G^\xi dx = z_\varepsilon(\xi),$$

and thus we deduce

$$z_\varepsilon(\xi) \geq -(\lambda+\bar{\lambda})\int_O G^\xi dx - \int_O z_\varepsilon G^\xi dx. \tag{44}$$

Noting, that

$$\Big|\int_O z_\varepsilon G^\xi dx\Big| \leq \|G^\xi\|_{L^\nu}\|z_\varepsilon\|_{L^\nu}, \qquad \forall \nu < \frac{n}{n-2},$$

and using (12), we deduce

$$\Big|\int_O z_\varepsilon G^\xi dx\Big| \leq C, \qquad \forall\varepsilon, \quad \forall\xi,$$

and thus we have proven the estimate from below

$$z_\varepsilon(\xi) \geq -C. \tag{45}$$

To obtain an estimate from above, we first write, using the assumption (18)

$$-\frac{1}{2}\Delta z_\varepsilon \leq (\lambda+\bar{\lambda}) + \bar{\lambda}_0|Dz_\varepsilon|^2. \tag{46}$$

We test (46) with $e^{2\lambda_0 z_\varepsilon}G^\xi$. We obtain

$$\frac{1}{2}\int_O Dz_\varepsilon DG^\xi e^{2\lambda_0 z_\varepsilon} dx \leq (\lambda+\bar{\lambda})\int_O e^{2\lambda_0 z_\varepsilon}G^\xi dx,$$

or also

$$\frac{1}{4\lambda_0}\int_O DG^\xi D\big(e^{2\lambda_0 z_\varepsilon}\big) dx \leq (\lambda+\bar{\lambda})\int_O e^{2\lambda_0 z_\varepsilon}G^\xi dx. \tag{47}$$

Testing now (2) with $e^{2\lambda_0 z_\varepsilon}$ yields

$$\frac{1}{2}\int\limits_O DG^\xi(e^{2\lambda_0 z_\varepsilon})dx + \int\limits_O G^\xi e^{2\lambda_0 z_\varepsilon} dx = e^{2\lambda_0 z_\varepsilon}. \tag{48}$$

So combining (47) and (48) yields

$$e^{2\lambda_0 z_\varepsilon(\xi)} \leq (1 + 2\lambda_0(\lambda + \bar{\lambda})) \int\limits_O G^\xi e^{2\lambda_0 z_\varepsilon} dx. \tag{49}$$

Suppose now ξ is a point of positive maximum of z_ε.
For any L, we have

$$\int\limits_O G^\xi e^{2\lambda_0 z_\varepsilon} dx = \int\limits_{O\cap\{x|z_\varepsilon > L\}} G^\xi e^{2\lambda_0 z_\varepsilon} dx + e^{2\lambda_0 L}\int\limits_O G^\xi dx \leq$$

$$\leq \|e^{2\lambda z_\varepsilon}\|_\infty \int\limits_{O\cap\{x|z_\varepsilon > L\}} G^\xi dx + e^{2\lambda_0 L}\int\limits_O G^\xi dx.$$

But

$$\int\limits_{O\cap\{x|z_\varepsilon > L\}} \leq \|G^\xi\|_{L^\nu} (\text{Meas}\{x|z_\varepsilon > L\})^{\frac{1}{\nu}} \leq$$

$$\leq \frac{1}{L}\|G^\xi\|_{L^\nu}\|z_\varepsilon\|_{L^{\nu'}} \leq \frac{C}{L}.$$

Then we deduce from (49)

$$\|e^{2\lambda_0 z_\varepsilon}\|_\infty \leq \frac{C}{L}\|e^{2\lambda z_\varepsilon}\|_\infty + Ce^{2\lambda_0 L},$$

and picking L sufficiently larger, we obtain

$$\|e^{2\lambda_0 z_\varepsilon}\|_\infty \leq C,$$

hence

$$z_\varepsilon \leq C,$$

which, in view of (45), completes the proof of (42).

3.3 Proof of Theorem 1

Since z_ε remains bounded in $H^1(O) \cap L^\infty(O)$, it is standard that picking a subsequence which converges weakly in $H^1(O)$ and $L^\infty(O)$, weak star, towards z, then in addition

$$z_\varepsilon \to z \quad \text{in} \quad H^1(O) \text{ strongly }. \tag{50}$$

It is then standard to pass to the limit in (35) and to prove that the limit (z,ρ) is a solution of (12). Moreover, regularity is standard since we know the solution is in $H^1(O) \cap L^\infty(O)$.

4 Uniqueness

In this section we prove the uniqueness of the solution z, ρ of (12) (z up to an additive constant) provided we make the additional assumption

$$|H(x,p) - H(x,p')| \leq (k + K|p| + K|p'|)|p - p'|. \tag{51}$$

We state the

Theorem 2. *We make the assumptions of Theorem 1, as well as (51), then there exists a unique solution z, ρ, where $z \in W^{2,s}(O)$, $\forall 2 \leq s < \infty$, ρ a constant (the uniqueness of z being up to an additive constant).*

Proof. Let z_1, ρ_1 and z_2, ρ_2 be two solutions. We assume

$$\rho_1 > \rho_2 \tag{52}$$

to fix the ideas. Let also

$$\xi = z_1 - z_2, \tag{53}$$

then ξ satisfies

$$\begin{aligned} -\frac{1}{2}\Delta\xi + \rho_1 - \rho_2 &= H(x, Dz_1) - H(x, Dz_2) \\ \frac{\partial \xi}{\partial n}|_\Gamma &= 0. \end{aligned} \tag{54}$$

We write

$$H(x, Dz_1) - H(x, Dz_2) = D\xi \cdot D\xi \frac{H(x, Dz_1) - H(x, Dz_2)}{|D\xi|^2}$$

and define

$$g(x) = D\xi \frac{H(x, Dz_1) - H(x, Dz_2)}{|D\xi|^2} \tag{55}$$

hence from the assumption (51), we have

$$\begin{aligned} |g(x)| &\leq \frac{|H(x, Dz_1) - H(x, Dz_2)|}{|D\xi|} \\ &\leq k + K|Dz_1(x)| + K|Dz_2(x)| \leq C \end{aligned} \tag{56}$$

from the regularity of z_1, z_2. Therefore (54) writes

$$\begin{aligned} -\frac{1}{2}\Delta\xi - g(x) \cdot D\xi + \rho_1 - \rho_2 &= 0 \\ \frac{\partial \xi}{\partial n}|_\Gamma &= 0. \end{aligned} \tag{57}$$

Suppose ξ is not a constant, it has a maximum x_0 on $\bar{O}$ which is a local maximum. Suppose first $x_0 \in O$, then necessarily $D\xi(x_0) = 0$, and by Bony maximum principle, see G. M. Troianiello [4], we have

$$\operatorname{ess} \liminf_{x \to x_0} \Delta\xi(x) \leq 0, \tag{58}$$

but from (57)

$$\operatorname{ess} \liminf_{x \to x_0} \Delta\xi(x) = 2(\rho_1 - \rho_2) > 0,$$

which contradicts (58).

Suppose now x_0 is located on Γ. By the smoothness of Γ, we may assume that there exists a ball $B_R(y)$, which is tangent to Γ at x_0, and since x_0 is a local maximum, R being small enough, so that

$$\xi(x) < \xi(x_0) \qquad \forall x \in B_R(y). \tag{59}$$

Define

$$L\xi = -\frac{1}{2}\Delta\xi - g(x) \cdot D\xi, \tag{60}$$

we have by (57)

$$L\xi \leq 0. \tag{61}$$

Set next

$$v(x) = e^{-\gamma|x-y|^2} - e^{-\gamma R^2}, \tag{62}$$

we can compute

$$Lv(x) = e^{-\gamma|x-y|^2}[-2\gamma^2|x-y|^2 + \gamma n + 2\gamma g \cdot (x-y)]. \tag{63}$$

Let $\rho < R$, we can define γ larger enough so that

$$Lv(x) \leq 0 \qquad \forall x \text{ with } \rho < |x-y| < R. \tag{64}$$

Thanks to (59), we can assert that

$$\max_{x \in \partial B_\rho(y)} \xi(x) < \xi(x_0), \tag{65}$$

hence we may find ε sufficiently small such that

$$\xi(x) - \xi(x_0) + \varepsilon v(x) \leq 0, \qquad x \in \partial B_\rho(y). \tag{66}$$

Noting that $v(x) = 0$ on $\partial B_R(y)$, we have also

$$\xi(x) - \xi(x_0) + \varepsilon v(x) \leq 0, \qquad x \in \partial B_R(y). \tag{67}$$

On the other hand, using (61) and (64), it follows

$$L(\xi - \xi(x_0) + \varepsilon v) = L\xi + \varepsilon Lv \leq 0, \tag{68}$$

on the annulus $\rho < |x - y| < R$.

Thanks to (66), (67), (68) and the maximum principle, we can assert that

$$\xi(x) - \xi(x_0) + \varepsilon v(x) \leq 0, \qquad \rho \leq |x - y| \leq R. \tag{69}$$

Apply (69) with $x = x_0 + \theta(y - x_0)$, with $0 < \theta < (R - \rho)/R$, which belongs to the annulus. We deduce, dividing by θ, and letting θ tend to 0,

$$D\xi(x_0)(y - x_0) + \varepsilon Dv(x_0)(y - x_0) \leq 0, \tag{70}$$

but $y - x_0 = -n|y - x_0|$, hence thanks to the boundary condition (57), we get

$$Dv(x_0) \cdot (y - x_0) \leq 0. \tag{71}$$

On the other hand, from (62), we have

$$Dv(x_0) \cdot (y - x_0) = 2\gamma |y - x_0|^2 e^{-\gamma |x_0 - y|^2} > 0,$$

which contradicts (71). Hence ξ is necessarily a constant, and $\rho_1 = \rho_2$. The proof has been completed.

5 Ergodic control in $\mathbb{R}^n$

Here we want to recover the result of the authors [1], concerning ergodic control in R^n. We assume that

$$f(x) - \bar{\lambda}(x) - \bar{\lambda}_0(x)|p|^2 \leq H(x,p) \leq \lambda(x) - \lambda_0(x)|p|^2 + f(x) \tag{72}$$

with

$$\begin{aligned} &\lambda_0, \frac{1}{\lambda_0}, \lambda, \bar{\lambda} \geq 0, \in L^\infty_{\text{loc}}(\mathbb{R}^n), \\ &\qquad \bar{\lambda}_0 \in L^\infty(\mathbb{R}^n), \qquad \bar{\lambda}_0 \geq 0, \\ &\qquad f \in L^\infty_{\text{loc}}(\mathbb{R}^n), \end{aligned} \tag{73}$$

$$f(x) - \bar{\lambda}(x) \to \infty \text{ as } |x| \to \infty. \tag{74}$$

Note, that for $x \in O$ any bounded domain $H(x,p)$ satisfies (20). We consider the problem

$$-\frac{1}{2}\Delta z + \rho = H(x, Dz), \qquad x \in R^n, \tag{75}$$

and we proved the following

Theorem 3. *We assume (72), (73), (74), then there exists a pair* (z, ρ) *with* $z \in W^{2,s}_{\text{loc}}$, ρ *a scalar solution of (75).*

We do not repeat the proof, which is rather long. We just claim that we can consider a sequence of smooth bounded domains, for instance

$$O_R = B_R(0)$$

the ball of center 0 and radius R, and solve

$$\begin{aligned} -\frac{1}{2}\Delta z^R + \rho^R &= H(x, Dz^R), \\ \frac{\partial z^R}{\partial n}|_{\partial B_R} &= 0, \end{aligned} \tag{76}$$

using Theorem 1. Then using the methods of [1], one can check that considering an appropriate pair z^R, ρ^R solution of (76), with $z^R \in W^{2,s}(B_R)$, then it converges in $H^s_{\text{loc}}(R^n)$ and $L^\infty_{\text{loc}(R^n)}$ weak star, towards a solution of (75).

References

1. Bensoussan, A. and Frehse, J. (1987) On Bellman equations of ergodic type with quadratic growth Hamiltonians, in Contributions to modern calculus of variations (Bologna, 1985), 13–25, Longman Sci. Tech., Harlow.
2. Bensoussan, A. and Frehse, J. (1992) On Bellman equations of ergodic control in R^n, J. Reine Angewandte Math., **429**, 125–160.
3. Ladyzhenskaya, O. A. and Ural'tseva, N. N. (1968) Linear and Quasilinear Elliptic Equations, Academic Press, N.Y.
4. Troianiello, G. M. (1987) Elliptic Differential Equations and Obstacle Problems, in The University series in Mathematics, Plenum Press, N.Y.

Regime Switching and European Options*

For Tyrone Duncan on his 60th birthday

John Buffington[1] and Robert J. Elliott[2]

[1] Deloitte-Touche, Houston, Texas, USA. jbuffington@deloitte.com
[2] University of Calgary, Calgary, Canada. relliott@ucalgary.ca

Abstract. We consider a Black-Scholes market in which the underlying economy, as modelled by the parameters and volatility of the processes, switches between a finite number of states. The switching is modelled by a hidden Markov chain. European options are priced and a Black-Scholes equation obtained.

Keywords: Option pricing, Black-Scholes equation.

1 Introduction

This paper discusses the pricing of European options in a standard Black-Scholes market comprising a bond and risky asset. However, we suppose the bank interest rate, and appreciation rate and volatility of the risky asset, depend on the state of the economy. This is modeled by a finite state Markov chain, X.

European options have been discussed in this framework by Di Masi et al. [2] and Guo [5]. However, our treatment of the European option and derivation of the characteristic function of occupation times is different.

We suppose the economic state of the world is described by a finite state Markov chain $X = \{X_t, t \geq 0\}$. In particular, there could be just two states for X representing 'good' and 'bad'.

As in [3], the state space of X can be taken to be, without loss of generality, the set of unit vectors $\{e_1, e_2, \dots, e_N\}$, $e_i = (0, \dots, 0, 1, 0, \dots, 0)' \in R^N$.

Suppose the process X is homogeneous in time and has a rate matrix A. Then if $p_t = E[X_t] \in R^N$

$$\frac{dp_t}{dt} = Ap_t.$$

As in [3] we can then show that

$$X_t = X_0 + \int_0^t AX_u du + M_t \tag{1}$$

* This work was partially supported by the Social Sciences and Humanities Research Council of Canada.

B. Pasik-Duncan: Stochastic Theory and Control, LNCIS 280, pp. 73–82, 2002.

where $M = \{M_t,\ t \geq 0\}$ is a martingale with respect to the filtration generated by X.

Our financial market has two underlying assets, a bank account and a risky asset.

We suppose the instantaneous interest rate in the bank depends on the state X of the economy, so that

$$r_t = \langle r, X_t \rangle$$

where $r = (r_1, r_2, \ldots, r_N)' \in R^N$.

Then \$1 invested at time zero in the bank becomes

$$S_t^0 = \exp\left(-\int_0^t r_u du\right) \tag{2}$$

at time t.

Similarly, suppose the rate of return $\mu = \{\mu_t,\ t \geq 0\}$ and volatility $\sigma = \{\sigma_t,\ t \geq 0\}$ depend on the state $X = \{X_t,\ t \geq 0\}$ of the economy. That is, there are vectors $\mu = (\mu_1, \ldots, \mu_N)'$ and $\sigma = (\sigma_1, \ldots, \sigma_N)' \in R^N$ such that

$$\begin{aligned} \mu_t &= \langle \mu, X_t \rangle \\ \sigma_t &= \langle \sigma, X_t \rangle. \end{aligned}$$

We suppose $\sigma_i > 0$, $i = 1, \ldots, N$. The price of the risky asset $S^1 = \{S_t^1,\ t \geq 0\}$ is then given by the dynamics

$$S_t^1 = S_0^1 + \int_0^t \mu_u S_u^1 du + \int_0^t \sigma_u S_u^1 dB_u. \tag{3}$$

Here $B = \{B_t,\ t \geq 0\}$ is a Brownian motion on $(\Omega, \mathcal{F}, P)$. Write $\{\mathcal{F}_t\}$ for the filtration generated by S.

The solution of (3) is

$$S_t^1 = S_0^1 \exp\left(\int_0^t \left(\mu_u - \frac{\sigma_u^2}{2}\right) du + \int_0^t \sigma_u dB_u^u\right).$$

The extra uncertainty introduced by the regime switching means our market is no longer complete. This question is discussed in Guo [6], where the market is completed by Arrow-Debreu securities related to the cost of switching. We shall not repeat this argument but assume we are already working under a risk neutral measure P so that the price of a European claim $g(S_T^1)$ with exercise time T is

$$E\left[\exp\left(-\int_0^T r_u du\right) g(S_T^1) \middle| \mathcal{F}_0\right].$$

In particular, the price of a European call at time $t \in [0, T]$ is

$$C(t, T, S_0^1, X_0) = E\left[\exp\left(-\int_t^T r_u du\right)(S_T^1 - K)^+ \middle| \mathcal{F}_t\right] \tag{4}$$

where

$$S_T^1 = S_t^1 \exp\left(\int_t^T \left(\mu_u - \frac{\sigma_u^2}{2}\right) du + \int_t^T \sigma_u dB_u\right). \tag{5}$$

If the processes r, μ, σ are deterministic and independent of X then

$$C(t, T, S_0) = S_t^1 N(d_1) - e^{-\int_t^T r_u du} K N(d_2) \tag{6}$$

where

$$d_1(t) = \left(\int_t^T \sigma_u^2 du\right)^{-1/2} \left[\log \frac{S^1}{K} + \left(\int_t^T r_u du + \frac{1}{2}\int_t^T \sigma_u^2 du\right)\right]$$

and

$$d_2(t) = d_1(t) - \left(\int_t^T \sigma_u^2 du\right)^{1/2}.$$

Suppose now the processes r, μ, σ depend on X so that

$$r_t = \langle r, X_t \rangle, \quad \mu_t = \langle \mu, X_t \rangle, \quad \sigma_t = \langle \sigma, X_t \rangle.$$

If we knew the trajectory of X between time 0 and time T we would know the values of

$$P_{t,T} = \int_t^T \langle r, X_u \rangle du, \quad L_{t,T} = \int_t^T \langle \mu, X_u \rangle du, \quad U_{t,T} = \int_t^T \langle \sigma, X_u \rangle^2 du.$$

That is, if $\mathcal{F}_t = \sigma\{X_u,\ 0 \le u \le t\}$

$$\begin{aligned} C(t, T, S_t^1, P_T, U_T) &= E\Big[\exp\big(-P_{t,T}\big)\big(S_T^1 - K\big)^+ \big| \mathcal{F}_T\Big] && (7)\\ &= S_t^1 N\big(d_1(t)\big) - \exp\big(-P_{t,T}\big) K N\big(d_2(t)\big) && (8) \end{aligned}$$

and

$$d_1(t) = (U_{t,T})^{-1/2}\Big[\log \frac{S_t^1}{K} + P_{t,T} + \frac{1}{2} U_{t,T}\Big]$$
$$d_2(t) = d_1(t) - U_{t,T}^{1/2}.$$

To determine the call option price in this model we must take a second expectation over the variables $P_{t,T}$ and $U_{t,T}$.

For now suppose $t = 0$. For $1 \le i \le N$ let

$$T_i = \int_0^T \langle e_i, X_u \rangle du$$

be the amount of time X has spent in state e_i up to time T. Then $T_1 + T_2 + \cdots + T_N = T$ so we can discuss just $T_1, T_2, \ldots, T_{N-1}$. Also

$$P_{0,T} = P_T = \int_0^T \langle r, X_u \rangle du = \sum_{i=1}^{N} r_i T_i = \sum_{i=1}^{N-1} (r_i - r_N) T_i + r_N T \tag{9}$$

and

$$U_{0,T} = U_T = \int_0^T \langle \sigma, X_u \rangle^2 du = \sum_{i=1}^{N} \sigma_i^2 T_i = \sum_{i=1}^{N-1} (\sigma_i^2 - \sigma_N^2) T_i + \sigma_N^2 T. \tag{10}$$

Suppose $\phi(\tau_1, \tau_2, \ldots, \tau_{N-1})$ is the density function of $(T_1, T_2, \ldots, T_{N-1})$. Then

$$\begin{aligned} &C(0, T, S_0^1, X_0) \\ &\quad = \int_0^T \cdots \int_0^T C(0, T, S_0^1, P_T, U_T) \phi(\tau_1, \tau_2, \ldots, \tau_{N-1}) d\tau_1 \ldots d\tau_{N-1}. \end{aligned}$$

We shall determine the Fourier transform of ϕ, that is, the characteristic function of $(T_1, T_2, \ldots, T_{N-1})$.

For any $\theta = (\theta_1, \theta_2, \ldots, \theta_{N-1}) \in R^{N-1}$ the characteristic function of $(T_1, T_2, \ldots, T_{N-1})$ is

$$E\left[\exp i \left(\sum_{j=1}^{N-1} \theta_j T_j \right)\right]_T = E\left[\exp i \left(\int_0^T \langle \boldsymbol{\theta}, X_u \rangle du \right)\right]$$

where $\boldsymbol{\theta} = (\theta_1, \theta_2, \ldots, \theta_{N-1}, 0) \in R^N$.

Lemma 1.

$$E\left[\exp i \left(\sum_{j=1}^{N-1} \theta_j T_j \right)\right] = \langle \exp[(A + i \operatorname{diag} \boldsymbol{\theta}) T] X_0, \mathbf{1} \rangle$$

where $\mathbf{1} = (1, 1, \ldots, 1)' \in R^N$.

Proof. Consider the R^N valued process

$$Z_t := \exp i \left(\int_0^t \langle \boldsymbol{\theta}, X_u \rangle \, du \right) X_t.$$

Then

$$dZ_t = \exp i\left(\int_0^t \langle \boldsymbol{\theta}, X_u\rangle du\right)\cdot dX_t + \exp i\left(\int_0^t \langle \boldsymbol{\theta}, X_u\rangle du\right)\cdot i\langle \boldsymbol{\theta}, X_t\rangle X_t dt$$
$$= (A + i\operatorname{diag}\boldsymbol{\theta})Z_u du + \exp i\left(\int_0^t \langle \boldsymbol{\theta}, X_u\rangle du\right) dM_t$$

where the dynamics of X are given by (1). Therefore,

$$Z_t = X_0 + \int_0^t (A + i\operatorname{diag}\boldsymbol{\theta})Z_u du + \int_0^t \exp\left(i\int_0^u \langle \boldsymbol{\theta}, X_u\rangle du\right) dM_u.$$

The final integral is a martingale, so taking expected values

$$E[Z_t] = X_0 + \int_0^t (A + i\operatorname{diag}\boldsymbol{\theta})E[Z_u]du$$

and

$$E[Z_t] = \exp[(A + i\operatorname{diag}\boldsymbol{\theta})t]\cdot X_0.$$

$E[Z_t]$ is an R^N valued process and the characteristic function

$$E\left[\exp i\left(\int_0^T \langle \boldsymbol{\theta}, X_u\rangle du\right)\right]$$

is obtained by summing its components. That is,

$$E\left[\exp i\left(\int_0^T \langle \boldsymbol{\theta}, X_u\rangle du\right)\right] = E\left[\left\langle \exp i\left(\int_0^T \langle \theta, X_u\rangle du\right) X_T, \mathbf{1}\right\rangle\right]$$
$$= \langle \exp[(A + i\operatorname{diag}\boldsymbol{\theta})T]X_0, \mathbf{1}\rangle.$$

Remark 1. When $N = 2$, that is X switches between only 2 states, the equation for the characteristic function of T_1 reduces to an ordinary differential equation. The inverse Fourier transform gives rise to a density function for T_1 in terms of Bessel processes. See Guo [5], [6].

In fact, for $X_t \in \{e_1, e_2\}$, $e_1 = (0,1)'$, $e_2 = (0,1)'$ write:

$$Z_t^1 := \exp\left(i\int_0^t \theta_1\langle e_1, X_u\rangle du\right)\langle e_1, X_t\rangle$$
$$Z_t^2 := \exp\left(i\int_0^t \theta_1\langle e_1, X_u\rangle du\right)\langle e_2, X_t\rangle.$$

Then:

$$Z_t^1 = \langle e_1, X_0\rangle + \int_0^t (i\theta_1 + a_{11})Z_u' du$$

$$- a_{22} \int_0^t Z_u^2 du + \int_0^t \exp\left(i \int_0^u \theta_1 \langle e_1, X_s \rangle ds\right) \langle e_1, dM_u \rangle$$

and

$$\begin{aligned} Z_t^2 = \langle e_2, X_0 \rangle &- \int_0^t (a_{11}) Z_u^1 du \\ &- a_{22} \int_0^t Z_u^2 du + \int_0^t \exp\left(i \int_0^u \theta_1 \langle e_1, X_s \rangle\right) \langle e_2, dM_u \rangle. \end{aligned}$$

The final integrals in Z^1 and Z^2 are martingales. Writing

$$\widehat{Z}_t^1 = E[Z_t^1], \quad \widehat{Z}_t^2 = E[Z_t^2]$$

and taking expectations we have

$$\widehat{Z}_t^1 = \langle e_1, X_0 \rangle + \int_0^t (i\theta_1 + a_{11}) \widehat{Z}_u^1 du - a_{22} \int_0^t \widehat{Z}_u^2 du \tag{11}$$

$$\widehat{Z}_t^2 = \langle e_2, X_0 \rangle - \int_0^t a_{11} \widehat{Z}_u^1 du + a_{22} \int_0^t \widehat{Z}_u^2 du. \tag{12}$$

Therefore,

$$\widehat{Z}_t^2 = e^{a_{22}t} \left(\langle e_2, X_0 \rangle - a_{11} \int_0^t e^{-a_{22}u} \widehat{Z}_u^1 du \right).$$

Substituting $\widehat{Z}_t^2$ in (11)

$$\begin{aligned} \widehat{Z}_t^1 = \langle e_1, X_0 \rangle &+ \int_0^t (i\theta_1 + a_{11}) \widehat{Z}_u^1 du - a_{22} \langle e_2, X_0 \rangle \int_0^t e^{a_{22}u} du \\ &+ a_{22} a_{11} \int_0^t e^{a_{22}u} \left(\int_0^u e^{-a_{22}s} \widehat{Z}_s^1 ds \right) du. \end{aligned} \tag{13}$$

Write $Y_t^1 = e^{-a_{22}t} \widehat{Z}_t^1$. Then from (13)

$$\frac{dY_t^1}{dt} = (a_{11} - a_{22} + i\theta_1) Y_t^1 - a_{22} \langle e_2, X_0 \rangle + a_{22} a_{11} \int_0^t Y_u^1 du.$$

Consequently

$$\frac{d^2 Y_t^1}{dt^2} - (a_{11} - a_{22} + i\theta_1) \frac{dY_t^1}{dt} - a_{22} a_{11} Y_t^1 = 0$$

with

$$Y_0^1 = \widehat{Z}_0^1 = \langle e_1, X_0 \rangle \tag{14}$$

and

$$\left. \frac{dY^1}{dt} \right|_{t=0} = (a_{11} - a_{22} + i\theta_1) \langle e_1, X_0 \rangle - a_{22} \langle e_2, X_0 \rangle. \tag{15}$$

Suppose y_1 and y_2 are the roots of the quadratic equation $y^2 - (a_{11} - a_{22} + i\theta_1)y - a_{22}a_{11}y = 0$. Then

$$Y_t^1 = c_1 e^{y_1 t} + c_2 e^{y_2 t}$$

where c_1 and c_2 are determined by the boundary conditions (14) and (15) so that $c_1 + c_2 = \langle e_1, X_0 \rangle$ and $c_1 y_1 + c_2 y_2 = (a_{11} - a_{22} + i\theta_1)\langle e_1, X_0\rangle - a_{22}\langle e_2, X_0\rangle$. Now

$$\widehat{Z}_t^1 = e^{a_{22}t} Y_t^1 = c_1 e^{(a_{22}+y_1)t} + c_2 e^{(a_{22}+y_2)t}$$

and

$$\begin{aligned} \widehat{Z}_t^2 &= e^{a_{22}t}\left(\langle e_2, X_0\rangle - a_{11}\int_0^t (c_1 e^{y_1 u} + c_2 e^{y_2 u})\, du\right) \\ &= e^{a_{22}t}\langle e_2, X_0\rangle - a_{11}e^{a_{22}t}\left[\frac{c_1}{y_1}(e^{y_1 t} - 1) + \frac{c_2}{y_2}(e^{y_2 t} - 1)\right]. \end{aligned}$$

Finally the characteristic function

$$\begin{aligned} E&\left[\left(i\int_0^t \theta_1 \langle e_1, X_u\rangle du\right)\right] \\ &= \widehat{Z}_t^1 + \widehat{Z}_t^2 \\ &= e^{a_{22}t}\Bigg[\langle e_2, X_0\rangle - a_{22}^{-1}\left((a_{11} - a_{22} + i\theta_1)\langle e_1, X_0\rangle - a_{22}\langle e_2, X_0\rangle\right) \\ &\qquad \times \frac{c_1}{y_1}(y_1 - a_{11})e^{y_1 t} + \frac{c_2}{y_2}(y_2 - a_{11})e^{y_2 t}\Bigg]. \end{aligned}$$

2 The Black Scholes Equation

Suppose, as above, the state of the economy is determined by the finite state Markov chain $X = \{X_t, t \geq 0\}$, $X_t \in \{e_1, \ldots, e_N\}$ with dynamics

$$X_t = X_0 + \int_0^t AX_s ds + M_t. \tag{16}$$

With $r_t = \langle r, X_t\rangle$, $\mu_t = \langle \mu, X_t\rangle$, and $\sigma_t = \langle \sigma, X_t\rangle$ our market includes the two underlying assets S^0 and S^1 where

$$S_t^0 = \exp\left(\int_0^t r_u du\right) \tag{17}$$

$$S_t^1 = S_0^1 \exp\left(\int_0^t \left(\mu_u - \frac{\sigma_u^2}{2}\right) du + \int_0^t \sigma_u dB_u\right). \tag{18}$$

If at time $t \in [0,T]$, $S_t^1 = S$ and $X_t = X$ the price of a European call option with exercise date T and strike K is

$$C(t,T,S,X) = E\left[\exp\left(-\int_t^T r_u du\right)\left(S_T^1 - K\right)^+ \middle| tS_t^1 = S,\, X_t = X\right]. \tag{19}$$

This is evaluated similarly to $C(0,T,S,X)$.

Write

$$\begin{aligned} V(t,S,X) &= \exp\left(-\int_0^t r_u du\right) C(t,T,S,X) \\ &= E\left[\exp\left(-\int_0^T r_u du\right)(S_T^1 - K)^+ \middle| S_t^1 = S, X_t = X\right] \\ &= E\left[\exp\left(-\int_0^T r_u du\right)(S_T^1 - K)^+ \middle| \mathcal{G}_t\right] \end{aligned} \tag{20}$$

where $\mathcal{G}_t = \sigma\{S_u^1,\, X_u : u \le t\}$. Consequently V is a $\mathcal{G}_t$-martingale. Write

$$\boldsymbol{V}(t,S) = \big(V(t,S,e_1),\ldots,V(t,S,e_2)\big),$$

so that $V(t,S_t^1,X_t) = \langle \boldsymbol{V}(t,S_t^1), X_t\rangle$. Applying the Itô rule to V we have

$$\begin{aligned} V(t,S_t^1,X_t) &= V(0,S_t^1,X_t) + \int_0^t \frac{\partial V}{\partial u} du \\ &\quad + \int_0^t \frac{\partial V}{\partial s}(\mu_u S_u^1 du + \sigma_u S_u^1 dB_u) \\ &\quad + \frac{1}{2}\int_0^t \frac{\partial^2 V}{\partial S^2}\sigma_u^2(S_u^1)du + \int_0^t \langle \boldsymbol{V}, dX_t\rangle \end{aligned} \tag{21}$$

and

$$dX_t = AX_t dt + dM_t.$$

By definition V is a martingale, so all the time integral terms in (21) must sum to zero identically. That is,

$$\frac{\partial V}{\partial t} + \mu_t S^1 \frac{\partial V}{\partial S} + \frac{1}{2}(\sigma_t S^1)^2 \frac{\partial^2 V}{\partial S^2} + \langle \boldsymbol{V}, AX\rangle = 0. \tag{22}$$

Now $V = \exp\big(-\int_0^t r_u du\big)C$, so with

$$\boldsymbol{C}(t,S) = \big(C(t,T,S,e_1),\ldots,C(t,T,S,e_N)\big)$$

we have

$$\left[\exp\left(-\int_0^t r_u du\right)\right]$$

$$\times \left(-r_t C + \frac{\partial C}{\partial t} + \mu_t S^1 \frac{\partial C}{\partial S} + \frac{1}{2}(\sigma_t S^1)^2 \frac{\partial^2 C}{\partial S^2} + \langle \boldsymbol{C}, AX \rangle \right) = 0 \quad (23)$$

with terminal condition

$$C(T,T,S,X) = (S-K)^+.$$

Equation (23) reduces to N equations with $X = e_1, e_2, \ldots, e_N$. That is, with $X = X_t = e_i$

$$r_t = \langle r, X_t \rangle = r_i$$
$$\mu_t = \langle \mu, X_t \rangle = \mu_i$$

and

$$\sigma_t = \langle \sigma, X_t \rangle = \sigma_i,$$

and, writing

$$C_i = C(t,T,S,e_i), \qquad \boldsymbol{C} = (C_1, C_2, \ldots, C_N)$$

we have that $\boldsymbol{C}$ satisfies the system of coupled Black-Scholes equations

$$-r_i C_i + \frac{\partial C_i}{\partial t} + \mu_i S \frac{\partial C_i}{\partial S} + \frac{1}{2}\sigma_i^2 S^2 \frac{\partial^2 C_i}{\partial S^2} + \langle \boldsymbol{C}, Ae_i \rangle = 0 \quad (24)$$

with terminal condition

$$C(T,T,S,e_i) = (S-K)^+.$$

3 Conclusion

We have considered an economy which can have finitely many states. These are identified with the standard unit vectors $e_i = (0, \ldots, 0, 1, 0, \ldots, 0)' \in R^N$ and the transition process is modeled as a Markov chain X. Black-Scholes pricing for European options is extended to the case where the parameters depend on X. The characteristic function of the occupation time is derived and used to price European options.

References

1. Barone-Adesi, G. and Whaley, R. (1987) Efficient analytic approximation of American option values, J. Finance **42**, 301–320.
2. Di Masi, G. B., Kabanov, Y. M., and Runggaldier, W. J. (1994) Mean variance hedging of options on stocks with Markov volatility, Theory of Probability and App. **39**, 173–181.
3. Elliott, R. J., Aggoun, L., and Moore, J. B. Hidden Markov Models: Estimation and Control, Springer-Verlag, New York.

4. Elliott, R. J. and Kopp, P. E. (1999) Mathematics of Financial Markets, Springer-Verlag, New York.
5. Guo, X. (1999) Insider Information and Stock Fluctuations, Ph.D. Thesis, Rutgers University, N.J.
6. Guo, X. (1999) An explicit solution to an optimal stopping problem with regime switching, Preprint, Rutgers University.

Equivalence of Two Kinds of Stability for Multi-dimensional ARMA Systems *

Xianbing Cao and Han-Fu Chen

Laboratory of Systems and Control, Institute of Systems Science, Academy of Mathematics and Systems Science, Chinese Academy of Sciences, Beijing, 100080, P. R. China

Abstract. Under some reasonable conditions imposed on the moving average part $C(z)w_k$ of the multi-dimensional system $A(z)y_k = C(z)w_k$ it is shown that for stability of $A(z)$, which means that all zeros of $\det A(z)$ are outside the closed unit disk, the necessary and sufficient condition is the stability of the system in the mean square sense, by which it is meant that the long run average of the squared output is bounded:

$$\limsup_{n\to\infty} \frac{1}{n}\sum_{k=1}^{n} \|y_k\|^2 < \infty \quad a.s.$$

1 Introduction

Consider the multi-dimensional ARMA system

$$A(z)y_k = C(z)w_k, \quad y_k = w_k = 0 \ \text{ for } \ k \le 0 \tag{1}$$

where both y_k and w_k are m-dimensional , z is the backward-shift operator $zy_k = y_{k-1}$, $A(z)$ and $C(z)$ are matrix polynomials:

$$A(z) = I + A_1 z + \cdots + A_p z^p, \quad \det A_p \neq 0, \tag{2}$$

$$C(z) = I + C_1 z + \cdots + C_r z^r, \quad \det C_r \neq 0. \tag{3}$$

In what follows we say that System (1) is stable if

$$\det A(z) \neq 0, \quad \forall z : |z| \le 1, \tag{4}$$

and stable in the mean square sense (MSS) if

$$\limsup_{n\to\infty} \frac{1}{n}\sum_{k=1}^{n} \|y_k\|^2 < \infty \quad a.s. \tag{5}$$

If System (1) is stable, then under appropriate conditions on w_i, it is natural to expect that the system is also stable in MSS. This paper mainly considers the converse problem: Does stability in MSS of System (1) imply stability?

* This work was supported by the National Natural Science Foundation of China.

B. Pasik-Duncan: Stochastic Theory and Control, LNCIS 280, pp. 83–96, 2002.

This problem arises from stochastic adaptive control for ARMAX systems (See, for example, [1]):

$$A(z)y_k = zB(z)u_k + D(z)w_k,$$

where $A(z)$, $B(z)$ and $D(z)$ are monic matrix polynomials with unknown coefficients. The control u_k depending upon the past output data $\{y_0, \cdots, y_{k-1}\}$ is designed to minimize some performance index and to stabilize the closed system. As a result, the limiting steady state system is of the form (1), for which only the stability in MSS is guaranteed. It is important to clarify if the limiting closed-loop system is truly stable.

For the single-input single-output (SISO) system, this problem is partly solved in [2], showing that the stability in MSS implies that no zero of $A(z)$ can be located inside the open unit disk without ruling out the possibility of zeros on the unit circle. The complete solution for SISO systems is provided in [3] having proved that for systems being stable in MSS, roots of $A(z)$ can neither be on the unit circle. Thus, in the SISO case stability of (1) is equivalent to its stability in MSS. However, the method of proof given in [2] cannot directly be extended to the multi-dimensional case. The purpose of this paper is to show that the equivalence of two kinds of stability mentioned above is also true for multi-dimensional systems. The result is presented for the special case $C(z) = I$ in Section 3, then it is extended to the general polynomial $C(z)$ with order $r \geq 1$ in Section 4. Some concluding remarks are given in Section 5.

2 Stability Implies Stability in MSS

In this section we prove that for System (1) stability implies stability in MSS. We present System (1) in the state-space form:

$$x_{k+1} = \Phi x_k + H^T C(z) w_{k+1}. \tag{6}$$

$$y_k = H x_k, \tag{7}$$

where

$$\Phi \triangleq \begin{bmatrix} -A_1 & I\ 0 \cdots 0 \\ -A_2 & 0\ I \cdots 0 \\ \vdots & \vdots\ \vdots \ddots \vdots \\ -A_{p-1} & 0\ 0 \cdots I \\ -A_p & 0\ 0 \cdots 0 \end{bmatrix}, \quad H \triangleq [\underbrace{I\ 0\ \cdots\ 0}_{mp}]\}\, m. \tag{8}$$

Lemma 1. Assume $\{w_i, \mathcal{F}_i\}$ is a martingale difference sequence with

$$\sup_i E(\|w_i\|^{2+\delta}|\mathcal{F}_i) < \infty$$

for some $\delta > 0$ or $\{w_i\}$ is uniformly bounded. If System (1) is stable, then it is also stable in MSS.

Proof. Since the roots of det $A(z)$ are the reciprocals of the eigenvalues of Φ (See, e.g., [4]), stability of (1) is equivalent to that the eigenvalues of Φ are inside the open unit disk. In this case there are constants $\alpha > 0, \beta \in (0, 1)$ such that

$$\|\Phi^k\| \leq \alpha\beta^k. \tag{9}$$

Note that

$$x_k = \Phi^k x_0 + \sum_{i=1}^{k} \Phi^{k-i} H^T C(z) w_i. \tag{10}$$

If w_i is uniformly bounded, $\|w_i\| \leq \zeta < \infty$ a.s. $\forall i$, then from (10) it follows that

$$\|x_k\| \leq \alpha\|x_0\|\beta^k + \sum_{i=1}^{k} \alpha\beta^{k-i} \sum_{j=0}^{r} \|C_j\|\zeta,$$

where $C_0 = I$ by definition, and

$$\|x_k\|^2 \leq 2\alpha^2\|x_0\|^2 + 2\alpha^2\zeta^2 \left(\sum_{j=0}^{r} \|C_j\|\right)^2 \frac{1}{(1-\beta)^2} < \infty, \ \forall k.$$

Therefore,

$$\limsup_{n\to\infty} \frac{1}{n} \sum_{k=1}^{n} \|x_k\|^2 < \infty \quad a.s. \tag{11}$$

and hence the system is stable in MSS.

Now, assume $\{w_i, \mathcal{F}_i\}$ is a martingale difference sequence with

$$\sup_i E(\|w_i\|^{2+\delta}|\mathcal{F}_i) < \infty$$

for some $\delta > 0$. From (9), (10) it follows that

$$\|x_k\| \leq \alpha\beta^k\|x_0\| + \sum_{i=1}^{k} \alpha\beta^{k-i}\|C(z)w_i\|,$$

$$\|x_k\|^2 \leq 2\alpha^2\|x_0\|^2\beta^{2k} + 2\alpha^2 \sum_{i=1}^{k} \beta^{k-i} \sum_{i=1}^{k} \beta^{k-i}\|C(z)w_i\|^2,$$

and

$$\limsup_{n\to\infty}\frac{1}{n}\sum_{k=1}^{n}\|x_k\|^2 = \frac{2\alpha^2}{1-\beta}\limsup_{n\to\infty}\frac{1}{n}\sum_{k=1}^{n}\sum_{i=1}^{k}\beta^{k-i}\|C(z)w_i\|^2$$
$$\leq \frac{2\alpha^2}{(1-\beta)^2}\limsup_{n\to\infty}\frac{1}{n}\sum_{i=1}^{n}\|C(z)w_i\|^2,$$

which is bounded if we can show that

$$\limsup_{n\to\infty}\frac{1}{n}\sum_{i=1}^{n}\|w_i\|^2 < \infty.$$

By the assumption, we have

$$\sup_i E\left(|\|w_{i+1}\|^2 - E(\|w_{i+1}\|^2|\mathcal{F}_i)|^{1+\frac{\delta}{2}}|\mathcal{F}_i\right) < \infty.$$

Then, by the convergence theorem for martingale difference sequences (See, e.g., [6], [7]) it follows that

$$\sum_{i=1}^{\infty}\frac{\|w_{i+1}\|^2 - E(\|w_{i+1}\|^2|\mathcal{F}_i)}{i} < \infty \quad a.s.$$

By the Kronecker lemma, this yields

$$\lim_{n\to\infty}\frac{1}{n}\sum_{i=1}^{n}\left(\|w_{i+1}\|^2 - E(\|w_{i+1}\|^2|\mathcal{F}_i)\right) = 0,$$

which implies (11) since $\sup_i E(\|w_{i+1}\|^2|\mathcal{F}_i) < \infty$. From (11) the stability of (1) in MSS immediately follows. □

3 Stability in MSS Implies Stability ($C(Z) = I$)

In order to show the essence of the proof, we present the main result first for the special case $C(z) = I$ in this section.

Let us list conditions to be used.

A1. $\{w_k\}$ is a sequence of mutually independent m-dimensional random vectors with following properties

$$E(w_k) = 0, \quad \forall k, \quad \sup_k E(\|w_k\|^\gamma) < \infty \quad \text{for some} \quad \gamma : \gamma > 2;$$

A2. $Ew_kw_k^T > 2\alpha_1 I, \ \ \forall k \geq i_0$ for some i_0, where $\alpha_1 > 0$.

Before describing Theorem 1 let us prove a lemma.

Lemma 2. Let

$$P^{-1}\Phi P = \begin{bmatrix} J_1 & & & \\ & J_2 & & \\ & & \ddots & \\ & & & J_d \end{bmatrix} \tag{12}$$

be the Jordan form of Φ, where $J_1,\ J_2,\ \cdots,\ J_d$ are Jordan blocks with

$$J_i = \begin{bmatrix} \lambda_i & 1 & & \\ & \ddots & \ddots & \\ & & \lambda_i & 1 \\ & & & \lambda_i \end{bmatrix}_{r_i \times r_i} \qquad r_i \geq 1, \quad i = 1, \cdots, d.$$

If $\lambda_1 \neq 0$, then the r_1th row $a_{r_1}^T$ of the matrix $P^{-1}H^T$ is nonzero.

Proof. Let us denote by B^* the matrix obtained by taking the complex conjugate and transpose of a matrix B and by $\bar{\lambda}$ the complex conjugate of a complex number λ.

Noticing that Φ is a real matrix, from (12) we have

$$\Phi^T (P^{-1})^* = (P^{-1})^* \begin{bmatrix} J_1^* & & & \\ & J_2^* & & \\ & & \ddots & \\ & & & J_d^* \end{bmatrix}. \tag{13}$$

Denote by

$$q_i = [(q_i^{(1)})^T \quad (q_i^{(2)})^T \quad \cdots \quad (q_i^{(p)})^T]^T$$

the r_1th column of $(P^{-1})^*, i = 1, \cdots, mp$, where $q_i^{(j)}$ are m-dimensional, $j = 1, \cdots, p$. Note that both sides of (13) are $mp \times mp$ matrices. Its r_1th column is expressed by

$$\begin{bmatrix} -A_1 & -A_2 & \cdots & -A_{p-1} & -A_p \\ I & 0 & \cdots & 0 & 0 \\ 0 & I & \cdots & 0 & 0 \\ \vdots & \vdots & \ddots & \vdots & \vdots \\ 0 & 0 & \cdots & I & 0 \end{bmatrix} \begin{bmatrix} q_{r_1}^{(1)} \\ q_{r_1}^{(2)} \\ \vdots \\ q_{r_1}^{(p-1)} \\ q_{r_1}^{(p)} \end{bmatrix} = \begin{bmatrix} q_{r_1}^{(1)} \\ q_{r_1}^{(2)} \\ \vdots \\ q_{r_1}^{(p-1)} \\ q_{r_1}^{(p)} \end{bmatrix} \bar{\lambda}_1. \tag{14}$$

By the definitions of a_{r_1} and $q_{r_1}^{(1)}$ we have

$$a_{r_1}^* = q_{r_1}^{(1)}. \tag{15}$$

If the Lemma were not true, then $q_{r_1}^{(1)} = 0$ and from (14) by noticing $\lambda_1 \neq 0$ we would have $q_{r_1}^{(2)} = \cdots q_{r_1}^{(p)} = 0$, i.e., $q_{r_1} = 0$. This contradicts the nondegeneracy of P, and the Lemma is proved. □

Theorem 1. Assume $A1$ and $A2$ hold and $C(z) = I$ in (1). Then System (1) is stable if and only if it is stable in MSS.

Proof. The necessity is given by Lemma 1. We now prove the sufficiency.

Assume the converse: $A(z)$ is unstable, i.e., there is at least one eigenvalue of Φ located outside or on the unit circle. Without loss of generality, we may assume this eigenvalue is $\lambda_1 \triangleq \rho e^{j\lambda}$ corresponding to the first Jordan block J_1 in (12), where $j = \sqrt{-1}$. By the converse assumption $\rho = |\lambda_1| \geq 1$.

Let us denote

$$P^{-1} x_{k+1} \triangleq \eta_{k+1} = \begin{bmatrix} \eta_{k+1}^{(1)} \\ \eta_{k+1}^{(2)} \\ \vdots \\ \eta_{k+1}^{(mp)} \end{bmatrix}. \tag{16}$$

Then, noticing $C(z) = I$ from (6) we obtain

$$P^{-1} x_{k+1} = P^{-1} \Phi P P^{-1} x_k + P^{-1} H^T w_{k+1}. \tag{17}$$

Using the vector a_{r_1} introduced in lemma 2, from (17) we find that

$$\eta_{k+1}^{(r_1)} = \lambda_1 \eta_k^{(r_1)} + a_{r_1}^T w_{k+1}, \tag{18}$$

and hence

$$\eta_{k+1}^{(r_1)} = \lambda_1^{k+1} \sum_{i=1}^{k+1} \lambda_1^{-i} a_{r_1}^T w_i. \tag{19}$$

Set

$$a_{r_1} = c + jd, \tag{20}$$

where c and d are real vectors, $j = \sqrt{-1}$.

By (19) we have

$$\begin{aligned}
|\eta_k^{(r_1)}|^2 &= \left(\lambda_1^k \sum_{i=1}^{k} \lambda_1^{-i} a_{r_1}^T w_i\right) (\bar{\lambda}_1)^k \sum_{i=1}^{k} (\bar{\lambda}_1)^{-i} \bar{a}_{r_1}^T w_i \\
&= |\lambda_1|^{2k} \left(\sum_{i=1}^{k} \lambda_1^{-i} (c^T w_i + jd^T w_i)\right) \sum_{i=1}^{k} (\bar{\lambda}_1)^{-i} (c^T w_i - jd^T w_i) \\
&= \rho^{2k} \left\{ \sum_{i=1}^{k} \rho^{-i} [\cos(i\lambda) c^T w_i + \sin(i\lambda) d^T w_i] \right. \\
&\qquad \left. + j \sum_{i=1}^{k} \rho^{-i} [\cos(i\lambda) d^T w_i - \sin(i\lambda) c^T w_i] \right\} \\
&\quad \times \left\{ \sum_{i=1}^{k} \rho^{-i} [\cos(i\lambda) c^T w_i + \sin(i\lambda) d^T w_i] \right. \\
&\qquad \left. - j \sum_{i=1}^{k} \rho^{-i} [\cos(i\lambda) d^T w_i - \sin(i\lambda) c^T w_i] \right\} \\
&= \rho^{2k} \left\{ \sum_{i=1}^{k} \rho^{-i} [\cos(i\lambda) c^T + \sin(i\lambda) d^T] w_i \right\}^2 \\
&\quad + \rho^{2k} \left\{ \sum_{i=1}^{k} \rho^{-i} [\cos(i\lambda) d^T - \sin(i\lambda) c^T] w_i \right\}^2,
\end{aligned} \tag{21}$$

which implies

$$|\eta_k^{(r_1)}|^2 \geq \rho^{2k} \left\{ \sum_{i=1}^{k} \rho^{-i} [\cos(i\lambda) c^T + \sin(i\lambda) d^T] w_i \right\}^2 \tag{21-1}$$

and

$$|\eta_k^{(r_1)}|^2 \geq \rho^{2k} \left\{ \sum_{i=1}^{k} \rho^{-i} [\cos(i\lambda) d^T - \sin(i\lambda) c^T] w_i \right\}^2. \tag{21-2}$$

By Lemma 2, $a_{r_1} \neq 0$. In other words, at least one of c and d does not equal 0. To be fixed, let us assume $c \neq 0$. In this case we will use (21-1). If $c = 0$, then $d \neq 0$ and in lieu of (21-1) we will use (21-2), but the proof is essentially the same.

Assume λ_1 is outside the unit circle, i.e., $\rho = |\lambda_1| > 1$.

By A1 we have

$$\sum_{i=1}^{\infty} E\{\rho^{-i}[\cos(i\lambda)c^T + \sin(i\lambda)d^T]w_i\}^2$$
$$\leq \sum_{i=1}^{\infty} \rho^{-2i} \| \cos(i\lambda)c^T + \sin(i\lambda)d^T \|^2 E\|w_i\|^2 < \infty.$$

Then by the convergence theorem for martingale difference sequences it follows that

$$\xi_n \triangleq \sum_{i=1}^{n} \rho^{-i}[\cos(i\lambda)c^T + \sin(i\lambda)d^T]w_i$$

converges a.s. to a limit ξ, $|\xi| < \infty$ a.s.

Notice that

$$E\xi_n^2 = \sum_{i=1}^{n} \rho^{-2i}[\cos(i\lambda)c^T + \sin(i\lambda)d^T]E(w_i w_i^T)[\cos(i\lambda)c + \sin(i\lambda)d],$$

and

$$\xi_n^2 \leq 2\sum_{i=1}^{n} \rho^{-i} \sum_{i=1}^{n} \rho^{-i}(\|c\|^2 + \|d\|^2)\|w_i\|^2$$
$$\leq \frac{2}{\rho - 1}(\|c\|^2 + \|d\|^2) \sum_{i=1}^{\infty} \rho^{-i}\|w_i\|^2 \triangleq \zeta.$$

By A1, it is clear that $E\zeta < \infty$. Therefore, applying the dominated convergence theorem (See, e.g., [6]) leads to

$$E\xi^2 = \lim_{n\to\infty} E\xi_n^2,$$

i.e.,

$$E\xi^2 = \sum_{i=1}^{\infty} \rho^{-2i}[\cos(i\lambda)c^T + \sin(i\lambda)d^T]E(w_i w_i^T)[\cos(i\lambda)c + \sin(i\lambda)d],$$

which implies

$$0 < 2\alpha_1 \sum_{i=i_0}^{\infty} \rho^{-2i} \| \cos(i\lambda)c + \sin(i\lambda)d\|^2 < E\xi^2 < \infty$$

by A2, and $c \neq 0$.

Therefore, there is an $\varepsilon > 0$ and an event E_0 with $P(E_0) > 0$ such that

$$\left(\sum_{i=1}^{k} \rho^{-i}[\cos(i\lambda)c + \sin(i\lambda)d]w_i\right)^2 > \varepsilon, \quad \forall\omega \in E_0,$$

for all sufficiently large k. Then by (21-1) it follows that

$$\begin{aligned}\limsup_{n} \frac{1}{n}\sum_{k=1}^{n} |\eta_k^{(r_1)}|^2 &\geq \limsup_{n\longrightarrow\infty} \frac{1}{n}\rho^{2n}\left\{\sum_{i=1}^{k} \rho^{-i}[\cos(i\lambda)c + \sin(i\lambda)d]w_i\right\}^2 \\ &> \limsup_{n\longrightarrow\infty} \frac{\varepsilon}{n}\rho^{2n} = \infty, \qquad \forall\omega \in E_0. \end{aligned} \tag{22}$$

By nondegeneracy of P, from (22) by (7), (16) it follows that

$$\limsup_{n\to\infty} \frac{1}{n}\sum_{k=1}^{n} \|y_k\|^2 = \infty \quad \forall\omega \in E_0,$$

which contradicts the assumption. Consequently, no zero of $A(z)$ can be inside the open unit disk.

We now rule out the possibility that $A(z)$ has zeros on the unit circle. Assume λ_1 is on the unit circle, i.e., $\rho = |\lambda_1| = 1$.

By (21-1) we have

$$|\eta_k^{(r_1)}| \geq \left\{\sum_{i=1}^{k}[\cos(i\lambda)c^T + \sin(i\lambda)d^T]w_i\right\}^2. \tag{23}$$

Notice that

$$\sin\lambda\sum_{k=1}^{n}\cos(2k\lambda) = \frac{1}{2}\sum_{k=1}^{n}[\sin(2k+1)\lambda - \sin(2k-1)\lambda] = \frac{1}{2}(\sin(2n+1)\lambda - \sin\lambda),$$

and hence

$$\sum_{k=1}^{n}\cos(2k\lambda) = \frac{\sin(2n+1)\lambda - \sin\lambda}{2\sin\lambda}, \qquad \lambda \neq 0, \quad \pi.$$

Similarly,

$$\begin{aligned}\sin\lambda\sum_{k=1}^{n}\sin(2k\lambda) &= -\frac{1}{2}\sum_{k=1}^{n}[\cos(2k+1)\lambda - \cos(2k-1)\lambda] \\ &= -\frac{1}{2}(\cos(2n+1)\lambda - \cos\lambda),\end{aligned}$$

and hence

$$\sum_{k=1}^{n}\sin(2k\lambda) = -\frac{\cos(2n+1)\lambda - \cos\lambda}{2\sin\lambda}, \qquad \lambda \neq 0, \quad \pi.$$

Then by A1, A2 we have

$$
\begin{aligned}
s_n^2 \triangleq & \sum_{i=1}^{n} E[(\cos(i\lambda)c^T + \sin(i\lambda)d^T)w_i]^2 \\
\geq & 2\alpha_1 \sum_{i=i_0}^{n} [\cos(i\lambda)c^T + \sin(i\lambda)d^T][\cos(i\lambda)c + \sin(i\lambda)d] \\
= & 2\alpha_1 \sum_{i=i_0}^{n} [\|c\|^2 \cos^2(i\lambda) + \|d\|^2 \sin^2(i\lambda) + c^T d \sin(2i\lambda)] \\
= & 2\alpha_1 \sum_{i=i_0}^{n} [\|c\|^2 \frac{1+\cos(2i\lambda)}{2} + \|d\|^2 \frac{1-\cos(2i\lambda)}{2} + c^T d \sin(2i\lambda)] \\
= & \begin{cases} \alpha_1(\|c\|^2 + \|d\|^2)(n - i_0 + 1)(1 + o(1)), & \lambda \neq 0, \quad \pi, \\ 2\alpha_1 \|c\|^2 (n - i_0 + 1), & \lambda = 0, \quad \pi. \end{cases}
\end{aligned}
$$

Similarly, we have

$$
s_n^2 \leq \sum_{i=1}^{n} 2(\|c\|^2 + \|d\|^2) E\|w_i\|^2 = O(n)
$$

by A1. Thus, s_n^2 has the following properties

1) $s_n^2 \to \infty$, as $n \to \infty$;

2) $\lim\limits_{n\to\infty} \frac{s_{n+1}^2}{s_n^2} = 1$;

$$
\begin{aligned}
3) \; & \sum_{k=1}^{\infty} (2s_k^2 \log\log s_k^2)^{-\gamma/2} E|[\cos(k\lambda)c^T + \sin(k\lambda)d^T w_k]|^\gamma] \\
& \leq \sum_{k=1}^{\infty} (2s_k^2 \log\log s_k^2)^{-\gamma/2} \|[\cos(k\lambda)c^T + \sin(k\lambda)d^T\|^\gamma E\|w_k]\|^\gamma] < \infty,
\end{aligned}
$$

Then by the iterated logarithm law (See, e.g., [5]) it follows that

$$
\limsup_{n\to\infty} \frac{\sum\limits_{i=1}^{n} [\cos(i\lambda)c^T + \sin(i\lambda)d^T] w_i}{(2s_n^2 \log\log s_n^2)^{1/2}} = 1 \quad a.s. \tag{24}
$$

Therefore there exists a subsequence $\{n_k\}$, which may depend on the sample point ω, such that

$$
\lim_{k\to\infty} \frac{\sum\limits_{i=1}^{n_k} [\cos(i\lambda)c^T + \sin(i\lambda)d^T] w_i}{(2s_{n_k}^2 \log\log s_{n_k}^2)^{1/2}} = 1 \quad a.s., \tag{25}
$$

i.e.,

$$\sum_{i=1}^{n_k}[\cos(i\lambda)c^T + \sin(i\lambda)d^T]w_i = (2s_{n_k}^2 \log\log s_{n_k}^2)^{1/2}(1+o(1))$$
$$\geq (2\alpha_1\|c\|^2 n_k \log\log \alpha_1\|c\|^2 n_k)^{1/2}(1+o(1)) \quad \text{a.s.} \tag{26}$$

From (23), (26) we obtain

$$|\eta_{n_k}^{(r_1)}|^2 \geq 2\alpha_1\|c\|^2 n_k(\log\log \alpha_1\|c\|^2 n_k)(1+o(1)) \quad a.s., \tag{27}$$

which implies

$$\frac{\|x_{n_k}\|^2}{n_k} \to \infty \quad (k\to\infty)$$

by (16) and the nondegeneracy of P.

Therefore

$$\limsup_{n\to\infty}\frac{1}{n}\sum_{k=1}^{n}\|y_k\|^2 = \infty \quad a.s.,$$

which contradicts the assumption that system (1) is stable in MSS. Thus A has no eigenvalue on the unit circle either. The proof is completed. □

4 General $C(z)$

We now prove that stability and stability in MSS are equivalent for the general case where the order of $C(z)$ may be greater or equal to 1, $r \geq 1$. We need the following condition.

A3 $C(z)$ is stable, i.e., $\det C(z) \neq 0, \quad \forall z : |z| \leq 1$.

Theorem 2. Assume $A1$, $A2$ and $A3$ hold. System (1) is stable if and only if it is stable in MSS.

Proof. The necessity part has been proved in Lemma 1. We need only to prove the sufficiency part. If the theorem were not true, then there would be at least one eigenvalue $\lambda_1 = \rho e^{j\lambda}$ of Φ with $\rho = |\lambda_1| \geq 1$.

In lieu of (18) we now have

$$\eta_{k+1}^{(r_1)} = \lambda_1\eta_k^{(r_1)} + a_{r_1}^T C(z)w_{k+1}. \tag{28}$$

Noticing $w_k = 0$ for $k \leq 0$ and setting $C_0 = I$, we have

$$\eta_k^{(r_1)} = a_{r_1}^T\lambda_1^k\sum_{i=1}^{k}\lambda_1^{-i}C(z)w_i$$
$$= a_{r_1}^T\lambda_1^k\sum_{i=1}^{k}\lambda_1^{-i}\sum_{s=0}^{r}C_s w_{i-s}$$

$$
\begin{aligned}
&= a_{r_1}^T \lambda_1^k \left[\sum_{s=0}^{r} \sum_{i=1}^{k-s} \lambda_1^{-i-s} C_s w_i \right] \\
&= a_{r_1}^T \lambda_1^k \sum_{i=1}^{k-r} \left(\sum_{s=0}^{r} C_s \lambda_1^{-s} \right) \lambda_1^{-i} w_i + a_{r_1}^T \lambda_1^k \sum_{s=0}^{r-1} \sum_{i=k-r+1}^{k-s} \lambda_1^{-i-s} C_s w_i \\
&= a_{r_1}^T \lambda_1^k C(\lambda_1^{-1}) \sum_{i=1}^{k-r} \lambda_1^{-i} w_i + a_{r_1}^T \lambda_1^k \sum_{s=0}^{r-1} \sum_{i=k-r+1}^{k-s} \lambda_1^{-i-s} C_s w_i \\
&= \lambda_1^k \sum_{i=1}^{k-r} \lambda_1^{-i} a_{r_1}^T C(\lambda_1^{-1}) w_i + a_{r_1}^T \lambda_1^k \sum_{i=k-r+1}^{k} \left[\sum_{s=0}^{k-i} \lambda_1^{-i-s} C_s \right] w_i \\
&= \lambda_1^k \sum_{i=1}^{k-r} \lambda_1^{-i} a_{r_1}^T C(\lambda_1^{-1}) w_i + \lambda_1^k \sum_{i=k-r+1}^{k} \lambda_1^{-i} a_{r_1}^T \left(\sum_{s=0}^{k-i} \lambda_1^{-s} C_s \right) w_i .
\end{aligned}
\tag{29}
$$

Define

$$
b^T = g^T + jh^T \triangleq a_{r_1}^T C(\lambda_1^{-1}), \quad q_i = u_i^T + jv_i^T \triangleq a_{r_1}^T \sum_{s=0}^{k-i} \lambda_1^{-s} C_s . \tag{30}
$$

Then we have

$$
\begin{aligned}
\eta_k^{(r_1)} &= \lambda_1^k \left[\sum_{i=1}^{k-r} \lambda_1^{-i} b^T w_i + \sum_{i=k-r+1}^{k} \lambda_1^{-i} q_i^T w_i \right] \\
&= \lambda_1^k \Bigg\{ \sum_{i=1}^{k-r} \rho^{-i} [\cos(i\lambda) g^T w_i + \sin(i\lambda) h^T w_i] \\
&\qquad + j \sum_{i=1}^{k-r} \rho^{-i} [\cos(i\lambda) h^T w_i - \sin(i\lambda) g^T w_i] \\
&\qquad + \sum_{i=k-r+1}^{k} \rho^{-i} [\cos(i\lambda) u_i^T w_i + \sin(i\lambda) v_i^T w_i] \\
&\qquad + j \sum_{i=k-r+1}^{k} \rho^{-i} [\cos(i\lambda) v_i^T w_i - \sin(i\lambda) u_i^T w_i] \Bigg\}
\end{aligned}
$$

Therefore,

$$
\begin{aligned}
|\eta_k^{(r_1)}|^2 \geq \rho^{2k} \Bigg\{ & \sum_{i=1}^{k-r} \rho^{-i} [\cos(i\lambda) g^T + \sin(i\lambda) h^T] w_i \\
& + \sum_{i=k-r+1}^{k} \rho^{-i} [\cos(i\lambda) u_i^T + \sin(i\lambda) v_i^T] w_i \Bigg\}^2 ,
\end{aligned}
\tag{31}
$$

and

$$|\eta_k^{(r_1)}|^2 \geq \rho^{2k}\left\{\sum_{i=1}^{k-r}\rho^{-i}[\cos(i\lambda)h^T - \sin(i\lambda)g^T]w_i \right.$$
$$\left. + \sum_{i=k-r+1}^{k}\rho^{-i}[\cos(i\lambda)v_i^T - \sin(i\lambda)u_i^T]w_i\right\}^2. \tag{32}$$

Since $|\lambda_1^{-1}| \leq 1$, by A3 $C(\lambda_1^{-1})$ is invertible, and hence b is a nonzero vector, which implies $g \neq 0$ or $h \neq 0$. Similar to Section 3, if $g \neq 0$, we use (31); otherwise we use (32). To be fixed, assume $g \neq 0$. Noticing $\|q_i\|$ is uniformly bounded in i, we see that

$$\lim_{n\to\infty}\left(\sum_{i=1}^{k-r}\rho^{-i}[\cos(i\lambda)g^T + \sin(i\lambda)h^T]w_i \right.$$
$$\left. + \sum_{i=k-r+1}^{k}\rho^{-i}[\cos(i\lambda)u_i^T + \sin(i\lambda)v_i^T]w_i\right)$$
$$= \sum_{i=1}^{\infty}\rho^{-i}[\cos(i\lambda)g^T + \sin(i\lambda)h^T]w_i.$$

Then, along the lines of the proof for (22), we conclude that ρ cannot be greater than 1. If $\rho = 1$, then from (31) it is seen that $|\eta_k^{(r_1)}|^2 \geq S_k^2$, where

$$S_k^2 = \left\{\sum_{i=1}^{k-r}[\cos(i\lambda)g^T + \sin(i\lambda)h^T]w_i + \sum_{i=k-r+1}^{k}[\cos(i\lambda)u_i^T + \sin(i\lambda)v_i^T]w_i\right\}^2,$$

and

$$s_k^2 \triangleq E(S_k^2)$$
$$= \sum_{i=1}^{k-r}E[(\cos(i\lambda)g^T + \sin(i\lambda)h^T)w_i]^2$$
$$+ \sum_{i=k-r+1}^{k}E[(\cos(i\lambda)u_i^T + \sin(i\lambda)v_i^T)w_i]^2.$$

Similar to the proof for (23)-(27), we are convinced that ρ can neither be equal to 1. Therefore System (1) is stable. □

5 Conclusion

In this paper we have shown that for m-dimensional ARMA systems, the boundedness of average of the squared output, which is a usual conclusion of stochastic adaptive controls for ARMAX systems, is equivalent to stability of the system. In other words, for stability of System (1) the necessary and sufficient condition is its stability in MSS. However, the conditions imposed on the driven noise for the sufficient part are stronger than those for the necessity part as shown by Lemma 1, A1, and A2 used in Theorems 1 and 2. It is of interest to minimize the gap between the noise conditions. Further, it is also of interest to weaken the condition on $C(z)$.

References

1. Chen, H. F. and Guo, L. (1991) Identification and Stachastic Adaptive Control, Birkhäuser, Boston.
2. Chen, H. F. Identification of both closed- and open-loop stochastic system while stabilizing it, Systems Science and Complexity, to appear.
3. Chen, H. F., Cao, X. B., and Fang, H. T. (2001) Stability of Adaptively Stabilized Stochastic Systems, IEEE Trans. Autom. Control **46**, 11, 1832–1836.
4. Chen, H. F. (1985) Recursive Estimation and Control for Stochastic System, Wiley, New York.
5. Wittmann, R. (1985) A general law of iterated logarithm, Z. Wahrs. Verw. Gebiete **68**, 521–543.
6. Chow, Y. S. and Teicher, H. (1988) Probabilty Theory, Springer-Verlag.
7. Stout, W. F. (1974) Almost Sure Convergence, Academic Press.

System Identification and Time Series Analysis: Past, Present, and Future

Manfred Deistler*

Institut für Ökonometrie, Operations Research und System Theorie, Abteilung für Ökonometrie und System Theorie

1 Introduction

The aim of this contribution is to describe main features in the development of system identification, in the sense of modelling from time series data. Given the restrictions in space, such an effort is necessarely fragmentary. Clearly, subjective judgements cannot be avoided. System identification has been developed in a number of different scientific communities, the most important of which are econometrics, statistics and system- and control theory.

The development of the field due to the requirements of applications and due to the intrinsic dynamics of its theories, and the interactions of the different communities in contributing to this development will be briefly described as well as the basic formal features of the problem. In addition some future perspectives are given.

System identification has attracted almost no interest from the part of the general public interested in the history or perspectives of other parts of science. This is explained not only be the relative importance of the subject, compared to subjects attracting a lot of attention, but also by its—in a certain sense—abstract scope and the fact that it provides an enabling technology, often hidden in wider problem solutions. What is more surprising to the author is how little interest the history of the subject has attracted for researchers in this area; a clear indication for this is the frequent lack of proper referencing to original results.

2 Time Series Analysis and System Identification

Time Series Analysis is concerned with the systematic approaches to extract information from *time series*, i.e. from observations ordered in time. We may distinguish two main basic questions in this context, namely system identification on the one hand and signal and feature extraction on the other hand. Here we are mainly concerned with system identification, i.e. with obtaining a (mathematical) model from or for a time series. Since often dependence relations in time series are of interest, the models are often dynamic and in many cases uncertainty is modeled by stochastic models.

* The author wants to thank D. Trummer for valuable comments.

B. Pasik-Duncan: Stochastic Theory and Control, LNCIS 280, pp. 97–109, 2002.

As has been mentioned already, system identification in the sense of data driven modeling is dealt with in different scientific communities and here we use the term system identification in a broad sense, i.e. we do not restrict it to the activities of the system and control community.

A system identification procedure is a set of rules for attaching a system (model) to time series data. In particular such rules may be algorithms.

Here the model is part of a (preliminary) model class, containing all candidate models, compatible with the a-priori specification of the phenomenon considered. Such an a-priori specification describes e.g. the list of variables of (potential) relevance or the functional forms to be considered.

In a next step data generation, data preprocessing and data transformations are considered. Typical data transformations are e.g. outlier removals.

The core part of system identification consists of the rules (in particular algorithms) of attaching a "good" model from the model class to the preprocessed data. Both construction and evaluation of the quality of such rules form the center part of the theory and of development of methods of system identification. In many cases two different steps can be distinguished, namely (data driven) model selection and estimation of real valued parameters. Typical problems of model selection are selection of inputs among a list of a-priori feasible inputs or order estimation. In many cases model selection leads to a new restricted model class resulting in a finite dimensional parameter space; this is what we call the semi-nonparametric approach to system identification.

In a stochastic setting system identification procedures often consist of estimators and tests, the evaluation of which is done in terms of e.g. asymptotic statistics. However also numerical properties of identification algorithms are important.

Model validation under different criteria often constitutes a final step in system identification.

3 The History

For the sake of brevitiy, we tried to keep the list of references short; for sections 3.1 and 3.2 the original references can be found in Davies (1941), Morgan (1990) and Hannan (1970), for section 3.3 we refer to Hannan and Deistler (1988).

3.1 The Early History of Time Series Analysis (1772–1920)

In the late 18th and early 19th century more accurate data from the orbits of the planets and the moon became available due to improvements in telescope building. This, and the fact that Kepler's laws result from a two body problem, whereas more than two bodies are involved in the orbits of the planets, triggered the interest in systematic deviations from Keplers' laws, in particular in the search for hidden periodicities and long term trends in those orbits. Harmonic analysis begins probably with a memoir published by J. L. Lagrange in 1772 on these problems. Subsequently the theory of Fourier series has been developed by L. Euler and J. Fourier.

The method of least squares for fitting a line into a scatter plot was introduced by A. M. Legendre and C. F. Gauss in the early 19th century.

Another major step in the history of time series was the introduction of the periodogram by Stokes (1879) and Schuster (1894). Schuster used the periodogram successfully to study e.g. sunspot numbers or the periodicitiy of earthquakes.

The statistical theory of regression and correlation analysis was developed at the turn of the 19th to the 20th century by F. Galton, K. Pearson, W.S. Gosset and others.

The empirical analysis of business cycles was another important area in early time series analysis. In the seventies and eighties of the 19th century, the British economist W.S. Jevons performed careful statistical analyses of periodic fluctuations in economic time series. Perhaps Jevons is known best by his much ridiculed, but methodologically interesting work on sunspot numbers and business cycles.

H. Moore published in 1914 his book "Economic Cycles: Their Law and Cause", where demand curves have been estimated by multiple regressions and where periodogram analysis was used. In 1915 W.M. Persons published his first business barometer. A very ambitions periodogram analysis was published in 1922 by W. Beveridge. This analysis was based on Beveridges' famous wheat price index ranging from 1545 till 1845.

3.2 The Formation of Modern Time Series Analysis (1920–1970)

A problem with "traditional" harmonic analysis for time series, were non exactly periodic fluctuations like business cycles. This lead G.U. Yule to propose moving average and autoregressive models (1921, 1927) for such time series. Closely related is the work of E. Slutzky "The Summation of Random Causes as a Source of a Cyclical Process", published in 1927 and R. Frisch's work "Propagation and Impulse Problems in Dynamic Economics", published in 1933.

In the thirties and fourties of the 20th century two extremely important developments took place: The formation of the theory of stationary processes as a main model class for time series and the development of Cowles Commission econometrics, which was a first systematic approach to modern system identification.

The theory of weakly stationary stochastic processes was developed in particular by A. Khinchin, A. N. Kolmogorov, H. Cramer, N. Wiener, H. Wold and K. Karhunen. Main results were the spectral representation of covariance functions and of stationary processes, Wold representation, factorization of spectra, linear least squares forecasting and filtering. These results were and still are of central importance for time series analysis, however are—in the narrow sense—not statistical in nature, since e.g. the forecasting problem commences from second order population moments, rather than from data. At about the same time the ergodic theory for strictly stationary processes has been developed in particular by G. D. Birkhoff, Khinchin and Kolmogorov, a special consequence of which is consistency of empirical first and second moments.

The great economic depression in 1929 triggered intensive economic research activities. Of particular importance in this context were the development of the Keynesian theory and its mathematical formulation, which yielded a description of the "macrodynamics" of an economy by difference equations on the one side and the development of national accounts for the construction of macrodata on the other side. Questions of quantitative economic policy directly led to problems of estimating parameters, such as the marginal propensity to consume, in such marcodynamic models. Tinbergen published his famous econometric model for the U.S. economy in 1939. A typical feature of Keynesian type macro models is instanteous feedback (simultanity) e.g. between income and consumption which leads to inconsistency of the ordinary least squares (OLS) estimators applied by Tinbergen. Subsequently the Cowles Commission approach for identifying simultaneous multi-input, multi-output (MIMO) ARX systems has been developed. The pioneering contributions there were the paper by Mann and Wald (1943), where the asymptotic properties of the OLS estimators were derived, the article by Haavelmo (1944), where a stochastic setting for macro models was proposed and the contribution by Koopmans et al (1950), where a rather complete theory of identification of linear static and ARX systems, including identifiability analysis and Gaussian maximum likelihood estimation (MLE), together with their asymptotic properties, was presented. Emphasis was laid on "structural" modeling, where restrictions, e.g. zerorestrictions on coefficients, from economic theory were available. Actual calculation of the MLE's was a huge problem at this time and for this reason subsequently, in order to reduce the computational burden, simpler, consistent, but not asymptotically efficient procedures, such as two stage least squares have been developed.The first econometric model estimated with Cowles Commission methods, was the famous Klein I model for the US.

In the fourties and fifties nonparametric estimation of spectral densities has been developed. The statistical properties of the periodogram were derived and Daniell proposed the first smoothed spectral estimator in 1946. Main steps in the development and evaluation of spectral estimators were the books by Blackman and Tukey, published in 1958, Grenander and Rosenblatt, published in 1957 and Hannan (1960).

Subsequent to a boom in nonparametric spectral- and transfer function estimation, parametric time domain methods became increasingly popular. Very important in the fifties and sixties were the developments in estimation for AR and ARMA models. Whereas for AR (and ARX) models least squares gave explicit solutions for (nonrestricted) reduced forms and the foundation of their asymptotic analysis was laid in the paper by Mann and Wald (1943), things turned out to be significantly more complicated in the ARMA (and ARMAX) case; this for two reasons: On the one hand, since in general no explicit form for the MLE exists in the ARMA case, special algorithms for (approximating) the MLE had to be developed. Many of these algorithms commence from a (consistent) initial estimator and perform one Gau-Newton type step (in order to obtain asymptotic efficiency). On the other hand the asymptotic theory for the MLE's turned out to be quite involved (see next section). The work in this time period has been summarized by the two main contributors to the area, namely in Hannan (1970) and Anderson (1971). In both books AR and ARMA modeling of regression residuals is also analysed in detail. Anderson (1971) also discusses multiple testing for AR orders, but order estimation was not considered in great detail at this time. To a large extent, the analysis was confined to the SISO (Single input, single output)-case.

3.3 The Recent Past (1970–1990)

In the seventies of the last century time series analysis and in particular identification of AR, ARMA and state space systems received substantial attention in statistics, econometrics and systems and control theory.

The book by Box and Jenkins (1970) triggered an enormous boom in applications, mainly because explicit instructions for actually performing identification (for SISO systems) from data in an integrated way were given. This included data transformation, rules for determining the AR and MA order from data, an algorithm for maximizing the likelihood and techniques for model validation.

Beginning with the sixties, to a good part started by R. E. Kalman's work on realization and parametrization, the structure of linear state space systems was carefully investigated. A major result in this time was the detection of the manifold structure of all systems of given order (Hazewinkel and Kalman 1976). Subsequently the structure theory of ARMA systems has been investigated. Although these results are not in a strict sense statistical in nature, they were essential for a deeper understanding of identification, in particular in the MIMO case.

The first correct proof for consistency of MLEs for (SISO) ARMA systems seems to be due to Hannan (1973). Subsequently consistency and asymptotic normality of the MLEs was established (under different assumptions) in a number of papers, also for the MIMO case (see e.g. Dunsmuir and Hannan 1976 and Caines and Ljung 1979).

Two major shortcomings of the Box-Jenkins approach were that order determination had to be done by an experienced modeller, e.g. by looking at the patterns of the autocorrelation and the partial autocorrelation function and that it was—to a certain degree at least—restricted to the single output case. A major step ahead was the development and evaluatin of automatic model selection (in particular order estimation) procedures based on information criteria like AIC or BIC by Akaike, Hannan, Rissanen, Schwartz and others.

In the eigthies of the last century the subject of identification of linear systems (including the MIMO case), reached a certain stage of maturity. This relates to the basic ideas, the mathematical instruments, theoretical results and algorithms, which, in particular for what we call main stream identification of linear systems, form a substantial and well connected body of theories and methods. This is also documented by a number of books appearing in this time (Ljung 1987, Caines 1988, Hannan and Deistler 1988, S"oderstr"om and Stoica 1989, Reinsel 1993). The seventies and eighties showed also a relatively high degree of exchange of information between different communities working on the subject.

4 Main Stream Identification of Linear Systems

Here again, we tried to keep the list of references short; here the reader is referred to the original references in Hannan and Deistler (1988).

In most applications still linear models are used; this is true even if e.g. physical theories suggest nonlinearities; but in many cases time constant of time varying linear systems give sufficient approximations. Also with respect to theory and methods identification of linear systems, in particular main stream identification, still forms the core part of system identification.

The main stream approach is characterized by the following five ingredients (see e.g. Deistler 2001):

- The model class consists of linear, time invariant, finite dimensional, causal and stable systems only. The classification of the variables into inputs and outputs is given a priori.
- Stochastic models for noise are used. In particular noise is assumed to be stationary ergodic with a rational spectral densitiy.
- The observed inputs are assumed to be free of noise and to be uncorrelated with the noise process.
- A seminonparametric approach is pursued in the sense that first a model selection procedure is applied such that then estimation is performed in a finite dimensional parameter space.
- In evaluation of identification procedures emphasis is placed on asymptotic statistical properties.

For the main stream case (and analogously for many other cases), the theory of identification may be decomposeed into three matching modules, namely structure theory, estimation of real valued parameters and model selection.

Let us consider linear state space systems of the form

$$x_{t+1} = Ax_t + B\varepsilon_t \tag{1}$$
$$y_t = Cx_t + \varepsilon_t \tag{2}$$

where y_t are the s-dimensional outputs, x_t the n-dimensional states, the ε_t form a white noise innovations process with $\Sigma = \mathbb{E}\varepsilon_t\varepsilon_t'$ and $A \in \mathbb{R}^{n\times n}, B \in \mathbb{R}^{n\times s}, C \in \mathbb{R}^{s\times n}$ are the parameter matrices. For simplicity of presentation we do not include observed inputs here. We assume the stability condition

$$|\lambda_{\max}(A) < 1| \tag{3}$$

and the miniphase condition

$$|\lambda_{\max}(A - BC) < 1| \tag{4}$$

to be satisfied (where $\lambda_{\max}$ denotes the eigenvalue of maximum modulus) and $\Sigma > 0$. The transfer function is of the form

$$k(z) = \sum_{j=1}^{\infty} K_j z^j + I; \qquad K_j = CA^{j-1}B. \tag{5}$$

For ARMA systems we have

$$a(z)y_t = b(z)\varepsilon_t \tag{6}$$

where $a(z)$ and $b(z)$ are polynomial matrices in the backward shift z, and and where the stability and miniphase condition are assumed to hold. Here the transfer function is given by

$$k(z) = a^{-1}(z)\dot{b}(z). \tag{7}$$

The spectral densitiy is given as

$$f(\lambda) = (2\pi)^{-1}k(e^{-i\lambda})\Sigma k^*(e^{-i\lambda}) \tag{8}$$

where $*$ denotes the conjugate transpose.

Under the assumptions mentioned above (where in the ARMA case in addtiion $a(0) = b(0)$ is required) $k(z)$ and Σ are uniquely determined by the (population) second moments of $(y_t|t \in \mathbb{Z})$, i.e. by f.

In *structure theory*, in general terms, the relation between external system behavior and internal parameters is analysed from the point of view of an idealized identification problem commencing from the extrenal behavior rather than from data. Here, for the linear main stream case, we commence from the transfer functions $k(z)$, and the internal parameters are contained in (A, B, C) or in $(a(z), b(z))$. For brevity, we will focus on the state space case, thus the relation defined by (6) will be considered: Let U_A denote the set of all transfer functions, defined by (6), for fixed s and variable n and satisfying our assumptions, let T_A denote the set of all state space systems (A, B, C), for fixed s but variable n, satisfying our assumptions and denote by $\pi : T_A \to U_A$ the mapping defined by (6) . If T_A (or U_A) is the original (preliminary) model class, then it is convenient to break down T_A and U_A into substes T_α and $U_\alpha, \alpha \in I$, respectively such that a parametrization, i.e. a mapping $\psi_\alpha : U_\alpha \to T_\alpha$ satisfying $\psi_\alpha(\pi(A, B, C)) = (A, B, C) \quad \forall (A, B, C) \in T_\alpha$ exists and that $\psi_\alpha$is continuous (w.r.t. a "reasonable" topology for U_α), and thus is a homeomorphism, and additionally has certain differentiability properties. For actual choices of parametrization see e.g. Deistler (2001) and the references there.

In most cases, for given T_α, not all entries in (A, B, C) will be free, then they will be described by free parameters $\tau \in \mathbb{R}^{d_\alpha}$ in a one-to-one homeomorphic way. We will use the symbol $\tau \subset \mathbb{R}^{d_\alpha}$ also for the set of free parameters.

Estimation for a given subclass: Here we assume that the subclass U_α, parametrized by a finite dimensional set $T_\alpha$of free parameters, is given or has been selected. In a certain sense, estimation here contains a model reduction step. Gau"sian maximum likelihood estimation is most important in this case. There $-2T^{-1}$ times the logarithm of the likelihood is given up to a constant by

$$\hat{L}_T(\theta) = T^{-1} \log \det \Gamma_T(\theta) + T^{-1} y'(T) \Gamma_T(\theta)^{-1} y(T). \tag{9}$$

Here T denotes the sample size, $y(T) = (y_1', \ldots, y_T')'$ is the stacked vector of observations, θ is the parameter vector consisting of τ and Σ and finally $\Gamma_T(\theta) = \mathbb{E} y(T; \theta) y'(T; \theta)$ is the variance matrix of a stacked vector $y(T; \theta)$ generated by a system with parameter θ. The MLE, $\hat{\theta}_T$ say, then is obtained by minimizing $\hat{L}_T$ over T_α and the set of all positive covariance matrices Σ. Note that $\hat{L}_T$ depends on τ only via k.

Consistency of the MLE's $\hat{k}_T$ and $\hat{\Sigma}$ can be shown under rather general conditions (see e.g. Hannan and Deistler 1988). For consistency proofs uniform laws of large numbers are used in order to show an appropriate mode of convergence of $\hat{L}_T$ towards the asymptotic likelihood function

$$\begin{aligned} L(k, \Sigma) &= \log \det \Sigma + (2\pi)^{-1} \\ &\times \int_{-\pi}^{\pi} \operatorname{tr} \left[\left(k\left(e^{-i\lambda}\right) \Sigma k^*\left(e^{-i\lambda}\right) \right)^{-1} \left(k_0\left(e^{-i\lambda}\right) \Sigma_0 k_0^*\left(e^{-i\lambda}\right) \right) \right] d\lambda, \end{aligned} \tag{10}$$

which guarantees that $\hat{k}_T, \hat{\Sigma}_T$ converge to the true values k_0, Σ_0 (and thus θ_0), which are the unique minimizers of L. The basic idea of this proof has been introduced by A. Wald for the i.i.d. case; the complications in this case arise from both, the dependence of observations and the properties (e.g noncompactness) of the parameter spaces.

Under further assumptions a central limit theorem for the MLE's $\hat{\theta}_T$ can be shown, namely that $T^{\frac{1}{2}}(\hat{\theta}_T - \theta_0)$ has a Gaussian limiting distribution with mean zero and covariance matrix given by the inverse of the Fisher information matrix. These results are valid for a rather large class of non Gaussian observations too.

A common way of *model selection*, in particular of order estimation is to minimize an information criterion of the form

$$A(\alpha) = \log\det \hat{\Sigma}_T(\alpha) + d_\alpha \cdot c(T) \cdot T^{-1} \tag{11}$$

where $\hat{\Sigma}_T(\alpha)$ is the MLE of Σ_0 corresponding to T_α, d_α is the dimension of the parameter space T_α and $c(T)$ is a prescribed function. The first term of the r.h.s. of (11) is a measure for goodness of fit of the best fitting system in T_α to the data. The second term is a measure of complexity of the model class used; thus the idea is to obtain an optimal tradeoff between fit and complexity. The parameter spaces T_α must have certain additional properties in order to make these procedures meaningful. By far the most important choices for $c(T)$ are $c(T) = 2$, in this case $A(\alpha)$ is called AIC criterion, and $c(T) = c \cdot \log(T), c \geq 1$ which gives the BIC criterion. Both criteria have been suggested by Akaike. Schwarz has given an important Bayesian derivation of BIC, showing that in this framework one has to commence from priors which are not absolutely continuous for the original overall model class, but give positive a priori probabilities to every T_α.

Consistency for BIC and AIC has been analysed by Hannan. It is important to note, that the uncertainty coming from model selection adds uncertainty to the estimators of real valued parameters. As shown in Leeb and P"otscher (2000) the statistical analysis of such post model selection estimators raises important questions and sheds doubts on the naive use and interpretation of such procedures.

5 Present State and Future Developments

As has been mentioned above, with the end of the eighties of the last century, the subject of identification of linear systems had reached a certain degree of maturity. Subsequentely, the development shows a much more fragmented picture.

The applications of system identification procedures have been drastically increasing over the last three decades.

Whereas thirty years ago the field was very much paradigm driven, the main driving forces today seem to be special requirements, model classes, and data structures arising in applications. As a consequence there is an increasing diversification corresponding to different fields of application, for instance major developments in time series econometrics such as cointegration and GARCH-type modeling barely received any attention in the engineering part of system identification.

The use of system identification in application has been drastically increasing over the last thirty years. On the econometric side here data driven modeling in finance, marketing or logistics should be mentioned; on the engineering side data driven modeling in speech processing or for chemical plants could serve as examples for a large and increasing number of applications in signal processing and process automation. System identification is also increasingly used in other areas, the scope ranging from metereology over telecommunication networks to physiology and genetics.

Perhabs the most active community in the nineties was the econometrics community. Triggered by the failures in the forecasting behavior of large macro models during the first oil crisis in the mid seventies and slightly later by the increasing interest in finance-, household- and firms data, a boom in the development of methods and theories which drastically changed econometrics at such, started with the late seventies. The two most important developments in time series econometrics are cointegration and nonlinear modeling for financial time series.

In cointegration (Engle and Granger 1987) observations are modeled by special classes of linear unstable systems. This allows to describe trends in means and variances which are often observed in economic time series. In addition, by such models linear combinations which make the nonstationary components stationary can be modeled. Such linear combinations represent the long term equilibrium of the economic system and therefore are of particular interest. A seminal paper on cointegration (in an ARX context), containing structure theory, estimation and specification is Johansen (1991). The asymptotic theory there is non standard, since e.g. nonnormal limiting laws of estimators are typical here.

In order to describe stylized facts exhibited by finance data, interesting new model classes and corresponding identification procedures have been developed. Of particular interest are GARCH—type models, the development of which was triggered by Engle (1982) and where volatility clustering often seen in finance data can be modeled.

Identification of linear systems is a mature, but not a saturated area. Novel research topics there are e.g.: Identification under control, identification of unstable and long memory systems, Bayesian time series analysis for systems with time varying parameters and identification of a priori highly structured systems (e.g. compartemental systems)

On the other hand in identification of non-linear systems there is still a large number of open problems. "Identification of nonlinear systems" is a statement like "non elephant zoology" as it has many different aspects.

Important parts are:

- For a rather general asymptotic theory for M-estimators in parametric classes of nonlinear systems, see e.g. P"otscher and Prucha (1997). Here the ideas from linear system identification (which is a nonlinear problem) are extended to the nonlinear case. Since there is no general structure

theory available in this case the results rest on assumptions which are "structural" and which have to be proved for the respective model class under consideration.
- Semi-parametric identification of (e.g. recurrent) neural networks or identification using wavelets
- Nonlinear, nonparametric modeling, e.g. by nonlinear autoregression, see e.g. Robinson (1983) for an early contribution in this direction
- Identification using special nonlinear model classes such as GARCH models

As far as future tendencies and developments are concerned, according to my subjective judgement the following will result:

- A further fragmentation according to different model classes, data structures and requirements from different fields of application. System identification will become more "multicultural" both from the methodological approach and with respect to applications and their feedback the methods and theory
- Intensified development in nonlinear system identification, both for black and grey box models. Here in particular structure theories for classes of nonlinear systems have to be developed
- Identification for spatio-temporal models e.g. of partial differential equations. This is clearly related to inverse problems
- Further automatization of identification procedures; this may serve e.g. for faciliteling experts or for complete automatization. Important problem classes here are: Removal of outliers, selection of inputs, specification of functional forms and dynamics.
- Identification in case of large data sets which are available for instance in data marts. Identification using data of different sources (data fusion)
- Increasing use of methods from computer algebra and symbolic computation
- Increasing use of nonstochastic approaches of different kind (e.g. Willems 1986)
- Increasing use of hybrid approaches mixing algorithms and logic modules for identification

System identification is not a new, but a still dynamic and interesting area. Part of its attraction is coming from the integration between data analysis and the analysis of (stochastic) dynamic systems. Whereas is the past, for several decades theoretical issues have received major attention, beginning with the seventies and eighties of the 20th century, and of course triggered by the enormous development of computers, aspects of application and calculation moved to the foreground. System identification as defined in this contribution is still treated in different scientific communities such as time series statistics, econometrics and system and control theory. The boundaries to other areas are flexible, many important ideas and methods come from different areas of applications where special tools for data analysis are developed. Of course this creates also permanent competition with other fields.

References

1. Anderson, T. W. (1971) The Statistical Analysis of Time Series, Wiley, New York.
2. Box, G. E. P. and Jenkins, G. M. (1970) Time Series Analysis: Forecasting and Control, San Francisco: Holden Day.
3. Caines P. E. (1988) Linear Stochastic Systems, John Wiley & Sons, New York.
4. Cramer, H. (1946) Mathematical Methods of Statistics, Princeton University Press, Princeton.
5. Davis, H. T. (1941) The Analysis of Economic Time Series, Principia press.
6. Deistler, M. (2001) System identification—General aspects and Structure. In: Goodwin G. C. (ed.), Model Identification and Adaptive Control, Springer, London.
7. Engle, R. F. (1982) Autoregressive Conditional Heteroskedasticity with Estimates of the Variance of the U.K. Inflation. Econometrica **50**, 987–1008
8. Engle, R. E. and Granger, C. W. J. (1987) Cointegration and error correction: Representation estimation and testing, Econometrica **55**, 251–270.
9. Haavelmo, T. (1944) The probability approach in econometrics. Econometrica **12** (Supplement), 1–115.
10. Hannan, E. J. (1960) Time Series Analysis, Methuen, New York.
11. Hannan, E. J. (1970) Multiple time series, New York: Wiley.
12. Hannan, E. J. and Deistler, M. (1998) The Statistical Theory of Linear Systems, John Wiley & Sons, New York.
13. Hazewinkel, M., and Kalman, R. E., (1976). Invariants, canonical forms and moduli for linear, constant, finite dimensional dynamical systems. In Marchesini G. and Mitter S. K. (eds.): Lecture Notes in Economics and Mathematical Systems, **131**, Springer Verlag, Berlin, 48–60.
14. Johansen, S. (1991) Estimation and hypothesis testing of cointegration in Gaussion vector autoregressive models, Econometrica **53**, 1551–1586.
15. Leeb, H. and Pötscher, B. M. (2000) The finite sample distribution of post-model-selection estimation, and uniform versus non-uniform approximations. Mimeo, Institute for Statistics and Decision Support Systems, University Vienna.
16. Ljung, L. (1987) System Identification: Theory for the User, Prentice Hall, Englewood Cliffs, New Jersey.
17. Koopmans, T. C., Rubin, H., and Leipnik, R. B. (1950). Measuring the Equation Systems of Dynamic Economics. In: Koopmans, T. C. (ed), Cowles Commission Monograph No. **10**, John Wiley, New York.
18. Mann, H. B. and Wald A. (1943) On the Statistical treatment of linear difference equations. Econometrica **11**, 173–220.
19. Morgan, M. S. (1990) The History of Econometric Ideas, Cambridge University Press, Cambridge.
20. Pötscher, B. M. and Prucha, I. (1997) Dynamic Nonlinear Econometric Models: Asymptotic Theory, Springer Verlag, Berlin.

21. Reinsel, G. C. (1993) Elements of Multivariate Time Series Analysis, Springer, New York.
22. Robinson, P. M. (1983) Nonparametric estimators for time series. Journal of Time Series Analysis **4**, 185–207
23. Söderström, T. and Stoica, P. (1989) System Identification, Prentice Hall, New York.
24. Willems, J. C. (1986) From time series to linear system. Part I: Finite dimensional linear time invariant systems, Automatica **22**, 561–580.

Max-Plus Stochastic Control

Wendell H. Fleming

Brown University, Providence, RI 02912 USA

Abstract. Max-plus stochastic processes are counterparts of Markov diffusion processes governed by Ito sense stochastic differential equations. In this framework, expectations are linear operations with respect to max-plus arithmetic. Max-plus stochastic control problems are considered, in which a minimizing control enters the state dynamics and running cost. The minimum max-plus expected cost is equal to the upper Elliott-Kalton value of an associated differential game.
Key words. Max-plus stochastic control, differential games.
AMS Subject Classification. 35F20, 60H10, 93E20.

1 Introduction

The Maslov idempotent calculus provides a framework in which a variety of asymptotic problems, including large deviations for stochastic processes, can be considered [15]. The asymptotic limit is typically described through a deterministic optimization problem. However, the limit still retains a "stochastic" interpretation, if probabilities are assigned which are additive with respect to "max-plus" addition and expectations are defined so as to be linear with respect to max-plus addition and scalar multiplication. Instead of "idempotent probability" we will use the term "max-plus probability". There is an extensive literature on max-plus probability and max-plus stochastic processes. See [1][2][14][15][17] and references cited there. The max-plus framework is also important for certain problems in discrete mathematics, and in computer science applications [2][14].

In [7] some elements of a max-plus stochastic calculus were considered, including max-plus martingales and stochastic differential equations. In the present paper, a brief introduction to max-plus stochastic control is given. As formulated in Section 2, the state x_s which is being controlled satisfies a differential equation (2). The state dynamics depend on a minimizing control u_s and also on a disturbance v_s. The max-plus probability law of disturbance trajectories is given by a likelihood function $Q(v_.)$ of the form (3). The objective is to choose a minimizing control, depending on past information, to minimize the max-plus expectation of a cost function Z of the form (5).

Associated with a max-plus stochastic control problem is a two controller, zero sum differential game in which the disturbance v_s is the maximizing control. The upper value function $\overline{V}$ satisfies the upper Isaacs PDE (14) in the viscosity sense. At least formally, an optimal feedback control $\mathbf{u}^*$ is obtained via (17). Under additional (rather restrictive) assumptions this is made precise in Section 3.

B. Pasik-Duncan: Stochastic Theory and Control, LNCIS 280, pp. 111–119, 2002.

In Section 4 strictly progressive strategies for the minimizing controller are defined. These are max-plus analogues of progressively measurable controls in the theory of controlled Markov diffusions. Theorem 4.1 states that $\overline{V}$ equals the max-plus optimal cost function W. The lower game value $\underline{V}$ would be obtained if the minimizing controller were allowed to use strategies which are progressive rather than strictly progressive. We call the difference $\overline{V} - \underline{V}$ an instantaneous information gap.

In Section 5 we recall a result according to which a max-plus stochastic control problem is the small-noise -intensity limit of risk sensitive stochastic control problems. This result is a control counterpart of results in the Freidlin-Wentzell theory of large deviations for small random perturbations of dynamical systems.

2 Max-Plus Control Problem

Let $I\!R$ denote the real numbers and $I\!R^- = I\!R \cup \{-\infty\}$. Let Ω be some "sample space", and let Q be a $I\!R^-$-valued function on Ω with

$$\sup_{\omega \in \Omega} Q(\omega) = 0.$$

We call $Q(\omega)$ the *likelihood* of ω and Q the *max-plus probability density function*. The max-plus expectation of a $I\!R^-$-valued Z on Ω is

$$E^+(Z) = \sup_{\omega \in \Omega} [Z(\omega) + Q(\omega)] \tag{1}$$

provided the right side is not $+\infty$. E^+ is a linear operation with respect to max-plus addition and scalar multiplication. The max-plus probability $P^+(A)$ of a set $A \subset \Omega$ is obtained when $Z(\omega) = 0$ for $\omega \in A$ and $Z(\omega) = -\infty$ for $\omega \in \Omega / A$.

We consider max-plus stochastic control problems which are formulated as follows. Let t denote an initial time and T a final time, with $0 \le t < T$. Let x_s denote the state and u_s the control at time s, with $x_s \in I\!R^n$, $u_s \in \mathcal{U}$, $t \le s \le T$. Moreover, let v_s denote a "disturbance" at times s, with $v_s \in \mathcal{V}$. The state dynamics are the differential equation

$$\frac{dx_s}{ds} = F(x_s, u_s, v_s) \ , \ t \le s \le T, \tag{2}$$

with initial data $x_t = x$. The likelihood of disturbance $v_.$ is

$$Q(v_.) = -\int_t^T q(v_s)\, ds. \tag{3}$$

We make the following assumptions:
(A1) $\mathcal{U} \subset I\!R^k$, $\mathcal{V} \subset I\!R^m$ and $\mathcal{U}, \mathcal{V}$ are compact.

(A2) F is of class C^1 on $\mathbb{R}^{\mathrm{n}} \times \mathcal{U} \times \mathcal{V}$, and F together with its first order partial derivatives is bounded.
(A3) q is continuous on $\mathcal{V}$ and $\max_{v \in \mathcal{V}} q(v) = 0$.

Note that in this model, "max-plus randomness" enters via the disturbances v. The sample space is $\Omega = L^\infty([t,T]\ ;\ \mathcal{V})$ with likelihood of $v_.$ given by (3).

In [7] the special case $F(x,u,v) = f(x,u) + \sigma(x,u)\,v$ is considered. In that case, (2) can be written equivalently as

$$dx_{\mathrm{s}} = f(x_{\mathrm{s}}, u_{\mathrm{s}})\,ds + \sigma(x_{\mathrm{s}}, u_{\mathrm{s}})\,dw_{\mathrm{s}}, \tag{4}$$
$$w_{\mathrm{s}} = \int_{\mathrm{t}}^{\mathrm{s}} v_{\mathrm{r}}\,dr.$$

In the terminology of [17], w_{s} is a max-plus independent increment process. In Section 5 we will take $\mathcal{V} = \mathbb{R}^{\mathrm{m}}$, $\Omega = L^2([t,T]\ ;\ \mathbb{R}^{\mathrm{m}})$ and $-Q(v_.)$ half the L^2 norm. In that case, (4) is the max-plus analogue of an Ito-sense stochastic differential equation in which w_{s} is replaced by a Brownian motion B_{s}.

In the terminology of [17], w_{s} is a max-plus independent increment process. In Section 5 we will take $\mathcal{V} = \mathbb{R}^{\mathrm{m}}$, $\Omega = L^2([t,T]\ ;\ \mathbb{R}^{\mathrm{m}})$ and $-Q(v_.)$ half the L^2 norm. In that case, (4) is the max-plus analogue of an Ito-sense stochastic differential equation in which w_{s} is replaced by a Brownian motion B_{s}.

Let $l(x,u)\,,g(x)$ denote running cost and terminal cost functions, such that:
(A4) l, g are of class C^1 and bounded together with their first order partial derivatives.

Let

$$Z = \int_{\mathrm{t}}^{\mathrm{T}} l(x_{\mathrm{s}}, u_{\mathrm{s}})\,ds + g(x_{\mathrm{T}})\,. \tag{5}$$

The control problem is to find a control u_{s} which minimizes the max-plus expectation $E^+_{\mathrm{t}x}(Z)$. (The subscripts t, x refer to the initial data $x_{\mathrm{t}} = x$ for (2).) The control u_{s} is chosen using "complete information" about the past up to time s, in a sense which must be made precise. In a way similar to the theory of controlled Markov diffusions [10], two kinds of control strategies are considered. In Section 3 feedback control policies based on current states are considered. Then in Section 4 strictly progressive control strategies are considered, which correspond to progressively measurable control processes in the theory of controlled Markov diffusions.

By (1), (5)

$$E^+_{\mathrm{t}x} Z = \sup_{v_.} P(t, x\ ;\ u_., v_.) \tag{6}$$

$$P(t, x\ ;\ u_., v_.) = \int_{\mathrm{t}}^{\mathrm{T}} [l(x_{\mathrm{s}}, u_{\mathrm{s}}) - q(v_{\mathrm{s}})]\,ds + g(x_{\mathrm{T}})\,. \tag{7}$$

This suggests that the max-plus stochastic control problem is closely related to a two-controller, zero sum differential game. In this differential game the minimizing controller chooses u_s. The maximizing controller chooses a disturbance v_s. The game dynamics are (2) and the game payoff is (7). Under assumptions (A.1)-(A.4) this game has upper and lower Elliott-Kalton sense values, $\overline{V}(t,x)$ and $\underline{V}(t,x)$. It will be seen that the infimum of $E^+_{tx}(Z)$ among strictly progressive strategies for the minimizing controller equals the upper game value $\overline{V}(t,x)$. See Theorem 4.1.

3 Feedback Control Policies

In this section we consider controls which are feedback functions of time and state:

$$u_s = \mathbf{u}(s, x_s) . \tag{8}$$

To avoid various technical difficulties, we consider only control policies which are Lipschitz continuous functions from $[0,T] \times I\!R^n$ into U. Let

$$J(t, x \; ; \; \mathbf{u}) = E^+_{tx} Z, \tag{9}$$

and substitute (8) in the state dynamics (2) and in (5). By (6), $J(t,x \; ; \; \mathbf{u})$ is the optimal cost function for the following control problem: choose a disturbance control $v_{\cdot}$ to maximize $P(t,x \; ; \; \mathbf{u}, v_{\cdot})$. Let

$$H^u(x,p) = \max_{v \in \mathcal{V}} \left[F(x,u,v) \cdot p + l(x,u) - q(v) \right] . \tag{10}$$

Then $J(t,x \; ; \; \mathbf{u})$ is the unique bounded, uniformly continuous viscosity solution to the dynamic programming PDE

$$0 = \frac{\partial J}{\partial t} + H^{\mathbf{u}}(x, D_x, J) \; , \; 0 \le t < T \tag{11}$$

with the terminal data

$$J(T, x \; ; \; \mathbf{u}) = g(x) \; , \; x \in I\!R^n . \tag{12}$$

See [10,Chapt. 2] [3,Chap. 3]. In (11), $D_x J$ denotes the gradient vector. Let

$$\bar{H}(x,p) = \min_{u \in \mathcal{U}} H^u(x,p) . \tag{13}$$

The upper value function $\overline{V}(t,x)$ is the unique bounded, uniformly continuous viscosity solution to

$$0 = \frac{\partial \overline{V}}{\partial t} + \bar{H}\left(x, D_x \overline{V}\right) \tag{14}$$

$$\overline{V}(T,x) = g(x). \tag{15}$$

See [6][3,Chap. 8]. Equation (14) is called the upper Isaacs PDE. Since $\bar{H} \leq H^{\mathbf{u}}$ a comparison principle for viscosity sub and supersolutions to (14)–(15)[10, p. 86][3,p. 152] gives that

$$\overline{V}(t,x) \leq J(t,x\,;\,\mathbf{u}). \tag{16}$$

If there exists $\mathbf{u}^*$ such that equality holds in (16), then $\mathbf{u}^*$ is an optimal feedback control policy.

The following is a sufficient condition that an optimal continuous $\mathbf{u}^*$ exists. Suppose that $\overline{V}$ is a classical solution to (14)–(15), such that $\overline{V}$ and $D_{\mathrm{x}}\overline{V}$ are continuous and bounded. Moreover, suppose that $\mathbf{u}^*$ is Lipschitz continuous, where

$$\mathbf{u}^*(t,x) \in \arg\min_{u\in\mathcal{U}} H^{\mathrm{u}}\left(x, D_{\mathrm{x}}\overline{V}(t,x)\right). \tag{17}$$

When $\mathbf{u} = \mathbf{u}^*$, $\overline{V}$ is a classical solution to (11)–(12), hence also a viscosity solution. Since $J(t,x\,;\,\mathbf{u}^*)$ is the unique bounded, uniformly continuous viscosity solution to (11)–(12)

$$\overline{V}(t,x) = J(t,x\,;\,\mathbf{u}^*). \tag{18}$$

These assumptions on $\overline{V}$ and $\mathbf{u}^*$ are quite restrictive. In many examples from control theory and differential games, value functions are not of class C^1 and optimal feedback control policies must be discontinuous.

4 Strictly Progressive Strategies

Let us recall the Elliott-Kalton definition of lower and upper game values $\underline{V}(t,x)$, $\overline{V}(t,x)$. See [5]. A strategy α for the minimizing controller maps $L^\infty([t,T]\,;\,\mathcal{V})$ into $L^\infty([t,T]\,;\,\mathcal{U})$. Moreover, $v_{\mathrm{r}} = \tilde{v}_{\mathrm{r}}$ for almost all $r \in [t,s]$ implies $\alpha(v_.)_{\mathrm{r}} = \alpha(\tilde{v}_.)_{\mathrm{r}}$ for almost all $r \in [t,s]$. We call such a strategy *progressive*. By definition

$$\underline{V}(t,x) = \inf_{\alpha}\sup_{v_.} P(t,x\,;\,\alpha(v_.), v_.) \tag{19}$$

where the inf is taken over progressive strategies α. By (6), this is the same as the infimum over progressive α of the max-plus expectation $E^+_{\mathrm{tx}}Z$. Similarly, a strategy β for the maximizing controller is a progressive map from $L^\infty([t,T]\,;\,\mathcal{U})$ into $L^\infty([t,T]\,;\,\mathcal{V})$, and

$$\overline{V}(t,x) = \sup_{\beta}\inf_{u_.} P(t,x\,;\,u_., \beta(u_.)). \tag{20}$$

A strategy α for the minimizing controller is called *strictly progressive* if: for every strategy β for the maximizing controller, the equations

$$u_{.} = \alpha\,(v_{.})\ ,\ v_{.} = \beta\,(u_{.}) \tag{21}$$

have a solution $\hat{u}_{.}, \hat{v}_{.}$. This implies that $\hat{u}_{.}$ is a fixed point of the composite map $\alpha \circ \beta$. At an intuitive level, a strictly progressive strategy α allows the minimizer to choose u_{s} knowing v_{r} for $t \le r < s$, while a progressive strategy allows knowledge of v_{s} also.

Examples 4.1–4.3 provide some useful classes of strictly progressive strategies.

Example 4.1. Let $t = t_0 < t_1 < \cdots < t_{\mathrm{N}} = T$ be a partition of the interval $[t, T]$. Suppose that $v_{\mathrm{r}} = \tilde{v}_{\mathrm{r}}$ for almost all $r \in [t, t_{\mathrm{j}}]$ implies $\alpha\,(v_{.})_{\mathrm{r}} = \alpha\,(\tilde{v}_{.})_{\mathrm{r}}$ for almost $r \in [t, t_{\mathrm{j}+1}]$, $j = 0, 1, \ldots, N-1$. Then α is strictly progressive. For $j = 0$, $u_{\mathrm{r}} = \hat{u}_{\mathrm{r}}$ is chosen open loop on the initial interval $[t, t_1)$ and $\hat{v}_{\mathrm{r}} = \beta\,(\hat{u}_{.})_{\mathrm{r}}$ on $[t, t_1)$. Then $\hat{u}_{\mathrm{r}}$ and $\hat{v}_{\mathrm{r}}$ are chosen on successive intervals $[t_{\mathrm{j}}, t_{\mathrm{j}+1})$ such that (21) holds.

Example 4.2. Suppose that there is a "time delay" $\delta > 0$ such that $\alpha\,(v_{.})_{\mathrm{s}}$ depends on v_{r} only for $r \le s - \delta$. We may assume that $\delta = N^{-1}\,(T - t)$ for some integer N and take $t_{\mathrm{j}} = jN^{-1}\,(T - t)$ in Example 4.1.

Example 4.3. Let $u_{\mathrm{j}} \colon I\!R^{\mathrm{n}} \to U$ and let

$$\alpha\,(v_{.})_{\mathrm{s}} = u_{\mathrm{j}}\,(x_{\mathrm{j}})\ \text{ for } t_{\mathrm{j}} \le s < t_{\mathrm{j}+1},\ j = 0, 1, \ldots, N-1, \tag{22}$$

where $x_{\mathrm{j}} = x_{t_{\mathrm{j}}}$ and x_{s} satisfies (2) on $[t, T]$ with $x_{\mathrm{t}} = x$. Then α is strictly progressive. On $[t, t_1)$, $u_{\mathrm{s}} = u_0\,(x)$ and u_{s} is defined on later intervals $[t_{\mathrm{j}}, t_{\mathrm{j}+1})$ by $u_{\mathrm{s}} = u_{\mathrm{j}}\,(x_{\mathrm{j}})$.

Example 4.4. In this example, α is progressive but not strictly progressive. Let $\alpha\,(v_{.})_{\mathrm{s}} = \phi\,(v_{\mathrm{s}})$ where ϕ is Borel measurable and not constant. There is a partition $\mathcal{U} = \mathcal{U}_1 \cup \mathcal{U}_2$ with $\mathcal{U}_1$ open, such that $\mathcal{V}_1 = \phi^{-1}\,(\mathcal{U}_1)$ and $\mathcal{V}_2 = \phi^{-1}\,(\mathcal{U}_2)$ are both nonempty. Choose $v_{\mathrm{i}} \in \mathcal{V}_{\mathrm{i}}$ for $i = 1, 2$ and ψ such that $\psi\,(u) = v_2$ if $u \in \mathcal{U}_1$, $\psi\,(u) = v_1$ if $u \in \mathcal{U}_2$. The composite $\phi \circ \psi$ has no fixed point. Let $\beta\,(u_{.})_{\mathrm{s}} = \psi\,(u_{\mathrm{s}})$. Then $(\alpha \circ \beta)\,(v_{.})_{\mathrm{s}} = (\phi \circ \psi)\,(v_{\mathrm{s}})$. If (21) holds, then $\hat{u}_{\mathrm{s}}$ is a fixed point of $\phi \circ \psi$ which is a contradiction.

Let

$$W\,(t, x) = \inf_{\alpha} E^{+}_{\mathrm{t}x} Z, \tag{23}$$

where the inf is taken among all strictly progressive strategies α.

Theorem 4.1. $W\,(t, x) = \overline{V}\,(t, x)$.

Let us only sketch a proof of Theorem 4.1, and refer to [7] for details. To show that $W \ge \overline{V}$, given $\theta > 0$ there exists a strategy β for the maximizing controller such that

$$P\,(t, x\,;\, u_{.}, \beta\,(u_{.})) \ge \overline{V}\,(t, x) - \theta \tag{24}$$

for all controls $u_{.}$ for the minimizer. Given any strictly progressive α, let $\hat{u}_{.}$, $\hat{v}_{.}$ be a solution of (21). Then

$$\begin{aligned} E^{+}_{tx} Z &= \sup_{v_{.}} P\left(t, x\,;\, \alpha\left(v_{.}\right), v_{.}\right) \\ &\geq P\left(t, x\,;\, \alpha\left(\hat{v}_{.}\right), \hat{v}_{.}\right) \\ &= P\left(t, x\,;\, \hat{u}_{.}, \beta\left(\hat{u}_{.}\right)\right) \geq \overline{V}\left(t, x\right) - \theta. \end{aligned}$$

Since θ and strictly progressive α are arbitrary, $W \geq \overline{V}$.

The inequality $W \leq \overline{V}$ is proved using time discretizations and viscosity solution arguments. Given a positive integer N, consider $t = t_j = Nj^{-1}T$, $j = 0, 1, \ldots, N$. Let $\overline{V}_N\left(t_j, x\right)$ denote the sup in (20) when u_s is required to be constant on each interval $[t_i, t_{i+1})$ with $i \geq j$. By a construction of Nisio,

$$\overline{V}_N\left(t_j, x\right) = F_N \overline{V}_N\left(t_{j+1}, \cdot\right)(x) \tag{25}$$

$$F_N \psi\left(x\right) = \inf_{u \in \mathcal{U}} \sup_{v_{.}} \left[\int_0^h \left[\ell\left(x_s, u\right) - q\left(v_s\right)\right] ds + \psi\left(x_h\right)\right],$$

where $h = h_N = N^{-1}T$ and x_s satisfies (2.2) with $u_s = u$ and $x_0 = x$. By a viscosity solution technique due to Souganidis [18,19][11], $\overline{V}_N\left(t_j, x\right)$ tends to $\overline{V}\left(t, x\right)$ uniformly on compact sets as $N \to \infty$, $t_j \to t$. Using (25), given $\theta > 0$ there is a strictly progressive α of the form in Example 4.3 such that

$$\sup_{v_{.}} P\left(0, x\,;\, \alpha\left(v_{.}\right), v_{.}\right) < \overline{V}_N\left(0, x\right) + \theta. \tag{26}$$

See [7,Section 7]. Let $N \to \infty$, $\theta \to 0$ to obtain $W\left(0, x\right) \leq \overline{V}\left(0, x\right)$. Since W and $\overline{V}$ depend on the time difference $T - t$, this implies $W \leq \overline{V}$.

Remark 4.1. We call the difference $\overline{V}\left(t, x\right) - \underline{V}\left(t, x\right)$ an *instantaneous information gap.* It equals the difference in the infimum over α of $E^{+}_{tx}(Z)$ depending on whether progressive or strictly progressive strategies α for the minimizing controller are used. If in (10)–(13) min max = max min, then $\overline{V} = \underline{V}$ and the instantaneous information gap is 0. The equality min max = max min in (10)–(13) is the Isaacs minimax condition.

5 Risk Sensitive Control Limits

Risk sensitive control theory provides a link between stochastic and deterministic (min − max) approaches to disturbances in control systems. Let us merely indicate this connection, in the context of controlled Markov diffusions subject to small noise intensity Brownian motion disturbances. For $\varepsilon > 0$, let X^{ε}_s be a controlled Markov diffusion satisfying the stochastic differential equation

$$dX^{\varepsilon}_s = f\left(X^{\varepsilon}_s, U_s\right) ds + \varepsilon^{1/2} dB_s, \quad t \leq s \leq T, \tag{27}$$

with initial data $X_t^\varepsilon = x$. In (27) the control process U_s is progressively measurable with respect to a family of σ-algebras to which the Brownian motion B_s is adapted, and $U_s \in \mathcal{U}$. Let

$$Z^\varepsilon = \int_t^T \ell\left(X_s^\varepsilon, U_s\right) ds + g\left(X_T^\varepsilon\right). \tag{28}$$

We again assume that $\mathcal{U}$ is compact. Moreover, f satisfies the same assumptions on $\mathbb{R}^n \times \mathcal{U}$ as in (A2) and ℓ, g satisfy (A4). The risk sensitive control problem is to choose $U_.$ to minimize

$$\mathcal{J}^\varepsilon\left(t, x\,;\, U_.\right) = E_{tx}\left[\exp\left(\varepsilon^{-1} Z^\varepsilon\right)\right] \tag{29}$$

where exp denotes the exponential function. The value function for this risk sensitive control problem is

$$\Phi^\varepsilon\left(t, x\right) = \inf_{U_.} \mathcal{J}^\varepsilon\left(t, x; U_.\right) \tag{30}$$

For the corresponding max-plus stochastic control problem we take $\mathcal{V} = \mathbb{R}^n$, $q(v) = 1/2|v|^2$ and $F\left(x, u, v\right) = f\left(x, u\right) + v$. This has the form (4), where for simplicity we now take σ to be the identity matrix. Since $\mathbb{R}^n$ is not compact, as required in assumption (A1), a "cut-off" argument in which $\mathbb{R}^n$ is replaced by $\{v : |v| \le M\}$ is needed to obtain Theorem 4.1 in this case. See [7,Sec. 7].

The upper differential game value $\overline{V}\left(t, x\right)$ (which equals $W\left(t, x\right)$ by Theorem 4.1) is obtained from the risk sensitive control value function in the following deterministic limit:

$$\overline{V}\left(t, x\right) = \lim_{\varepsilon \to 0} \varepsilon \log \Phi^\varepsilon\left(t, x\right). \tag{31}$$

See [8][12] and also related results on finite and infinite time horizons in [4][9][13][16]

References

1. Akian, M., Quadrat, J.-P., and Viot, M. (1994) Bellman processes, in Lecture Notes in Control and Info. Sci. No. 199, eds. G. Cohen and J.-P. Quadrat, Springer Verlag.
2. Baccelli, F., Cohen, G. Olsder, G. J., and Quadrat, J.-P. (1992) Synchronization and Linearity: an Algebra for Discrete Event Systems, John Wiley and Sons.
3. Bardi, M. and Capuzzo-Dolcetta, I. (1997) Optimal Control and Viscosity Solutions of Hamilton-Jacobi-Bellman Equations, Birkäuser.
4. Bensoussan, A. and Nagai, H. (1997) Min-max characterization of a small noise limit of risk-sensitive control, SIAM J. Control Optim. **35**, 1093–1115.
5. Elliott, R. J. and Kalton, N. J. (1972) The existence of value in differential games, Mem. Amer. Math. Soc. **126**.

6. Evans, L. C. and Souganidis, P. E. (1984) Differential games and representation formulas for solutions of Hamilton-Jacobi-Isaacs equations, Indiana Univ. Math. J. **33**, 773–797.
7. Fleming, W. H. Max-plus stochastic processes and control, preprint.
8. Fleming, W. H. and McEneaney, W. M. (1992) Risk sensitive control and differential games, Lecture Notes in Control and Inform. Sci. **184**, Springer-Verlag, New York, 185–197.
9. Fleming, W. H. and McEneaney, W. M. (1995) Risk sensitive control on an infinite time horizon, SIAM J. Control Optim. **33**, 1881–1915.
10. Fleming, W. H. and Soner, H. M. (1993) Controlled Markov Processes and Viscosity Solutions, Springer-Verlag.
11. Fleming, W. H. and Souganidis, P. E. (1989) On the existence of value functions of two-player, zero-sum stochastic differential games, Indiana Univ. Math. J. **38**, 293–314.
12. James, M. R. (1992) Asymptotic analysis of nonlinear risk-sensitive control and differential games, Math. Control Signals Systems **5**, 401–417.
13. Kaise, H. and Nagai, H. (1999) Ergodic type Bellman equations of risk-sensitive control with large parameters and their singular limits, Asymptotic Analysis **20**, 279–299.
14. Litvinov, G. L. and Maslov, V. P. (1998) Correspondence Principle for Indempotent Calculus and Some Computer Applications, in Idempotency, J. Gunawardena ed, Publ. Newton Inst. **11**, Cambridge Univ. Press, pp. 420–443.
15. Maslov, V. P. and Samborskii, S. M., eds. (1992) Idempotent Analysis, Advances in Soviet Math. No. **13**, Amer. Math. Soc.
16. Nagai, H. (1996) Bellman equations of risk-sensitive control, SIAM J. Control Optimiz. **34**, 74–101.
17. Quadrat, J.-P. (1998) Min-plus probability calculus, Actes 26 eme École de Printemps d'Informatique Theorique, Noirmoutier.
18. Souganidis, P. E. (1985) Approximation schemes for viscosity solutions of Hamilton-Jacobi equations, J. Differential Eqns. **59**, 1–43.
19. Souganidis, P. E. (1985) Approximation schemes for viscosity solutions of Hamilton-Jacobi equations with applications to differential games, J. Nonlinear Anal., T.M.A.**9**, 217–257.

An Optimal Consumption–Investment Problem for Factor-Dependent Models

Wendell H. Fleming[1] and Daniel Hernández-Hernández[2]*

[1] Division of Applied Mathematics, Brown University, Providence, RI 02912, USA. whf@cfm.brown.edu
[2] Centro de Investigación en Matemáticas, Apartado Postal 402, Guanajuato, Gto. 36000, México. dher@cimat.mx

Abstract. An extension of the classical Merton model with consumption is considered when the diffusion coefficient of the asset prices depends on some economic factor. The objective is to maximize total expected discounted HARA utility of consumption. Optimal controls are provided as well as a characterization of the value function in terms of the associated Hamilton-Jacobi-Bellman equation.

Keywords: Stochastic volatility, portfolio optimization, factor modeling, mean reverting.

AMS Subject Classification: 60F17, 60J20, 65U05, 90A09, 90C40, 93E20

1 Introduction

In this paper we are concerned with an optimal consumption–investment model, where the volatility of the risky asset is explicitly affected by some economic factor. Our goal is to find consumption and investment strategies which maximize total expected discounted HARA utility of consumption. We consider only the case when the HARA parameter γ is negative, however analogous arguments can be used to study the case $0 < \gamma < 1$. The problem is formulated in Section 2.

At each period of time the investor's wealth is divided between a risky asset and a riskless asset. The controls in this problem are the fraction u_t of wealth X_t in the risky asset and the consumption rate c_t as a fraction of X_t. The dynamics of the risky asset are governed by the stochastic differential equation (2). We assume that the riskless interest rate and the mean rate of return of the risky asset are constant. The factor process Y_t represents an underlying economic variable, which is modelled as an ergodic Markov diffusion process, and affects the risky asset price dynamics via the diffusion coefficient $\sigma(Y_t)$.

It is assumed that the Brownian motions driving the risky asset price evolution and the factor process are independent. In [5] we studied the case when noises are correlated, but it relies on different methods from those used in the present paper. In this case the arguments are based in viscosity solutions

* Partially supported by Conacyt under grant 37643-E

B. Pasik-Duncan: Stochastic Theory and Control, LNCIS 280, pp. 121–130, 2002.

for partial differential equations and dynamic programming methods. By a change of probability measure argument, this portfolio optimization problem is reduced to a stochastic control problem with state Y_t. It is shown that the value function $W(y)$ associated with this control problem is a classical solution of the dynamic programming equation (13); see Theorem 3.1.

Several other authors have considered portfolio optimization models in which the interest rate r, the mean return μ or the volatility coefficient σ in (2) may depend on a Markov diffusion process Y_t. The model (2)–(2) for stochastic volatility is of the kind considered by Fouque-Papanicolaou-Sircar [11]. References [14], [15], [8], [16] and [22] consider the problem of maximizing the expected HARA utility of wealth X_T at a final time T, with consumption omitted from the model. See also [3], where the discrete time version of this problem is studied. In his forthcoming Brown University PhD thesis, T. Pang considers investment-consumption problems in which the factor Y_t is the "riskless" interest rate, with μ and σ constant.

2 Problem Formulation

We begin describing the model for which we want to find optimal consumption–investment strategies. In this model the investor consumes from his wealth, denoted by X_t, generated by his investment strategy, subject to the restriction that he must be solvent all the time, i.e. $X_t > 0$ for all $t \geq 0$. At each time t, the investor divides his wealth into two kinds of assets, one called "risky" and the other "riskless". The dynamics of the risky asset are governed by the Ito sense stochastic differential equation

$$dP_t = \mu P_t dt + \sigma(Y_t) P_t dW_t^1. \tag{1}$$

In this model the diffusion coefficient of the process P_t depends on some diffusion process Y_t of underlying economic factors. That is, the volatility is stochastic, while the mean rate of return μ is constant. See [14], [11]. The dynamics of Y_t are given by

$$dY_t = g(Y_t)dt + \beta dW_t^2, \qquad Y_0 = y \in \mathbb{R}, \tag{2}$$

where the processes W^1 and W^2 are independent Browinan motions with respect to the σ-algebra $\mathcal{F}_t$ defined on a probability space $(\Omega, \mathcal{F}, \mathbf{P})$. This means that the Brownian fluctuations in the factor process are independent of those in the risky asset. The riskless asset pays a fixed rate of return r called interest rate, with $\mu > r$.

Throughout we assume the following: The functions σ and g belong to $C^1(\mathbb{R})$ and

$$\begin{cases} \text{(i) } \sigma_y \text{ is bounded and } \sigma_l \le \sigma(\cdot) \le \sigma_u \text{ for some constants } \sigma_u > \sigma_l > 0; \\ \text{(ii) } g_y \text{ is bounded and there exists k} > 0 \text{ such that } g_y \le -\text{k}. \end{cases} \tag{3}$$

Let u_t be the fraction of wealth held in the risky asset and $c_t X_t$ be the rate at which wealth is consumed. We will consider the case without restrictions, that is u_t takes values in $\mathbb{R}$, while $c_t \ge 0$. The set of admissible strategies (u, c), denoted by $\mathcal{A}$, consists of bounded $\mathcal{F}_t$-progressively measurable processes. The probability space and family $\{\mathcal{F}_t\}$ of σ-algebras is given. Since there are optimal control policies which are feedback functions of Y_t, the particular choice of probability space and $\{\mathcal{F}_t\}$ turns out not to be important.

Then, from the self-financing property of our portfolio, for $(u, c) \in \mathcal{A}$, X_t satisfies

$$dX_t = X_t[(r + (\mu - r)u_t - c_t)dt + u_t\sigma(Y_t)dW_t^1], \tag{4}$$

with initial condition $X_0 = x > 0$. Furthermore, $X_t > 0$ for all $t \ge 0$ with solution given by

$$X_t = x \exp\left\{rt + \int_0^t \left[(\mu - r)u_s - c_s - \frac{1}{2}u_s^2\sigma^2(Y_s)\right]ds + \int_0^t u_s\sigma(Y_s)dW_s^1\right\}. \tag{5}$$

The objective is to maximize total expected discounted HARA utility of consumption on the infinite horizon

$$J(x, y; c, u) = \mathbf{E}\int_0^\infty \frac{1}{\gamma}e^{-\alpha t}(c_t X_t)^\gamma dt, \quad \alpha > 0, \ \gamma < 0, \tag{6}$$

over the set of addmisible strategies $\mathcal{A}$. We anticipate that the value function has the form $V(x, y) = \frac{x^\gamma}{\gamma}\tilde{W}(y)$ for some function $\tilde{W}$. Under Condition (3) we shall characterize $\tilde{W}$ as the value function of another optimal control problem. The corresponding dynamic programming equation associated with V is

$$\begin{aligned} &-\alpha V + \frac{1}{2}\beta^2 V_{yy} + g(y)V_y \\ &\quad + \sup_{u\in\mathbb{R}, c\ge 0}\left\{[r + (\mu - r)u - c]xV_x + \frac{1}{2}u^2\sigma^2(y)x^2V_{xx} + \frac{1}{\gamma}c^\gamma x^\gamma\right\} = 0. \end{aligned} \tag{7}$$

Given $T > 0$ and $(u, c) \in \mathcal{A}$, using (5) the form of the corresponding finite horizon functional can be derived

$$\begin{aligned}
&\mathbf{E} \int_0^T \frac{1}{\gamma} e^{-\alpha t} (c_t X_t)^\gamma dt \\
&\quad = \mathbf{E} \int_0^T \frac{x^\gamma}{\gamma} e^{-\alpha t} c_t^\gamma e^{r\gamma t + \gamma \int_0^t [(\mu - r) u_s - c_s - \frac{1}{2} u_s^2 \sigma^2(Y_s)] ds + \gamma \int_0^t u_s \sigma(Y_s) dW_s^1} dt \\
&\quad = \frac{x^\gamma}{\gamma} \mathbf{E}T \int_0^T e^{-\alpha t} c_t^\gamma e^{r\gamma t + \gamma \int_0^t [(\mu - r) u_s - c_s] ds + \frac{\gamma(\gamma-1)}{2} \int_0^t u_s^2 \sigma^2(Y_s) ds} dt \\
&\quad = \frac{x^\gamma}{\gamma} \tilde{J}(y; c, u, T),
\end{aligned} \tag{8}$$

with $\tilde{J}$ defined implicitly from (8). The second equality follows changing the probability measure, via a Girsanov transformation, with Radon-Nikodym derivative

$$\frac{d\tilde{\mathbf{P}}}{d\mathbf{P}}|_{\mathcal{F}_T} = \exp\{\gamma \int_0^T u_s \sigma(Y_s) dW_s^1 - \frac{1}{2}\gamma^2 \int_0^T u_s^2 \sigma^2(Y_s) ds\}.$$

Under the new measure $\tilde{\mathbf{P}}$ the process $B_t := W_t^1 - \gamma \int_0^t u_s \sigma(Y_s) ds$ is a Brownian motion adapted to $\{\mathcal{F}_t\}$. This argument is valid since u is bounded. This transformation has been used previously in [7] and [8], where the criterion to be optimized is expected HARA utility of an investor's final wealth.

Given (u, c) admissible, defining the process

$$z_t := -\alpha t + r\gamma t + \gamma \int_0^t [(\mu - r) u_s - c_s] ds + \frac{\gamma(\gamma - 1)}{2} \int_0^t u_s^2 \sigma^2(Y_s) ds,$$

write

$$\tilde{J}(y; c, u, T) = \mathbf{E}T \int_0^T c_t^\gamma e^{z_t} dt$$

and define the value function

$$W(y, T) := \inf_{c,u} \tilde{J}(y; c, u, T).$$

Observe that independence of Brownian motions W^1 and W^2 implies that the dynamics of the factor process Y_t remains the same after the change of measure $\tilde{\mathbf{P}}$. Now, writing the z_t process as

$$\begin{aligned}
z_t = &-\alpha t + r\gamma t - \gamma \int_0^t c_s ds + \int_0^t \frac{\gamma}{2(1-\gamma)} \frac{(\mu - r)^2}{\sigma^2(Y_s)} ds \\
&+ \frac{\gamma(\gamma - 1)}{2} \int_0^t \sigma^2(Y_s)(u_s - \frac{\mu - r}{(1-\gamma)\sigma^2(Y_s)})^2 ds,
\end{aligned}$$

we conclude, since u_t does not affect the dynamics of the state process Y_t and $\gamma < 0$, that

$$u_t^* = \frac{\mu - r}{(1-\gamma)\sigma^2(Y_t)} \tag{9}$$

is an optimal control for $\tilde{J}(y; c, u, T)$ for y and T fixed. Note that u^* is also optimal for the infinite horizon functional $\tilde{J}(y; c, u) := \tilde{\mathbf{E}} \int_0^\infty c_t^\gamma e^{z_t} dt$. The value function associated with this functional is denoted by $W(y)$.

Then, evaluating $\tilde{J}(y; c, u)$ in the optimal control u^*, we get a new functional $\bar{J}$, which we want to minimize, defined by

$$\begin{aligned} \bar{J}(y; c) &:= \tilde{J}(y; c, u^*) \\ &= \tilde{\mathbf{E}} \int_0^\infty e^{-\alpha t} c_t^\gamma e^{\gamma \int_0^t (Q(Y_s) - c_s) ds} dt \\ &= \tilde{\mathbf{E}} \int_0^\infty c_t^\gamma e^{\bar{z}_t} dt, \end{aligned}$$

with

$$Q(y) := r + \frac{1}{2} \frac{(\mu - r)^2}{(1-\gamma)\sigma^2(y)}$$

and

$$\bar{z}_t = -\alpha t + r\gamma t - \gamma \int_0^t c_s ds + \int_0^t \frac{\gamma}{2(1-\gamma)} \frac{(\mu - r)^2}{\sigma^2(Y_s)} ds.$$

From (3) we get

$$\begin{aligned} &\tilde{\mathbf{E}} \int_0^T e^{-\alpha t} c_t^\gamma e^{\gamma \int_0^t (r + \frac{1}{2} \frac{(\mu - r)^2}{(1-\gamma)\sigma_l^2} - c_s) ds} dt \\ &\qquad \leq \tilde{J}(y; c, u^*, T) \leq \tilde{\mathbf{E}} \int_0^T e^{-\alpha t} c_t^\gamma e^{\gamma \int_0^t (r + \frac{1}{2} \frac{(\mu - r)^2}{(1-\gamma)\sigma_u^2} - c_s) ds} dt. \end{aligned}$$

Let

$$\bar{\alpha} := \alpha - \gamma r - \frac{\gamma(\mu - r)^2}{2(1-\gamma)\sigma^2}.$$

Then, taking the infimum with respect to addmisible c, the value of the left and rigth side correspond to the constant

$$C(\sigma, T) = e^{-\alpha T} \left[\frac{1-\gamma}{\bar{\alpha}} (1 - e^{-\frac{\bar{\alpha} T}{1-\gamma}}) \right]^{1-\gamma},$$

with σ equal to σ_l and σ_u, respectively. See [6], p. 161. Hence, fixing $T_1 > 0$, there exist positive constants $C(\sigma_l)$ and $C(\sigma_u)$ such that for $T > T_1$

$$C(\sigma_l) \leq W(y), \ W(y, T) \leq C(\sigma_u). \tag{10}$$

Given $y, \hat{y}$ initial conditions for the factor process and $(u, c) \in \mathcal{A}$, denote by Y_t, $\hat{Y}_t$ the corresponding solutions of (2). Define also $\hat{\bar{z}}_t$ as above when $\hat{Y}_t$ replaces Y_t. Then

$$\mathbf{E}T \int_0^T c_t^\gamma e^{\hat{\bar{z}}_t} - \mathbf{E}T \int_0^T c_t^\gamma e^{\bar{z}_t} dt = \mathbf{E}T \int_0^T c_t^\gamma e^{\hat{\bar{z}}_t}(1 - e^{\bar{z}_t - \hat{\bar{z}}_t}) dt$$
$$\leq \mathbf{E}T \int_0^T c_t^\gamma e^{\hat{\bar{z}}_t}(\hat{\bar{z}}_t - \bar{z}_t) dt. \tag{11}$$

On the other hand, using (3) and Gronwall's inequality, we obtain

$$\begin{aligned} \hat{\bar{z}}_t - \bar{z}_t &= \frac{\gamma(\mu - r)^2}{2(1-\gamma)} \int_0^t \left[\frac{1}{\sigma^2(\hat{Y}_s)} - \frac{1}{\sigma^2(Y_s)} \right] ds \\ &\leq \frac{\gamma|\sigma_y|\sigma_u(\mu - r)^2}{(\gamma - 1)\sigma_l^4} \int_0^t |\hat{Y}_s - Y_s| ds \\ &\leq \frac{\gamma|\sigma_y|\sigma_u(\mu - r)^2}{(\gamma - 1)\mathrm{k}\sigma_l^4} |\hat{y} - y|, \end{aligned}$$

which, together with (11), implies

$$|W(\hat{y}) - W(y)| \leq \frac{\gamma|\sigma_y|\sigma_u(\mu - r)^2}{(\gamma - 1)\mathrm{k}\sigma_l^4} C(\sigma_u)|\hat{y} - y|.$$

The same arguments can be used to get the Lipschitz property of $W(\cdot, T)$, with Lipschitz constant independent of T. The next theorem summarizes some implications of these properties. For the definition and basic properties of viscosity solutions of partial differential equations see, for instance, [4], [9].

Theorem 1. *Let $\tilde{W}(y) := \lim_{T\to\infty} W(y, T)$. Then, $W(y, T)$ and $\tilde{W}(y)$ are solutions in the viscosity sense to the Hamilton-Jacobi-Bellman equations*

$$W_T + \alpha W = g W_y + \frac{\beta^2}{2} W_{yy} + \inf_{c \in C}[-c\gamma W + c^\gamma] + \gamma Q W \tag{12}$$

and

$$\alpha W = g W_y + \frac{\beta^2}{2} W_{yy} + \inf_{c \in C}[-c\gamma W + c^\gamma] + \gamma Q W, \tag{13}$$

respectively.

In view of (10), in (12) and (13) the consumption control set $\mathcal{C}$ can be chosen as a closed interval $[c_l, c_u]$, with $0 < c_l < c_u < \infty$ and

$$c_l \leq C(\sigma_u)^{\frac{1}{\gamma-1}}, \quad C(\sigma_l)^{\frac{1}{\gamma-1}} \leq c_u.$$

Proof. Note first that $W(y, T)$ converges to $\tilde{W}(y)$ uniformly on compact sets because both, $W(\cdot, T)$ and $\tilde{W}(\cdot)$, are continuous. To prove the theorem it is enough to show that $W(y,T)$ is a viscosity solution of (12), in view of the stability results of viscosity solutions. See for instance Lemma 2.6.2 in [9]. This follows from the dynamic programming principle, which in this case holds because $Q(\cdot)$ is continuous and bounded, and hence assumptions of Theorem 4.7.1 of [9] hold. The theorem follows from Theorem 2.5.1 of [9]. □

3 Main Result

In this section we shall prove that the the value function $W(y)$ is smooth and that it is the unique classical solution of (13) in a suitable class of functions. Note that the infimum in (13) is achieved at $W(y)^{\frac{1}{\gamma-1}} \in (c_l, c_u)$, and substituiting it, we can rewrite this equation as

$$\alpha W = gW_y + \frac{\beta^2}{2}W_{yy} + (1-\gamma)W^{\frac{\gamma}{\gamma-1}} + \gamma QW. \tag{14}$$

Theorem 2. *$\tilde{W}(y)$ is a classical solution of (14) and it is unique in the class of positive, bounded Lipschitz functions. Moreover, $\tilde{W}(y) = W(y)$ for each $y \in \mathbb{R}$.*

Proof. To simplify notation let us write equation (14) as

$$w_{yy} = \Phi(y, w, w_y),$$

with $\Phi(y,p,q) = (2/\beta^2)[\alpha p - g(y)q - (1-\gamma)p^{\frac{\gamma}{\gamma-1}} - \gamma Q(y)p]$, and defining $v = w_y$ write it down as a first order ode

$$\begin{cases} v_y = \Phi(y, w, v) \\ w_y = v. \end{cases} \tag{15}$$

The basic idea consists to prove that locally there exists a smooth solution to (5), which coincides with $\tilde{W}$. Given $y_0 \in \mathbb{R}$, let $v_0 \in \mathbb{R}$ and $w_0 := \tilde{W}(y_0) > 0$ be initial conditions of (5), and define the rectangle $R = \{(y,p,q) \in \mathbb{R}^3 : y_0 \le y \le y_0 + 1, \quad |p - w_0| \le L, |q - v_0| \le L\}$, with $L < w_0$. Note that Φ is continuously differentiable on R. Let $M = \sup\{|\Phi(y,p,q)| : (y,p,q) \in R\}$ and $a := \min\{1, \frac{L}{M}\}$. Then, there exists a unique solution $w(y; v_0)$ to (5) in $[y_0, y_0 + a]$, and from its Taylor expansion around (y_0, v_0),

$$\begin{aligned} w(y;v_0) &= w_0 + w_y(y_0;v_0)(y-y_0) + \frac{1}{2}w_{yy}(y_0;v_0)(y-y_0)^2 + o((y-y_0)^2) \\ &= w_0 + v_0(y-y_0) + \frac{1}{2}\Phi(y_0,w_0,v_0)(y-y_0)^2 + o((y-y_0)^2). \end{aligned}$$

Therefore,

$$\frac{\partial w}{\partial v_0}(y;v_0) = (y-y_0) + O((y-y_0)^2).$$

Let $y_1 \in (y_0, y_0 + a]$ and $w_1 := \tilde{W}(y_1)$ and define $\psi(y_1, v_0) = w_1 - w(y_1; v_0)$ with $v_0 = (w_1 - w_0)/(y_1 - y_0)$. Hence,

$$|\psi(y_1, v_0)| = O((y_1 - y_0)^2).$$

Furthermore, for $\theta > 0$ fixed and $|y_1 - y_0|$ sufficiently small,

$$\psi(y_1, v_0 - \theta) > 0 \quad \text{and} \quad \psi(y_1, v_0 + \theta) < 0.$$

From the implicit function theorem, there exists $\delta > 0$ such that if $|y_1 - y_0| < \delta$, then

$$\psi(y_1, \cdot) = 0$$

has a unique solution $\hat{v}$. This proves that for each $y_0 \in \mathbb{R}$ there exists $\delta > 0$ such that for $y_1 \in (y_0, y_0 + \delta]$ fixed, there is a solution w of (14) in $[y_0, y_1]$ such that $w(y_0) = \tilde{W}(y_0)$ and $w(y_1) = \tilde{W}(y_1)$.

Now we shall prove that $\tilde{W}$ and w coincide in $[y_0, y_1]$. We will prove only that $w \leq \tilde{W}$, since the same arguments can be used to obtain the reverse inequality. Suppose that

$$\tau := \min_{y \in [y_0, y_1]} \frac{\tilde{W}(y)}{w(y)} < 1$$

and let $x_0 \in (y_0, y_1)$ such that the minimum is achieved. Then, $y \to \tilde{W}(y) - \tau w(y)$ has a local minimum at x_0 and $\tilde{W}(x_0) = \tau w(x_0)$. Since $\tilde{W}$ is a viscosity supersolution of (14) by Theorem 2.1,

$$\tau[\alpha - \gamma Q(x_0)]w(x_0) - (1-\gamma)\tau^{\frac{\gamma}{\gamma-1}} w^{\frac{\gamma}{\gamma-1}}(x_0) - \tau g(x_0) w_y(x_0) - \tau \frac{\beta^2}{2} w_{yy}(x_0) \geq 0,$$

and, using the fact that w is a classical solution of (14), we get

$$(1-\gamma) w^{\frac{\gamma}{\gamma-1}}(x_0)(\tau - \tau^{\frac{\gamma}{\gamma-1}}) \geq 0,$$

which is a contradiction, since $\frac{\gamma}{\gamma-1} < 1$. Thus, $w \leq \tilde{W}$.

Finally we shall prove uniqueness. Define $\tilde{Z}(y) := \log \tilde{W}(y)$, which is bounded, Lipschitz, and solves the ode

$$\frac{1}{2}\beta^2 Z_{yy} + g(y) Z_y + \frac{1}{2}\beta^2 Z_y^2 + \gamma Q(y) - \alpha + F(Z) = 0, \tag{16}$$

with $F(z) = (1-\gamma)e^{-\frac{z}{1-\gamma}}$. Observe that $F_z(z) = -[1/(1-\gamma)]F(z) < 0$. Suppose that $\hat{Z}(y) = \log \hat{W}$ is another such solution to (16) and let $R(y) = \tilde{Z}(y) - \hat{Z}(y)$. Then

$$\frac{1}{2}\beta^2 R_{yy} + G(y) R_y + \Lambda(y) R = 0$$

where $G(y) = g(y) + (1/2)\beta^2(\tilde{Z}_y(y) + \hat{Z}_y(y))$ and Λ is defined implicitly, such that $\Lambda(y) \leq -\delta < 0$ by the mean value theorem. Then, Feynman-Kac formula (see [12], pp. 366-368) applied to solutions η_t of the stochastic differential equation

$$d\eta_t = G(\eta_t)dt + \beta dW_t,$$

with initial condition $\eta_0 = y$, gives

$$R(y) = \mathbf{E}[\exp\{\int_0^T \Lambda(\eta_t)dt\}R(\eta_T)] \leq e^{-\delta T}\|R\|.$$

Letting $T \to \infty$ we get $R(y) \leq 0$. Similarly, $R(y) \geq 0$. Thus, $\hat{Z} = \tilde{Z}$ and then $\hat{W} = \tilde{W}$.

The last part of the theorem follows from standard verification arguments. See [9], p. 146. □

The same verification arguments mentioned before can be used to check that the value function $V(x, y)$ has the form $(x^\gamma/\gamma)W(y)$, which is a classical solution of (7). Also, $u_t^* := (\mu - r)/(1-\gamma)\sigma^2(Y_t)$ and $c_t^* := W(Y_t)^{\frac{1}{\gamma-1}}$ are optimal strategies.

References

1. Bielecki, T. R. and Pliska, S. R. (1999) Risk sensitive dynamic asset management, Applied Math. Optim. **39**, 337–360.
2. Bielecki, T. R. and Pliska, S. R. Risk sensitive intertemporal CAPM, with application to fixed income management, preprint.
3. Bielecki, T. R., Hernández-Hernández, D., and Pliska, S. R. (1999) Risk sensitive control of finite state Markov chains in discrete time, with applications to portfolio management, Math. Meth. Oper. Res. **50**, 167–188.
4. Crandall, M. G., Ishii, H., and Lions, P. L. (1992) A user's guide to viscosity solutions, Bulletin A.M.S., N.S. **27**, 1–67.
5. Fleming, W. H. and Hernández-Hernández, D. An optimal consumption model with stochastic volatility, submitted for publication.
6. Fleming, W. H. and Rishel, R. W. (1975) Deterministic and Stochastic Optimal Control, Springer-Verlag.
7. Fleming, W. H. and Sheu, S. J. (1999) Optimal long term growth rate of expected utlity wealth, Ann. Applied Prob. **9**, 871-903.
8. Fleming, W. H. and Sheu, S. J. (2000) Risk sensitive control and an optimal investment model, Mathematical Finance **10**, 197–213.
9. Fleming, W. H. and Soner, H. M. (1993) Controlled Markov Processes and Viscosity Solutions, Springer-Verlag.
10. Fleming, W. H. and Zariphopoulou, T. (1991) An optimal investment/consumption model with borrowing, Mathematics of Operations Research **16**, 802–822.
11. Fouque, J. P., Papanicolaou, G., and Sircar, R. (2000) Derivatives in Financial Markets with Stochastic Volatility, Cambridge Univ. Press.

12. Karatzas, I. and Shreve, S. E. (1988) Brownian Motion and Stochastic Calculus, Springer-Verlag.
13. Nagai, H. Optimal strategies for risk sensitive portfolio optimization problems for general factor models, preprint.
14. Zariphopoulou, T. (2001) A solution approach to valuation with unhedgeable risks, Finance and Stochastics **5**, 61–88.

Adaptation of a Real-Time Seizure Detection Algorithm

Mark G. Frei[1], Shane M. Haas[1,2], and Ivan Osorio[1,3]

[1] Flint Hills Scientific, L.L.C.; 5020 W. 15th St., Suite A, Lawrence, KS 66049, U.S.A.
[2] Massachusetts Institute of Technology
[3] Department of Neurology, Kansas University Medical Center

Abstract. The time-varying dynamics and non-stationarity of epileptic seizures makes their detection difficult. Osorio et. al. in ([1]) proposed an adaptable seizure detection algorithm ('SDA'), however, that has had great success. In this presentation, we begin with an overview of the original detection algorithm's architecture, describing its degrees of freedom that provide flexibility and outline a procedure to adapt the method to improve performance.

The adaptation consists of generating multiple candidate digital filters using various techniques from signal processing, defining a practical optimization criteria, and using this criteria to select the best filter candidate. Coupled within the procedure is the selection of a corresponding optimal percentile value for use in the nonlinear (order statistic) filtering step that follows in the algorithm.

Finally, we discuss how the algorithm has been utilized for closed-loop therapy, in which seizure detections are used to trigger electrical stimulations in the brain designed to prevent the development of a seizure before its disabling effects occur.

1 Background

Epilepsy is the most prevalent serious neurological disease which occurs across all age groups and genders ([2]).

1.1 Epilepsy Numbers

- 2.7 million in the U. S. have epilepsy.
- 1 to 2% in industrialized countries have epilepsy
- 5 to 10% in non-industrialized countries have epilepsy
- 60 million worldwide have epilepsy

1.2 Epilepsy and Disability

- Intermitency and criticality conditions manifest as unpredictability
- UNPREDICTABILITY = DISABILITY
- ACCURATE SEIZURE PREDICTION = PARTIAL REHABILITATION
- AUTOMATED SEIZURE BLOCKAGE = FULL REHABILITATION

B. Pasik-Duncan: Stochastic Theory and Control, LNCIS 280, pp. 131–136, 2002.

2 Overview of the Seizure Detection Algorithm

Figure 1 shows a block diagram of the algorithm. The primary steps of the algorithm are:

(1) Decomposition of the raw signals into components which conceptually represent the seizure or epileptiform component and the non-seizure or residual component. One means for accomplishing this step is via digital filtering. (Figure 2).

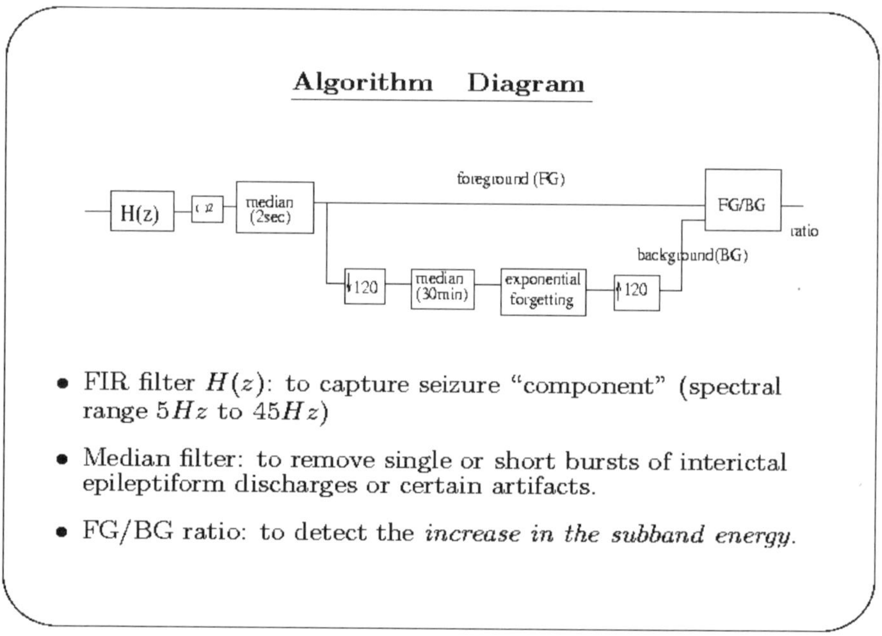

Fig. 1.

(2) After the signal is filtered and squared, the next step of the method is designed to distinguish between isolated 'outlier' spikes that correspond to epileptiform activity but do not constitute a seizure. This is accomplished using an order statistic filter (e.g., a median filter).

(3) The output of the order statistic filter ('foreground') is normalized via division by background levels (constructed using a longer time-scale median filter applied to the foreground signal) to produce a dimensionless ratio.

(4) The maximum ratio over all individual signals being analyzed is then compared to threshold and duration constraints, and a detection is logged if the ratio remains above the threshold level, T, for longer than the duration constraint, D.

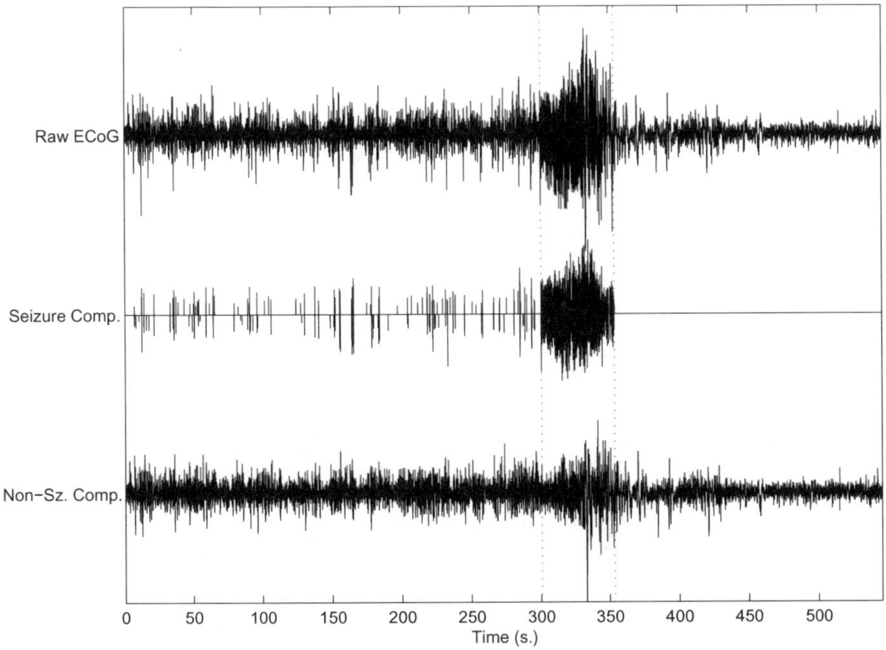

Fig. 2. Decomposition of a Data Segment of ECoG Containing a Seizure

The first two algorithm steps are illustrated in the upper panel, and the latter two in the lower panel of Figure 3. Figure 4 illustrates the onset of a seizure in multi-channel electrocorticogram signals recorded bilaterally from the temporal lobes of a subject (upper panel). The lower panel shows the output of the seizure detection algorithm during the corresponding time interval and 10 minutes centered on the detection (inset).

3 Adaptation of the Seizure Detection Algorithm

The purposes for adapting the algorithm include:

- Improve sensitivity
  - ♦ Utilize "seizure fingerprint(s)" to eliminate any false negatives
- Improve specificity
  - ♦ Eliminate undesirable detections
- Speed up detection/Increase warning
- Customize algorithm to user preference

The various algorithm parameters that may be adapted to improve algorithm performance are shown in Figure 5. The adaptation procedure for simultaneously optimizing the digital filter used in the first step of the algorithm, along with the percentile utilized in the second step's order statistic filter were presented. Details of the procedure are available in ([3]).

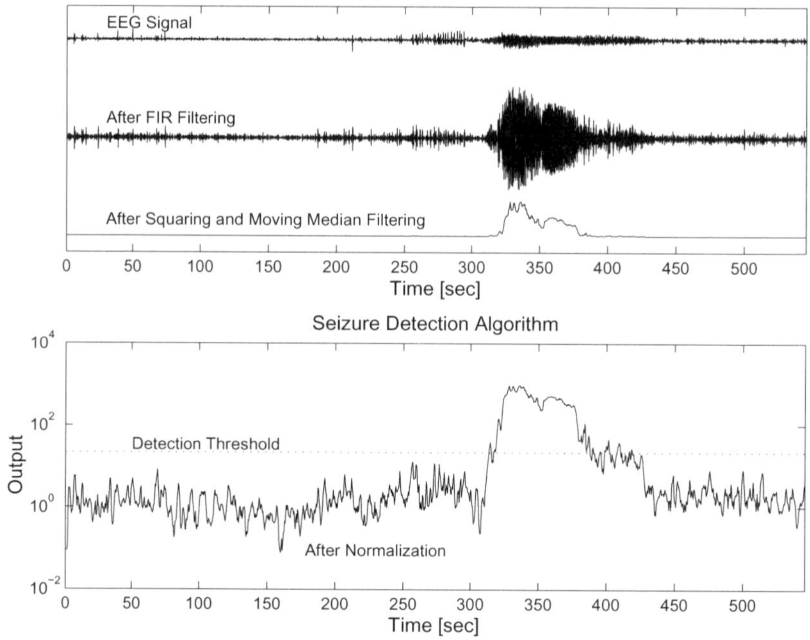
EEG Signal
After FIR Filtering
After Squaring and Moving Median Filtering
0 50 100 150 200 250 300 350 400 450 500
Time [sec]
Seizure Detection Algorithm
10^4
10^2
10^0
10^{-2}
Output
Detection Threshold
After Normalization
0 50 100 150 200 250 300 350 400 450 500
Time [sec]

Fig. 3.

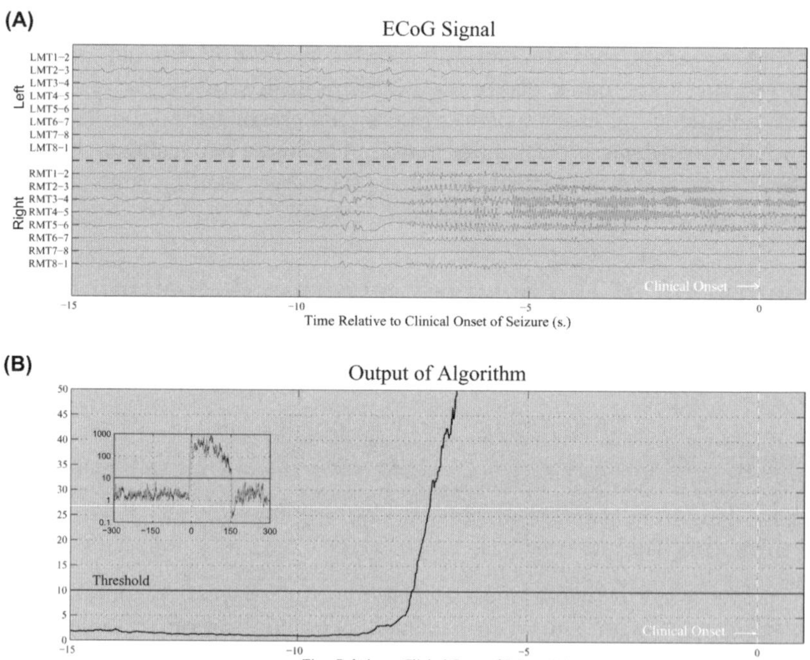
(A)
ECoG Signal
Left
LMT1-2
LMT2-3
LMT3-4
LMT4-5
LMT5-6
LMT6-7
LMT7-8
LMT8-1
Right
RMT1-2
RMT2-3
RMT3-4
RMT4-5
RMT5-6
RMT6-7
RMT7-8
RMT8-1
Clinical Onset
-15 -10 -5 0
Time Relative to Clinical Onset of Seizure (s.)
(B)
Output of Algorithm
Threshold
Clinical Onset
-15 -10 -5 0
Time Relative to Clinical Onset of Seizure (s.)

Fig. 4.

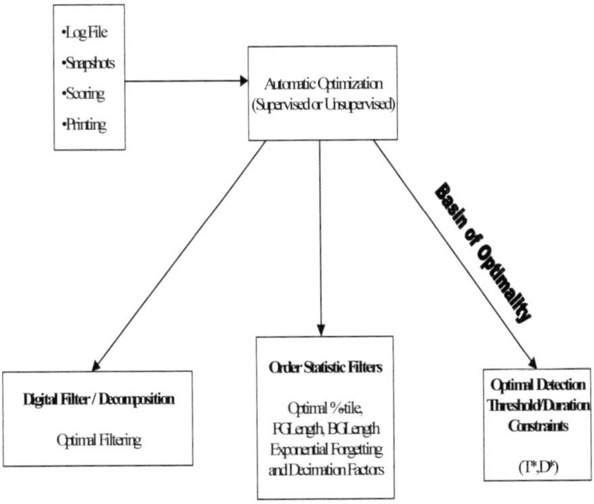

Fig. 5. Individualize Adaptation of the Algorithm

4 Using the SDA for Closed-Loop Stimulation

The successful performance of the SDA outlined in ([1]), along with its adaptability as described here, make it an ideal choice for use in a device for automated therapy in which electrical brain stimulation or other therapeutic intervention means is triggered by the automated real-time detection of an epileptic seizure. The benefits of this form of closedloop therapy were reviewed as compared to open-loop therapy in which stimulation are delivered periodically without knowledge of when seizures are occurring (Figure 6). Preliminary results of closed-loop seizure therapy have been successful ([4]). The next stage of development we are pursuing is the development of an implantable device for closed-loop seizure therapy (see, e.g., Figure 7).

References

1. Osorio, I., Frei, M. G., and Wilkinson, S. B. (1998) Real-time automated detection and quantitative analysis of seizures and short-term prediction of clinical onset. Epilepsia **39**(6), 615–627.
2. Hauser, W. A., Annegers, J. F., and Rocca, W. A. (1996) Descriptive epidemiology of epilepsy: Contributions of population-based studies from Rochester, Minnesota. Mayo Clin Proc. **71**, 576–586.
3. Haas, S. M., Frei, M. G., and Osorio, I. (2002) Adaptation strategies for automated seizure detection. Preprint.
4. Osorio, I., Frei, M. G., Manly, B. F. J., Sunderam, S., Bhavaraju, N. C., and Wilkinson, S. B. (2001) An introduction to contingent (closed-loop) brain electrical stimulation for seizure blockage, to ultra-short term clinical trials and to multidimensional statistical analysis of therapeutic efficacy. J Clin. Neurophysiol. **18**(6), 533–544.

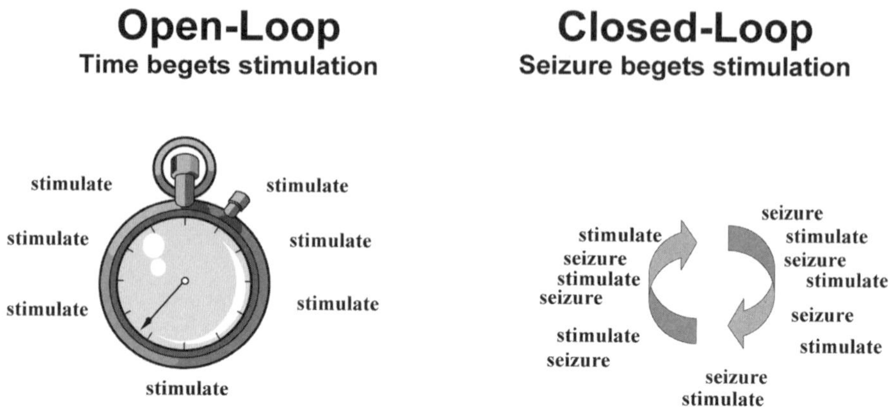

Fig. 6. Brain Stimulation: Open Loop vs Closed Loop

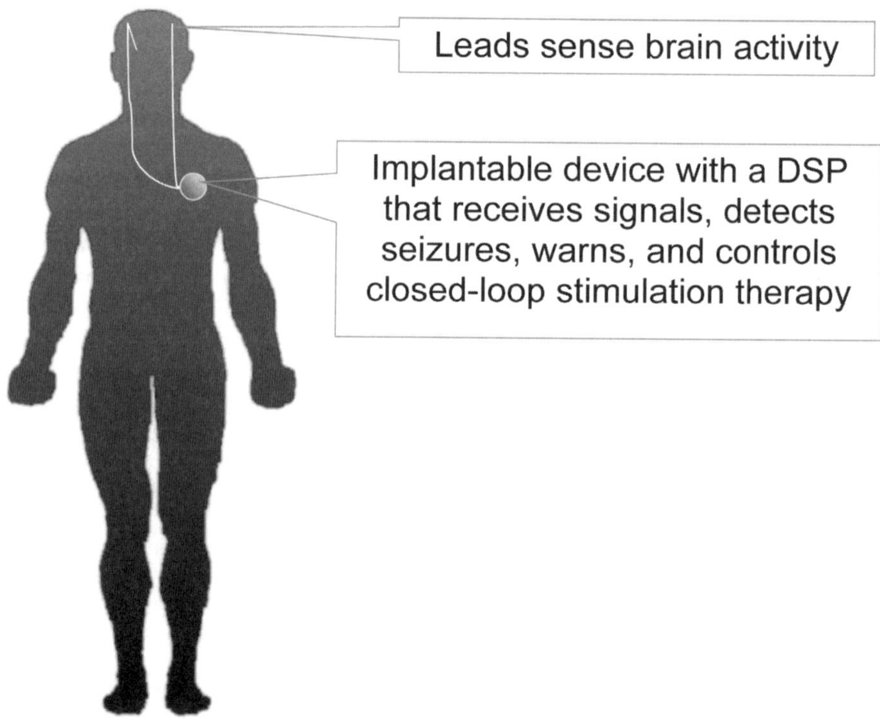

Fig. 7.

Randomization Methods in Optimization and Adaptive Control

Dedicated to Tyrone Duncan on occasion of his 60th birthday

László Gerencsér[1], Zsuzsanna Vágó[1,3], and H. Hjalmarsson[2]

[1] Computer and Automation Institute, Hungarian Academy of Sciences, 13-17 Kende u., H-1111 Budapest, Hungary
[2] Dept. of Signals, Sensors and Systems, The Royal Institute of Technology, S-100 44 Stockholm, Sweden
[3] Pázmány Péter Catholic University, Budapest, 28 Szentkirályi u., H-1088 Budapest, Hungary

Abstract. We consider simultaneous perturbation stochastic approximation (SPSA) methods applied to noise-free problems in optimization and adaptive control. More generally, we consider discrete-time fixed gain stochastic approximation processes that are defined in terms of a random field that is identically zero at some point θ^*. The boundedness of the estimator process is enforced by a resetting mechanism. Under appropriate technical conditions the estimator sequence converges to θ^* with geometric rate almost surely. This result is in striking contrast to classical stochastic approximation theory where the typical convergence rate is $n^{-1/2}$. For the proof a discrete-time version of the ODE-method is used and the techniques of [10] are extended. A simple variant of noise free-SPSA is applied to extend a direct controller tuning method named Iterative Feedback Tuning (IFT), see [16]. Using randomization, the number of experiments required to obtain an unbiased estimate of the gradient of the cost function can be reduced significantly for multi-input multi-output systems.

1 Introduction

Consider the following problem: minimize a function $L(\theta)$ defined for $\theta \in \mathbb{R}^p$, such that it is three-times continuously differentiable with respect to θ, and $L(\cdot)$ has a unique global minimizing value θ^*. Assume that the computation of $L(\cdot)$ is expensive and the gradient of $L(\cdot)$ is not computable at all by direct experiment. It is our objective to minimize L using as few function evaluation as possible.

In principle we can approximate the gradient of L using some numerical technique, but then the number of function evaluations would be proportional to the dimension of the problem. A much more economical procedure seems to be to use a randomization technique proposed in [29], and using only 2 function evaluations per iteration compute an approximately unbiased esti-

B. Pasik-Duncan: Stochastic Theory and Control, LNCIS 280, pp. 137–153, 2002.

mate of the gradient, which will be denoted by

$$G(\theta) = L_\theta(\theta).$$

This gradient estimator is obtained by considering *random* simultaneous perturbations of the components of $\theta \in \mathbb{R}^p$ as follows: take a sequence of independent, identically distributed (i.i.d.) random variables, with time-index n, $\Delta_{ni} = \Delta_{ni}(\omega)$, $i = 1, \dots, p$ defined over some probability space $(\Omega, \mathcal{F}, \mathcal{P})$ satisfying certain weak technical conditions given in [29]. A standard choice is to take a Bernoulli-sequence with

$$P(\Delta_{ni}(\omega) = +1) = 1/2 \qquad P(\Delta_{ni}(\omega) = -1) = 1/2.$$

Let $c > 0$ be a small positive number, the step-size of the perturbation, and evaluate $L(\cdot)$ at two randomly, symmetrically chosen points, $\theta + c\Delta_n(\omega)$ and $\theta - c\Delta_n(\omega)$, respectively. Then

$$\frac{1}{2c}(L(\theta + c\Delta_n) - L(\theta - c\Delta_n))$$

is an approximate value of the directional derivative of L at the point θ in the random direction Δ_n. From this the gradient can be approximately reconstructed as follows: set

$$\Delta_n^{-1}(\omega) = \left[\Delta_{n1}^{-1}(\omega), \dots, \Delta_{np}^{-1}(\omega)\right]^T.$$

Then we can write

$$H(n, \theta, \omega) = \Delta_n^{-1}(\omega)\frac{1}{2c}(L(\theta + c\Delta_n(\omega)) - L(\theta - c\Delta_n(\omega))).$$

The remarkable property of this estimator is that it is approximately unbiased, in fact, for quadratic functions it is exactly unbiased.

Using this gradient estimator with a decreasing $c = c_n$, where c_n tends to zero at a rate $1/n^\gamma$ with some $\gamma > 0$, typically with $\gamma = 1/6$, an iterative minimization procedure, called the simultaneous perturbations stochastic approximation (SPSA) method has been developed in [29]. It is a stochastic approximation procedure with decreasing gain $1/n^a$ with some $0 < a \leq 1$, typically with $a = 1$:

$$\theta_{n+1} = \theta_n + \frac{1}{n+1}H(n+1, \theta_n, \omega) \qquad \theta_0 = \xi. \tag{1}$$

This procedure was originally designed for cases when the function evaluation is noisy: instead of $L(\theta)$ we measure $L(\theta) + \varepsilon_n$ where ε_n is a zero mean, typically state-independent noise. It has been studied in a number of papers [3,11,24,29–31]

Noise-free SPSA. In the case when L is evaluated without noise, the SPSA procedure (1) is in fact a Robbins-Monroe type procedure, which has been studied extensively in the statistical and system-identification literature, see e.g. [1,27,25]. The asymptotic covariance matrix of the estimator process has been determined under various conditions in the above references, see also [2,7,28]. A special feature of noise-free SPSA when applied to the minimization of quadratic-functions is that

$$H(n,\theta^*,\omega) = \Delta_n^{-1}(\omega)\Delta_n^T(\omega)G(\theta^*) = 0$$

identically. In more generality, we will consider the following problem: let $H(n,\theta,\omega)$ be a random field defined over some probability space $(\Omega, \mathcal{F}, P)$ for $n \geq 1$ and $\theta \in D \subset \mathbb{R}^p$, where D is a bounded open domain. Assume that for some $\theta^* \in \mathbb{R}^p$ the random field identically vanishes, i.e. we have for all ω

$$H(n,\theta^*,\omega) = 0. \tag{2}$$

The problem is to determine θ^* via a stochastic approximation procedure based on observed values of $H(n,\theta,\omega)$.

Now it is easy to see that $H(n,\theta^*,\omega) = 0$ implies the asymptotic covariance matrix of the estimator process is 0. Hence the convergence rate for Robbins-Monroe type stochastic approximation procedures with vanishing field will be better than the standard rate $n^{-1/2}$. But how much better can it be? Following the analysis of [8] it can be seen that the convergence rate can be better than n^{-k} for any $k > 0$. This observation induces us to try to improve the convergence rate by increasing the stepsize, and to consider fixed gain stochastic approximation processes of the form

$$\theta_{n+1} = \theta_n + \lambda H(n+1,\theta_n,\omega) \qquad \theta_0 = \xi. \tag{3}$$

Fixed gain recursive estimation processes of this general form have been widely used and studied in the engineering literature assuming some form of stationarity, see e.g. the papers [1,9,10,15,21,23,25]. Fixed gain SPSA methods for noisy problems have been first considered in [13]. Using fixed gain we can not expect convergence, but the tracking error of the estimator is can be shown to be bounded in an appropriate sense.

The additional assumption of the present paper, namely that $H(n,\theta^*,\omega) = 0$ will radically change convergence properties. We have the surprising result that θ_n does converge to θ^* almost surely, and the rate of convergence is *geometric*. This result is proved under the condition that, following [8], we enforce θ_n to stay in a prescribed compact domain $D_0 \subset D$ by using a resetting mechanism.

Resetting. A main technical difficulty in recursive estimation is the analysis of the effect of enforced boundednes that is ensured by a resetting mechanism (cf. [8]). Let D_θ be a compact truncation domain to be specified later and let D_ξ be a compact domain of initial estimates such that $\xi = \widehat{\theta}_0 \in D_\xi$. At time n we first define a tentative value θ_{n+1-} following (3) as

$$\theta_{n+1-} = \theta_n + \lambda H(n+1,\theta_n,\omega)$$

and then we set

$$\begin{aligned} \theta_{n+1} &= \theta_{n+1-} \quad \text{if} \quad \theta_{n+1-} \in D_\theta \\ \theta_{n+1} &= \theta_0 \qquad \text{if} \quad \theta_{n+1-} \notin D_\theta. \end{aligned} \tag{4}$$

Intuitively it is clear that the domain D_ξ should be inside the interior of the domain D_θ, and exact technical conditions are described below in Condition 6.

The main technical result presented in this paper is Theorem 1 of Section 3. The proof is based on a discretized version of the ODE-method, developed in [12], and techniques used in [10].

Computable gradients. In contrast to the standard setup that is used in Kiefer-Wolfowitz and SPSA-theory there are interesting applications when the directional derivatives of the cost function *can* be computed experimentally exactly. But these experiments can be costly and hence the use of randomization is justified. Thus consider the problem of solving

$$L_\theta(\theta) = G(\theta) = 0$$

under the assumption that for any fixed θ and any direction $v \in \mathbb{R}^p$ the directional derivative $v^T L_\theta(\theta)$ can be computed experimentally.

An unbiased estimate of the gradient can be obtained from a directional derivative in a random direction, which in this case simplifies to the following construction: we shall perform an experiment to compute the random directional derivative $\Delta_n^T(\omega) L_\theta(\theta)$ and set

$$H(n, \theta, \omega) = \Delta_n^{-1}(\omega) \Delta_n^T(\omega) G(\theta). \tag{5}$$

Then $H(n, \theta, \omega)$ is an unbiased estimator of $G(\theta)$ and, since $G(\theta^*) = 0$, we have exactly

$$H(n, \theta^*, \omega) = 0.$$

The algorithm for solving $G(\theta) = 0$ will take the form

$$\theta_{n+1} = \theta_n - \lambda \Delta_n^{-1}(\omega) \Delta_n^T(\omega) G(\theta) \qquad \theta_0 = \xi. \tag{6}$$

Now the general result is applicable and we find that θ_n converges to θ^* at geometric rate almost surely.

Iterative Feedback Tuning. A nice application of noise-free SPSA is an extension of a new, and interesting idea of direct adaptive control called Iterative Feedback Tuning (IFT), [18]. This is a model-free method for tuning the controller parameters of a linear control system. In this paper we consider systems without disturbances. The objective is to minimize a design criterion with respect to the controller parameters using measurements from the plant collected during (essentially) normal operating conditions. No modeling of the eventually highly complex plant-dynamics is required. Based on the experiments, the gradient of the design criterion is obtained with high accuracy and is subsequently used to update the parameters of the controller.

For a single-input/single-output (SISO) system, two experiments are required in each iteration regardless of the number of parameters in the controller. For a multi-input/multi-output (MIMO) system the situation is quite different, however. A MIMO system using a controller having n_w inputs and n_u outputs requires $1 + n_u \times n_w$ experiments. This clearly limits the applicability of the method for multivariable systems.

A possible way of reducing the number of experiments is to use a simultaneous perturbation randomization technique. Using these ideas we get a randomized IFT where an estimate of gradient is obtained from a physical experiment that gives an approximate value of the directional derivative in a *random direction*. Based on this single experiment an approximately unbiased estimate of the complete gradient can be obtained regardless of the dimension of the controller. In this contribution the focus is on the problem where the reference signal is *periodic*. It is shown that in this case randomized IFT method inherits the geometric convergence rate of noise-free SPSA.

2 Technical Assumptions

The Euclidean-norm of a vector x will be denoted by $|x|$, the operator norm of a matrix A will be denoted be $\|A\|$. Assume $\theta^* = 0$ and assume that $H(n, \theta, \omega)$ can be written in the form

$$H(n, \theta, \omega) = A(n, \theta, \omega)\theta$$

where the $p \times p$ matrix-valued random-field $A(n, \theta, \omega)$ satisfies the conditions below. If $H(n, \theta, \omega)$ is continuously differentiable with respect θ then, taking into account that $H(n, \theta^*, \omega) = 0$, an exact Taylor-series expansion gives a representation of the above form with some $A(n, \theta, \omega)$.

First we formulate a simple analytical condition for $A(n, \theta, \omega)$. For this purpose we define the $p \times p$ matrix-valued random-field $\Delta A/\Delta\theta$, for $\theta, \theta + h$ $h \neq 0$, by

$$(\Delta A/\Delta\theta)(n, \theta, \theta + h, \omega) = |A(n, \theta + h, \omega) - A(n, \theta, \omega)|/|h|.$$

Condition 1 *The matrix-valued random-fields A and $\Delta A/\Delta\theta$ are defined and bounded for $n \geq 1$, θ, $\theta + h \in D$, $h \neq 0$, where D is a bounded domain, say*

$$||A(n, \theta, \omega)|| \leq K', \quad ||(\Delta A/\Delta\theta)(n, \theta, \theta + h, \omega)|| \leq L'.$$

To ensure a stochastic averaging effect we use the following condition (for the the concept of L-mixing see [5]):

Condition 2 *A and $\Delta A/\Delta\theta$ are L-mixing uniformly in θ for $\theta \in D$ and in $\theta, \theta+h$ for $\theta, \theta+h \in D$, $h \neq 0$, respectively, with respect to a pair of families of σ-algebras $(\mathcal{F}_n, \mathcal{F}_n^+), n \geq 1$.*

In particular, if $H(n,\theta,\omega)$ is given by (5), $G(\theta)$ is twice continuously differentiable, and θ is restricted to some compact domain, and the perturbations Δ_k are Bernoulli, then Conditions 1 and 2 are satisfied.

For the sake of convenience we assume that the mean field $\mathrm{E}H(n,\theta,\omega)$ is independent of n, i.e. we can write

$$G(\theta) = \mathrm{E}H(n,\theta,\omega). \tag{7}$$

This implies the following condition:

Condition 3 *$B(\theta) = \mathrm{E}A(n,\theta,\omega)$ is independent of n.*

Write

$$\overline{A}(n,\theta,\omega) = A(n,\theta,\omega) - B(\theta).$$

Thus (3) can be written as a quasi-linear random iterative process of the form

$$\theta_{n+1} = \theta_n + \lambda A(n+1,\theta_n,\omega)\theta_n. \tag{8}$$

There is an extensive literature on products of random mappings, see e.g. [4,32,22]. A main advance that is presented here is that the assumption on stationarity is removed, and completely new techniques are used to establish geometric rate of convergence.

An important technical tool in recursive identification is the associated ordinary differential equation (ODE) defined as

$$\dot{y}_t = \lambda G(y_t), \qquad y_s = \xi = z \in D_z, \qquad s \geq 0 \tag{9}$$

where $D_z \subset D$ is a compact domain.

Condition 4 *The function G defined on D is continuous and bounded in y together with its first and second partial derivatives, say*

$$|G(y)| \leq K, \quad \|\partial G(y)/\partial y\| \leq L \quad \|\partial^2 G(y)/\partial y^2\| \leq L. \tag{10}$$

The solution of (9) will be denoted by $y(t,s,\xi)$. The time-homogeneous flow associated with (9) is defined as the mapping $\xi \to y_t(\xi) = y(t,0,\xi)$. Let D_z be such that for $z \in D_z$ we have $y_t(z) \in D$ for any $t \geq 0$. For any fixed t the image of D_z under y_t will be denoted as $y_t(D_z)$, i.e. $y_t(D_z) = \{y : y = y(t,0,z),\ z \in D_z\}$. The union of these sets will be denoted by $y(D_z)$, i.e.

$$y(D_z) = \{y : y = y(t,0,z) \quad \text{for some} \quad t \geq 0,\ z \in D_z\}.$$

Condition 5 *The ordinary differential equation (9) is globally exponentially stable in D_z: for some $C_0 > 0$ and $\alpha > 0$ we have for all $0 \leq s \leq t,\ z \in D_z$ $|y(t,s,z)| \leq C_0 e^{-\alpha(t-s)}$. In addition we assume that*

$$\|\frac{\partial}{\partial z} y(t,s,z)\| \leq C_0 e^{-\alpha(t-s)}. \tag{11}$$

This condition is not particularly restrictive: if the Jacobian-matrix of the right hand side of (9), i.e. $G_\theta(\theta^*)$ has all its eigenvalues on the open left-half plane, then this condition is certainly satisfied with some small domain D_z. In our application, when $G(\theta)$ is the gradient of some function $L(\theta)$, D_z can be an arbitrary compact set, assuming that L is a so-called α-convex function.

To connect the discrete-time algorithm given for (3) and the continuous-time procedure given by (9), the standard procedure would be a piecewise-linear extension of the discrete-time procedure. However the error-term due to this imbedding seems to be too big for geometric convergence rate. Therefore a new auxiliary device will be introduced: a discrete-time version of the ODE will be developed, following [12].

Define the discrete-time deterministic process (z_n) by

$$z_{n+1} = z_n + \lambda G(z_n), \qquad z_0 = \xi = \theta \in D_\theta, \tag{12}$$

Let $z(n,m,\xi)$ denote the solution of (12) with initial condition $z_m = \xi$. Define the time-homogeneous mapping associated with (12) $\xi \to z_n(\xi) = z(n,0,\xi)$. Let $D_\theta \subset D$ be such that for $\theta \in D_\theta$ we have $z_n(\theta) \in D$ for any $n \geq 0$. For any fixed n the image of D_θ under z_n will be denoted as $z_n(D_\theta)$ i.e. $z_n(D_\theta) = \{z : z = z(n,0,\theta),\ \theta \in D_\theta\}$. The union of these sets will be denoted by $z(D_\theta)$, i.e.

$$z(D_\theta) = \{z : z = z(n,0,\theta) \quad \text{for some} \quad n \geq 0,\ \theta \in D_\theta\}.$$

It can be proved that, under suitable technical conditions, $z(D_\theta) \subset D_z \subset D$ where D_z is some compact domain.

For any set D_0 write $S(D_0, \varepsilon) = \{\theta : |\theta - z| < \varepsilon \text{ for some } z \in D_0\}$. Finally the interior of a compact domain D_0 is denoted by int D_0. Using the above auxiliary processes and flows, the following technical conditions are imposed on the domains that show up in the resetting mechanism:

Condition 6 *There exist compact domains $D_\theta \subset D_z \subset D_y \subset D$ and $d > 0$ such that $0 \in \mathrm{int} D_\theta$ and*

$$S(y(D_\theta), d) \subset D_z \quad \text{and} \quad y(D_z) \subset D_y \subset D. \tag{13}$$

It is also assumed that $0 \in \mathrm{int} D_\xi$ and for some $0 < r < R$ we have

$$D_\xi \subset S(0,r) \subset S(0,R) \subset D_\theta. \tag{14}$$

3 The Main Results

In this section we state the main result and give an outline of the proof.

Theorem 1 *Consider the algorithm (3). Assume that Conditions 1-6 are satisfied. Let* $\xi = \theta_0 \in D_\xi$ *and assume that* $C_0^3 r/R < 1$. *Then there exists a* γ *with* $0 < \gamma < 1$ *and a positive random variable* $C(\omega)$ *such that for sufficiently small* λ *we have*

$$|\theta_N| \leq C(\omega)\gamma^{\lambda N}.$$

For the proof we exploit stability properties of the discrete flow defined by (12), which has been established in [12]. The above theorem is directly applicable to derive geometric rate of convergence for algorithm (6).

A discrete-time local ODE principle: In what follows the discrete-time parameter n will be replaced by t and n will stand for a rescaled discrete time-index. Let T be a fixed positive integer. Let us subdivide the set of integers into intervals of length T. Let n be a non-negative integer and let $\tau(nT)$ denote the first integer $t > nT$ for which $\theta_t \notin D_\theta$. In the interval $[nT, (n+1)T-1]$ we consider the solution of (12) starting from θ_{nT} at time nT. This will be denoted by $\overline{z}_t$, i.e. $\overline{z}_t$ is defined by

$$\overline{z}_{t+1} = \overline{z}_t + \lambda G(\overline{z}_t), \qquad \overline{z}_{nT} = \theta_{nT}.$$

We can also write $\overline{z}_t = z(t, nT, \theta_{nT})$ for $nT \leq t \leq (n+1)T$. The definition of $\overline{z}_t$ is non-unique for $t = (n+1)T$, therefore we use the notation $\overline{z}_{(n+1)T-} = z((n+1)T, nT, \theta_{nT})$ and $\overline{z}_{(n+1)T} = \theta_{(n+1)T}$. A key step in the derivation is to get an upper bound for $|\theta_t - \overline{z}_t|$. In the lemma below the definition of $\theta_{\tau(nT)}$ will be temporarily changed to denote the value of θ_t at time $\tau(nT)$ prior to resetting.

Lemma 1. *For any* T *we have*

$$\sup_{nT \leq t \leq (n+1)T \wedge \tau(nT)} |\theta_t - \overline{z}_t| \leq c^* \eta_n^* |\theta_{nT}| \tag{15}$$

where (η_n^*) *is L-mixing with respect to* $(\mathcal{F}_{nT}, \mathcal{F}_{nT}^+)$. *Choosing* $T = [(\lambda\alpha)^{-1}]+1$ *we get for any* $2 < q < \infty$ *and* $r > p$ *the inequality* $M_q(\eta^*) \leq C_q \lambda^{1/2}$ *where* C_q *is independent of* λ.

The presence of the multiplicative term $|\theta_{nT}|$ on the right hand side is a key feature that ensures convergence with exponential rate.

A technically challenging problem is to deal with resetting. The key idea here is to find an appropriate variant of (15) that would relate $|\theta_{(n+1)T}|$ to $|\theta_{nT}|$ in a multiplicative form. This is indeed possible and then the geometric decay of $|\theta_{nT}|$ is obtained by applying fairly straightforward estimations for random products as in [6]

4 A Direct Adaptive Control Problem

We consider a system described by the discrete-time linear time-invariant, multi-input-multi-output (LTI MIMO) model

$$\begin{pmatrix} y_n \\ w_n \end{pmatrix} = G \begin{pmatrix} r_n \\ u_n \end{pmatrix} \tag{16}$$

where G is the generalized plant consisting of the true plant and possibly some frequency weighting filters and a reference model and is represented by a transfer function matrix, $r_n \in \mathbf{R}^{n_r}$ represents external signals such as set-points or reference signals, and $u_n \in \mathbf{R}^{n_u}$ represents the control signals. Furthermore $w_n \in \mathbf{R}^{n_w}$ represents the measurements, and $y_n \in \mathbf{R}^{n_y}$ represents the variables that will be included in the control criterion, which may include also some of the control signals. By proper definition of the generalized plant, these signals can be frequency-weighted filtered versions of measured signals in the real system. A key technical assumption in this paper is that the reference signal r is periodic.

The system is controlled by the controller

$$u_n = C(\rho) w_n \tag{17}$$

where $C(\rho)$ is a $n_u \times n_w$ transfer function matrix parametrized by some parameter vector $\rho \in \mathbf{R}^{n_\rho}$. The feedback system is shown in Figure 1. The external signal $z \in \mathbf{R}^{n_u}$ in this figure will be used later.

We shall in this paper assume that each channel in the controller is independently parameterized i.e. each block C_{ij} of the controller is parameterized by an independent parameter vector ρ_{ij}.

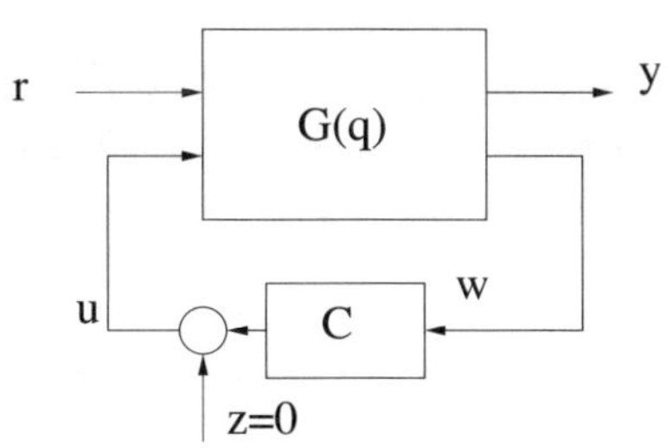

Fig. 1. Feedback system.

Signals obtained from the closed loop system with the controller $C(\rho)$ operating, will be marked by the argument ρ. In particular, if the initial state of the system is ξ then let $\hat{y}_n(\rho, \xi)$ be the corresponding output-process.

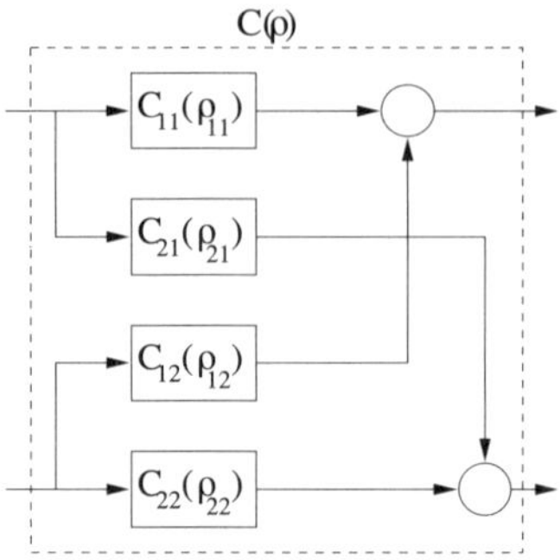

Fig. 2. A controller where each block C_{ij} is independently parameterized.

Let the period of the reference signal (r_t) be N_0. For any fixed ρ let $x_t^*(\rho)$ and $y_t^*(\rho)$ denote the steady-state state-process and output-process, respectively. Then the cost function will be defined as

$$J(\rho) = \frac{1}{2N_0}\left[\sum_{n=1}^{N_0} \left[y_n^*(\rho)\right]^T W_n^y y_n^*(\rho)\right]$$

where W_n^y is a weight-matrix. The optimal controller parameter ρ^* is defined as the minimizing value ρ. To carry out the minimization it is necessary to compute the gradient of this criterion with respect to the controller parameters. The novel contribution of the IFT approach [19] was to show that, in contrast to previous approaches, for LTI SISO systems and controllers, both with arbitrary complexity, an unbiased approximation of the gradient can be obtained directly from input-output data collected on the actual closed loop system, by performing one special experiment.

5 Iterative Feedback Tuning

During the last years quite some experience has been gained with IFT e.g. to robust control of a simulation model of a flexible transmission system [20] and to vibration attenuation [26]. It has also been applied by the chemical multinational S.A. Solvay to tune PID controllers for temperature control in furnaces, in distillation columns, flow control in evaporators etc., see [18].

In the standard setup we have a single plant which is to be controlled and at some fraction of the real time used for experimenting to get the gradient. In some applications, such as car-manufacturing, it is possible to *replicate* the plant, and use one plant for experimenting with the closed loop behaviour of the system and use a duplicate for running an experiment to get the gradient. To get mathematically clean results we shall use the latter setup. To facilitate the notation we shall from now on assume that the weighting matrix W_n^y equals the identity matrix which we denote by I.

To minimize $J(\rho)$ we need its gradient, and for this we need to get an approximate value of

$$\frac{\partial y}{\partial \rho}(\rho^k).$$

The computation of this quantity has always been the stumbling block in solving this direct optimal controller parameter tuning problem. The key contribution in the original derivation of IFT [19] was to show that for single input/single output (SISO) systems these quantities can indeed be obtained by performing experiments on the closed loop system formed by the actual system in feedback with the controller $C(\rho)$.

To obtain the derivative of y w.r.t. the parameter vector ρ_{ij} which parameterizes the ijth entry C_{ij} of the controller we proceed as follows. We can write the relation between controller block C_{ij} and an arbitrary output y_l as

$$\begin{pmatrix} y_l \\ \\ w_j \end{pmatrix} = \bar{G}_{ijl} \begin{pmatrix} r \\ \\ u_{ij} \end{pmatrix} \tag{18}$$

$$u_{ij} = C_{ij} w_j \tag{19}$$

where $\bar{G}_{ijl}$ is obtained from the interconnection between the generalized plant G and all controller blocks except C_{ij}, see Figure 3. Here, u_{ij} denotes the contribution from controller block C_{ij} to the ith input u_i. Obviously $\bar{G}_{ijl}$ does not depend on ρ_{ij}. The feedback system in Figure 3 has the structure of a SISO system and we can proceed as in [19]. Differentiating (18)–(19) w.r.t. an arbitrary element of ρ_{ij} gives, denoting the derivatives by $'$,

$$\begin{pmatrix} y_l' \\ \\ w_j' \end{pmatrix} = \bar{G}_{ijl} \begin{pmatrix} 0 \\ \\ u_{ij}' \end{pmatrix} \tag{20}$$

$$u_{ij}' = C_{ij}' w_j + C_{ij} w_j' \tag{21}$$

These equations represent the same feedback system as in Figure 3 but with $r = 0$ and the external input $z_{ij} = C_{ij}' w_j$. Hence, the gradient y_l' is obtained by performing a closed loop experiment with all external signals zero except for $z_{ij} = C_{ij}' w_j$ where w_j is first obtained from a closed loop experiment corresponding to (18)–(19).

Now, using the linearity of the system, the derivatives w.r.t. all parameters in C_{ij} can be obtained from these two experiments. The procedure is as follows: first perform a normal experiment with the desired reference signal r, c.f. Figure 1. Measure w and y. Then for each controller block C_{ij}, cf. Figure 2, perform an experiment where $r = 0$ and where w_j is injected at the output of controller block C_{ij}, i.e. $z_{ij} = w_j$ in the block diagram in Figure 3. By filtering the lth output element, y_l, of y from this experiment through C_{ij}', the derivative of y_l w.r.t. any parameter in the vector ρ_{ij} can be obtained. To compute the derivatives w.r.t. to all possible parameters in a full block controller, $n_u \times n_w$ experiments are necessary, giving a total of $1 + n_u \times n_w$ experiments in order to be able to compute all gradients.

The above procedure was originally designed in such a way that the real time is split in alternating periods of control and gradient experiment. The feedback loop is tuned iteratively, hence the name Iterative Feedback Tuning. This a practically useful approach, but there is a drawback: namely, with alternating reference signals the transient effects will not become negligible even over a very long time-horizon. A first step to remove the effect of the transients is to set up the experiments differently: a key novel assumption is that the system can *replicated*, and thus one control system is purely used for evaluating the controller performance, while a copy of it is used to evaluate a directional derivative.

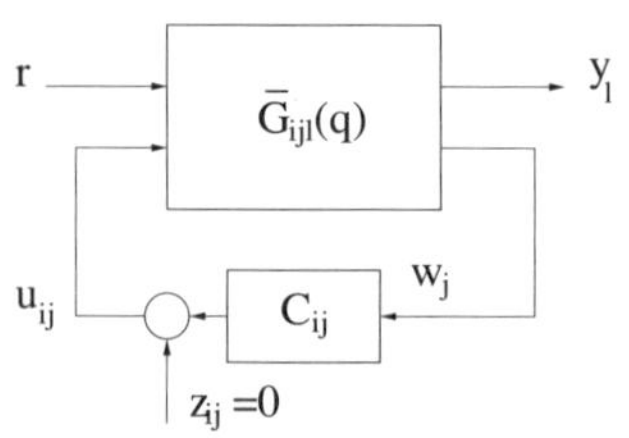

Fig. 3. Feedback system where each channel in the controller is independently parameterized. The generalized plant $\bar{G}_{ijl}$ is independent of the parameter vector ρ_{ij}

6 Randomized IFT

We now present a randomized version of the IFT gradient estimation procedure. A second device to remove the effect of the transients is to introduce a fixed *delay* or settling time τ. Then a single experiment to evaluate the controller performance will last $\tau + N_0$ time-units, and we consider a computable approximation of $J(\rho)$, denote by $J(\rho, m, \xi) =$, corresponding to an initial state ξ at time m:

$$\frac{1}{2N_0} \sum_{n=\tau+1}^{\tau+N_0} y_n(\rho, m, \xi)^T y_n(\rho, m, \xi). \tag{22}$$

Then it is easy to see that

$$|J(\rho, m, \xi) - J(\rho)| \le Ce^{-\alpha\tau} |x^*_m(\rho) - \xi| \tag{23}$$

with some $\alpha > 0$. The same technique and estimate apply for the computation of the gradient processes.

Let now ρ_n be a sequence of controller parameters that are obtained by some iterative procedure and used in successive experiments. Let the corresponding state-sequence in the original control system be $\bar{x}_n = x_{n(\tau+N_0)} = x_{\overline{n}}$. Then under mild conditions we have

$$|x_n - x_n^*(\rho_n)| \leq c|\rho_n - \rho_{n-1}| \leq c|(\partial/\partial\theta)J(\rho_{n-1}, \overline{n}-1, \bar{x}_n - 1|.$$

Taking into account (23) and $|(\partial/\partial\theta)J(\rho_{n-1}| = O(\rho_{n-1} - \rho^*)$ we get $|x_n - x_n^*(\rho_n)| \leq c|\rho_n - \rho^*|$ with some constant c.

Ultimately we come to consider the perturbed version of (6) where $G(\theta)$ is replaced by $G(\theta) + \delta G_n(\theta)$, where the norm of the latter is bounded by $c|\theta_n - \theta^*|$. Note that the perturbation is multiplicative. Taking τ large enough c can be made arbitrarily small, and thus the analysis of Section 3 can be carried over, and geometric convergence takes place. This suggests that randomized IFT can greatly reduce the experimentation time as compared to MIMO IFT as outlined above.

7 Simulation Results

Fixed gain SPSA for noise-free problems has been tested for randomly generated quadratic functions up to dimension 100. In this case the estimation of the gradient is exactly unbiased for any c and the algorithm (6) applies. It is instructive to consider first the SPSA method without resetting. Let the quadratic function L be $L(\theta) = \frac{1}{2}\theta^T A\theta$, with some symmetric positive definite A. Then the gradient estimate is $H(k,\theta) = \Delta_k^{-1}\Delta_k^T G(\theta)$. Taking into account $G(\theta) = A\theta$ we get the following recursion for θ_k:

$$\theta_{k+1} = (I - \lambda\Delta_k^{-1}\Delta_k^T A)\theta_k. \tag{24}$$

Since $A_k = \Delta_k^{-1}\Delta_k^T$ is an i.i.d. sequence, hence it is stationary and ergodic, we can directly apply Oseledec's multiplicative ergodic theorem, see [32]. Thus we get that

$$\lim_{k\to\infty} \frac{1}{k} \log|\theta_k - \theta^*| = \mu = \text{const.} \tag{25}$$

exists with probability 1 for Lebesgue-almost all initial conditions θ_0. The number μ is called the top Lyapunov-exponent. It can also be shown that $\mu < 0$ for sufficiently small λ, e.g. by using the techniques of [6].

In practice we do use resetting. Let B_k denote the event that a resetting takes place at time k:

$$B_k = \{\omega : (I - \lambda\Delta_k^{-1}\Delta_k^T A)\theta_k \notin D_\theta\}.$$

Then the algorithm becomes:

$$\theta_{k+1} = \theta_k - (1 - \chi_{B_k})\lambda \Delta_k^{-1} \Delta_k^T A)\theta_k + \chi_{B_k}(\theta_0 - \theta_k).$$

For this non-linear procedure Theorem 1 guarantees that the Lyapunov exponent

$$\overline{\lim}_{k \to \infty} \frac{1}{k} \log |\theta_k - \theta^*| = \mu < 0$$

with probability 1 for all initial conditions. The value of μ is called a Lyapunov-exponent, it may depend on the initial condition, but it can be easily shown that for each fixed ω at most p different μ may occur.

In Figures 4 and 5 we plot simulation results for the following problem. We generated a quadratic function $L(\theta) = \frac{1}{2}\theta^T A \theta$ with randomly chosen the eigenvalues of A by exponential distribution, and randomly chosen rotations in dimension 20, i.e. $p = 20$. Then we made iterations using the algorithm (6) with different stepsizes λ. In each experiment we had $N = 500$ iterations. The Lyapunov-exponent is approximated in real-time by $\log |\theta_n - \theta^*|/n$, which becomes fairly stable after cca 100 iterations, underscoring that geometric rate of convergence does take place. In Figures 4 and 5 the approximate Lyapunov exponent using the latest values, i.e. $\log |\theta_N - \theta^*|/N$ is plotted against the step-size λ for the SPSA algorithm and the deterministic gradient method, respectively.

Since the Lyapunov-exponent for the SPSA method is more sensitive to the stepsize than the deterministic gradient method we use different scales in Figures 4 an 5.A key indicator of the efficiency of the SPSA method is the *number of measurements* that is needed to achieve a given accuracy. Here a measurement means the evaluation of a single directional derivative by some experiment. This is in turn essentially determined by the Lyapunov exponent the control of which via the stepsize is therefore a key issue. An adaptive procedure for choosing the optimal λ has been proposed in [14].

Consider now the cases when the SPSA and the gradient method both operate under their respective optimal stepsize for a problem that we described above. The optimal Lyapunov exponent is found to be $\mu = -0.0158$ for the SPSA method and $\mu = -0.1734$ for the gradient method. A simple calculation gives that the theoretical number of experiments to achieve a precision $\varepsilon = 0.01$ is approximately $n = \log\ \varepsilon/\mu = 290$ in the SPSA case and $n = \log\ \varepsilon/\mu \cdot p = 540$ using gradient method. These results are supported by experimental data: the required accuracy is indeed achieved after the indicated number of steps. Thus we conclude that SPSA is more economical than similar deterministic methods.

We have also considered non-quadratic problems of the form

$$L(\theta) = \frac{1}{2}\theta^T A(\theta)\theta \quad \text{with} \quad A(\theta) = A + u(\theta)u(\theta)^T$$

where $u(\theta) = D\theta$ with some fixed matrix D where the elements of D were chosen uniformly in the range of $[0, c_p]$, where c_p was chosen in $[0.01,\ 1]$.

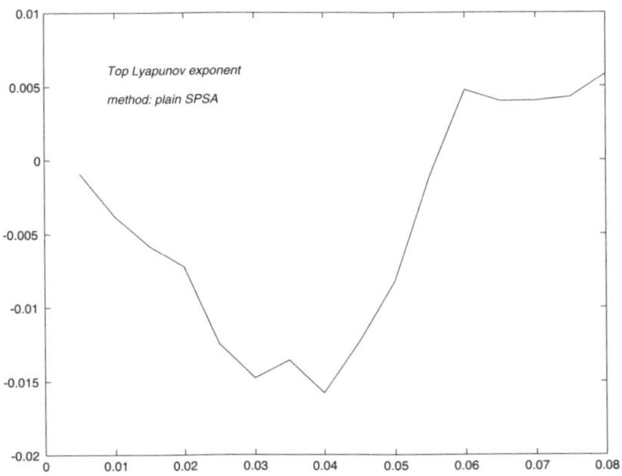

Fig. 4. The Lyapunov-exponent vs. step-size for the SPSA method.

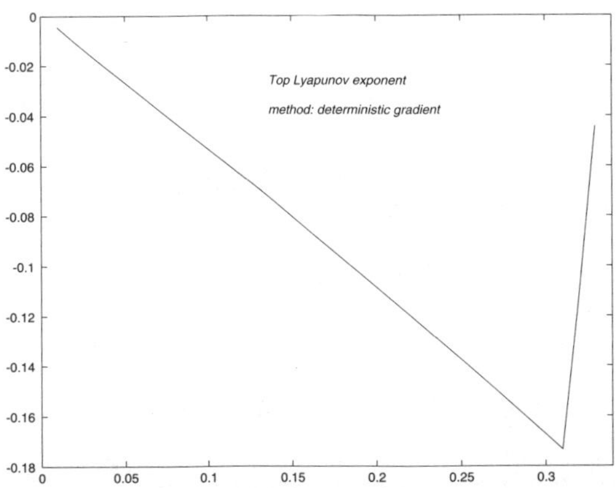

Fig. 5. The Lyapunov-exponent vs. step-size for the gradient method.

Geometric rate of convergence in this case is ensured by the results of this paper and indeed confirmed by simulation results similar to the quadratic case. Experiments for the randomized IFT are in progress.

8 Conclusion

We have introduced a fixed-gain version of the SPSA method and have demonstrated that for noise-free optimization it gives geometric rate of convergence almost surely. These theoretical results are supported by simulation, and the superiority of SPSA compared to deterministic gradient method has been shown experimentally for both quadratic and perturbed quadratic problems of dimension 20. Fixed gain SPSA has been used to derive a randomized version of Iterative Feedback Tuning (IFT) for MIMO systems, where a good approximation of the directional derivative of the cost function in a random direction is computed by a single physical experiment.

Acknowledgement

The first author gratefully acknowledges the support of Kansas University, Lawrence and the NSF to attend the Stochastic Theory and Control Workshop, and expresses his thanks to Bozenna Pasik-Duncan for her invitation. A number of useful comments by James C. Spall are also gratefully acknowledged. His work was also partially supported by the National Research Foundation of Hungary (OTKA) under Grant no. T 032932. The second author gratefully acknowledges the support of the Bolyai János Research Fellowship of the Hungarian Academy of Sciences. The third author gratefully acknowledges the support by MTA SZTAKI's EU Center of Excellence program.

References

1. Benveniste, A., Métivier, M., and Priouret, P. (1990) Adaptive Algorithms and Stochastic Approximations. Springer-Verlag, Berlin.
2. Borodin, A. N. (1979) A stochastic approximation procedure in the case of weakly dependent observations, Theory of Probability and Appl. **24**, 34–52.
3. Chen, H. F., Duncan, T. E. and Pasik-Duncan, B. (1999) A Kiefer-Wolfowitz algorithm with randomized differences, IEEE Trans. Automat. Contr. **44**, 442–453.
4. Furstenberg, H. and Kesten, H. (1960) Products of random matrices, Ann. Math. Statist. **31**, 457–469.
5. Gerencsér. L. (1989) On a class of mixing processes, Stochastics **26**, 165–191.
6. Gerencsér. L. (1991) Almost sure exponential stability of random linear differential equations, Stochastics **36**, 411–416.
7. Gerencsér. L. (1991) A representation theorem for the error of recursive estimators. Submitted to SIAM J. Control and Optimization in 1991. Revised in 1997.
8. Gerencsér. L. (1992) Rate of convergence of recursive estimators, SIAM J. Control and Optimization **30**(5), 1200–1227.
9. Gerencsér. L. (1995) Rate of convergence of the LMS algorithm, Systems and Control Letters **24**, 385–388.
10. Gerencsér. L. (1996) On fixed gain recursive estimation processes, J. of Mathematical Systems, Estimation and Control **6**, 355–358. Retrieval code for full electronic manuscript: 56854.
11. Gerencsér. L. (1999) Rate of convergence of moments for a simultaneuous perturbation stochastic approximation method for function minimization, IEEE Trans. Automat. Contr. **44**, 894–906.
12. Gerencsér. L. (2002) Stability of random iterative mappings. In M. Dror, P. L'ecuyer, and F. Szidarovszky, editors, Modeling Uncertainty. An Examination of its Theory, Methods, and Applications, to appear. Dordrecht, Kluwer, 2002.

13. Gerencsér, L., Hill, S. D., and Vágó, Zs. (1999) Optimization over discrete sets via SPSA. In Proceedings of the 38-th Conference on Decision and Control, CDC'99, 1791–1794.
14. Gerencsér, L. and Vágó, Zs. (2001) A stochastic approximation method for noise free optimization. In Proceedings of the European Control Conference, ECC'01, Porto, 1496–1500.
15. Györfi, L. and Walk, H. (1996) On the average stochastic approximation for linear regression, SIAM J. Control and Optimization **34(1)**, 31–61.
16. Hjalmarsson, H. (1999) Efficient tuning of linear multivariable controllers using iterative feedback tuning, Int. J. Adapt.Control Signal Process. **13**, 553–572.
17. Hjalmarsson, H. (1998) Control of nonlinear systems using Iterative Feedback Tuning. In Proc. 1998 American Control Conference, Philadelphia, 2083–2087.
18. Hjalmarsson, H., Gevers, M., Gunnarsson, S., and Lequin, O. (1998) Iterative Feedback Tuning: theory and applications. IEEE Control Systems Magazine **18**(4), 26–41.
19. Hjalmarsson, H., Gunnarsson, S., and Gevers, M. (1994) A convergent iterative restricted complexity control design scheme. In Proc. 33rd IEEE CDC, Orlando, Florida, 1735–1740.
20. Hjalmarsson, H., Gunnarsson, S., and Gevers, M. (1995) Model-free tuning of a robust regulator for a flexible transmission system. European Journal of Control **1**, 148–156.
21. Joslin, J. A. and Heunis, A. J. (2000) Law of the iterated logarithm for a constant-gain linear stochastic gradient algorithm, SIAM J. on Control and Optimization **39**, 533–570.
22. Kifer, Y. (1986) Ergodic Theory of Random Transformations. Birkhäuser, 1986.
23. Kushner, H. J. and Shwartz, A. (1984) Weak convergence and asymptotic properties of adaptive filters with constant gains, IEEE Trans. Informat. Theory **30**(2), 177–182.
24. Kushner, H. J. and Yin, G. (1997) Stochastic Approximation Algorithms and Applications. Springer Verlag. New York.
25. Ljung, L. and Söderström, T. (1983) Theory and Practice of Recursive Identification. The MIT Press.
26. Meurers, T. and Veres, S. M. (1999) Iterative design for vibration attenuation. International Journal of Acoustics and Vibration **4**(2), 79–83.
27. Nevelson, M. B. and Khasminskii, R. Z. (1972) Stochastic Approximation and Recursive Estimation. Nauka, Moscow (in Russian).
28. Solo, V. (1981) The second order properties of a time series recursion, Ann. Stat. **9**, 307–317.
29. Spall, J. C. (1992) Multivariate stochastic approximation using a simultaneous perturbation gradient approximation, IEEE Trans. Automat. Contr. **37**, 332–341.
30. Spall, J. C. (1997) A one-measurement form of simultaneous perturbation stochastic approximation, Automatica **33**, 109–112.
31. Spall, J. C. (2000) Adaptive stochastic approximation by the simultaneous perturbation method, IEEE Trans. Automat. Contr. **45**, 1839–1853.
32. Oseledec, V. I. (1968) A multiplicative ergodic theorem. Lyapunov charasteristic numbers for dynamical systems, Trans. Moscow Math. Soc. **19**, 197–231.

Capacity of the Multiple-Input, Multiple-Output Poisson Channel

Shane M. Haas and Jeffrey H. Shapiro

Massachusetts Institute of Technology, Laboratory for Information and Decision Systems, Research Laboratory of Electronics, Cambridge MA 02139, USA

Abstract. This paper examines the Shannon capacity of the single-user, multiple-input, multiple-output (MIMO) Poisson channel with peak and average transmit power constraints. The MIMO Poisson channel is a good model for the physical layer of a multi-aperture optical communication system that operates in the shot-noise-limited regime. We derive upper and lower bounds on the capacity that coincide in a number of special cases. The capacity is bounded below by that of the MIMO channel with an additional on-off keying (OOK) transmitter constraint, and it is bounded above by that of parallel, independent multiple-input, single-output (MISO) channels.

1 Introduction

The multiple-input, multiple-output (MIMO) Poisson channel serves as a good model for a shot-noise-limited, direct detection optical communication system that uses multiple transmit and receive apertures. Each receive aperture focuses its incident light onto a photodetector, producing an output that can be modelled as a Poisson counting process (PCP). The rate of this PCP is proportional to the power of the impinging light [1].

The capacity of the single-input, single-output (SISO) Poisson channel is well understood. Kabanov [5] derived the information capacity of the SISO Poisson channel with a peak transmit power constraint using martingale techniques. Davis [6] considered the addition of an average transmit power constraint. Wyner [8] derived the capacity and error exponent from first principles, using a discrete memoryless channel approximation. Shamai (Shitz) [9,10] derived the capacity with constraints on the transmitted pulse width. Frey [12] allowed for time-varying peak and average power constraints, as well as for random noise intensities. Shamai (Shitz) and Lapidoth [13] considered general spectral constraints on the PCP rate process.

Recently, the multiple-user Poisson channel has received attention. Lapidoth and Shamai (Shitz) [14] computed the two-user capacity region of the multiple-access channel. Their concluding section included a comment on the capacity of the multiple-input, single-output channel in the absence of background noise and average power constraints. We will formalize their comment later, in section 2. Bross, et. al. [15] calculated the error exponent for the two-user Poisson multiple access channel. Lapidoth discusses the Poisson broadcast channel in [16].

B. Pasik-Duncan: Stochastic Theory and Control, LNCIS 280, pp. 155–168, 2002.

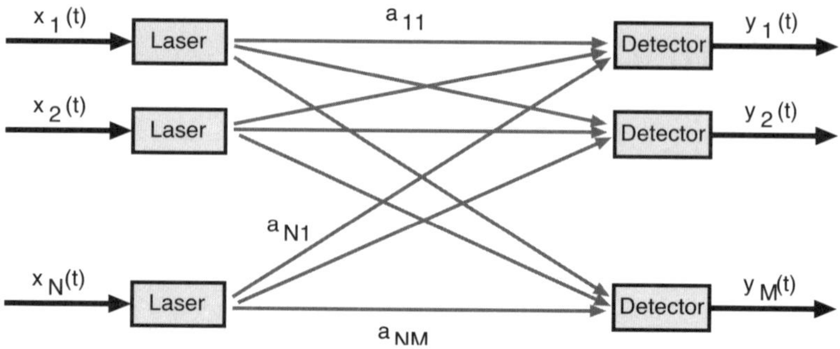

Fig. 1. The N inputs of the MIMO Poisson channel $\{x_n(t),\ 1 \leq n \leq N, 0 \leq t < T\}$ produce M Poisson counting processes $\{y_m(t),\ 1 \leq m \leq M,\ 0 \leq t < T\}$ whose rates when conditioned on the inputs are $\{\mu_m(t) = \sum_{n=1}^{N} \alpha_{nm} x_n(t) + \lambda_m,\ 1 \leq m \leq M, 0 \leq t < T\}$, where λ_m is the background-noise rate for the m-th aperture.

In what follows, we shall derive upper and lower bounds on the capacity of the MIMO Poisson channel. We will show that our bounds coincide in a number of interesting special cases. These include the limits of low and high signal-to-background ratio, and the multiple-input, single-output (MISO) channel. We will also show that our lower bound gives the MIMO capacity for the single-input, multiple-output (SIMO) channel, and that, in general, our bounds are quite close.

2 The MIMO Poisson Channel

In this paper, we will use the incoherent channel model as in [14]. For every receiver, we will assume that the fields received from the multiple transmitters are sufficiently separated in frequency or angle of arrival to make the received power equal to the sum of the powers from the individual transmitters. Under shot-noise-limited operation, the output of the m-th $(1 \leq m \leq M)$ receiver, $\{y_m(t),\ 0 \leq t < T\}$, conditioned on the knowledge of the transmitted codeword, can be taken to be a PCP with rate

$$\mu_m(t) = \sum_{n=1}^{N} \alpha_{nm} x_n(t) + \lambda_m,$$

where N is the number of transmit apertures, α_{nm} is the path gain from the n-th transmit aperture to the m-th receive aperture, and λ_m is the background-noise rate for the m-th aperture. In this paper, we assume that the transmitter and receiver know the non-random path gains and background-noise rates. The (non-negative) codeword waveform $x_n(t)$, which is proportional to the power waveform sent from the n-th transmitter, will be forced to satisfy the peak power constraint,

$$0 \leq x_n(t) \leq A_n, \tag{1}$$

and the average power constraint,

$$\frac{1}{T}\int_0^T E[x_n(t)]dt \le \sigma A_n, \quad \text{where } 0 \le \sigma \le 1. \tag{2}$$

Figure 1 shows an illustrative block diagram of the MIMO Poisson channel.

2.1 Capacity and Mutual Information

The m-th output process path up to time t, $Y_m(t) \equiv \{y_m(\tau), 0 \le \tau < t \le T\}$, is completely described by the total number of arrivals occurring prior to time t, denoted by $y_m(t)$, and its ordered arrival times, $0 \le t_{m1} \le t_{m2} \le \dots \le t_{my_m(t)} < t \le T$. Note that $Y_m(t)$ is continuous from the left, viz., it includes all arrivals up to, but not including, time t. We can then equivalently define $Y_m(t) \equiv \{y_m(t), t_{m1}, \dots, t_{my_m(t)}\}$. Let $Y_m \equiv Y_m(T)$ represent the path of the m-th output process on the interval $[0, T)$. Denote the M channel output process paths up to time t as $Y(t) \equiv \{Y_1(t), \dots, Y_M(t)\}$, and the entire path as $Y \equiv Y(T)$. Let $X \equiv \{x_1(t), \dots, x_N(t);\, 0 \le t < T\}$ represent the channel input.

Define $\mathcal{X}$ as all distributions on X that satisfy the transmitter constraints in (1) and (2). The capacity of the MIMO Poisson channel in nats[1] per second is then

$$C_{\text{MIMO}} = \sup_{p_X \in \mathcal{X}} \frac{1}{T} I(X;Y), \tag{3}$$

where the mutual information, $I(X;Y)$, is given by [4],

$$\begin{aligned} I(X;Y) &= E\left[\log\left(\frac{p_{Y|X}}{p_Y}\right)\right] = E\left[\log p_{Y|X}\right] - E\left[\log p_Y\right] \\ &= h(Y) - h(Y|X). \end{aligned} \tag{4}$$

Here: $h(Y) \equiv -E[\log p_Y]$ and $h(Y|X) \equiv -E[\log p_{Y|X}]$ are the unconditional and conditional entropies, respectively, of the output processes' ordered arrival times described by the densities p_Y and $p_{Y|X}$. A more general derivation of the mutual information for point processes is given in [7] and [2] using martingale techniques. In what follows, we present a specialized derivation for doubly-stochastic Poisson processes [11], sufficient for our application, using ordered arrival time densities.

First consider the conditional entropy, $h(Y|X)$, of the output processes' ordered arrival times. Conditioned on the channel input X, the output processes $Y_1, \dots, Y_M$ are independent, each with density ([11],Th. 2.3.2),

$$p_{Y_m|X} = \exp\left(\int_0^T \log[\mu_m(t)]dY_m(t) - \int_0^T \mu_m(t)dt\right), \tag{5}$$

[1] All logarithms in this paper are natural logarithms.

where

$$\int_0^T \log[\mu(t)]dY_m(t) \equiv \begin{cases} 0, & y_m(T) = 0, \\ \sum_{k=1}^{y_m(T)} \log[\mu_m(t_{mk})], & y_m(T) > 0, \end{cases} \tag{6}$$

and $0 \leq t_{m1} \leq t_{m2} \leq \ldots \leq t_{m,y_m(T)} < T$ are the ordered arrival times of the m-th output process, Y_m. We will use the following lemma in both the conditional and unconditional entropy derivations.

Lemma 1. *Let $\{P(t), 0 \leq t < T\}$ be an inhomogeneous PCP with rate function $m(t)$. Then,*

$$E\left[\int_0^T \log[m(t)]dP(t)\right] = \int_0^T m(t)\log[m(t)]dt. \tag{7}$$

Proof. Using the arrival time density in (5), symmetry about the $k!$ permutations of the k variables of integration, and the unity sum of the Poisson mass function:

$$\begin{aligned}
E&\left[\int_0^T \log[m(t)]dP(t)\right] \\
&= \sum_{k=1}^{\infty}\int_0^T\int_0^{t_k}\cdots\int_0^{t_2}\left(\prod_{n=1}^{k} m(t_n)\right)\exp\left(-\int_0^T m(t)dt\right) \\
&\qquad\qquad \times \sum_{j=1}^{k}\log\left[m(t_j)\right]dt_1\cdots dt_k \\
&= \sum_{k=1}^{\infty}\frac{\exp\left(-\int_0^T m(t)dt\right)}{k!} \\
&\qquad \times \int_0^T\cdots\int_0^T\left(\prod_{n=1}^{k} m(t_n)\right)\sum_{j=1}^{k}\log\left[m(t_j)\right]dt_1\cdots dt_k \\
&= \sum_{k=1}^{\infty}\frac{\exp\left(-\int_0^T m(t)dt\right)}{(k-1)!}\left(\int_0^T m(t)dt\right)^{k-1}\int_0^T m(t)\log[m(t)]dt \\
&= \int_0^T m(t)\log[m(t)]dt,
\end{aligned}$$

where $0 \leq t_1 \leq \cdots \leq t_{P(t)} < T$ are the ordered arrival times up to time t.

Using Lm. 1 and (5), the conditional entropy of the ordered arrival times is

$$h(Y|X) = -E\left[\log p_{Y|X}\right]$$

$$
\begin{aligned}
&= -E\left[\sum_{m=1}^{M} \log p_{Y_m|X}\right] \\
&= -\sum_{m=1}^{M} E\left[\int_0^T \log[\mu_m(t)]dY_m(t) - \int_0^T \mu_m(t)dt\right] \\
&= -\sum_{m=1}^{M} \left(E\left[E\left[\int_0^T \log[\mu_m(t)]dY_m(t)\middle|\, \mu_m(t), 0 \le t < T\right]\right]\right. \\
&\qquad \left. - E\left[\int_0^T \mu_m(t)dt\right]\right) \\
&= -\sum_{m=1}^{M} \left(E\left[\int_0^T \mu_m(t)\log[\mu_m(t)]dt\right] - \int_0^T E[\mu_m(t)]dt\right). \qquad (8)
\end{aligned}
$$

We now derive the unconditional entropy, $h(Y)$, of the ordered arrival times. The output Y is a multi-channel, doubly-stochastic, Poisson process with density given by ([11],pg. 425):

$$
p_Y = \exp\left(\sum_{m=1}^{M}\left(\int_0^T \log[\widehat{\mu}_m(t)]dY_m(t) - \int_0^T \widehat{\mu}_m(t)dt\right)\right), \qquad (9)
$$

where

$$
\widehat{\mu}_m(t) \equiv E\left[\mu_m(t) \mid y_1(\tau), \ldots, y_M(\tau), 0 \le \tau < t\right] = E\left[\mu_m(t) \mid Y(t)\right], \qquad (10)
$$

is the causal least-squares estimator of $\mu_m(t)$ based on the M output processes.

Using Lm. 1, (9), and noting that $E[\widehat{\mu}(t)] = E[\mu(t)]$, the unconditional entropy is

$$
\begin{aligned}
h(Y) &= -E\left[\log p_Y\right] \\
&= -\sum_{m=1}^{M} E\left[\int_0^T \log[\widehat{\mu}_m(t)]dY_m(t) - \int_0^T \widehat{\mu}_m(t)dt\right] \\
&= -\sum_{m=1}^{M} \left(E\left[E\left[\int_0^T \log[\widehat{\mu}_m(t)]dY_m(t)\middle|\, \widehat{\mu}_m(t), 0 \le t < T\right]\right]\right. \\
&\qquad \left. - E\left[\int_0^T \widehat{\mu}_m(t)dt\right]\right) \\
&= -\sum_{m=1}^{M} \left(E\left[\int_0^T \widehat{\mu}_m(t)\log[\widehat{\mu}_m(t)]dt\right] - \int_0^T E[\mu_m(t)]dt\right). \qquad (11)
\end{aligned}
$$

Substituting (8) and (11) into (4) gives the mutual information for the MIMO Poisson channel,

$$I(X;Y) = \sum_{m=1}^{M} E\left[\int_0^T \{\mu_m(t)\log[\mu_m(t)] - \widehat{\mu}_m(t)\log[\widehat{\mu}_m(t)]\}\, dt\right]. \tag{12}$$

Paralleling the SISO capacity derivation by Davis [6], we manipulate the mutual information as follows. Define

$$\phi_m(z) \equiv (\lambda_m + z)\log(\lambda_m + z) - \lambda_m \log \lambda_m, \tag{13}$$

and note that

$$\widehat{\mu}_m(t) = E\left[\sum_{n=1}^{N} \alpha_{nm} x_n(t) + \lambda_m \,\middle|\, Y(t)\right] = \sum_{n=1}^{N} \alpha_{nm}\widehat{x}_n(t) + \lambda_m, \tag{14}$$

where $\widehat{x}_n(t) \equiv E\,[\,x_n(t) \mid Y(t)]$. The mutual information is then

$$\begin{aligned}
I(X;Y)& \\
&= \sum_{m=1}^{M} E\left[\int_0^T \left\{\phi_m\left(\sum_{n=1}^{N}\alpha_{nm}x_n(t)\right) - \phi_m\left(\sum_{n=1}^{N}\alpha_{nm}\widehat{x}_n(t)\right)\right\} dt\right] \\
&= \sum_{m=1}^{M} E\left[\int_0^T \left\{\frac{\sum_{n=1}^{N}\alpha_{nm}\widehat{x}_n(t)}{R_m}\phi_m(R_m) - \phi_m\left(\sum_{n=1}^{N}\alpha_{nm}\widehat{x}_n(t)\right)\right.\right. \\
&\qquad \left.\left. - \frac{\sum_{n=1}^{N}\alpha_{nm}x_n(t)}{R_m}\phi_m(R_m) + \phi_m\left(\sum_{n=1}^{N}\alpha_{nm}x_n(t)\right)\right\} dt\right] \\
&= \sum_{m=1}^{M} E\left[\int_0^T \left\{h_m\left(\sum_{n=1}^{N}\alpha_{nm}\widehat{x}_n(t)\right) - h_m\left(\sum_{n=1}^{N}\alpha_{nm}x_n(t)\right)\right\} dt\right],
\end{aligned} \tag{15}$$

with

$$h_m(z) \equiv \frac{z}{R_m}\phi_m(R_m) - \phi_m(z), \tag{16}$$

$$R_m \equiv \sum_{n=1}^{N} \alpha_{nm} A_n. \tag{17}$$

Equation (15) is the MIMO generalization of the SISO mutual information expression derived in [5] and [6]. The function $h_m(R_m p)$ is shown in Fig. 2 as a function of p for illustrative high and low background noise rates.

We have not been able to find the supremum in (3), but instead have derived upper and lower bounds. The derivations of these bounds, given below, exploit the following rewritten form of (15):

$$
\begin{aligned}
I(X;Y)& \\
= \sum_{m=1}^{M} \int_0^T & \left\{ h_m\left(\sum_{n=1}^{N} \alpha_{nm} E[x_n(t)] \right) - E\left[h_m\left(\sum_{n=1}^{N} \alpha_{nm} x_n(t) \right) \right] \right\} dt \\
- \sum_{m=1}^{M} \int_0^T & \left\{ h_m\left(\sum_{n=1}^{N} \alpha_{nm} E[x_n(t)] \right) - E\left[h_m\left(\sum_{n=1}^{N} \alpha_{nm} \widehat{x}_n(t) \right) \right] \right\} dt,
\end{aligned}
\tag{18}
$$

and the fact that $h_m(z)$ is concave, which makes both summations positive with the first being greater than the second.

2.2 The Parallel-Channel Upper Bound

An upper bound on the capacity can be found by maximizing the first term in (18) and ignoring the second term. Because the supremum of a sum cannot exceed the sum of the supremums, we will upper bound the mutual information by maximizing each term of the summation, i.e.,

$$
\begin{aligned}
C_{\text{MIMO}} \leq \sum_{m=1}^{M} \sup_{p_X(x) \in \mathcal{X}} \frac{1}{T} \int_0^T & \left\{ h_m \left(\sum_{n=1}^{N} \alpha_{nm} E[x_n(t)] \right) \right. \\
& \left. - E\left[h_m\left(\sum_{n=1}^{N} \alpha_{nm} x_n(t) \right) \right] \right\} dt.
\end{aligned}
\tag{19}
$$

To maximize each term, fix the argument $\sum_{n=1}^{N} \alpha_{nm} E[x_n(t)]$ and consider the distribution that minimizes $E\left[h_m\left(\sum_{n=1}^{N} \alpha_{nm} x_n(t)\right)\right]$. Because $h_m(0) = h_m(R_m) = 0$, this distribution is concentrated at $x_1(t) = \cdots = x_N(t) = 0$ and $x_1(t) = A_1, \cdots, x_N(t) = A_N$ for all t. Notice that this distribution by construction satisfies the peak transmit power constraint, and makes the expectation term zero.

Let $p_m(t) = \Pr(x_1(t) = A_1, \cdots, x_N(t) = A_N)$ for the m-th term. Noting that $E[x_n(t)] = A_n p_m(t)$, we now find the argument,

$$
\sum_{n=1}^{N} \alpha_{nm} E[x_n(t)] = R_m p_m(t), \tag{20}
$$

or equivalently the function $p_m(t)$, that maximizes the m-th term

$$
\frac{1}{T} \int_0^T h_m[R_m p_m(t)]\, dt. \tag{21}
$$

The average power constraint requires that $p_m(t)$ obey

$$\frac{1}{T}\int_0^T p_m(t)dt = \frac{1}{T}\int_0^T \frac{E[x_n(t)]}{A_n}dt \leq \sigma. \tag{22}$$

Using the concavity of $h_m[R_m p(t)]$, the Taylor series expansion with respect to $R_m p(t)$ about a point $R_m p_m^*$, $0 \leq p_m^* \leq 1$, yields

$$h_m[R_m p(t)] - h_m(R_m p_m^*) \leq h_m'(R_m p_m^*)\left[R_m p_m(t) - R_m p_m^*\right], \tag{23}$$

where

$$\begin{aligned} h_m'(z) &\equiv \frac{d}{dz}h_m(z) \\ &= \frac{\phi_m(R_m)}{R_m} - \log\left[e(z+\lambda_m)\right] \\ &= \frac{(\lambda_m + R_m)\log(\lambda_m + R_m) - \lambda_m \log \lambda_m}{R_m} - \log\left[e(z+\lambda_m)\right] \\ &= (1+s_m)\log(1+s_m) - s_m \log s_m - \log\left[e\left(\frac{z}{R_m} + s_m\right)\right], \end{aligned} \tag{24}$$

and $s_m \equiv \frac{\lambda_m}{R_m}$ is the ratio of background noise to aggregate received power, i.e., it is the reciprocal signal-to-background ratio. Averaging over the interval $[0, T)$ yields

$$\begin{aligned} &\frac{1}{T}\int_0^T h_m\left[R_m p_m(t)\right]dt - h_m(R_m p_m^*) \\ &\quad \leq R_m h_m'(R_m p_m^*)\left(\frac{1}{T}\int_0^T p_m(t)dt - p_m^*\right). \end{aligned} \tag{25}$$

Let $p_m^* = \min(p_m^{\max}, \sigma)$ where

$$p_m^{\max} = \frac{(1+s_m)^{(1+s_m)}}{e s_m^{s_m}} - s_m, \tag{26}$$

is the value of p that maximizes $h_m(R_m p)$ found by solving $h_m'(R_m p_m^{\max}) = 0$ in (24). When $p_m^{\max} \leq \sigma$, then $h_m'(R_m p_m^*) = 0$, and

$$\frac{1}{T}\int_0^T h_m\left[R_m p_m(t)\right]dt \leq h_m(R_m p_m^*), \tag{27}$$

with equality when $p_m(t) \equiv p_m^*$. Similarly, when $\sigma < p_m^{\max}$, then $h_m'(R_m p_m^*) > 0$ due to strict concavity, and

$$\frac{1}{T}\int_0^T p_m(t)dt - p_m^* \leq 0 \tag{28}$$

due to the average energy constraint (22). Consequently, (27) is also true, holding with equality when $p_m(t) \equiv p_m^*$.

We now have our parallel-channel upper bound on C_{MIMO}, viz.,

$$C_{\mathrm{MIMO}} \leq C_{\mathrm{PC-UB}} = \sum_{m=1}^{M} h_m \left(\sum_{n=1}^{N} \alpha_{nm} A_n p_m^* \right), \tag{29}$$

where our parallel-channel terminology is due to this expression's being the sum of the capacities of M independent N-by-1 MISO channels. This terminology becomes clearer by expressing the mutual information via entropy chain rule and conditioning inequalities [4] as,

$$\begin{aligned} I(Y_1, \ldots, Y_M; X) &= h(Y_1, \ldots, Y_M) - h(Y_1, \ldots, Y_M \mid X) \\ &= h(Y_1, \ldots, Y_M) - \sum_{m=1}^{M} h(Y_m \mid X) \\ &= \sum_{m=1}^{M} \left[h(Y_m \mid Y_1, \ldots, Y_{m-1}) - h(Y_m \mid X) \right] \\ &\leq \sum_{m=1}^{M} \left[h(Y_m) - h(Y_m \mid X) \right] \\ &= \sum_{m=1}^{M} I(Y_m; X). \end{aligned} \tag{30}$$

2.3 The On-Off Keying Lower Bound

A lower bound on the channel capacity can be found by restricting the supremum in (3) to a subset of $\mathcal{X}$, namely those input distributions that, in addition to the peak and average power constraints, also satisfy $x_1(t) = A_1, \cdots, x_N(t) = A_N$ or $x_1(t) = \cdots = x_N(t) = 0$ for each time t. As in [5] and [6], switching between these "on" and "off" states arbitrarily fast causes the causal least-squares estimator, $\widehat{x}_n(t)$, to reduce to the unconditional mean, $E[x_n(t)]$, which makes the second term in (18) vanish. For details, see [3].

Because $h_m(0) = h_m(R_m) = 0$, the mutual information for rapidly switching on-off keying (OOK) inputs reduces (18) to

$$I(X;Y) = \int_0^T \left(\sum_{m=1}^{M} h_m \left[R_m p(t) \right] \right) dt, \tag{31}$$

where $p(t) = \Pr(x_1(t) = A_1, \cdots, x_N(t) = A_N)$ is the probability that the inputs are in the "on" state at time t. We now find the $p(t)$ that maximizes (31).

Define

$$g[p(t)] = \sum_{m=1}^{M} h_m \left[R_m p(t) \right]. \tag{32}$$

Expanding $g[p(t)]$ with respect to $p(t)$ about a point p^*, $0 \le p^* \le 1$, and using the concavity of $g[p(t)]$ gives

$$g[p(t)] - g(p^*) \le g'(p^*)[p(t) - p^*], \tag{33}$$

where

$$\begin{aligned} g'(p) &= \frac{d}{dp} g(p) \\ &= \sum_{m=1}^{M} R_m h'_m \left(R_m p \right) \\ &= \sum_{m=1}^{M} R_m \log \left(\frac{(1+s_m)^{(1+s_m)}}{e s_m^{s_m} (p + s_m)} \right) \\ &= \log \left\{ \prod_{m=1}^{M} \left[\frac{(1+s_m)^{(1+s_m)}}{e s_m^{s_m} (p + s_m)} \right]^{R_m} \right\}. \end{aligned} \tag{34}$$

Let $p^* = \min(p^{\max}, \sigma)$ where $p^{\max}$ is the value of p that maximizes $g(p)$, found by solving

$$\prod_{m=1}^{M} \left[\frac{(1+s_m)^{(1+s_m)}}{e s_m^{s_m} (p^{\max} + s_m)} \right]^{R_m} = 1. \tag{35}$$

Similar to the derivation of the parallel-channel upper bound, we obtain

$$\frac{1}{T} \int_0^T \left(\sum_{m=1}^{M} h_m [R_m p(t)] \right) dt \le \sum_{m=1}^{M} h_m (R_m p^*), \tag{36}$$

with equality when $p(t) \equiv p^*$.

We now have a lower bound on the MIMO capacity C_{MIMO}:

$$C_{\text{MIMO}} \ge C_{\text{OOK-LB}} = \sum_{m=1}^{M} h_m \left(\sum_{n=1}^{N} \alpha_{nm} A_n p^* \right), \tag{37}$$

where $C_{\text{OOK-LB}}$ is the capacity of using OOK signaling with arbitrarily fast toggling.

3 Comparison of Bounds

Figure 2 shows the relationship between the upper and lower bounds on the MIMO channel capacity. Assuming no average energy constraint ($\sigma = 1$), the PC-UB is the sum of the concave functions $h_m(R_m p)$ evaluated at their respective maxima, whereas the OOK-LB is the maximum of their sum. Davis [6] and Wyner [8] have shown that the maximum of $h_m(R_m p)$ occurs between $1/e$ for high signal-to-background ratio ($s_m \to 0$), and $1/2$ for low signal-to-background ratio ($s_m \to \infty$). It follows that,

$$\frac{1}{e} \leq \min_{1 \leq m \leq M} p_m^{\max} \leq p^{\max} \leq \max_{1 \leq m \leq M} p_m^{\max} \leq \frac{1}{2}, \tag{38}$$

which, in turn, implies that the PC-UB and OOK-LB are equal in both the high and low signal-to-background limits. See Fig. 3 for a plot of $p_m^{\max}$ as a function of s_m.

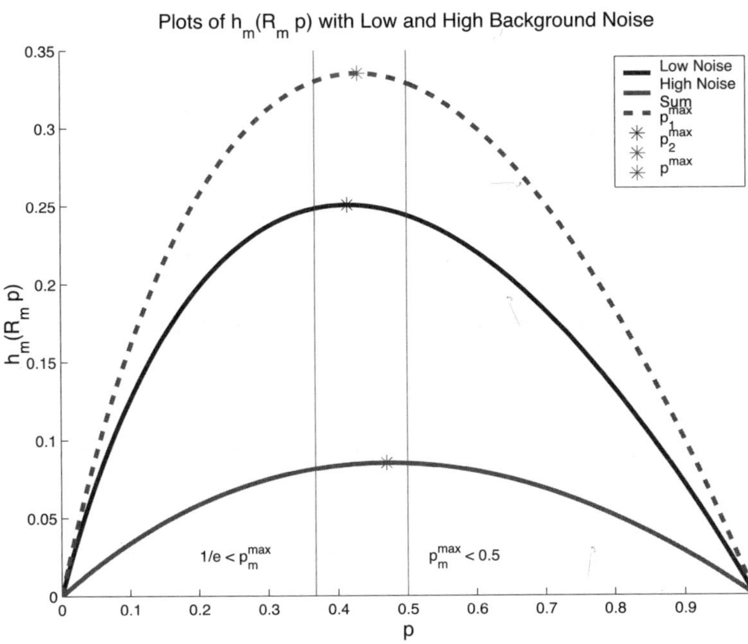

Fig. 2. The the parallel-channel upper bound (PC-UB) is the sum of concave functions evaluated at their respective maxima, whereas the the on-off keying lower bound (OOK-LB) is the maximum of the their sum. In this two receive aperture ($M = 2$) example, $C_{PC-UB} = h_1(R_1 p_1^{\max}) + h_2(R_2 p_2^{\max})$, and $C_{OOK-LB} = h_1(R_1 p^{\max}) + h_2(R_2 p^{\max})$.

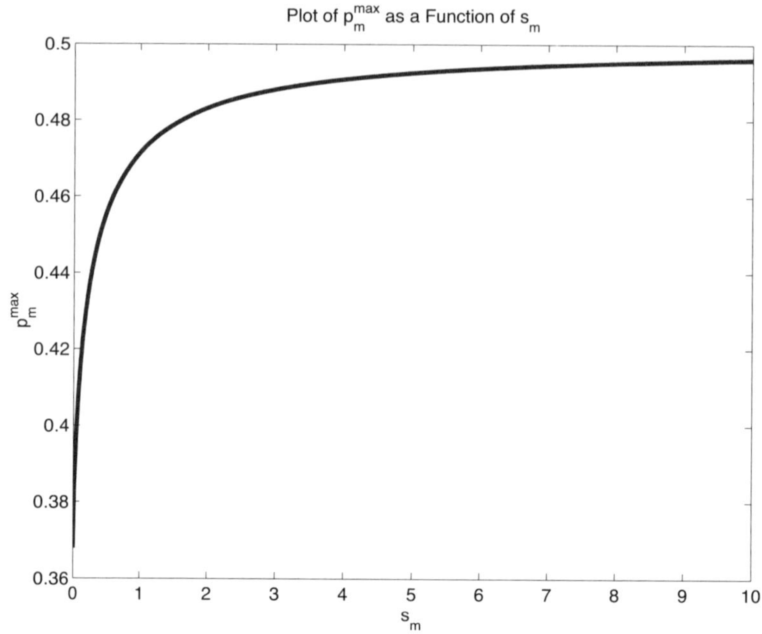

Fig. 3. The maximum of $h_m(R_m p)$ as a function of s_m lies between $e^{-1} \approx 0.3679$ for low noise ($s_m \to 0$) and 0.5 for high noise ($s_m \to \infty$).

In addition to the preceding high and low signal-to-background regimes, there are several other special cases for which $C_{\mathrm{OOK-LB}} = C_{\mathrm{MIMO}} = C_{\mathrm{PC-UB}}$ prevails:

- the multiple-input, single-output (MISO) channel ($M = 1$ case), in which case $p_1^* = p^*$;
- the flat signal-to-background ratio case ($s_1 = \cdots = s_M$), in which case $p_1^* = \cdots = p_M^* = p^*$;
- and the low average-power case ($\sigma \leq \min_{1 \leq m \leq M} p_m^{\max}$), in which case $p_1^* = \cdots = p_M^* = p^* = \sigma$.

For the single-input, multiple-output (SIMO) Poisson channel (the $N = 1$ case), we have that $C_{\mathrm{MIMO}} = C_{\mathrm{OOK-LB}}$, because it can be shown that the optimal distribution at any given time satisfies the OOK constraint.

In general, when the bounds are not equal, they are usually quite close, as can be seen from Fig. 4, which was computed for the $N = 2$, $M = 3$ special case using the arbitrarily-chosen path gains given in Table 1.

4 Conclusions

In this paper we have introduced the MIMO Poisson channel with peak and average transmit power constraints. We have derived upper and lower bounds

Table 1. Path gains for the upper and lower bound comparison in Fig. 4

α_{nm}	$m = 1$	$m = 2$	$m = 3$
$n = 1$	0.2977	0.5760	0.1279
$n = 2$	0.0692	0.6979	2.0322

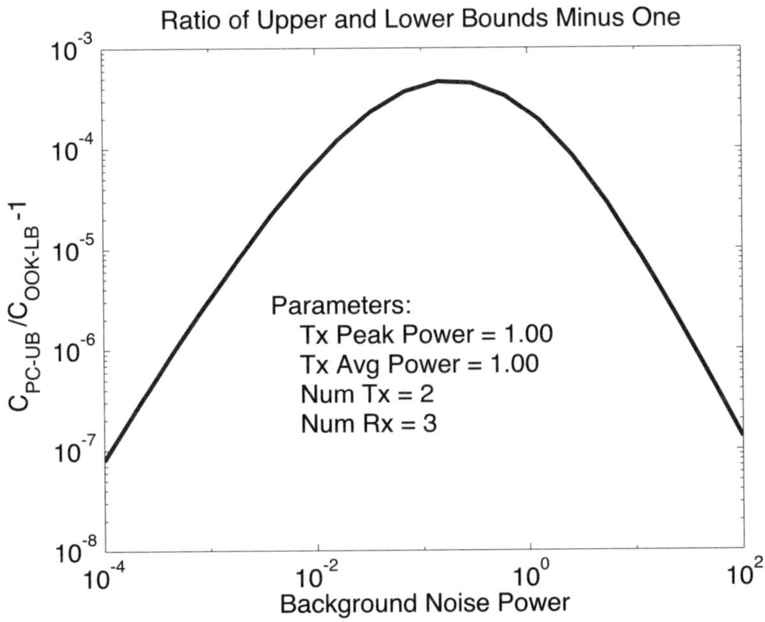

Fig. 4. In general, the upper and lower bounds on channel capacity are quite close. This figure shows the fractional difference between the PC-UB and OOK-LB for the $N = 2$, $M = 3$ special case whose path gains are given in Table 1.

on the channel capacity that are equal in a number of special cases, such as for channels with low or high background noise, a single receiver (the MISO channel), or low average input power constraints. The lower bound is equal to the MIMO capacity for the single transmitter case (the SIMO channel).

References

1. Gagliardi, R. M. and Karp, S. (1995) Optical Communications. Wiley, New York.
2. Boel, R., Varaiya, P., and Wong, E. (1975) Martingales on jump processes. II: Applications, SIAM J. Control **13**(5), 1022–1061.
3. Bremaud, P. (1988) An averaging principle for filtering a jump process with point observations, IEEE Trans. Inform. Theory **34**(3), 582–586.

4. Cover, T. M. and Thomas, J. A. (1991) Elements of Information Theory. John Wiley& Sons, Inc., 231.
5. Kabanov, Y. M. (1978) The capacity of a channel of the Poisson type, Theory Probab. Appl. **23**, 143–147.
6. Davis, M. A. (1980) Capacity and cutoff rate for Poisson-type channels, IEEE Trans. Inform. Theory **26**, 710–715.
7. Lipster, R. S. and Shiryaev, A. N. (2000) Statistics of Random Processes, Vol. 2, Springer, Berlin Heidelberg, New York, 2nd Ed., 346.
8. Wyner, A. D. (1988) Capacity and error exponent for the direct detection photon channel. Part 1, IEEE Trans. Inform. Theory **34**(6), 1449–1461.
9. Shamai (Shitz), S. (1991) On the capacity of a direct-detection photon channel with intertransition-constrained binary input, IEEE Trans. on Inform. Theory **37**(6), 1540–1550.
10. Shamai (Shitz), S. (1990) Capacity of a pulse-amplitude modulated direct detection photon channel, Proc. IEEE **137**(6), 424–436.
11. Snyder, D. L. and Miller, M. I. (1991) Random Point Processes in Time and Space., 2nd ed., Springer-Verlag New York, Inc..
12. Frey, M. R. (1991) Information capacity of the Poisson channel, IEEE Trans. on Inform. Theory **37**(2), 244–256.
13. Shamai (Shitz), S. and Lapidoth, A. (1993) Bounds on the capacity of a spectrally constrained Poisson channel, IEEE Trans. on Inform. Theory **39**(1), 19–29.
14. Lapidoth, A. and Shamai (Shitz), S. (1998) The Poisson multiple-access channel, IEEE Trans. on Inform. Theory **44**(2), 488–501.
15. Bross, S. I., Brunashev, M. V., and Shamai (Shitz), S. (2001) Error exponents for the two-user Poisson multiple-access channel, IEEE Trans. on Inform. Theory **47**(5), 1999–2016.
16. Lapidoth, A., Telatar, I. E., and Urbanke, R. (1998) On wideband broadcast channels. In IEEE Intl. Symp. on Inform. Theory, August 16–21, 1998, Cambridge, MA, 188.

Stochastic Analysis of Jump–Diffusions for Financial Log–Return Processes

Floyd B. Hanson[1] and John J. Westman[2]

[1] University of Illinois, Laboratory for Advanced Computing, Chicago IL 60607–7045, USA
[2] University of California, Department of Mathematics, Los Angeles CA 90095–1555, USA

Abstract. A jump–diffusion log–return process with log–normal jump amplitudes is presented. The probability density and other properties of the theoretical model are rigorously derived. This theoretical density is fit to empirical log–returns of Standard & Poor's 500 stock index data. The model repairs some failures found from the log–normal distribution of geometric Brownian motion to model features of realistic financial instruments: (1) No large jumps or extreme outliers, (2) Not negatively skewed such that the negative tail is thicker than the positive tail, and (3) Non–leptokurtic due to the lack of thicker tails and higher mode.

1 Introduction

Encouraged by the long term success of the Black–Scholes [3] option pricing model in spite of its deficiencies, many financial modelers have tried to improve on this basic model. Merton [18] applied discontinuous sample path Poisson processes, along with Brownian motion processes, i.e., jump–diffusions, to the problem of pricing options. Merton derived several extensions of the already classical diffusion theory of Black–Scholes minimizing the portfolio variance for jump–diffusion models using techniques similar to those used to derive the Black–Scholes formulae.

Before Black–Scholes, Merton [17] analyzed the optimal consumption and investment portfolio with either geometric Brownian motion or jump Poisson noise, illustrated explicit solutions for constant risk–aversion utility. In [16], Merton also examined constant risk–aversion problems. In [12], Karatzas, Lehoczky, Sethi and Shreve pointed out that it is necessary to enforce non–negativity feasibility conditions on both wealth and consumption. They formally derive explicit solutions from a consumption–investment dynamic programming models with a time–to–bankruptcy horizon, qualitatively correcting the results of Merton [17]. Sethi and Taksar [23] directly present corrections to certain formulae Merton's finite horizon consumption–investment model [17]. Merton [19] revisited the problem in the sixth chapter of his continuous–time finance book, correcting his earlier work by adding a simpler absorbing boundary condition at zero wealth and using other techniques.

Wilmott [25] presents a good discussion on difficulty of hedging with jump–diffusion models in finance, in fact, that *perfect risk–free hedging is*

B. Pasik-Duncan: Stochastic Theory and Control, LNCIS 280, pp. 169–183, 2002.

impossible when there are jumps in the underlying, using a single option. Lipton–Lifshitz [15] presents a good discussion of predictability and unpredictability, mainly for foreign exchange applications, but is applicable to other financial applications as well.

In the computational finance paper of Hanson and Westman [9], they solved an optimal portfolio and consumption policies model modified from a theoretical important event model proposed by Rishel [22]. The model is an optimal portfolio and consumption model for a portfolio of stocks and a bond. The stock prices are dependent on both deterministic (scheduled) and stochastic (unscheduled) jump external events in an environment of geometric jump–diffusion processes. The jumps can affect both the stock prices directly or indirectly through parameters. The deterministic jumps are quasi–deterministic, in that the timing of the scheduled events is deterministic, but the magnitude of the jumps is random. The computations were illustrated with a simple discrete jump model, such that both stochastic and quasi–deterministic jump magnitudes were heuristically estimated to be discretely distributed as single negative or positive jumps. A partial motivation for this quasi–deterministic are the more or less monthly announcements of the Federal Open Market Committee [7], but the response of the market to changes in Federal Funds Rate or Federal Discount Rate is difficult to predict. This quasi–deterministic process might be called the "Greenspan Process". The current paper focuses more on the stock log–return distribution and the estimating the parameters of this log–normal jump–diffusion distribution for a more basic stock process.

The empirical distribution of daily log–returns for actual financial instruments differ in many ways from the ideal log–normal diffusion process as assumed in the Black–Scholes model and other financial models. The most notable difference is that actual log–returns suffer occasional large jumps in value. The negative large jumps are called crashes and buying frenzies lead to positive large jumps. Another difference is that the empirical log–return will typically be negatively skewed, since The negative jumps are usually larger than the positive jumps. Thus, the coefficient of skew [6] is negative,

$$\eta_3 \equiv M_3/(M_2)^{1.5} < 0 \,, \tag{1}$$

where M_2 and M_3 are the 2nd and 3rd central moments. Still another difference is that the empirical distribution is leptokurtic since the coefficient of kurtosis [6] satisfies

$$\eta_4 \equiv M_4/M_2^2 > 3 \,, \tag{2}$$

where the value 3 is the normal distribution kurtosis value and M_4 is the fourth central moment. Qualitatively, this means that the tails are fatter than a normal with the same mean and standard deviation, and consequently the distribution is also more slender about the mode (local maximum).

In Merton's discontinuous returns paper[18] (see also Chapter 9 along with background in Chapter 3 of [19]) treated the case of option pricing modeled by a jump–diffusion process where the jumps have a log–normal distribution in one important example. Andersen and Andreasen [1] treat the log–normal jump–diffusion option pricing problem in much more detail, both analytically through forward partial integral–differential equations and numerically mainly though alternating direction implicit methods.

Kou [14] has developed a Laplacian double exponential jump–diffusion model to account of for the leptokurtic, negative skew and other properties in option pricing. In the model, jumps occur in time according to a Poisson process and the amplitude of the jump are distributed as a double exponential with mean κ and variance 2η, i.e., a shifted exponential depending on the absolute value of the deviation. Many probability properties are developed along with required special functions.

There are many other approaches, such as Poisson random measure related Lévy distributions (see [2], for example) and stochastic volatility (see [25], for instance).

In Section 2, the explicit form of the log–return density is shown to be a infinite sum of log–normal distributions weighted by Poisson counting discrete distribution when both the diffusion and Poisson jump amplitudes are both log–normal. In Section 3, the five log–normal jump–diffusion parameters are estimated for the empirical log–returns of the Standard & Poor's 500 (S&P500) stock index closing under the constraints that theoretical jump–diffusion distribution has the same mean and variance as the empirical distribution.

2 Density for Log–Normal Jump–Diffusions

Let S(t) be the price of a single stock or mutual fund that satisfies the Markov, geometric, log–normal jump–diffusion stochastic differential equation (SDE),

$$dS(t) = S(t)\left[\mu_d dt + \sigma_d dZ(t) + J dP(t)\right] \ , \tag{3}$$

$S(0) = S_0$ and $S(t) > 0$, where μ_d is the mean appreciation return rate, σ_d is the diffusive volatility, $Z(t)$ is a continuous, one–dimensional Brownian motion process, J is a random jump amplitude with log–return mean μ_j and variance σ_j^2 defined below, and $P(t)$ is a discontinuous, one–dimensional, standard Poisson process with jump rate λ. Here, we will assume that the jump–diffusion parameters μ_d, σ_d, μ_j, σ_j and λ are constant. The stochastic processes $Z(t)$ and $P(t)$ are Markov and pairwise independent. The jump amplitude process J, given a Poisson jump in time, is also independently distributed.

The continuous, diffusion process $Z(t)$ is standard and is specified by the two infinitesimal moments,

$$E[dZ(t)] = 0$$

and

$$\text{Var}[dZ(t)] = dt ,$$

since the process is normally distributed. The discontinuous space–time jump process $JdP(t)$ can be more generally expressed in terms of Poisson random measure $\mathcal{P}(dt, dq)$ through the stochastic integral,

$$JdP(t) = \int_{\mathcal{Q}} J(q)\mathcal{P}(dt, dq) , \tag{4}$$

where $Q = q$ is the mark for the jump amplitude process, i.e., the underlying random variable. The differential Poisson process is basically a counting process with the probability of the jump count given by the usual Poisson distribution,

$$p_k(\lambda dt) = \text{Prob}[dP(t) = k] = \exp(-\lambda dt)(\lambda dt)^k/k!, \quad k = 0, 1, \dots ,$$

with parameter $u = \lambda dt > 0$. The jump in the stock price corresponding to the jump of the space–time Poisson process is

$$[S](t_j) \equiv S(t_j^+) - S(t_j^-) = J(Q)S(t_j^-)$$

at some jump time t_j. Hence, it is assumed that $-1 < J(Q) < \infty$ so that one jump does not make the underlying stock worthless. The infinitesimal moments of the jump process are

$$E[J(Q)dP(t)] = \lambda dt \int_{\mathcal{Q}} J(q)\phi_Q(q)dq = E[J(Q)]\lambda dt$$

and

$$\text{Var}[J(Q)dP(t)] = \lambda dt \int_{\mathcal{Q}} J^2(q)\phi_Q(q)dq = E[J^2(Q)]\lambda dt ,$$

neglecting here $O^2(dt)$ to obtain infinitesimal moments, where $\phi_Q(q)$ is the Poisson mark density, providing it exists on the mark space $\mathcal{Q}$.

Before describing the jump amplitude distribution in more detail, the stock price SDE (9) is first transformed to the SDE of the instantaneous stock log–returns using the stochastic chain rule for Markov processes in continuous time [11,4],

$$d[\ln(S(t))] = \mu_{ld}dt + \sigma_d dZ(t) + \ln(1 + J(Q))dP(t) , \tag{5}$$

where the log–diffusion drift $\mu_{ld} \equiv \mu_d - \sigma_d^2/2$ includes the Itô calculus shift of the mean appreciation rate by the diffusion coefficient and the log–return jump amplitude is the logarithm of the relative post–jump amplitude $\ln(1 + J)$. This log–return SDE (5) will be the model that will be compared to the S&P500 log–returns, since the log–returns are the preferred financial investment metric measuring the relative changes in investment value, as opposed to the absolute change of the stock price represented by the geometric jump–diffusion SDE in (9).

Since $J > -1$, it is convenient to select the mark process to be the log–return jump amplitude

$$Q = \ln(1 + J) ,$$

which has the inverse

$$J(Q) = \exp(Q) - 1 ,$$

on the mark space $\mathcal{Q} = (-\infty, +\infty)$. On this fully infinite domain, the ideal choice for the mark density is the normal density,

$$\phi_Q(q) = \phi_n(q; \mu_j, \sigma_j^2) \equiv \frac{\exp\left(-(q-\mu_j)^2 / \left(2\sigma_j^2\right)\right)}{\sqrt{2\pi\sigma_j^2}} , \tag{6}$$

having a mean $E[Q] = \mu_j$ and variance $\text{Var}[Q] = \sigma_j^2$, which define the log–return jump amplitude moments through $Q = \ln(1 + J(Q))$. Hence, $J(Q)+1$ is log–normally distributed, with mean

$$E[J(Q)] = E\left[e^Q\right] - 1 = \exp(\mu_j + \sigma_j^2/2) - 1 ,$$

and variance

$$\text{Var}[J(Q)] = E^2\left[e^Q\right] \cdot \left(\exp(\sigma_j^2) - 1\right) .$$

The basic moments of the stock log–return differential are

$$M_1^{(\mathrm{jd})} \equiv E[d[\ln(S(t))]] = (\mu_{ld} + \lambda\mu_j)dt , \tag{7}$$

$$\begin{aligned} M_2^{(\mathrm{jd})} &\equiv \text{Var}[d[\ln(S(t))]] = \text{Var}[\sigma_d dW(t)] \\ &\quad + \left(\text{Var}[Q] + E^2[Q]\right)\text{Var}[dP(t)] + \text{Var}[Q]E^2[dP(t)] \\ &= \left(\sigma_d^2 + \lambda\left(\sigma_j^2(1 + \lambda dt) + \mu_j^2\right)\right) dt , \end{aligned} \tag{8}$$

where the $O^2(dt)$ term has been retained in the variance rather than being neglected as usual, since the return time, dt, the daily fraction of one trading year, will be small, but not negligibly small. The log–return is the primary model independent variable of interest in this paper, as the investor is interested in the percent or relative change in a portfolio and the log–return is the continuous limit of the relative change.

Next the log–normal density will be found by basic probabilistic methods and the results are summarized in the following theorem.

Theorem. The probability density for the log–normal jump–diffusion log–return differential $d[\ln(S(t))]$ specified in the SDE (5) is given by

$$\phi_{d\ln(S(t))}(z) = \sum_{k=0}^{\infty} p_k(\lambda dt)\phi_n\left(z; \mu_{ld}dt + \mu_j k, \sigma_d^2 dt + \sigma_j^2 k^2\right) , \tag{9}$$

$-\infty < z < +\infty$, where the Poisson distribution $p_k(\lambda dt)$ is specified in (5) and the normal density ϕ_n is specified in (6).

Proof. The basic idea of the proof follows from finding the density of a triad of independent random variables, $X + Y \cdot Z$ given the densities of the three components X, Y, and Z. Here, the process $X = \mu_{ld}dt + \sigma_d dZ(t)$ which is the diffusion plus log–drift term, normally distributed with density $\phi_n(z; \mu_{ld}dt, \sigma_d^2 dt)$, the process $Y = Q = \ln(1 + J(Q))$ which is the jump amplitude of the log–return, normally distributed with density $\phi_n(y; \mu_j, \sigma_j^2)$ and $Z = dP(t)$ which is the differential Poisson process. The discrete distribution of the Poisson process given in (5).

According to Feller [8], the density of a sum of independent random variables is given by a convolution of densities,

$$\phi_{X+YZ}(z) = (\phi_X * \phi_{YZ})(z) \equiv \int_{-\infty}^{+\infty} \phi_X(z-y)\phi_{YZ}(y)dy \,, \tag{10}$$

but before calculating the convolution the density for the product of two random variables $\phi_{YZ}(x)$, the density of the *Poisson–Normal* process, is needed. Fortunately, the Z process is a non–negative process while the Y process has fully infinite range, otherwise there would be more cases of passage through zero to treat.

Feller's argument for sum of two distributions will be modified for the product distribution, i.e., $\Phi_{YZ}(x) \equiv \text{Prob}[YZ \le x]$, in the special case that when the non–negative random variable Z vanishes, i.e., in the case when the Poisson jump counter $k = 0$, so that normal variate Y is indeterminate in the inequality $Y \cdot Z \le x$. In terms of realized, observed variables, $Y = y$ and $Z = z$, the inequality $yz \le x$ implies that $y \le x/z$ if $z \ne 0$, else if $z = 0$ then y is unbounded in $(-\infty, +\infty)$, but when $z = 0$ there is an implied constraint on x that $x \ge 0$. The case $z < 0$ is empty since the simple Poisson process is non–negative. Thus, the conditional probability

$$\text{Prob}[YZ \le x | Z = z] = \left\{ \begin{array}{ll} \text{Prob}[Y \le x/z], & z > 0 \\ H(x), & z = 0 \\ 0, & z < 0 \end{array} \right\}$$

$$= \left\{ \begin{array}{ll} \Phi_Y(x/k), & z = k = 1, 2, \ldots \\ H(x), & z = k = 0 \\ 0, & \text{otherwise} \end{array} \right\} ,$$

since the process Z is discretely distributed Poisson process with non-negative values k. Here, $H(x)$ is the Heaviside unit step function serving to handle the situation when $Z = 0$ in $YZ \leq x$, so that $x \geq 0$ becomes a consistency constraint. Hence, the discrete analog of the average over the second density $\phi_{YZ}(y)$ in (10) is the sum over the Poisson jump count probabilities $p_k(\lambda dt)$ for $z = k = 0, 1, 2, \ldots,$

$$\Phi_{YZ}(x) \equiv \text{Prob}[YZ \leq x] = p_0(\lambda dt)H(x) + \sum_{k=1}^{\infty} p_k(\lambda dt)\Phi_Y(x/k) \ ,$$

Thus, the Poisson–Normal density of the product YZ, assuming differentiability in the sense of generalized functions, is

$$\phi_{YZ}(x) = \Phi'_{YZ}(x) = p_0(\lambda dt)\delta(x) + \sum_{k=1}^{\infty} p_k(\lambda dt)\phi_Y(x/k)/k \tag{11}$$

$$= p_0(\lambda dt)\delta(x) + \sum_{k=1}^{\infty} p_k(\lambda dt)\phi_n\left(x/k; \mu_j, \sigma_j^2\right)/k \tag{12}$$

$$= p_0(\lambda dt)\delta(x) + \sum_{k=1}^{\infty} p_k(\lambda dt)\phi_n\left(x; \mu_j k, \sigma_j^2 k^2\right) \ , \tag{13}$$

where the fact that Y has a normal density and the easily derived normal density scaling property

$$k^{-1}\phi_n(x/k; a, b^2) = \phi_n(x; ak, b^2k^2) \tag{14}$$

have been used. Note that probability is conserved since

$$E_{YZ}[1] = \int_{-\infty}^{\infty} \phi_{YZ}(x)dx = 1$$

by conservation of probability for both normal and Poisson processes. The probabilistic mass at $x = 0$, represented by the Dirac $\delta(x)$, is essential.

Finally, using the convolution formula (10) for density of the sum reduces the target density to a Poisson distribution weighted sum of the convolution of two normal densities,

$$\begin{aligned}
\phi_{d\ln(S(t))}(x) = \phi_{X+YZ}(x) &= \int_{-\infty}^{\infty} \phi_X(x-z)\phi_{YZ}(z)dz \\
&= e^{-\lambda dt}\phi_n(x; \mu_{ld}dt, \sigma_d^2 dt) \\
&+ \sum_{k=1}^{\infty} p_k(\lambda dt) \int_{-\infty}^{\infty} \phi_n(x-z; \mu_{ld}dt, \sigma_d^2 dt)\phi_n(z; \mu_j k, \sigma_j^2 k^2)dz \\
&= e^{-\lambda dt}\phi_n(x; \mu_{ld}dt, \sigma_d^2 dt) \\
&+ \sum_{k=1}^{\infty} p_k(\lambda dt) \int_{-\infty}^{\infty} \phi_n(z; x-\mu_{ld}dt, \sigma_d^2 dt)\phi_n(z; \mu_j k, \sigma_j^2 k^2)dz \\
&= e^{-\lambda dt}\phi_n(x; \mu_{ld}dt, \sigma_d^2 dt)
\end{aligned}$$

$$+\sum_{k=1}^{\infty} p_k(\lambda dt)\phi_n(x;\mu_{ld}dt+\mu_j k,\sigma_d^2 dt+\sigma_j^2 k^2)$$
$$=\sum_{k=0}^{\infty} p_k(\lambda dt)\phi_n(x;\mu_{ld}dt+\mu_j k,\sigma_d^2 dt+\sigma_j^2 k^2)\ ,$$

where first the convolution argument x of the left density was shifted to the drift so that both densities have the same argument z, and then an identity, derivable by the *completing the square* technique, combining a product of two normal densities into one,

$$\phi_n(x;\mu_1,\nu_1)\cdot\phi_n(x;\mu_2,\nu_2)=\phi_n\left(x;\frac{\mu_1\nu_2+\mu_2\nu_1}{\nu_1+\nu_2},\frac{\nu_1\nu_2}{\nu_1+\nu_2}\right)\cdot\frac{1}{\sqrt{2\pi(\nu_1+\nu_2)}}\exp\left(-\frac{(\mu_1-\mu_2)^2}{\nu_1+\nu_2}\right)\ , \tag{15}$$

has been used. This completes the proof of the Theorem. □

Note that the log–return density is the Poisson mean over a normal density since (9) can be rewritten as

$$\phi_{d\ln(S(t))}(z)=E_{dP(t)}\left[\phi_n(z;\mu_{ld}dt+\mu_j dP(t),\sigma_d^2 dt+\sigma_j^2(dP(t))^2)\right]\ ,$$

where the Poisson jump counter k has been replaced by the simple differential Poisson process $dP(t)$ and the normal density ϕ_n is specified in (6). Thus the density is the expectation over a normal distribution with jumping mean and variance, the jump in the mean scaled by the mark jump mean μ_j and the jump in the variance scaled by the mark jump variance σ_j^2, respectively.

In the 2001 thesis of Düvelmeyer [5], the probability distribution function for the log of the process $\ln(S(t))$, rather than the density of the log–returns $d\ln(S(t))$ needed here, is given, but there is an error in the kth variance, in that $\sigma_j^2 k$ appears rather than the $\sigma_j^2 k^2$ which arises from the normal density scaling property (14), using the notation here. This thesis work appears to be the basis for Kluge's jump–diffusion version of the *Share Simulator* [13]. Also, the intended optimal portfolio and consumption application for the present paper is very different, not generalizations of the Black–Scholes option pricing as in the thesis [5].

Using the log–normal jump–diffusion log–return density in (9), the third and fourth central moments with finite return time dt are computed and verified with MapleV^{TM} [24] symbolic computation yielding

$$\begin{aligned}M_3^{(\mathrm{jd})}&=E\left[(d[\ln(S(t))]-M_1^{(\mathrm{jd})})^3\right]\\&=\int_{-\infty}^{+\infty}\left(z-M_1^{(\mathrm{jd})}\right)^3\phi_{d\ln(S(t))}(z)dz\\&=6\mu_j(\lambda dt)^2\sigma_j^2+(3\mu_j\sigma_j^2+\mu_j^3)\lambda dt\ ,\end{aligned} \tag{16}$$

$$\begin{aligned}
M_4^{(\mathrm{jd})} &= E\left[(d[\ln(S(t))] - M_1^{(\mathrm{jd})})^4\right] \\
&= \int_{-\infty}^{+\infty} \left(z - M_1^{(\mathrm{jd})}\right)^4 \phi_{d\ln(S(t))}(z)dz \\
&= 3(\sigma_j^2)^2(\lambda dt)^4 + (6\mu_j^2\sigma_j^2 + 18(\sigma_j^2)^2)(\lambda dt)^3 \\
&\quad +(3\mu_j^4 + 30\mu_j^2\sigma_j^2 + 21(\sigma_j^2)^2 + 6\sigma_j^2 dt\sigma_j^2)(\lambda dt)^2 \\
&\quad +(\mu_j^4 + 6\sigma_j^2 dt\sigma_j^2 + 6\mu_j^2\sigma_j^2 + 6\mu_j^2\sigma_j^2 dt + 3(\sigma_j^2)^2)\lambda dt + 3(\sigma_j^2)^2 dt^2 \ ,
\end{aligned} \tag{17}$$

which are used to compute the theoretical coefficients of skew (1) and kurtosis (2), respectively.

3 Log–Normal Jump–Diffusion Model Parameter Estimation

Now the log–normal jump–diffusion density (9) is available for fitting to realistic data to obtain some of the parameters of the log–normal diffusion model (5) for $d[\ln(S(t))]$. For realistic data, the daily closings of the Standard and Poor's 500 (S&P500) stock index from 1995 to July 2001 will be used from data available on–line [26]. The data consists of $nsp = 1657$ points. The S&P500 data is an example of one large mutual fund rather than a single stock but has the advantage of not being biased severely to any one stock. The data has been transformed into the discrete analog of the continuous log–return, i.e., into changes in the natural logarithm of the index closings, $\Delta[\ln(SP_i)] \equiv \ln(SP_{i+1}) - \ln(SP_i)$ for $i = 1, \ldots, nsp - 1$ points. The scatter for the $nsp - 1 = 1656$ points of $\Delta[\ln(SP_i)]$ is shown in Figure 1 versus time in years, along with confidence intervals for one, two and three standard deviations. A slight, but noticeable, time dependence of the local mean and volatility is seen, but the time–dependent behavior is the topic of another paper and the constant coefficient case needs to be treated here. In different view, the histogram of the data using 50 equally spaced data bins is given in Figure 2. The mean is $M_1^{(sp)} \simeq 5.754 \times 10^{-4}$ and the variance is $M_2^{(sp)} \simeq 1.241 \times 10^{-4}$, the coefficient of skew is

$$\eta_3^{(sp)} \equiv M_3^{(sp)}/(M_2^{(sp)})^{1.5} \simeq -0.2867$$

which is negative and the coefficient of kurtosis or normalized fourth central moment is

$$\eta_4^{(sp)} \equiv M_4^{(sp)}/(M_2^{(sp)})^2 \simeq 6.862 \ .$$

Compared to the normal distribution, the empirical distribution has negative skew while the normal distribution has zero skew. Also, the empirical kurtosis is 2.3 times the normal distribution kurtosis of 3. The S&P500 log–return skew and kurtosis are characteristic of log–returns of many market instruments as noted in the Introduction.

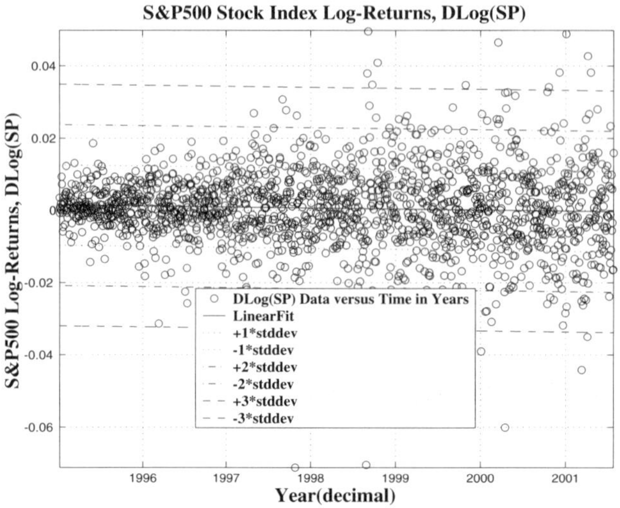

Fig. 1. Daily changes in the logarithm of the S&P500 stock index. Linear fit (light solid line) is nearly zero and horizontal. The confidence intervals for one (68%), two (95%) and three (99%) standard deviations are presented (light dashed lines)

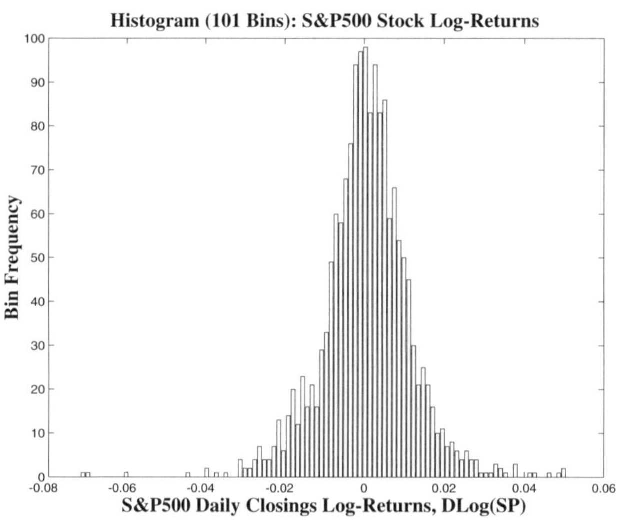

Fig. 2. Histogram of daily changes in the logarithm of the S&P500 stock index

Using MATLAB™ [20], the theoretical log–normal jump–diffusion density $\phi_{d\ln(S(t))}$ in (9) is compared to the 50 bin histogram shown in Figure 2 by discretizing the theoretical density using the same 50 bin data structure as for the histogram. However, the five parameter set $\{\mu_d, \sigma_d^2, \mu_j, \sigma_j^2, \lambda dt\}$ had to be reduced to a more manageable set to avoid large fitting errors and to preserve the *Principle of Modeling Parsimony* (striving for economies or simplicity of the model). The empirical return time is taken as the reciprocal average number of trading days per year or $1/252.3 \simeq 0.003964 = dt$ for the data used (250 trading days seems to be a standard value [14]), so the empirical dt is small, but not infinitesimal. Two parameters, μ_d and σ_d^2, were eliminated by forcing the theoretical means and variances to be the same as the mean and variance of the empirical data, respectively,

$$\sigma_d^2 = (M_2 - \lambda dt((1 + \lambda dt)\sigma_j^2 + \mu_j^2))/dt \ , \tag{18}$$

$$\mu_d = (M_1 - \lambda dt \mu_j)/dt \ , \tag{19}$$

using the log–return moment formulas (7,8). The objective is to select the reduced set $\{\mu_j, \sigma_j^2, \lambda dt\}$ then to minimize the variance (i.e., histogram least squares) of the difference between the empirical and theoretical model distribution. The search stopping criteria was that the final maximum relative change in successive values of the parameters μ_j, σ_j^2 and λdt plus the relative change in the variance of the histogram difference was less than a 0.5e-3 tolerance. Due to the complexity of the jump–diffusion density and the need to keep finance methods simple, a multi–dimensional modification of *Golden Section Search* that needs no derivatives and searches beyond the current range when a local minimum is not found in the current search hypercube [10]. In addition, hypercube constraints were implemented so that the free model parameters $\{-\mu_j, \sigma_j^2, \lambda dt\}$ would remain non–negative and be bounded. The final parameter results are

$$\begin{aligned} &\mu_d \simeq 0.2712 \ , \quad \sigma_d^2 \simeq 0.01048 \ , \\ &\mu_j \simeq -0.0007474 \ , \quad \sigma_j^2 \simeq 0.00007812 \ , \quad \lambda \simeq 161.7 \ , \end{aligned} \tag{20}$$

with a final variance of the deviation of 11.45 and an mean deviation of 2.68×10^{-2}, with a total frequency count of 1656 over all bins. Also, note that the diffusion mean and variance have dimension per year, while the jump mean and variance are dimensionless so the jump values should be weighted by the jump intensity λ for a jump to diffusion comparison (i.e., the jump values are not as relatively small as they seem). For a better empirical to theoretical comparison, the moments of the deviation of the empirical S&P500 histograms from the normal density histogram, where the normal density matches only the empirical mean and variance since there are only two parameters to match given dt, with a mean deviation of -6.015×10^{-5}, but a deviation variance of 56.32, almost five times deviation variance of the fit jump–diffusion. The corresponding jump–diffusion normalized higher moments are the coefficients

$$\eta_3^{(\mathrm{jd})} \equiv M_3^{(\mathrm{jd})} / \left(M_2^{(\mathrm{jd})}\right)^{1.5} \simeq -0.2114$$

for skew and

$$\eta_4^{(\mathrm{jd})} \equiv M_4^{(\mathrm{jd})} / \left(M_2^{(\mathrm{jd})}\right)^{2} \simeq 8.082$$

for kurtosis, whose values are qualitatively similar to the empirical values with somewhat larger (+18%) leptokurtosis and lesser (-26%) negative skew, but much more realistic than the normal density model with zero skew and kurtosis of three.

Note that the single log–normal jump amplitude distribution and the minimum variance comparison yields only a very slender distribution (very small σ_j^2) around a negative mean (μ_j), so only the dominant negative tail is represented and not the subdominant positive tail. The quality of the fit may be due to the simple minimum variance technique used, the single log–normal jump amplitude distribution and the fully infinite theoretical domain. A better model would use a log–bi–normal distribution similar to the log–bi–discrete distribution used in [9] with a positive as well as a negative discrete jump, but enhanced with a slight spread. Such a log–bi–normal distribution would add three more unknown parameters to search for, i.e., additional mean, variance and the probability of positive jump relative to a negative jump. The multidimensional Golden Section Search could be used, although it has the slowness of a general method, one that needs no derivatives. If more speed and accuracy were required, a nonlinear least squares method such as that of Levenberg–Marquardt [21] hybrid method can be used, but the parameter gradient of the log–normal jump–diffusion density would be needed, which could be facilitated by symbolic computation like MapleV™ [24]. The histogram of the final discretized theoretical density is displayed in Figure 3 and the deviation of the empirical S&P500 from the theoretical jump–diffusion histogram data is displayed in Figure 4. The discrepancy between the empirical and theoretical is best seen in the difference histogram in Figure 4, considering that the frequency scale being ten times finer for the positive deviations than in full histogram Figure 3, with significant frequency deviations of $(-15, +10)$ around the mode and adjacent shoulders fairly well–distributed within the deviation range of $(-0.03, +0.02)$.

Conclusions

The probability density for a jump–diffusion whose jump amplitudes are distributed log–normally has been found and rigorously justified using basic probabilistic theory. This density is a discrete weighted sum of normal densities whose parameters depend on the $\{\mu_d, \sigma_d^2\}$ mean–variance parameters of the continuous drift–diffusion subsystem, the $\{\mu_j, \sigma_j^2\}$ mean–variance parameters of the log–normal jump mark distribution and the Poisson jump rate λ, with the weights being the Poisson discrete distribution of the jump counts. The log–normal jump–diffusion should be useful for fitting data for real investment markets with a distribution of random jumps, negative skew and leptokurtic properties that are not present in the standard log–normal or geometric diffusion model alone.

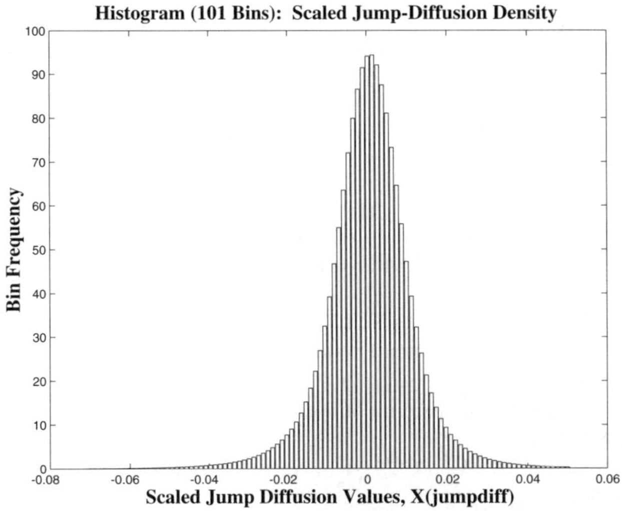

Fig. 3. Histogram of log–normal jump–diffusion as in Figure 2 with same bin structure

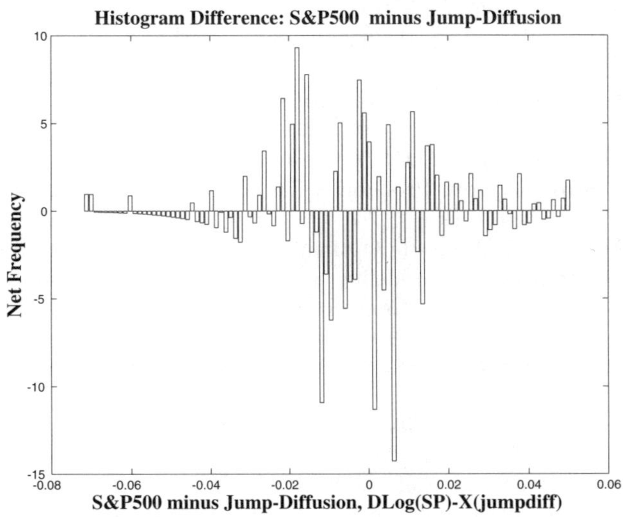

Fig. 4. Difference of histograms with daily changes in the logarithm of the S&P500 stock index minus the log–normal jump–diffusion bin values (Note: scale in this figure is one tenth the scale of the previous figure)

Acknowledgements

The authors thank Professor Bozenna Pasik–Duncan for the invitation to this splendid Kansas Workshop on Stochastic Theory in honor of Professor Tyrone Duncan's 60th birthday, providing local support via the University of Kansas and the National Science Foundation. The first author (FBH) acknowledges that this work was supported in part by the National Science Foundation Grant DMS–99–73231.

References

1. Andersen, L. and Andreasen, J. (2000) Jump–diffusion processes: Volatility smile fitting and numerical methods of option pricing, Review of Derivatives Research **4**, 231–262.
2. Barndorff-Neilsen, O. E. and Shephard, N. (2001) Non–Gaussian OU based models and Ssome of their uses in financial economics, J. Roy. Stat. Soc. Ser. B **63**, 167–241.
3. Black, F. and Scholes, M. (1973) The pricing of options and corporate liabilities, J. Political Economy **81**, 637–659.
4. Gihman, I. I. and Skorohod, A. V. (1979) Controlled Stochastic Processes, Springer-Verlag, New York.
5. Düvelmeyer D. (2001) Untersuchungen zu Chancen und Risiken von Anti–Trend–Strategien am Beispiel des DAX–Futures. Thesis, Facultät für Mathematik, Technische Universität Chemnitz, Chemnitz, *URL: http://www-usercgi.tu-chemnitz.de/~dana/diplom_pdf_dd.zip*
6. Evans, M., Hastings, N., and Peacock, B. (2000) Statistical Distributions, 3rd ed. John Wiley, New York.
7. Federal Open Market Committee (2001) Press Releases. Federal Reserve System,
URL: http://www.federalreserve.gov/FOMC/PreviousCalendars.htm
8. Feller, W. (1971) An Introduction to Probability Theory and Its Application, Vol 2, 2nd ed. John Wiley, New York.
9. Hanson, F. B. and Westman, J. J. (2001) Optimal consumption and portfolio policies for important jump events: Modeling and computational considerations. In Proceedings of 2001 American Control Conference, 4456–4661.
10. Hanson, F. B. and Westman, J. J. (2002) Golden Super Search: Multidimensional Modification of Golden Section Search Unrestricted by Initial Domain. In preparation
11. Kushner, H. J. (1967) Stochastic Stability and Control. Academic Press, New York.
12. Karatzas, I., Lehoczky, J. P., Sethi, S. P., and Shreve, S. E. (1986) Explicit solution of a general consumption/investment problem, Math Oper Res **11**, 261–294.
13. Kluge, T. (2000) Share Simulator: Black–Scholes, Heston and Jump–Diffusion Models. *URL: http://www.mathfinance.de/TinoKluge/*
14. Kou, S. G. (2000) A Jump Diffusion Model for Option Pricing with Three Properties: Leptokurtic Feature, Volatility Smile, and Analytical Tractability. *URL: http://www.ieor.columbia.edu/~kou/expo.pdf*
15. Lipton-Lifshitz, A. (1999) Predictability and unpredictability in financial markets, Physica D **133**, 321–347.
16. Merton, R. C. (1969) Lifetime portfolio selection under uncertainty: The continuous-time case, Rev Economics and Statistics **51**, 247–257.

17. Merton, R. C. (1971) Optimum consumption and portfolio rules in a continuous-time model, J Economic Theory bf3, 373–413.
18. Merton, R. C. (1976) Option pricing when underlying stock returns are discontinuous, J Financial Economics **3**, 125–144.
19. Merton, R. C. (1990) Continuous-Time Finance. Basil Blackwell, Cambridge, MA.
20. Moler, C., et al. (2000) Using MATLAB, vers 6. Mathworks, Natick, MA.
21. Press, W. H., et al. (1992) Numerical Recipes in C: The Art of Scientific Computing. Cambridge University Press, Cambridge, UK.
22. Rishel, R. (1999) Modeling and portfolio optimization for stock prices dependent on external events. In Proceedings of 38th IEEE Conference on Decision and Control, 2788–2793.
23. Sethi, S. P. and Taksar, M. (1988) A note on Merton's optimum consumption and portfolio rules in a continuous-time model. J Economic Theory **46**, 395–401.
24. Waterloo Maple (1998) Maple V. Waterloo Maple Inc, Waterloo, CA.
25. Wilmott, P. (2000) Paul Wilmott on Quantitative Finance, Vol 1. John Wiley, New York
26. Yahoo! Finance (August 2001) Historical Quotes, S&P 500, Symbol ^SPC. *URL: http://chart.yahoo.com/*

Numerical Methods for Optimal Stopping Using Linear and Non-linear Programming *

Kurt Helmes

Humboldt University of Berlin, Institute of Operations Research, Germany

Abstract. Computational methods for optimal stopping problems are presented. The first method to be described is based on a linear programming approach to exit time problems of Markov processes and is applicable whenever the objective function is a unimodal function of a threshhold parameter which specifies a stopping time. The second method, using linear and non-linear programming techniques, is a modification of a general linear programming approach to optimal stopping problems recently proposed by S. Röhl. Both methods are illustrated by solving Shiryaev's quickest detection problem for Brownian motion.

Keywords: Optimal stopping, linear programming, numerical methods for exit time problems, detection problem.

Subject Classification: *Primary* 60G40. *Secondary* 90C05, 90C30, 60G35, 93E20, 65C20.

1 Introduction

The purpose of the paper is to describe numerical methods based on linear programming and on non-linear optimization techniques for solving optimal stopping problems of Markov processes. We shall illustrate the power of these methods by (numerically) analyzing Shiryaev's quickest detection problem for a Wiener process.

The linear programming approach to optimal stopping and to stochastic control in general is an extension of work by Manne [13] who initiated the formulation of stochastic control problems as linear programs over a space of stationary distributions for the long-term average control of finite-state Markov chains, see Hernandez-Lerma et. al. [10] for details, generalizations and additional references. The generalization of the LP formulation for continuous time, general state and control spaces, and different objective functions has been established by Stockbridge [20], Kurtz and Stockbridge [6], [12], and Bhatt and Borkar [1]. The LP-formulation for stopping time problems and numerical methods for the solution of such problems have been presented by Cho [2], Cho and Stockbridge [3], and Röhl [16].

The basic idea of the LP-approach to stochastic control of Markov processes is to formulate such control problems as linear programs over a space of

* This research is partially supported under grant DMS 9803490

B. Pasik-Duncan: Stochastic Theory and Control, LNCIS 280, pp. 185–203, 2002.

stationary distributions. Specifically, the variables in these infinite-dimensional linear programs are measures on the product of the state and control spaces and in the case of exit problems, each such variable is augmented by a second measure on the exterior of the state space. These variables are tied together by an adjoint equation involving the generator A of the Markov process and a family of test functions. Different numerical methods are determined by a judious choice of a finite set of test functions combined with a selection of a finite number of variables and/or restrictions imposed on the support of the occupation measure and the exterior measure. Such choices determine approximations of the infinite dimensional optimization problem by a finite dimensional one. The viability of these numerical methods has been demonstrated in an uncontrolled setting by Helmes et. al. [6] and Helmes and Stockbridge [8] by analyzing the distribution of the exit time for a variety of processes evolving on bounded intervals in one- and two-dimensions. In [7], [9] Helmes and Stockbridge have applied the method of moments, i. e. the test functions chosen are the monomials up to a given order, on a stochastic control problem and on a particular class of diffusion processes defined on a higher dimensional state space. In [14], [15] Mendiondo and Stockbridge have applied the discretization method, i. e. the support of all measures is restricted to a finite set of gridpoints, to the analysis of some stochastic control problems, while Cho [2] uses the discretization method to analyze a stopping time problem. Röhl [16] uses the method of moments and proposes an iterative scheme for the numerical analysis of one- and two-dimensional stopping problems.

In this note we shall built on [6] and [16] and exploit an advantage which the method of moments has over other variants of the LP technique. For exit time problems, cf. [6], the method naturally provides bounds on quantities of interest, e. g. upper and lower bounds on the mean exit time from a bounded interval, etc. So for optimal stopping problems whose objective function is a unimodal function of a threshhold value which specifies a stopping time, we can apply line search techniques, f. i. the Fibonacci or Golden Section rule, to the optimal values of such parametrized LP-problems. This way we obtain a range for the optimal stopping rule parameter and get bounds for the optimal value, see Section 3 below. Ideally, if the bounds converge to the same value we shall find an optimal solution to the original problem.

The second method which we propose is applicable to more general stopping problems and replaces Röhl's iteration technique by a pair of optimization problems, one being linear the other one being non-linear, and another LP-problem which constitutes the verification step of the procedure.

2 Formulation and Fundamental Theorems

We formulate optimal stopping problems in a restricted setting which fits the numerical methods which we propose and are adequate for the example to be analyzed. In order to keep the notation simple we shall also restrict the formulation to bounded intervals $I \subset \mathbb{R}$. Moreover, the Markov processes to be considered are diffusions with polynomial coefficients, i. e. the generator A has the form, $f \in \mathcal{D} := \text{domain}(A)$, $x \in I$,

$$Af(x) = \frac{a(x)}{2} f''(x) + b(x)f'(x), \tag{1}$$

where $a(x)$ and $b(x)$ are polynomials on I. For the quickest detection problem to be analyzed in Sections 3 and 4, see also Shiryaev [19], $I = [0,1]$, $a(x) = \text{const} \cdot x^2(1-x)^2$, $b(x) = \tilde{\text{const}} \cdot (1-x)$, while for a stopping problem related to the valuation of a perpetual Russian option, cf. Shepp and Shiryaev [17], [18], $I = [1, K)$, $K \gg 1$, $a(x) = \text{const} \cdot x^2$, $b(x) = \tilde{\text{const}} \cdot x$, and $\mathcal{D} = \{g \in \mathcal{C}^2((1,\infty)) \mid g'(1+) = 0\}$.

The process $(X_t)_{t \geq 0}$ to be stopped is characterized as a solution to the martingale problem for the generator A and an initial position x, i. e. there exists a filtration $\{\mathcal{F}_t\}_{t \geq 0}$ such that $(X_t)_t$ is $\{\mathcal{F}_t\}$-progressively measurable, $X_0 = x$, and for every $f \in \mathcal{D}$, $t \geq 0$, the expression

$$f(X_t) - f(X_0) - \int_0^t Af(X_s)ds$$

defines an $\{\mathcal{F}_t\}$-martingale.

The objective of the decision maker is to minimize (or maximize) an expected pay-off,

$$E_x \left[R(X_\tau) + \int_0^\tau \ell(X_s)ds \right], \tag{2}$$

over all $\{\mathcal{F}_t\}$-stopping times τ for which $E_x[\tau] < \infty$, where R and ℓ are polynomial functions on I. For instance, for the detection problem $R(x) = 1 - x$ and $\ell(x)$ is proportional to x, and (30) is to be minimized.

There are two well known methods which can be employed to solve the optimal stopping problem

$$\inf_{\tau, E_x[\tau]<\infty} \left\{ E_x \left[R(X_\tau) + \int_0^\tau \ell(X_s)ds \right] \right\} =: v^*(x), \tag{3}$$

viz. the supermartingale characterization or the variational inequality approach, cf. Shiryaev [19]. The LP-approach to exit time problems and the LP-approach to stopping provide an alternative to these methods and are particularly important from the point of view of numerical computations.

In the sequel we shall repeatedly use the following shorthand writing

$$\langle \mu, f \rangle := \int_I f(x)\mu(dx),$$

where μ denotes a non-negative ($\mu \geq 0$) measure on I and f is any (Borel)-measurable and μ-integrable function defined on the interval. We formally define the adjoint operator A^* of A, i. e. A^* is applied to measures μ, by the equation

$$\langle A^*\mu, f\rangle := \langle \mu, Af\rangle$$

for all $f \in \mathcal{D}$. We let δ_x denote the Dirac measure at x. So the equation

$$\mu_1 - \delta_x - A^*\mu_0 = 0, \tag{4}$$

where μ_0 and μ_1 are non-negative measures on I, is to be understood as shorthand writing for the family of equations

$$\forall\, f \in \mathcal{D}, \qquad \langle \mu_1, f\rangle - f(x) - \langle \mu_0, Af\rangle = 0.$$

The symbol $\mathbb{1}$ stands for the constant function identical to one.

2.1 The exit time approach (cf. Method I)

Let $(X_t)_{t\geq 0}$ denote the Markov process to be stopped so to minimize the expected pay-off (30). By definition, the quantity

$$f(X_t) - f(x) - \int_0^t Af(X_s)ds$$

is a martingale for each $f \in \mathcal{D}$ and thus it follows by the optional sampling theorem that for each admissible stopping time τ (note that $E_x[\tau] < \infty$)

$$E_x[f(X_\tau)] - f(x) - E_x\left[\int_0^\tau Af(X_s)ds\right] = 0. \tag{5}$$

Define the occupation measure μ_0 and exit distribution μ_1 by

$$\mu_0(\Gamma) = E_x\left[\int_0^\tau I_\Gamma(X_s)ds\right] \qquad \text{and} \qquad \mu_1(\Gamma) = P_x[X_\tau \in \Gamma]$$

for Borel sets $\Gamma \subset I$; I_Γ denotes the indicator function of Γ. It then follows that (6) can be written as (5).

We refer to (5) as the basic adjoint equation. In Kurtz and Stockbridge [6] it is shown for very general (controlled) models which include our model as a special case that for each μ_0 and μ_1 satisfying (5) there is a process $X = (X_t)_t$ and a stopping time τ for which (6) is satisfied, and τ is essentially the first exit time of X. Thus the basic adjoint equation characterizes the occupation measure μ_0 and the exit distribution μ_1 of a Markov process defined on I having generator A. Applying this reasoning to any subinterval $B \subset I$ we obtain the following result.

Theorem 1. For each measure μ_0 and μ_1 restricted to $B = [a, b] \subset I$ and satisfying the basic adjoint equation, the expression

$$\langle \mu_1, R \rangle + \langle \mu_0, \ell \rangle =: \Psi(a, b)$$

equals the expected pay-off of a Markov process with generator A which is stopped when hitting a or b.

2.2 A general LP-approach to optimal stopping (cf. Method II)

The exit time approach will lead to a numerical method which is only applicable to a restricted class of stopping problems. Moreover, it is an indirect approach by which optimal solutions can be found. A direct method based on linear programming has recently been proposed in two theses, cf. Cho [2] and Röhl [16]. The following result is a special case of general theorems proved by these authors and provides the analytical underpinning for Method II, see Section 4.

Theorem 2. Consider the optimal stopping time problem (3). Then $v^*(x)$ equals the optimal value of the infinite-dimensional linear program

$$\inf_{\mu_\tau, \mu \geq 0} \left\{ \langle \mu_\tau, R \rangle + \langle \mu, \ell \rangle \mid \langle \mu_\tau, \mathbb{1} \rangle = 1, \;\; \mu_\tau - \delta_x - A^* \mu = 0 \right\}. \tag{6}$$

3 Method I: The Exit Time Approach

To further simplify the exposition we shall – without loss of generality – assume from now on that I equals the unit interval $[0\,1]$; the change of variable $x \mapsto (x-a)/(b-a)$ will transform general cases to this special one.

In Section 2.1 we have seen that each exit time problem is equivalent to a particular infinite-dimensional linear program. Since measures on bounded intervals are determined by their moments and since the generator A is assumed to have polynomial coefficients, choosing finitely many moments as variables we can associate with each stopping time $\tau_b := \inf\{t \mid X_t \geq b\}$, $0 \leq b \leq 1$, i. e. we define $B = [0, b]$ in Theorem 1, two linear programs whose optimal values sandwhich the expected pay-off when using τ_b, $0 \leq x \leq b$,

$$\varphi(b) := \Psi(0, b) = E_x \left[R(X_{\tau_b}) + \int_0^{\tau_b} \ell(X_s) ds \right].$$

To formulate these programs we let $a(x) = \sum_{i=0}^{N_a} \alpha_i x^i$, $b(x) = \sum_{i=0}^{N_b} \beta_i x^i$, $\ell(x) = \sum_{i=0}^{N_\ell} \gamma_i x^i$ and $R(x) = \sum_{i=0}^{N_R} \delta_i x^i$, where N_a, N_b, N_ℓ and N_R are integers, and $\alpha_0, \dots, \alpha_{N_a}, \beta_0, \dots, \delta_1, \dots, \delta_{N_R} \in \mathbb{R}$. For any integer $M \geq N := \max\{N_a, N_b, N_\ell, N_R\}$ we define for $\boldsymbol{\mu} = (\mu_0, \mu_1, \dots, \mu_M) \in \mathbb{R}^{M+1}$ the i-th iterated differences of $\boldsymbol{\mu}$, $(i,n) \in \mathfrak{M} := \{(i,n) \mid 0 \leq i \leq M,\ 0 \leq n \leq M-i\}$,

$$(-1)^i \Delta^i \boldsymbol{\mu}(n) = \sum_{k=0}^{i} (-1)^k \binom{i}{k} \mu_{k+n}, \tag{7}$$

and call

$$\mathcal{H}^M := \left\{ \boldsymbol{\mu} \in \mathbb{R}^{M+1} \mid \Delta^i \boldsymbol{\mu}(n) \geq 0,\ (i,n) \in \mathfrak{M} \right\} \subset \mathbb{R}^{M+1}$$

the Hausdorff polytope of order M, cf. Helmes [5] and, for the generalization of this concept to higher dimensions, Röhl [16]. For vectors $\boldsymbol{\alpha} = (\alpha_0, \dots, \alpha_{N_a}, 0, \dots, 0) \in \mathbb{R}^{M+1}$, $\boldsymbol{\beta}$, $\boldsymbol{\gamma}$ and $\boldsymbol{\delta}$ similarly defined, we denote their scalar product with vectors $\boldsymbol{\mu} \in \mathcal{H}^M$ by

$$\langle \boldsymbol{\mu}, \boldsymbol{\alpha} \rangle := \sum_{k=0}^{M} \mu_k \alpha_k;$$

for further use we shall also introduce the following abbreviations, $k \in \mathcal{K} := \{k \in \mathbb{N} \mid 0 \leq k \leq M - N\}$, $\eta_k := \eta_k(x, \boldsymbol{\alpha}, \boldsymbol{\beta}, \boldsymbol{\mu}^{(1)}, \boldsymbol{\mu}^{(0)})$ and

$$\eta_k := \frac{k(k-1)}{2} \sum_{i=0}^{N_a} \alpha_i \mu_{i+k-2}^{(0)} + k \sum_{i=0}^{N_b} \beta_i \mu_{i+k-1}^{(1)} + x^k - \mu_k^{(1)}.$$

Next we define two linear programming problems, $\mathcal{P}_{\min}$ and $\mathcal{P}_{\max}$:

$$\underline{\varphi}(b) := \min_{\boldsymbol{\mu}^{(0)}, \boldsymbol{\mu}^{(1)}} \left\{ \langle \boldsymbol{\mu}^{(1)}, \boldsymbol{\delta} \rangle + \langle \boldsymbol{\mu}^{(0)}, \boldsymbol{\gamma} \rangle \;\middle|\; \begin{array}{l} \boldsymbol{\mu}^{(0)}, \boldsymbol{\mu}^{(1)} \in \mathcal{H}^M,\ \mu_0^{(1)} = 1, \\ \\ \eta_k = 0,\ k \in \mathcal{K} \end{array} \right\}$$

and

$$\bar{\varphi}(b) := \max_{\boldsymbol{\mu}^{(0)}, \boldsymbol{\mu}^{(1)}} \left\{ \langle \boldsymbol{\mu}^{(1)}, \boldsymbol{\delta} \rangle + \langle \boldsymbol{\mu}^{(0)}, \boldsymbol{\gamma} \rangle \;\middle|\; \begin{array}{l} \boldsymbol{\mu}^{(0)}, \boldsymbol{\mu}^{(1)} \in \mathcal{H}^M,\ \mu_0^{(1)} = 1, \\ \\ \eta_k = 0,\ k \in \mathcal{K} \end{array} \right\}.$$

Since the Hausdorff polytope includes the set of all moment sequences up to order M, we obtain the inequalities

$$\min_{0 \leq b \leq 1} \underline{\varphi}(b) \leq \min_{0 \leq b \leq 1} \varphi(b) \leq \min_{0 \leq b \leq 1} \bar{\varphi}(b).$$

Furthermore, if there is a b^* such that

$$\min_{0 \leq b \leq 1} \underline{\varphi}(b) = \underline{\varphi}(b^*) = \bar{\varphi}(b^*) = \min_{0 \leq b \leq 1} \bar{\varphi}(b) \tag{8}$$

then τ_{b^*} is the optimal stopping time in the class of all stopping rules $\{\tau_b\}_{0\leq b\leq 1}$. If

$$\varepsilon^* := \min_{0\leq b\leq 1} \bar{\varphi}(b) - \min_{0\leq b\leq 1} \underline{\varphi}(b) > 0, \tag{9}$$

and $\bar{b}^*$, $\underline{b}^*$ resp., is a solution of $\mathcal{P}_{\max}$, $\mathcal{P}_{\min}$ resp., then $\tau_{\bar{b}^*}$ and $\tau_{\underline{b}^*}$ are ε^*-optimal stopping times within the class $\{\tau_b\}_b$.

These observations underly the following numerical procedure, where the tacit assumption about $\underline{\varphi}(b)$ and $\bar{\varphi}(b)$ is that unimodality of φ might ensure unimodality of $\underline{\varphi}$ and $\bar{\varphi}$, at least for large values of M.

Method I.

> *Assume that $\varphi(b)$ is a unimodal function of b. Apply a line search technique, e. g. the Golden Section rule, with either a specified number of iterations or a prescribed tolerance level, etc. to $\underline{\varphi}(b)$ and $\bar{\varphi}(b)$. If equality (8), inequality (9) resp., holds then the line search will determine an optimal, ε^*-optimal resp., stopping rule within the class of all stopping times $\{\tau_b\}_{0\leq b\leq 1}$.*

We shall illustrate Method I by analyzing Shiryaev's quickest detection problem.

Example (The detection problem for a Wiener process, part I).

The detection problem, sometimes called the disruption problem, for Brownian motion is to detect the onset, assumed to be conditionally exponentially distributed and independent of the noise, of a drift value r. The decision is to choose a random variable τ, the time at which an "alarm signal" is given, such that a linear combination of the *probability of false alarm* and the *average delay of detecting the occurence of disruption* is minimized. The disruption problem for a Wiener process is equivalent to an optimal stopping time problem of a diffusion. This problem has been solved by Shiryaev using a variational inequality approach, see [19] for more details. The stopping problem has the form

$$\inf_{\tau, E_x[\tau]<\infty} E_x\left[(1-X_\tau) + c\int_0^\tau X_s\, ds\right], \tag{10}$$

where $c > 0$ is a given number, and $(X_t)_t$ satisfies the stochastic differential equation, x, r, σ, λ positive parameters, $0 < x < 1$,

$$dX_t = \lambda(1-X_t)dt + \frac{r}{\sigma}X_t(1-X_t)d\bar{W}_t, \quad X(0) = x, \tag{11}$$

where $(\bar{W}_t)_t$, a Brownian motion, is the innovation process determined by the original noise process and accumulated estimates up to time t. The process

$(X_t)_{t\geq 0}$ represents the conditional probability of the events $\{\theta \geq t\}$, θ the time of disruption, given the observations $\mathfrak{F}_t^Y = \sigma(Y_s, 0 \leq s \leq t)$, where the data are described by

$$Y_t = r(t-\theta)^+ + \sigma W_t, \quad 0 \leq t,$$

and $(W_t)_t$ is a Wiener process.

The random variable θ is assumed to be distributed according to

$$P[\theta = 0] = x \quad \text{and} \quad P[\theta \geq t \mid \theta > 0] = e^{-\lambda t}.$$

Since the Markov process X satisfies Equation (16), the generator A of this process equals, $f \in \mathfrak{C}^2(\mathbb{R})$, cf. (1),

$$Af(x) = \lambda(1-x)f'(x) + \frac{r^2}{2\sigma^2}\, x^2(1-x)^2 f''(x).$$

The following (infinite dimensional) LP-problem, cf. Section 2, solves (19):

$$v^*(x) = \min_{\mu_\tau,\mu} \left\{ \int_0^1 (1-\xi)\mu_\tau(d\xi) + c\int_0^1 \xi\mu(d\xi) \;\middle|\; \begin{array}{l} \mu_\tau - \delta_x - A^*\mu = 0, \\ \langle \mu_\tau, \mathbb{1}\rangle = 1, \\ \mu_\tau, \mu \geq 0 \end{array} \right\}.$$

Below we compare the numerical results for the optimal value (as a function of x) using Method I with the exact values based on Shiryaev's formula, viz.

$$v^*(x) = \begin{cases} (1-A^*) - \displaystyle\int_\xi^{A^*} \psi^*(z)\,dz, & x \in [0, A^*) \\ 1-x, & x \in [A^*, 1] \end{cases} \tag{12}$$

where, $\Lambda = 2\sigma^2\lambda/r^2$, $C = 2\sigma^2 c/r^2$, $H(y) = \ln(y/(1+y)) - 1/y$,

$$\psi^*(z) = -C\int_0^z \exp\Big[-\Lambda[H(z) - H(y)]\Big]\frac{dy}{y(1-y)^2},$$

and A^* is the root of the equation

$$\psi^*(A^*) = -1. \tag{13}$$

Figures 1 and 2 show the graphs of the functions $A^*(c)$ and $v^*(x)$ as defined by (12) and (13) for a particular set of parameters.

For the same parameters Table 1 reports the numerical results for $A^*(c)$ and v^* as a function of c using Method I, while Table 2 reports the results for $v^*(x)$ as a function of the initial position x. In each case we have applied

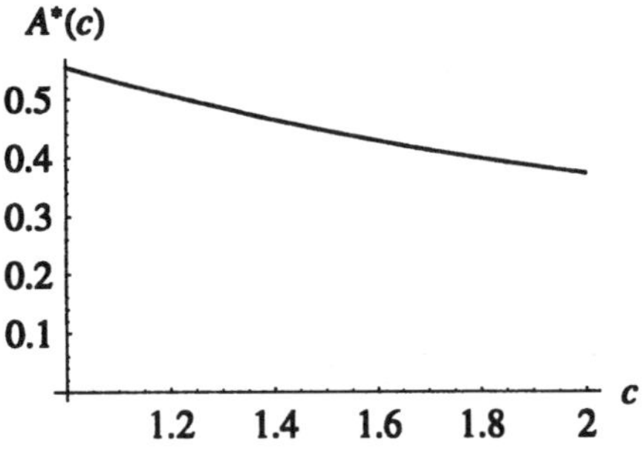

Fig. 1. The optimal stopping point A^* as a function of c ($r = \sigma = \lambda = 1$).

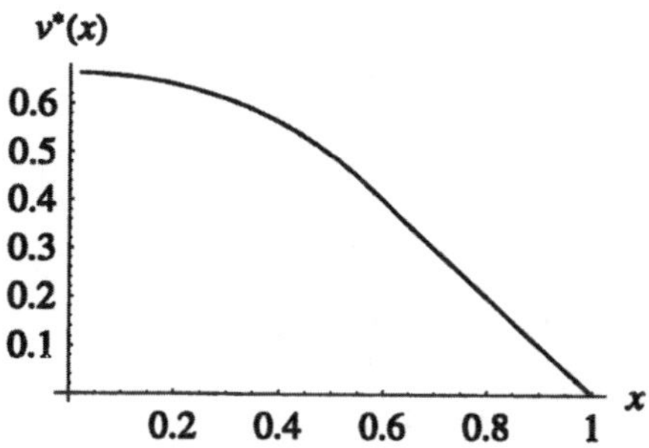

Fig. 2. The value function v^* on $[0.02, 1]$ ($r = \sigma = \lambda = c = 1$).

the Golden Section rule to the min and max LP-problems using moments up to order 30 and terminated the line search after 40 iterations. Note the excellent agreement of the numerical results with the exact values which were obtained employing Mathematica to evaluate formulas (12) and (13). Table 3 illustrates the limitations of the method if more extreme parameter settings, e. g. $r/\sigma \gg 1$, for instance $r/\sigma = 10$, are analyzed. In Table 3 we display the computed quantities $v^*(0.3)$ and A^* (for $\lambda = 1$, $r = 10$ and $\sigma = 1$) as a function of the number of moments used.

The parameter setting $\lambda = \sigma = 1$ and $r = 10$ provides an example where $\underline{\varphi}(b)$ and $\bar{\varphi}(b)$ are not unimodal functions. Using numerical integration we found the optimal stopping point to be $A^* \doteq 0.977968$ and the optimal value $v^*(0.3) \doteq 0.129128$.

4 Method II: A General LP-Approach

According to Theorem 2, choosing as variables the first $M + 1$ moments of measures defined on I we can associate with each optimal stopping problem (3) *one* finite dimensional linear *minimization* problem $\underline{\mathcal{P}}_M$ whose optimal value $\underline{v}_M(x)$ is a lower bound on $v^*(x)$. Let $a(x)$, $b(x)$, $\ell(x)$, $R(x)$, $\boldsymbol{\alpha}$, $\boldsymbol{\beta}$, $\boldsymbol{\gamma}$, $\boldsymbol{\delta}$, $\boldsymbol{\mu}$, $\mathfrak{M}$, $\mathcal{K}$ and $(\eta_k)_{k \in \mathcal{K}}$ be defined as in Section 3, then we have the following

Table 1. The optimal stopping point A^* and the optimal value v^* as a function of c using Method I ($r = \sigma = \lambda = 1$, $x = 0.3$ and $M = 30$)

c value	objective of LPs	optimal stopping point	exact value for A^*	optimal value	exact value for v^*
1.0	min	0.556066	0.556066	0.609534	0.609534
	max	0.556066		0.609534	
1.2	min	0.506103	0.506093	0.637820	0.637820
	max	0.506091		0.637820	
1.4	min	0.463688	0.463688	0.658360	0.658360
	max	0.463687		0.658360	
1.6	min	0.427376	0.427384	0.673251	0.673254
	max	0.427376		0.673251	
1.8	min	0.396020	0.396014	0.683900	0.683910
	max	0.396015		0.683900	
2.0	min	0.368711	0.368709	0.691282	0.691308
	max	0.368709		0.691282	

inequality:

$$\begin{aligned} v^*(x) &= \inf_{\mu_\tau, \mu \geq 0} \{ \langle \mu_\tau, R \rangle + \langle \mu, \ell \rangle \mid \langle \mu_\tau, \mathbb{1} \rangle = 1, \quad \mu_\tau - \delta_x - A^* \mu = 0 \} \\ &\geq \min_{\boldsymbol{\mu}^{(\tau)}, \boldsymbol{\mu}} \left\{ \langle \boldsymbol{\mu}^{(\tau)}, \delta \rangle + \langle \boldsymbol{\mu}, \boldsymbol{\gamma} \rangle \;\middle|\; \begin{array}{l} \Delta^i \boldsymbol{\mu}^{(\tau)}(n), \ \Delta^i \boldsymbol{\mu}(n) \geq 0, \\ (i,n) \in \mathfrak{M}, \\ \mu_0^{(\tau)} = 1, \ \eta_k = 0, \ k \in \mathcal{K} \end{array} \right\} \qquad (14) \\ &=: \underline{v}_M(x). \end{aligned}$$

Note that for a general stopping problem the method of moments for the LP-approach does not determine a linear optimization problem which bounds $v^*(x)$ from above. However, as will be illustrated below, the transformation

Table 2. The optimal stopping point A^* and the optimal value v^* as a function of the initial position using Method I ($r = \sigma = \lambda = c = 1$ and $M = 30$)

initial position	objective of LPs	optimal stopping point	optimal value	exact value for v^*
0.1	min	0.556064	0.656103	0.656103
	max	0.556075	0.656103	
0.2	min	0.556066	0.639540	0.639540
	max	0.556067	0.639540	
0.3	min	0.556066	0.609534	0.609534
	max	0.556066	0.609534	
0.4	min	0.556065	0.562906	0.562906
	max	0.556064	0.562906	
0.5	min	0.556065	0.494628	0.494628
	max	0.556066	0.494628	
0.6	min	0.600000	0.400000	0.400000
	max	0.600000	0.400000	
0.7	min	0.700000	0.300000	0.300000
	max	0.700000	0.300000	
0.8	min	0.800000	0.200000	0.200000
	max	0.800000	0.200000	
0.9	min	0.900000	0.100000	0.100000
	max	0.900000	0.100000	

Table 3. The computed values v^* and A^* as functions of M ($\lambda = \sigma = 1$, $r = 10$)

number of moments used M	objective of LPs	optimal stopping point	optimal value
30	min	0.975893	0.125080
	max	0.977355	0.134995
40	min	0.979529	0.126333
	max	0.976450	0.132733
50	min	0.973717	0.126144
	max	0.979016	0.131450
60	min	0.979407	0.126074
	max	0.978503	0.130838
70	min	0.976796	0.126813
	max	0.977235	0.130920
80	min	0.978065	0.126408
	max	0.982095	0.130700
90	min	0.979725	0.126207
	max	0.978065	0.130427
100	min	0.975573	0.126339
	max	0.979408	0.130085

T_M, to be defined next, cf. Feller [4] and Röhl [16], when applied to a solution $\underline{\boldsymbol{\mu}}^{(\tau)}$ yields valuable information which allows to improve the lower bound $\underline{v}_M(x)$:

$$\underline{\boldsymbol{\mu}}^{(\tau)} \stackrel{T_M}{\longmapsto} \left(q_{\frac{k}{M}}\right)_{0\le k\le M} \qquad \text{and} \qquad q_{\frac{k}{M}} := \binom{M}{k}(-1)^{M-k}\Delta^{M-k}\underline{\boldsymbol{\mu}}^{(\tau)}(k).$$

Along with the optimization problem $\underline{\mathcal{P}}_M$, see (14), we shall consider a non-linear optimization problem $\hat{\mathcal{P}}_M$. The problem $\hat{\mathcal{P}}_M$ differs from $\underline{\mathcal{P}}_M$ in that convex combinations of moments (up to M) of a (fixed) finite set of Dirac measures are substituted for the variables $\left(\mu_k^{(\tau)}\right)_{0\le k\le M}$. For instance, deciding on N_p Dirac measures at (variable) points $0 \le b_1, b_2, \dots, b_{N_p} \le 1$ and a (variable) non-negative vector $\boldsymbol{p} \in \mathbb{R}^{N_p}$, $\sum_{j=1}^{N_p} p_j = 1$, we put

$$\boldsymbol{\mu}_m^{(\tau)} = \sum_{j=1}^{N_p} p_j b_j^m, \qquad 0 \le m \le M. \tag{15}$$

The non-linear problem $\hat{\mathcal{P}}_M$ is defined as

$$\hat{v}_M(x) := \min_{\boldsymbol{\mu},\boldsymbol{p},\boldsymbol{b}} \left\{ \langle \boldsymbol{\mu}^{(\tau)}, \boldsymbol{\delta} \rangle + \langle \boldsymbol{\mu}, \boldsymbol{\gamma} \rangle \;\middle|\; \begin{array}{l} \boldsymbol{\mu} \in \mathcal{H}^M, \ \boldsymbol{\mu}^{(\tau)} \text{ satisfies (15)} \\ \text{and } \eta_k = 0, \ k \in \mathcal{K} \end{array} \right\}$$

In general, nothing can be said about the relative size of $v^*(x)$ and $\hat{v}_M(x)$. In many applications, however, $\hat{v}_M(x)$ equals $v^*(x)$ up to numerical accuracy. A benefit of computing $\hat{v}_M(x)$ is that an optimal solution of $\hat{\mathcal{P}}_M$ yields a refinement of $\underline{\mathcal{P}}_M$ which yields a better lower bound than $\underline{v}_M(x)$. Again, to simplify the exposition, let us assume that an optimal solution of $\hat{\mathcal{P}}_M$ assigns weights to only two points, $\hat{b}_1$ and $\hat{b}_2$; the general case is a straight forward extension of this special one. Then we cover the unit interval by five subintervals, $0 \le \varepsilon_1, \varepsilon_2 \ll 1$,

$$\begin{aligned}[0,1] &= [0, \hat{b}_1 - \varepsilon_1] \cup [\hat{b}_1 - \varepsilon_1, \hat{b}_1 + \varepsilon_1] \cup [\hat{b}_1 + \varepsilon_1, \hat{b}_2 - \varepsilon_2] \\ &\qquad \cup [\hat{b}_2 - \varepsilon_2, \hat{b}_2 + \varepsilon_2] \cup [\hat{b}_2 + \varepsilon_2, 1] \\ &= \bigcup_{j=1}^{5} I_j.\end{aligned}$$

For any such covering of $[0,1]$ the infinite-dimensional linear program (6) can be written as

$$\inf_{\mu^{(1)},\dots,\mu^{(5},\mu\ge 0} \left(\sum_{i=1}^{5} \langle \mu^{(j)}, R \rangle + \langle \mu, \ell \rangle \right)$$

subject to

$$\sum_{i=1}^{5} \langle \mu^{(j)}, \mathbb{1} \rangle = 1, \qquad \text{support}\left(\mu^{(j)}\right) \subset I_j, \qquad 0 \le j \le 5,$$

and

$$\sum_{i=1}^{5} \mu^{(i)} - \delta_k - A^* \mu = 0.$$

Switching from measures $\mu^{(j)}$ to finite sequences $\boldsymbol{\mu}^{(j)} \in \mathbb{R}^{M+1}$ such that each vector $\boldsymbol{\mu}^{(j)}$ satisfies the analogue of the "Hausdorff conditions" (7) for measures defined on a general interval $[a,b]$, $0 \le a < b \le 1$, i. e.

$$\boldsymbol{\mu}_n^{(j)} = \sum_{k=0}^{n} \binom{n}{k} (b-a)^k a^{n-k} \boldsymbol{\nu}_k^{(j)},$$

where each vector $\boldsymbol{\nu}^{(j)}$ satisfies (7), we obtain a refinement of $\underline{\mathcal{P}}_M$. The value of the refined problem will be denoted by $\underline{v}_M^*(x)$. By construction the following inequalities hold:

$$\underline{v}_M(x) \le \underline{v}_M^*(x) \le v^*(x).$$

A refinement of $\underline{\mathcal{P}}_M$ typically yields a much improved lower bound. The following procedure, Method II, formalizes the ideas described above.

Method II.

Step 1. *Solve* $\underline{\mathcal{P}}_M$.

Step 2. *Solve* $\hat{\mathcal{P}}_M$.

Use the solution of $\underline{\mathcal{P}}_M$ *and the transformtion* T_M *to specify an initial value for a non-linear solver.*

Step 3. *Use the solution of* $\hat{\mathcal{P}}_M$ *to determine a refinement of* $\underline{\mathcal{P}}_M$*; choose* ε_i *"small", e. g.* 10^{-4} *or* 10^{-5}.

If $\underline{v}_M^*(x) \approx \hat{v}_M(x)$ *take* $\underline{v}_M^*(x)$ *as an estimate (lower bound) for* $v^*(x)$.

Remark. Whenever the solution of $\hat{\mathcal{P}}_M$ involves but one Dirac measure δ_{b^*} a further heuristic is to combine methods I and II and to compare the values $\underline{v}_M^*(x)$ and $\min_{0 \le b \le 1} \bar{\varphi}(b)$. Should the numbers be close then these numbers determine a (reasonable) range for the optimal value of a general stopping problem.

Example (The quickest detection problem for a Wiener process, part II).

Tables 4–7 display the results of our analysis of the quickest detection problem using Method II for the parameters $r = \sigma = \lambda = 1$; $c = 1$ and $x = 0.3$ if fixed. In Table 4 we compare the values $\underline{v}_M(x)$, $\hat{v}_M(x)$ and $\underline{v}_M^*(x)$, $M = 25$, with $v^*(x)$ as functions of x, and in Table 5 we compare the values of $A^*(c)$, as a function of c, with the values for A^* derived from the optimization problems $\underline{\mathcal{P}}_M$, $\hat{\mathcal{P}}_M$ and a refinement of $\underline{\mathcal{P}}_M$. For the verification step (Step 3 of Method II) we used the covering

$$[0.0556] \cup [0.556, 0.55612] \cup [0.55612, 1] \tag{16}$$

which is suggested by the numbers in Table 6.

Table 4. The values $\underline{v}_M(x)$, $\hat{v}_M(x)$, $\underline{v}^*_M(x)$ and $v^*(x)$ as functions of x; $r = c = \sigma = \lambda = 1$, $M = 25$

initial position	$\underline{v}_M(x)$	$\hat{v}_M(x)$	$\underline{v}^*_M(x)$	exact value $v^*(x)$
0.1	0.63958	0.65610	0.656101	0.656103
0.2	0.62301	0.63954	0.639538	0.639540
0.3	0.59301	0.60953	0.609533	0.609534
0.4	0.54643	0.56291	0.562904	0.562906
0.5	0.47995	0.49463	0.494630	0.494628
0.6	0.39497	0.4	0.4	0.4
0.7	0.29941	0.3	0.3	0.3
0.8	0.19997	0.2	0.2	0.2
0.9	0.09999	0.1	0.1	0.1

In Table 6 we illustrate how the transformation T_M is used. The numbers shown, $(q_{k/M})_{0 \le k \le M}$, are the image of $\boldsymbol{\mu}^{(\tau)}$, an optimal solution of $\underline{\mathcal{P}}_M$, under T_M; $x = 0.3$ in this case. From these numbers we can infer that the non-linear solver should be initialized at a point nearby $14/25 = 0.56$; $14/25$ is our first estimate of the optimal stopping point.

The same idea is applied in Step 3 of Method II. Table 7 illustrates this part of the procedure. The solution of $\hat{\mathcal{P}}_M$, $b^* = 0.55607194$, is an estimate of the optimal stopping time and b^* suggests the covering (32). Applying transformation T_M to each solution vector $\boldsymbol{\mu}^{(i)}$ associated with covering (32) we obtain Table 7. It shows that the solution of the refined LP concentrates "all" its mass on $[0.556, 0.55612]$; the spurious mass $q_0^{(3)} = 0.0242408$ actually disappears for larger values of M. So we use $0.556 + 14/25 \cdot 0.00012 = 0.550672$ and $0.556 + 16/25 \cdot 0.00012 = 0.550738$ to specify a range for the optimal stopping point.

Table 5. Approximating values of $A^*(c)$ based on $\underline{\mathcal{P}}_M$, $\hat{\mathcal{P}}_M$ and a refinement of $\underline{\mathcal{P}}_M$ ($r = \sigma = \lambda = 1$, $x = 0.3$, $M = 25$)

c value	estimate of A^* based on Step 1	estimate of A^* based on Step 2	estimate of A^* based on Step 3	exact value $A^*(c)$
1	0.56	0.55607194	0.5561	0.556066
1.2	0.52	0.50609462	0.5065	0.506093
1.4	0.47	0.46368731	0.4640	0.463688
1.6	0.44	0.42737578	0.4270	0.427384
1.8	0.40	0.39601437	0.3960	0.396014
2	0.40	0.36870895	0.3687	0.368709

Since the detection problem can be analyzed by Method I as well as Method II we can combine the values $\underline{v}^*_M(x)$ and $\bar{\varphi}(b^*)$, see Table 2 for the latter value, to get an estimate of $v^*(x)$ and to obtain a (numerical) error bound, viz.

$$0.609533 + 10^{-6}.$$

Table 6. The values $q_{k/M}$, $0 \le k \le M = 25$; $r = c = \sigma = \lambda = 1$, $x = 0.3$ for an optimal solution of $\underline{\mathcal{P}}_M$

k	$q_{k/M}$	k	$q_{k/M}$	k	$q_{k/M}$	k	$q_{k/M}$	k	$q_{k/M}$
0	0	6	0	12	0	18	0	24	0
1	0	7	0	13	0.299145	19	0	25	0
2	0	8	0	14	0.512821	20	0		
3	0	9	0	15	0.188034	21	0		
4	0	10	0	16	0	22	0		
5	0	11	0	17	0	23	0		

Table 7. The values $q^{(i)}_{k/M}$, $0 \le k \le M = 25$, $1 \le i \le 3$, for an optimal solution of a refinement of $\underline{\mathcal{P}}_M$, cf. page 199; $r = c = \sigma = \lambda = 1$, $x = 0.3$

k	$q^{(1)}_{k/M}$	$q^{(2)}_{k/M}$	$q^{(3)}_{k/M}$
0	0	0	0.0242408
1	0	0	0
2	0	0	0
3	0	0	0
4	0	0	0
5	0	0	0
6	0	0	0
7	0	0	0
8	0	0	0
9	0	0	0
10	0	0	0
11	0	0	0
12	0	0	0
13	0	0	0
14	0	0.333593	0
15	0	0.48927	0
16	0	0.152897	0
17	0	0	0
18	0	0	0
19	0	0	0
20	0	0	0
21	0	0	0
22	0	0	0
23	0	0	0
24	0	0	0
25	0	0	0

5 Concluding Remarks

We have described numerical procedures for analyzing optimal stopping problems of Markov processes. Both methods are based on a linear programming approach to such kind of decision problems. We have illustrated these methods by numerically analyzing Shiryaev's quickest detection problem for a Wiener process. This example was chosen for its importance and for the fact that the numerical results can be compared with analytical ones.

We conclude by commenting on some of our computational experiences. We used AMPL as a convenient interface and employed the CPLEX solver. When using Method I to analyze the quickest detection problem we ran LP problems with M up to 140 accepting unscaled infeasibilities. If $M = 50$ an individual LP-run typically requires ~ 500 iterations, most of them in phase I of the Simplex algorithm. To be able to check computations within loops, etc. we set AMPL- and CPLEX-options in such way that previously computed bases were not used. Large values of M and N naturally increase the run time of Method I. For different parameter settings we made 'ad hoc' decisions to strike a compromise between accuracy and run time.

When using Method II we used smaller values for M, $M \leq 40$; for larger values we very often experienced the program to exit because of detected (numerical) infeasibilities. For the larger linear and non-linear problems of Method II the solvers, CPLEX for the linear problems and MINOS for the non-linear ones, typically required 1000 – 1500 iterations, half of them during phase I of the Simplex algorithm. For the detection problem, as far as refinements of $\underline{\mathcal{P}}_M$ are concerned, we had to use but one Dirac measure δ_b (cf. Method II); depending on the size of M and the parameters given we used $\varepsilon = 10^{-4}$ or 10^{-5}.

In light of the excellent agreement between the numerical and analytical results for a large set of different parameters we consider the LP-techniques a convenient and easy to use tool for analyzing the detection problem as well as similarly structured ones, e. g. pricing perpetual Russian options, etc.

References

1. Bhatt, A. G. and Borkar, V. S. (1996) Occupation Measures for Controlled Markov Processes: Characterization and Optimality. Ann. Probab. **24**, 1531–1562.
2. Cho, M. J. (2000) Linear Programming Formulation for Optmal Stopping. PhD thesis, The Graduate School, University of Kentucky, Lexington.
3. Cho, M. J. and Stockbridge, R. H. Linear Programming Formulation for Optimal Stopping Problems. to appear in SIAM J. Control Optim.
4. Feller, W. (1965) An Introduction to Probability Theory and its Applications. Vol. 2, Wiley, New York.
5. Helmes, K. (1999) Computing Moments of the Exit Distribution for Diffusion Processes Using Linear Programming. In: Kall P., Lüthi J.-H. (Eds.) Operations Research Proceedings 1998. Springer, Berlin, 231–240.

6. Helmes, K., Röhl, S., and Stockbridge, R. H. (2001) Computing Moments of the Exit Time Distribution for Markov Processes by Linear Programming, Operations Research **49**, 516–530.
7. Helmes K. and Stockbridge R. H. (2000) Numerical Comparison of Controls and Verification of Optimality for Stochastic Control Problems. J. Optim. Th. Appl. **106**, 107–127.
8. Helmes, K. and Stockbridge, R. H. (2001) Numerical Evaluation of Resolvents and Laplace Transforms of Markov Processes. Mathematical Methods of Operations Research **53**, 309–331.
9. Helmes, K. and Stockbridge, R. H. Analyzing Diffusion Approximations of the Wright-Fisher Model Using Linear Programming. Submitted for publication.
10. Hernandez-Lerma, O., Hennet, J. C., and Lasserre, J. B. (1991) Average Cost Markov Decision Processes: Optimality Conditions. J. Math. Anal. Appl. **158**, 396–406.
11. Kurtz, T. G. and Stockbridge, R. H. (1998) Existence of Markov Controls and Characterization of Optimal Markov Controls. SIAM J. Control Optim. **36**, 609–653.
12. Kurtz, T. G. and Stockbridge, R. H. (1999) Martingale Problems and Linear Programs for Singular Control. Thirty-Seventh Annual Allerton Conference on Communication, Control, and Computing (Monticello, Ill.), 11-20, Univ. Illinois, Urbana-Champaign, Ill.
13. Manne, A. S. (1960) Linear Programming and Sequential Decisions, Management Sci. **6**, 259–267.
14. Mendiondo, M. S. and Stockbridge, R. H. (1998) Approximation of Infinite-Dimensional Linear Programming Problems which Arise in Stochastic Control. SIAM J. Control Optim. **36**, 1448–1472.
15. Mendiondo, M. S. and Stockbridge, R. H. Long-Term Average Control of a Local Time Process. To appear in Markov Processes and Controlled Markov Chains, Kluwer.
16. Röhl, S. (2001) Ein linearer Programmierungsansatz zur Lösung von Stopp- und Steuerungsproblemen. Ph. D. Dissertation, Humboldt-Universität zu Berlin, Berlin, Germany.
17. Shepp, L. and Shiryaev, A. N. (1993) The Russian Option: Reduced Regret. Ann. Appl. Probab. **3**, 631–640.
18. Shepp, L. and Shiryaev, A. N. (1993) A New Look at Pricing of the "Russion Option". Theory Probab. Appl. **39**, 103–119.
19. Shiryaev, A. N. (1978) Optimal Stopping Rules. Springer, New York.
20. Stockbridge, R. H. (1990) Time-Average Control of Martingale Problems: A Linear Programming Formulation. Ann. Probab. **18**, 206–217.

The ODE Method and Spectral Theory of Markov Operators

Jianyi Huang[1], Ioannis Kontoyiannis[2], and Sean P. Meyn[1]

[1] University of Illinois at Urbana-Champaign, Urbana IL 61801, USA
[2] Brown University, Box F, 182 George St., Providence, RI 02912, USA

Abstract. We give a development of the ODE method for the analysis of recursive algorithms described by a stochastic recursion. With variability modeled via an underlying Markov process, and under general assumptions, the following results are obtained:

(i) Stability of an associated ODE implies that the stochastic recursion is stable in a strong sense when a gain parameter is small.
(ii) The range of gain-values is quantified through a spectral analysis of an associated linear operator, providing a non-local theory, even for nonlinear systems.
(iii) A second-order analysis shows precisely how variability leads to sensitivity of the algorithm with respect to the gain parameter.

All results are obtained within the natural operator-theoretic framework of geometrically ergodic Markov processes.

1 Introduction

Stochastic approximation algorithms and their variants are commonly found in control, communication and related fields. Popularity has grown due to increased computing power, and the interest in various 'machine learning' algorithms [6,7,12]. When the algorithm is linear, then the error equations take the following linear recursive form,

$$X_{t+1} = [I - \alpha M_t]\, X_t + W_{t+1}, \tag{1}$$

where $\boldsymbol{X} = \{X_t\}$ is an error sequence, $\boldsymbol{M} = \{M_t\}$ is a sequence of $k \times k$ random matrices, $\boldsymbol{W} = \{W_t\}$ is a random "disturbance" or "noise", $\alpha \geq 0$ is a fixed constant, and I is the $k \times k$ identity matrix.

An important example is the LMS (least mean square) algorithm. Consider the discrete linear time-varying model,

$$y(t) = \theta(t)^T \phi(t) + n(t), \quad t \geq 0, \tag{2}$$

where $\{y(t)\}$ and $\{n(t)\}$ are the sequences of (scalar) observations and noise, respectively. The vector-valued processes $\theta(t) = [\theta_1(t), \dots, \theta_k(t)]^T$ and $\phi(t) = [\phi_1(t), \dots, \phi_k(t)]^T$, $t \geq 0$, denote the k-dimensional regression vector and

B. Pasik-Duncan: Stochastic Theory and Control, LNCIS 280, pp. 205–221, 2002.

time varying parameters, respectively. These will be taken to be functions of an underlying Markov chain in the analysis that follows.

The LMS algorithm is given by the recursion

$$\hat{\theta}(t+1) = \hat{\theta}(t) + \alpha\phi(t)e(t), \tag{3}$$

where $e(t) := y(t) - \hat{\theta}(t)^T\phi(t)$, and the parameter $\alpha \in (0,1]$ is the *step size.* Hence,

$$\tilde{\theta}(t+1) = (I - \alpha\phi(t)\phi(t)^T)\tilde{\theta}(t) + [\theta(t+1) - \theta(t) - \alpha\phi(t)n(t)], \tag{4}$$

where $\tilde{\theta}(t) \triangleq \theta(t) - \hat{\theta}(t)$. This is of the form (1) with $M_t = \phi(t)\phi(t)^T$, $W_{t+1} = \theta(t+1) - \theta(t) - \alpha\phi(t)n(t)$, and $X_t = \tilde{\theta}(t)$.

On iterating (1) we obtain the representation,

$$\begin{aligned} X_{t+1} &= (I - \alpha M_t)\, X_t + W_{t+1} \\ &= (I - \alpha M_t)\, [(I - \alpha M_{t-1})\, X_{t-1} + W_t] + W_{t+1} \\ &= \prod_{i=t}^{0}(I - \alpha M_i)X_0 + \prod_{i=t}^{1}(I - \alpha M_i)W_1 + \cdots + (I - \alpha M_t)W_t + W_{t+1}. \end{aligned} \tag{5}$$

From the last expression it is clear that the matrix products $\prod_{i=t}^{s}(I - \alpha M_i)$ play an important role in the behavior of (1).

Properties of products of random matrices are of interest in a wide range of fields. Application areas include numerical analysis [15,34], statistical physics [9,10], recursive algorithms [11,5,27,17], perturbation theory for dynamical systems [1], queueing theory [23], and even botany [30]. Seminal results are contained in [3,13,29,28].

A complementary and popular research area concerns the eigenstructure of *large* random matrices (see e.g. [33,16] for a recent application to capacity of communication channels). Although the results of the present paper do not address these issues, they provide justification for simplified models in communication theory, leading to bounds on the capacity for time-varying communication channels [24].

The relationship with dynamical systems theory is particularly relevant to the issues addressed here. Consider a *nonlinear* dynamical system described by the equations,

$$X_{t+1} = X_t - f(X_t, \Phi_{t+1}) + W_{t+1}\,, \tag{6}$$

where $\boldsymbol{\Phi} = \{\Phi_t\}$ is an ergodic Markov process, evolving on a state space X, and $f\colon \mathbb{R}^k \times \mathsf{X} \to \mathbb{R}^k$ is smooth and Lipschitz continuous. For this nonlinear model we can construct a random linear model of the form (1) to address many interesting issues. Viewing the initial condition $\gamma = X_0 \in \mathbb{R}^k$ as a continuous variable, we write $X_t(\gamma)$ as the resulting state trajectory, and consider the sensitivity matrix,

$$S_t = \frac{\partial}{\partial\gamma} X_t(\gamma), \qquad t \geq 0\,.$$

From (6) we have the linear recursion,

$$S_{t+1} = [I - M_{t+1}]S_t, \tag{7}$$

where $M_{t+1} = \nabla_x f(X_t, \Phi_{t+1})$, $t \geq 0$. If $\boldsymbol{S} = \{S_t\}$ is suitably stable then the same is true for the nonlinear model, and we find that trajectories couple to a steady state process $\boldsymbol{X}^* = \{X_t^*\}$:

$$\lim_{t\to\infty} \|X_t(\gamma) - X_t^*\| = 0\,.$$

These ideas are related to issues developed in Section 3.

The traditional analytic technique for addressing the stability of (6) or of (1) is the *ODE method* of [22]. For linear models, the basic idea is that, for small values of α, the behavior of (1) should mimic that of the linear ODE,

$$\frac{d}{dt}\gamma_t = -\alpha\overline{M}\gamma_t + \overline{W}\,, \tag{8}$$

where $\overline{M}$ and $\overline{W}$ are steady-state means of M_t and W_t, respectively. To obtain a finer performance analysis one can instead compare (1) to the linear diffusion model,

$$d\Gamma_t = -\alpha\overline{M}\Gamma_t + dB_t, \tag{9}$$

where $\boldsymbol{B} = \{B_t\}$ is a Brownian Motion.

Under certain assumptions one may show that, if the ODE (8) is stable, then the stochastic model (1) is stable in a statistical sense for a range of small α, and comparisons with (9) are possible under still stronger assumptions (see e.g. [4,8,21,20,14] for results concerning both linear and nonlinear recursions).

In [27] an alternative point of view was proposed where the stability verification problem for (1) is cast in terms of the spectral radius of an associated discrete-time semigroup of linear operators. This approach is based on the functional analytic setting of [26], and analogous techniques are used in the treatment of multiplicative ergodic theory and spectral theory in [2,18,19]. The main results of [27] may be interpreted as a significant extension of the ODE method for linear recursions.

Our present results give a unified treatment of both the linear and nonlinear models treated in [27] and [8], respectively.[1] Utilizing the operator-theoretic framework developed in [18] also makes it possible to offer a transparent treatment, and also significantly weaken the assumptions used in earlier results.

We provide answers to the following questions:

(i) For what range of $\alpha > 0$ is the random linear system (1) L_2-stable, in the sense that $\mathrm{E}_x[\|X_t\|^2]$ is bounded in t for any initial condition $\Phi_0 = x \in \mathrm{X}$?

[1] Our results are given here with only brief proof outlines; a more detailed and complete account is in preparation.

(ii) What does the averaged model (8) tell us about the behavior of the original stochastic model?

(iii) What is the impact of variability on *performance* of recursive algorithms?

2 Linear Theory

In this section we develop stability theory and structural results for the linear model (1) where $\alpha \geq 0$ is a fixed constant.

It is assumed that an underlying Markov chain $\boldsymbol{\Phi}$, with general state-space X, governs the statistics of (1) in the sense that $\boldsymbol{M}$ and $\boldsymbol{W}$ are functions of the Markov chain:

$$M_t = m(\Phi_t), \quad W_t = w(\Phi_t), \qquad t \geq 0\,. \tag{10}$$

We assume that the entries of the $k \times k$-matrix valued function m are bounded functions of $x \in \mathrm{X}$. Conditions on the vector-valued function w are given below.

We begin with some basic assumptions on $\boldsymbol{\Phi}$, required to construct a linear operator with useful properties.

2.1 Some spectral theory

We assume throughout that the Markov chain $\boldsymbol{\Phi}$ is *geometrically ergodic* [25,18]. This is equivalent to assuming the validity of the following conditions:

Irreducibility & aperiodicity: There exists a σ-finite measure ψ on the state space X such that, for any $x \in \mathrm{X}$ and any measurable $A \subset \mathrm{X}$ with $\psi(A) > 0$,

$$P^t(x, A) := \mathrm{P}\{\Phi_t \in A \mid \Phi(0) = x\} > 0, \quad \text{for all sufficiently large } t > 0.$$

Minorization: There exists a non-empty set $C \in \mathcal{B}(\mathrm{X})$, a non-zero, positive measure ν on $\mathcal{B}(\mathrm{X})$, and $t_0 \geq 1$ satisfying

$$P^{t_0}(x, A) \geq \nu(A) \qquad x \in C,\ A \in \mathcal{B}(\mathrm{X}).$$

In this case, the set C and the measure ν are called *small.*

Geometric drift: There exists a *Lyapunov function* $V : \mathrm{X} \to [1, \infty)$, $\gamma < 1$, $b < \infty$, a small set C, and small measure ν, satisfying

$$PV(x) := \int P(x, dy)V(y) \leq \gamma V(x) + b\mathbb{I}_C(x), \qquad x \in \mathrm{X} \tag{11}$$

Under these assumptions it is known that there is a unique invariant probability measure π, and the underlying distributions of $\boldsymbol{\Phi}$ converge to π geometrically fast, in total-variation norm. Moreover, in (11) we may assume without loss of generality that $\pi(V^2) := \int V^2(x)\pi(dx) < \infty$. For a detailed development of geometrically ergodic Markov processes see [25,26,18].

Let L_∞^V denote the set of measurable *vector-valued* functions $g\colon \mathsf{X} \to \mathbb{C}^k$ satisfying

$$\|g\|_V := \sup_{x\in\mathsf{X}} \frac{\|g(x)\|}{V(x)} < \infty\,,$$

where $\|\cdot\|$ is the Euclidean norm on $\mathbb{C}^k$, and $V : \mathsf{X} \to [1,\infty)$ is the Lyapunov function as above. For a linear operator $\mathcal{L}\colon L_\infty^V \to L_\infty^V$ we define the induced operator norm via

$$|||\mathcal{L}|||_V := \sup \|\mathcal{L}f\|_V/\|f\|_V$$

where the supremum is over all non-zero $f \in L_\infty^V$. We say that $\mathcal{L}$ is a bounded linear operator if $|||\mathcal{L}|||_V < \infty$, and its spectral radius is then given by

$$\xi := \lim_{t\to\infty} \left(|||\mathcal{L}^t|||\right)^{1/t} \tag{12}$$

The *spectrum* $S(\mathcal{L})$ of the linear operator $\mathcal{L}$ is

$$S(\mathcal{L}) := \{z \in \mathbb{C} : (Iz - \mathcal{L})^{-1} \text{ does not exist as a bdd linear operator on } L_\infty^V\}.$$

If $\mathcal{L}$ is a finite matrix, its spectrum is just the collection of all its eigenvalues. Generally, for the linear operators considered in this paper, the dimension of $\mathcal{L}$ and its spectrum will be infinite.

The family of linear operators $\mathcal{L}_\alpha\colon L_\infty^V \to L_\infty^V$, $\alpha \in \mathbb{R}$, that will be used to analyze the recursion (1) are defined by

$$\begin{aligned}\mathcal{L}_\alpha f(x) &:= \mathrm{E}\left[(I - \alpha m(\Phi_1))^{\tau} f(\Phi_1) \mid \Phi_0 = x\right] \\ &= \mathrm{E}_x\left[(I - \alpha M_1)^{\tau} f(\Phi_1)\right]\,,\end{aligned} \tag{13}$$

and we let ξ_α denote the spectral radius of $\mathcal{L}_\alpha$. The motivation for (13) comes from the representation (5), and the following expression for the iterates of this semigroup:

$$\mathcal{L}_\alpha^t f\,(x) = \mathrm{E}_x\left[(I - \alpha M_1)^{\tau} \cdots (I - \alpha M_t)^{\tau} f(\Phi_t)\right]\,, \qquad t \geq 1\,. \tag{14}$$

The transpose ensures that the matrices are multiplied in order consistent with (5).

We assume throughout the paper that $m\colon \mathsf{X} \to \mathbb{R}^{k\times k}$ is a bounded function. Under these conditions we obtain the following result as in [27].

Theorem 1. *There exists* $\alpha_0 > 0$ *such that for* $\alpha \in (0, \alpha_0)$, $\xi_\alpha < \infty$, *and* $\xi_\alpha \in S(\mathcal{L}_\alpha)$. □

To ensure that the recursion (1) is stable in the mean, it is sufficient that the spectral radius satisfy $\xi_\alpha < 1$. Under this condition it is obvious that the mean $\mathrm{E}[X_t]$ is uniformly bounded in t (see (14)). The following result summarizes additional conclusions obtained below.

Theorem 2. *Suppose that the eigenvalues of* $\overline{M} := \int m(x)\, \pi(dx)$ *are all positive, and that* $w^2 \in L_\infty^V$, *where the square is interpreted component-wise. Then, there exists a bounded open set* $O \in \mathbb{R}$ *containing* $(0, \alpha_0)$, *where* α_0 *is given in Theorem 1, such that:*

(i) *For all* $\alpha \in O$ *we have* $\xi_\alpha < 1$, *and for any initial condition* $\Phi_0 = x \in \mathrm{X}$, $X_0 = \gamma \in \mathbb{R}^k$,

$$\mathrm{E}_x[\|X_t\|^2] \to \tau_\alpha^2, \quad \textit{geometrically fast, as } t \to \infty,$$

for a finite constant τ_α^2.

(ii) *If* $\boldsymbol{\Phi}$ *is stationary, then for* $\alpha \in O$ *there exists a stationary process* $\boldsymbol{X}^\alpha$ *such that for any initial condition* $\Phi_0 = x \in \mathrm{X}$, $X_0 = \gamma \in \mathbb{R}^k$,

$$\mathrm{E}_x[\|X_t(\gamma) - X_t^\alpha\|^2] \to 0, \quad \textit{geometrically fast, as } t \to \infty.$$

(iii) *If* $\alpha \notin \overline{O}$ *and the noise* $\boldsymbol{W}$ *is i.i.d. with a positive definite covariance matrix, then* $\mathrm{E}_x[\|X_t\|^2]$ *is unbounded.*

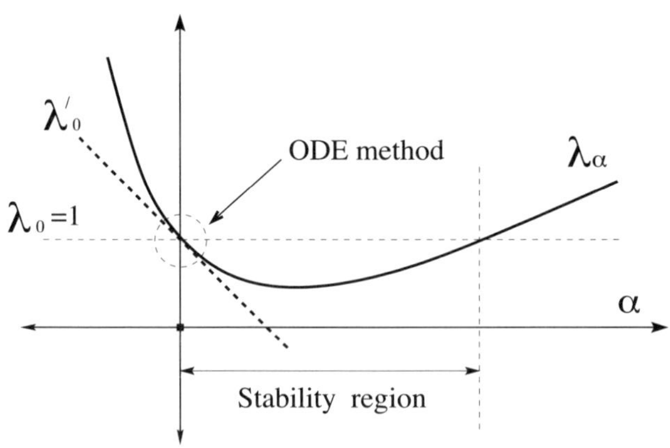

Fig. 1. The graph shows how $\lambda_\alpha := \xi_\alpha$ varies with α. When α is close to 0, Theorem 4 below implies that the ODE (8) determines stability of the algorithm since it determines whether or not $\xi_\alpha < 1$. A second-derivative formula is also given in Theorem 4: If λ_0'' is large, then the range of α for stability will be correspondingly small.

PROOF OUTLINE FOR THEOREM 2 Starting with (5), we may express the expectation $\mathrm{E}_x\left[X_{t+1}^T X_{t+1}\right]$ as a sum of terms of the form,

$$\mathrm{E}_x\left[W_j^T\left(\prod_{i=t}^{j}(I-\alpha M_i)\right)^T\left(\prod_{i=t}^{k}(I-\alpha M_i)\right)W_k\right], \quad j,k=0,\dots,t. \tag{15}$$

For simplicity consider the case $j = k$. Taking conditional expectations at time j, one can then express the expectation (15) as

$$\operatorname{trace}\left(\mathrm{E}_x\left[(\mathcal{Q}_\alpha^{t-j}h\,(\Phi_j))w(\Phi_j)w(\Phi_j)^T\right]\right)$$

where $\mathcal{Q}_\alpha$ is defined in (20), and $h \equiv I_{k\times k}$. We define O as the set of α such that the spectral radius of this linear operator is strictly less than unity. Thus, for $\alpha \in O$ we have, for some $\eta_\alpha < 1$,

$$\operatorname{trace}\left((\mathcal{Q}_\alpha^{t-j}h\,(y))w(y)w(y)^T\right) = O(V(y)^2 e^{-\eta_\alpha(t-j)}), \qquad \Phi_j = y \in \mathsf{X}.$$

Similar reasoning may be applied for arbitrary k, j, and this shows that $\mathrm{E}_x[\|X_t\|^2]$ is bounded in $t \geq 0$ for any deterministic initial conditions $\Phi_0 = x \in \mathsf{X}$, $X_0 = \gamma \in \mathbb{R}^k$.

To construct the stationary process X^α we apply backward coupling as presented in [32]. Consider the system starting at time $-n$, initialized at $\gamma = 0$, and let $X_t^{\alpha,n}$, $t \geq -n$, denote the resulting state trajectory. We then have for all $n, m \geq 1$,

$$X_t^{\alpha,m} - X_t^{\alpha,n} = \left(\prod_{i=t}^{0}(I-\alpha M_i)\right)[X_0^{\alpha,m} - X_0^{\alpha,n}], \qquad t \geq 0,$$

which implies convergence in L_2 to a stationary process: $X_t^\alpha := \lim_{n\to\infty} X_t^{\alpha,n}$, $t \geq 0$. We can then compare to the process initialized at $t = 0$,

$$X_t^\alpha - X_t(\gamma) = \left(\prod_{i=t}^{0}(I-\alpha M_i)\right)[X_0^\alpha - X_0(\gamma)], \qquad t \geq 0,$$

and the same reasoning as before gives (ii). □

2.2 Spectral decompositions

Next we show that $\lambda_\alpha := \xi_\alpha$ is in fact an eigenvalue of $\mathcal{L}_\alpha$ for a range of $\alpha \sim 0$, and we use this fact to obtain a multiplicative ergodic theorem. The maximal eigenvalue λ_α in Theorem 3 is a generalization of the Perron-Frobenius eigenvalue; c.f. [31,18].

Theorem 3. *Suppose that the eigenvalues $\{\lambda_i(\overline{M})\}$ of $\overline{M}$ are distinct, then*

(i) *There exists* $\varepsilon_0 > 0, \delta_0 > 0$ *such that the linear operator* $\mathcal{L}_z$ *has exactly* k *distinct eigenvalues* $\{\lambda_{1,z}, \dots, \lambda_{k,z}\} \subset S(\mathcal{L}_z)$ *within the restricted range,*

$$B_1(\delta_0) = \{\lambda \in S(\mathcal{L}_z) : |\lambda - 1| < \delta_0\},$$

whenever z *lies in the open ball* $B_0(\varepsilon_0) := \{z \in \mathbb{C} : |z| < \varepsilon_0\}$*. The ith eigenvalue* $\lambda_{i,z}$ *is an analytic function of* z *on this domain for each* i*.*

(ii) *For* $z \in B(\varepsilon_0)$ *there are associated eigenfunctions* $\{h_{1,z}, \dots, h_{k,z}\} \subset L_\infty^V$ *and eigenmeasures* $\{\mu_{1,z}, \dots, \mu_{k,z}\}$ *satisfying*

$$\mathcal{L}_z h_{i,z} = \lambda_{i,z} h_{i,z}, \qquad \mu_{i,z}\mathcal{L}_z = \lambda_{i,z}\mu_{i,z}\,.$$

Moreover, for each i, $x \in \mathrm{X}$, $A \in \mathcal{B}(\mathrm{X})$, $\{h_{i,z}(x), \mu_{i,z}(A)\}$ *are analytic on* $B(\varepsilon_0)$.

(iii) *Suppose moreover that the eigenvalues* $\{\lambda_i(\overline{M})\}$ *are real. Then we may take* $\varepsilon_0 > 0$ *sufficiently small so that* $\{\lambda_{i,\alpha}, h_{i,\alpha}, \mu_{i,\alpha}\}$ *are real for* $\alpha \in (0, \varepsilon_0)$*. The maximal eigenvalue* $\lambda_\alpha := \max_i \lambda_{i,\alpha}$ *is equal to* ξ_α*, and the corresponding eigenfunction and eigenmeasure may be scaled so that the following limit holds,*

$$\lambda_\alpha^{-t}\mathcal{L}_\alpha^t \to h_\alpha \otimes \mu_\alpha, \qquad t \to \infty,$$

where the convergence is in the V*-norm.*
In fact, there exists $\delta_0 > 0$ *and* $b_0 < \infty$ *such that for any* $f \in L_\infty^V$ *the following limit holds:*

$$\left| \lambda_\alpha^{-t} \mathrm{E}_x\Big[\Big(\prod_{i=1}^{t}(I - \alpha M_i)\Big)^{T} f(\Phi_t)\Big] - h_\alpha(x)\mu_\alpha(f)\right| \le b_0 e^{-\delta_0 t} V(x)\,.$$

PROOF. The linear operator $\mathcal{L}_0$ possesses a k-dimensional eigenspace corresponding to the eigenvalue $\lambda_0 = 1$. This eigenspace is precisely the set of constant functions, with a corresponding basis of eigenfunctions given by $\{e^i\}$, where e^i is the ith basis element in $\mathbb{R}^k$. The k-dimensional set of vector-valued eigenmeasures $\{\pi^i\}$ given by $\pi^i = {e^i}^T\pi$ spans the set of all eigenmeasures with eigenvalue $\lambda_{0,i} = 1$.

Consider the rank-k linear operator $\Pi\colon L_\infty^V \to L_\infty^V$ defined by $\Pi f := \pi(f)$. This is equivalently expressed as

$$\Pi f\,(x) = (\pi(f_1), \dots, \pi(f_k))^T = \Big[\sum e^i \otimes \pi^i\Big] f, \qquad f \in L_\infty^V.$$

It is obvious that $\Pi\colon L_\infty^V \to L_\infty^V$ is a rank-k linear operator, and for $\alpha = 0$ we have from the V-uniform ergodic theorem of [25],

$$\mathcal{L}_0^t - \Pi = [\mathcal{L}_0 - \Pi]^t \to 0, \qquad t \to \infty,$$

where the convergence is in norm, and hence takes place exponentially fast. It follows that the spectral radius of $(\mathcal{L}_0 - \Pi)$ is strictly less than unity. By standard arguments it follows that, for some $\varepsilon_0 > 0$, the spectral radius of $\mathcal{L}_z - \Pi$ is also strictly less than unity. The results then follow as in Theorem 3 of [19]. □

Conditions under which the bound $\xi_\alpha < 1$ is satisfied are given in Theorem 4, where we also provide formulae for the derivatives of λ_α:

Theorem 4. *Suppose that the eigenvalues $\{\lambda_i(\overline{M})\}$ are real and distinct, then the maximal eigenvalue $\lambda_\alpha = \xi_\alpha$ satisfies:*

(i) $\frac{d}{d\alpha}\lambda_\alpha\Big|_{\alpha=0} = -\lambda_{\min}(\overline{M})$.

(ii) *The second derivative is given by,*

$$\frac{d^2}{d\alpha^2}\lambda_\alpha\Big|_{\alpha=0} = 2\sum_{l=0}^{\infty} v_0^T \mathrm{E}_\pi[(M_0 - \overline{M})(M_{l+1} - \overline{M})]r_0\,,$$

where r_0 is a right eigenvector of $\overline{M}$ corresponding to $\lambda_{\min}(\overline{M})$, and v_0 is the left eigenvector, normalized so that $v_0^T r_0 = 1$.

(iii) *Suppose that $m(x) = m^T(x)$, $x \in \mathrm{X}$, then we may take $v_0 = r_0$ in (ii), and the second derivative may be expressed,*

$$\frac{d^2}{d\alpha^2}\lambda_\alpha\Big|_{\alpha=0} = \mathrm{trace}\,(\Gamma - \Sigma),$$

where an Γ is the Central Limit Theorem covariance for the stationary vector-valued stochastic process $F_k = [M_k - \overline{M}]v_0$, and $\Sigma = \mathrm{E}_\pi[F_k F_k^T]$ is its variance [25].

Proof. To prove (i), we differentiate the eigenfunction equation $\mathcal{L}_\alpha h_\alpha = \lambda_\alpha h_\alpha$ to obtain

$$\mathcal{L}_\alpha{}' h_\alpha + \mathcal{L}_\alpha h'_\alpha = \lambda'_\alpha h_\alpha + \lambda_\alpha h'_\alpha. \tag{16}$$

Setting $\alpha = 0$ then gives a version of *Poisson's equation*,

$$\mathcal{L}'_0 h_0 + P h'_0 = \lambda'_0 h_0 + h'_0, \tag{17}$$

where $\mathcal{L}'_0 h_0 = \mathrm{E}_x\left[-m(\Phi_1)^T h_0(\Phi_1)\right]$. An application of Theorem 3 (ii) shows that $h'_0 \in L^V_\infty$, which justifies an integration of both sides of (17) with respect to the invariant probability π to obtain

$$\mathrm{E}_\pi\left[-m(\Phi_1)^T\right] h_0 = -\overline{M}^T h_0 = \lambda'_0 h_0.$$

This shows that λ'_0 is an eigenvalue of $-\overline{M}$, and h_0 is an associated eigenvector for $\overline{M}^T$. It follows that $\lambda'_0 = -\lambda_{\min}(\overline{M})$ by maximality of λ_α.

We note that Poisson's equation (17) combined with equation (17.39) of [25] implies the formula,

$$h'_0(x) = \mathrm{E}_\pi[h'_0(\Phi(0))] - \sum_{l=0}^{\infty} \mathrm{E}_x[(M_{l+1} - \overline{M})^T] h_0\,. \tag{18}$$

To prove (ii) we consider the second-derivative formula,

$$\mathcal{L}''_\alpha h_\alpha + 2\mathcal{L}'_\alpha h'_\alpha + \mathcal{L}_\alpha h''_\alpha = \lambda_\alpha h''_\alpha + 2\lambda'_\alpha h'_\alpha + \lambda''_\alpha h_\alpha .$$

Evaluating these expressions at $\alpha = 0$ and integrating with respect to π then gives the steady state expression,

$$\lambda''_0 h_0 = -2\mathrm{E}_\pi[(M_1 + \lambda'_0)h'_0(\Phi_1)]. \tag{19}$$

In deriving this identity we have used the expressions,

$$\mathcal{L}'_0 f(x) = \mathrm{E}_x[M_1 f(\Phi_1)], \quad \mathcal{L}''_0 f(x) = 0, \qquad f \in L^V_\infty,\ x \in \mathrm{X}.$$

This combined with (19) gives the desired formula since we may take $v_0 = h_0$ in (ii).

To prove (iii) we simply note that in the symmetric case the formula in (ii) becomes,

$$\lambda''_0 = \sum_{k\neq 0} \mathrm{E}_\pi[\|F_k\|^2] = \operatorname{trace}(\Gamma - \Sigma). \qquad \square$$

2.3 Second-order statistics

In order to understand the second-order statistics of $\boldsymbol{X}$ it is convenient to introduce another linear operator $\mathcal{Q}_\alpha$ as follows,

$$\begin{aligned}\mathcal{Q}_\alpha f(x) &= \mathrm{E}\left[(I - \alpha m(\Phi_1))^T f(\Phi_1)(I - \alpha m(\Phi_1)) | \Phi_0 = x\right] \\ &= \mathrm{E}_x\left[(I - \alpha M_1)^T f(\Phi_1)(I - \alpha M_1)\right],\end{aligned} \tag{20}$$

where the domain of $\mathcal{Q}_\alpha$ is the collection of matrix-valued functions $f\colon \mathrm{X} \to \mathbb{C}^{k\times k}$. When considering $\mathcal{Q}_\alpha$ we redefine L^V_∞ accordingly. It is clear that $\mathcal{Q}_\alpha : L^V_\infty \to L^V_\infty$ is a bounded linear operator under the geometric drift condition and the boundedness assumption on m.

Let ξ^Q_z denote the spectral radius of $\mathcal{Q}_\alpha$. We can again argue that ξ^Q_z is smooth in a neighborhood of the origin, and the following follows as in Theorem 4:

Theorem 5. *Assume that the eigenvalues of $\overline{M}$ are real and distinct. Then there exists $\varepsilon_0 > 0$ such that for each $z \in B(\varepsilon_0)$ there exists an eigenvalue $\eta_z \in \mathbb{C}$ for $\mathcal{Q}_z$ satisfying $|\eta_z| = \xi^Q_z$, and η_α is real for real $\alpha \in (0, \varepsilon_0)$. The eigenvalue η_z is smooth on $B(\varepsilon_0)$ and satisfies,*

$$\eta'_0(\mathcal{Q}) = -2\lambda_{\min}(\overline{M}).$$

PROOF. This is again based on differentiation of the eigenfunction equation given by $\mathcal{Q}_\alpha h_\alpha = \eta_\alpha h_\alpha$, where η_α and h_α are the eigenvalue and matrix-valued eigenfunction, respectively. Taking derivatives on both sides gives

$$\mathcal{Q}_\alpha{}' h_\alpha + \mathcal{Q}_\alpha h'_\alpha = \eta'_\alpha h_\alpha + \eta_\alpha h'_\alpha \tag{21}$$

where $\mathcal{Q}'_0 h_0 = \mathrm{E}_x\left[-m(\Phi_1)^T h_0(\Phi_1) - h_0(\Phi_1) m(\Phi_1)\right]$. As before, we then obtain the steady-state expression,

$$\mathrm{E}_\pi\left[-m(\Phi_1)^T h_0 - h_0 m(\Phi_1)\right] = -\overline{M}^T h_0 - h_0 \overline{M} = \eta'_0 h_0. \tag{22}$$

And, as before, we may conclude that $\eta'_0 = 2\lambda'_0 = -2\lambda_{\min}(\overline{M})$. □

2.4 An illustrative example

Consider the discrete-time, linear time-varying model

$$y_t = \theta_t^T \phi_t + n_t, \qquad t \geq 0\,, \tag{23}$$

where $\boldsymbol{y} = \{y_t\}$ is a sequence of scalar observations, $\boldsymbol{n} = \{n_t\}$ is a noise process, $\boldsymbol{\theta} = \{\theta_t\}$ is the sequence of k-dimensional regression vectors, and $\boldsymbol{\phi}=\{\phi_t\}$ are k-dimensional time-varying parameters. In this section we illustrate the results above using the LMS (least mean square) parameter estimation algorithm,

$$\widehat{\theta}_{t+1} = \widehat{\theta}_t + \alpha\phi_t e_t,$$

where $\boldsymbol{e} = \{e_t\}$ is the error sequence, $e_t := y_t - \widehat{\theta}_t^T \phi_t$, $t \geq 0$.

As in the Introduction, writing $\widetilde{\theta}_t = \theta_t - \widehat{\theta}_t$ we obtain

$$\tilde{\theta}_{t+1} = (I - \alpha\phi_t\phi_t^T)\tilde{\theta}_t + \left[\theta_{t+1} - \theta_t - \alpha\phi_t n_t\right]\,.$$

This is of the form (1) with $X_t = \tilde{\theta}_t$, $M_t = \phi_t\phi_t^T$ and $W_{t+1} = \theta_{t+1} - \theta_t - \alpha\phi_t n_t$.

For the sake of simplicity and to facilitate explicit numerical calculations, we consider the following special case: We assume that $\boldsymbol{\phi}$ is of the form $\phi_t = (s_t, s_{t-1})^T$, where the sequence $\boldsymbol{s}$ is Bernoulli ($s_t = \pm 1$ with equal probability) and take $\boldsymbol{n}$ to be an i.i.d. noise sequence.

In analyzing the random linear system we may ignore the noise $\boldsymbol{n}$ and take $\boldsymbol{\Phi} = \boldsymbol{\phi}$. This is clearly geometrically ergodic since it is an ergodic, finite state space Markov chain, with four possible states. In fact, $\boldsymbol{\Phi}$ is geometrically ergodic with Lyapunov function $V \equiv 1$. In the case $k = 2$, viewing $h \in L_\infty^V$ as a real vector, the eigenfunction equation for $\mathcal{L}_\alpha$ becomes

$$\mathcal{L}_\alpha h_\alpha = \frac{1}{2}\begin{bmatrix} A_1 \; A_0 \; A_2 \; A_0 \\ \\ A_1 \; A_0 \; A_2 \; A_0 \\ \\ A_0 \; A_2 \; A_0 \; A_1 \\ \\ A_0 \; A_2 \; A_0 \; A_1 \end{bmatrix} h_\alpha = \lambda_\alpha h_\alpha \tag{24}$$

where $A_0 = \begin{bmatrix} 0 & 0 \\ 0 & 0 \end{bmatrix}$, $A_1 = \begin{bmatrix} 1-\alpha & -\alpha \\ -\alpha & 1-\alpha \end{bmatrix}$, $A_2 = \begin{bmatrix} 1-\alpha & \alpha \\ \alpha & 1-\alpha \end{bmatrix}$.

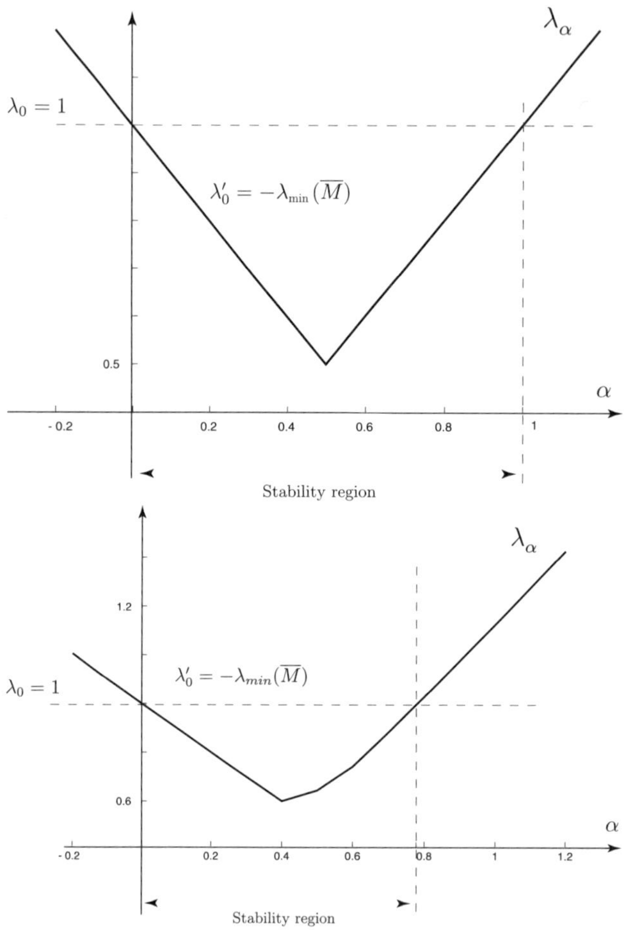

Fig. 2. The figure on the left shows the Perron-Frobenius eigenvalue $\lambda_\alpha = \xi_\alpha$ for the LMS model with $\phi_t = (s_t, s_{t-1})^T$. The figure on the right shows the case where $\phi_t = (s_t, s_{t-1}, s_{t-2})^T$. In both cases, the sequence $\boldsymbol{s}$ is i.i.d. Bernoulli.

In this case, we have the following local behavior:

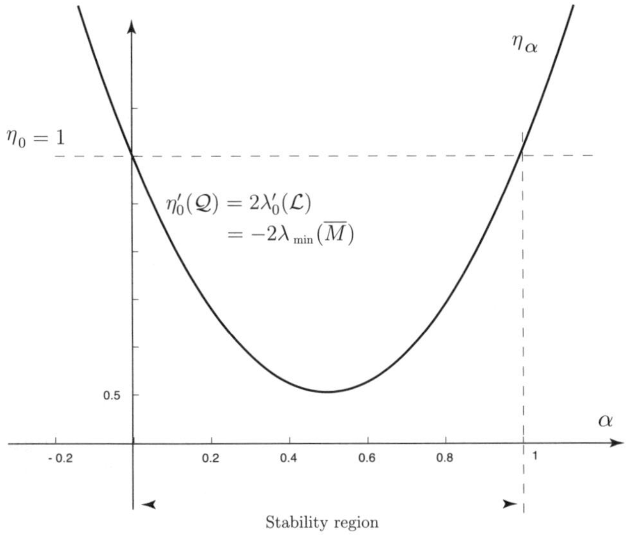

Fig. 3. The maximal eigenvalues $\eta_\alpha = \xi_\alpha^Q$ are piecewise quadratic in α in the case where $\phi_t = (s_t, s_{t-1})^T$ with $\boldsymbol{s}$ as above.

Theorem 6. *In a neighborhood of* 0*, the spectral radii of* $\mathcal{L}_\alpha$*,* $\mathcal{Q}_\alpha$ *satisfy*

$$\frac{d}{d\alpha}\xi_\alpha\Big|_{\alpha=0} = -\lambda_{\min}(\overline{M}) = -1; \qquad \frac{d}{d\alpha}\xi_\alpha^Q\Big|_{\alpha=0} = -2\lambda_{\min}(\overline{M}) = -2$$

$$\frac{d^n}{d\alpha^n}\xi_\alpha\Big|_{\alpha=0} = 0, n \geq 2; \qquad \frac{d^n}{d\alpha^n}\xi_\alpha^Q\Big|_{\alpha=0} = 0, n \geq 3.$$

So λ_α *and* η_α *are linear and quadratic around* 0*, respectively.*

Proof. This follows from differentiating the respective eigenfunction equations. Here we only show the proof for operator $\mathcal{Q}$; the proof for operator $\mathcal{L}$ is similar.

Taking derivatives on both sides of the eigenfunction equation for $\mathcal{Q}_\alpha$ gives,

$$\mathcal{Q}_\alpha{}' h_\alpha + \mathcal{Q}_\alpha h_\alpha' = \eta_\alpha' h_\alpha + \eta_\alpha h_\alpha' \tag{25}$$

Setting $\alpha = 0$ gives a version of *Poisson's equation*,

$$\mathcal{Q}_0' h_0 + \mathcal{Q} h_0' = \eta_0' h_0 + \eta_0 h_0' \tag{26}$$

Using the identities of h_0 and $\mathcal{Q}_0' h_0 = \mathrm{E}_x[-M_1^T h_0 - h_0 M_1]$, we obtain the steady state expression

$$\overline{M}^T h_0 + h_0 \overline{M} = -\eta_0' h_0. \tag{27}$$

Since $\overline{M} = I$, we have $\eta_0' = -2$. Now, taking the 2nd derivatives on both sides of (25) gives,

$$\mathcal{Q}_\alpha'' h_\alpha + 2\mathcal{Q}_\alpha' h_\alpha' + \mathcal{Q}_\alpha h_\alpha'' = \eta_\alpha'' h_\alpha + 2\eta_\alpha' h_\alpha' + \eta_\alpha h_\alpha''. \tag{28}$$

Letting $\alpha = 0$ and considering the steady state, we obtain

$$2\overline{M}^T h_0 \overline{M} - 2\mathrm{E}_\pi[M_1^T h_0' + h_0' M_1] = \eta_0'' h_0 + 2\eta_0' \mathrm{E}_\pi[h_0']. \tag{29}$$

Poisson's equation (26) combined with equation (27) and equation (17.39) of [25] implies the formula,

$$\begin{aligned} h_0'(x) &= \mathrm{E}_\pi(h_0') + \textstyle\sum_{l=0}^\infty \mathrm{E}_x[-M_{l+1}^T h_0 - h_0 M_{l+1} - \eta_0' h_0] \\ &= \mathrm{E}_\pi(h_0') + \textstyle\sum_{l=0}^\infty \mathrm{E}_x[(\overline{M} - M_{l+1})^T h_0 + h_0(\overline{M} - M_{l+1})]. \end{aligned} \tag{30}$$

So, from $\overline{M} = I$, $\eta_0' = -2$ and (29) we have $\eta_0'' = 2$. In order to show η_α is quadratic near zero, we take the 3rd derivative on both sides of (28) and consider the steady state at $\alpha = 0$,

$$\mathcal{Q}_0''' h_0 + 3\mathcal{Q}_0'' h_0' + 3\mathcal{Q}_0' h_0'' + \mathcal{Q}_0 h_0''' = \eta_0''' h_0 + 3\eta_0'' h_0' + 3\eta_0' h_0'' + \eta_0 h_0'''. \tag{31}$$

With equation (17.39) of [25] and $\eta_0' = -2$ and $\eta_0'' = 2$, we can show $\eta_0''' = 0$ and $\eta_0^{(n)} = 0$ for $n > 3$, hence η_α is quadratic around 0. □

3 Nonlinear Models

We now turn to the nonlinear model shown in (6). We take the special form,

$$X_{t+1} = X_t - \alpha[f(X_t, \Phi_{t+1}) + W_{t+1}], \tag{32}$$

We continue to assume that $\boldsymbol{\Phi}$ is geometrically ergodic, and that $W_t = w(\Phi_t)$, $t \geq 0$, with $w^2 \in L_\infty^V$. The associated ODE is given by

$$\frac{d}{dt}\gamma_t = \overline{f}(\gamma_t), \tag{33}$$

where $\overline{f}(\gamma) = \int f(\gamma, x)\, \pi(dx)$, $\gamma \in \mathbb{R}^k$.

We assume that $\overline{W} = \mathrm{E}_\pi[W_1] = 0$, and the following conditions are imposed on f. The function $\overline{f}_\infty$ appearing in Condition (N1) may be used to construct an ODE that approximates the behavior of (32) when the initial condition is very large.

(N1) The function f is Lipschitz, and there exists a function $\overline{f}_\infty : \mathbb{R}^k \to \mathbb{R}^k$ such that

$$\lim_{r\to\infty} r^{-1}\overline{f}(r\gamma) = \overline{f}_\infty(\gamma), \qquad \gamma \in \mathbb{R}^k.$$

Furthermore, the origin in $\mathbb{R}^k$ is an asymptotically stable equilibrium point for the ODE,

$$\frac{d}{dt}\gamma_t^\infty = \overline{f}_\infty(\gamma_t^\infty). \tag{34}$$

(N2) There exists $b_f < \infty$ such that $\sup_{\gamma\in\mathbb{R}^k} \|f(\gamma,x)-\overline{f}(\gamma)\|^2 \le b_f V(x),\ x \in \mathrm{X}$.

(N3) There exists a unique stationary point x^* for the ODE (33) that is a globally asymptotically stable equilibrium.

Define the absolute error by

$$\varepsilon_t := \|X_t - x^*\|, \qquad t \ge 0. \tag{35}$$

The following result is an extension of Theorem 1 of [8] to Markov models:

Theorem 7. *Assume that (N1)–(N3) hold. Then there exists $\varepsilon_0 > 0$ such that for any $0 < \alpha < \varepsilon_0$:*

(i) *For any $\delta > 0$, there exists $b_1 = b_1(\delta) < \infty$ such that*

$$\limsup_{n\to\infty} \mathrm{P}(\varepsilon_n \ge \delta) \le b_1\alpha.$$

(ii) *If the origin is a globally exponentially asymptotically stable equilibrium for the ODE (33), then there exists $b_2 < \infty$ such that for every initial condition $\Phi_0 = x \in \mathrm{X}$, $X_0 = \gamma \in \mathbb{R}^k$,*

$$\limsup_{n\to\infty} \mathrm{E}[\varepsilon_n^2] \le b_2\alpha.$$

Proof Outline for Theorem 7 The continuous-time process $\{x_t^o : t \ge 0\}$ is defined to be the interpolated version of $\boldsymbol{X}$ given as follows: Let $T_j = j\alpha$, $j \ge 0$, and define $x^o(T_j) = \alpha X_j$, with $\boldsymbol{x}^o$ defined by linear interpolation on the remainder of $[T_j, T_{j+1}]$ to form a piecewise linear function. Using geometric ergodicity we can bound the error between $\boldsymbol{x}^o$ and solutions to the ODE (33) as in [8], and we may conclude that the joint process $(\boldsymbol{X}, \boldsymbol{\Phi})$ is geometrically ergodic with Lyapunov function $V_2(\gamma, x) = \|\gamma\|^2 + V(x)$. □

We conclude with an extension of Theorem 2 describing the behavior of the sensitivity process $\boldsymbol{S}$.

Theorem 8. *Assume that (N1)–(N3) hold, and that the eigenvalues of the matrix $\overline{M}$ have strictly positive real part, where*

$$\overline{M} := \nabla\overline{f}\,(x^*)\,.$$

Then there exists $\varepsilon_1 > 0$ such that for any $0 < \alpha < \varepsilon_1$, the conclusions of Theorem 7 (ii) hold, and, in addition:

(i) *The spectral radius ξ_α of the random linear system (7) describing the evolution of the sensitivity process is strictly less than one.*

(ii) *There exists a stationary process $\boldsymbol{X}^\alpha$ such that for any initial condition $\Phi_0 = x \in \mathrm{X}$, $X_0 = \gamma \in \mathbb{R}^k$,*

$$\mathrm{E}_x[\|X_t - X_t^\alpha\|^2] \to 0, \qquad t \to \infty .$$

References

1. L. Arnold. *Random dynamical systems.* Springer-Verlag, Berlin, 1998.
2. S. Balaji and S.P. Meyn. Multiplicative ergodicity and large deviations for an irreducible Markov chain. *Stochastic Process. Appl.*, 90(1):123–144, 2000.
3. R. Bellman. Limit theorems for non-commutative operations. I. *Duke Math. J.*, 21, 1954.
4. Michel Benaïm. Dynamics of stochastic approximation algorithms. In *Séminaire de Probabilités, XXXIII*, pages 1–68. Springer, Berlin, 1999.
5. Albert Benveniste, Michel Métivier, and Pierre Priouret. *Adaptive algorithms and stochastic approximations.* Springer-Verlag, Berlin, 1990. Translated from the French by Stephen S. Wilson.
6. D.P. Bertsekas and J. Tsitsiklis. *Neuro-Dynamic Programming.* Atena Scientific, Cambridge, Mass, 1996.
7. B. Bharath and V. S. Borkar. Stochastic approximation algorithms: overview and recent trends. *Sādhanā*, 24(4-5):425–452, 1999. Chance as necessity.
8. V.S. Borkar and S.P. Meyn. The O.D.E. Method for Convergence of Stochastic Approximation and Reinforcement Learning. *SIAM J. Control Optim.*, 38:447–69, 2000.
9. P. Bougerol. Limit theorem for products of random matrices with Markovian dependence. In *Proceedings of the 1st World Congress of the Bernoulli Society, Vol. 1 (Tashkent, 1986)*, pages 767–770, Utrecht, 1987. VNU Sci. Press.
10. A. Crisanti, G. Paladin, and A. Vulpiani. *Products of random matrices in statistical physics.* Springer-Verlag, Berlin, 1993.
11. O. Dabeer and E. Masry. The LMS adaptive algorithm: Asymptotic error analysis. In *Proceedings of the 34th Annual Conference on Information Sciences and Systems, CISS 2000*, pages WP1–6 – WP1–7, Princeton, NJ, March 2000.
12. Paul Fischer and Hans Ulrich Simon, editors. *Computational learning theory*, Berlin, 1999. Springer-Verlag. Lecture Notes in Artificial Intelligence.
13. H. Furstenberg and H. Kesten. Products of random matrices. *Ann. Math. Statist*, 31:457–469, 1960.
14. László Gerencsér. Almost sure exponential stability of random linear differential equations. *Stochastics Stochastics Rep.*, 36(2):91–107, 1991.
15. R. Gharavi and V. Anantharam. Structure theorems for partially asynchronous iterations of a nonnegative matrix with random delays. *Sādhanā*, 24(4-5):369–423, 1999. Chance as necessity.
16. S.V. Hanly and D. Tse. Multiaccess fading channels. II. Delay-limited capacities. *IEEE Trans. Inform. Theory*, 44(7):2816–2831, 1998.
17. J. A. Joslin and A. J. Heunis. Law of the iterated logarithm for a constant-gain linear stochastic gradient algorithm. *SIAM J. Control Optim.*, 39(2):533–570 (electronic), 2000.

18. I. Kontoyiannis and S.P. Meyn. Spectral theory and limit theorems for geometrically ergodic Markov processes. *Submitted*, 2001. Also presented at the 2001 INFORMS Applied Probability Conference, NY, July, 2001.
19. I. Kontoyiannis and S.P. Meyn. Spectral theory and limit theorems for geometrically ergodic Markov processes. Part II: Empirical measures & unbounded functionals. *Preprint*, 2001.
20. H. J. Kushner. *Approximation and weak convergence methods for random processes, with applications to stochastic systems theory.* MIT Press, Cambridge, MA, 1984.
21. H.J. Kushner and G. Yin. *Stochastic approximation algorithms and applications.* Springer-Verlag, New York, 1997.
22. L. Ljung. On positive real transfer functions and the convergence of some recursive schemes. *IEEE Trans. Automatic Control*, AC-22(4):539–551, 1977.
23. Jean Mairesse. Products of irreducible random matrices in the $(\max, +)$ algebra. *Adv. in Appl. Probab.*, 29(2):444–477, 1997.
24. M. Medard, S.P. Meyn, and J. Huang. Capacity benefits from channel sounding in Rayleigh fading channels. *INFOCOM (submitted)*, 2001.
25. S. P. Meyn and R. L. Tweedie. *Markov Chains and Stochastic Stability.* Springer-Verlag, London, 1993.
26. S.P. Meyn and R.L. Tweedie. Computable bounds for geometric convergence rates of Markov chains. *Ann. Appl. Probab.*, 4(4):981–1011, 1994.
27. G.V. Moustakides. Exponential convergence of products of random matrices, application to the study of adaptive algorithms. *International Journal of Adaptive Control and Signal Processing*, 2(12):579–597, 1998.
28. V. I. Oseledec. Markov chains, skew products and ergodic theorems for "general" dynamic systems. *Teor. Verojatnost. i Primenen.*, 10:551–557, 1965.
29. V. I. Oseledec. A multiplicative ergodic theorem. Characteristic Ljapunov, exponents of dynamical systems. *Trudy Moskov. Mat. Obšč.*, 19:179–210, 1968.
30. J. B. T. M Roerdink. Products of random matrices or "why do biennials live longer than two years? *CWI Quarterly*, 2:37–44, 1989.
31. E. Seneta. *Non-negative Matrices and Markov Chains.* Springer-Verlag, New York, Second edition, 1980.
32. H. Thorisson. *Coupling, stationarity, and regeneration.* Springer-Verlag, New York, 2000.
33. D. Tse and S.V. Hanly. Multiaccess fading channels. I. Polymatroid structure, optimal resource allocation and throughput capacities. *IEEE Trans. Inform. Theory*, 44(7):2796–2815, 1998.
34. Divakar Viswanath. Random Fibonacci sequences and the number 1.13198824 .... *Math. Comp.*, 69(231):1131–1155, 2000.

Sign-Regressor Adaptive Filtering Algorithms Using Averaged Iterates and Observations

Dedicated to Professor Tyrone Duncan on the Occasion of His 60th Birthday

C. Ion[1], G. Yin[1], and V. Krishnamurthy[2]

[1] Wayne State University, Detroit, MI 48202, USA
[2] University of Melbourne, Parkville, Victoria 3052, Australia

Abstract. Motivated by the resurgent interest in efficient adaptive signal processing algorithms for interference suppression in wireless CDMA (Code Division Multiple Access) communication networks, this paper is concerned with asymptotic properties of adaptive filtering algorithms. Our focus is on improving efficiency of sign-regressor procedures, which are known to have reduced complexity compared with the usual LMS algorithms and better performance compared with the sign-error procedures. In view of the recent developments in iterate averaging for stochastic approximation methods, algorithms that include both iterate and observation averaging are suggested. It is shown that such algorithms converge to the true parameter and the convergence rate is optimal.

1 Introduction

This work is concerned with adaptive filtering problems. By concentrating on sign-regressor algorithms, we develop a procedure that uses recursive estimates with larger than $O(1/n)$ step sizes and averages of the iterates and observations. Our focus is on the improvement of the performance of the underlying algorithms. We aim to obtain asymptotic efficiency; we show that the averaging algorithms for adaptive filtering algorithms are asymptotically optimal.

Owing to its importance, adaptive filtering algorithms have received much attention; see for example, [2,4,6,13,16,22,23], among others. Recently due to the applications in adaptive multiuser detection in CDMA wireless communication networks, there is increasing interest in further our understanding of adaptive filtering algorithms; see [3,7–10,21,25] and the references therein.

Suppose that $X_n \in \mathbb{R}^r$ and $s_n \in \mathbb{R}$ are sequences of measured output and reference signals, respectively. Assume that the sequence $\{s_n, X_n\}$ is stationary. By adjusting the system parameter H adaptively, one aims to make the weighted output $H'X_n$ best match the reference signal s_n as well as possible in the sense that a cost function is minimized. [Throughout the paper, we use z' to denote the transpose of $z \in \mathbb{R}^{\ell\times r}$ for $\ell, r \geq 1$, and use $|z|$ to denote the norm of z. For notational simplicity, K denotes a generic

B. Pasik-Duncan: Stochastic Theory and Control, LNCIS 280, pp. 223–238, 2002.

positive constant whose values may vary for different usage. For a square matrix B, by $B > 0$ we mean that it is positive definite.] If a mean squares cost $L(H) = E|s_1 - H'X_1|^2$ is used, the gradient of $L(H)$ is then given by $L_H(H) = -2EX_1(s_1 - H'X_1)$, and the recursive algorithm is of the form

$$H_{n+1} = H_n + a_n X_n (s_n - H_n' X_n), \tag{1}$$

which is known as an LMS algorithm. If the cost $L(H) = E|s_n - H'X_n| = E|s_1 - H'X_1|$ is used, then $L_H(H) = -E(X_1 \text{sign}(s_1 - H'X_1)) := -f(H)$, and a recursive algorithm of gradient descent type takes the form

$$H_{n+1} = H_n + a_n X_n \text{sign}(s_n - H_n' X_n), \tag{2}$$

where sign(0) $= 0$ and sign (y)$= |y|$ if $y \neq 0$ for any $y \in \mathbb{R}$. In both (1) and (2), $\{a_n\}$ is a sequence of step sizes satisfying $a_n \geq 0$, $a_n \to 0$ as $n \to \infty$ and $\sum_n a_n = \infty$. Algorithm (1) is commonly referred to as LMS algorithm, whereas (2) is called a sign-error algorithm. Compared with (1), algorithm (2) has reduced complexity. The use of the sign operator makes the algorithm be easily implementable so it is appealing in various applications, see [4,6] and the references therein. However, sign$X_n(s_n - H'X_n)$ as a function of H is not continuous. Thus to analyze such an algorithm is more difficult than that of (1). Much effort has been devoted to the improvement of sufficient conditions for convergence of such algorithms. Recently, in [3], by treating an algorithm with randomly generated truncation bounds, we obtained with probability (w.p.1) convergence of the recursive algorithm by assuming only the stationarity and finite second moments of the signals. The condition used is close to the minimal requirement needed. In addition, we also examined rate of convergence of the algorithm by weak convergence methods, which was identified as an open problem in the previous work. Local analysis was carried out in the rates of convergence study. Although $X_n \text{sign}(s_n - H'X_n)$ is not continuous in H, $L_H(H)$ can be a smooth function, thanks to the smoothing effect provided by taking expectation.

This line of work is continued in [25], in which we have suggested an averaging algorithm using iterate averaging ideas of [18,19] (see also [14]). Asymptotic optimality is obtained; applications to multiuser detection has also been investigated. In this paper, we will examine sign-regressor algorithms with averaging in both iterates and observations. The extra averages of the observations make the resulting trajectories more smooth; see also related work [1,20,26].

The sign-regressor algorithm is also used frequently in applications, which have reduced complexity compared with the LMS algorithm and better performance compared with the sign-error algorithm. To take advantages of both LMS and sign-error algorithms, in lieu of (2), use sign operator only for the regressor X_n componentwise.

Our plan is as follows. Section 2 presents the algorithm. Section 3 proceeds with the convergence of the algorithm. Then asymptotic optimality is treated in Section 4. Further discussions are provided in Section 5. Finally, Section 6 presents an application in adaptive multiuser detection in CDMA wireless networks.

2 Algorithm

Let $H \in \mathbb{R}^r$, $X_n \in \mathbb{R}^r$ and $s_n \in \mathbb{R}$ be system parameter, measured output, and reference signals, respectively. Consider the following algorithm

$$\begin{aligned} H_{n+1} &= \overline{H}_n + \frac{1}{n^\gamma} \sum_{j=1}^{n} \mathrm{Sgn}(X_j)(s_j - X_j' H_j), \quad 2/3 < \gamma < 1, \\ \overline{H}_n &= \frac{1}{n} \sum_{j=1}^{n} H_j. \end{aligned} \tag{3}$$

Note that the averaging of the iterates can be executed recursively as

$$\overline{H}_{n+1} = \overline{H}_n - \frac{1}{n+1} \overline{H}_n + \frac{1}{n+1} H_{n+1}, \tag{4}$$

where $\mathrm{Sgn}(\cdot)$ denotes $\mathrm{Sgn}(X) = (\mathrm{sign}(X_1), \ldots, \mathrm{sign}(X_r))'$, i.e., the sign operator applied componentwise to a vector $X \in \mathbb{R}^r$.

Remark 7. As can be seen that larger than $O(1/n)$ step sizes are used to force the iterates getting to a neighborhood of the true parameter H_* faster. Then averages are taken for the iterates as well as observations.

To carry out the asymptotic analysis, we need the following assumptions:

(**A**) The sequence $\{s_n, X_n\}$ is stationary such that
- (a) $E\mathrm{Sgn}(X_1)X_1' = D$, $E\mathrm{Sgn}(X_1)s_1 = b$;
- (b) $-D$ is Hurwitz (all eigenvalues have negative real part);
- (c) the sequence $\{s_n, X_n\}$ is bounded and uniformly mixing with mixing measures satisfying $\sum_k \varphi^{1/2}(k) < \infty$.

3 Convergence

Theorem 8. *Under* (A), *both* H_n *and* $\overline{H}_n$ *converges to* $H_* = D^{-1}b$ *w.p.1.*

Idea of Proof. To prove the convergence, we use Theorem 7.1 of Kushner and Yin [15, p. 163]. The recurrence condition, namely, "for each $0 < \rho < 1$ let R_ρ be a compact set such that $H_n \in R_\rho$ i.o. with probability at least ρ" needs to be verified. A sufficient condition that guarantees the recurrence is: $\{H_n\}$ is bounded in probability (tight). That is, we need only verify for any $\varepsilon > 0$, there is a K_ε such that $P(|H_n| \geq K_\varepsilon) \leq \varepsilon$. By Tchebyshev's inequality, this can be achieved if $E|H_n|^2 < \infty$. Thus, we need only verify the finite second moment condition. Since we will prove a sharper bound on $H_n - H_*$ in Theorem 9, we will omit the details here. Once the recurrence condition is verified, $\limsup_n |H_n| < \infty$ w.p.1. As a result, using the ODE method, we construct a sequence of piecewise constant interpolation of the stochastic approximation iterates H_n as:

$$H^n(t) = H_{n+i},\ t \in [t_{n+i} - t_n, t_{n+i+1} - t_n), \qquad t_n = \sum_{i=0}^{n-1} \frac{1}{i^\gamma},\ i \geq 0.$$

We show that $\{H^n(\cdot)\}$ is uniformly bounded and equicontinuous in the extended sense (an extension of the usual equicontinuity to a class of measurable but noncontinuous functions (see [15] for a definition)). Then the Arzela-Ascoli theorem still holds and implies $\{H^n(\cdot)\}$ has a convergent subsequence with limit $H(\cdot)$ satisfying the asymptotically stable ODE,

$$\dot{H}(t) = b - DH(t), \tag{5}$$

which has as a unique stationary point H_*. Finally, we obtain $H_n \to H_*$ w.p.1.

4 Asymptotic Normality

Next, we consider the asymptotic distribution of the smoothed sign-regressor algorithm. Following the approach of [15, Chapter 10], our analysis consists of two parts. In the first part, we derive an upper bound on the estimation error $E|H_n - H_*|^2$. In the second part, we show that a suitably scaled sequence converges to the solution of a stochastic differential equation.

4.1 Estimate of $E|H_n - H_*|^2$

For future use, define $\widetilde{H}_n = H_n - H_*$. By virtue of (3),

$$H_n = \overline{H}_{n-1} + \frac{1}{(n-1)^\gamma} \sum_{j=1}^{n-1} \mathrm{Sgn}(X_j)(s_j - X_j' H_j). \tag{6}$$

Then

$$\begin{aligned} H_{n+1} - H_n = {} & \overline{H}_n - \overline{H}_{n-1} + \frac{1}{n^\gamma} \mathrm{Sgn}(X_n)(s_n - X_n' H_n) \\ & + \left(\frac{1}{n^\gamma} - \frac{1}{(n-1)^\gamma} \right) \sum_{j=1}^{n-1} \mathrm{Sgn}(X_j)(s_j - X_j' H_j). \end{aligned}$$

Since $\frac{1}{n^\gamma} - \frac{1}{(n-1)^\gamma} = \frac{-(\gamma/n)+\eta_n}{(n-1)^\gamma}$, where $\eta_n = O(1/n^2)$ and in view of (6),

$$\begin{aligned}\overline{H}_n - \overline{H}_{n-1} &= -\frac{1}{n}\overline{H}_{n-1} + \frac{1}{n}H_n \\ &= \frac{1}{n(n-1)^\gamma}\sum_{j=1}^{n-1}\mathrm{Sgn}(X_j)(s_j - X_j'H_j),\end{aligned}$$

by (3), we deduce that

$$\begin{aligned}H_{n+1} &= H_n + \frac{1}{n^\gamma}\mathrm{Sgn}(X_n)(s_n - X_n'H_n) \\ &+ \frac{1}{(n-1)^\gamma}\left(\frac{1-\gamma}{n} + \eta_n\right)\sum_{j=1}^{n-1}\mathrm{Sgn}(X_j)(s_j - X_j'H_j),\end{aligned}$$

$$\widetilde{H}_{n+1} = \widetilde{H}_n - \frac{1}{n^\gamma}D\widetilde{H}_n + \frac{1}{n^\gamma}(D - \mathrm{Sgn}(X_n)X_n')\widetilde{H}_n + \frac{1}{n^\gamma}\xi_n + \frac{1}{n^\gamma}\zeta_n \tag{7}$$

where

$$\begin{aligned}\xi_n &= \mathrm{Sgn}(X_n)s_n - \mathrm{Sgn}(X_n)X_n'H_* \quad \text{and} \quad \zeta_n = \widehat{\zeta}_n(1 + O(1/n)), \\ \widehat{\zeta}_n &= \frac{(1-\gamma)n^\gamma}{n(n-1)^\gamma}\sum_{j=1}^{n-1}\mathrm{Sgn}(X_j)(s_j - X_j'H_j) \\ &+ \frac{\eta_n n^\gamma}{(n-1)^\gamma}\sum_{j=1}^{n-1}\mathrm{Sgn}(X_j)(s_j - X_j'H_j), \ \eta_n = O(1/n^2).\end{aligned} \tag{8}$$

To obtain the desired estimate, define $V(\widetilde{H}) = (\widetilde{H}'\widetilde{H})/2$.

Theorem 9. *Assume* (A). *Then* $EV(\widetilde{H}_n) = O(1/n^\gamma)$ *for* n *sufficiently large.*

Idea of Proof. As in Yin and Yin [26], we use the techniques of perturbed test function (see [15]) to obtain the estimate. The result is essentially a stability statement. The rationale is that we add a small perturbation to the Liapunov so that it will result in needed cancellations. Nevertheless, a direct adoption of the usual approach does not work due to the use of the averages in the noisy observations. We thus use a stopping time argument to circumvent the difficulty. Note that H_n is $\mathfrak{F}_n$ measurable so is $\widetilde{H}_n$. For some $\delta > 0$, let $p(j,\delta) = \max\{k : \sum_{l=j}^k E|\zeta_l| \le \delta\}$. For some $M > 0$ sufficiently large, consider the following partition into subintervals $M = \tau_0 < \tau_1 = p(\tau_0,\delta) < \cdots < \tau_\nu = p(\tau_{\nu-1},\delta)$. Then we have

$$\sum_{l=\tau_i}^{\tau_{i+1}} E|\zeta_l| < \delta, \quad \text{for each} \quad i \le \nu.$$

Let us work with the subintervals $[\tau_i, \tau_{i+1})$. In what follows, we use E_n to denote the conditional expectation with respect to $\mathfrak{F}_n$, the σ-algebra generated by $\{H_1, s_k, X_k,\ k < n\}$. For any i and n satisfying $\tau_i \le n < \tau_{i+1})$,

$$\begin{aligned}E_n V(\widetilde{H}_{n+1}) - V(\widetilde{H}_n) &= V'_H(\widetilde{H}_n)(\widetilde{H}_{n+1} - \widetilde{H}_n) \\ &+ \int_0^1 (\widetilde{H}_{n+1} - \widetilde{H}_n)' V_{HH}(H_n + s(H_{n+1} - H_n))(\widetilde{H}_{n+1} - \widetilde{H}_n) ds,\end{aligned}$$

and

$$\begin{aligned}E_n V(\widetilde{H}_{n+1}) = & \widetilde{H}'_n \left[-\frac{1}{n^\gamma} D\widetilde{H}_n + \frac{1}{n^\gamma}(D - \mathrm{Sgn}(X_n)X'_n)\widetilde{H}_n + \frac{1}{n^\gamma}\xi_n + \frac{1}{n^\gamma}\zeta_n \right] \\ & + O(n^{-2\gamma})(1 + V(\widetilde{H}_n)) + \rho_n,\end{aligned}$$

where $E|\rho_n| = O(n^{-2\gamma})$. We first derive the desired order estimates on a subsequence, and then establish the result for n large enough. To proceed, we introduce a number of perturbed Liapunov functions that are small in magnitude and result in the desired cancellations as

$$\begin{aligned} V_1(\widetilde{H}, n) &= \sum_{j=n}^{\tau_{i+1}} \frac{1}{j^\gamma} E_n \widetilde{H} \xi_j, \\ V_2(\widetilde{H}, n) &= \sum_{j=n}^{\tau_{i+1}} \frac{1}{j^\gamma} E_n \widetilde{H}'[D - \mathrm{Sgn}(X_j)X'_j]\widetilde{H}, \\ V_3(\widetilde{H}, n) &= \sum_{j=n}^{\tau_{i+1}} \frac{1}{j^\gamma} E_n \widetilde{H}' \zeta_j. \end{aligned} \tag{9}$$

First, by virtue of the ϕ-mixing conditions

$$\begin{aligned} |V_1(\widetilde{H}, n)| &\le K \frac{|\widetilde{H}|}{n^\gamma} \sum_{j=n}^{\tau_{i+1}} \varphi^{1/2}(j-n) \le O\left(\frac{1}{n^\gamma}\right)(1 + V(\widetilde{H})), \\ |V_2(\widetilde{H}, n)| &\le O\left(\frac{1}{n^\gamma}\right)(1 + V(\widetilde{H})), \\ |V_3(\widetilde{H}, n)| &\le O\left(\frac{1}{n^\gamma}\right)(1 + V(\widetilde{H})). \end{aligned} \tag{10}$$

Thus, the perturbations are small. We next show that the desired cancellations take place. In fact

$$\begin{aligned} E_n V_1(\widetilde{H}_{n+1}, n+1) &- V_1(\widetilde{H}_n, n) \\ &= E_n V_1(\widetilde{H}_{n+1}, n+1) - E_n V_1(\widetilde{H}_n, n+1) \\ &\quad + E_n V_1(\widetilde{H}_n, n+1) - V_1(\widetilde{H}_n, n) \\ &= \rho_n - \frac{1}{n^\gamma} \widetilde{H}'_n \xi_n, \end{aligned} \tag{11}$$

where $E|\rho_n| = O(n^{-2\gamma})$. Similarly,

$$E_n V_2(\widetilde{H}_{n+1}, n+1) - V_2(\widetilde{H}_n, n) = \omega_n - \frac{1}{n^\gamma} \widetilde{H}'_n (D - \operatorname{Sgn}(X_n) X_n) \widetilde{H}_n,$$

$$E_n V_3(\widetilde{H}_{n+1}, n+1) - V_3(\widetilde{H}_n, n) = \widetilde{\varrho}_n - \frac{1}{n^\gamma} \widetilde{H}'_n \zeta_n,$$

where $E|\omega_n| = O(n^{-2\gamma})$ and $E|\widetilde{\varrho}_n| = O(n^{-2\gamma})$.

By using (A) (b), it can be verified that $\widetilde{H}'_n D \widetilde{H}_n \leq -\lambda V(\widetilde{H}_n)$ for some $\lambda > 0$. It follows that there is a λ_0 with $0 < \lambda_0 < \lambda$ such that

$$(1/n^\gamma) \widetilde{H}'_n D \widetilde{H}_n + (1/n^\gamma) o(|\widetilde{H}_n|^2) + O(1/n^{2\gamma}) V(\widetilde{H}_n) \leq -\frac{\lambda_0}{n^\gamma} V(\widetilde{H}_n). \tag{12}$$

Define

$$W(n) = V(\widetilde{H}_n) + V_1(\widetilde{H}_n, n) + V_2(\widetilde{H}_n, n) + V_2(\widetilde{H}_n, n).$$

Using the estimates obtained thus far,

$$\begin{aligned} E_n W(n+1) &- W(n) \\ &= E_n V(\widetilde{H}_{n+1}) - V(\widetilde{H}_n) + E_n V_1(\widetilde{H}_{n+1}) - V_1(\widetilde{H}_n) \\ &\qquad + E_n V_2(\widetilde{H}_{n+1}) - V_2(\widetilde{H}_n) + E_n V_3(\widetilde{H}_{n+1}) - V_3(\widetilde{H}_n) \\ &\leq -\frac{\lambda_0}{n^\gamma} V(\widetilde{H}_n) + O\left(\frac{1}{n^{2\gamma}}\right) (1 + V(\widetilde{H}_n) + E_n (s_n - X'_n \widetilde{H}_n)(s_n - X'_n \widetilde{H}_n). \end{aligned}$$

By (10), we obtain

$$\begin{aligned} E_n W(n+1) &\leq (1 - \lambda_0/n^\gamma) W(n) + O(1/n^{2\gamma})(1 + W(n)) \\ &\qquad + E_n (s_n - X'_n \widetilde{H}_n)(s_n - X'_n \widetilde{H}_n)) \\ &\leq (1 - \lambda_1/n^\gamma) W(n) + O(1/n^{2\gamma}) \\ &\qquad + E_n (s_n - X'_n \widetilde{H}_n)(s_n - X'_n \widetilde{H}_n), \end{aligned} \tag{13}$$

for some $0 < \lambda_1 < \lambda_0$. Taking expectation and iterating on (13), for all $\kappa > N$,

$$EW(n+1) \le \prod_{j=\kappa}^{n}(1-\lambda_1/j^\gamma)EW(\kappa)+\sum_{j=\kappa}^{n}\prod_{i=j}^{n}(1-\lambda_1/i^\gamma)O(1/j^{2\gamma}) = O(1/n^\gamma).$$

Furthermore, by using (10) again, we also have $EV(\widetilde{H}_n) = O(1/n^\gamma)$. This concludes the proof.

4.2 Asymptotic Normality

This section is devoted to the asymptotic normality of the averaging sign-regressor algorithm. We first derive the asymptotic equivalence and then obtain an invariance theorem. To proceed, note that $\overline{H}_n$ can be rewritten as

$$\overline{H}_{n+1} = \overline{H}_n + \frac{1}{n^\gamma(n+1)}\sum_{j=1}^{n}\mathrm{Sgn}(X_j)(s_j - X_j'H_j). \tag{14}$$

Define $\widehat{H}_n = \overline{H}_n - H_*$. Then, with $\{\xi_n\}$ given by (8), rewrite (14) as

$$\widehat{H}_{n+1} = \widehat{H}_n + \frac{1}{n^\gamma(n+1)}\sum_{j=1}^{n}\xi_j - \frac{1}{n^\gamma(n+1)}\sum_{j=1}^{n}\mathrm{Sgn}(X_j)X_j'\widetilde{H}_j. \tag{15}$$

Note that (15) can be written as

$$\begin{aligned}\widehat{H}_{n+1} =& \widehat{H}_n - \frac{D}{n^\gamma}\widehat{H}_n + \frac{D}{n^\gamma(n+1)}\widehat{H}_n \\ &+\frac{1}{n^\gamma(n+1)}\sum_{j=1}^{n}(D-\mathrm{Sgn}(X_j)X_j')\widetilde{H}_j + \frac{1}{n^\gamma(n+1)}\sum_{j=1}^{n}\xi_j, \\ =& (I - \frac{D}{n^\gamma})\widehat{H}_n + \frac{D}{n^\gamma(n+1)}\widehat{H}_n \\ &+\frac{1}{n^\gamma(n+1)}\sum_{j=1}^{n}(D-\mathrm{Sgn}(X_j)X_j')\widetilde{H}_j + \frac{1}{n^\gamma(n+1)}\sum_{j=1}^{n}\xi_j.\end{aligned} \tag{16}$$

Define

$$D_{nk} = \begin{cases}\prod_{j=k+1}^{n}(I - D/j^\gamma) & k < n, \\ I, & k = n.\end{cases}$$

Using (15) and multiplying by $\sqrt{n}$ give us

$$\begin{aligned}\sqrt{n}\widehat{H}_{n+1} = & \sqrt{n}D_{n0}\widehat{H}_1 + \sqrt{n}\sum_{k=1}^{n}\frac{1}{k^\gamma(k+1)}D_{nk}D\widehat{H}_k \\ & +\sqrt{n}\sum_{k=1}^{n}\frac{1}{k^\gamma(k+1)}D_{nk}\sum_{j=1}^{k}(D-\mathrm{Sgn}(X_j)X_j')\widetilde{H}_j \\ & +\sqrt{n}\sum_{k=1}^{n}\frac{1}{k^\gamma(k+1)}D_{nk}\sum_{j=1}^{k}\xi_j.\end{aligned} \tag{17}$$

We shall show that asymptotically, the first three terms on the right-hand side are unimportant and the last term is equivalent to $(D^{-1}/\sqrt{n})\sum_{j=1}^{n}\xi_j$.

Theorem 10. *Under the assumptions of Theorem 9,*

$$\sqrt{n}\widehat{H}_{n+1} = \sqrt{n}(\overline{H}_{n+1} - H_*) = \frac{D^{-1}}{\sqrt{n}}\sum_{j=1}^{n}\xi_j + o(1),$$

where $o(1) \to 0$ in probability as $n \to \infty$.

<u>Idea of Proof.</u> In what follows, we examine each term in (17) separately. First, since H_n is bounded w.p.1 and $\gamma < 1$,

$$\sqrt{n}|D_{n0}\widehat{H}_1| \le |\sqrt{n}||D_{n0}||\widehat{H}_1| \to 0 \text{ w.p.1.} \tag{18}$$

As for the second term, by virtue of (9) and a partial summation,

$$\begin{aligned} & E\left|\sqrt{n}\sum_{k=1}^{n}\frac{1}{k^\gamma(k+1)}D_{nk}D\widehat{H}_k\right| \\ & \le \sqrt{n}\sum_{k=1}^{n}\frac{1}{k^\gamma(k+1)}|D_{nk}||D|E|\widehat{H}_k| \\ & \le Kn^{1/2-\gamma} \to 0 \text{ as } n \to \infty.\end{aligned}$$

Therefore, this term also tends to 0 in probability. We proceed to examine the last term in (17). A partial summation leads to

$$\begin{aligned}\sqrt{n}\sum_{k=1}^{n}\frac{1}{k^\gamma}D_{nk}\frac{1}{k}\sum_{j=1}^{k}\xi_j = & \left(\sum_{k=1}^{n}\frac{1}{k^\gamma}D_{nk}\right)\left(\frac{1}{\sqrt{n}}\sum_{j=1}^{n}\xi_j\right) \\ & +\sqrt{n}\sum_{k=1}^{n-1}\left(\sum_{j=1}^{k}\frac{1}{j^\gamma}D_{nj}\right)\left(\frac{1}{k}\sum_{j=1}^{k}\xi_j - \frac{1}{k+1}\sum_{j=1}^{k+1}\xi_j\right).\end{aligned} \tag{19}$$

Note that

$$
\begin{aligned}
&D_{nk} - D_{n,k-1} = \frac{D}{k^\gamma} D_{nk}, \\
&\sum_{k=1}^{n} \frac{1}{k^\gamma} D_{nk} = D^{-1}(I - D_{n0}) \ \text{ and } \ \sum_{j=1}^{k} \frac{1}{j^\gamma} D_{nj} = D^{-1}(D_{nk} - D_{n0}).
\end{aligned}
$$

Furthermore, since $D_{n0} \to 0$ as $n \to \infty$ and $(1/\sqrt{n}) \sum_{j=1}^{n} \xi_j$ is bounded in probability,

$$
\left(\sum_{k=1}^{n} \frac{1}{k^\gamma} D_{nk} \right) \left(\frac{1}{\sqrt{n}} \sum_{j=1}^{n} \xi_j \right) = \frac{D^{-1}}{\sqrt{n}} \sum_{j=1}^{n} \xi_j + o(1), \tag{20}
$$

where $o(1) \to 0$ in probability. Likewise, we obtain

$$
\begin{aligned}
&\sqrt{n} \sum_{k=1}^{n-1} \left(\sum_{j=1}^{k} \frac{1}{j^\gamma} D_{nj} \right) \left(\frac{1}{k} \sum_{j=1}^{k} \xi_j - \frac{1}{k+1} \sum_{j=1}^{k+1} \xi_j \right) \\
&\quad = \sqrt{n} D^{-1} \sum_{k=1}^{n-1} (D_{nk} - D_{n0}) \left(\frac{1}{k} \sum_{j=1}^{k} \xi_j - \frac{1}{k+1} \sum_{j=1}^{k} \xi_j - \frac{1}{k+1} \xi_{k+1} \right) \\
&\quad = \sqrt{n} D^{-1} \sum_{k=1}^{n-1} (D_{nk} - D_{n0}) \left(\frac{1}{(k+1)k} \sum_{j=1}^{k} \xi_j - \frac{1}{k+1} \xi_{k+1} \right)
\end{aligned}
$$

tends to 0 in probability.

For the next to the last term in (17), define $\psi_j = D - \text{Sgn}(X_j) X_j'$. Similar to the derivation of (20),

$$
\sqrt{n} \sum_{k=1}^{n} \frac{1}{k^\gamma (k+1)} D_{nk} \sum_{j=1}^{k} \psi_j \widetilde{H}_j = \frac{D^{-1}}{\sqrt{n}} \sum_{j=1}^{n} \psi_j \widetilde{H}_j + o(1), \tag{21}
$$

where $o(1) \to 0$ in probability. Choose $m = m(n)$ such that $m(n) \to \infty$ as $n \to \infty$, but $m(n)/\sqrt{n} \to 0$. It then follows

$$
\frac{D^{-1}}{\sqrt{n}} \sum_{j=1}^{n} \psi_j \widetilde{H}_j = \frac{D^{-1}}{\sqrt{n}} \sum_{j=m}^{n} \psi_j \widetilde{H}_j + o(1), \tag{22}
$$

where $o(1) \to 0$ in probability. Using (7) with

$$
\pi_n := -D \widetilde{H}_n + \psi_n \widetilde{H}_n + \xi_n + \zeta_n,
$$

noting the boundedness of the signals and the mixing property of $\{\psi_j\}$, and applying Theorem 9 yield that

$$\begin{aligned}
&E\left|\frac{1}{\sqrt{n}}\sum_{j=m}^{n}\psi_j\widetilde{H}_j\right|^2 \\
&=\frac{1}{n}\sum_{j=m}^{n}E\widetilde{H}_j'\psi_j'\left[\sum_{k=j}^{n}E_j\psi_k(\widetilde{H}_k-\widetilde{H}_m)+\sum_{k=j}^{n}E_j\psi_k\widetilde{H}_m\right] \\
&\le\frac{K}{n}\sum_{j=m}^{n}E|\widetilde{H}_j'||\psi_j'|\sum_{l=m}^{n}\left|\sum_{k=l}^{n}E_l\psi_k\right|\frac{1}{l^\gamma}|\pi_l|+o(1) \\
&\le Kn^{1-(3\gamma/2)}\to 0 \quad\text{as}\quad n\to\infty.
\end{aligned}$$

Thus the left-hand side of (21) also tends to 0 in probability.

Next, consider the asymptotic distribution of a suitable scaled sequence. Noting that $\sqrt{n}(\overline{H}_n - H_*) = \sqrt{n}(\overline{H}_{n+1} - H_*) + o(1)$, where $o(1) \to 0$ as $n \to \infty$, we obtain the following result on the limiting distribution.

Theorem 11. *Under the conditions of Theorem* 10, $\sqrt{n}(\overline{H}_n - H_*)$ *converges in distribution to a normal random variable with mean* 0, *and covariance matrix* $D^{-1}S_0D^{-1}$ *as* $n \to \infty$, *where* S_0 *is given by*

$$S_0 = E\xi_1\xi_1 + \sum_{i=2}^{m} E\xi_1\xi_i' + \sum_{i=2}^{m} E\xi_i\xi_1'. \tag{23}$$

5 Discussion

So far, we have established convergence and rates of convergence of the adaptive filtering algorithms with averaging. In this section, first, we exploit the issue of asymptotic optimality and efficiency. Then we generalize the results obtained to functional limit theorems, which give far reaching results and explain fully the stochastic aspects of the problems. We also mention possible future topics of investigation.

Asymptotic Optimality. Consider an algorithm of the form

$$H_{n+1} = H_n + a_n \mathrm{Sgn}(X_n)(s_n - H_n'X_n),$$

with step sizes $a_n = 1/n^\gamma$ for some $2/3 < \gamma \le 1$. Similar as in the previous development, it can be shown that $n^{\gamma/2}(H_n - H^*)$ is asymptotically normal. It is clear that among the γ's given above, the best one is $\gamma = 1$. If one uses a matrix-valued parameter Γ and $a_n = \Gamma/n$, then it can be demonstrated that the best choice of Γ is the D^{-1}. Nevertheless, D is usually not available, so additional estimates on D are needed. From a computation point of view, one may not wish to use a rapidly decreasing sequence of step sizes such as $a_n = O(1/n)$ since this results in a very slow movement in the initial stage. For further discussions, see [15, Chapter 11]. Inspired by the iterate averaging method [18,19], in conjunction with the double averaging procedure of [1], we have used a sequence of step sizes larger than $O(1/n)$ in the estimation together with taking averages of both iterates and observations. This yields the best scaling factor $\sqrt{n}$ as well as the smallest possible covariance. Thus the algorithm is asymptotically optimal.

Generalization and Further Remarks. Bounded mixing signals are treated here. For martingale difference sequences or m-dependent signals, the ideas used in [24] can be adopted. The asymptotic normality obtained in Theorem 11 is of central limit type. Such a result can be generalized to functional central limit theorem. The main idea is to replace the definition of $\sqrt{n}(\overline{H}_n - H_*)$ by $B^n(t) := \frac{\lfloor nt \rfloor}{\sqrt{n}} (\overline{H}_{\lfloor nt \rfloor +1} - H_*)$, $t \in [0,1]$, $\lfloor z \rfloor$ denotes the integer part of a real number z. Using essentially the same idea as in the previous section, we will be able to show that $B^n(t) = \frac{D^{-1}}{\sqrt{n}} \sum_{k=0}^{\lfloor nt \rfloor -1} \xi_k + o(1)$, where $o(1) \to 0$ in probability uniformly in $t \in [0,1]$. The conclusion of Theorem 11 then changes to: $B^n(\cdot)$ converges weakly to a Brownian motion with covariance $D^{-1} S_0 D^{-1} t$. This is far reaching result since it gives the evolution of the random process involved and delineate its behavior fully.

Further work may be directed to study almost sure properties associated with the averaging procedures; see for instance the recent work [17]. Another interesting problem is to study algorithms with constant step size. One may also consider L_p moment bounds (see [5] and the references therein) associated with the averaging procedures.

6 Adaptive Multiuser Detection in CDMA Wireless Networks

Code-division multiple-access (CDMA) implemented with direct-sequence (DS) spread-spectrum signaling will be widely used in 3G cellular telecommunications services such as personal communications, mobile telephony, and indoor wireless networks. The advantages of DS/CDMA include superior operation in multi-path environments, flexibility in the allocation of channels and the ability to operate asynchronously, increased capacity in bursty or fading networks [8]. Demodulating a given user in a DS/CDMA network requires processing the received signal to minimize multiple access interference (MAI) caused by other spread-spectrum users in the channel–as well as ambient channel noise. MAI arises in DS/CDMA systems due to the fact that all users communicate through the same physical channel using *non-orthogonal* multiplexing. The use of non-orthogonal signature waveforms (resulting in MAI) has many advantages in wireless CDMA systems such as greater bandwidth utilization under conditions of channel fading and bursty traffic.

The conventional (single-user) detector, which suppresses the MAI using code-matched filtering, is very sensitive to the "near-far problem," i.e., differences between the received energies from the desired user and interfering users. Recently, *blind multiuser detection* techniques [7,8] have been developed that allow one to use a linear multiuser detector for a given user, with no knowledge beyond that required for implementation of the conventional detector for that user. Blind multiuser detection is useful in mobile wireless channels when the desired user can experience a deep fade or if a strong interferer suddenly appears. The advantage of these methods is that they require no knowledge of system-parameters; the only knowledge required is the signature-sequence of the user of interest and that the channel general properties fit that of the model. In [7], a blind least mean square (LMS) algorithm is given for linear minimum mean square error (MMSE) detection. In [8] a code-aided blind recursive least squares algorithm for jointly suppressing MAI and NBI is given. More recently, in [9] a blind averaged LMS algorithm is presented while in [11] adaptive step size LMS algorithms are presented. The objective of this section is to outline the use of the sign regressor algorithm described above in designing an adaptive linear blind mulituser detector.

Consider a synchronous K-user binary DS/CDMA communication system which transmits through an additive white Gaussian noise channel. After preprocessing and sampling at the CDMA receiver the resulting discrete time received signal at time n, denoted by r_n, is given by (see [9] for details)

$$r_n = \sum_{k=1}^{K} \sqrt{P_k} b_k(n) \xi_k + \sigma w_n. \tag{24}$$

Here r_n is an N-dimensional vector; N is called the processing (spreading) gain; ξ_k is an N-vector denoting the normalized signature sequence of the kth user, i.e., each element $\xi_{ki} \in \{-1/\sqrt{N}, +1/\sqrt{N}\}$ for $i = 1, 2, \ldots, N$, so that $\xi_k' \xi_k = 1$; $b_k(n)$ denotes the data bit of the kth user transmitted at time n; $P_k = A_k^2$ is the received power of the kth user; σ is the standard deviation of the noise samples; w_n is a white Gaussian vector with mean zero and covariance matrix I, where I denotes the $N \times N$ identity matrix. It is assumed that the discrete-time stochastic processes $\{b_k(n)\}$, and $\{w_n\}$ are mutually independent, and that $\{b_k(n)\}$ is a collection of independent equi-probable ± 1 random variables.

Assuming user 1 being the user of interest, for user 1, $\sum_{k=2}^{K} \sqrt{P_k} b_k(n) \xi_k$ in (24) is termed multiple access interference (MAI). The aim of a multiuser detector is to suppress the MAI and adaptively estimate (demodulate) the bit sequence $b_1(n)$ given the observation sequence r_n. A linear blind multiuser detector demodulates the bits of user 1 according to $\widehat{b}_1(n) = \text{sgn}(c_*' r_n)$, where $\widehat{b}_1(n)$ denotes the estimate of the transmitted bit $b_1(n)$ at time n, and c_* denotes an appropriately chosen "weight vector." We consider here the widely used code-aided blind linear mean output error (MOE) detector [7], which chooses the "weight vector" c so as to minimize the MOE cost function

$$\zeta_n = E\{(c' r_n)^2\} \text{ subject to the constraint } c'\xi_1 = 1. \tag{25}$$

The constraint ensures that the received energy from the user of interest is equal to 1 (see [7] for further insights and motivation). Thus the above is a minimization of the energy from the interferers. Furthermore, as shown in [7], the MOE cost function has a unique global minimum (w.r.t. c). The blind MOE detector yields the following estimate $\hat{b}_1(n)$ of the transmitted signal

$$\hat{b}_1(n) = \operatorname{sgn}(c_*{}' r_n) \quad \text{where} \quad c_* = \frac{R^{-1}\xi_1}{\xi_1' R^{-1}\xi_1}, \tag{26}$$

where $R = E\{rr'\}$ denotes the autocorrelation matrix of the received signal r. In the above equation, c_* is the optimal linear MOE "weight vector." Such a detector is "blind" since it does not assume any knowledge of the data symbols $b_1(n)$ and signature sequences of other users.

In *adaptive* blind multiuser detection problems, we are interested in recursively adapting the weight vector c_n to minimize ζ_n the MOE given by (25). Note that it is often necessary to use a constant step size tracking algorithm due to the time-varying nature of c^* caused by the birth and death of users (MAI interferers).

In presenting the sign algorithms for blind adaptive multiuser detection, it is convenient to work with an unconstrained optimization problem rather than (25). Let $c_{n,i}$, for $i = 1, \dots, N$ denote the elements of c_n. The constrained optimization problem (25) may be transformed into an unconstrained one by solving for one of the elements $c_{n,i}$, $i \in [1, \dots, N]$ using the constraint (25). With no loss of generality, we solve for the first element $c_{n,1}$ and obtain

$$c_{n,1} = \frac{1}{\xi_{1,1}} \left(1 - \sum_{i=2}^{N} \xi_{1,i} c_{n,i} \right). \tag{27}$$

By defining the $(N-1)$-dimensional vector $H_n = (c_{n,2}, \dots, c_{n,N})'$, we obtain the equivalent unconstrained optimization problem:

$$\text{Minimize } L(H) \quad \text{with} \quad L(H) = E(s_n - H' X_n)^2. \tag{28}$$

Here $s_n = -r_{n,1}/\xi_{1,1}$ and X_n denotes the $(N-1)$-dimensional vector

$$X_n = (r_{n,2} - r_{n,1}\xi_{1,2}/\xi_{1,1}, \dots, r_{n,N} - r_{n,1}\xi_{1,N}/\xi_{1,1})'.$$

As in Section 2, (3) and (4) result in an adaptive blind sign regressor multiuser detection algorithm with iterate and observation averaging. Our numerical studies on iterate averaged algorithms with constant step size in [25] and [9] have demonstrated significant improvements in convergence of the linear multiuser detector. These lead us to believe that significant improvements in convergence also hold for the sign regressor multiuser detector algorithm developed here. Moreover, as in [11], adaptive step size algorithms that cope with time varying user arrival and departure statistics can also be developed for the sign regressor algorithm.

Acknowledgement. The research of C. Ion and G. Yin was supported in part by the National Science Foundation. The research of V. Krishnamurthy was supported in part by the ARC Special Research Center in Ultra-Broadband Information Networks (CUBIN).

References

1. J.A. Bather, Stochastic approximation: A generalization of the Robbins-Monro procedure, *Proc. Fourth Prague Symp. Asymptotic Statist.*, P. Mandl and M. Hušková Eds., 1989, 13-27.
2. A. Benveniste, M. Gooursat, and G. Ruget, Analysis of stochastic approximation schemes with discontinuous and dependent forcing terms with applications to data communication algorithms, *IEEE Trans. Automatic Control*, **AC-25** (1980), 1042-1058.
3. H.F. Chen and G. Yin, Asymptotic properties of sign algorithms for adaptive filtering, preprint, 2000.
4. E. Eweda, Convergence analysis of the sign algorithm without the independence and Gaussian assumptions, *IEEE Trans. Signal Processing*, **48** (2000), 2535-2544.
5. L. Gerencsér, Rate of convergence of recursive estimators, *SIAM J. Control Optim.*, **30** (1992), 1200-1227.
6. A. Gersho, Adaptive filtering with binary reinforcement, *IEEE Trans. Inform. Theory*, **IT-30** (1984), 191-199.
7. M.L. Honig, U. Madhow, and S. Verdu, Adaptive blind multiuser detection. *IEEE Transactions on Information Theory*, **41** (1995), 944–960.
8. M.L. Honig and H.V. Poor, Adaptive interference suppression in wireless communication systems, in H.V. Poor and G.W. Wornell, editors, *Wireless Communications: Signal Processing Perspectives*, Prentice Hall, 1998.
9. V. Krishnamurthy, Averaged stochastic gradient algorithms for adaptive blind multiuser detection in DS/CDMA systems, *IEEE Trans. Comm.*, **48** (2000), 125-134.
10. V. Krishnamurthy and A. Logothetis, Adaptive nonlinear filters for narrowband interference suppression in spread spectrum CDMA systems, *IEEE Trans. Comm.*, **47** (1999), 742-753.
11. V. Krishnamurthy, G. Yin, and S. Singh, Adaptive step size algorithms for blind interference suppression in DS/CDMA systems, *IEEE Trans. Signal Proc.*, **49** (2001), 190-201.
12. H. J. Kushner, *Approximation and Weak Convergence Methods for Random Processes, with Applications to Stochastic Systems Theory*, MIT Press, Cambridge, MA, 1984.
13. H.J. Kushner and A. Shwartz, Weak convergence and asymptotic properties of adaptive filters with constant gains, *IEEE Trans. Inform. Theory* **IT-30** (1984), 177-182.

14. H.J. Kushner and J. Yang, Stochastic approximation with averaging of the iterates: optimal asymptotic rate of convergence for general processes, *SIAM J. Control Optim.* **31** (1993), 1045-1062.
15. H.J. Kushner and G. Yin, *Stochastic Approximation Algorithms and Applications*, Springer-Verlag, New York, 1997.
16. O. Macchi and E. Eweda, Convergence analysis of self-adaptive equalizers, *IEEE Trans. Information Theory*, **IT-30** (1984), 161-176.
17. M. Pelletier, Asymptotic almost sure efficiency of averaged stochastic algorithms, *SIAM J. Control Optim.* **39** (2000), 49-72.
18. B.T. Polyak, New method of stochastic approximation type, *Automat. Remote Control* **51** (1990), 937-946.
19. D. Ruppert, Stochastic approximation, in *Handbook in Sequential Analysis*, B.K. Ghosh and P.K. Sen Eds., Marcel Dekker, 503-529, New York, 1991.
20. R. Schwabe, Stability results for smoothed stochastic approximation procedures, *Z. angew. Math. Mech.* **73** (1993), 639-644.
21. S. Verdu, Multiuser detection, in H.V. Poor and J.B Thomas, Eds., *Advances in Statistical Signal Processing.* JAI Press, Greenwich, CT., 1993.
22. B. Widrow and S.D. Stearns, *Adaptive Signal Processing*, Prentice-Hall, Englewood, Cliffs, NJ, 1985.
23. G. Yin, Asymptotic properties of an adaptive beam former algorithm, *IEEE Trans. Inform. Theory* **IT-35** (1989), 859-867.
24. G. Yin, Adaptive filtering with averaging, In G. Goodwin K. Aström and P.R. Kumar, editors, *IMA Volumes in Mathematics and Its Applications*, Vol. 74, 375-396, Springer-Verlag, 1995.
25. G. Yin, V. Krishnamurthy, and C. Ion, Iterate-Averaging sign algorithms for adaptive filtering with applications to blind multiuser detection, preprint, 2000.
26. G. Yin and K. Yin, Asymptotically optimal rate of convergence of smoothed stochastic recursive algorithms, *Stochastics and Stochastic Reports*, **47** (1994), 21-46.

Kalman-Type Filters Approach for Some Nonparametric Estimation Problems

R. Khasminskii[1]

Wayne State University, Detroit, MI 48202, USA

Abstract. Some results on Kalman-type filters for nonparametric estimation problems are presented. On-line recursive filters are proposed for an estimation of a signal and it's derivatives observed in Gaussian white noise and for a regression estimation with equidistant observation design.

Keywords: Kalman filter; nonparametric estimation; regression; equidistant design; Gaussian White Noise.

1 Introduction

Let $\Sigma(\beta, L)$ be a class of functions $S(t), t \in [0,T]$, having k derivatives for $t \in (0,T)$ with the kth derivative $S^{(k)}(t)$ satisfying the Hőlder condition with the exponent α: for $t, t+h \in [0,T]$, $0 < \alpha \leq 1$, $\beta = k + \alpha$

$$|S^{(k)}(t+h) - S^{(k)}(t)| \leq L|h|^{\alpha}$$

(We write sometimes below for brevity that S has the smoothness β in t). Assume that the observation process $X_\varepsilon(t)$ has the form

$$X_\varepsilon(t) = \int_0^t S(s)ds + \varepsilon W(t), \tag{1}$$

ε is a small parameter, $W(t)$ is a standard Wiener process. The following problem was considered in [7],[5]. What is the rate of convergence to 0 for $\varepsilon \to 0$ of the risk of a best estimators of $S, S^{(1)}, \dots, S^{(k)}$, based on $X_\varepsilon(t), t \in [0,T]$, and how to create these estimators? It was proven in [7] and [5] that the kernel estimator (see [10,12] and other) for the suitable choice of the kernel and the bandwidth δ (see below) for any $t \in (0,T)$ has a property

$$\sup_{S \in \Sigma(\beta,L)} E\Big(\Big[\frac{\widehat{S}(t) - S(t)}{\varepsilon^{2\beta/(2\beta+1)}}\Big]^2 + \sum_{j=1}^{k}\Big[\frac{\widehat{S}^{(j)}(t) - S^{(j)}(t)}{\varepsilon^{2(\beta-j)/(2\beta+1)}}\Big]^2\Big) \leq C. \tag{2}$$

Here and below we denote by C, C_i generic constants, do not depending on ε. In more details, the estimators $\widehat{S}(t) = \widehat{S}^{(0)}(t), \dots, \widehat{S}^{(k)}(t)$ can be chosen by the following way:

$$\widehat{S}^{(j)}(t) = \frac{1}{\delta^{j+1}} \int_{-\infty}^{\infty} K^{(j)}\Big(\frac{t-s}{\delta}\Big) dX_\varepsilon(s), \ j = 0, ..., k, \tag{3}$$

B. Pasik-Duncan: Stochastic Theory and Control, LNCIS 280, pp. 239–250, 2002.

where $\delta = \delta(\varepsilon) = \varepsilon^{\frac{2}{2\beta+1}}$, and $K(t)$ is a compactly supported smooth kernel having properties

$$\int_{\mathbb{R}} K(t)dt = 1; \int_{\mathbb{R}} t^p K(t)dt = 0, p = 1, \dots, k. \tag{4}$$

Moreover, it was proven also that better (uniformly in an open subset of $\Sigma(\beta, L)$) rate of convergence of risks to 0 is unavailable.

The estimator (3) has an essential inconvenience: Even for small h the estimators for $S(t+h), S^{(1)}(t+h), \dots, S^{(k)}(t+h)$ (in spite of a known smoothness of S) are *not* a small corrections of estimators for $S(t), S^{(1)}(t), \dots, S^{(k)}(t)$, based on new observations at the time interval $[t, t+s]$. So these estimators are not recursive. Another inconvenience follows from the condition (4). Due to this condition the kernel K becomes more and more cumbersome with the growth of a priori known smoothness of S. The goal of this paper is to present some recent results concerning recursive nonparametric estimation for the model (1) and analogous multidimensional and regression models.

2 Kalman-Type Filter for the Model (1)

Note first of all that the observation model (1) admit the equivalent form

$$dX_\varepsilon(t) = S(t)dt + \varepsilon dW(t); \; X_\varepsilon(0) = 0. \tag{5}$$

So it can be treated as the observation of an one-dimensional signal S in a Gaussian White Noise (GWN) with the spectral density ε^2. In [1] the following recursive estimator inspired by the Kalman's ideas, was proposed:

$$\begin{aligned} d\widehat{S}(t) &= \widehat{S}^{(1)}(t)dt + \frac{q_1}{\varepsilon^{2/(2\beta+1)}}\big(dX_t - \widehat{S}(t)dt\big), \\ d\widehat{S}^{(j)}(t) &= \widehat{S}^{(j+1)}(t)dt + \frac{q_{j+1}}{\varepsilon^{2(j+1)/(2\beta+1)}}\big(dX_t - \widehat{S}(t)dt\big), j = 1, \dots, k-1, \\ d\widehat{S}^{(k)}(t) &= \frac{q_{k+1}}{\varepsilon^{2k/(2\beta+1)}}\big(dX_t - \widehat{S}(t)dt\big), \end{aligned} \tag{6}$$

subject to initial conditions $\widehat{S}^{(j)}(0) = S_\circ^{(j)}$, $j = 0, ..., k$, where parameters $q_1, \dots, q_{k+1}$ are chosen so that all roots λ_i of the polynomial

$$p^k(\lambda) = \lambda^{k+1} + q_1\lambda^k + ... + q_k\lambda + q_{k+1} \tag{7}$$

have negative real parts:

$$\Re\lambda_i < 0. \tag{8}$$

The form (6) of the filter is motivated by the fact that it is the form of Kalman filter for the signal S observed in the GWN with the spectral density ε^2, the kth derivative of which is also GWN with suitably chosen small intensity, see details in [1].

The following result was proven in [1].

Theorem 1. *For every vector* $(q_1, ..., q_k)$ *such that (8) is valid and every signal* $S(t)$ *from* $\Sigma(\beta, L)$*, the tracking filter (6) with arbitrary initial conditions* $S_\circ^{(j)}$, $j = 0, ..., k$ *has the property: there exist constants* C_1 *and* C_2*, and an initial boundary layer*

$$\Delta^\varepsilon = C_1 \varepsilon^{2/(2\beta+1)} \log(1/\varepsilon)$$

such that for $t \geq \Delta^\varepsilon$

$$E\Big(\Big[\frac{\widehat{S}(t) - S(t)}{\varepsilon^{2\beta/(2\beta+1)}}\Big]^2 + \sum_{j=1}^{k} \Big[\frac{\widehat{S}^{(j)}(t) - S^{(j)}(t)}{\varepsilon^{2(\beta-j)/(2\beta+1)}}\Big]^2\Big) \leq C_2. \tag{9}$$

It follows from aforementioned results in [7,5] that the rate of convergence of estimators $\widehat{S}^{(j)}(t), j = 0, \ldots, k$ is optimal. It was proven in [1] that the initial boundary layer of the order $\varepsilon^{2/(2\beta+1)}$ is also inevitable for on-line estimators.

3 Recursive Estimation of a Smooth Regression Function

The model (1) and (5) and an analogous multidimensional models can be considered as continuous approximations for the regression estimation models with the equidistant design. So the natural problem is to construct also on-line tracking filters for the regression estimation.

Remind first of all the statement of the regression estimation problem. Let (t, X) be a pair of random variables, $t \in [0,1], X \in \mathbb{R}^1$, and $f(t) = E(X|t)$ be a regression function. There are two the most popular setting for a regression estimation problem. For the first setting (*random design*), statistician estimates $f(t)$ on the base of a sample $(t_1, X_1), \ldots, (t_n, X_n)$, where $(t_1, X_1), \ldots, (t_n, X_n)$ are independent copies of (t, X), while for the second one (*equidistant design*) $t_{in} = i/[n^\gamma], i = 1, \ldots, , [n^\gamma]$ are fixed, and the statistician uses a sample $X_{1i}, \ldots, X_{ki}, k = [n/[n^\gamma]]$ for each i, and the distribution of all elements of this sample coincides with the conditional distribution $P\{X \in A | t = t_{in}\}, A \subset \mathbb{R}^1$.

For *nonparametric* regression estimation problem the function f is assumed to be belong to a collection of functions Σ which cannot be specified by a finite number of parameters. Here, following to [13], [14], [4] and [7], we consider the class $\Sigma(\beta, L)$ again. It is well known from these citations that for both designs there are estimators $\widehat{f^{(j,n)}}(t)$ for $f^{(j)}(t) = \frac{d^j}{dt^j} f(t)$, $j = 0, 1, \ldots, k$ such that for a wide class of loss functions $\ell(*)$ and any $t \in [0,1]$

$$\sup_{f \in \Sigma(\beta, L)} E\ell\Big(n^{\frac{\beta-j}{2\beta+1}} |\widehat{f^{(j,n)}(t)} - f^{(j)}(t)|\Big) < C, \; j = 0, 1, \ldots, k \tag{10}$$

and do not exist estimators with better rate of the convergence to zero of estimation risks in n uniformly in a nonempty open subset of $\sum(\beta, L)$.

In [8] a simplest case (with $\gamma = 1$) of afore-mentioned design was analyzed: a statistician has one 'measurement' X_{in} for each $t_{in} = \frac{i}{n}, i = 1, \ldots, n$, so that X_{in} can be written as

$$X_{in} = f(t_{in}) + \sigma(t_{in})\xi_{in}, \tag{11}$$

here $(\xi_{in})_{i \le n}$ is a sequence of i.i.d. random variables, $E\xi_{in} \equiv 0, \quad E\xi_{in}^2 = 1$. The natural discrete analogy of the filter (6) is the on-line estimator (we write X_i and t_i instead of X_{in} and t_{in} for brevity)

$$\begin{aligned} \widehat{f^{(j,n)}}(t_i) &= \widehat{f^{(j,n)}}(t_{i-1}) + \frac{1}{n}\widehat{f^{(j+1,n)}}(t_{i-1}) + \frac{q_{j+1}}{n^{\frac{(2\beta-j)}{2\beta+1}}}\big(X_i - \widehat{f^{(0,n)}}(t_{i-1})\big) \\ j &= 0, 1, \ldots, k-1 \\ \widehat{f^{(k,n)}}(t_i) &= \widehat{f^{(k,n)}}(t_{i-1}) + \frac{q_{k+1}}{n^{\frac{(2\beta-k)}{2\beta+1}}}\big(X_i - \widehat{f^{(0,n)}}(t_{i-1})\big) \end{aligned} \tag{12}$$

subject to some initial conditions $\widehat{f^{(0,n)}}(0), \widehat{f^{(1,n)}}(0), \ldots, \widehat{f^{(k,n)}}(0)$. It is shown in [8] that the estimator (12) has the optimal rate of convergence of risks to 0 for $n \to \infty$. More precisely the following result is proven there.

Theorem 2. *Let $q_1, \ldots, q_{k+1}$ are chosen such that all roots of the polynomial (7) are different and have negative real parts. Let the observation model has the form (11), $f \in \Sigma(\beta, L)$ and $\sigma^2(t) < C$. Then the estimator (12) with arbitrary bounded initial conditions $\widehat{f^{(0,n)}}(0), \widehat{f^{(1,n)}}(0), \ldots, \widehat{f^{(k,n)}}(0)$ possesses the property: for $t_l > Cn^{-\frac{1}{2\beta+1}} \log n = \Delta_n$*

$$\sup_{f \in \sum(\beta,L)} \sum_{j=0}^{k} E\big(f^{(j)}(t_\ell) - \widehat{f^{(j,n)}}(t_\ell)\big)^2 n^{\frac{2(\beta-j)}{2\beta+1}} \le C. \tag{13}$$

Remark 1. It follows from (10) that the rate of convergence in (13) is unimprovable, and analogously to the model (5) it is easy to prove that the initial layer of the order $n^{-\frac{1}{2\beta+1}}$ is inevitable for the on-line estimators.

Remark 2. The estimator (12) estimates f only at $t_i = i/n$, but taking use of smoothness f it can be applied for all $t \in [\Delta_n, 1]$. For instance, the optimal in n rate of convergence of risks gives the estimator

$$\widehat{f^{(j,n)}}(t) = \widehat{f^{(j,n)}}(t_\ell), \; t_\ell \le t < t_{\ell+1}.$$

The same estimator can be easy applied for the extrapolation problems. For instance, assume that $f(t_\ell + h)$ has to be estimated, and only observations from (11) for $i \le \ell$ are available. Then consider the estimator

$$\widehat{f_n}(t_\ell + h) = \sum_{j=0}^{k} \frac{h^j}{j!}\widehat{f^{(j,n)}}(t_\ell).$$

It is not hard to prove, using (13) that for this estimator

$$E\left(\frac{\widetilde{f}_n(t_\ell+h)-f(t_\ell+h)}{(\max\{h,n^{-\frac{1}{2\beta+1}}\})^\beta}\right)^2 < C$$

and the rate of convergence $\left(\max\{h,n^{-\frac{1}{2\beta+1}}\}\right)^\beta$ cannot be exceeded uniformly in $\sum(\beta,L)$ by any other extrapolator.

4 Estimation of Time Dependent Spatial Signal Observed in GWN

It was found in [6],[13], [14] that for a regression function $f(x), x \in \mathcal{K} \subset \mathbb{R}^d, \mathcal{K}$ is a compact set, having a priori known smoothness β_k in $x_k, k = 1, \dots, d$, the estimation performance can be characterized by the parameter

$$\widetilde{\beta} = \Big(\sum_{q=1}^{n} \frac{1}{\beta_q}\Big)^{-1} \tag{14}$$

In more details, it was proven that for the sample of size n there exist such estimators $\hat{f}_n$, that uniformly in $\mathcal{K}$ the inequality

$$E|\hat{f}_n(x) - f(x)|^2 \le Cn^{-\frac{2\widetilde{\beta}}{2\widetilde{\beta}+1}}$$

is valid, and there is no estimators with uniformly better rate of convergence of risks, as $n \to \infty$. The analogous upper bound was established also for estimation of the partial derivatives of f.

The precise upper bounds for an estimation problem of a signal $S(x), x \in [0,1]^n$ observable in a presence of GWN of a small intensity ε were firstly obtained by N. Nussbaum [9]. Under assumptions that the signal is 1-periodic in each argument and possesses partial Sobolev's derivatives $(D^{\beta_q}_{x_q} S(x))_{1\le q\le n}$ with $\|S\|^2_\beta := \sum\limits_{q=1}^{n} \|D^{\beta_q}_{x_q} S\|^2 \le P$ (P is a prescribed constant) and applying Pinsker's results, [11], he proved in [9] that the best in the mimimax sense for $\varepsilon \to 0$ quadratic risk is $C(P,\beta)\varepsilon^{4\widetilde{\beta}/(2\widetilde{\beta}+1)}(1+o(1))$ with $\widetilde{\beta}$ from (14) and even the best in mimimax sense constant $C(P,\beta)$ was found.

In [3] an analogous problem was considered for the case of time dependent spacial signal $S(t,x)$, $t \in [0,T]$, $x \in [0,1]^n$. Similar [9] it was assumed there that $S(t,x)$ is 1-periodic in each argument of the vector $x = (x_1, \dots, x_n)$ and obeys the above-mentioned Sobolev's smoothness. It was assumed also that $S(t,x)$ obeys Sobolev's smoothness in t with known parameter β_0 . It is not hard to create, using kernel or projection estimators approach (see, e.g. [7]), estimators with the rate of convergence of quadratic risks not exceeding

$C\varepsilon^{4\beta/(2\beta+1)}(1+o(1))$ $(\varepsilon \to 0)$ with β defined by the equation

$$\beta = \Big(\sum_{q=0}^{n} \frac{1}{\beta_q}\Big)^{-1} \tag{15}$$

The main problem of [3] was the construction of a recursive with respect to the time variable t on-line estimator with optimal rate of convergence of risks to 0 for $\varepsilon \to 0$.

The observation model for this problem can described by the following way:

$$Y^\varepsilon(dt, dx) = S(t,x)dtdx + \varepsilon W(dt, dx),\ t \in [0,T],\ x \in [0,1]^n, \tag{16}$$

where ε is a small parameter, $W(dt, dx)$ is a cylindrical orthogonal Gaussian random measure, such that for any time intervals and sets Γ, Γ_1, Γ_2 from the Borel σ-algebra on $[0,1]^n$

$$\begin{aligned} &EW([0,t]\times\Gamma) = 0,\\ &EW([t_1',t_1'']\times\Gamma_1)W([t_2',t_2'']\times\Gamma_2) = \big([t_1',t_1'']\cap[t_2',t_2'']\big)\Lambda(\Gamma_1\cap\Gamma_2); \end{aligned} \tag{17}$$

here Λ is a Lebesgue measure on $[0,1]^n$. As was mentioned above, $S(t,x)$ is a 1-periodic function in each component of $x = (x_1, \dots, x_n)$ and so $S(t,x)$ can be presented in a form of the Fourier series ($\mathbf{i} = \sqrt{-1}$)

$$S(t,x) = \sum_j S_j(t) e^{2\pi\mathbf{i}\langle j,x\rangle}, \tag{18}$$

where $j = (j_1, \dots, j_n)$ is a multi-index and $\langle j,x\rangle = \sum_{q=1}^{n} j_q x_q$. Further, denoting dY_j^ε an *observable* random measure

$$dY_j^\varepsilon(t) = \int_{[0,1]^n} e^{-2\pi\mathbf{i}\langle j,x\rangle} Y^\varepsilon(dt,dx)$$

and taking use of well known properties of $W(dt,dx)$, we can rewrite the observation model (16) at the form

$$dY_j^\varepsilon(t) = S_j(t)dt + \varepsilon dW_j(t),\ 0 \le t \le T,\ j \in J, \tag{19}$$

J is the set of all multi-indices, $W_j(t)$ are independent standard Wiener processes. Thus it is possible to use approach of the Section **2** and to create the on-line estimator $\widetilde{S}_j(t)$ for each Fourier component $S_j(t)$:

$$\begin{aligned} d\widetilde{S}_j(t) &= \widetilde{S}_j^{(1)}(t)dt + \frac{q_1}{\varepsilon^{2/(2\beta'+1)}}\Big(dY_j^\varepsilon(t) - \widetilde{S}_j(t)dt\Big)\\ d\widetilde{S}_j^{(p)}(t) &= \widetilde{S}_j^{(p+1)}(t)dt + \frac{q_{p+1}}{\varepsilon^{2(p+1)/(2\beta'+1)}}\Big(dY_j^\varepsilon(t) - \widetilde{S}_j(t)dt\Big), \end{aligned}$$

$$p = 1, \dots, k$$
$$d\widetilde{S}_j^{(k)}(t) = \frac{q_{k+1}}{\varepsilon^{2(k+1)/(2\beta'+1)}} \left(dY_j^\varepsilon(t) - \widetilde{S}_j(t)dt \right), \tag{20}$$

subject to deterministic initial conditions $\widetilde{S}_j(0)$, $\widetilde{S}_j^{(p)}(0)$, $p = 1, \dots, k$ such that for some $c > 0$

$$\sum_j \left(\widetilde{S}_j^2(0) + \sum_{p=1}^k (\widetilde{S}_j^{(p)}(0))^2 \right) \leq c. \tag{21}$$

Here $\widetilde{S}_j^{(p)}(t)$ is estimator for $S_j^{(p)}(t)$, $p = 1, \dots, k$. Then we can create estimator for $S(t, x)$ as follows:

$$\tilde{S_N}(t, x) = \sum_{j \in \mathcal{N}} \tilde{S}_j(t) e^{2\pi \mathbf{i} \langle j, x \rangle}, \tag{22}$$

where $\mathcal{N}$ is a finite set of multi-indices. We characterize the set $\mathcal{N}$ by multi-index N with positive components $N_1, \dots, N_n$ so that

$$\mathcal{N} = \{ j : |j_q| \leq N_q, q = 1, \dots, n \}. \tag{23}$$

Estimator (22) is therefore a symbiosis of Kalman-type and Chentsov's projection (see, e.g., [7]) estimators. The only question is: if it is possible to choose multi-index $N = N^\varepsilon$ and the parameter β' of the Kalman-type filter for to reach an optimal rate of convergence to 0 risks, as $\varepsilon \to 0$? It is proven in [3] that the answer is positive if the following natural conditions concerning smoothness of S are fulfilled.

Condition **A** (Smoothness in t).

$S(t, x)$ is k times differentiable in t and $D_t^k S(t, x) = \frac{\partial^k S(t,x)}{\partial t^k}$ is Hölder continuous in $L^2_{[0,1]^n}$ norm with exponent $\alpha (\leq 1)$: for $t, t + h \in [0, T]$

$$\| D_t^k S(t + h, .) - D_t^k S(t, .) \|^2_{L^2_{[0,1]^n}} \leq L_0 |h|^{2\alpha}.$$

The parameter $\beta_0 = k + \alpha$ characterizes a smoothness in t of $S(t, x)$.

Condition **B** (Smoothness in x).

For $q = 1, \dots, n$

$$\sup_{t \leq T} \| D_{x_q}^{\beta_q} S(t, x) \|^2_{L^2_{[0,1]^n}} = \sup_{t \leq T} \sum_j |2\pi j_q|^{2\beta_q} \left| S_j(t) \right|^2 \leq L_q < \infty.$$

Theorem 3. *Assume* (**A**) *and* (**B**) *and let parameters* $q_1, \dots, q_{k+1}$ *are chosen so that* (8) *is valid. Choose* $N_q^\varepsilon \asymp \varepsilon^{-\frac{2\beta}{\beta_q(2\beta+1)}}$, $q = 1, \dots, n$, *and* $\beta' = \frac{(1+2\beta)\beta_0-\beta}{2\beta}$. *Then estimator (22) with* $N = N^\varepsilon$ *has for* ε *small enough property*

$$E\big\|S(t,*) - \widetilde{S}_{N^\varepsilon}(t,*)\big\|^2_{L^2_{[0,1]^n}} \leq C\varepsilon^{\frac{4\beta}{2\beta+1}}$$

for $t \in \left[C\varepsilon^{\frac{2\beta}{(2\beta+1)\beta_0}}\log(1/\varepsilon),\ T\right]$.

Remark 3. It is proven in [3] also that under suitable conditions the time derivatives of $\widetilde{S}_{N^\varepsilon}(t,*)$ are estimators of the time derivatives of $S(t,x)$ with the best possible rate of convergence risks to 0 for $\varepsilon \to 0$.

Condition **B** means the global smoothness of S with respect to x. So Theorem 3 does not give the answer to the question: if it is possible to create recursive (in time) estimator for $S(t,x^{(0)})$ with optimal rate of convergence of risk to 0 if only information on smoothness S in $x \in O(x^{(0)})$ is available (here and below $O(x^{(0)})$ is some ball in $\mathbb{R}^n$ with center at $x^{(0)}$)? For this case it is more natural to use symbiosis of kernel estimator in x and Kalman-type one in t.

Assume that instead of conditions **A** and **B** the following conditions are fulfilled.

Condition **A1**. The function $S(t,x) = S(t,x_1,\dots,x_n)$ has smoothness $\beta_i = k_i + \alpha_i$ in $x_i, i = 1,\dots,n$ uniformly in $t \in [0,T], x \in O(x^{(0)})$:

$$|D^{k_i}_{x_i}S(t,x_1,\dots,x_i+h,x_{i+1},\dots,x_n)$$
$$-D^{k_i}_{x_i}S(t,x_1,\dots,x_i,x_{i+1},\dots,x_n)| \leq L|h|^{\alpha_i}.$$

Condition **B1**. $S(t,x)$ has the smoothness $\beta_0 = k + \alpha$ in t also uniformly in $t \in [0,T], x \in O(x^{(0)})$:

$$|D^k_t S(t+h,x) - D^k_t S(t,x)| \leq L|h|^\alpha$$

Let $S(t,x)$ satisfy the conditions **A1** ,**B1** and $K(x), x \in \mathbb{R}^n$ be a compactly supported function with properties (cp. (4)):

$$\int_{\mathbb{R}^n} K(x)dx = 1; \int_{\mathbb{R}^n} x_i^l K(x)dx = 0; l = 1,...,k_i; i = 1,...n.$$

(Note (cp.[2]) that such a function can be found at the form

$$K(x) = \prod_{i=1}^n K_i(x_i),$$

where

$$\int_{\mathbb{R}^1} K_i(x)dx = 1; \int_{\mathbb{R}^1} x^l K_i(x)dx = 0;\ l = 1, \dots, k_i.)$$

Consider the following symbiosis of a kernel in x and on-line in t estimators for the model (16).Denote

$$\tilde{Y_\varepsilon}(dt,x) = \int_{\mathbb{R}^n} K_{\bar{\delta}}(x-y) Y^\varepsilon(dt,dy) \tag{24}$$

Here

$$\bar{\delta} = (\delta_1, ..., \delta_n), K_{\bar{\delta}}(x) = (\delta_1 ... \delta_n)^{-1} K(\frac{x_1}{\delta_1}, ..., \frac{x_n}{\delta_n}), \delta_i \asymp \varepsilon^{\frac{2\beta}{\beta_i(2\beta+1)}}, \tag{25}$$

β is defined in (15). Consider now the estimator

$$\begin{aligned} d\tilde{S}_\varepsilon(t,x) &= \tilde{S}^{(1)}_\varepsilon(t,x)dt + \frac{q_1}{\varepsilon^{2/(2\beta+1)}}\big(\tilde{Y_\varepsilon}(dt,x) - \tilde{S}_\varepsilon(t,x)dt\big), \\ d\tilde{S}^{(j)}_\varepsilon(t,x) &= \tilde{S}^{(j+1)}(t,x)dt + \frac{q_{j+1}}{\varepsilon^{2(j+1)/(2\beta+1)}}\big(\tilde{Y_\varepsilon}(dt,x) - \tilde{S}_\varepsilon(t,x)dt\big), \\ &\quad j = 1, \dots, k-1, \\ d\tilde{S}^{(k)}_\varepsilon(t,x) &= \frac{q_{k+1}}{\varepsilon^{2k/(2\beta+1)}}\big(\tilde{Y_\varepsilon}(dt,x) - \tilde{S}_\varepsilon(t,x)dt\big) \end{aligned} \tag{26}$$

with initial conditions $\tilde{S}^{(j)}_\varepsilon(0,x) = S^{(j)}_\circ(x)$, $j = 0, \dots, k$. We assume again that parameters $q_1, \dots, q_{k+1}$ are chosen so that all roots λ_i of the polynomial (7) have strictly negative real parts.

Theorem 4. *Let $S(t,x)$ satisfy conditions (**A1**) and (**B1**). Then the estimator given by (24), (25), and (26) has the property: for*

$$t \in [C\varepsilon^{\frac{2\beta}{\beta_0(2\beta+1)}} \log(1/\varepsilon), T],\ x \in O(x^{(0)})$$

and ε small enough

$$E(|\frac{\tilde{S}_\varepsilon(t,x) - S(t,x)}{\varepsilon^{2\beta/(2\beta+1)}}|^2 + \sum_{j=1}^k |\frac{\tilde{S}^{(j)}_\varepsilon(t,x) - D^j_t S(t,x)}{\varepsilon^{\frac{2\beta}{2\beta+1}(1-\frac{j}{\beta_0})}}|^2) \leq C$$

Proof. Denote

$$S_{\bar{\delta}}(t,x) = \int_{\mathbb{R}^n} K_{\bar{\delta}}(x-y) S(t,y))dy. \tag{27}$$

Then we can rewrite (24) at the form

$$\tilde{Y_\varepsilon}(dt,x) = S_{\bar{\delta}}(t,x)dt + \varepsilon \int_{\mathbb{R}^n} K_{\bar{\delta}}(x-y) W(dt,dy) \tag{28}$$

It is well known (see, e.g., Lemma 3.1 in [2]) that under condition **A1** the upper bound

$$|S_{\bar{\delta}}(t,x) - S(t,x)| < C(\delta_1^{\beta_1} + ... + \delta_n^{\beta_n}) \tag{29}$$

is valid for $x \in O(x^{(0)})$.

Lemma 1. *For any fixed* x *the expression*

$$\int_A \zeta_{\bar{\delta}}(dt) = \int_A \int_{\mathbb{R}^n} K_{\bar{\delta}}(x-y) W(dt, dy), \ A \subset [0,T]$$

is a Gaussian orthogonal random measure on $\mathbb{R}^1$ *, and*

$$E[\zeta_{\bar{\delta}}(dt)] = 0; E[\zeta_{\bar{\delta}}(dt)]^2 = (\delta_1 ... \delta_n)^{-1} \|K\|^2_{L^2_{[0,1]^n}} dt.$$

Proof of this lemma is straightforward consequence of (17) and properties of stochastic integral.

Due to Lemma 1 the equation (28) can be rewritten as

$$\tilde{Y}_\varepsilon(dt,x) = S_{\bar{\delta}}(t,x)dt + \varepsilon_\delta \|K\|_{L^2_{[0,1]^n}} d\tilde{w}(t),$$

where $\tilde{w}(t)$ is a standard Wiener process, $\varepsilon_\delta = \varepsilon(\delta_1 ... \delta_n)^{-1/2}$. It follows from (27) and condition **B1** that the function $S_{\bar{\delta}}(t,x)$ also has the smoothness β_0 in t. Thus, applying Theorem 1, we can assert that for $t \in [C_1 \varepsilon_\delta^{\frac{2}{2\beta+1}} \log(1/\varepsilon_\delta), T]$ and ε_δ small enough the upper bound

$$E\Big(\Big|\frac{\tilde{S}_\varepsilon(t,x) - S(t,x)}{\varepsilon_\delta^{2\beta_0/(2\beta_0+1)}}\Big|^2 + \sum_{j=1}^{k} \Big|\frac{\tilde{S}_\varepsilon^{(j)}(t,x) - D_t^j S(t,x)}{\varepsilon_\delta^{2(\beta_0-j)/(2\beta_0+1)}}\Big|^2\Big) \le C \tag{30}$$

is valid. It is easy to check that for parameters δ_i chosen in accordance with (25) the equation

$$\varepsilon_\delta \asymp \varepsilon^{\frac{\beta(2\beta_0+1)}{\beta_0(2\beta+1)}} \tag{31}$$

holds. The assertion of theorem follows from (31) and (30).

5 Concluding Remarks

1. It follows from [2] that the rate of convergence of risks in Theorem 4 cannot be exceeded uniformly in an open set $K \subset \Sigma(\beta, L)$. It can be proven analogously to [1] that the initial boundary layer of the order $\varepsilon^{\frac{2\beta}{\beta_0(2\beta+1)}}$ is also inevitable for on-line estimators.

2. The natural question arises: if it is possible to create an analogous recursive on-line estimators for the space derivatives of $S(t,x)$? The answer to this question is positive if some additional information on the smoothness of $D^{(j)}_{x_i}S(t,x)$ in t is available. One way to construct this sort of estimators is the making use of the approach at the Theorem 4. We outline it restricting ourselves only by estimation of $D_{x_i}S(t,x)$. The Kalman-type estimator (26) can be used again but instead of $\tilde{Y}_\varepsilon(dt,x)$ one have to use statistics

$$D_{x_i}\tilde{Y}_\varepsilon(dt,x) := \int_{\mathbb{R}^n} D_{x_i}K_{\bar{\delta}}(x-y)Y^\varepsilon(dt,dy).$$

Thus one can use the estimator

$$\begin{aligned} dD_{x_i}\tilde{S}_\varepsilon(t,x) &= D_{x_i}\tilde{S}^{(1)}_\varepsilon(t,x)dt \\ &\quad + \frac{q_1}{\varepsilon^{2/(2\beta+1)}}\big(D_{x_i}\tilde{Y}_\varepsilon(dt,x) - D_{x_i}\tilde{S}_\varepsilon(t,x)dt\big), \\ dD_{x_i}\tilde{S}^{(j)}_\varepsilon(t,x) &= D_{x_i}\tilde{S}^{(j+1)}(t,x)dt \\ &\quad + + \frac{q_{j+1}}{\varepsilon^{2(j+1)/(2\beta+1)}}\big(D_{x_i}\tilde{Y}_\varepsilon(dt,x) - D_{x_i}\tilde{S}_\varepsilon(t,x)dt\big), \\ j &= 1,\dots,k-1, \\ dD_{x_i}\tilde{S}^{(k)}_\varepsilon(t,x) &= \frac{q_{k+1}}{\varepsilon^{2k/(2\beta+1)}}\big(D_{x_i}\tilde{Y}_\varepsilon(dt,x) - D_{x_i}\tilde{S}_\varepsilon(t,x)dt\big) \end{aligned} \tag{32}$$

with arbitrary initial conditions. If the a'priori information $D_{x_i}S(t,x) \in \Sigma(\beta^*, L^*)$ is available, one can prove completely analogously to Theorem 4 that $D_{x_i}\tilde{S}(t,x), D_{x_i}\tilde{S}^{(1)}(t,x),\dots,D_{x_i}\tilde{S}^{(k)}(t,x)$ are 'good' (in the sense of Theorem 4) estimators for $D_{x_i}S(t,x),\dots,D_{x_i}D^{(k)}_t S(t,x)$.

Acknowledgement. The research of this author was partially supported by NSF under Grant DMS 9971608.

References

1. Chow, P.-L., Khasminskii, R., and Liptser, R. Sh. (1977) Tracking of signal and its derivatives in Gaussian white noise, *Stochastic Processes and Applications* **69**, pp. 259-273.
2. Chow, P.-L.,Ibragimov I., and Khasminskii, R. (1999) Statistical approach to some ill-posed problems for linear partial differential equations. *Probab. Theory Relat.Fields*, **113**, pp. 421-441.
3. Chow, P.-L., Khasminskii, R., and Liptser, R. Sh. (2001) On estimation of time dependent spatial signal in Gaussian white noise, *Stochastic Processes and Applications* ,**96**, pp.161-175.
4. Ibragimov, I. and Khasminskii, R. (1980) On nonparametrical regression estimation, *Soviet Math.Doklady*, **252**, No 4, pp.780-784.
5. Ibragimov, I. and Khasminskii, R. (1980) Estimation of signal, its derivatives, and point of maximum for Gaussian distributions. *Theory of Probab. Appl.*, **15**, pp. 703–720.
6. Ibragimov, I. and Khasminskii, R. (1981) Asymptotic quality bounds for regression estimation in L_p, *Zap. Nauchn. Sem. LOMI*, **97**, pp.88-101 (In Russian).

7. Ibragimov, I. and Khasminskii, R. *Statistical estimation: Asymptotic theory.* Springer Verlag, 1981.
8. Khasminskii, R. and Liptser, R. Sh. (2001) On-line estimation of a smooth regression function, submitted to *Theory of Probab. Appl.*
9. Nussbaum, M. (1983) Optimal filtration of a function of many variables in white Gaussian noise. *Problems Inform. Transmission*, **19**, 2, pp. 23-29.
10. Parzen, E. (1962) On estimation of a probability density function and mode. *Ann. Math. Statist.*, **33**, No 3, pp. 1065–1073.
11. Pinsker, M. (1980) Optimal filtration of square-integrable signals in Gaussian noise,*Problems Inform. Transmission*, **16**, pp. 52-68.
12. Rosenblatt, M. (1956) Remarks on some nonparametric estimates of a density function. *Ann. Math. Statist.*, **27**, No 3, pp. 832–837.
13. Stone, C. (1980) Optimal rates of convergence for nonparametric estimators, *Ann. Statist.*, **8**, pp. 1348–1360.
14. Stone, C. (1982) Optimal global rates of convergence for nonparametric regression, *Ann. Statist.*, **10**, pp. 1040–1053.

Detection and Estimation in Stochastic Systems with Time-Varying Parameters

Tze Leung Lai

Department of Statistics, Stanford University, Stanford, CA 94305, USA

Abstract. We give a brief survey of recent developments in change-point detection and diagnosis and in estimation of parameters that may undergo occasional changes. There is a large variety of detection and estimation procedures widely scattered in the engineering, economics, statistics and biomedical literature. These procedures can be broadly classified as sequential (or on-line) and fixed sample (or off-line). We focus on detection and estimation procedures that strike a suitable balance between computational complexity and statistical efficiency, and present some of their asymptotically optimal properties.

1 Introduction

The subject of change-point detection in stochastic systems has many applications, including statistical quality control, fault detection and diagnosis in complex dynamical systems, biomolecular sequence analysis and modeling structural changes in economic systems. There is a large literature on the subject widely scattered in statistics, engineering, bioinformatics and econometrics. We can broadly classify change-point detection problems as sequential (or "on-line") and fixed sample (or "off-line"). Section 2 gives a review of sequential change-point detection theory. Diagnosis of the nature of a change, or "fault isolation," is often an important sequel to fault detection in applications, and Section 3 describes some recent developments in the theory of on-line fault detection and isolation. Section 4 considers fixed sample change-point problems.

Closely related to the problem of detection of parameter changes in stochastic systems is the problem of estimating the time-varying parameters and predicting future states of the system. In state-space models, if the parameters change with time, it is more convenient to regard them as states. A Bayesian approach to do this is to take a stochastic process as the prior distribution for the time-varying parameters, whose posterior distribution then provides an estimate of the current parameter value given the observations. This reduces the estimation problem to a filtering or smoothing problem, which is typically nonlinear except for some particular models of parameter variations, and which is usually very complicated to compute and analyze. In Section 5 we describe some bounded-complexity approximations to these infinite-dimensional filters/smoothers and show that despite the much lower computational complexity which enables them to be implemented on-line,

B. Pasik-Duncan: Stochastic Theory and Control, LNCIS 280, pp. 251–265, 2002.

they can be asymptotically as efficient as the computationally intensive Bayes procedures.

2 Theory of Sequential Change-Point Detection

When the observations X_t are independent with a common density function f_0 for $t < \nu$ and with another common density function f_1 for $t \geq \nu$, Shiryayev [38] formulated the problem of optimal sequential detection of the change-time ν in a Bayesian framework, by putting a geometric prior distribution on ν and assuming a loss of c for each observation taken after ν and a loss of 1 for a false alarm before ν. He used optimal stopping theory to show that the Bayes rule triggers an alarm as soon as the posterior probability that a change has occurred exceeds some fixed level. Instead of the Bayesian approach, Lorden [29] used the minimax approach of minimizing the worst-case expected delay

$$\bar{E}_1(T) = \sup_{\nu \geq 1} \text{ ess } \sup E[(T - \nu + 1)^+ | X_1, \ldots, X_{\nu-1}] \tag{1}$$

over the class $\mathcal{F}_\gamma$ of all rules T satisfying the lower bound constraint $E_0(T) \geq \gamma$ on the expected duration to false alarm. He showed that as $\gamma \to \infty$, (1) is asymptotically minimized by the CUSUM (cumulative sum) rule

$$N = \inf\{n : \max_{1 \leq k \leq n} \sum_{t=k}^{n} \log(f_1(X_t)/f_0(X_t)) \geq c\} \tag{2}$$

proposed by Page [34], where c is so chosen that $E_0(N) = \gamma$. Specifically he showed that as $\gamma \to \infty$,

$$\bar{E}_1(N) \sim \inf_{T \in \mathcal{F}_\gamma} \bar{E}_1(T) \sim (\log \gamma)/I(f_1, f_0), \tag{3}$$

where $I(f_1, f_0) = E_1\{\log(f_1(X_t)/f_0(X_t))\}$ is the Kullback-Leibler information number. Lorden's method is to relate the CUSUM procedure to certain one-sided sequential probability ratio tests which are optimal for testing f_0 versus f_1. Instead of studying the optimal detection problem via sequential testing theory, Moustakides [30] was able to formulate the worst-case detection delay problem subject to a false alarm constraint as an optimal stopping problem and to prove that the CUSUM rule is a solution to the optimal stopping problem. Bansal and Papantoni-Kazakos [3] extended Lorden's asymptotic theory to the case of stationary ergodic X_t before, and after, ν, and Yakir [41] generalized Shiryayev's Bayesian approach to finite-state Markov chains.

The CUSUM rule (2) can be easily extended to non-independent observations by replacing $f_j(X_t)$ in (2) by the conditional density $f_j(X_t | X_1, \ldots, X_{t-1})$ for $j = 0, 1$. By using a change-of-measure argument and the strong law for log-likelihood ratio statistics, Lai [22] has shown that (3) holds quite generally for such extensions of the CUSUM rule to general stochastic systems. Moreover, alternative performance criteria are proposed in [21] and [22] and the CUSUM rule is shown to be asymptotically optimal under these performance criteria.

In practice, the post-change distributions are usually modeled by parametric families with unknown parameters, and the preceding theory that assumes completely specified $f_1(\cdot|x_1, \dots, x_{t-1})$ is too restrictive. Nevertheless the asymptotic theory assuming known f_1 provides benchmarks that we try to attain even when unknown parameters are present. An obvious way to modify the CUSUM rule for the case of $f_\theta(\cdot|x_1, \dots, x_{t-1})$ with unknown post-change parameter θ is to estimate it by maximum likelihood, leading to the generalized likelihood ratio (GLR) rule

$$N_G = \inf\left\{n : \max_{1\le k\le n} \sup_{\theta\in\Theta} \sum_{t=k}^{n} \log \frac{f_\theta(X_t|X_1, \dots, X_{t-1})}{f_{\theta_0}(X_t|X_1, \dots, X_{t-1})} \ge c\right\}. \tag{4}$$

For the problem of detecting shifts in the mean θ of independent normal observations with known variance, this idea was proposed by Barnard [4], but the statistical properties of the procedure remained a long-standing problem that was recently solved by Siegmund and Venkatraman [39], whose asymptotic approximations for the average run lengths $E_\theta(N_G)$ of the GLR rule under $\theta = \theta_0$ and under $\theta \neq \theta_0$ show that the GLR rule is asymptotically optimal in the sense of (3).

For practical implementation, the CUSUM rule (2) can be written in the recursive form $N = \inf\{n : \ell_n \ge c\}$, where $\ell_n = \{\ell_{n-1}+\log(f_1(X_n)/f_0(X_n))\}^+$ with $\ell_0 = 0$. The GLR rule (4) does not have such convenient recursive forms and the memory requirements and number of computations at time n grow to infinity with n. A natural modification to get around this difficulty is to replace $\max_{1\le k\le n}$ in (4) by $\max_{n-M\le k\le n}$. Such window-limited GLR rules were first introduced by Willsky and Jones [40] in the context of detecting additive changes in linear state-space models. Consider the stochastic system described by the state-space representation of the observed signals y_t:

$$x_{t+1} = F_t x_t + G_t u_t + w_t \tag{5a}$$

$$y_t = H_t x_t + J_t u_t + \epsilon_t \tag{5b}$$

in which the unobservable state vector x_t, the input vector u_t and the measurement vector y_t have dimensions p, q and r, respectively, and w_t, ϵ_t are independent Gaussian vectors with zero means and $\mathrm{Cov}(w_t) = Q_t, \mathrm{Cov}(\epsilon_t) = R_t$. The Kalman filter provides a recursive algorithm to compute the conditional expectation $\widehat{x}_{t|t-1}$ of the state x_t given the past observations $y_{t-1}, u_{t-1}, y_{t-2}, u_{t-2}, \dots$ The innovations $e_t := y_t - H_t\widehat{x}_{t|t-1} - J_t u_t$ are independent Gaussian vectors with covariance matrices V_t, and their means $m_t = E(e_t)$ for $t \ge \nu$ are of the form $m_t = \rho(t, \nu)\theta$ instead of the baseline values $m_t = 0$

for $t < \nu$, where $\rho(t,k)$ is a matrix that can be computed recursively and θ is an unknown parameter vector, cf. [5], [40]. Willsky and Jones [40] proposed the window-limited GLR detector of the form

$$\begin{aligned} N_w &= \inf\left\{n : \max_{n-M\le k\le n-\tilde M} \sup_\theta \sum_{i=k}^n \log\left[\frac{f(V_i^{-1/2}(e_i - \rho(i,k)\theta))}{f(V_i^{-1/2}e_i)}\right] \ge c\right\} \\ &= \inf\left\{n : \max_{n-M\le k\le n-\tilde M} \left(\sum_{i=k}^n \rho^T(i,k)V_i^{-1}e_i\right)^T\right. \\ &\qquad \times \left(\sum_{i=k}^n \rho^T(i,k)V_i^{-1}\rho(i,k)\right)^{-1} \\ &\qquad \times \left.\left(\sum_{i=k}^n \rho^T(i,k)V_i^{-1}e_i\right)/2 \ge c\right\}, \end{aligned} \tag{6}$$

where f denotes the standard d-dimensional normal density function and $d = \dim(\theta)$.

Although window-limited GLR rules of the type (6) found many successful applications in fault detection of navigation and other control systems and in tracking of maneuvering targets and signal processing, it remained a difficult open problem concerning how the window size and threshold should be chosen in (6) and whether the corresponding rule has any optimal properties (cf. [5]). To address this problem, Lai [21] began by considering the simpler situation of detecting changes in the mean θ of independent normal observations $X_1, X_2, \ldots$. Here the window-limited GLR rule has the form

$$N_W = \inf\left\{n : \max_{n-M\le k\le n} (X_k + \cdots + X_n)^2/2(n-k+1) \ge c\right\}, \tag{7}$$

and the methods of Siegmund and Venkatraman [39] to analyze the GLR rule (4) in this independent normal case can be extended to the window-limited modification (7). In particular, if we choose $M \sim \gamma$, then we have $E_0 N_W \sim E_0 N_G \sim Kc^{-1/2}e^c$ as $c \to \infty$, where an explicit formula for K is given in [39]. Therefore, choosing $c = \log\gamma + \frac{1}{2}\log\log\gamma - \log K + o(1)$ gives $E_0 N_W \sim E_0 N_G \sim \gamma$. With this choice of c, we also have $E_\theta N_W \sim E_\theta N_G \sim \min\{\gamma, (2\log\gamma)/\theta^2\}$ uniformly in $|\theta| \le (\log\gamma)^{\frac{1}{2}-\epsilon}$ for every $\epsilon > 0$. The choice $M = \gamma$ for the window size in (7) requires computation of $\gamma + 1$ quantities $(X_{n-i} + \cdots + X_n)^2/(i+1)$, $i = 0, \ldots, \gamma$, at every stage $n > \gamma$, and it is desirable to reduce the computational burden for large γ by using a smaller window size. To develop efficient detection schemes that involve $O(\log\gamma)$ computations at every stage n, Lai [21] replaced $\max_{0\le n-k\le M}$ in (7) by $\max_{n-k+1\in\mathcal{N}}$, where $\mathcal{N} = \{1, \ldots, M\} \cup \{[b^j M] : 1 \le j \le J\}$, with $M \sim a\log\gamma, b > 1$ and $J = \min\{j : [b^j M] \ge \gamma\} \sim (\log\gamma)/(\log b)$. Specifically, replacing N_W by

$$\widetilde{N}_W = \inf\left\{N : \max_{k:n-k+1\in\mathcal{N}} (X_k + \cdots X_n)^2/2(n-k+1) \geq c\right\}, \tag{8}$$

it is shown in [21] that $E_0(\widetilde{N}_W) \sim \widetilde{K}c^{-1/2}e^c \sim \gamma$ if $c = \log\gamma + \frac{1}{2}\log\log\gamma - \log\widetilde{K} + o(1)$, and that $E_\theta(\widetilde{N}_W) \sim (2\log\gamma)/\theta^2$ if $|\theta| > \sqrt{2/a}$ while $E_\theta(\widetilde{N}_W) \leq (1+o(1))\min\{\gamma, (2b\log\gamma)/\theta^2\}$ uniformly in $|\theta| \leq \sqrt{2/a}$. Hence, choosing b close to 1 (say $b = 1.1$), there is little loss of efficiency in reducing the computational complexity of N_G by its window-limited modification (8).

Lai and Shan [25] extended the preceding idea to address the long-standing problem concerning how the window size and the threshold should be chosen in the Willsky-Jones rule (6). As pointed out by Basseville and Nikiforov [5], the main difficulty of this problem lies in the coupling effect between the threshold and window size on the performance of the rule. The basic idea of Lai and Shan is to decouple the effects of the threshold and window size. A threshold of the order of $\log\gamma$ is needed to ensure a false alarm duration of γ, as in the simpler problem of detecting changes in a normal mean. With the threshold thus chosen to control the false alarm rate, the choice of the windows is targeted towards making the rule as efficient as possible for detecting the unknown change. Putting a complexity constraint of the order of $O(\log\gamma)$ on the number of elements of the window, the Willsky-Jones window in (6) is enlarged to the form $\widetilde{\mathcal{N}} = \{\widetilde{M}, \ldots, M\} \cup \{[b^j M] : 1 \leq j \leq J\}$ as in (8). Here we need a minimal delay $\widetilde{M} \geq \dim(\theta)$ to avoid difficulties with GLR statistics when $n - k < \dim(\theta)$. Under certain stability assumptions on the Kalman filter, such window-limited GLR rules with $\max_{n-M\leq k\leq n-\widetilde{M}}$ in (6) replaced by $\max_{k:n-k+1\in\widetilde{\mathcal{N}}}$ can be shown to be asymptotically optimal, under different performance criteria, for detecting changes with $I(\theta, 0) > a^{-1}$, and to be within b times the asymptotic lower bound for expected delay in detecting smaller changes. Moreover, Lai and Shan [25] have also shown that for these window-limited GLR rules T,

$$\sup_{k\geq 1} P_0(k \leq T < k+m) \sim P_0(T \leq m) \sim m/E_0(T),$$

as $E_0(T) \sim \gamma \to \infty$ and $m/\log\gamma \to \infty$ but $\log m = o(\log\gamma)$. Hence to determine the threshold c by Monte Carlo simulations, the constraint $E_0(T) \doteq \gamma$ can be replaced by the probability constraint $P_0(T \leq m) \doteq m/\gamma$, which is much more tractable since simulating $P_0(T \leq m)$ involves many fewer random variables (no more than m in each simulation run) than directly simulating $E_0(T)$. Importance sampling methods have also been developed in [25] for the Monte Carlo evaluation of $P_0(T \leq m)$.

By using weak convergence theory the CUSUM rule has been extended to non-likelihood-based detection statistics by Benveniste et al. [6], [7], whose "asymptotic local approach" can be summarized as follows. Suppose the detection statistics Y_i are such that for every fixed μ,

$$\left\{\gamma^{-1/2}\sum_{i=1}^{[\gamma t]} Y_i, t \geq 0\right\} \text{ converges weakly under } P_{\mu/\sqrt{\gamma}} \text{ to}$$
$$\{W_\mu(t), t \geq 1\} \text{ as } \gamma \to \infty, \tag{9}$$

where $\{W_\mu(t), t \geq 0\}$ is a multivariate Gaussian process with independent increments such that $EW_\mu(t) = \mu t$ and $\mathrm{Cov}(W_\mu(t)) = tV$. The baseline probability measure is P_0 that corresponds to $\mu = 0$. Let $S_{n,k} = \sum_{i=k}^{n} Y_i$ and $\lambda_\gamma = \lambda/\sqrt{\gamma}$. Consider the CUSUM rule

$$T_\gamma = \inf\left\{n : \max_{1\leq k\leq n} [\lambda_\gamma^T S_{n,k} - (n-k+1)\lambda_\gamma^T V \lambda_\gamma/2] \geq c\right\}. \tag{10}$$

Then the weak convergence property (9) implies that for fixed c, T_γ/γ converges in distribution under $P_{\mu/\sqrt{\gamma}}$ to $\tau_\mu(c) = \inf\{t : \max_{0\leq u\leq t}[\lambda^T(W_\mu(t) - W_\mu(u)) - (t-u)\lambda^T V\lambda/2] \geq c\}$, and therefore $E_{\mu/\sqrt{\gamma}}(T_\gamma) \sim \gamma E\tau_\mu(c)$. How should (10) be modified when λ is not specified in advance?

Maximizing the CUSUM statistics $\lambda_\gamma^T S_{n,k} - (n-k+1)\lambda_\gamma^T V\lambda_\gamma/2$ over λ_γ yields $S_{n,k}^T V^{-1} S_{n,k}/2(n-k+1)$, which leads to the following window-limited rule suggested in [7] and [43]:

$$\widetilde{T}_\gamma = \inf\left\{n > b_1\gamma : \max_{n-b_2\gamma\leq k\leq n-b_1\gamma} S_{n,k} V^{-1} S_{n,k}/2(n-k+1) \geq c\right\}, \tag{11}$$

with $b_2 > b_1 > 0$. For fixed c and μ, (9) implies that $\widetilde{T}_\gamma/\gamma$ converges in distribution under $P_{\mu/\sqrt{\gamma}}$ to $\widetilde{\tau}_\mu = \inf\{t > b_1 : \max_{t-b_2\leq u\leq t-b_1}(W_\mu(t) - W_\mu(u))^T V^{-1}(W_\mu(t) - W_\mu(u))/2(t-u) \geq c\}$. The preceding asymptotic approach has been called "local" because it is based on weak convergence of T_γ/γ or $\widetilde{T}_\gamma/\gamma$ under P_0 to the same limiting distribution as that in the canonical setting of independent standard normal $V^{-1/2}Y_t$, when λ_γ is of the order $\lambda/\sqrt{\gamma}$ in (10) or when the window is of the form $b_1\gamma \leq n-k \leq b_2\gamma$ in (11). Such choice of window size (or λ_γ) makes $\widetilde{T}_\gamma$ (or T_γ) very inefficient for detecting changes that are considerably larger than the $O(\gamma^{-1/2})$ order of magnitude for the "local" changes. Chan and Lai [10] have developed another approach based on moderate deviations theory (instead of weak convergence approximations) to extend the window-limited GLR rule (4) to non-likelihood-based detection statistics, which are widely used in monitoring adaptive algorithms for system identification and control (cf. [5], [7], [43]), leading to detection rules of the general form

$$N_W^* = \inf\left\{n > \widetilde{M} : \max_{k:n-k+1\in\mathcal{N}_{\widetilde{M}}} \left(\sum_{i=k}^{n} Y_i\right)^T V_{n,k}^{-1}\left(\sum_{i=k}^{n} Y_i\right)/2 \geq c\right\}, \tag{12}$$

where $c \sim \log\gamma$ (instead of bounded c in the asymptotic local approach), $\mathcal{N}_{\widetilde{M}} = \{\widetilde{M}j : j \in \mathcal{N}\}$, $\mathcal{N} = \{1, \ldots, M\} \cup \{[b^j M] : 1 \le j \le J\}, b > 1, M \sim ac, J = \min\{j : [b^j M] \ge \gamma\} \sim c/\log b$, $\widetilde{M}/c \to \infty$ (to ensure that the sums $\sum_{i=k}^{n} Y_i$ become approximately Gaussian in the moderate deviations sense), and $V_{n,k}$ is an estimate of the asymptotic covariance matrix of $S_{n,k}$ under P_0. Note that $V^{-1}/(n-k+1)$ in (11) corresponds to the inverse of $V_{n,k} = (n-k+1)V$. In many applications V is unknown and needs to be estimated. Even when V is known, using $V_{n,k}$ instead of $(n-k+1)V$ offers the flexibility of making adjustments for non-normality in treating (12) under P_0 as if it were a window-limited GLR rule with independent standard normal $V^{-1/2}Y_t$, where the Y_t are actually dependent random vectors with unknown distributions. Grouping the observations into batches as in (12) arises naturally in quality control applications in which samples of size $\widetilde{M}$ are taken at regular intervals of time.

3 On-Line Fault Detection and Isolation

The change diagnosis (or fault isolation) problem is to determine, upon detection of change in a system, which one in a set of possible changes has actually occurred, cf. [5], [14], [31], [36]. It is of particular importance in integrity monitoring of a navigation system whose goal is to detect and isolate faulty measurement sources so that they can be removed from the navigation solution as soon as possible (cf. [32], [35]). Although it is closely related to the subject of sequential multihypothesis testing, the unknown time of change makes it difficult to apply the sequential multihypothesis tests in the literature to infer the kind of change. Nikiforov [31] recently addressed the problem of jointly detecting and isolating changes subject to a certain constraint on false alarms/isolations, in the case where the observations X_t are independent with a common density function f_0 for $t < \nu$ and with another common density function belonging to the set $\{f_1, \ldots, f_J\}$ for $t \ge \nu$. Without assuming independence, suppose that under P_0, the conditional density of X_n given $X_1, \ldots, X_{n-1}$ is $f_0(\cdot\,|X_1, \ldots, X_{n-1})$ for every $n \ge 1$, and that under $P_j^{(\nu)}$, the conditional density is $f_0(\cdot \mid X_1, \ldots, X_{n-1})$ for $n < \nu$ and is $f_j(\cdot \mid X_1, \ldots, X_{n-1})$ for $n \ge \nu$ $(j = 1, \ldots, J)$. A detection-isolation rule can be represented by the pair $(T, \widehat{j})$, in which T is a stopping rule and $\widehat{j}$ is a terminal decision rule taking values in $\{1, \ldots, J\}$. Thus $\{T = t, \widehat{j} = j\}$, which represents the event that a change is detected at time t and determined to be of type j, belongs to the σ-field generated by $X_1, \ldots, X_t$. For the case of independent X_t, Nikiforov [31] established an asymptotic lower bound for the worst-case detection-isolation delay

$$\bar{E}(T) = \sup_{\nu \ge 1, 1 \le j \le J} \text{ess sup}\, E_j^{(\nu)} \left[(T - \nu + 1)^+ \,|X_1, \ldots, X_{\nu-1}\right], \tag{13}$$

subject to the constraint

$$E_0(T) \geq \gamma \quad \text{and} \quad \infty > E_j^{(1)}(T_{n(h)}) \geq \gamma \quad \text{for} \quad 1 \leq h \neq j \leq J\ , \tag{14}$$

where $(T_1, \widehat{j}_1)$, $(T_2 - T_1, \widehat{j}_2), \ldots$ are i.i.d. copies of $(T, \widehat{j})$ and $n(h) = \inf\{r \geq 1 : \widehat{j}_r = h\}$.

Without this independence assumption, Lai [23] recently showed that (14) can be replaced by the following more general constraint on false alarms and false isolations:

$$\begin{aligned} &E_0(T) \geq \gamma \text{ and } P_j^{(1)}\{T < \infty,\ \widehat{j} \neq j\} \leq a_\gamma E_j^{(1)}(T) \text{ with} \\ &\qquad \log a_\gamma^{-1} \sim \log \gamma \text{ as } \gamma \to \infty. \end{aligned} \tag{15}$$

By extending the methods developed in [22], he generalized Nikiforov's asymptotic lower bound for the worst-case detection-isolation delay far beyond the independent model considered by Nikiforov, subject to the constraint (15) on false alarms and false isolations. He also generalized Nikiforov's construction of asymptotically optimal detection-isolation rules that attain the lower bound. The constraint on $P_j^{(1)}\{T < \infty, \widehat{j} \neq j\}$ in (15) assumes that the change occurs at the outset ($t = 1$) to avoid false alarms (for which correctness of the change-type determination following the alarm becomes meaningless) under $P_j^{(\nu)}$. A more natural constraint is the error probability of the terminal decision rule under $P_j^{(\nu)}$, which is $P_j^{(\nu)}(T < \nu) + P_j^{(\nu)}(\nu \leq T < \infty, \widehat{j} \neq j)$, consisting of the false alarm probability and the probability of false isolation when the alarm is correct. Using a Bayesian approach that assumes a prior distribution p on the change-type j and an independent prior distribution π_α on the change-time ν, the error probability constraint can be formulated as

$$\sum_{j=1}^{J} \sum_{\nu=1}^{\infty} p(j)\pi_\alpha(\nu) P_j^{(\nu)}(\infty > T \geq \nu, \widehat{j} \neq j) + \sum_{\nu=1}^{\infty} \pi_\alpha(\nu) P_0(T < \nu) \leq \alpha, \tag{16}$$

noting that $P_j^{(\nu)}(T < \nu) = P_0(T < \nu)$. Subject to this probability constraint on false alarms and false isolations, an asymptotic lower bound for the Bayesian expected delay $\sum_{\nu=1}^{\infty} \pi_\alpha(\nu) E_j^{(\nu)}(T - \nu + 1)^+$ is derived in [23], where it is also shown that this asymptotic lower bound, subject to (4), can be attained by relatively simple rules involving likelihood ratio statistics.

In practice, the post-change distribution corresponding to the jth type of change often involves unknown parameters. Instead of a known conditional density function $f_j(\cdot|X_1, \ldots, X_{t-1})$ for $t \geq \nu$, suppose that one has a parametric family of density functions $f_{j,\theta}(\cdot|X_1, \ldots, X_{t-1})$, $\theta \in \Theta_j$. As an alternative to (15) and (16), Lai [23] also introduced the following constraint on the false detection-isolation probability per unit time:

$$\sup_{\nu\ge 1} P_0(\nu \le T < \nu + m_\alpha)/m_\alpha \le \alpha,$$

$$\sup_{\nu\ge 1} P_{\theta_j}^{(\nu)}(\nu \le T < \nu + m_\alpha,\ \widehat{j} \ne j)/m_\alpha \le \alpha\ \forall \theta_j \in \Theta_j,\ 1 \le j \le J. \tag{17}$$

He developed an asymptotic lower bound for the detection-isolation delay subject to this constraint, and showed that this bound can be attained by the window-limited GLR rule

$$\widehat{\tau} = \inf\left\{ n > \widetilde{M} : \exists\, 1 \le j \le J \text{ such that } \max_{n-M\le k\le n-\widetilde{M}} \sup_{\theta_j\in\Theta_j} \sum_{t=k}^{n} Z_t(\theta_j, 0) \right. \tag{18}$$

$$\left. \ge c + \left(\max_{0\le h\le J, h\ne j} \max_{n-M\le k\le n-\widetilde{M}} \sup_{\theta\in\Theta_h} \sum_{t=k}^{n} Z_t(\theta_h, 0) \right)^{+} \right\}, \tag{19}$$

where $Z_t(\theta_j, 0) = \log\{f_{j,\theta_j}(X_t|X_1, \ldots, X_{t-1})/f_0(X_t|X_1, \ldots, X_{t-1})\}$, $\theta_j \in \Theta_j$. This kind of rules is an extension of the window-limited GLR rules for sequential change-point detection in Section 2, where how M and $\widetilde{M}$ can be chosen is also discussed.

4 Fixed Sample Change-point Problems

Unlike sequential change-point detection in which one observes a stochastic system sequentially and the detection rule can be represented by a stopping rule that depends on the current and past observations, fixed sample change-point problems deal with testing for changes, to be discussed in this section, and estimating the change-points and other parameters from a given set of data, to be considered in Section 5. Such problems arise naturally in applications to econometrics, biomolecular sequence analysis and gene mapping.

Beginning with Quandt's pioneering paper [37] on testing for a change in a linear regression model, an active area of research in econometrics is testing and modeling structural changes. Some recent directions in this area include likelihood ratio tests of multiple structural changes in regression models (e.g. [1], [2]) and modeling and testing for structural changes in volatilities of stock returns (e.g. [20], [26]). There is also an extensive literature in statistics on tests of the hypothesis $H_0 : f_i = g$ for $1 \le i \le n$ versus the change-point alternative $H_1 : \exists\, \nu$ such that $f_i = g$ for $i < \nu$ and $f_i = \widetilde{g}(\ne g)$ for $i \le \nu \le n$, based on independent observations $Y_1, \ldots, Y_n$ such that Y_i is a d-dimensional random vector with density function f_i. Here g and $\widetilde{g}$ are density functions belonging to a given parametric family $\{g_\theta, \theta \in \Theta\}$. Some representative works include [8], [12] and [18]. The GLR or score statistics for testing these hypotheses can be expressed in the form

$$\mathcal{M}_n = \max_{1\le i<j\le n: j-i\in J_n} (j-i)h((S_j - S_i)/(j-i)), \tag{20}$$

where $S_j = \Sigma_{t=1}^n Y_t$. Statistics of the type (19) also arise in biomolecular sequence analysis under the rubric of "scan statistics." Under certain conditions, Chan and Lai [9] have recently shown that there exist $q \in \{0, \dots, d\}$ and $r > 0$ (depending on h) such that as $n \to \infty$,

$$\mathcal{M}_n - r\left\{\log n + \left(\frac{q}{2}\right)\log\log n\right\}$$
$$\text{has a limiting extreme-value distribution.} \tag{21}$$

This generalizes an earlier result of Karlin, Dembo and Kawabata [19] who consider the case $d = 1$ and $h(x) = x$ (for which $q = 0$) in the context of high-scoring segments in a DNA sequence.

As pointed out in [9], the asymptotic distribution theory of the scan statistic $\mathcal{M}_n$ based on a fixed sample of size n is closely related to that of the sequential detection rule

$$T_c = \inf\left\{n : \max_{k<n: n-k\in J(c)} (n-k)h((S_n - S_k)/(n-k)) > c\right\}. \tag{22}$$

Analogous to (20),

$$e^{-c/r}(c/r)^{q/2}T_c \quad \text{has a limiting exponential distribution as} \quad c \to \infty\ ; \tag{23}$$

see [9]. The special case of this result for univariate normal Y_i with zero means and $h(x) = x^2/2$ (for which $q = r = 1$) was obtained by Siegmund and Venkatraman [39], who used (22) to derive $E_0 N_G \sim Kc^{-1/2}e^c$ that we have referred to in Section 2, where K is the mean of the limiting exponential distribution in (22).

Gene mapping is another area of applications of fixed sample change-point theory. An approximating Gaussian model for mapping genes on the basis of whole genome scans, developed by Lander and Botstein [27] and Feingold, Brown and Siegmund [13], can be described as follows. Moving along a chromosome of the organism being studied, one observes at the locus t (defined by its genetic distance in recombination units from a particular end of the chromosome) a random process $X(t)$. If there are no loci contributing to the trait under consideration, $X(t)$ is a mean 0 stationary Gaussian process with covariance function $\sigma^2 R(t)$. On a chromosome on which a single locus t_0 contributes susceptibility to the trait, there is superimposed a mean value function $\xi R(t - t_0)$. Typical examples of R are a two-sided exponential function, for which $X(t)$ is an Ornstein-Uhlenbeck process, or a linear combination of such functions. There can be several loci contributing to the trait, on the same or on different chromosomes, and then the models may involve interactions between loci. The statistical problems studied in [13] and [27] are to test the hypothesis that $\xi = 0$, for which the likelihood ratio test in the Gaussian model has the rejection region $\max_t X(t) \geq b$ for an appropriate threshold b, and to estimate the change-point parameter t_0 when $\xi \neq 0$.

5 Estimation of Time-Varying Parameters

To begin with, consider the simple mean shift model $X_t = \theta_t + \epsilon_t$, $t = 1, 2, \ldots$, in which the ϵ_t are i.i.d. zero-mean normal random variables, the sequence of change points of $\{\theta_t\}$ forms a discrete renewal process with geometric interarrival times with parameter p, and the post-change values of $\{\theta_t\}$ are i.i.d. normal. Chernoff and Zacks [11] gave an expression of the Bayes estimate $\widehat{\theta}_n = E(\theta_n | X_1, \ldots, X_n)$, which requires $O(2^n)$ operations to compute. Yao [42] later found another representation of the estimate that requires only $O(n^2)$ operations. Thus, even in this simple example, the memory required and the number of operations needed to compute the Bayes estimate grow to ∞ with n. Although in practice mean shifts typically occur very infrequently (i.e., p is very small), the unknown times of their occurrence leads to the great complexity of the Bayes estimate $\widehat{\theta}_n$. By extending the "window limiting" idea in the detection problem, Lai, Liu and Liu [24] have developed alternative estimators $\widetilde{\theta}_n$ which involve no more than a fixed number (depending on p) of parallel recursions such that the Bayes risk is asymptotically equivalent to that of $\widehat{\theta}_t$ as $p \to 0$ and $np \to \infty$:

$$\sum_{t=1}^{n} E(\widetilde{\theta}_t - \theta_t)^2 \sim \sum_{t=1}^{n} E(\widehat{\theta}_t - \theta_t)^2 \sim 3np \log(1/p) \ . \tag{24}$$

One such estimator uses window-limited GLR detection schemes to segment the data and estimates θ_t by some suitably weighted mean of the observations in the segment containing time t. Instead of using "hard thresholding" to segment the data, an alternative approximation to the optimal filter is to use a bounded-complexity mixture (BCMIX). The BCMIX estimator is close in spirit to Yao's method for computing the exact Bayes procedure but only keeps a fixed number of linear filters at every stage. Let $\mathbf{X}_{s,t} = (X_s, \ldots, X_t)$ for $s < t$. Let $K(t)$ be the most recent change-point prior to t. The conditional density f_t of θ_t given $\mathbf{X}_{1,t}$ is a mixture of $t-1$ normal densities $f(\mu | \mathbf{X}_{1,t}, K(t))$:

$$f_t(\mu | \mathbf{X}_{1,t}) = \sum_{j=2}^{t} f(\mu | \mathbf{X}_{1,t}, K(t) = j) P(K(t) = j | \mathbf{X}_{1,t}),$$

so that $E(\theta_t | \mathbf{X}_{1,t})$ is a weighted average of $t-1$ linear filters. When p is very small, most of the mass for $P(K(t) = j | \mathbf{X}_{1,t})$ should be concentrated in a neighborhood around the true change-point. This inspires the main idea of the BCMIX approximation: One can approximate $f_t(\mu|\mathbf{X}_{1,t})$ with a much smaller number of filters than $t-1$, using a fixed number m (depending on p) of filters to approximate $E(\theta_t|\mathbf{X}_{1,t})$. Each filter i is characterized by

- $K_i(t)$: the candidate for the most recent change-point position;
- $p_i(t)$: the weight which is related to but not equal to $P(K(t)=K_i(t)|\mathbf{X}_{1,t})$.

The approximation is given by

$$f_t(\mu|\mathbf{X}_{1,t}) \approx \sum_{i=1}^{m} p_i(t) f(\mu|K_i(t), \mathbf{X}_{K_i(t),t}).$$

With the arrival of a new data point at $t+1$, we update the conditional density as follows:

$$f_{t+1}(\mu|\mathbf{X}_{1,t+1}) \propto \varphi(X_{t+1}-\mu)[p f_0(\mu) + (1-p)\sum_{i=1}^{m} p_i(t) f(\mu|K_i(t), \mathbf{X}_{K_i(t),t})],$$

where φ denotes the common normal density function of the ϵ_t and f_0 is the prior normal density function of θ_t immediately after a change-point. Note that we only need to keep track of the weight and the last change-time of each the m filters to update them. Hence, if $f_t(\mu|\mathbf{X}_{1,t})$ can be approximated by a mixture of m normals, then $f_{t+1}(\mu|\mathbf{X}_{1,t+1})$ can be approximated by a mixture of $m+1$ normals. The added filter represents the possibility that a new change occurs at $t+1$. In order to keep the number of filters fixed at m, [24] deletes the filter with the smallest probability. Simulation studies show that both the BCMIX and the segmentation filters perform well in terms of mean squared errors, and that the segmentation filter has larger mean squared errors than BCMIX when p is not sufficiently small. Moreover, BCMIX compares favorably with $\hat{\theta}_t$ even in moderately sized samples for which it is feasible to compute the Bayes procedure $\hat{\theta}_t$ by using Yao's algorithm.

The BCMIX procedure can be readily extended to the smoothing problem of estimating θ_t from the fixed sample $\mathbf{X}_{1,n}$ for $1 \le t \le n$. Yao [42] developed an algorithm involving $O(n^3)$ operations to compute the Bayes estimator $E(\theta_t|\mathbf{X}_{1,n})$ by combining forward and backward filters. For the BCMIX approximation to $E(\theta_t|\mathbf{X}_{1,n})$, [24] approximates the backward (time-reversed) filter by

$$f_t(\mu|\mathbf{X}_{t,n}) \approx \sum_{j=1}^{m} q_j(t) f(\mu|J_j(t), \mathbf{X}_{t,J_j(t)}),$$

where $J_j(t)$ is the analogue of $K_i(t)$ and $q_j(t)$ is the counterpart of $p_i(t)$ for the jth backward filter. The m forward and m backward filters are then combined to yield a BCMIX estimator of θ_t.

These methods have been extended in [24] to the problem of estimating time-varying parameters in exponential families. We are currently working on extending these ideas to the time-varying stochastic regression model

$X_t = \theta_t^T \phi_t + \epsilon_t$, in which X_t is the observed output at time t and ϕ_t is a p-dimensional vector consisting of past outputs and inputs. The θ_t are unknown parameters that may undergo occasional changes. Instead of imposing a stochastic model on the parameter variations as we have done here, an alternative approach in the engineering literature is to forgo modeling of θ_t and to simply weight the data so that those around time t get the most weights to estimate θ_t. This is the idea behind the recursive least squares algorithm with "forgetting" (discount) factor $\lambda \in (0,1)$, which estimates θ_{n+1} by minimizing $\sum_{i=1}^n \lambda^{n-i}(X_i - \theta^T \phi_i)^2$ and which has the recursive form

$$\begin{aligned} \widehat{\theta}_{t+1} &= \widehat{\theta}_t + \lambda P_t \phi_t (X_t - \phi_t^T \widehat{\theta}_t)/[\lambda + (1-\lambda)\phi_t^T P_t \phi_t], \\ P_{t+1} &= \lambda^{-1}\{P_t - (1-\lambda)P_t \phi_t \phi_t^T P_t/[\lambda + (1-\lambda)\phi_t^T P_t \phi_t]\}. \end{aligned} \tag{25}$$

Instead of using matrix gains, an alternative idea is to use scalar gains, leading to the LMS (least mean squares) algorithm

$$\widehat{\widehat{\theta}}_{t+1} = \widehat{\widehat{\theta}}_t + \lambda \phi_t (X_t - \phi_t^T \widehat{\widehat{\theta}}_t)/(1 + \|\phi_t\|^2) \ , \tag{26}$$

where $\lambda \in (0,1]$ is called the step size or adaptation rate. The past decade has witnessed substantial progress in the analysis of (i) the statistical properties of (24) and (25) and their variants, and (ii) their associated adaptive prediction and control schemes of certainty-equivalence type; see e.g. [15], [16], [17]. However, there is substantial loss in efficiency in using these estimators when θ_t changes only infrequently, and a comprehensive statistical theory of estimation in time-varying stochastic regression models is still lacking.

The Bayesian approach of putting a stochastic model on the parameter variations reduces the estimation problem to a smoothing (or filtering) problem for hidden Markov models. Markov chain Monte Carlo or sequential importance sampling methods can often be used to implement the smoothers off-line. For applications of this simulation-based approach to bioinformatics, see [28].

References

1. Bai, J. (1999) LIkelihood ratio tests for multiple structural changes, J. Econometrics **91**, 299–323.
2. Bai, J. and Perron, P. (1998) Estimating and testing for multiple structural changes, Econometrica **66**, 47–78.
3. Bansal, R. K. and Papantoni-Kazakos, P. (1986) An algorithm for detecting a change in a stochastic process, IEEE Trans. Information Theory **32**, 227–235.
4. Barnard, G. A. (1959) Control charts and stochastic processes (with discussion), J. Roy. Statist. Soc. Ser. B **21**, 239–271.
5. Basseville, M. and Nikiforov, I. V. (1993) Detection of Abrupt Changes: Theory and Applications, Prentice-Hall, Englewood Cliffs.

6. Benveniste, A., Basseville, M., and Moustakides, G. (1987) The asymptotic local approach to change detection and model validation. IEEE Trans. Automatic Control **32**, 583–592.
7. Benveniste, A., Metivier, M., and Priouret, P. (1987) Adaptive Algorithms and Stochastic Approximations, Springer-Verlag, New York.
8. Brown, R. L., Durbin, J., and Evans, J. M. (1975) Techniques for testing the constancy of regression relationships over time (with discussion), J. Roy. Statist. Soc. Ser. B **37**, 149–192.
9. H.P. Chan and T.L. Lai (2001) Saddlepoint approximations and nonlinear boundary crossing probabilities for Markov random walks. Tech. Report, Dept. Statistics, Stanford Univ.
10. Chan, H. P. and Lai, T. L. (2001) Maxima of Gaussian random fields and moderate deviation approximations to boundary crossing probabilities. Tech. Report, Dept. Statistics, Stanford Univ.
11. Chernoff, H. and Zacks, S. (1964) Estimating the current mean of a normal distribution which is subjected to change in time, Ann. Math. Statist. **35**, 999–1018.
12. Davis, R. A., Huang, D., and Yao, Y. C. (1995) Testing for a change in parameter values and order of an autoregressive model, Ann. Statist. **23**, 282–304.
13. Feingold, E., Brown, P. O., and Siegmund, D. (1993) Gaussian models for genetic linkage analysis using high resolution maps of identity-by-descent, Amer. J. Human Genetics **53**, 234–251.
14. Frank, P. M. (1991) Fault diagnosis in dynamic systems using analytical and knowledge-based redundancy—a survey and some new results, Automatica **26**, 459–474.
15. Guo, L. (1994) Stability of recursive tracking algorithms, SIAM J. Control and Optimization **32**, 1195–1225.
16. Guo, L. and Ljung, L. (1995) Performance analysis of general tracking algorithms, IEEE Trans. Automatic Control **40**, 1388–1402.
17. Guo, L., Ljung, L., and Priouret, P. (1993) Performance analysis of the forgetting factor RLS algorithm, Internat. J. Adaptive Control and Signal Processing **7**, 525–537.
18. James, B., James, K. L., and Siegmund, D. (1987) Tests for a change-point, Biometrika **74**, 71–84.
19. Karlin, S., Dembo, A., and Kawabata, T. (1990) Statistical composition of high scoring segments from molecular sequences, Ann. Statist. **18**, 571–581.
20. Kim, D. and Kon, S. J. (1999) Structural change and time dependence in models of stock returns, J. Empirical Finance **6**, 283–308.
21. Lai, T. L. (1995) Sequential change-point detection in quality control and dynamical systems (with discussion), J. Roy. Statist. Soc. Ser. B **57**, 613–658.
22. Lai, T. L. (1998) Information bounds and quick detection of parameter changes in stochastic systems, IEEE Trans. Information Theory **44**, 2917–2929.
23. Lai, T. L. (2000) Sequential multiple hypothesis testing and efficient fault detection-isolation in stochastic systems, IEEE Trans. Information Theory **46**, 595–608.
24. Lai, T. L., Liu, H., and Liu, T. (2001) Change-point detection and bounded-complexity estimation of parameters undergoing occasional changes, Tech. Report, Dept. Statistics, Stanford Univ.

25. Lai, T. L. and Shan, J. Z. (1999) Efficient recursive algorithms for detection of abrupt changes in signals and control systems, IEEE Trans. Automatic Control **44**, 952–966.
26. Lamoureux, C. G. and Lastrapes, W. D. (1990) Persistence in variance, structural change, and the GARCH model, J. Business and Econ. Statist. **8**, 225–234.
27. Lander, E. S. and Botstein, D. (1989) Mapping Mendelian factors underlying quantitative traits using RFLP linkage maps, Genetics **121**, 185–199.
28. Liu, J. and Lawrence, C. (1999) Bayesian inference on biopolymer models, Bioinformatics **15**, 38–52.
29. Lorden, G. (1971) Procedures for reacting to a change in distribution, Ann. Math. Statist. **42**, 1897–1908.
30. Moustakides, G. (1986) Optimal procedures for detecting changes in distributions, Ann. Statist. **14**, 1379–1387.
31. Nikiforov, I. V. (1995) A generalized change detection problem. IEEE Trans, Information Theory **41**, 171–187.
32. Nikiforov, I. V. (1996) New optimal approach to global positioning system/differential global positioning system integrity monitoring, J. Guidance, Control and Dynamics **19**, 1023–1033.
33. Nikiforov, I. V., Varavva, V., and Kireichikov, V. (1993) Application of statistical fault detection algorithms to navigation systems monitoring, Automatica **29**, 1275–1290.
34. Page, E. S. (1954) Continuous inspection schemes, Biometrika **41**, 100–115.
35. Parkinson, B. W. and Axelrad, P. (1988) Autonomous GPS integrity monitoring using the pseudorange residual, Navigation **35**, 255–274.
36. Patton, R. J., Frank, P. M. and Clark, R. N. (1989) Fault Diagnosis in Dynamic Systems: Theory and Applications. Prentice-Hall, Englewood Cliffs.
37. Quandt, R. E. (1960) Tests of the hypothesis that a linear regression system obeys two separate regimes, J. Amer. Statist. Assoc. **55**, 324–330.
38. Shiryayev, A. N. (1978) Optimal Stopping Rules. Springer-Verlag, New York.
39. Siegmund, D. and Venkatraman, E. S. (1995) Using the generalized likelihood ratio statistics for sequential detection of a change-point, Ann. Statist. **23**, 255–271.
40. Willsky, A. S. and Jones, H. L. (1976) A generalized likelihood ratio approach to detection and estimation of jumps in linear systems, IEEE Trans. Automatic Control **21**, 108–112.
41. Yakir, B. (1994) Optimal detection of a change in distribution when the observations form a Markov chain with a finite state space. In Change-point Problems, E. Carlstein et al., eds., IMS Monograph series **23**, 346–358.
42. Yao, Y. C. (1984) Estimation of a noisy discrete-time step function: Bayes and empirical Bayes approaches, Ann. Statist. **12**, 1434–1447.
43. Zhang, Q., Basseville, M., and Benveniste, A. (1994) Early warning of slight changes in systems and plants with application to condition based maintenance, Automatica **30**, 95–114.

Asymptotic Normality in Partially Observed Diffusions with Small Noise: Application to FDI

This paper is dedicated to Tyrone E. Duncan on the occasion of his 60th birthday

François LeGland and Bo Wang

IRISA / INRIA, Campus de Beaulieu, 35042 RENNES Cedex, France

Abstract. The problem of residual evaluation for fault detection in partially observed diffusions is investigated, using the local asymptotic approach, under the small noise asymptotics. The score function (i.e. the gradient of the log–likelihood function) evaluated at the nominal value of the parameter, and suitably normalized, is used as residual. It is proved that this residual is asymptotically Gaussian, with mean zero under the null hypothesis, with a different mean (depending linearly on the parameter change) and the same covariance matrix under the contiguous alternative hypothesis. This result relies on the local asymptotic normality (LAN) property for the family of probability distributions of the observation process, which is also proved.

Keywords: fault detection, residual evaluation, local approach, asymptotic normality, LAN

1 Introduction

The problem of fault detection and isolation (FDI) in dynamical properties of signals and systems has received growing attention, and has been theoretically and experimentally investigated with different types of approaches. Among these, the local approach is known to be very powerful and has been successfully applied in many practical situations. Some early and fundamental works on the subject include Benveniste, Basseville and Moustakides [2], Benveniste, Métivier and Priouret [3], Basseville [1], etc. As explained there, the local approach consists of the following two steps. The first step, called residual generation, is to propose a statistics, called the residual, which depends only on the observations and on the nominal value of the parameter, and which ideally should be close to zero under the null hypothesis, and significantly different from zero under the alternative hypothesis. The second step, called residual evaluation, is to study the asymptotic behaviour of the residual under the null hypothesis and under a contiguous alternative hypothesis, and to design a simple test based on the residual.

In the case of completely observed stationary Markov processes (controlled by the parameter), Basseville [1], Delyon, Juditsky and Benveniste [7]

B. Pasik-Duncan: Stochastic Theory and Control, LNCIS 280, pp. 267–282, 2002.

have considered some residuals (such as the efficient score, the quasi–score, etc.) and have investigated their asymptotic behaviour when the size N of the sample goes to infinity : typically, the residual ζ_N satisfies the following central limit theorem as $N \uparrow \infty$

$$\zeta_N \Longrightarrow \begin{cases} \mathcal{N}(0, \Sigma(\alpha)) \ , & \text{under } \mathbb{P}_\alpha, \\ \mathcal{N}(M(\alpha)\, \Delta, \Sigma(\alpha)) \ , & \text{under } \mathbb{P}_{\alpha+\Delta/\sqrt{N}}, \end{cases}$$

where α is the nominal value of the parameter. ¿From this result, one can design the optimum test by reducing the original problem of detecting a change in a dynamical system, to the simple and universal static problem of detecting a change in the mean of a Gaussian r.v., see [1].

In the case of partially observed diffusions, Cérou, LeGland and Newton [5] have addressed the problem of residual generation. They considered the score function, i.e. the gradient of the log-likelihood function w.r.t. the parameter, evaluated at the nominal value, as the residual, and proposed an efficient numerical approximation using particle filters. In this paper, the corresponding residual evaluation problem is studied. Instead of the large time asymptotics, the small noise asymptotics is used here, which usually is simpler to deal with, see Kutoyants [10].

The paper is organized as follows. Section 2 presents the model and gives the expression of the residual. Some preliminary computations based on small noise expansion, are presented in Section 3. Asymptotic normality of the residual under the nominal hypothesis is studied in Section 4. In Section 5, we prove the local asymptotic normality of the family of probability distributions of the observation process, and in Section 6 we prove the asymptotic normality of the residual under the contiguous alternative hypothesis, and we apply these results to the FDI problem.

2 Statistical Model and Residual Definition

On a measurable space $(\Omega, \mathcal{F}, P)$ are given

- for each $\varepsilon > 0$, a family $\mathcal{M}^\varepsilon = \{\mathbb{P}_{\theta,\varepsilon}, \theta \in \Theta\}$ of probability measures,
- two stochastic processes $X = \{X_t \, , \, 0 \le t \le T\}$ and $Y = \{Y_t \, , \, 0 \le t \le T\}$, taking values in $\mathbb{R}^m$ and $\mathbb{R}^d$ respectively,

such that under $\mathbb{P}_{\theta,\varepsilon}$

$$dX_t = b(\theta, X_t)dt + \varepsilon\, dW_t^\theta \ , \quad X_0 = \varepsilon\, \xi \ ,$$

$$dY_t = h(X_t)dt + \varepsilon\, dV_t^\theta \ , \quad Y_0 = 0 \ ,$$

where $\{W_t^\theta, 0 \leq t \leq T\}$ and $\{V_t^\theta, 0 \leq t \leq T\}$ are independent standard Wiener processes, and ξ is an $\mathbb{R}^m$–valued r.v. independent of the Wiener processes, which we assume to be a standard Gaussian r.v., without loss of generality.

Assume that Ω is the canonical space $C([0,T];\mathbb{R}^{m+d})$, in which case X and Y are the canonical processes on $C([0,T];\mathbb{R}^m)$ and $C([0,T];\mathbb{R}^d)$, respectively, and $\mathbb{P}_{\theta,\varepsilon}$ is the probability distribution of (X,Y). The set of parameters $\Theta \subset \mathbb{R}^p$ is compact, and the coefficients satisfy the following hypotheses

- the mapping $(\theta,x) \longmapsto b(\theta,x)$ is bounded, with bounded and Lipschitz continuous derivatives $\partial b(\theta,x)$ and $b'(\theta,x)$ (w.r.t. the parameter and the state variable respectively),
- the mapping $x \longmapsto h(x)$ is bounded, with bounded and Lipschitz continuous derivative $h'(x)$.

Our purpose is to design statistical tests to decide whether $\theta = \alpha$, corresponding to a nominal behaviour of the system, or $\theta \neq \alpha$, on the basis of a given observation path $\{Y_t, 0 \leq t \leq T\}$. Let $\mathcal{Y}_T$ denote the σ–algebra generated by the process Y. The hypotheses ensure that the probability measures in $\mathcal{M}^\varepsilon$ are mutually absolutely continuous, thus, according to Campillo and LeGland [4], the likelihood function for estimating the parameter θ given $\mathcal{Y}_T$ can be expressed as

$$L_\varepsilon(\theta) = \mathbb{E}^\dagger_{\theta,\varepsilon}[Z_T^\varepsilon \mid \mathcal{Y}_T] \; ,$$

where

$$Z_t^\varepsilon = \exp\{\frac{1}{\varepsilon^2} \int_0^t h^*(X_s)\, dY_s - \frac{1}{2\varepsilon^2} \int_0^t |h(X_s)|^2\, ds\} \; ,$$

for any $0 \leq t \leq T$, and where $\mathbb{P}^\dagger_{\theta,\varepsilon}$ is a probability measure on Ω, equivalent to $\mathbb{P}_{\theta,\varepsilon}$ with Radon–Nikodym derivative

$$\frac{d\mathbb{P}_{\theta,\varepsilon}}{d\mathbb{P}^\dagger_{\theta,\varepsilon}} = Z_T^\varepsilon \; ,$$

independent of $\theta \in \Theta$, so that under $\mathbb{P}^\dagger_{\theta,\varepsilon}$

$$dX_t = b(\theta, X_t)dt + \varepsilon\, dW_t^\theta, \qquad X_0 = \varepsilon\, \xi \; ,$$

and $\{Y_t, 0 \leq t \leq T\}$ is a Wiener process independent of $\{W_t^\theta, 0 \leq t \leq T\}$, with covariance matrix $\varepsilon^2\, I_d$. Alternatively

$$L_\varepsilon(\theta) = \mathbb{E}^\dagger_\varepsilon[\Lambda_T^{\theta,\varepsilon}\, Z_T^\varepsilon \mid \mathcal{Y}_T] \; , \tag{1}$$

where

$$\Lambda_t^{\theta,\varepsilon} = \exp\{\frac{1}{\varepsilon^2} \int_0^t b^*(\theta, X_s)\, dX_s - \frac{1}{2\varepsilon^2} \int_0^t |b(\theta, X_s)|^2\, ds\} \; ,$$

for any $0 \leq t \leq T$, and where $\mathbb{P}^\dagger_\varepsilon$ is a probability measure on Ω, equivalent to $\mathbb{P}^\dagger_{\theta,\varepsilon}$ with Radon–Nikodym derivative

$$\frac{d\mathbb{P}^\dagger_{\theta,\varepsilon}}{d\mathbb{P}^\dagger_\varepsilon} = \Lambda_T^{\theta,\varepsilon} ,$$

so that under $\mathbb{P}^\dagger_\varepsilon$, $\{X_t, 0 \leq t \leq T\}$ and $\{Y_t, 0 \leq t \leq T\}$ are two independent Wiener processes, with covariance matrices $\varepsilon^2 I_m$ and $\varepsilon^2 I_d$ respectively.

By definition, the residual ζ_ε is the gradient (column–vector) of the log-likelihood function w.r.t. the parameter θ, evaluated at the nominal value $\theta = \alpha$, and suitably normalized, i.e.

$$\zeta_\varepsilon = (\, \varepsilon\, \partial \log L_\varepsilon(\alpha)\,)^* .$$

It follows from (1) that

$$\partial\, L_\varepsilon(\theta) = \mathbb{E}^\dagger_\varepsilon[\partial \log \Lambda_T^{\theta,\varepsilon}\, \Lambda_T^{\theta,\varepsilon}\, Z_T^\varepsilon \mid \mathcal{Y}_T] = \mathbb{E}^\dagger_{\theta,\varepsilon}[\partial \log \Lambda_T^{\theta,\varepsilon}\, Z_T^\varepsilon \mid \mathcal{Y}_T] ,$$

hence

$$\begin{aligned} \varepsilon\, \partial\, \log L_\varepsilon(\theta) = \frac{\varepsilon\, \partial\, L_\varepsilon(\theta)}{L_\varepsilon(\theta)} &= \frac{\varepsilon\, \mathbb{E}^\dagger_{\theta,\varepsilon}[\partial\, \log \Lambda_T^{\theta,\varepsilon}\, Z_T^\varepsilon \mid \mathcal{Y}_T]}{\mathbb{E}^\dagger_{\theta,\varepsilon}[Z_T^\varepsilon \mid \mathcal{Y}_T]} \\ &= \varepsilon\, \mathbb{E}_{\theta,\varepsilon}[\partial\, \log \Lambda_T^{\theta,\varepsilon} \mid \mathcal{Y}_T] = (\, \mathbb{E}_{\theta,\varepsilon}[\Xi_T^{\theta,\varepsilon} \mid \mathcal{Y}_T]\,)^* , \end{aligned}$$

where

$$\begin{aligned} \Xi_t^{\theta,\varepsilon} &= \int_0^t \partial\, b^*(\theta, X_s)\, \frac{1}{\varepsilon}\, [dX_s - b(\theta, X_s)\, ds] \\ &= \int_0^t \partial\, b^*(\theta, X_s)\, dW_s^\theta , \end{aligned}$$

for any $0 \leq t \leq T$.

Let $p_t^{\theta,\varepsilon}$ denote the unnormalized conditional density of the r.v. X_t given $\mathcal{Y}_t$ under $\mathbb{P}_{\theta,\varepsilon}$, and let $w_t^{\theta,\varepsilon}$ denote its derivative w.r.t. the parameter θ, i.e.

$$\int \phi(x)\, p_t^{\theta,\varepsilon}(x)\, dx = \mathbb{E}^\dagger_{\theta,\varepsilon}[\phi(X_t)\, Z_t^\varepsilon \mid \mathcal{Y}_t] ,$$

and

$$\int \phi(x)\, w_t^{\theta,\varepsilon}(x)\, dx = \mathbb{E}^\dagger_{\theta,\varepsilon}[\phi(X_t)\, \partial \log \Lambda_t^{\theta,\varepsilon}\, Z_t^\varepsilon \mid \mathcal{Y}_t] ,$$

for any test function ϕ defined on $\mathbb{R}^m$. Then the likelihood function can be expressed as

$$L_\varepsilon(\theta) = \mathbb{E}^\dagger_{\theta,\varepsilon}[Z_T^\varepsilon \mid \mathcal{Y}_T] = \int p_T^{\theta,\varepsilon}(x)\, dx , \qquad (2)$$

and it follows from Campillo and LeGland [4, Section 4] that the residual can be expressed as

$$\zeta_\varepsilon = \mathbb{E}_{\alpha,\varepsilon}[\Xi_T^{\alpha,\varepsilon} \mid \mathcal{Y}_T] = \frac{[\displaystyle\int \varepsilon\, w_T^{\alpha,\varepsilon}(x)\, dx\,]^*}{\displaystyle\int p_T^{\alpha,\varepsilon}(x)\, dx} \; . \tag{3}$$

Moreover, $\{p_t^{\theta,\varepsilon}\,,\, 0 \leq t \leq T\}$ satisfies the Duncan–Mortensen–Zakai (DMZ) equation

$$\begin{cases} dp_t^{\theta,\varepsilon}(x) = \frac{1}{2}\,\varepsilon^2 \displaystyle\sum_{i,j=1}^m \frac{\partial^2}{\partial x_i \partial x_j}\, p_t^{\theta,\varepsilon}(x)\, dt - \sum_{i=1}^m \frac{\partial}{\partial x_i}\,[\, b^i(\theta,x)\, p_t^{\theta,\varepsilon}(x)\,]\, dt \\ \qquad\qquad + \dfrac{1}{\varepsilon^2} h^*(x)\, p_t^{\theta,\varepsilon}(x)\, dY_t \;, \\ p_0^{\theta,\varepsilon}(x) = \dfrac{1}{\varepsilon^m (2\pi)^{m/2}}\, \exp\{-\dfrac{|x|^2}{2\,\varepsilon^2}\} \;, \end{cases}$$

and $\{w_t^{\theta,\varepsilon}\,,\, 0 \leq t \leq T\}$ satisfies the stochastic partial differential equation

$$\begin{cases} dw_t^{\theta,\varepsilon}(x) = \frac{1}{2}\,\varepsilon^2 \displaystyle\sum_{i,j=1}^m \frac{\partial^2}{\partial x_i \partial x_j}\, w_t^{\theta,\varepsilon}(x)\, dt - \sum_{i=1}^m \frac{\partial}{\partial x_i}\,[\, b^i(\theta,x)\, w_t^{\theta,\varepsilon}(x)\,]\, dt \\ \qquad\qquad + \dfrac{1}{\varepsilon^2}\, h^*(x)\, w_t^{\theta,\varepsilon}(x)\, dY_t \\ \qquad\qquad - \displaystyle\sum_{i=1}^m \frac{\partial}{\partial x_i}\,[\, \partial b^i(\theta,x)\, p_t^{\theta,\varepsilon}(x)\,]\, dt \;, \\ w_0^{\theta,\varepsilon}(x) = 0 \;. \end{cases}$$

3 Small Noise Expansion

Let $(X^{(0)}, Y^{(0)})$ and $(X^{(1)}, Y^{(1)})$ denote the solutions of the following limiting ordinary (deterministic) differential system

$$\begin{cases} \dot{X}_t^{(0)} = b(\alpha, X_t^{(0)}) \;, \quad X_0^{(0)} = 0 \;, \\ \dot{Y}_t^{(0)} = h(X_t^{(0)}) \;, \quad Y_0^{(0)} = 0 \;, \end{cases}$$

and *linear tangent* stochastic differential system

$$\begin{cases} dX_t^{(1)} = b'(\alpha, X_t^{(0)})\, X_t^{(1)}\, dt + dW_t^\alpha \;, \quad X_0^{(1)} = \xi \;, \\ dY_t^{(1)} = h'(X_t^{(0)})\, X_t^{(1)}\, dt + dV_t^\alpha \;, \quad Y_0^{(1)} = 0 \;, \end{cases} \tag{4}$$

respectively. Define also

$$\bar{X}_t^\varepsilon = \frac{1}{\varepsilon}\,(X_t - X_t^{(0)})\;, \quad \text{and} \quad \bar{Y}_t^\varepsilon = \frac{1}{\varepsilon}\,(Y_t - Y_t^{(0)})\;,$$

for any $0 \leq t \leq T$ and any $\varepsilon > 0$. Then $(\bar{X}^\varepsilon, \bar{Y}^\varepsilon)$ is the solution of the following *contaminated* stochastic differential system under $\mathbb{P}_{\theta,\varepsilon}$

$$\left\{ \begin{aligned} d\bar{X}_t^\varepsilon &= \frac{1}{\varepsilon}\,[b(\theta, X_t^{(0)} + \varepsilon\,\bar{X}_t^\varepsilon) - b(\alpha, X_t^{(0)})]\,dt + dW_t^\theta\;, \quad \bar{X}_0^\varepsilon = \xi\;, \\ d\bar{Y}_t^\varepsilon &= \frac{1}{\varepsilon}\,[h(X_t^{(0)} + \varepsilon\,\bar{X}_t^\varepsilon) - h(X_t^{(0)})]\,dt + dV_t^\theta\;, \quad \bar{Y}_0^\varepsilon = 0\;, \end{aligned} \right. \tag{5}$$

Notice that the mapping $u \longmapsto \dfrac{1}{\varepsilon}\,[b(\theta, X_t^{(0)} + \varepsilon\,u) - b(\alpha, X_t^{(0)})]$ is bounded and Lipschitz continuous, with a Lipschitz constant independent of $\varepsilon > 0$, hence the r.v. $\bar{X}_t^\varepsilon$ has moments of any order, which are uniform in $\varepsilon > 0$.

Under the nominal behaviour of the system, we have the following well–known result, see Freidlin and Wentzell [8].

Lemma 1. *As* $\varepsilon \downarrow 0$

$$\sup_{0 \leq t \leq T} |\bar{X}_t^\varepsilon - X_t^{(1)}| \longrightarrow 0\;, \quad \textit{and} \quad \sup_{0 \leq t \leq T} |\bar{Y}_t^\varepsilon - Y_t^{(1)}| \longrightarrow 0\;,$$

in $\mathbb{P}_{\alpha,\varepsilon}$*–probability.*

Set

$$\phi_t = \frac{1}{\varepsilon^2} \int_0^t h^*(X_s^{(0)})\,dY_s - \frac{1}{\varepsilon^2} \int_0^t |h(X_s^{(0)})|^2\,ds\;,$$

and define

$$\bar{p}_t^{\theta,\varepsilon}(u) = \varepsilon^m\, \exp\{-\phi_t\}\; p_t^{\theta,\varepsilon}(X_t^{(0)} + \varepsilon\,u)\;,$$

$$\bar{w}_t^{\theta,\varepsilon}(u) = \varepsilon^{m+1}\, \exp\{-\phi_t\}\; w_t^{\theta,\varepsilon}(X_t^{(0)} + \varepsilon\,u)\;,$$

Then by the Itô lemma, $\{\bar{p}_t^{\theta,\varepsilon}\;, 0 \leq t \leq T\}$ and $\{\bar{w}_t^{\theta,\varepsilon}\;, 0 \leq t \leq T\}$ satisfy the stochastic partial differential equations

$$\left\{ \begin{aligned} d\bar{p}_t^{\theta,\varepsilon}(u) &= \tfrac{1}{2} \sum_{i,j=1}^m \frac{\partial^2}{\partial u_i \partial u_j}\, \bar{p}_t^{\theta,\varepsilon}(u)\,dt \\ &\quad - \sum_{i=1}^m \frac{\partial}{\partial u_i}(\,\frac{1}{\varepsilon}\,[b^i(\theta, X_t^{(0)} + \varepsilon\,u) - b^i(\alpha, X_t^{(0)})]\,\bar{p}_t^{\theta,\varepsilon}(u)\,)\,dt \\ &\quad + \frac{1}{\varepsilon^2}\,[h(X_t^{(0)} + \varepsilon\,u) - h(X_t^{(0)})]^*\,\bar{p}_t^{\theta,\varepsilon}(u)\,[dY_t - h(X_t^{(0)})\,dt]\;, \\ \bar{p}_0^{\theta,\varepsilon}(u) &= \frac{1}{(2\pi)^{m/2}}\, \exp\{-\tfrac{1}{2}\,|u|^2\}\;, \end{aligned} \right.$$

and

$$
\left\{
\begin{aligned}
d\bar{w}_t^{\theta,\varepsilon}(u) &= \tfrac{1}{2} \sum_{i,j=1}^{m} \frac{\partial^2}{\partial u_i \partial u_j}\, \bar{w}_t^{\theta,\varepsilon}(u)\, dt \\
&\quad - \sum_{i=1}^{m} \frac{\partial}{\partial u_i} (\frac{1}{\varepsilon} [b^i(\theta, X_t^{(0)} + \varepsilon\, u) - b^i(\alpha, X_t^{(0)})]\, \bar{w}_t^{\theta,\varepsilon}(u))\, dt \\
&\quad + \frac{1}{\varepsilon^2} [h(X_t^{(0)} + \varepsilon\, u) - h(X_t^{(0)})]^*\, \bar{w}_t^{\theta,\varepsilon}(u)\, [dY_t - h(X_t^{(0)})\, dt]\ , \\
&\quad - \sum_{i=1}^{m} \frac{\partial}{\partial u_i} [\partial b^i(\theta, X_t^{(0)} + \varepsilon\, u)\, \bar{p}_t^{\theta,\varepsilon}(u)\,]\, dt\ , \\
\bar{w}_0^{\theta,\varepsilon}(u) &= 0\ ,
\end{aligned}
\right.
$$

respectively. Let $\bar{\mathcal{Y}}_T^\varepsilon$ denote the σ–algebra generated by the process $\bar{Y}^\varepsilon$. It follows from Theorem 3.7 in Charalambous, Elliott and Krishnamurthy [6], that

$$
\frac{\displaystyle\int \bar{p}_T^{\theta,\varepsilon}(u)\, du}{\displaystyle\int \bar{p}_T^{\alpha,\varepsilon}(u)\, du} = \mathbb{E}_{\alpha,\varepsilon}[\bar{\Upsilon}_T^{\theta,\varepsilon} \mid \bar{\mathcal{Y}}_T^\varepsilon]\ , \tag{6}
$$

where

$$
\begin{aligned}
\bar{\Upsilon}_t^{\theta,\varepsilon} = \exp\{ \int_0^t \frac{1}{\varepsilon} [b(\theta, X_s^{(0)} + \varepsilon\, \bar{X}_s^\varepsilon) - b(\alpha, X_s^{(0)} + \varepsilon\, \bar{X}_s^\varepsilon)]^*\, dW_s^\alpha \\
- \tfrac{1}{2} \int_0^t |\frac{1}{\varepsilon} [b(\theta, X_s^{(0)} + \varepsilon\, \bar{X}_s^\varepsilon) - b(\alpha, X_s^{(0)} + \varepsilon\, \bar{X}_s^\varepsilon)|^2\, ds\}\ ,
\end{aligned}
$$

for any $0 \le t \le T$. Similarly, it follows from Theorem 3.2 and Remark 3.3 in Charalambous, Elliott and Krishnamurthy [6], that

$$
\frac{\displaystyle[\int \bar{w}_T^{\alpha,\varepsilon}(u)\, du]^*}{\displaystyle\int \bar{p}_T^{\alpha,\varepsilon}(u)\, du} = \mathbb{E}_{\alpha,\varepsilon}[\bar{\Xi}_T^{\alpha,\varepsilon} \mid \bar{\mathcal{Y}}_T^\varepsilon]\ , \tag{7}
$$

where

$$
\bar{\Xi}_t^{\alpha,\varepsilon} = \int_0^T \partial\, b^*(\alpha, X_s^{(0)} + \varepsilon\, \bar{X}_s^\varepsilon)\, dW_s^\alpha = \Xi_t^{\alpha,\varepsilon}\ ,
$$

for any $0 \le t \le T$.

4 Asymptotic Normality of the Residual Under the Nominal Hypothesis

Using (7) and the change of variable $x = X_T^{(0)} + \varepsilon u$ in (3), yields

$$\zeta_\varepsilon = \frac{[\int \varepsilon\, w_T^{\alpha,\varepsilon}(x)\, dx]^*}{\int p_T^{\alpha,\varepsilon}(x)\, dx} = \frac{[\int \varepsilon^{m+1}\, w_T^{\alpha,\varepsilon}(X_T^{(0)} + \varepsilon\, u)\, du]^*}{\int \varepsilon^m\, p_T^{\alpha,\varepsilon}(X_T^{(0)} + \varepsilon\, u)\, du}$$

$$= \frac{[\int \bar{w}_T^{\alpha,\varepsilon}(u)\, du]^*}{\int \bar{p}_T^{\alpha,\varepsilon}(u)\, du} = \mathbb{E}_{\alpha,\varepsilon}[\bar{\Xi}_T^{\alpha,\varepsilon} \mid \bar{\mathcal{Y}}_T^\varepsilon] \; .$$

To study the asymptotic behaviour of ζ_ε under $\mathbb{P}_{\alpha,\varepsilon}$ as $\varepsilon \downarrow 0$, we extend the contaminated model (5) and the linear tangent model (4) as follows

$$\begin{cases} d\bar{X}_t^\varepsilon = \dfrac{1}{\varepsilon}\, [\, b(\alpha, X_t^{(0)} + \varepsilon\, \bar{X}_t^\varepsilon) - b(\alpha, X_t^{(0)})\,]\, dt + dW_t^\alpha \; , \quad \bar{X}_0 = \xi \; , \\ d\bar{\Xi}_t^{\alpha,\varepsilon} = \partial b^*(\alpha, X_t^{(0)} + \varepsilon\, \bar{X}_t^\varepsilon)\, dW_t^\alpha \; , \quad \bar{\Xi}_0^{\alpha,\varepsilon} = 0 \; , \\ d\bar{Y}_t^\varepsilon = \dfrac{1}{\varepsilon}\, [\, h(X_t^{(0)} + \varepsilon\, \bar{X}_t^\varepsilon) - h(X_t^{(0)})\,]\, dt + dV_t^\alpha \; , \quad \bar{Y}_0^\varepsilon = 0 \; , \end{cases}$$

and

$$\begin{cases} dX_t^{(1)} = b'(\alpha, X_t^{(0)}) X_t^{(1)} dt + dW_t^\alpha \; , \quad X_0^{(1)} = \xi \; , \\ d\Xi_t^{(1)} = \partial b^*(\alpha, X_t^{(0)})\, dW_t^\alpha \; , \quad \Xi_0^{(1)} = 0 \; , \\ dY_t^{(1)} = h'(X_t^{(0)}) X_t^{(1)} dt + dV_t^\alpha \; , \quad Y_0^{(1)} = 0 \; , \end{cases} \tag{8}$$

respectively.

Lemma 2. *As* $\varepsilon \downarrow 0$

$$\sup_{0 \le t \le T} |\bar{\Xi}_t^{\alpha,\varepsilon} - \Xi_t^{(1)}| \longrightarrow 0 \; ,$$

in $\mathbb{P}_{\alpha,\varepsilon}$*–probability.*

PROOF. Notice that

$$|\partial b(\alpha, X_t^{(0)} + \varepsilon\, u) - \partial b(\alpha, X_t^{(0)})| \le K\, \varepsilon\, |u| \; ,$$

for any $u \in \mathbb{R}^m$, and the Doob inequality yields

$$\mathbb{E}_{\alpha,\varepsilon}[\sup_{0 \le t \le T} |\bar{\Xi}_t^{\alpha,\varepsilon} - \Xi_t^{(1)}|^2\,] \le 4\, K^2\, \varepsilon^2\, \mathbb{E}_{\alpha,\varepsilon}[\int_0^T |\bar{X}_t^\varepsilon|^2\, dt] \; ,$$

which goes to zero as $\varepsilon \downarrow 0$, since

$$\sup_{0<\varepsilon\leq\varepsilon_0} \sup_{0\leq t\leq T} \mathbb{E}_{\alpha,\varepsilon}|\bar{X}^\varepsilon_t|^2 < \infty \ . \quad \square$$

Let $\mathcal{G}$ denote the σ–algebra generated by $(X^{(1)}, \Xi^{(1)}, \bar{Y}^\varepsilon)$, and let $\bar{\mathbb{Q}}^{(1)}_\varepsilon$ and $\mathbb{Q}^{(1)}$ denote the probability distribution of the r.v.'s $(X^{(1)}, \Xi^{(1)}, \bar{Y}^\varepsilon)$ and $(X^{(1)}, \Xi^{(1)}, Y^{(1)})$ respectively, on the canonical space $C([0,T]\,;\,\mathbb{R}^{m+p+d})$.

Lemma 3. *As* $\varepsilon \downarrow 0$

$$\|\bar{\mathbb{Q}}^{(1)}_\varepsilon - \mathbb{Q}^{(1)}\|_{\mathrm{TV}} \longrightarrow 0 \ .$$

Proof. It holds

$$\frac{d\mathbb{Q}^{(1)}}{d\bar{\mathbb{Q}}^{(1)}_\varepsilon} = \mathbb{E}_{\alpha,\varepsilon}[\exp\{\bar{M}^\varepsilon_T - \tfrac{1}{2}\,\langle \bar{M}^\varepsilon\rangle_T\} \mid \mathcal{G}] \ ,$$

where the martingale $\{\bar{M}^\varepsilon_t\,,\, 0\leq t\leq T\}$ is defined by

$$\begin{aligned}
\bar{M}^\varepsilon_t &= \int_0^t (\, h'(X^{(0)}_s)\, X^{(1)}_s - \frac{1}{\varepsilon}\,[\, h(X^{(0)}_s + \varepsilon\, \bar{X}^\varepsilon_s) - h(X^{(0)}_s)\,]\,)^*\, dV^\alpha_s \\
&= -\int_0^t (\,\frac{1}{\varepsilon}\,[\, h(X^{(0)}_s + \varepsilon\, \bar{X}^\varepsilon_s) - h(X^{(0)}_s)\,] - h'(X^{(0)}_s)\, \bar{X}^\varepsilon_s\,)^*\, dV^\alpha_s \\
&\qquad - \int_0^t (\, h'(X^{(0)}_s)\, (\bar{X}^\varepsilon_s - X^{(1)}_s)\,)^*\, dV^\alpha_s \ ,
\end{aligned}$$

for any $0\leq t\leq T$. Notice that

$$|\frac{1}{\varepsilon}\,[\, h(X^{(0)}_t + \varepsilon\, u) - h(X^{(0)}_t)\,] - h'(X^{(0)}_t)\, u| \leq K\, \varepsilon\, |u|^2 \ ,$$

for any $u\in\mathbb{R}^m$, and

$$|h'(X^{(0)}_t)\, (\bar{X}^\varepsilon_t - X^{(1)}_t)\,| \leq K\, |\bar{X}^\varepsilon_t - X^{(1)}_t| \ ,$$

hence

$$\langle \bar{M}^\varepsilon\rangle_T \leq 2\, K^2\, \varepsilon^2 \int_0^T |\bar{X}^\varepsilon_t|^4\, dt + 2\, K^2\, T \sup_{0\leq t\leq T} |\bar{X}^\varepsilon_t - X^{(1)}_t|^2 \ ,$$

which goes to zero in $\mathbb{P}_{\alpha,\varepsilon}$–probability as $\varepsilon \downarrow 0$, since

$$\sup_{0<\varepsilon\leq\varepsilon_0} \sup_{0\leq t\leq T} \mathbb{E}_{\alpha,\varepsilon}|\bar{X}^\varepsilon_t|^4 < \infty \ ,$$

and using Lemma 1. It follows from Problem 1.9.2 in Liptser and Shiryayev [12] that the same convergence holds for $\bar{M}^\varepsilon_T$, hence

$$\bar{Z}^\varepsilon_T = \exp\{\bar{M}^\varepsilon_T - \tfrac{1}{2}\,\langle \bar{M}^\varepsilon\rangle_T\} \longrightarrow 1 \ ,$$

in $\mathbb{P}_{\alpha,\varepsilon}$–probability as $\varepsilon \downarrow 0$. Therefore, using the Jensen inequality and the Scheffé theorem yields

$$\begin{aligned}\|\bar{\mathbb{Q}}^{(1)}_\varepsilon - \mathbb{Q}^{(1)}\|_{\mathrm{TV}} &= \mathbb{E}_{\alpha,\varepsilon}|1 - \frac{d\mathbb{Q}^{(1)}}{d\bar{\mathbb{Q}}^{(1)}_\varepsilon}| \\ &= \mathbb{E}_{\alpha,\varepsilon}|\mathbb{E}_{\alpha,\varepsilon}[1 - \bar{Z}^\varepsilon_T \mid \mathcal{G}]\,| \le \mathbb{E}_{\alpha,\varepsilon}|1 - \bar{Z}^\varepsilon_T| \longrightarrow 0 \ . \quad \square\end{aligned}$$

The convergence in distribution of ζ_ε follows from Lemmas 2 and 3, and from Theorem 1.2 in Kleptsina, Liptser and Serebrovski [9].

Theorem 1. *As* $\varepsilon \downarrow 0$

$$\zeta_\varepsilon = \mathbb{E}_{\alpha,\varepsilon}[\bar{\Xi}^{\alpha,\varepsilon}_T \mid \bar{\mathcal{Y}}^\varepsilon_T] \Longrightarrow \mathbb{E}_\alpha[\Xi^{(1)}_T \mid \mathcal{Y}^{(1)}_T] = \zeta \ ,$$

under $\mathbb{P}_{\alpha,\varepsilon}$.

To characterize the limit r.v. ζ, the extended linear tangent model (8) is rewritten as

$$\begin{cases} d\begin{pmatrix} X^{(1)}_t \\ \Xi^{(1)}_t \end{pmatrix} = \begin{pmatrix} F_t & 0 \\ 0 & 0 \end{pmatrix}\begin{pmatrix} X^{(1)}_t \\ \Xi^{(1)}_t \end{pmatrix} dt + \begin{pmatrix} I \\ J^*_t \end{pmatrix} dW^\alpha_t, & \begin{pmatrix} X^{(1)}_0 \\ \Xi^{(1)}_0 \end{pmatrix} = \begin{pmatrix} \xi \\ 0 \end{pmatrix}, \\ dY^{(1)}_t = \begin{pmatrix} H_t & 0 \end{pmatrix}\begin{pmatrix} X^{(1)}_t \\ \Xi_t \end{pmatrix} dt + dV^\alpha_t, \quad Y^{(1)}_0 = 0, \end{cases}$$

with the notations $F_t = b'(\alpha, X^{(0)}_t)$, $H_t = h'(X^{(0)}_t)$ and $J_t = \partial b(\alpha, X^{(0)}_t)$, and the solution of the above filtering problem yields

$$\zeta = \mathbb{E}_\alpha[\Xi^{(1)}_T \mid \mathcal{Y}^{(1)}_T] = \int_0^T r^*_t\, H^*_t\; [dY^{(1)}_t - H_t\, \hat{X}^{(1)}_t\, dt] \ , \tag{9}$$

and

$$\mathbb{E}_\alpha[(\Xi^{(1)}_T - \zeta)\, (\Xi^{(1)}_T - \zeta)^* \mid \mathcal{Y}^{(1)}_T] = \int_0^T (J^*_t\, J_t - r^*_t\, H^*_t\, H_t\, r_t)\, dt \ , \tag{10}$$

where $\{r_t\, ,\, 0 \le t \le T\}$ satisfies the ODE

$$\dot{r}_t = F_t\, r_t - P_t\, H^*_t\, H_t\, r_t + J_t \ , \quad r_0 = 0 \ ,$$

and where $\hat{X}^{(1)}_t = \mathbb{E}_\alpha[X^{(1)}_t \mid \mathcal{Y}^{(1)}_t]$ and P_t are respectively the Kalman filter estimate and its asociated error covariance matrix in the linear tangent model (4), for any $0 \le t \le T$. Furthermore, the innovation process is a standard Wiener process, hence the r.v. ζ is Gaussian, with mean 0 and covariance matrix

$$\Sigma = \int_0^T r^*_t\, H^*_t\, H_t\, r_t\, dt \ .$$

Theorem 2. *The residual ζ_ε is asymptotically normal under the nominal hypothesis, i.e. as $\varepsilon \downarrow 0$*

$$\zeta_\varepsilon \Longrightarrow \mathcal{N}(0, \Sigma) \ , \quad \text{under } \mathbb{P}_{\alpha,\varepsilon} .$$

5 Local Asymptotic Normality (LAN)

To study the asymptotic normality of the residual under the contiguous alternative hypothesis, we first prove the local asymptotic normality property of the family of probability distributions of the observation process, a result which is of interest by itself.

As is well known [11], we can express the likelihood ratio as

$$\mathbb{E}_{\alpha,\varepsilon}[\frac{d\mathbb{P}_{\theta,\varepsilon}}{d\mathbb{P}_{\alpha,\varepsilon}} \mid \mathcal{Y}_T] = \frac{L_\varepsilon(\theta)}{L_\varepsilon(\alpha)} = \frac{\displaystyle\int p_T^{\theta,\varepsilon}(x)\, dx}{\displaystyle\int p_T^{\alpha,\varepsilon}(x)\, dx} \ . \tag{11}$$

Taking $\theta = \alpha + \varepsilon \Delta$, using (6) and the change of variable $x = X_T^{(0)} + \varepsilon\, u$ in the expression (11), yields

$$\begin{aligned} Z_\varepsilon(\Delta) = \frac{L_\varepsilon(\alpha + \varepsilon\, \Delta)}{L_\varepsilon(\alpha)} &= \frac{\displaystyle\int p_T^{\alpha+\varepsilon\, \Delta,\varepsilon}(x)\, dx}{\displaystyle\int p_T^{\alpha,\varepsilon}(x)\, dx} \\ &= \frac{\displaystyle\int \varepsilon^m\, p_T^{\alpha+\varepsilon\, \Delta,\varepsilon}(X_T^{(0)} + \varepsilon\, u)\, du}{\displaystyle\int \varepsilon^m\, p_T^{\alpha,\varepsilon}(X_T^{(0)} + \varepsilon u)\, du} \\ &= \frac{\displaystyle\int \bar{p}_T^{\alpha+\varepsilon\, \Delta,\varepsilon}(u)\, du}{\displaystyle\int \bar{p}_T^{\alpha,\varepsilon}(u)\, du} = \mathbb{E}_{\alpha,\varepsilon}[\bar{\Upsilon}_T^{\alpha+\varepsilon\, \Delta,\varepsilon} \mid \bar{\mathcal{Y}}_T^\varepsilon] \ . \end{aligned}$$

To study the joint asymptotic behaviour of $(\zeta_\varepsilon, Z_\varepsilon(\Delta))$ under $\mathbb{P}_{\alpha,\varepsilon}$ as $\varepsilon \downarrow 0$, we further extend the contaminated model (5) and the linear tangent model (4) as follows

$$\left\{ \begin{aligned} d\bar{X}_t^\varepsilon &= \frac{1}{\varepsilon}\, [\, b(\alpha, X_t^{(0)} + \varepsilon\, \bar{X}_t^\varepsilon) - b(\alpha, X_t^{(0)})\,]\, dt + dW_t^\alpha \ , \quad \bar{X}_0^\varepsilon = \xi \ , \\ d\bar{\Xi}_t^{\alpha,\varepsilon} &= \partial b^*(\alpha, X_t^{(0)} + \varepsilon\, \bar{X}_t^\varepsilon)\, dW_t^\alpha \ , \quad \bar{\Xi}_0^{\alpha,\varepsilon} = 0 \ , \\ d\bar{\Upsilon}_t^{\alpha+\varepsilon\, \Delta,\varepsilon} &= \frac{1}{\varepsilon}\, [b(\alpha + \varepsilon\, \Delta, X_t^{(0)} + \varepsilon\, \bar{X}_t^\varepsilon) \\ &\qquad - b(\alpha, X_t^{(0)} + \varepsilon\, \bar{X}_t^\varepsilon)]^*\, \bar{\Upsilon}_t^{\alpha+\varepsilon\, \Delta,\varepsilon}\, dW_t^\alpha \ , \quad \bar{\Upsilon}_0^{\alpha,\varepsilon} = 0 \ , \\ d\bar{Y}_t^\varepsilon &= \frac{1}{\varepsilon}\, [\, h(X_t^{(0)} + \varepsilon\, \bar{X}_t^\varepsilon) - h(X_t^{(0)})\,]\, dt + dV_t^\alpha \ , \quad \bar{Y}_0^\varepsilon = 0 \ , \end{aligned} \right.$$

and

$$\begin{cases} dX_t^{(1)} = b'(\alpha, X_t^{(0)}) X_t^{(1)} dt + dW_t^\alpha \ , \quad X_0^{(1)} = \xi \ , \\ d\Xi_t^{(1)} = \partial b^*(\alpha, X_t^{(0)})\, dW_t^\alpha \ , \quad \Xi_0^{(1)} = 0 \ , \\ d\Upsilon_t^{(1)} = \Delta^* \, \partial b^*(\alpha, X_t^{(0)})\, \Upsilon_t^{(1)}\, dW_t^\alpha \ , \quad \Upsilon_0^{(1)} = 0 \ , \\ dY_t^{(1)} = h'(X_t^{(0)}) X_t^{(1)} dt + dV_t^\alpha \ , \quad Y_0^{(1)} = 0 \ , \end{cases}$$

respectively.

Lemma 4. *As* $\varepsilon \downarrow 0$

$$\sup_{0 \le t \le T} |\bar{\Upsilon}_t^{\alpha + \varepsilon\, \Delta, \varepsilon} - \Upsilon_t^{(1)}| \longrightarrow 0 \ ,$$

in $\mathbb{P}_{\alpha,\varepsilon}$*–probability.*

PROOF. It holds

$$\begin{aligned} d(&\bar{\Upsilon}_t^{\alpha + \varepsilon\, \Delta, \varepsilon} - \Upsilon_t^{(1)}) \\ &= (\frac{1}{\varepsilon}\, [b(\alpha + \varepsilon\, \Delta, X_t^{(0)} + \varepsilon\, \bar{X}_t^\varepsilon) - b(\alpha, X_t^{(0)} + \varepsilon\, \bar{X}_t^\varepsilon)] \\ &\qquad\qquad - \partial b(\alpha, X_t^{(0)})\, \Delta\,)^*\, \Upsilon_t^{(1)}\, dW_t^\alpha \\ &\quad + \frac{1}{\varepsilon}\, [b(\alpha + \varepsilon\, \Delta, X_t^{(0)} + \varepsilon\, \bar{X}_t^\varepsilon) - b(\alpha, X_t^{(0)} + \varepsilon\, \bar{X}_t^\varepsilon)]^* \\ &\qquad\qquad (\bar{\Upsilon}_t^{\alpha + \varepsilon\, \Delta, \varepsilon} - \Upsilon_t^{(1)})\, dW_t^\alpha \ . \end{aligned}$$

Notice that

$$\begin{aligned} |&\frac{1}{\varepsilon}\, [b(\alpha + \varepsilon\, \Delta, X_t^{(0)} + \varepsilon\, u) - b(\alpha, X_t^{(0)} + \varepsilon\, u)] - \partial b(\alpha, X_t^{(0)})\, \Delta| \\ &\le |\frac{1}{\varepsilon}\, [b(\alpha + \varepsilon\, \Delta, X_t^{(0)} + \varepsilon\, u) - b(\alpha, X_t^{(0)} + \varepsilon\, u)] - \partial b(\alpha, X_t^{(0)} + \varepsilon\, u)\, \Delta| \\ &\quad + |\, [\partial b(\alpha, X_t^{(0)} + \varepsilon\, u) - \partial b(\alpha, X_t^{(0)})]\, \Delta| \\ &\le K\, \varepsilon\, |\Delta|^2 + K\, \varepsilon\, |u|\, |\Delta| \ , \end{aligned}$$

and

$$|\frac{1}{\varepsilon}\, [b(\alpha + \varepsilon\, \Delta, X_t^{(0)} + \varepsilon\, u) - b(\alpha, X_t^{(0)} + \varepsilon\, u)]\, | \le K\, |\Delta| \ ,$$

for any $u \in \mathbb{R}^m$, and the Doob inequality yields

$$\mathbb{E}_{\alpha,\varepsilon}[\sup_{0 \le t \le T} |\bar{\Upsilon}_t^{\alpha + \varepsilon\, \Delta, \varepsilon} - \Upsilon_t^{(1)}|^2\,]$$

$$
\begin{aligned}
&\le 12\, K^2\, \varepsilon^2\, |\Delta|^4\, \mathbb{E}_{\alpha,\varepsilon}[\int_0^T |\Upsilon_t^{(1)}|^2\, dt] \\
&+ 12\, K^2\, \varepsilon^2\, |\Delta|^2\, \mathbb{E}_{\alpha,\varepsilon}[\int_0^T |\bar{X}_t^\varepsilon|^2\, |\Upsilon_t^{(1)}|^2\, dt] \\
&+ 12\, K^2\, |\Delta|^2\, \mathbb{E}_{\alpha,\varepsilon}[\int_0^T |\bar{\Upsilon}_t^{\alpha+\varepsilon\,\Delta,\varepsilon} - \Upsilon_t^{(1)}|^2\, dt] \;,
\end{aligned}
$$

which goes to zero as $\varepsilon \downarrow 0$, using the Gronwall lemma, since

$$
\sup_{0<\varepsilon\le\varepsilon_0}\ \sup_{0\le t\le T}\ \mathbb{E}_{\alpha,\varepsilon}|\bar{X}_t^\varepsilon|^4 < \infty \;,
$$

and

$$
\sup_{0<\varepsilon\le\varepsilon_0}\ \sup_{0\le t\le T}\ \mathbb{E}_{\alpha,\varepsilon}|\Upsilon_t^{(1)}|^4 < \infty \;. \quad \square
$$

The joint convergence in distribution of $(\zeta_\varepsilon, Z_\varepsilon(\Delta))$ follows from Lemmas 2 and 4, from another instance of Lemma 3, and from Theorem 1.2 in Kleptsina, Liptser and Serebrovski [9].

Theorem 3. *As* $\varepsilon \downarrow 0$

$$
\begin{aligned}
(\zeta_\varepsilon, Z_\varepsilon(\Delta)) \;&=\; (\mathbb{E}_{\alpha,\varepsilon}[\bar{\Xi}_T^{\alpha,\varepsilon} \mid \bar{\mathcal{Y}}_T^\varepsilon], \mathbb{E}_{\alpha,\varepsilon}[\bar{\Upsilon}_T^{\alpha+\varepsilon\,\Delta,\varepsilon} \mid \bar{\mathcal{Y}}_T^\varepsilon]) \\
&\Longrightarrow (\mathbb{E}_\alpha[\Xi_T^{(1)} \mid \mathcal{Y}_T^{(1)}], \mathbb{E}_\alpha[\Upsilon_T^{(1)} \mid \mathcal{Y}_T^{(1)}]) = (\zeta, Z(\Delta)) \;,
\end{aligned}
$$

under $\mathbb{P}_{\alpha,\varepsilon}$.

To characterize the limit r.v. $Z(\Delta)$, notice that

$$
\Upsilon_T^{(1)} = \exp\{\Delta^* \int_0^T J_t^*\, dW_t^\alpha - \tfrac{1}{2}\, \Delta^* \int_0^T J_t^*\, J_t\, dt\; \Delta\} \;,
$$

where $J_t = \partial b(\alpha, X_t^{(0)})$, hence

$$
\begin{aligned}
Z(\Delta) &= \mathbb{E}_\alpha[\Upsilon_T^{(1)} \mid \mathcal{Y}_T^{(1)}] \\
&= \exp\{-\tfrac{1}{2}\, \Delta^* \int_0^T J_t^*\, J_t\, dt\; \Delta\}\; \mathbb{E}_\alpha[\exp\{\Delta^*\, \Xi_T^{(1)}\} \mid \mathcal{Y}_T^{(1)}] \;,
\end{aligned}
$$

and it follows from (9) and (10) that

$$
\mathbb{E}_\alpha[\exp\{\Delta^*\, \Xi_T^{(1)}\} \mid \mathcal{Y}_T^{(1)}] = \exp\{\Delta^*\, \zeta + \tfrac{1}{2}\, \Delta^* \int_0^T J_t^*\, J_t\, dt\, \Delta - \tfrac{1}{2}\, \Delta^*\, \Sigma\, \Delta\} \;.
$$

Therefore

$$
Z(\Delta) = \exp\{\Delta^*\, \zeta - \tfrac{1}{2}\, \Delta^*\, \Sigma\, \Delta\} \;.
$$

Thus, we have proved the following local asymptotic normality property.

Theorem 4. *The family of probability distributions of the observation process is locally asymptotically normal (LAN) at the point α as $\varepsilon \downarrow 0$, i.e. for any $\Delta \in \mathbb{R}^p$*

$$Z_\varepsilon(\Delta) = \mathbb{E}_{\alpha,\varepsilon}[\frac{d\mathbb{P}_{\theta,\varepsilon}}{d\mathbb{P}_{\alpha,\varepsilon}} \mid \mathcal{Y}_T] \Longrightarrow \exp\{\Delta^* \zeta - \tfrac{1}{2}\,\Delta^*\,\Sigma\,\Delta\} = Z(\Delta) \;,$$

where $\zeta \sim \mathcal{N}(0,\Sigma)$, under $\mathbb{P}_{\alpha,\varepsilon}$.

6 Application to FDI

Under the contiguous alternative hypothesis $\theta = \alpha + \varepsilon\Delta$, for any test function ϕ defined on $\mathbb{R}^p$, it holds

$$\begin{aligned}
\mathbb{E}_{\alpha+\varepsilon\Delta,\varepsilon}[\phi(\zeta_\varepsilon)] &= \mathbb{E}_{\alpha,\varepsilon}[\phi(\zeta_\varepsilon)\,\frac{d\mathbb{P}_{\alpha+\varepsilon\Delta,\varepsilon}}{d\mathbb{P}_{\alpha,\varepsilon}}\,] \\
&= \mathbb{E}_{\alpha,\varepsilon}[\phi(\zeta_\varepsilon)\,Z_\varepsilon(\Delta)\,] \\
&\longrightarrow \mathbb{E}_\alpha[\phi(\zeta)\,\exp\{\Delta^*\,\zeta - \tfrac{1}{2}\,\Delta^*\,\Sigma\,\Delta\}\,] \;,
\end{aligned}$$

where the convergence follows from Theorems 3 and 4. Let F and G denote the Gaussian distributions on $\mathbb{R}^p$ with (not necessarily invertible) covariance matrix Σ, and mean vector 0 and $\Sigma\,\Delta$ respectively. Since G is absolutely continuous w.r.t. F, with Radon–Nikodym derivative

$$\frac{dG}{dF}(x) = \exp\{\Delta^*\,x - \tfrac{1}{2}\,\Delta^*\,\Sigma\,\Delta\} \;,$$

it holds

$$\begin{aligned}
&\mathbb{E}_\alpha[\phi(\zeta)\,\exp\{\Delta^*\,\zeta - \tfrac{1}{2}\,\Delta^*\,\Sigma\,\Delta\}\,] \\
&\quad = \int \phi(x)\,\exp\{\Delta^*\,x - \tfrac{1}{2}\,\Delta^*\,\Sigma\,\Delta\}\;F(dx) = \int \phi(x)\;G(dx) \;,
\end{aligned}$$

i.e. the following result has been proved.

Theorem 5. *The residual ζ_ε is asymptotically normal under the contiguous alternative hypothesis, i.e. as $\varepsilon \downarrow 0$*

$$\zeta_\varepsilon \Longrightarrow \mathcal{N}(\Sigma\,\Delta, \Sigma) \;, \quad \textit{under } \mathbb{P}_{\alpha+\varepsilon\Delta,\varepsilon}.$$

Combining the results of Theorems 2 and 5, yields

$$\zeta_\varepsilon \Longrightarrow \zeta \sim \begin{cases} \mathcal{N}(0,\Sigma) \;, & \text{under } \mathbb{P}_{\alpha,\varepsilon}, \\ \mathcal{N}(\Delta\,\Sigma, \Sigma) \;, & \text{under } \mathbb{P}_{\alpha+\varepsilon\,\Delta,\varepsilon}, \end{cases}$$

where

$$\Sigma = \int_0^T r_t^* \, H_t^* \, H_t \, r_t \, dt \ ,$$

depends only on the nominal value α of the parameter. To detect a change in the mean vector of the limiting Gaussian r.v., one can use a simple χ^2 test

$$\zeta^* \, \Sigma^{-1} \, \zeta \lesseqgtr c \ , \tag{12}$$

where $c > 0$ is the threshold, provided the covariance matrix Σ is invertible. Plugging the residual ζ_ε in, which depends only on the nominal value α of the parameter and on observations $\{Y_t \, , \, 0 \leq t \leq T\}$ produced by the original dynamical system, yields the following test

$$\zeta_\varepsilon^* \, \Sigma^{-1} \, \zeta_\varepsilon \lesseqgtr c \ ,$$

where $c > 0$ is the threshold. To select the threshold, error probabilities for this test can be approximated by those of the test (12).

Acknowledgments. This work was completed while the second author was visiting IRISA, with a post–doctoral fellowship from Ministère de la Recherche, and further support from INRIA. It was partially supported by the Commission of the European Communities, under the TMR project *System Identification*, project number FMRX–CT98–0206, and under the IHP research training network *Statistical Methods for Dynamical Stochastic Models*, network number HPRN–CT2000–00100.

References

1. M. Basseville. On–board component fault detection and isolation using the statistical local approach. *Automatica*, 34(11):1391–1416, Nov. 1998.
2. A. Benveniste, M. Basseville, and G. V. Moustakides. The asymptotic local approach to change detection and model validation. *IEEE Transactions on Automatic Control*, AC–32(7):583–592, July 1987.
3. A. Benveniste, M. Métivier, and P. Priouret. *Adaptive Algorithms and Stochastic Approximations*, volume 22 of *Applications of Mathematics*. Springer–Verlag, New York, 1990.
4. F. Campillo and F. Le Gland. MLE for partially observed diffusions : direct maximization vs. the EM algorithm. *Stochastic Processes and their Applications*, 33(2):245–274, 1989.
5. F. Cérou, F. Le Gland, and N. J. Newton. Stochastic particle methods for linear tangent filtering equations. In J.-L. Menaldi, E. Rofman, and A. Sulem, editors, *Optimal Control and Partial Differential Equations. In honour of professor Alain Bensoussan's 60th birthday*, pages 231–240. IOS Press, Amsterdam, 2001.
6. C. D. Charalambous, R. J. Elliott, and V. Krishnamurthy. Conditional moment generating functions for integrals and stochastic integrals. In *Proceedings of the 36th Conference on Decision and Control, San Diego 1997*, pages 3944–3949. IEEE–CSS, Dec. 1997.

7. B. Delyon, A. Juditsky, and A. Benveniste. On the relationship between identification and local tests. Publication Interne 1104, IRISA, Rennes, May 1997. `ftp://ftp.irisa.fr/techreports/1997/PI-1104.ps.gz`.
8. M. I. Freidlin and A. D. Wentzell. *Random Perturbations of Dynamical Systems*, volume 260 of *Grundlehren der mathematischen Wissenschaften*. Springer–Verlag, New York, 1984.
9. M. L. Kleptsina, R. S. Liptser, and A. P. Serebrovski. Nonlinear filtering problem with contamination. *The Annals of Applied Probability*, 7(4):917–934, 1997.
10. Y. A. Kutoyants. *Identification of Dynamical Systems with Small Noise*, volume 300 of *Mathematics and its Applications*. Kluwer Academic Publisher, Dordrecht, 1994.
11. R. S. Liptser and A. N. Shiryayev. *Statistics of Random Processes I. General Theory*, volume 5 of *Applications of Mathematics*. Springer–Verlag, New York, 1977.
12. R. S. Liptser and A. N. Shiryayev. *Theory of Martingales*, volume 49 of *Mathematics and its Applications (Soviet Series)*. Kluwer Academic Publishers, Dordrecht, 1989.

Stochastic Lagrangian Adaptive LQG Control

David Levanony[1] and Peter E. Caines[2]

[1] Dept. of Electrical and Computer Engineering, Ben Gurion University, Beer Sheva 84105, Israel. levanony@ee.bgu.ac.il.

[2] Dept. of Electrical Engineering, McGill University, Montreal, Quebec H2A 2A7, Canada, and The Canadian Institute for Advanced Research. peterc@cim.mcgill.ca

Abstract. This paper presents a continuous time stochastic adaptive control algorithm for completely observed linear stochastic systems with unknown parameters. The adaptive estimation algorithm is designed so that, first, it drives the estimate into a neighbourhood I^δ of a set I of parameters corresponding to the true closed loop dynamics and then, second, by activating a performance monitoring feature in I^δ, the estimate converges to the true system parameter and the resulting control yields optimal long run LQ closed loop performance.

1 Introduction

The work of Kumar and Becker [1982] and Kumar and Lin [1982] introduced the technique of biasing standard parameter estimation schemes when used in adaptive control algorithms for controlled Markov chains. This biasing is a function of the system performance that would result if the true system were described by the current parameter estimate. The basic idea is to direct the process of parameter estimates in a way that takes account of both identification and steady state system performance. This line of research was continued for controlled diffusion processes by Borkar [1991]. One notable advantage of this approach is that it permits a significant weakening of the persistent excitation (PE) requirement that is to be found in the work on consistency based stochastic adaptive control and which is a particular difficulty in the adaptive stabilization in both the continuous and discrete time cases (see Lai and Wei [1982], Caines [1988, 1992], Chen and Guo [1991]).

One contribution of this paper lies in the fact that the adaptive control algorithm described is completely recursive, in contrast to that of Borkar [1991], which implicitly involves an instantaneous optimization procedure.

A particular feature of our approach is the use of the geometric analysis of Polderman [1986, 1989] who studied the structure of the sets of parameters corresponding to (i) indistinguishable closed loop dynamical behaviour and (ii) optimal LQ closed loop performance. This analysis was specifically carried out with the analysis of LQ adaptive control schemes in view.

B. Pasik-Duncan: Stochastic Theory and Control, LNCIS 280, pp. 283–300, 2002.

Methodology and Principal Results. Consider a class of completely observed LTI systems whose states evolve according to the Itô equations

$$dx_t = Ax_t dt + Bu_t dt + Cdw_t,$$

where x, u, w take values in $\mathbb{R}^n, \mathbb{R}^m, \mathbb{R}^r$ (respectively) and w is a standard Brownian process independent of x. For a given system, parameterized by unknown $(A^*, B^*) \triangleq \theta^*$, our objective is to recursively generate estimates $\{\theta_t; t \geq t_o\}$ to be used in an LQ certainty equivalence control to obtain the optimal LQ (long run) cost which would be obtained if θ^* were known. In the solution to this problem provided in this paper we consider the same control policy as described in Caines [1992]. The basic difference lies in the parameter estimation algorithm. In Caines [1992] a standard RLS algorithm has been employed under a (sample-wise) PE condition which requires independent verification. In the adaptive parameter estimation and control work of Duncan and Pasik-Duncan [1986,1991], the role of this PE condition is effectively replaced by the condition that a certain determinant $det\widetilde{A}_t$ shall be almost surely bounded away from zero. This condition is verifiable in certain cases of interest. Alternatively, the required PE property can be *created* by the injection of a diminishing excitation signal, as has been shown by Duncan, Guo and Pasik–Duncan [1999]. In this work, no external dither is utilized.

Let $\phi_t^T = (x_t^T, -x_t^T K_t^T)$ denote the regression vector where $K_t = K(\theta_t)$ is computed via the control law; then, roughly speaking, one class of PE conditions is equivalent to assuming that the matrix $\int_0^t \phi_s \phi_s^T ds$ (properly normalized) converges to a strictly positive definite limit. Another important class is that involving the comparative rates of growth of eigenvalues of this matrix (see Lai and Wei [1982], Chen and Guo [1991]). Note, however, that if $K_t \to K$, the existence of such a positive definite limit is questionable. PE conditions are almost invariably used to ensure consistency; an alternative formulation, which we pursue in this work, is that where one simply considers the natural convergence $\theta_t \to I$, as $t \to \infty$, where I is some limit set, which occurs without further conditions for essentially all parameter estimation algorithms. (In our case, I corresponds to parameters yielding indistinguishable system trajectories.)

Let $J(\theta)$ be the long run optimal performance cost for a system with $(A, B) = \theta$. As has been established by Polderman [1986], the limit set of indistinguishable dynamics

$$I = \{(A, B) | A - BK(A, B) = A^* - B^* K(A, B)\}$$

is a smooth manifold upon which J has a unique minimum θ^* (= the true parameter) which, in addition, we show has no local stationary points other than θ^* on I. With those properties in mind, the intuition behind the performance biased adaptive control algorithm is as follows: A conventional certainty equivalence adaptive controller, using an MLE-type parameter estimation scheme, forms Stage 1 of the adaptation process. It uses current parameter estimates to generate approximately optimal (in a certain sense) predictions of the state trajectories together with a feedback control law. This has the standard implicit effect that prediction errors are used to adjust parameter estimates so that asymptotically the estimated and the actual closed loop system dynamics will match exactly. By the discussion above, estimated and actual system dynamics for the closed loop system will match exactly on the set I, but performance will be suboptimal for any $\theta \epsilon I$, $\theta \neq \theta^*$. Therefore, the refining Stage 2 is introduced into the adaptation: A performance biased cost is embedded within a constrained optimization problem whose recursive solution constitutes the Stage 2, or Lagrangian mode, algorithm. As a result, the overall algorithm eventually drives the parameter estimate in the direction of $-\nabla J(\theta)$; this, together with the unique minimum property of J on I results in the desired consistency which in turn implies the optimality of the long run performance. Since the trajectory of the estimate resulting from Stage 1 is eventually within I^δ, the second stage eventually becomes dominant. It should be noted that, since there are no means to determine the "right time" to activate Stage 2, one has in fact to use a two-phase algorithm from some predetermined time T onwards. More precisely, the overall algorithm switches to the performance-monitored Lagrangian mode algorithm whenever the parameter estimate lies in a certain set $\mathcal{A}(t)$, while, on the other hand, it activates the MLE-type scheme when this condition fails to hold.

A switching policy is used in order to avoid infinitely many switchings (for an example, see Levanony, Shwartz and Zeitouni [1994]). Moreover, the performance monitored parameter sets $\mathcal{A}(t)$ should be defined in such a way that will ensure that indeed $\theta_t \to I^\delta$, as $t \to \infty$, and once in I^δ, the estimator will remain in I^δ whilst operating in the performance monitored mode, to obtain $\theta_t \to \theta^*$, as $t \to \infty$.

Our results are organized as follows: The adaptive control problem is formulated in Section 2. In Section 3 a class of AML (Asymptotic Maximum Likelihood) estimates is defined and further, is shown to have a common limit parameter set $\mathcal{I}$. Section 4 is devoted to the investigation of the geometric properties of the limit set $\mathcal{I}$. In Section 5, after a formal derivation of the algorithm, the following statements are made: (1) The SDE defined by the proposed algorithm, has a unique strong solution. (2) There exists a $\theta_\infty(\omega)$ such that the parameter estimates $\theta_t \to \theta_\infty$, *a.s.* as $t \to \infty$. (3) $\theta_\infty \in \mathcal{I}$, *a.s.* (4) There exists a $\tau(\omega) < \infty$ (*a.s.*) such that $\theta_t \in \mathcal{A}(t)$, for all $t \geq \tau$. (5) $\theta_\infty = \theta^*$, *a.s.*, and optimal, long–run LQ performance is obtained.

Remark. Due to obvious constrains on the paper's length, all proofs are omitted. These appear in [16].

2 Problem Statement

We consider the system

$$dx_t = Ax_t dt + Bu_t dt + C dw_t, \tag{1}$$

where $x_t \in \mathbb{R}^n, u_t \in \mathbb{R}^m$ and $w_t \in \mathbb{R}^p$ for all $t \geq 0, x_0$ is a non-random initial condition and w is a standard Brownian motion. Let the process $w = \{w_t,\ t \geq 0\}$ be measurable with respect to the increasing family of σ-fields $\mathcal{F}_t$ for all $t \geq 0$. The solution process $\{x_t,\ t \geq 0\}$ for (1) is dependent upon the values taken by A, B, C when these are treated as non–random, time independent variables.

Let x generate the increasing family of σ-fields $\mathcal{F}_t^x, t \geq 0$. Then the process u is assumed to satisfy the (adaptive) non-anticipating control condition that u_t is only an $\mathcal{F}_t^x$ measurable function for $t \geq 0$, and hence is not an explicit function of A, B, C. We shall denote this condition by $u \in \mathcal{U}$.

The objective in the application of adaptive control to the system (1) is to achieve, along almost all sample paths,

$$J_\infty^o \triangleq \inf_{u \in \mathcal{U}} \lim_{t \to \infty} \frac{1}{t} \int_0^t (\|x_s\|^2 + \|u_s\|^2) ds \tag{2}$$

where the control laws employed are such that the inner limit exists, along almost all paths.

Let us define the matrix parameter $\Theta = [A, B]^T$ and the corresponding vector parameter $\theta = [col(A, B)] \in \mathbb{R}^{n(n+m)}$, where $[col(A, B)]^T = [(A_1^T, B_1^T) \cdots, (A_n^T, B_n^T)]$, with A_i (respectively B_i) denoting the i-th row of A (respectively B). Further define the $n \times (n+m)n$ matrix Ψ_t by

$$\Psi_t = \begin{bmatrix} \phi_t^T & 0 & \cdots & 0 \\ 0 & \phi_t^T & 0 & 0 \\ 0 & \cdots & \cdots & \phi_t^T \end{bmatrix},$$

where $\phi_t^T \triangleq (x_t^T, u_t^T)$. Assuming a full rank noise (i.e. $CC^T > 0$), we take for simplicity $C = I$. The system equation (1) may be conveniently re-expressed as

$$dx_t = \Psi_t \theta dt + dw_t, \quad t \geq 0. \tag{3}$$

For convenience, when $u \in \mathcal{U}$, we shall refer to both (1) and (3) as the system $\Xi(\theta)$.

We use the notation θ^* to denote the value of the deterministic parameter of the system (3) generating a given set of observations for a given control law $u \in \mathcal{U}$. This parameter θ^* will be referred to as the true system parameter and we assume that

$$\theta^* \in \mathcal{S} \triangleq \{\theta = col(A, B) : (A, B) \text{ stabilizable}\}.$$

The adaptive control algorithms we study in this paper are based upon the class of certainty equivalence (CE) algorithms which have the following form for the adaptive LQ problem: for each $t \geq 0$,

(i) Compute an estimate $\theta_t \in \mathcal{S}$ of $\theta^* \in \mathcal{S}$.
(ii) Use the feedback control law

$$u_t = -K(\theta_t)x_t,$$

where

$$K(\theta) = K(A, B) = B^T V(A, B) \tag{4}$$

where $V(A, B)$ is the (unique) positive definite symmetric solution to the algebraic Riccati equation (ARE):

$$A^T V + VA - VBB^T V + I = 0. \tag{5}$$

It is well known that for any given system, parameterized by a stabilizable pair (A, B), the optimal achievable performance (2) equals to

$$J(A, B) \triangleq \operatorname{Tr} V(A, B). \tag{6}$$

Therefore, (with a slight abuse of notation) we shall refer to $J(\theta) \triangleq \operatorname{Tr} V(\theta)$ as the (synthetic) cost function. As will be shown below, a minimization of (6), which takes place during the adaptation procedure, leads to the desired optimal performance, that is, along almost all sample paths, one obtains,

$$\lim_{t\to\infty} \frac{1}{t} \int_0^t (\|x_s\|^2 + \|u_s\|^2) ds = J(\theta^*) = J^o_\infty. \tag{7}$$

One of the motives for the use of the gradient search maximum likelihood (ML) class of parameter estimation schemes in the adaptive control algorithm introduced in this paper is that such schemes are sufficiently flexible to permit various modifications to the algorithm while retaining consistency; another motive is that they are comparatively efficient numerically. To facilitate the analysis in this paper, we reduce the proof of the consistency of the ML scheme (for systems subject to feedback control which is dependent upon current parameter estimates) to a convergence analysis of recursive least squares schemes (for feedback systems) and this, in turn, is reduced to the study of minimum variance estimates (i.e. conditional expectations) within a Bayesian framework. As a consequence, after the presentation of an introductory result (Lemma 3.1) in Section 3 concerning ML estimates, we proceed to establish a Bayesian convergence result (Theorem 3.2) and then an RLS convergence result (Theorem 3.3).

3 Maximum Likelihood Identification

Let $\{(u_t, x_t),\ t \geq 0\}$ denote an observed input-state sample path of the system $\Xi(\theta^*)$, and let $L_t(\theta) \triangleq L(\theta, (u_0^t, x_0^t))$ denote the log-likelihood function of (u_0^t, x_0^t) at any $\theta \in \mathcal{S}$ where its gradient $\nabla L_t(\theta)$ is given by

$$\nabla L_t(\theta) = \int_0^t \Psi_s^T dx_s - \int_0^t \Psi_s^T \Psi_s ds\theta$$
$$= \int_0^t \Psi_s^T dw_s - \int_0^t \Psi_s^T \Psi_s ds(\theta - \theta^*) \triangleq m_t - \Phi_t \widetilde{theta} \quad (8)$$

where $\{m_t, t \geq 0\}$ is an $n(n+m)$ dimensional martingale, $\widetilde{theta} \triangleq \theta - \theta^*$ and

$$\Phi_t \triangleq \int_0^t \Psi_s^T \Psi_s ds.$$

In terms of (8) we may present the following preliminary asymptotic maximum likelihood estimation (MLE) result, which we note does not constitute a consistency result without the addition of further hypotheses.

Lemma 3.1 For the system $\Xi(\theta^*)$, and the observed process $\{(u_t, x_t),\ t \geq 0\}$, let $\{\theta_t = \theta_t(\theta^*, \omega),\ t \geq 0\}$, be a process which is progressively measurable with respect to the σ-fields $\mathcal{F}_t^x, t \geq 0$. Assume that $\{\theta_t,\ t \geq 0\}$ belongs to the class of *Asymptotic ML estimates* $(\mathcal{AML})$ in the sense that

$$\nabla L_t(\theta_t) \to 0 \qquad \text{a.s. as } t \to \infty, \quad (9)$$

which is denoted by $\{\theta_t,\ t \geq 0\} \in \mathcal{AML}$. Further assume that almost surely Φ_t is non-singular for all t sufficiently large. Then there exists an a.s. finite random variable $\theta_\infty = \theta_\infty(\theta^*, \omega)$ for which

$$\theta_t \to \theta_\infty \qquad \text{a.s. as } t \to \infty \quad (10)$$

for all $\theta^* \notin \mathcal{N}$, where $\mathcal{N}$ is a Lebesgue null set in $\mathbb{R}^{n(n+m)}$ independent of ω. □

The characterization of the limit set of $\{\theta_t\} \in \mathcal{AML}$ is the purpose of the final phase of this section. Following the Bayesian embedding approach (Kumar [1989]), we begin with a consistency proof in a Gaussian setting:

Theorem 3.2 Consider the system (1) where it is assumed that $C = I$, (A^*, B^*, x_0) have a joint Gaussian distribution and $w = \{w_t, t \geq 0\}$ is a standard Brownian vector process, independent of (A^*, B^*, x_0).

Suppose that the system is controlled by $u_t = -K_t x_t,\ t \geq 0$, where K_t is a causal, $\mathcal{F}_t^x$-measurable, continuous and bounded semi-martingale (matrix) which converges a.s. to a finite, possibly random limit K_∞. Let

$$\widehat{\Theta}_t \triangleq \mathrm{E}[\Theta^* | \mathcal{F}_t^x] = \mathrm{E}([A^*, B^*]^T | \mathcal{F}_t^x) \quad (11)$$

be the MV (minimum variance) estimate of $\Theta^* \triangleq [A^*, B^*]^T$. Then

$$\widehat{\Theta}_t \to \{[A, B]^T : A - BK_\infty = A^* - B^* K_\infty\} \quad \text{a.s. as } t \to \infty \quad (12)$$

Corollary 3.3 Let Θ_t^{RLS} be the matrix RLS estimate corresponding to θ_t^{RLS} of Lemma 3.1 (i.e. $\theta_t^{RLS} = col\{\Theta_t^{RLS}\}$). Then, under the conditions of Theorem 3.2 but with $[A^*, B^*]$ deterministic and stabilizable, Θ_t^{RLS} is a.s. convergent to some Θ_∞^{RLS} with

$$\Theta_\infty^{RLS} \in \{[A, B]^T : A - BK_\infty = A^* - B^* K_\infty\}, \text{ a.s.}$$

for all $[A^*, B^*]^T = \Theta^* \notin \mathcal{N}$ where $\mathcal{N}$ is a Lebesgue null set in $\mathbb{R}^{(n+m)\times n}$ independent of ω. □

To conclude this section, the next theorem shows that ML estimates posses the same limit set as RLS estimates do:

Theorem 3.4 Consider the system (1) with $C = I$ and $[A^*, B^*]$ a stabilizable (deterministic) pair. Suppose that $\Phi_t > 0$ a.s. for all $t \geq t_0$, for some $t_0 < \infty$, and let $\{\theta_t, t \geq t_0\} \in \mathcal{AML}$ in the sense of (9) with $\theta_t \in \mathcal{S}$ for all $t \in [t_0, \infty]$, a.s. Then $\theta_\infty = \lim_{t\to\infty} \theta_t$ exists and is finite, w.p.1, and $\theta_\infty \in \mathcal{I}$ a.s., where

$$\theta_\infty \in \mathcal{I} \triangleq \{\theta = col(A, B) : A - BK(\theta) = A^* - B^* K(\theta)\}, \; a.s. \tag{13}$$

This holds for all stabilizable pairs $[A^*, B^*]$ outside a Lebesgue null set $\mathcal{N}$ in $\mathbb{R}^{n\times n} \times \mathbb{R}^{n\times m}$. □

4 Geometric Results

In this section we examine the geometric characteristics of the limit set $\mathcal{I}$, the set of systems with indistinguishable closed-loop dynamics. As is apparent from Theorem 3.2, standard CE LQ schemes may only lead to parameter estimate convergence into $\mathcal{I}$ and producing suboptimal performance. To deal with this situation, the geometric study in this section provides information which facilitates the construction of an adaptive control scheme which will result in the desired optimal performance.

Let

$$\mathcal{C} = \{\theta \in \mathcal{S}; K(\theta) = K(\theta^*)\}$$

Since $\theta_t \to \mathcal{I}$, as $t \to \infty$, one would achieve the optimal performance (7) if, further, $\mathcal{I} \subset \mathcal{C}$. However, Polderman [1986] showed that

(i) $\mathcal{I} \cap \mathcal{C} = \theta^*$
(ii) $V(\theta) \geq V(\theta^*) \qquad \forall \theta \in \mathcal{I}$.

(Recall that V is the solution of ARE (5).) We show below that without consistent identification (i.e. $\theta_t \to \theta^*$) only suboptimal performance can be achieved. Due to (i) on the other hand, by using standard $\mathcal{AML}$ estimates one may encounter

$$\theta_t \not\to \theta^* \tag{14}$$

thus getting $J(\theta_\infty) \geq J(\theta^*)$ (where, as define in (6), $J(\theta)$ is the optimal achievable performance for a system parameterized by θ). We now examine the first order derivatives of V and J on $\mathcal{I}$. We show that, in addition of being the unique minimum of J over $\mathcal{I}, \theta^*$ is also a unique stationary point for the gradient of J with respect to B (Lemma 4.1 below). Such a result is important as a key tool in establishing the consistency of various gradient and Newton type algorithms including the one used in this paper.

First note that for any $\theta \in \mathcal{I}$, the calculation of $V(\theta)$ can be made either by the ARE (5) or by

$$[A^* - B^*K(\theta)]^T V + V[A^* - B^*K(\theta)] + K(\theta)K^T(\theta) + I = 0 \tag{15}$$

where $K(\theta) = B(\theta)^T V(\theta) = B^T V(\theta)$. Note that by (15) $V(\theta) = V(A, B)$ is actually a function of B only (where Definition (13) determines the corresponding A for all $\theta \in \mathcal{I}$).

Let $dJ(\theta)/dB$ be an $n \times m$ matrix whose typical (i, j) entry is $dJ(\theta)/dB_{ij}$. Note that since for all $\theta \in \mathcal{I}$, J is a function of B only, one has,

$$\frac{dJ(\theta)}{dB_{ij}} = \frac{\partial J(\theta)}{\partial B_{ij}} + \sum_{p,q=1}^{n} \frac{\partial A_{pq}}{\partial B_{ij}} \frac{\partial J(\theta)}{\partial A_{pq}}.$$

Lemma 4.1 There exists an open path-wise connected subset of $\mathcal{I}' \subset \mathcal{I}$ with $\theta^* \in \mathcal{I}'$ such that θ^* is the only stationary point of dJ/dB over $\mathcal{I}'$, that is,

$$\frac{dJ(\theta)}{dB} = 0 \Leftrightarrow \theta = \theta^*, \quad \theta \in \mathcal{I}'. \tag{16}$$

□

Finally, let

$$\nabla J(\theta) = \left[\left(\frac{\partial J(\theta)}{\partial col A} \right)^T, \left(\frac{\partial J(\theta)}{\partial col B} \right)^T \right]^T. \tag{17}$$

We end this section with the presentation of the *projection* of $\nabla J(\theta)$ on the tangent space to $\mathcal{I}$ at a point $\theta \in \mathcal{I}$.

Recall that for any $\theta \in \mathcal{I}, A = A(B)$ is determined by (13). Let $D = D(\theta)$ be an $n^2 \times nm$ matrix whose entries are

$$D_{ij}(\theta) = \frac{\partial (col A)_i}{\partial (col B)_j}$$

Then, $\nabla^{\mathcal{I}} J(\theta)$ is defined to be the projection of $\nabla J(\theta)$ on the tangent space to $\mathcal{I}$ at θ. It may be written as

$$\nabla^{\mathcal{I}} J(\theta) = \begin{bmatrix} D(\theta) \\ I \end{bmatrix} [D^T(\theta)D(\theta) + I]^{-1} \left[D^T(\theta) \frac{\partial J(\theta)}{\partial col A} + \frac{\partial J(\theta)}{\partial col B} \right] \tag{18}$$

where I is an $nm \times nm$ identity matrix.

Note that

$$[D^T(\theta), I]\nabla^{\mathcal{I}} J(\theta) = D^T(\theta)\frac{\partial J(\theta)}{\partial col A} + \frac{\partial J(\theta)}{\partial col B} = col \frac{dJ(\theta)}{dB} \tag{19}$$

Hence,

$$\nabla^{\mathcal{I}} J(\theta) = 0 \Leftrightarrow \frac{dJ(\theta)}{dB} = 0 \tag{20}$$

where, with $\theta \in \mathcal{I}'$, (16) implies that $\theta = \theta^*$.

5 Lagrangian Adaptation

The results quoted in the previous section show that for optimal adaptive LQ performance it is necessary to generate consistent parameter estimates. This leads us to adopt a methodology related to the biased ML approach of Kumar [1983] and Borkar [1991], (see also Kumar and Becker [1982], Kumar and Lin [1982]). The techniques of the aforementioned authors invoke the minimization of a weighted sum of the log-likelihood function and the computed performance J of the controlled system.

In light of Theorem 3.2, the point of view adapted in this paper is that, in the limiting case of an infinite observation sample, the control task is to minimize J over the parameterized system descriptions and parameterized controllers that satisfy the *constraint* given by the vanishing of the gradient of the log-likelihood function. Consequently, on the finite time interval, a suitable approximation to this constrained optimization task is sought after.

In Kumar [1983] (who considers finite parameter sets) and Borkar [1991], the minimization of the relevant weighted sum is assumed to occur instantaneously. Since our problem formulation involves uncountable parameter spaces this is highly impractical in the present case. Consequently the adaptive algorithm presented in this paper involves the solution of stochastic differential equations whose solutions approximate the solution of the constrained optimization problems referred to above.

Motivation and Sketch of the Conceptual Adaptive Algorithm. Suppose that a positive scalar stochastic process $\{\delta_t; t \geq 0\}$ is given with δ_t monotonically decreasing to zero. Then we formulate the adaptive optimization problem as follows:

$$\begin{aligned} &\text{minimize} \quad && J(\theta, \theta), \\ &\text{subject to} \quad && \begin{cases} \theta \in \mathcal{S} \\ \|\nabla L_t(\theta)\| \leq \delta_t, \end{cases} \end{aligned} \tag{21}$$

where, with a slight change of notation, the first parameter, θ, appearing in the function $J(\cdot, \cdot)$ denotes the system $\Xi(\cdot)$, while the second refers to the parameterization of the feedback control law $K(\theta)$. Hence $J(\theta,\theta')$ denotes the LQ performance of the system $\Xi(\theta)$ subject to the control law $u_t = -K(\theta')x_t$, and we adopt the convention that $\nabla J(\theta,\theta)$ denotes the gradient of $J(\theta,\theta)$ with respect to the θ parameter in both entries.

The conceptual adaptive algorithm is based upon necessary conditions for the solution of (21). Suppose that $\theta^o \in \mathcal{S}$ is a solution to (21), then there exists an adjoint variable $\lambda^o \geq 0$ such that

$$\nabla J(\theta^o,\theta^o) - \lambda^o \Phi_t \nabla L_t(\theta^o) = 0, \tag{22}$$

$$\|\nabla L_t(\theta^o)\| \leq \delta_t, \tag{23}$$

where we note that $\frac{1}{2}\frac{\partial}{\partial\theta}\|\nabla L_t(\theta)\|^2 = -\Phi_t \nabla L_t(\theta)$.

The scalar case shows ∇J and $\Phi_t \nabla L_t$ to be asymptotically parallel for *any* $\theta \in \mathcal{S}$, and this may lead to an asymptotically trivial satisfaction of the necessary condition (22). Specifically, we observe that in the scalar case $(\theta = (a,b)^T)$ (17) reduces to

$$\frac{\partial v}{\partial b} = -k\frac{\partial v}{\partial a}; \quad \frac{\partial v}{\partial a} \neq 0 \quad \forall \theta = (a,b). \tag{24}$$

In this case $J = v$ and so $\nabla J(\theta,\theta)$ is parallel to the vector $(1,-k)^T$. However, by a simple calculation, $\Phi_t \nabla L_t(\theta)$ converges to a vector parallel to $(1,-k)^T$, as is shown by

$$\Phi_t \nabla L_t(\theta) = r_t^2(1+k^2(\theta)) \in_t \begin{bmatrix} 1 \\ -k \end{bmatrix} + o(1/r_t), \tag{25}$$

where $\in_t \to 0$ and $r_t = \operatorname{Tr}\Phi_t \to \infty$. Assuming that $\lambda_t, t \to \infty$, may be chosen to balance the growth of r_t and the decrease of $\in_t$ to zero, one may obtain

$$\lim_{t\to\infty} (\nabla J(\theta,\theta) - \lambda_t \Phi_t \nabla L_t(\theta)) = \lim_{t\to\infty} \left(\frac{\partial J}{\partial a} \begin{bmatrix} 1 \\ -k \end{bmatrix} - \lambda_t \Phi_t \nabla L_t(\theta) \right) = 0 \tag{26}$$

for *any* $\theta \in \mathcal{S}$. This triviality essentially results from the fact that we have shown both terms in (26) are parallel to the vector $(1,-k)^T$ throughout $\mathcal{S}$ (as $t \to \infty$). Furthermore, if the minimization of $J(\theta^*,\theta)$ is approximated for θ^* by adopting a certainty equivalence approach, that is, assuming $\theta^* = \theta_t$, the current estimate, we obtain the triviality

$$\theta_t = \operatorname*{argmin}_{\theta \in \mathcal{S}} J(\theta_t,\theta).$$

The property above leads to the formulation of the following set of equations which are to be solved for (θ,λ) during the adaptation process:

$$\nabla^{\mathcal{I}} J(\theta) - \lambda \Phi_t \nabla L_t(\theta) = 0, \tag{27}$$

$$\|\nabla L_t(\theta)\| \leq \delta_t, \tag{28}$$

where the definition (18) of $\nabla^{\mathcal{I}} J(\theta)$ is extended to apply to *all* θ's in $\mathcal{S}$ and $\nabla^{\mathcal{I}} J(\theta)$ denotes the projection of $\nabla J(\theta, \theta)$ onto the tangent space T'_θ to $\mathcal{I}$ at θ', where θ' is the nearest point to θ in $\mathcal{I}$. We note that since $\mathcal{I}$ is unknown, $\nabla^{\mathcal{I}}$ is not computable. Therefore a consistent approximation of $\nabla^{\mathcal{I}} J$ is introduced below.

We begin with a statement (for which there is substantial evidence [16] and whose proof will be given in a future publication) concerning the consistent approximation of the (full) derivative of the closed loop dynamic matrix $A^* - B^* K(\theta)$ with respect to B when $\theta \in \mathcal{I}$.

Hypothesis 5.1 Let $\{\theta_t\}$ be an $\mathcal{AML}$ estimate incorporated in a CE LQ adaptive control, and suppose that there exists $\theta_\infty = \theta_\infty(\omega) \in \mathcal{I}$ such that $\theta_t \to \theta_\infty$ a.s. Let $P_t \triangleq \int_0^t x_s x_s^T ds, \Gamma_t \triangleq P_t^{-1} \int_0^t x_s dx_s^T$. For a fixed (i,j) $i = 1, \cdots n; j = 1, \cdots m$, define the matrix $\Pi_t(i,j)$ and the vector $\zeta_t(i,j)$ by

$$\Pi_t(i,j) = P_t^{-1}\{\int_0^t \zeta_s(i,j) dx_s^T + \int_0^t x_s d\zeta_s^T(i,j)$$
$$-[\int_0^t (\zeta_s(i,j) x_s^T + x_s \zeta_s^T(i,j)) ds] \Gamma_t\} \tag{29}$$

$$\dot{\zeta}_t(i,j) = F(\theta_t)\zeta_t(i,j) + \Pi_t^T(i,j) x_t, \quad \zeta_{t_0} \tag{30}$$

where $F(\theta) = A - BK(\theta), \theta = col(A, B)^T$. Then

$$\Pi_t(i,j) \to \frac{d}{dB_{ij}} F^T(\theta_\infty) \qquad \text{a.s. as } t \to \infty \tag{31}$$

□

Now note that for any $\theta \in \mathcal{I}, \lim_{t\to\infty} \Pi_t(i,j,\theta)$ satisfies

$$\lim_{t\to\infty} \Pi_t(i,j,\theta) = \frac{dA}{dB_{ij}} - \left[\frac{dB}{dB_{ij}} B^T + B \frac{dB^T}{dB_{ij}}\right] V(\theta) - BB^T \frac{dV(\theta)}{dB_{ij}} \tag{32}$$

where

$$\frac{dV_{pq}}{dB_{ij}} = \sum_{k=1}^n \sum_{\ell=1}^n \frac{\partial A_{k\ell}}{dB_{ij}} + \frac{\partial V_{pq}}{\partial B_{ij}} \tag{33}$$

for $p, q = 1, 2 \cdots n$ $\;i = 1, 2 \cdots n$ and $j = 1, 2 \cdots m$.

By definition $D = dcol A / dcol B$, hence its $n^3 m$ entries correspond to the elements of the tensor $\{dA_{k\ell}/dB_{ij}\}$ (where $k, \ell, i = 1, \cdots n$ and $j = 1, 2 \cdots m$). Therefore equations (32) and (33) (with $1 \leq i \leq n, 1 \leq j \leq m$) form a set of $n^3 m$ algebraic equations for the $n^3 m$ unknown entries of D.

We claim the following:

Lemma 5.2 Consider a fixed $\theta \in S$ and an arbitrary $n \times n$ matrix Π. Then the set of n^2 equations

$$\Pi = \frac{dA}{dB_{ij}} - \left[\frac{dB}{dB_{ij}} B^T + B\frac{dB^T}{dB_{ij}}\right] V(\theta) - BB^T \frac{dV(\theta)}{dB_{ij}} \tag{34}$$

$$\frac{dV_{pq}}{dB_{ij}} = \sum_{k=1}^{n}\sum_{\ell=1}^{n} \frac{\partial V_{pq}}{\partial A_{k\ell}} \frac{dA_{k\ell}}{dB_{ij}} + \frac{\partial V_{pq}}{\partial B_{ij}} \tag{35}$$

has a unique solution $\{dA_{k\ell}/dB_{ij}\}$ $k, \ell = 1, 2 \cdots n$, depending analytically on θ. □

Corollary 5.3 Assume that Hypothesis 5.1 holds, then given an estimate $\{\theta_t\}$, let $\widehat{\Delta}_t(\theta_t)$ be the unique solution of the equations (34), (35) (with (i, j) ranging over $\{1, 2 \cdots n\}$ and $\{1, 2 \cdots m\}$ respectively), where $\Pi = \Pi_t(i, j, \theta_t)$ (see (29, 30)).

Then, if there exists a $\theta_\infty \in \mathcal{I}$ such that

$$\theta_t \to \theta_\infty \text{ a.s. as } t \to \infty,$$

it is the case that

$$\widehat{\Delta}_t(\theta_t) \to D(\theta_\infty) \text{ a.s. as } t \to \infty, \tag{36}$$

and, with $\widehat{\nabla^{\mathcal{I}}} J_t(\theta)$ defined by (18) with $D(\theta)$ replaced by $\widehat{\Delta}_t(\theta)$, it follows that

$$\widehat{\nabla^{\mathcal{I}}} J_t(\theta_t) \to \nabla^{\mathcal{I}} J(\theta_\infty) \text{ a.s. as } t \to \infty. \tag{37}$$

□

The last result leads to the following set of Lagrange-type equations whose solution (θ_t, λ_t) forms the foundation of the Lagrangian Adaptation procedure: First choose an $\mathcal{F}_t^x$ measurable processes $0 < \delta_t \downarrow 0$ such that with $r_t \triangleq \operatorname{Tr} \Phi_t, r_t\delta_t \to 0$ as $t \to \infty$.

Second, let (θ_t, λ_t) denote the solutions to the following set of equations:

$$\widehat{\nabla^{\mathcal{I}}} J_t(\theta_t) - \lambda_t \Phi_t \nabla L_t(\theta_t) = 0, \tag{38}$$

$$\|\nabla L_t(\theta_t)\| \leq \delta_t, \tag{39}$$

where,

$$\widehat{\nabla^{\mathcal{I}}} J_t(\theta) \triangleq \begin{bmatrix} \widehat{\Delta}_t(\theta) \\ I \end{bmatrix} \left[\widehat{\Delta}_t^T(\theta)\widehat{\Delta}_t(\theta) + I\right]^{-1} \left[\widehat{\Delta}_t^T(\theta)\frac{\partial J(\theta)}{\partial col A} + \frac{\partial J(\theta)}{\partial col B}\right] \tag{40}$$

and the entries of $\widehat{\Delta}_t(\theta), \theta = col(A,B)^T$ are given by $\{dA_{k\ell}/dB_{ij}\}$, which are the solutions of these quantities

$$\Pi_t(i,j) = \frac{dA}{dB_{ij}} - \left[\frac{dB}{dB_{ij}}B^T + B\frac{dB^T}{dB_{ij}}\right]V(\theta) - BB^T\frac{dV(\theta)}{dB_{ij}} \tag{41}$$

$$\frac{dV_{pq}(\theta)}{dB_{ij}} = \sum_{k=1}^{n}\sum_{\ell=1}^{n}\frac{\partial V_{pq}}{\partial A_{k\ell}}\frac{dA_{k\ell}}{dB_{ij}} + \frac{\partial V_{pq}(\theta)}{\partial B_{ij}}, \tag{42}$$

where $i = 1,2\cdots n, j = 1,2\cdots m, p,q = 1,2\cdots n$. As a direct derivation from (29), (30), Π_t evolves according to the SDE:

$$\begin{aligned} d\Pi_t(i,j) = &-P_t^{-1}x_tx_t^T\Pi_t(i,j)dt + P_t^{-1}\{\zeta_t(i,j)dx_t^T + x_t\zeta_t^T(i,j)F^T(\theta_t)dt \\ &+x_tx_t^T\Pi_t(i,j)dt - (\zeta_t(i,j)x_t^T + x_t\zeta_t^T(i,j))\Gamma_t dt \\ &+\int_0^t(\zeta_t(i,j)x_s^T + x_s\zeta_t^T(i,j))dsP_t^{-1}x_t\big[dx_t^T - x_t^T\Gamma_t ds\big]\} \end{aligned} \tag{43}$$

$$\dot{\zeta}_t(i,j) = F(\theta_t)\zeta_t(i,j) + \Pi_t^T(i,j)x_t \tag{44}$$

with arbitrary non-zero initial conditions $\Pi_{t_0}(i,j), \zeta_{t_0}(i,j)$ at $t_0 > 0$ for which $P_{t_0} = \int_0^{t_0} x_sx_s^T ds > 0$.

Obviously, a real-time application would require a *recursive* algorithm. Therefore, equations (38), (39) are to be used as a basis for the derivation of SDEs for the evolution of solutions (θ_t, λ_t) in time. As an asymptotic solution would evidently be sufficient, we shall not require (θ_t, λ_t) to satisfy (38) at any given finite time, but instead, to drive the RHS of (38) to zero as $t \to \infty$ (with a similar procedure applied for (39)). The resulting adaptation scheme, termed the Lagrangian Mode, is derived below.

Derivation of the SDEs of the Lagrangian Mode. The basic idea behind the derivation of the recursive, constrained optimization phase of the adaptive algorithm is as follows: Define

$$h_t(\theta,\lambda) = \widehat{\nabla^{\mathcal{I}}}J_t(\theta) - \lambda\Phi_t\nabla L_t(\theta), \tag{45}$$

and the slack variable process

$$q_t(\theta) = \frac{1}{2}(\|\nabla L_t(\theta)\|^2 - \delta_t^2). \tag{46}$$

Fix $\alpha > 0$ and assume that there exist continuous semimartingales $(\theta,\lambda)_t$ such that

$$dh_t(\theta_t,\lambda_t) = -\alpha h_t(\theta_t,\lambda_t)dt \tag{47}$$

$$dq_t(\theta_t) = -\alpha q_t(\theta_t)dt \tag{48}$$

Then, a *formal* application of a generalized Itô rule (Kunita [1990], Theorem 3.3.1), for $h_t(\cdot)$ and $q_t(\cdot)$, enables to obtain SDE's for (θ_t,λ_t), via an explicit presentation of the LHS of (47, 48), by which, the SDEs for h_t and q_t, driven by (θ_t, λ_t), are constructed. (The detailed derivation is omitted. See a similar application in Levanony et.al. [1994].)

In order to simplify the equations to follow, we omit the explicit dependence upon $(\theta, \lambda)_t$ in most of the functions involved.

We first define:

$$\begin{aligned} M_t(\theta,\lambda) &= \Phi_t H_t^{-1}(\theta,\lambda)\Phi_t \quad (\text{an } n(n+m)\times n(n+m) \text{ matrix}) \\ \mu_t(\theta,\lambda) &= [\nabla L_t^T(\theta) M_t(\theta,\lambda)\nabla L_t(\theta)]^{-1} \quad (\text{a scalar}) \\ S_t(\theta,\lambda) &= \lambda I + \mu_t(\theta,\lambda)\nabla L_t(\theta)\nabla L_t^T(\theta)[I-\lambda M_t(\theta,\lambda)] \\ &\qquad (\text{an } n(n+m)\times n(n+m) \text{ matrix}) \end{aligned}$$

From the SDEs for h_t and q_t, equated to the RHS of (47), (48) (respectively), the SDEs for (θ_t, λ_t) are obtained in the following form:

$$\begin{aligned} d\theta_t &= H_t^{-1}\Phi_t S_t \Psi_t^T\{dx_t - \Psi_t\theta_t dt\} + \mu_t H_t^{-1}\Phi_t \nabla L_t(\theta_t)\{\alpha q_t dt - \delta_t d\delta_t\} \\ &\quad + H_t^{-1}[I - \mu_t \Phi_t \nabla L_t(\theta_t)\nabla L_t(\theta_t)^T \Phi_t H^{-1}]\Big\{\lambda_t \dot{\Phi}_t \nabla L_t(\theta_t) dt \\ &\quad - \alpha h_t dt - \frac{1}{2}\sum_{i=1}^{n(n+m)} \frac{\partial}{\partial\theta_i}\nabla\widehat{\nabla^{\mathcal{I}}} J_t(\theta_t) d\langle\theta,\theta_i\rangle_t \\ &\quad + \Phi_t[\mu_t M_t S_t - I]\Psi_t^T d\langle x\rangle_t \Psi_t [I - \lambda_t M_t]\nabla L_t(\theta_t)\Big\} \\ &\quad + \frac{1}{2}\mu_t H_t^{-1}\Phi_t \nabla L_t(\theta_t)\,\mathrm{Tr}\left\{[M_t S_t - I]\Psi_t^T d\langle x\rangle_t \Psi_t [M_t S_t - I]\right\} \end{aligned} \tag{49}$$

and

$$\begin{aligned} d\lambda_t &= \mu_t \nabla L_t^T(\theta_t)[I - \lambda_t M_t]\Psi_t^T\{dx_t - \Psi_t^T\theta_t dt\} + \mu_t\{\alpha q_t dt - \delta_t d\delta_t\} \\ &\quad + \mu_t \nabla L_t^T(\theta_t)\Phi_t H_t^{-1}\Big\{\frac{1}{2}\sum_{i=1}^{n(n+m)} \frac{\partial}{\partial\theta_i}\nabla\widehat{\nabla^{\mathcal{I}}} J_t(\theta_i) d\langle\theta,\theta_i\rangle_t \\ &\qquad + \alpha h_t dt - \lambda_t \dot{\Phi}_t \nabla L_t(\theta_t) dt\Big\} \\ &\quad + \mu_t \nabla L_t(\theta_t)^T[\mu_t M_t S_t - I]\Psi_t^T d\langle x\rangle_t \Psi_t[\lambda_t M_t - I]\nabla L_t(\theta_t) \\ &\quad + \frac{1}{2}\mu_t\,\mathrm{Tr}\{[M_t S_t - I]\Psi_t^T d\langle x\rangle_t \Psi_t [M_t S_t - I]^T\} \end{aligned} \tag{50}$$

where $d\langle x\rangle_t = dt$ and

$$d\Phi_t = \Psi_t^T \Psi_t dt \tag{51}$$

$$d\langle\theta\rangle_t = H_t^{-1}\Phi_t S_t \Psi_t^T d\langle x\rangle_t \Psi_t S_t^T \Phi_t H_t^{-1} \tag{52}$$

The recursive solution of (49, 50) forms what we refer to as the Lagrangian mode.

The control law we use on the Lagrangian mode is the standard CE LQ control, that is $u_t = -K(\theta_t)x_t$. It is shown below that the parameter estimate at the Lagrangian phase is naturally kept bounded away from the boundary of $\mathcal{S}$ (on which K might not exist) thus, making such a control law valid.

We note that since H_t may become singular, (49, 50) cannot always be expected to have a solution (θ_t, λ_t) which guarantees (47, 48) hold. Therefore, one has to introduce another algorithm to be active when the Lagrangian equations are unsolvable. Such an algorithm, which we call the ML mode, should ensure that indeed $\theta_t \to \mathcal{I}$ as $t \to \infty$ independently of the Lagrangian mode.

The ML Mode. This consists of the recursive solution of

$$d\theta_t = \Phi_t^{-1}[\Psi_t^T dx_t - (\dot{\Phi}_t\theta_t + \alpha\nabla L_t(\theta_t))dt], \quad t \geq t_0 > 0, \tag{53}$$

where it is assumed that $\Phi_{t_0} > 0$. Note that (53), combined with the Itô rule leads to $d\nabla L_t(\theta_t) = -\alpha\nabla L_t(\theta_t)dt$ and thus ensures an exponential decay of ∇L, and hence the desired approach to $\mathcal{I}$.

Let $\mathcal{S}' \subset \mathcal{S}$ be the largest open *connected* set in $\mathcal{S}$ in which θ^* is assumed to lie. We shall restrict the search for θ^* in the ML mode to $\mathcal{S}'$. Furthermore, unlike the Lagrangian mode, in the ML mode the estimate might reach the boundary of $\mathcal{S}$ on which the standard LQ feedback law doesn't exist. We therefore modify the CE LQ control law in the ML mode, following Caines [1993]:

Let $\{\mathcal{S}'_\ell, \ell \geq 1\}$ be a sequence of compact subsets of $\mathcal{S}'$, which are strictly increasing on $\mathcal{S}'$, that is,

$$\mathcal{S}'_\ell \uparrow \mathcal{S}' \text{ as } \ell \to \infty \text{ and } \mathcal{S}'_\ell \subset \mathcal{S}'_k \quad \forall \ell < k.$$

Fix $\mathcal{S}_\ell$ for some large $\ell < \infty$ and choose *a* $\theta_0 \in$ the interior of $\mathcal{S}_\ell$ = the interior of $\mathcal{S}_\ell$. Then, the state feedback control law (only in the ML mode) is defined by

$$u_t = -K^\ell(\theta_t)x_t \tag{54}$$

where $K^\ell(\theta)$ constructed such that it is a Lipschitz continuous function which equals to $K(\theta)$ on the closed subset $\mathcal{S}'_\ell$ and equals $K(\theta_0)$ on the boundary $\partial\mathcal{S}'$.

Remark. Following Caines [1993], we use below a mechanism to reset the parameter estimate to θ_0, once it reaches the boundary $\partial\mathcal{S}'$. Hence, the resulting control law becomes continuous. Unlike Caines [1993], we are not required here to assume that $\theta^* \in \mathcal{S}'_\ell$.

The Adaptive Algorithm. Recall that $r_t = \mathrm{Tr}(\Phi_t)$. Choose δ_t such that $\delta_t r_t \to 0$ with $\delta_t \leq \lambda^2_{\min}(\Phi_t)/r_t^2$ for all large t's (e.g. $\delta_t = 1/r_t^2$). Define $\eta_t = 1 + 2(\|\nabla\widehat{\nabla^{\mathcal{I}}} J_t(\theta_t)\| \wedge r_t)/\lambda^2_{\min}(\Phi_t)$ and take $\theta_0 \in \mathcal{S}'_\ell$.

The algorithm which combines the ML and Lagrangian modes is as follows:

(1) Use the ML mode equation (53) until either
   $(i)\quad \theta \in \partial\mathcal{S}$ or,
   $(ii)\quad H_t(\theta_t, \frac{3}{2}\eta_t) \geq 2\delta_t I.$
   Where the computation of the modified certainty equivalence feedback matrix $K_t^\ell = K^\ell(\theta_t)$ is done as above, that is, the control law is given by: $u_t = -K^\ell(\theta_t)x_t$, where K^ℓ is Lipschitz continuous, uniformly over $\mathcal{S}'$ with $K^\ell(\theta) = K(\theta)\ \forall\theta \in \mathcal{S}_\ell$. and $K^\ell(\theta) = K(\theta_o)\ \forall\theta \in \partial\mathcal{S}$.
(2) In the case of $(1-i)$, reset $\theta_t = \theta_0$ and return to (1).
(3) In the case of $(1 - ii)$, apply the Lagrangian mode equations (49, 50) with initial conditions $(\theta_t, \frac{3}{2}\eta_t)$ and equations (41)–(44) (for the recursive computation of $\widehat{\nabla^{\mathcal{I}}} J_t(\theta_t))$, until either
   $(i)\quad H_t(\theta_t, \lambda_t) \leq \delta_t I$ or,
   $(ii)\quad \lambda_t \notin (\eta_t, 2\eta_t).$
   where the control law is a standard CE LQ control that is, $u_t = -K(\theta_t)x_t$.
(4) In the case of $(3-i)$, return to (1).
(5) In the cases of $(3-ii)$, reset $\lambda_t = \frac{3}{2}\eta_t$ and return to (3).

Remark. The choice of two levels for activating and deactivating the Lagrangian mode is made to provide the sufficient delay (between switchings) in order to avoid the possibility of dithering i.e. switching between the two phases infinitely many times over bounded intervals. (See e.g. Levanony et.al. [1994].)

In the analysis to follow, we adopt the convention that whenever a continuous modification of a stochastic process exists, it is this modification that is considered. Furthermore, it is assumed throughout that $\theta^* \notin \mathcal{N}$, $\mathcal{N}$ being a Lebesgue null set cited in Theorem 3.2.

Theorem 5.4 Fix $t_0 > 0$, choose $(\theta_0, \lambda_0) \in \mathcal{S}' \times (0, \infty)$ and consider the following stopping times related to either (49, 50) or (53), respectively:

$$\sigma = \inf\{t \geq t_o | \lambda_{\min}(H_t) = 0 \text{ or } \nabla L_t(\theta_t) = 0 \text{ or } \lambda_t \notin (0, \eta_t)\},$$
$$\tau = \inf\{t \geq t_o | \theta_t \in \partial\mathcal{S}\}$$

(i) Suppose that both Φ_{t_0} and $H_{t_0}(\theta_0, \lambda_0)$ are strictly positive definite w.p.1. Then (1), together with (49, 50) have a unique strong solution $(x, \theta, \lambda)_t$ over $[t_0, \sigma)$.
(ii) With $\Phi_{t_0} > 0$ a.s., (1) and (53) have a unique strong solution $(x, \theta)_t, t \in [t_0, \tau)$.

□

As each of the two modes has been shown to possess a unique strong solution (until the appropriate stopping time), it remains to show that the combination of the two modes, together with the particular resetting policy on each mode, is valid. To this end, it suffices to show that only a finite number of resettings and/or switchings may take place within any bounded time interval. Actually, as we show in Theorem 5.7 below, only a finite number of such changes may take place over $[t_0, \infty)$.

Lemma 5.5 Assume that $\Phi_{t_0} > 0$ a.s. Then, with probability 1, only a finite number of switchings and resettings may take place on finite time intervals. Hence, the state equation (1) together with the algorithm equations (49, 50) or (53) with the switching and resetting policy above have a unique strong solution on $[t_0, \infty)$. □

Corollary 5.6 Under the conditions of Lemma 5.5,

$$\theta_t \to \theta_\infty(\omega) \in \mathcal{I} \quad \text{as } t \to \infty, \text{ a.s.}$$

□

Theorem 5.7 Assume that Hypothesis 5.1 holds and suppose that $\Phi_{t_0} > 0$. Then with probability 1, $\theta_\infty(\omega) = \theta^*$, that is, the estimate θ_t is strongly consistent, and the long–run LQ performance cost equals $J(\theta^*)$. Moreover, along almost all sample paths, only a finite number of resettings and switching may take place over $[t_0, \infty)$ where there exists a $\tau_0 = \tau_0(\omega) < \infty$ (a.s.) s.t. the Lagrangian mode remains active on $[\tau_0, \infty)$. □

References

1. Borkar, V. S. (1991) Self-Tuning Control of Diffusions without the Identifiability Condition, J. Opt. Theory and Appl **68**, 117–138.
2. Caines, P. E. (1988) Linear Stochastic Systems, John Wiley.
3. Caines, P. E. (1992) Continuous-Time Stochastic Adaptive Control: Non-explosion, $\in$-Consistency and Stability, Systems and Control Letters **19**, 169–176.
4. Chen, H. F. and Guo, L. (1991) Identification and Stochastic Adaptive Control, Birkhäuser.
5. Duncan, T. E. and Pasik-Duncan, B. (1986) A Parameter Estimate Associated with the Adaptive Control of Stochastic Systems. In Analysis and Optimization of Systems, L. N. Control and Inf. Sc. **83**, Springer, 508–514.
6. Duncan, T. E. and Pasik-Duncan, B. (1991) Some Methods for the Adaptive Control of Continuous Time Linear Stochastic Systems. In: Topics in Stochastic Systems: Modeling, Estimation and Adaptive Control, Gerencsér, L. and Caines, P. E., Eds. L. N. Control and Info. Sc. **161**, Springer.
7. Duncan, T. E., Guo, L., and Pasik-Duncan, B. (1999) Adaptive Continuous-Time Linear Quadratic Gaussian Control, IEEE Trans. Automatic Control **44**, 1653–1662.
8. Kumar, P. R. and Becker, A. (1982) A New Family of Optimal Adaptive Controllers for Markov Chains, IEEE Trans. Automatic Control **AC-27**, 137–146.
9. Kumar, P. R. and Lin, W. (1982) Optimal Adaptive Controllers for Unknown Markov Chains, IEEE Trans. Automatic Control **AC-27**, 765–774.
10. Kumar, P. R. (1983) Optimal Adaptive Control of Linear-Quadratic-Gaussian Systems, SIAM J. Control and Optimization **21**, 163–178.

11. Kumar, P. R. (1989) Convergence of Adaptive Control Schemes Using Least Squares Estimates. in Proceedings of the 28th IEEE Conference on Decision and Control, 727–731.
12. Kunita, H. (1990) Stochastic Flows and Stochastic Differential Equations, Cambridge University Press.
13. Lai, T. L. and Wei, C. Z. (1982) Least Squares Estimates in Stochastic Regression Models with Application to Identification and Control of Dynamic Systems, Ann. Stat. **10**, 154–166.
14. Levanony, D., Shwartz, A., and Zeitouni, O. (1994) Recursive Identification in Continuous-Time Stochastic Processes, Stochastic Proc. and Appl. **49**, 245–275.
15. Levanony D. and Caines, P. E. (2001) On Persistent Excitation for Linear Systems with Stochastic Coefficients, SIAM J. Control and Optimization, to appear.
16. Levanony, D. and Caines, P. E. (2001) Stochastic Lagrangian Adaptation, McGill University Research Report.
17. Polderman, J. W. (1986) A Note on the Structure of Two Subsets of the Parameter Space in Adaptive Control Problems, Systems and Control Letters **7**, 25–34.
18. Polderman, J. W. (1989) Adaptive LQ Control: Conflict Between Identification and Control, Linear Algebra and Applications, 219–244.

Optimal Control of Linear Backward Stochastic Differential Equations with a Quadratic Cost Criterion

Andrew E.B. Lim[1] and Xun Yu Zhou[2]*

[1] Department of Industrial Engineering and Operations Research, Columbia University, New York, NY 10027, USA
[2] Department of Systems Engineering and Engineering Management, The Chinese University of Hong Kong, Shatin, N.T., Hong Kong

Abstract. Backward Stochastic Differential Equations (BSDEs) are Ito SDEs with a random *terminal* condition. While it is the case that *uncontrolled* BSDEs have been the topic of extensive research for a number of years, little has been done on optimal control of BSDEs. In this paper, we consider the problem of linear–quadratic control of a BSDE. A complete solution to this problem is obtained, in terms of a pair of Riccati type equations and an uncontrolled BSDE, using an approach that is based on the completion of squares technique.

1 Introduction

A backward stochastic differential equation (BSDE) is an Ito stochastic differential equation for which a *random terminal condition* on the state has been specified. The linear version of this type of equations was first introduced by Bismut [2] as the adjoint equation in the stochastic maximum principle (see also [3,19,22]). General nonlinear BSDEs, introduced independently by Pardoux and Peng [18] and Duffie and Epstein [9], have received considerable research attention in recent years due to their nice structure and wide applicability in a number of different areas, especially in mathematical finance (see, e.g., [7,10,11,10,15,17,21]). For example, the Black–Scholes formula for options pricing can be recovered via a system of forward–backward stochastic differential equations. In this case, the random terminal condition is related to the price of the underlying stock at a given terminal date. Unlike a (forward) stochastic differential equation (SDE), the solution of a BSDE is a *pair* of adapted processes $(x(\cdot), z(\cdot))$. The additional term $z(\cdot)$ may be interpreted as a risk-adjustment factor and is required for the equation to have *adapted* solutions. This restriction of solutions to the class of *adapted processes* is necessary if the insights gained from the study of BSDEs are to be useful in applications. Adapted processes depend on *past* and *present* information but do *not* rely (clairvoyantly) on future knowledge. This is natural in virtually

* The research of this author was supported by the RGC Earmarked Grants CUHK 4435/99E.

B. Pasik-Duncan: Stochastic Theory and Control, LNCIS 280, pp. 301–317, 2002.

all applications; for example, the replicating portfolio for a contingent claim may depend at any particular time on past and present stock prices but not, quite naturally, on future stock prices. For recent accounts on BSDE theory and applications, the reader is referred to the books [17,21].

Since a BSDE is a well-defined dynamic system, it is very natural and appealing, first at the theoretical level, to consider the optimal control of the BSDE. As for applications, optimally controlled BSDEs promise to have a great potential. For example, an optimal control problem of a linear BSDE comes out in the process of solving a *forward* stochastic linear–quadratic (LQ) control problem in [6]. Moreover, controlled BSDEs are expected to have important applications in mathematical finance. For instance, a situation in which funds may be injected or withdrawn from the replication process of a contingent claim so as to achieve some other goal may be viewed quite naturally as an optimal BSDE control problem. However, the study on controlled BSDEs is quite lacking in literature. To our best knowledge there are only a few papers dealing with optimal control of BSDEs, including [20] and [8] which established local and global maximum principles, respectively, and [11] in which a controlled BSDE with linear state drift is studied.

This paper is concerned with optimal control of a linear BSDE with a quadratic cost criteria, namely, a stochastic backward linear–quadratic (BLQ) problem. It is well-known that LQ control is one of the most important classes of optimal control and together with the maximum principle and dynamic programming is one of the cornerstones of optimal control theory. Stochastic forward LQ theory has been well established, especially with the recent development on the so-called *indefinite* stochastic LQ control ([1,5,6,14]). However, stochastic BLQ control remains an almost completely unexplored area.

The main contribution of this paper is a complete solution of a general BLQ problem. As it turns out, the optimal control can no longer be expressed as a linear feedback of the *current* state as in the deterministic or stochastic forward case. Rather it depends, in general, on the entire past history of the state pair $(x(\cdot), z(\cdot))$. It will be shown that this dependence is linear, and explicit formulas for the optimal control and the optimal cost in terms of a pair of Riccati equations, a Lyapunov equation, an uncontrolled BSDE and an uncontrolled SDE are established. The basic idea is to first establish a lower bound to the optimal cost via the completion-of-squares technique, and then to construct a control that achieves exactly this lower bound.

It is interesting to remark that our original approach to solving the BLQ problem was inspired by [12] where an (uncontrolled) BSDE is viewed as a *controlled forward* SDE. Extending this idea, we can show that the optimal control of the BLQ problem is the limit of a sequence of square integrable processes, obtained by solving a family of *forward* LQ problems. During this procedure, the key Riccati equations, along with other related equations, come out very naturally. What is more interesting is that once these equations are in place, one may forget about the forward formulation and limiting procedure, which is rather complicated, and instead use these equations *directly* along with the completion-of-square technique to obtain the optimal control for the original BLQ problem. Nevertheless, the forward formulation still represents an alternative, and insightful, approach to the backward control problem and for this reason, an outline of this procedure is also presented in this paper.

The outline of this paper is as follows. In Section 2, we formulate the BLQ problem. In Section 3, we present the main result of the paper. In addition, we compare the solution of the stochastic BLQ problem with that of the deterministic case. An outline of the proof of the main result is carried out in Section 4. In Section 5, we explain, in a rather informal way, an alternative approach to the BLQ problem. In particular, we show that the optimal BLQ control, established in Section 3, coincides with the limit of the solutions of a family of forward LQ problems. Finally, Section 6 concludes the paper. Due to limitations on space, we have not included any proofs in this paper. The interested reader may consult [14]

2 Problem Formulation

We assume throughout that $(\Omega, \mathcal{F}, \{\mathcal{F}\}_{t\geq 0}, P)$ is a given and fixed complete filtered probability space and that $W(\cdot)$ is a scalar-valued Brownian motion on this space. (Our assumption that $W(\cdot)$ is scalar-valued is for the sake of simplicity. No essential difficulties are encountered when extending our analysis to the case of vector-valued Brownian motions). In addition, we assume that $\mathcal{F}_t$ is the augmentation of $\sigma\{W(s) \mid 0 \leq s \leq t\}$ by all the P-null sets of $\mathcal{F}$.

Throughout this paper, we denote the set of symmetric $n\times n$ matrices with real elements by S^n. If $M \in S^n$ is positive (semi-)definite, we write $M > (\geq)$ 0. Let X be a given Hilbert space. The set of X-valued continuous functions is denoted by $C(0,T;X)$. If $N(\cdot) \in C(0,T;S^n)$ and $N(t) > (\geq)\ 0$ for every $t \in [0,T]$, we say that $N(\cdot)$ is positive (semi-)definite, which is denoted by $N(\cdot) > (\geq)\ 0$. Suppose $\eta : \Omega \to \mathbb{R}^n$ is an $\mathcal{F}_T$-random variable. We write $\eta \in L^2_{\mathcal{F}_T}(\Omega;\mathbb{R}^n)$ if η is square integrable (i.e. $E|\eta|^2 < \infty$). Consider now the case when $f : [0,T] \times \Omega \to \mathbb{R}^n$ is an $\{\mathcal{F}_t\}_{t\geq 0}$ adapted process. If $f(\cdot)$ is square integrable (i.e. $E\int_0^T |f(t)|^2 dt < \infty$) we shall write $f(\cdot) \in L^2_{\mathcal{F}}(0,T;\mathbb{R}^n)$; if $f(\cdot)$ is uniformly bounded (i.e. $E\operatorname{ess\,sup}_{t\in[0,T]} |f(t)| < \infty$) then $f(\cdot) \in L^\infty_{\mathcal{F}}(0,T;\mathbb{R}^n)$. If $f(\cdot)$ has (P-a.s.) continuous sample paths and $E\sup_{t\in[0,T]} |f(t)|^2 < \infty$ we write $f(\cdot) \in L^2_{\mathcal{F}}(0,T;C(0,T;\mathbb{R}^n))$; if $E\sup_{t\in[0,T]} |f(t)| < \infty$ then $f(\cdot) \in L^\infty_{\mathcal{F}}(0,T;$
$C(0,T;\mathbb{R}^n))$. These definitions generalize in the obvious way to the case when $f(\cdot)$ is $\mathbb{R}^{n\times m}$- or S^n-valued. Finally, in cases where we are restricting ourselves to deterministic Borel measurable functions $f : [0,T] \to \mathbb{R}^n$, we shall drop the subscript $\mathcal{F}$ in the notation; for example $L^\infty(0,T;\mathbb{R}^n)$.

Consider the BSDE:

$$\begin{cases} dx(t) = \Big\{A(t)\,x(t) + B(t)\,u(t) + C(t)z(t)\Big\}\,dt + z(t)\,dW(t), \\ x(T) = \xi, \end{cases} \tag{1}$$

where $u(\cdot)$ is the control process. The class of *admissible controls* for (1) is:

$$\mathcal{U} = L^2_{\mathcal{F}}(0,\, T;\, \mathbb{R}^m). \tag{2}$$

Later, we shall state assumptions on the coefficients $A(\cdot)$, $B(\cdot)$, $C(\cdot)$ and the terminal condition ξ so as to guarantee the existence of a unique solution pair $(x(\cdot),\, z(\cdot)) \in L^2_{\mathcal{F}}(\Omega; C(0,\, T;\, \mathbb{R}^n)) \times L^2_{\mathcal{F}}(0,\, T;\, \mathbb{R}^n)$ of the BSDE (1) for every admissible control $u(\cdot) \in \mathcal{U}$. We refer to such a 3-tuple $(x(\cdot),\, z(\cdot);\, u(\cdot))$ as an *admissible triple*. The cost associated with an admissible triple $(x(\cdot),\, z(\cdot);\, u(\cdot))$ is given by:

$$\begin{aligned} J(\xi; u(\cdot)) := E\,\frac{1}{2}\Big[&x(0)'Hx(0) \\ &+ \int_0^T \Big(x(t)'Q(t)x(t) + z(t)'S(t)z(t) + u(t)'R(t)u(t)\Big)\,dt\Big]. \end{aligned} \tag{3}$$

The *backward linear–quadratic (BLQ) control problem* can be stated as follows:

$$\begin{cases} \min J(\xi; u(\cdot)) \\ \text{subject to:} \\ u(\cdot) \in \mathcal{U}, \qquad (x(\cdot),\, z(\cdot);\, u(\cdot)) \text{ satisfies (1).} \end{cases} \tag{4}$$

Throughout this paper, we shall assume the following:

Assumption:

(**A1**)

$$\begin{cases} A,\, C \in L^\infty(0,\, T;\, \mathbb{R}^{n\times n}), \\ \quad B \in L^\infty(0,\, T;\, \mathbb{R}^{n\times m}), \\ Q,\, S \in L^\infty(0,\, T;\, S^n),\;\; Q,\, S\, \geq\, 0, \\ \quad R \in L^\infty(0,\, T; S^m),\;\; R\, >\, 0, \\ \quad H \in S^n,\;\; H\, \geq\, 0, \\ \quad \xi \in L^2_{\mathcal{F}_T}(\Omega;\, \mathbb{R}^n). \end{cases}$$

In particular, Assumption (**A1**) is sufficient to guarantee the existence of a unique solution pair $(x(\cdot), z(\cdot)) \in L^2_{\mathcal{F}}(\Omega; C(0, T; \mathbb{R}^n)) \times L^2_{\mathcal{F}}(0, T; \mathbb{R}^n)$ of (1) for every admissible control $u(\cdot) \in \mathcal{U}$; see [21, Chapter 7].

3 Main Result

Before we present the main result of the paper, which gives a complete solution to the above BLQ problem, let us see how one would solve the *deterministic* BLQ problem. This corresponds to $\xi \in \mathbb{R}^n$ being deterministic, $C = 0$, $S = 0$, and an admissible class $\mathcal{U}_d = L^2(0, T; \mathbb{R}^n)$. The other parameters satisfy (**A1**) while the cost and dynamics are given by:

$$J_d(\xi; u(\cdot)) := \frac{1}{2} x(0)' H x(0) + \frac{1}{2} \int_0^T \Big(x(t)' Q(t) x(t) + u(t)' R(t) u(t) \Big)\, dt,$$

$$\begin{cases} \dot{x}(t) = A(t)\, x(t) + B(t)\, u(t), \\ x(T) = \xi, \end{cases}$$

respectively. By reversing time, $\tau = T - t, \quad t \in [0, T]$, we obtain an equivalent forward LQ problem that can be solved using standard (Riccati) approach (see, e.g., [21, Chapter 6, Section 2]). In particular, this gives us the following result:

Proposition 1 (Deterministic BLQ Problem). *The optimal cost and optimal feedback control for the deterministic BLQ problem are*

$$J_d^*(\xi) = \frac{1}{2} \xi'\, Z(T)\, \xi, \tag{5}$$

$$u(t) = R(t)^{-1} B(t)' Z(t) x(t) \tag{6}$$

respectively, where $Z(\cdot)$ is the unique solution of the Riccati equation

$$\begin{cases} \dot{Z}(t) + Z(t)A(t) + A(t)'Z(t) + Z(t)B(t)R(t)^{-1}B(t)'Z(t) - Q(t) = 0, \\ Z(0) = H, \end{cases} \tag{7}$$

and $x(\cdot)$ is the unique solution of the differential equation:

$$\begin{cases} \dot{x}(t) = (A(t) + B(t)\, R(t)^{-1} B(t)' Z(t))\, x(t), \\ x(T) = \xi. \end{cases}$$

It is important to recognize that the above time reversal technique can*not* be extended to the stochastic BLQ problem, (4), as it would destroy the adaptiveness which is essential in the model. In particular, a control obtained in this way will not, in general, be $\{\mathcal{F}_t\}_{t \geq 0}$-adapted and hence is not admissible.

It turns out that the solution to (4) is more involved. In the remainder of this section, we present two alternative expressions (which are later shown to be equivalent) for the optimal BLQ control. The first one is analogous to the solution to the deterministic BLQ problem just presented. It gives an explicit formula via a pair of Riccati equations, an uncontrolled BSDE and an uncontrolled SDE.

First, consider the following Riccati-type equation:

$$\begin{cases} \dot{\Sigma}(t) - A(t)\Sigma(t) - \Sigma(t)A(t)' - \Sigma(t)Q(t)\Sigma(t) \\ \quad + B(t)R(t)^{-1}B(t)' + C(t)\Sigma(t)\Big(S(t)\Sigma(t) + I\Big)^{-1}C(t)' = 0, \\ \Sigma(T) = 0. \end{cases} \tag{8}$$

The existence and uniqueness of solution to this equation is addressed in [14]. With $\Sigma(\cdot)$ denoting the solution to (8), we define the following equations:

$$\begin{cases} \dot{Z}(t) + Z(t)A(t) + A(t)'Z(t) \\ \quad + Z(t)\Big[B(t)R(t)^{-1}B(t)' + C(t)\Sigma(t)(I + S(t)\Sigma(t))^{-1}C(t)'\Big]Z(t) \\ \quad - Q(t) = 0, \\ Z(0) = H, \end{cases} \tag{9}$$

$$\begin{cases} \dot{N}(t) + N(t)(A(t) + \Sigma(t)Q(t)) + (A(t) + \Sigma(t)Q(t))'N(t) - Q(t) = 0, \\ N(0) = \frac{1}{2}\Big\{H(I + \Sigma(0)H)^{-1} + (I + H\Sigma(0))^{-1}H\Big\}, \end{cases} \tag{10}$$

$$\begin{cases} dh(t) = \Big\{\Big(A(t) + \Sigma(t)Q(t)\Big)h(t) + C(t)\Big(I + \Sigma(t)S(t)\Big)^{-1}\eta(t)\Big\}\,dt \\ \quad + \eta(t)\,dW(t), \\ h(T) = -\xi. \end{cases} \tag{11}$$

The first equation (9) is again a Riccati-type equation. It is a generalization of the Riccati equation (7) associated with the deterministic problem. The second equation is a Lyapunov equation while the third is a linear BSDE. Based on the solutions $Z(\cdot)$ and $(h(\cdot), \eta(\cdot))$ to (9) and (11), respectively, we finally introduce

$$\begin{cases} dq(t) = \Big\{ - \Big[A(t) + B(t)R(t)^{-1}B(t)'Z(t) \\ +C(t)(I + \Sigma(t)S(t))^{-1}\Sigma(t)C(t)'Z(t)\Big]' q(t) \\ \quad +Z(t)C(t)(I + \Sigma(t)S(t))^{-1}\eta(t)\Big\}dt \\ \quad +\Big\{(Z(t) - S(t))(I + \Sigma(t)S(t))^{-1}\eta(t) \\ \quad +(I + Z(t)\Sigma(t))(I + S(t)\Sigma(t))^{-1} \\ \quad \times C(t)'(I + Z(t)\Sigma(t))^{-1}(Z(t)h(t) - q(t))\Big\} dW(t), \\ q(0) = 0. \end{cases} \tag{12}$$

Existence and uniqueness of solutions of equations (9)–(12) is discussed in [14]. It should be noted that the equations (8), (10), (11) and (12) play no role in the solution of the deterministic BLQ problem.

Theorem 1. *The BLQ problem (4) is uniquely solvable. Moreover, the following control*

$$u(t) = R(t)^{-1}B(t)'(Z(t)x(t) + q(t)) \tag{13}$$

is optimal, where $Z(\cdot)$ and $q(\cdot)$ are the solutions of (9) and (12), respectively. The optimal state trajectory $(x(\cdot), z(\cdot))$ is the unique solution of the BSDE:

$$\begin{cases} dx(t) = \Big\{\Big(A(t) + B(t)\, R(t)^{-1}B(t)'Z(t)\Big)\, x(t) \\ \qquad\qquad +C(t)z(t) + B(t)\, R(t)^{-1}\, B(t)'\, q(t)\Big\}\, dt \\ \qquad\qquad +z(t)\, dW(t), \\ x(T) = \xi, \end{cases} \tag{14}$$

and the optimal cost is

$$\begin{aligned} J^*(\xi) := E\frac{1}{2}\Big\{\xi' N(T)\xi + \int_0^T \{\eta(t)'\Big[(S(t)\Sigma(t) + I)^{-1}S(t) - N(t)\Big]\eta(t) \\ -2\eta(t)'(I + S(t)\Sigma(t))^{-1}C(t)'N(t)h(t)\}\, dt\Big\} \end{aligned} \tag{15}$$

where $N(\cdot)$ is the unique solution of (10).

Remark 1. If we compare the two optimal controls, (6) and (13), for the deterministic and stochastic BLQ problems respectively, we see that the latter involves an additional random non-homogeneous term $q(\cdot)$. This addition disqualifies (13) from a feedback control of the *current* state, contrary to the deterministic BLQ (see Proposition 1) or stochastic forward LQ (see [5]) cases. The reason is because $q(\cdot)$ depends on $(h(\cdot), \eta(\cdot))$ which in turn depends on ξ, the terminal condition of part of the state variable, $x(\cdot)$. This is one of the major distinctive features of the stochastic BLQ problem. On the other hand, when ξ is non-random, $C = 0$ and $S = 0$, the optimal control (13) reduces to the solution (6) of the deterministic problem: In this case, it is easy to see (by the uniqueness of solutions of (11)) that $\eta(t) \equiv 0$. This implies, in turn, that $q(t) \equiv 0$ and hence, the optimal control (13) agrees with the solution (6) of the deterministic problem. In addition, since

$$N(t) = \frac{1}{2}\Big[Z(t)(I + \Sigma(t)Z(t))^{-1} + (I + Z(t)\Sigma(t))^{-1}Z(t)\Big]$$

(see [14]), it follows that $N(T) = Z(T)$ and the optimal cost (15) reduces to (5) for the deterministic problem. Through the above comparison, we can also see that the fundamental difference between the solutions to the deterministic and stochastic BLQ problems lies in the introduction of the equation (8).

Although for the stochastic BLQ problem the optimal control is no longer a feedback of the current state, it is indeed a linear state feedback of the *entire past history* of the state process $(x(\cdot),\, z(\cdot))$. This conclusion is a consequence of the second form of the optimal control we will present, which is in terms of the Hamiltonian system:

$$\begin{cases} dx(t) = \Big\{A(t)x(t) - B(t)R(t)^{-1}B(t)'y(t) + C(t)z(t)\Big\}dt \\ \qquad\qquad +z(t)dW(t), \\ x(T) = \xi, \end{cases} \tag{16}$$

$$\begin{cases} dy(t) = \Big\{ - A(t)'y(t) - Q(t)x(t)\Big\}dt \\ \qquad\qquad +\Big\{ - C(t)'y(t) - S(t)z(t)\Big\}dW(t), \\ y(0) = -Hx(0). \end{cases} \tag{17}$$

Notice that the combination of (16)–(17) does not qualify as a conventional forward–backward stochastic differential equation (FBSDE) as defined in, say, [21,17]. The subtle difference is that the forward and backward variables in (16)–(17) are directly related at the *initial* time, while those in the FBSDE are related at the *terminal* time. Moreover, one cannot transform between these two types of equations by reversing the time, due to the required adaptiveness. In the sequel, we shall refer to any three-tuple of processes:

$$(x(\cdot),\, z(\cdot),\, y(\cdot)) \in L^2_{\mathcal{F}}(\Omega;\, C(0,\, T;\, \mathbb{R}^n))$$

$$\times L^2_{\mathcal{F}}(0, T; \mathbb{R}^n) \times L^2_{\mathcal{F}}(\Omega; C(0, T; \mathbb{R}^n))$$

which satisfies the equations (16)–(17) as a *solution of the Hamiltonian system (16)–(17)*.

Theorem 2. *The system (16)–(17) has a unique solution* $(x(\cdot), z(\cdot), y(\cdot))$. *Moreover, the BLQ problem (4) is uniquely solvable, with the optimal control*

$$u(t) = -R(t)^{-1}B(t)'y(t), \tag{18}$$

and $(x(\cdot), z(\cdot))$ *the corresponding optimal state process. The optimal cost is (15).*

Remark 2. If (18) is optimal, then (16)–(17) are exactly the corresponding state equation and adjoint equation; see [8]. This is the reason why we call (16)–(17) the Hamiltonian system.

Theorem 2 shows that the optimal control is linear in the process $y(\cdot)$. The following simple result further reveals that the optimal control is a linear feedback of the past and current values of the state process $(x(\cdot), z(\cdot))$.

Proposition 2. *Let* $y(\cdot)$ *be the process obtained from the Hamiltonian system (16)–(17). Then:*

$$\begin{aligned} y(t) = \Phi(t)\Big\{ - Hx(0) + \int_0^t \Phi(s)^{-1}\Big[Q(s)x(s) + C(s)'S(s)z(s)\Big]\, ds \\ - \int_0^t \Phi(s)^{-1}S(s)z(s)dW(s)\Big\} \end{aligned}$$

where $\Phi(\cdot)$ *is the unique solution of the matrix SDE:*

$$\begin{cases} d\Phi(t) = -A(t)'\Phi(t)dt - C(t)'\Phi(t)dW(t), \\ \Phi(0) = I. \end{cases}$$

Proof. This is an immediate consequence of the variation-of-constant formula; see [21, p. 47, Theorem 6.14].

4 Proofs of Theorems 1 and 2: Outline

In this section we give an outline of the proofs of the main results of the paper, Theorems 1 and 2. For further details, we refer the reader to [14]. The basic idea is first to find a lower bound of the cost function (3) (see Lemma 1), and then to identify a control which achieves *exactly* this lower bound (see Proposition 4).

Lemma 1. *For every $u(\cdot) \in \mathcal{U}$, we have*

$$J(\xi; u(\cdot)) \geq h(0)'\Big[H\Sigma(0) + I\Big]^{-1} Hh(0) + E\int_0^T \Big\{h'Qh + \eta'(S\Sigma + I)^{-1}S\eta\Big\}dt, \tag{19}$$

where $\Sigma(\cdot)$ and $(h(\cdot), \eta(\cdot))$ are the solutions of (8) and (11), respectively.

Our next step involves finding a control that achieves this lower bound in Lemma 1. To this end, recall the Hamiltonian system (16)–(17).

Proposition 3. *The Hamiltonian system (16)–(17) has a unique solution $(x(\cdot),\, z(\cdot),\, y(\cdot))$. Moreover, the following relations are satisfied:*

$$\begin{cases} x(t) = \Sigma(t)y(t) - h(t), \\ z(t) = -\Sigma(t)(S(t)\Sigma(t) + I)^{-1}C(t)'y(t) \\ \quad -(\Sigma(t)S(t) + I)^{-1}\eta(t), \\ x(0) = -(\Sigma(0)H + I)^{-1}h(0), \end{cases} \tag{20}$$

where $\Sigma(\cdot)$ and $(h(\cdot), \eta(\cdot))$ are the solutions of (8) and (11), respectively.

The following result gives us a control which achieves the lower bound in Lemma 1.

Proposition 4. *Let $(x(\cdot),\, z(\cdot),\, y(\cdot))$ be the solution of the Hamiltonian system (16)–(17) and $u(\cdot)$ be given by*

$$u(t) = -R(t)^{-1}B(t)'y(t). \tag{21}$$

Then $(x(\cdot),\, z(\cdot))$ is the solution of the BSDE (1) corresponding to (21) and

$$J(\xi; u(\cdot)) = h(0)'\Big[H\Sigma(0) + I\Big]^{-1} Hh(0) + E\int_0^T \Big\{h'Qh + \eta'(S\Sigma + I)^{-1}S\eta\Big\}\, dt. \tag{22}$$

is the associated cost.

Proof of Theorem 2:
The unique solvability of the Hamiltonian system (16)–(17) has been proved in Proposition 3. The optimality of (21) follows from the fact that the cost (22) associated with the control (21) is equal to a lower bound to the optimal cost; see Lemma 1. The expression (15) for the optimal cost can be obtained by applying Ito's formula to $h(t)'N(t)h(t)$. Finally, we are able to conclude that the control (21) is unique because the BLQ problem (4) is a (strictly) convex optimization problem: The set of admissible triples $(x(\cdot),\, z(\cdot),\, u(\cdot))$

associated with (1) is a convex set, and the cost (3) is a strictly convex function on this set. This proves Theorem 2.

The following lemma is important in the proof of Theorem 1:

Lemma 2. *Let $(x(\cdot), z(\cdot), y(\cdot))$ be the solution of the Hamiltonian system (16)–(17) and $q(\cdot)$ be the solution of the SDE (12). Then*

$$y(t) = -Z(t)x(t) - q(t). \tag{23}$$

Proof of Theorem 1:
It follows immediately from Proposition 4 and Lemma 2.

5 Alternative Derivation: Forward Formulation

In Section 4, we obtained the solution of the BLQ problem (4) by showing that the control (18) or (13) achieves a lower bound to the cost function. In showing this result, equations (8)–(12), especially the Riccati equations (8) and (9), play a crucial role. In other words, once these equations are in place, then the whole derivation, albeit quite tedious, is essentially in the same spirit as the completion-of-square technique commonly used in tackling forward LQ problems. However, the reader may be puzzled about how these (rather complicated) equations were obtained in the first place. This section serves to unfold the origin of those equations by presenting an alternative, and intuitively appealing, approach to the BLQ problem (4). The idea is basically inspired by [12] where an (uncontrolled) BSDE is regarded as a *controlled forward* SDE. Here we go one step further to show that the BLQ problem can also be viewed as a (constrained) forward LQ problem, and that the solution (21) of the BLQ problem and the relationships (20) coincide with the limiting solution of a sequence of unconstrained forward LQ problems. In this process, the Riccati equations (8) and (9), along with other related equations, come out very naturally. It should be noted that our aim in this section is to highlight the origin of the equations (8)–(12) as well as (20), and hence the material in this section will be presented in an informal way. For this reason, certain convergence results required in this derivation, for example, are taken for granted, although they can be verified rigorously using standard techniques from stochastic analysis, the details of which are left to the interested reader. Finally, for the sake of notational convenience, we shall assume throughout this section that $S = 0$. The extension to the case $S \geq 0$ can be obtained in a similar way.

Forward LQ Problem:

Consider the following SDE:

$$\begin{cases} dx(t) = \Big\{A(t)x(t) + B(t)u(t) + C(t)v(t)\Big\}dt + v(t)\,dW(t), \\ x(0) = x^0, \end{cases} \tag{24}$$

We assume throughout that $x^0 \in \mathbb{R}^n$ and $(u(\cdot),\, v(\cdot)) \in \bar{\mathcal{U}}$ where

$$\bar{\mathcal{U}} = L^2_{\mathcal{F}}(0,\, T;\, \mathbb{R}^m) \times L^2_{\mathcal{F}}(0,\, T;\, \mathbb{R}^n).$$

For every $i \in \mathbb{Z}^+$ let

$$\begin{aligned} &J(x^0,\, u(\cdot),\, v(\cdot);\, i) \\ &:= E\frac{1}{2}\Big\{x^{0\prime} H x^0 + \int_0^T \Big(x(t)'Q(t)x(t) + u(t)'R(t)u(t)\Big)\, dt \\ &+ i\, |x(T) - \xi|^2\Big\}. \end{aligned} \tag{25}$$

The family of LQ problems, parameterized by i, is defined by:

$$\begin{cases} \min_{x^0,\, (u(\cdot),\, v(\cdot))} J(x^0,\, u(\cdot),\, v(\cdot);\, i), \\ \text{Subject to:} \\ \quad (u(\cdot),\, v(\cdot)) \in \bar{\mathcal{U}},\ x^0 \in \mathbb{R}^n, \\ \quad (x^0,\, x(\cdot),\, u(\cdot),\, v(\cdot)) \text{ satisfies (24).} \end{cases} \tag{26}$$

Comparing (26) with the BLQ problem (4) it is clear that the control $v(\cdot)$ replaces the process $z(\cdot)$ in the BSDE, while the terminal condition $x(T) = \xi$ in (4) is replaced by a penalty term in the cost of the forward problem (26). One fundamental differences between (4) and (26) should be recognized. In the BLQ problem (4), the initial condition $x(0)$ and the process $z(\cdot)$ are part of the state process $(x(\cdot),\, z(\cdot))$; that is, once $u(\cdot)$ has been chosen, the pair $(x(\cdot),\, z(\cdot))$ (and hence, $x(0)$) is uniquely determined. On the other hand, the pair $(u(\cdot),\, v(\cdot))$ and the initial condition $x(0)$ are decision variables in the forward problem (26). This additional degree of freedom is possible because the forward problem (26) does not involve a terminal condition on the state $x(\cdot)$. We shall show that the optimal solution of the BLQ problem (4), as stated in Theorems 2 and 1, can be obtained by solving the forward problem (26) and letting $i \uparrow \infty$.

Completion of Squares:
The solution of the forward problem (26) can be obtained by using a completion-of-square approach via the Riccati equation studied in [5]. In particular, let $P_i(\cdot)$ and $(h_i(\cdot),\, \eta_i(\cdot))$ be the unique solutions of the following equations:

$$\begin{cases} \dot{P}_i(t) + P_i(t)A(t) + A(t)'P_i(t) \\ \quad -P_i(t)\Big[B(t)R(t)^{-1}B(t)' + C(t)P_i(t)^{-1}C(t)'\Big]P_i(t) + Q(t) = 0, \\ P_i(T) = i\, I, \end{cases} \tag{27}$$

$$\begin{cases} dh_i(t) = \Big\{(A(t) + P_i(t)^{-1}Q(t))h_i(t) + C(t)\,\eta_i(t)\Big\}\,dt + \eta_i(t)\,dW(t), \\ h_i(T) = -\xi. \end{cases} \tag{28}$$

Note that (28) is introduced to cope with the linear term $E\{\,i\xi x(T)\}$ in the terminal cost part of (25). Evaluating $\Sigma_i(t) := P_i(t)^{-1}$, it turns out that $\Sigma_i(\cdot)$ is a solution of the Riccati equation:

$$\begin{cases} \dot{\Sigma}_i(t) = A(t)\Sigma_i(t) + \Sigma_i(t)A(t)' - C(t)\Sigma_i(t)C(t)' \\ \qquad +\Sigma_i(t)Q(t)\Sigma_i(t) - B(t)R(t)^{-1}B(t)', \\ \Sigma_i(T) = \frac{1}{i}I. \end{cases} \tag{29}$$

(The above explains the origin of the key equations (8) and (11).) It can be shown (see [14]) that:

$$\begin{aligned} &J(x^0,\, u(\cdot),\, v(\cdot);\, i) \\ &= \frac{1}{2}h_i(0)'(H\Sigma(0)+I)^{-1}Hh_i(0) + E\frac{1}{2}\int_0^T h_i(t)'Q(t)h_i(t)\,dt \\ &\quad +\frac{1}{2}\Big[x^0 + \Big(I+\Sigma_i(0)H\Big)^{-1}h(0)\Big]'\Big[H + P_i(0)\Big] \\ &\quad \times\Big[x^0 + \Big(I+\Sigma_i(0)H\Big)^{-1}h_i(0)\Big] \\ &\quad +E\frac{1}{2}\int_0^T \Big[\Big(u + R^{-1}B'P_i(x+h_i)\Big)'R\Big(u + R^{-1}B'P_i(x+h_i)\Big) \\ &\quad +\Big(v + \Sigma_iC'P_i(x+h_i) + \eta\Big)'P_i\Big(v + \Sigma_iC'P_i(x+h_i) + \eta_i\Big)\Big]\,dt. \end{aligned} \tag{30}$$

Since $(u(\cdot),\, v(\cdot))$ and $x(0)$ in (30) are free to be chosen, it follows that the optimal cost for the forward LQ problem (26) is

$$J_i^*(\xi) = \frac{1}{2}h_i(0)'(H\Sigma(0)+I)^{-1}Hh_i(0) + E\frac{1}{2}\int_0^T h_i(t)'Q(t)h_i(t)\,dt, \tag{31}$$

which is obtained when

$$\begin{cases} u_i(t) = -R(t)^{-1}B(t)'P_i(t)(x_i(t)+h_i(t)), \\ v_i(t) = -\Sigma(t)C(t)'P(t)(x_i(t)+h_i(t)) - \eta_i(t), \\ x_i^0 = -(I+\Sigma_i(0)H)^{-1}h_i(0), \end{cases} \tag{32}$$

where $x_i(\cdot) \in L^2_{\mathcal{F}}(\Omega;\, C(0,\, T;\, \mathbb{R}^n))$, the optimal state trajectory, is the unique solution of the SDE:

$$\begin{cases} dx_i(t) = \Big\{A(t)x_i(t) - B(t)R(t)^{-1}B(t)'P_i(t)\Big(x_i(t) + h_i(t)\Big) \\ \qquad\qquad -C(t)P_i(t)^{-1}C(t)'P_i(t)\Big(x_i(t) + h_i(t)\Big) - C(t)\eta_i(t)\Big\}\,dt \\ \qquad\qquad -\Big\{P_i(t)^{-1}C(t)'P_i(t)\Big(x_i(t) + h_i(t)\Big) + \eta_i(t)\Big\}\,dW(t), \\ x_i(0) = x_i^0. \end{cases} \tag{33}$$

Limiting Solution: $i \uparrow \infty$

Let $y_i(\cdot)$ be defined by the relation:

$$y_i(t) := P_i(t)(x_i(t) + h_i(t)). \tag{34}$$

It follows that

$$x_i(t) = \Sigma_i(t)y_i(t) - h_i(t) \tag{35}$$

where $x_i(\cdot)$ the solution of (33). It is easy to show that

$$\begin{cases} dx_i(t) = \Big\{A(t)x_i(t) - B(t)R(t)^{-1}B(t)'y_i(t) \\ \quad +C(t)\Big(- \Sigma_i(t)C(t)'y_i(t) - \eta_i(t)\Big)\Big\}\,dt \\ \quad +\Big\{ - \Sigma_i(t)C(t)'y_i(t) - \eta_i(t)\Big\}\,dW(t), \\ x_i(0) = x_i^0 \end{cases} \tag{36}$$

$$\begin{cases} dy_i(t) = \Big\{ - (A(t) + \Sigma_i(t)Q(t))'y_i(t) + Q(t)h_i(t)\Big\}\,dt \\ \quad -C(t)'y_i(t)\,dW(t), \\ y_i(0) = H(\Sigma_i(0)H + I)^{-1}h_i(0). \end{cases} \tag{37}$$

Substituting (34) into (29), (31)-(33) and letting $i \uparrow \infty$, we obtain:

$$\begin{cases} u(t) = -R(t)^{-1}B(t)'y(t), \\ x(t) = \Sigma(t)y(t) - h(t), \\ v(t) = -\Sigma(t)C(t)'y(t) - \eta(t), \\ x^0 = -\Big(I + \Sigma(0)H\Big)^{-1}h(0), \end{cases} \tag{38}$$

and:

$$J^*(\xi) := \frac{1}{2}h(0)'(H\Sigma(0)+I)^{-1}Hh(0) + E\frac{1}{2}\int_0^T h(t)'Q(t)h(t)\,dt$$
$$= E\frac{1}{2}\Big\{\xi' N(T)\xi - \int_0^T \Big(\eta' N\eta + 2h'NC\eta\Big)\,dt\Big\}, \tag{39}$$

where

$$\begin{cases} \dot\Sigma(t) = A(t)\Sigma(t) + \Sigma(t)A(t)' \\ -C(t)\Sigma(t)C(t)' + \Sigma(t)Q(t)\Sigma(t) - B(t)R(t)^{-1}B(t)', \\ \Sigma(T) = 0, \end{cases} \tag{40}$$

$$\begin{cases} \dot N(t) + N(t)(A(t)+\Sigma(t)Q(t)) + (A(t)+\Sigma(t)Q(t))'N(t) - Q(t) = 0, \\ N(0) = \frac{1}{2}\Big\{H(I+\Sigma(0)H)^{-1} + (I+H\Sigma(0))^{-1}H\Big\}, \end{cases} \tag{41}$$

$$\begin{cases} dh(t) = \Big\{(A(t)+\Sigma(t)Q(t))h(t) + C(t)\,\eta(t)\Big\}\,dt + \eta(t)\,dW(t), \\ h(T) = -\xi, \end{cases} \tag{42}$$

$$\begin{cases} dx(t) = \Big\{A(t)x(t) - B(t)R(t)^{-1}B(t)'y(t) \\ \quad +C(t)\Big(-\Sigma_i(t)C(t)'y(t) - \eta(t)\Big)\Big\}\,dt \\ \quad +\Big\{-\Sigma(t)C(t)'y(t) - \eta(t)\Big\}\,dW(t), \\ x(0) = x^0, \end{cases} \tag{43}$$

$$\begin{cases} dy(t) = \Big\{-(A(t)+\Sigma(t)Q(t))'y(t) + Q(t)h(t)\Big\}\,dt \\ \quad -C(t)'y(t)\,dW(t), \\ y(0) = H(\Sigma(0)H+I)^{-1}h(0). \end{cases} \tag{44}$$

The Hamiltonian system (16)–(17) is obtained by substituting (38) into (43)-(44) together with the observation that

$$x(T) = \Sigma(T)y(T) - h(T) = \xi.$$

The optimal control (18), the optimal cost (15), and the relations (20) are recovered in (38)-(39). Hence, the solution of the optimal BLQ control problem (4) as outlined in Theorem 2 coincides with the limiting solution of a family (26) of forward LQ problems. Theorem 1 can be recovered simply by applying the transformation as outlined in Lemma 2.

6 Conclusion

Uncontrolled BSDEs have been a topic of active research for many years. Fundamental issues such as existence and uniqueness of solutions of BSDEs are well established (see [18]), and for this reason, a natural 'next step' is the development of a stochastic optimal control theory. In the case of optimal control of (forward) Ito diffusions, it is well accepted that Kalman's linear–quadratic control, Pontryagin's maximum principle, and Bellman's dynamic programming are the cornerstones of the theory. In developing a control theory for backwards equations, it makes sense, therefore, to begin with the maximum principle, linear–quadratic control, and dynamic programming. In this paper, we solve the linear–quadratic problem. A maximum principle has been derived in [8].

The optimal solution of the linear–quadratic problem is derived explicitly in terms of a pair of Riccati equations, a forward SDE and a BSDE. Moreover, this optimal control coincides with the solution of a constrained forward LQ problem, and is the limiting solution of a family of unconstrained forward LQ problems.

An outstanding open problem is to study the BLQ problem where all the coefficients are random. In this case, the Riccati equations (8) and (9) both become (nonlinear) BSDEs (rather than ODEs as in this paper), the solvability of which is very challenging to prove.

References

1. M. Ait Rami and X. Y. Zhou. Linear matrix inequalities, Riccati equations, and indefinite stochastic linear quadratic control, *IEEE Trans. Automat. Contr.*, **45**, 2000, pp. 1131 – 1143.
2. B.D.O. Anderson and J.B. Moore. *Optimal Control - Linear Quadratic Methods*, Prentice-Hall, New Jersey, 1989.
3. A. Bensoussan. Lecture on stochastic control, part I, *Lecture Notes in Math.*, **972**, 1983, pp 1 – 39.
4. J.M. Bismut. An introductory approach to duality in optimal stochastic control, *SIAM Rev.*, **20**, 1978, pp 62 – 78.
5. S. Chen, X.J. Li and X.Y. Zhou. Stochastic linear quadratic regulators with indefinite control weight costs, *SIAM J. Contr. Optim.*, **36**, 1998, pp 1685 – 1702.
6. S. Chen and X.Y. Zhou. Stochastic linear quadratic regulators with indefinite control weight costs. II, *SIAM Journal on Control and Optimization*, **39**, 2000, pp. 1065 – 1081.
7. J. Cvitanić and J. Ma. Hedging options for a large investor and forward-backward SDEs, *Ann. Appl. Probab.*, **6**, 1996, pp 370 – 398.

8. N.G. Dokuchaev and X.Y. Zhou. Stochastic control problems with terminal contingent conditions, *J. Math. Anal. Appl.* **238**, 1999, pp 143 – 165.
9. D. Duffie and L. Epstein. Stochastic differential utility, *Econometrica*, **60**, 1992, pp 353–394.
10. D. Duffie, J. Ma and J. Yong. Black's consol rate conjecture, *Ann. Appl. Prob.*, **5**, 1995, pp 356 – 382.
11. N. El Karoui, S. Peng and M.C. Quenez. Backward stochastic differential equations in finance, *Math. Finance*, **7**, 1997, pp 1 – 71.
12. M. Kohlmann and X.Y. Zhou. Relationship between backward stochastic differential equations and stochastic controls: A linear–quadratic approach, *SIAM J. Contr. Optim.*, **38**, 2000, pp 1392 – 1407.
13. A.E.B. Lim. Quadratic hedging and mean–variance portfolio selection with random parameters in an incomplete market. (Preprint).
14. A.E.B. Lim and X.Y. Zhou. Linear-quadratic control of backward stochastic differential equations. *SIAM J. Contr. Optim.*, Vol 40 No. 2, 2001, pp 450 – 474.
15. A.E.B.Lim and X.Y. Zhou. Mean-variance portfolio selection with random parameters. To appear in *Math. Oper. Res*, 2002.
16. A.E.B. Lim and X.Y. Zhou. Stochastic optimal LQR control with integral quadratic constraints and indefinite control weights. *IEEE Trans. Automat. Contr.*, **44(7)**, 1999, pp 359 – 369.
17. J. Ma and J. Yong. *Forward-Backward Stochastic Differential Equations and Their Applications*, Lect. Notes Math., Vol. 1702, Springer-Verlag, New York, 1999.
18. E. Pardoux and S. Peng. Adapted solution of backward stochastic differential equation, *Syst. & Contr. Lett.*, **14**, 1990, pp 55 – 61.
19. S. Peng. A general stochastic maximum principle for optimal control problems, *SIAM J. Contr. Optim.*, **28**, 1990, pp. 966 – 979.
20. S. Peng. Backward stochastic differential equations and applications to optimal control, *Appl. Math. Optim.*, **27**, 1993, pp 125 – 144.
21. J. Yong and X.Y. Zhou. *Stochastic Controls: Hamiltonian Systems and HJB Equations*, Springer-Verlag, New York, 1999.
22. X.Y. Zhou. A unified treatment of maximum principle and dynamic programming in stochastic controls, *Stoch. & Stoch. Rep.*, **36**, 1991, pp 137 – 161.

Hilbert Spaces Induced by Toeplitz Covariance Kernels

Mihaela T. Matache and Valentin Matache

Department of Mathematics, University of Nebraska, Omaha, NE 68182-0243, USA. dmatache@mail.unomaha.edu; vmatache@mail.unomaha.edu

Abstract. We consider the reproducing kernel Hilbert space $\mathcal{H}_\mu$ induced by a kernel which is obtained using the Fourier-Stieltjes transform of a regular, positive, finite Borel measure μ on a locally compact abelian topological group Γ. Denote by G the dual of Γ. We determine $\mathcal{H}_\mu$ as a certain subspace of the space $\mathcal{C}_0(G)$ of all continuous function on G vanishing at infinity. Our main application is calculating the reproducing kernel Hilbert spaces induced by the Toeplitz covariance kernels of some well-known stochastic processes.

Keywords: Covariance Kernel, Fourier Transform, Reproducing Kernel.

AMS Subj. Class. Primary: 60B15, Secondary: 60G10, 46C15.

1 Introduction

Let K denote a reproducing kernel on a nonempty set X. Such a kernel is called a Toeplitz kernel if X is an abelian group and there exists a function $\Phi : X \to \mathbb{C}$ such that

$$K(x, y) = \Phi(x - y), \qquad \forall x, y \in X.$$

The covariance kernels associated to wide-sense stationary stochastic processes (see Definition 2 in Section 3 of this paper) are Toeplitz reproducing kernels. Let G be a locally compact, abelian, topological group and K a continuous Toeplitz reproducing kernel on G. A well known theorem of Bochner, [23, 1.4.3], states that K is necessarily induced by a positive, finite, regular Borel measure μ on Γ, the dual of G, in the sense that

$$K(x, y) = \hat{\mu}(y - x) \qquad \forall x, y \in G$$

where $\hat{\mu}$ is the Fourier-Stieltjes transform of μ. For that reason the reproducing kernel Hilbert space (RKHS) induced by such a kernel K is denoted by $\mathcal{H}_\mu$. The main result of this paper is Theorem 2 in which we describe $\mathcal{H}_\mu$ as follows.

If $\hat{\mu} \in L^1_G(dx)$, then μ is absolutely continuous with respect to the Haar measure $d\gamma$ of Γ, there is a continuous function φ such that $\varphi = d\mu/d\gamma$,

B. Pasik-Duncan: Stochastic Theory and Control, LNCIS 280, pp. 319–333, 2002.

$d\gamma$-a.e., and $f \in \mathcal{H}_\mu$ if and only if $f \in \mathcal{C}_0(G) \cap L^1_G(dx)$, $\hat{f}(\gamma) = 0$ on $\{\gamma \in \Gamma : \varphi(\gamma) = 0\}$ and

$$\int_{\{\gamma \in \Gamma : \varphi(\gamma) \neq 0\}} \frac{|\hat{f}(\gamma)|^2}{\varphi(\gamma)} d\gamma < \infty.$$

In the text above $\hat{f}$ denotes the Fourier transform of f and $\mathcal{C}_0(G)$ is the algebra of all continuous functions vanishing at infinity on G. The space $L^1_G(dx)$ is calculated with respect to the Haar measure dx of G.

Let $\mathbb{T}$ be the unit circle in the complex plane and $\mathbb{Z}$ the set of all integers. The representation of $\mathcal{H}_\mu$ contained by Theorem 2 generalizes similar results which describe the RKHS induced by a continuous, Toeplitz reproducing kernel in the particular cases $G = \mathbb{T}$, $\Gamma = \mathbb{Z}$, [19, page 84], respectively $G = \Gamma = \mathbb{R}$, [13], [21].

Section 2 of this paper is dedicated to introducing in detail the notions mentioned without many details or explanations in this section. We also prove some preliminary results and technical lemmas necessary to the proof of the main result. Section 3 contains that result (Theorem 2), its particular form for compact, abelian, topological groups (Theorem 3), and several examples of RKHS spaces induced by Toeplitz covariance kernels associated to well known stochastic processes, (such as the increment-process associated to a Poisson process, the Ornstein-Uhlenbeck process, some discrete first order autoregressive processes, and some moving average processes). Theorems 2 and 3 are used to determine those spaces.

There are many applications of the theory of RKHS spaces in various fields. To give an example, T. Kailath, E. Parzen, and some of their co-workers developed techniques of RKHS spaces to solve detection and estimation problems, [4], [5], [6], [13], [14], [21]. RKHS techniques play a central role in these papers. They are used to solve problems of extraction, detection, and prediction of signals in the presence of noise. The main message in [21] is that it is important to know if the signal belongs or not to the RKHS induced by the covariance kernel of the noise. For more on the importance of RKHS theory and its applications in this area of mathematics we refer to [2], [3], [11], [12], [16], [17], to quote only few of many eligible references.

2 Preliminary Results

Let X denote a nonempty set. A reproducing kernel on X is any function $K : X \times X \to \mathbb{C}$ with the property that

$$\sum_{i,j=1}^{n} K(x_i, x_j) c_i \bar{c}_j \geq 0 \qquad \forall x_1, \ldots, x_n \in X, \quad \forall c_1, \ldots, c_n \in \mathbb{C}.$$

Each reproducing kernel K on X induces in a unique way a Hilbert space $\mathcal{H}_K$ consisting of complex valued functions on X called the Reproducing Kernel Hilbert Space (RKHS) induced by K.

For each $y \in X$ denote by k_y the kernel function associated to y, i.e. the function

$$k_y(x) = K(x, y), \qquad x \in X.$$

Denote by $\mathcal{K}$ the set of all kernel functions, i.e.

$$\mathcal{K} = \{k_y : y \in X\}.$$

$\mathcal{H}_K$ is the completion of the linear space Span$\mathcal{K}$ spanned by the functions in $\mathcal{K}$ endowed with the inner product $\langle \cdot, \cdot \rangle$ determined by the following relation

$$\langle k_y, k_x \rangle = K(x, y) \qquad \forall x, y \in X.$$

The reason why $\mathcal{H}_K$ is called an RKHS with kernel K is the so called reproducing property

$$f(x) = \langle f, k_x \rangle \qquad \forall x \in X, \quad \forall f \in \mathcal{H}_K.$$

All these facts are well known. We refer the reader to [1] and [8] for the basics on RKHS.

For each $f \in \mathcal{H}_K$ we denote by $\|f\|$ the $\mathcal{H}_K$-norm of f, while $\|f\|_\infty$ denotes the following quantity

$$\|f\|_\infty := \sup\{f(x) : x \in X\}.$$

Clearly $\|f\|_\infty \in [0, \infty]$, $\forall f \in \mathcal{H}_K$. The following lemma contains an elementary remark.

Lemma 1. *Under the assumptions above, if $\mathcal{K}$ is a norm bounded set then there is $M > 0$ such that*

$$\|f\|_\infty \leq M\|f\| \qquad \forall f \in \mathcal{H}_K. \tag{1}$$

Therefore a norm-convergent sequence in $\mathcal{H}_K$ must also be uniformly convergent on X toward the same limit.

Proof. Since $\mathcal{K}$ is norm bounded there is $M > 0$ such that

$$\|k_y\| \leq M \qquad \forall y \in X.$$

Therefore for each $x \in X$ one can write

$$|f(x)| = |\langle f, k_x \rangle| \leq \|f\| \|k_x\| \leq M\|f\|.$$

□

Let G be an abelian, locally compact, topological group having the dual group denoted by Γ. It is well known that the reproducing kernels $K(x, y)$ on G of the form $K(x, y) = \Phi(x - y)$ for some $\Phi : G \to \mathbf{C}$ are necessarily induced by a finite, positive, regular Borel measure μ on Γ in the sense that such a μ with the following property always exists

$$\Phi(x) = \int_\Gamma (x, \gamma) d\mu(\gamma) \qquad x \in G. \tag{2}$$

In (2) (x, γ) denotes $\gamma(x)$. We will use this notation all over this paper. Let $\hat{\mu}$ denote the Fourier-Stieltjes transform of μ. Identifying as usual G to the dual group of Γ, observe that (2) can be rewritten in the following form

$$\Phi(x) = \hat{\mu}(-x) \qquad x \in G. \tag{3}$$

In (3) and all over this paper we use additive notation for the group law of G. Since $\hat{\mu}$ is the Fourier-Stieltjes transform of a complex regular Borel measure on a locally compact abelian group, Φ must be uniformly continuous and bounded on G and hence

$$k_y \in \mathcal{U}_b(G) \qquad \forall y \in G$$

where for each y, k_y is the kernel-function associated to the reproducing kernel $K(x, y) = \Phi(x - y)$, $x, y \in G$, Φ and μ are related as in (2), and $\mathcal{U}_b(G)$ is the space of all bounded, uniformly continuous, complex functions on G. Recall also the notation $\mathcal{H}_K = \mathcal{H}_\mu$ for the RKHS with the previously described reproducing kernel. Our first observation on $\mathcal{H}_\mu$ is the following.

Theorem 1. *Under the assumptions above one has that*

$$\mathcal{H}_\mu \subseteq \mathcal{U}_b(G).$$

Proof. Denote as before by $\mathcal{K}$ the set of all kernel-functions of $\mathcal{H}_\mu$. Clearly the linear space spanned by $\mathcal{K}$, Span$\mathcal{K}$ is a dense subset of $\mathcal{H}_\mu$. On the other hand, Span$\mathcal{K} \subseteq \mathcal{U}_b(G)$ and $\mathcal{K}$ is a norm bounded set since one can write

$$\|k_y\|^2 = \langle k_y, k_y \rangle = \Phi(y - y) = \Phi(0) \qquad \forall y \in G.$$

By the density of Span$\mathcal{K}$ in $\mathcal{H}_\mu$ and Lemma 1, we deduce that

$$\mathcal{H}_\mu \subseteq \mathcal{U}_b(G).$$

□

Corollary 1. *If G is separable, then $\mathcal{H}_\mu$ is also separable.*

Proof. Let $S = \{x_n : n \in I\}$ be a countable dense subset of G. Observe that if $f \in \mathcal{H}_\mu$ is perpendicular to k_{x_n}, $\forall n \in I$, then $f(x_n) = 0$, $\forall n \in I$. Since f is a continuous function and S a dense subset of G, it follows the f is the null function. The immediate consequence of this fact is that Span$\{k_y : y \in S\}$ is dense in $\mathcal{H}_\mu$. Therefore, since S is countable, if one considers the linear span of the vectors in $\{k_y : y \in S\}$ with coefficients chosen in the set $\mathbb{Q}[i]$ of all complex numbers with rational real and imaginary parts, one obtains a countable, dense subset of $\mathcal{H}_\mu$. □

The Haar measures on G and Γ will be denoted by dx and $d\gamma$ respectively. It will always be assumed that they are normalized in such a way that the inversion theorem ([23, 1.5.1]) holds i.e. the following formula holds

$$f(x) = \int_\Gamma \hat{f}(\gamma)(x,\gamma)\, d\gamma \qquad x \in G. \tag{4}$$

In order that (4) holds f must be an $L^1_G(dx)$-function which belongs to the class $B(G)$ of all functions on G which are the Fourier-Stieltjes transforms of complex, Borel measures on Γ. Assume now that μ is absolutely continuous with respect to the Haar measure of Γ, $\varphi \in B(\Gamma)$ and

$$d\mu = \varphi(\gamma)\, d\gamma.$$

Since μ is a finite, positive measure, it follows that the Radon-Nykodim derivative φ is a nonnegative $L^1_\Gamma(d\gamma)$-function. The following lemma shows that the kernel-functions of $\mathcal{H}_\mu$ are complex conjugates of shifted Fourier transforms of φ, more precisely we can prove the following.

Lemma 2. *Let $y \in G$ be arbitrary and fixed. The kernel-function k_y can be calculated by the following formula*

$$k_y(x) = \overline{\widehat{(y,\gamma)\varphi(\gamma)}(x)} \qquad x \in G. \tag{5}$$

Proof. Indeed, one can write the following.

$$\begin{aligned} k_y(x) = \Phi(x-y) &= \int_\Gamma (x-y,\gamma)\, d\mu(\gamma) = \int_\Gamma (x-y,\gamma)\varphi(\gamma)\, d\gamma \\ &= \int_\Gamma \overline{\gamma(y-x)}\varphi(\gamma)\, d\gamma = \overline{\int_\Gamma (-x,\gamma)(y,\gamma)\varphi(\gamma)\, d\gamma} = \overline{\widehat{(y,\gamma)\varphi(\gamma)}(x)}. \end{aligned}$$

□

Corollary 2. *All the kernel-functions of $\mathcal{H}_\mu$ are $L^1_G(dx)$-functions if and only if $\hat{\varphi} \in L^1_G(dx)$.*

Proof. Observe that $k_0 = \overline{\hat{\varphi}}$, hence $k_0 \in L^1_G(dx)$ if and only if $\hat{\varphi} \in L^1_G(dx)$. Given that $k_y(x) = \Phi(x-y) = k_0(x-y)$ and dx is translation-invariant, it follows that $k_y \in L^1_G(dx)$, $\forall y \in G$ if and only if $k_0 \in L^1_G(dx)$. □

Under the assumptions $d\mu/d\gamma = \varphi$ $d\gamma$-a.e and $\varphi \in B(\Gamma) \cap L^1_\Gamma(d\gamma)$ one has the following useful formula for the Fourier transform of a kernel-function.

Lemma 3. *If $\varphi \in B(\Gamma) \cap L^1_\Gamma(d\gamma)$ the Fourier transform $\hat{k}_y$ of an arbitrary, fixed kernel-function k_y is given by the following formula*

$$\hat{k}_y(\gamma) = \overline{(y,\gamma)\varphi(\gamma)} = (-y,\gamma)\varphi(y) \qquad \gamma \in \Gamma. \tag{6}$$

Proof. The equality (6) is a direct consequence of Lemma 2 and the inversion formula (4). Indeed

$$\begin{aligned}\hat{k}_y(\gamma) &= \int_G (-x,\gamma)k_y(x)\,dx = \overline{\int_G (x,\gamma)\overline{k_y(x)}\,dx} \\ &= \overline{\int_G (x,\gamma)\widehat{(y,\gamma)\varphi}(\gamma)\,dx} = \overline{(y,\gamma)\varphi(\gamma)} = (-y,\gamma)\varphi(y).\end{aligned}$$

Above we were able to apply formula (4) to the function $(y,\gamma)\varphi(\gamma)$ because $B(\Gamma)$ is invariant under multiplication by (y,γ), [23, 1.3.3]. □

From now on, we will work under the assumptions in Lemma 3, namely assume that μ is a finite, positive Borel measure on Γ such that $d\mu \ll d\gamma$, there is a finite, positive Borel measure λ on G such that $d\mu/d\gamma$ is equal $d\gamma$-a.e to φ, the Fourier-Stieltjes transform of λ i.e. such that

$$\varphi(\gamma) = \int_G (-x,\gamma)\,d\lambda(x) \qquad \gamma \in \Gamma, \tag{7}$$

and $\varphi \in L^1_\Gamma(d\gamma)$.

Remark 1. The assumptions above hold if and only if $\hat{\mu} \in L^1_G(dx)$.

Proof. By [23, 1.7.3], if $\hat{\mu} \in L^1_G(dx)$, then $d\mu \ll d\gamma$, $d\mu/d\gamma \in L^1_\Gamma(d\gamma)$, and $d\mu/d\gamma$ is equal $d\gamma$-a.e. to the Fourier transform of $g(x) = \hat{\mu}(-x)$, an $L^1_G(dx)$-function. The converse implication is a direct consequence of [23, 1.5.1]. □

Under these assumptions we introduce a positive, not necessarily finite measure $\tilde{\mu}$ associated to μ as follows.

Definition 1. Let S denote the following open subset of Γ, $S := \{\gamma \in \Gamma : \varphi(\gamma) \neq 0\}$. Let E denote an arbitrary, fixed, Borel subset of Γ. The measure $\tilde{\mu}$ is the Borel measure on Γ given by the following equality

$$\tilde{\mu}(E) := \int_{S\cap E} \frac{1}{\varphi(\gamma)}\,d\gamma.$$

From now on we will use the notation $S = \text{supp}\tilde{\mu}$. The following is the last technical lemma we need prior to proving the theorem containing the description of $\mathcal{H}_\mu$.

Lemma 4. *Let μ, $\tilde{\mu}$, λ, and φ be as described above. The following inequality holds.*

$$\frac{\lambda(G)}{\varphi(\gamma)} \geq 1 \qquad \forall\gamma \in \text{supp}\tilde{\mu}. \tag{8}$$

Let f be any function in $B(G) \cap L^1_G(dx)$ such that $\hat{f}(\gamma) = 0$ $d\gamma$-a.e. on $\Gamma \setminus \text{supp}\tilde{\mu}$. For such f the following equality holds.

$$f(x) = \int_\Gamma \hat{f}(\gamma)\bar{\hat{k}}_y(\gamma)\, d\tilde{\mu}(\gamma). \tag{9}$$

Proof. Relation (8) is an immediate consequence of (7), as for equality (9), observe that one can write

$$\begin{aligned}\int_\Gamma \hat{f}(\gamma)\bar{\hat{k}}_x(\gamma)\, d\tilde{\mu}(\gamma) &= \int_{\text{supp}\tilde{\mu}} \hat{f}(\gamma)(x,\gamma)\varphi(\gamma)\frac{1}{\varphi(\gamma)}\, d\gamma \\ &= \int_\Gamma \hat{f}(\gamma)(x,\gamma)\, d\gamma = f(x).\end{aligned}$$

Above we made use of both Lemma 3 and the inversion formula (4). □

3 The Main Results

Let $\mathcal{C}_0(G)$ denote the space of all continuous, complex functions on G which vanish at infinity. We are ready to characterize the space $\mathcal{H}_\mu$ now.

Theorem 2. *If $\hat{\mu} \in L^1_G(dx)$ then the space $\mathcal{H}_\mu$ consists of those functions $f \in L^1_G(dx) \cap \mathcal{C}_0(G)$ which satisfy the following two conditions*

$$\hat{f}(\gamma) = 0 \qquad \forall\gamma \in \Gamma \setminus \text{supp}\tilde{\mu} \tag{10}$$

$$\int_\Gamma |\hat{f}(\gamma)|^2\, d\tilde{\mu}(\gamma) < \infty. \tag{11}$$

Any function $f \in \mathcal{H}_\mu$ has the property $\|f\|_2 < \infty$ where $\|\cdot\|_2$ is the norm of $L^2_G(dx)$.

Proof. Let $\mathcal{H}_0$ denote the space of all functions $f \in L^1_G(dx) \cap \mathcal{C}_0(G)$ satisfying conditions (10) and (11). First we will show that $\mathcal{H}_0$ is complete under the norm induced by the inner product

$$\langle f, g\rangle = \int_\Gamma \hat{f}(\gamma)\bar{\hat{g}}(\gamma)d\tilde{\mu}(\gamma)$$

$\|f\|_2 < \infty$, $\forall f \in \mathcal{H}_0$, and $\mathcal{H}_0 \subseteq B(G)$. First, observe that Span$\mathcal{K} \subseteq \mathcal{H}_0$. This is a direct consequence of Lemma 3 and the following computation

$$\int_\Gamma |\hat{k_y}(\gamma)|^2 d\tilde{\mu}(\gamma) = \int_{\text{supp}\tilde{\mu}} \varphi(\gamma)\, d\gamma = \mu(\Gamma) < \infty.$$

Denote by $\|\cdot\|$ the norm of $\mathcal{H}_0$. If $f \in \mathcal{H}_0$ then

$$\int_\Gamma |\hat{f}(\gamma)| d\gamma < \infty \quad \text{and} \quad \int_\Gamma |\hat{f}(\gamma)|^2 d\gamma < \infty.$$

Indeed, by Lemma 4

$$\frac{1}{\varphi(\gamma)} \geq \frac{1}{\lambda(G)} \qquad \forall \gamma \in \operatorname{supp}\tilde{\mu}.$$

Therefore

$$\int_\Gamma |\hat{f}(\gamma)|^2 d\gamma = \int_{\operatorname{supp}\tilde{\mu}} |\hat{f}(\gamma)|^2 d\gamma \leq \lambda(G) \int_{\operatorname{supp}\tilde{\mu}} \frac{|\hat{f}(\gamma)|^2}{\varphi(\gamma)} d\gamma = \lambda(G)\|f\|^2 < \infty.$$

Also

$$\int_\Gamma |\hat{f}(\gamma)| d\gamma = \int_{\operatorname{supp}\tilde{\mu}} |\hat{f}(\gamma)| d\gamma \leq \sqrt{\int_{\operatorname{supp}\tilde{\mu}} \frac{|\hat{f}(\gamma)|^2}{\varphi(\gamma)} d\gamma} \sqrt{\int_{\operatorname{supp}\tilde{\mu}} \varphi(\gamma) d\gamma}$$
$$= \|f\| \sqrt{\mu(\Gamma)} < \infty.$$

Observe that we established the inequalities

$$\|\hat{f}\|_1 \leq \sqrt{\mu(\Gamma)}\|f\| \qquad \forall f \in \mathcal{H}_0$$

where $\|\cdot\|_1$ is the norm of $L^1_\Gamma(d\gamma)$, and

$$\|\hat{f}\|_2 \leq \sqrt{\lambda(G)}\|f\| \qquad \forall f \in \mathcal{H}_0.$$

We will also denote by $\|\cdot\|_1$ and $\|\cdot\|_2$ the norms of $L^1_G(dx)$ and $L^2_G(dx)$ respectively.

Now assume that $(f_n)_n$ is a Cauchy sequence in $\mathcal{H}_0$ endowed with the norm $\|\cdot\|$. Then $(\hat{f}_n)_n$ will be Cauchy in both $L^1_\Gamma(d\gamma)$ and $L^2_\Gamma(d\gamma)$. Since both these spaces are complete, there is a $g \in L^1_\Gamma(d\gamma) \cap L^2_\Gamma(d\gamma)$ such that $\|\hat{f}_n - g\|_1 \to 0$ and $\|\hat{f}_n - g\|_2 \to 0$. The reason why the limit is the same modulo equality $d\gamma$-a.e. is the fact that convergent sequences in L^p-spaces have subsequences converging a.e. toward the limit-function, [24, 3.12]. Also, the sequence $(\hat{f}_n)_n$ is Cauchy in the norm $\|\cdot\|_\infty$ because $\|\hat{f}_n\|_\infty \leq \|f_n\|_1$, $\forall n$, [23, 1.2.4]. Denote by h the uniform limit of $(\hat{f}_n)_n$. Clearly $h = g$ $d\gamma$-a.e. and $h \in B(\Gamma)$ because, by Bochner's theorem [23, 1.4.3], $B(\Gamma)$ is closed with respect to uniform convergence on Γ. So we established that $h \in B(\Gamma) \cap L^1_\Gamma(d\gamma) \cap L^2_\Gamma(d\gamma)$, $\|\hat{f}_n - h\|_1 \to 0$, $\|\hat{f}_n - h\|_\infty \to 0$, and hence $h(\gamma) = 0$ $\forall \gamma \in \Gamma \setminus \operatorname{supp}\tilde{\mu}$. Let

$$f(x) := \int_\Gamma h(\gamma)(x,\gamma)\, d\gamma.$$

By the inversion theorem $f \in L^1_G(dx) \cap B(G)$ and for each $x \in G$ one can write

$$|f(x) - f_n(x)| = \left|\int_\Gamma (h(\gamma) - \hat{f}_n(x))(x,\gamma)\, d\gamma\right| \leq \|h - \hat{f}_n\|_1 \to 0.$$

Again by the inversion theorem one can see that $\hat{f} = h$. Since $h \in L^1_\Gamma(d\gamma) \cap L^2_\Gamma(d\gamma)$ it follows that $\|f\|_2 < \infty$, as a consequence of the Plancherel theorem, [23, 1.6.1].

Let us prove now that f is the $\|\cdot\|$ - limit of $(f_n)_n$. For arbitrary fixed $\epsilon > 0$ consider n_0 a positive integer such that

$$\|f_m - f_n\| < \frac{\epsilon}{2} \qquad \forall m, n \geq n_0.$$

One has that

$$\begin{aligned}\|f - f_n\|^2 &= \int_\Gamma \liminf_{k\to\infty} |\hat{f}_{m_k}(\gamma) - \hat{f}_n(\gamma)|^2 d\tilde{\mu}(\gamma) \\ &\leq \liminf_{k\to\infty} \int_\Gamma |\hat{f}_{m_k}(\gamma) - \hat{f}_n(\gamma)|^2 d\tilde{\mu}(\gamma) \leq \left(\frac{\epsilon}{2}\right)^2 < \epsilon^2\end{aligned}$$

whenever $n \geq n_0$. Above we used Fatou's lemmma [24, 1.28] and the existence of a subsequence $(\hat{f}_{m_k})_k$ of $(\hat{f}_n)_n$ convergent to $h = \hat{f}$ $d\gamma$-a.e. on Γ. So $\mathcal{H}_0$ is a Hilbert space and since $\mathcal{H}_0 \subseteq L^1_G(dx) \cap B(G)$ one deduces that the reproducing property holds on $\mathcal{H}_0$, i.e.

$$\langle f, k_x \rangle = f(x) \qquad x \in G.$$

The above equality is a direct consequence of (9). Thus $\mathcal{H}_0$ is an RKHS with kernel

$$K(x,y) = \langle k_y, k_x \rangle = \int_{\text{supp}\tilde{\mu}} (-y,\gamma)\varphi(\gamma)(x,\gamma)\varphi(\gamma)\frac{1}{\varphi(\gamma)}\, d\gamma = \hat{\mu}(y-x)$$

by Lemma 3. Given the uniqueness of the RKHS associated to a given reproducing kernel it follows that $\mathcal{H}_\mu = \mathcal{H}_0$. □

The statement in the Theorem 2 in the particular case $G = \Gamma = \mathbb{R}$ appears in [13]. Theorem 2 can be formulated in a special way if G is compact. Before stating it in that context we need to make some simple observations and introduce more notations. Recall that if G is compact, then Γ is a complete orthonormal subset of $L^2_G(dx)$, [18]. For each $f \in L^1_G(dx)$ and each $\gamma \in \Gamma$ we denote by $c_\gamma(f)$ the Fourier coefficient of f of index γ, i.e.

$$c_\gamma(f) = \int_G f(x)(-x,\gamma)dx.$$

Denote by $\mathcal{C}(G)$ the algebra of all complex-valued continuous functions on G. One can give the following characterization to the space $\mathcal{H}_\mu$.

Theorem 3. *Let G be a compact, abelian topological group. Let Γ denote its dual group, and let μ be a finite, positive, regular Borel measure on Γ. Then $f \in \mathcal{H}_\mu$ if and only if $f \in \mathcal{C}(G)$,*

$$c_\gamma(f) = 0 \qquad \forall \gamma \in \Gamma \setminus \mathit{supp}\tilde{\mu} \tag{12}$$

and

$$\sum_{\gamma \in \mathrm{supp}\tilde{\mu}} \frac{|c_\gamma(f)|^2}{\mu(\{\gamma\})} < \infty. \tag{13}$$

Proof. Since dx is a finite measure any function in $\mathcal{C}(G)$ is an $L^1_G(dx)$-function. On the other hand, conditions (12) and (13) are exactly (10) and (11) in our context, since Γ is a discrete topological group, [23, 1.7.3]. Also, $\hat{\mu}$ is automatically in $L^1_G(dx)$ when G is compact because $\hat{\mu}$ is continuous on G and hence bounded. □

Theorem 3, in the particular case $G = \mathbb{T}$ and $\Gamma = \mathbb{Z}$ appears in [19, page 84]. In the following we will illustrate the utility of Theorem 2 by calculating the reproducing kernel space associated to some stochastic processes. Let $(X_t)_{t\in I}$ be a stochastic process. The RKHS generated by the covariance kernel of a stochastic process is often a valuable instrument. We will designate the aforementioned reproducing kernel Hilbert space as the RKHS associated to the stochastic process.

Definition 2. A process $(X(t))_{t\in S}$ is called wide-sense stationary, if it has constant mean and the autocorrelation function $K(s,t) = \mathbf{E}[X(s)X(t)]$, $s, t \in S$, depends only on the difference $t - s$.

The index set S is assumed to be the subset of a group. Theorem 2 can be used to calculate the RKHS associated to wide-sense stationary processes. We give several examples in the following. Recall that if K is a reproducing kernel on X, then for each nonempty subset E of X, the restriction of K to $E \times E$ is a reproducing kernel on E and the RKHS induced by this second kernel is simply the space of the restrictions to E of all functions in $\mathcal{H}_K$. We will use this fact in some of the examples without mentioning it each time.

Example 1. Let $(N(t))_{t\geq 0}$ be a Poisson process, and define its increment process as follows $X(t) := N(t+1) - N(t), t \geq 0$. Denote by dx the Lebesgue measure on the real line. The RKHS associated to $(X(t))_{t\geq 0}$ is the space of all functions g which are restrictions to $[0,\infty)$ of functions $f \in \mathcal{C}_0(\mathbb{R}) \cap L^1_{\mathbb{R}}(dx)$ with the following properties

$$\hat{f}(2k\pi) = 0 \qquad \forall k \in \mathbb{Z}, k \neq 0 \tag{14}$$

and

$$\int_{-\infty}^{\infty} \frac{|\hat{f}(x)|^2 x^2}{\sin^2(x/2)}\, dx < \infty. \tag{15}$$

Proof. Let $G = \mathbb{R}$ and $\Gamma = \mathbb{R}$, (see [23] for the fact that they are duals of each other). Let $d\mu = \sin^2(x/2)/(\sqrt{2\pi}(x/2)^2)dx$. Clearly $(\sin^2(x/2)/(x/2)^2) \in$

$L^1_{\mathbb{R}}(dx)$. Working as usual with the normalized Haar measure $dx/\sqrt{2\pi}$, one obtains by a straightforward calculation

$$\hat{\mu}(x) = \begin{cases} 1 - |x| & \text{if} \quad |x| \leq 1 \\ 0 & \text{if} \quad |x| > 1 \end{cases} .$$

This is a continuous, compactly supported function and hence belongs to $L^1_{\mathbb{R}}(dx)$. Therefore Theorem 2 can be applied to the kernel $K(t,s) = \nu\hat{\mu}(s-t)$ whose restriction to $[0,\infty)$ is the covariance kernel of the process $(X(t))_{t\geq 0}$, [20]. The positive constant ν is sometimes called the intensity of the Poisson process, [20]. Since the function φ in Theorem 2 is in our case $\varphi(x) = \sin^2(x/2)/(\sqrt{2\pi}(x/2)^2)$ whose set of zeros is $\{2k\pi : k \in \mathbb{Z}, k \neq 0\}$, one gets the characterization above. □

Our next example is concerned with the following stochastic process.

Definition 3. The stationary Ornstein-Uhlenbeck process is the unique Gaussian process with mean zero and covariance kernel $K(t,s) = (\sigma^2/(2\beta))e^{-\beta|t-s|}$ $t,s \in \mathbb{R}$.

Its associated RKHS is described in the next example.

Example 2. The *RKHS* associated to the stationary Ornstein-Uhlenbeck process consists of all functions $f \in \mathcal{C}_0(\mathbb{R}) \cap L^1_{\mathbb{R}}(dx)$ which satisfy the following condition

$$\int_{-\infty}^{\infty} |\hat{f}(x)|^2(\beta^2 + x^2)\, dx < \infty. \tag{16}$$

Proof. Using again the notations of Theorem 2, consider $\varphi(x) = \sigma^2/(\sqrt{2\pi}(\beta^2 + x^2))$. It is easy to check that $\hat{\varphi}(x) = (\sigma^2/(2\beta))e^{-\beta|x|}$, [23, 1.5.3]. Clearly φ satisfies the assumptions in Theorem 2 and condition (10) is vacuously satisfied by all functions $f \in \mathcal{C}_0(\mathbb{R}) \cap L^1_{\mathbb{R}}(dx)$ since φ has no zeros. □

Let us consider some examples of discrete processes now.

Example 3. Let r and σ be real constants, $0 < r < 1$, $(A_n)_{n\geq 0}$ a sequence of zero mean uncorrelated random variables such that

$$\operatorname{Var}(A_0) = \frac{\sigma^2}{1-r^2} \quad \text{and} \quad \operatorname{Var}(A_n) = \sigma^2 \qquad \forall n > 0.$$

The first order autoregressive process AR(1) $(X_n)_{n\geq 0}$ is defined as follows. $X_0 = A_0$, $X_n = rX_{n-1} + A_n$. The RKHS $\mathcal{H}$ associated to this process has the following description. $\mathcal{H}$ consists of all absolutely-summable sequences $(w_n)_{n\geq 0}$ of complex numbers which are restrictions to the set $\mathbb{N}$ of non-negative integers of sequences $(z_n)_{n\in\mathbb{Z}}$ of complex numbers with the following property

$$\sum_{n=-\infty}^{\infty} \left((1+r^2)|z_n|^2 - r(\bar{z}_{n+1} + \bar{z}_{n-1})z_n\right) < \infty.$$

Proof. A straightforward computation leads to the formula

$$\mathrm{Cov}(X_n, X_m) = \frac{\sigma^2}{1-r^2} r^{|m-n|}.$$

Let

$$\varphi(e^{i\theta}) := \frac{\sigma^2}{1-r^2} \sum_{n=-\infty}^{\infty} r^{|n|} e^{in\theta}.$$

Let $G = \mathbb{Z}$ and $\Gamma = \mathbb{T}$ (see [23] for the fact that they are each other' s dual). The Fourier transform of φ is the sequence of its Fourier coefficients calculated with respect to the standard orthonormal basis $\{e^{in\theta} : n \in \mathbb{Z}\}$ of $\Gamma = \mathbb{T}$. Obviously, if $d\mu := \varphi(e^{i\theta}) \frac{d\theta}{2\pi}$ (where $d\theta$ is the arc-length measure on $\mathbb{T}$), one has that

$$\hat{\varphi} = \left(\frac{r^{|n|}\sigma^2}{1-r^2} \right)_{n\in\mathbb{Z}}.$$

Therefore $\hat{\varphi}(m-n) = \mathrm{Cov}(X_n, X_m)\ \forall m, n \in \mathbb{Z}$ and hence Theorem 2 can be used to calculate the RKHS induced by the covariance kernel of $(X_n)_{n\geq 0}$. Indeed, since $0 < r < 1$, the sequence $\hat{\varphi}$ is absolutely summable, i.e. $\hat{\mu} \in L^1_G(dx)$. A straightforward computation leads to the following simpler representation of φ

$$\varphi(e^{i\theta}) = \frac{\sigma^2}{|1 - re^{i\theta}|^2}.$$

Applying Theorem 2 to the groups $G = \mathbb{Z}$, $\Gamma = \mathbb{T}$ and the measure $d\mu = \varphi(e^{i\theta}) \frac{d\theta}{2\pi}$ one obtains that $\mathcal{H}_\mu$ consists of those absolutely summable sequences $(z_n)_{n\in\mathbb{Z}}$ of complex numbers with the property that

$$\int_{-\pi}^{\pi} \left| \sum_{n=-\infty}^{\infty} z_n e^{in\theta} \right|^2 |1 - re^{i\theta}|^2 d\theta < \infty. \tag{17}$$

Note that $\varphi(e^{i\theta})$ is never zero, so (17) is the only condition $(z_n)_{n\in\mathbb{Z}}$ must satisfy besides being absolutely summable. The absolute summability of $(z_n)_{n\in\mathbb{Z}}$ and a straightforward computation lead to the following alternative expression of (17)

$$\sum_{n=-\infty}^{\infty} ((1+r^2)|z_n|^2 - r(\bar{z}_{n-1} + \bar{z}_{n+1})z_n) < \infty.$$

□

Example 4. Let q be a positive integer. Consider a sequence $(A_n)_{n\geq 0}$ of random variables with the properties $E[A_n] = E[A_0]$, $\forall n \geq 0$ and $\mathrm{Var}(A_n) = \sigma^2 > 0$, $\forall n \geq 0$. Let $(X_n)_{n\geq q}$ be the following moving average process of order q, MA(q)

$$X_n = \frac{1}{q+1}\sum_{k=0}^{q} A_{n-k} \quad \forall n \geq q.$$

The *RKHS* $\mathcal{H}$ associated to this process has the following description. $\mathcal{H}$ consists of those absolutely summable sequences $(w_n)_{n\geq q}$ of complex numbers which are restrictions to the set $\{n \in \mathbb{Z} : n \geq q\}$ of sequences $(z_n)_{n\in\mathbb{Z}}$ for which the following conditions hold

$$\sum_{n=-\infty}^{\infty} z_n e^{\frac{2nk\pi i}{q+1}} = 0 \qquad \forall k \in \mathbb{Z}, 0 < |k| \leq \frac{q+1}{2} \tag{18}$$

and

$$\int_{-\pi}^{\pi} \frac{\left|\sum_{n=-\infty}^{\infty} z_n e^{in\theta}\right|^2 \sin^2\frac{\theta}{2}}{\sin^2\frac{\theta}{2} + \sin\frac{(q+2)\theta}{2}\sin\frac{q\theta}{2}} d\theta < \infty. \tag{19}$$

Proof. The covariance kernel of $(X_n)_{n\geq q}$ is described by the following

$$\mathrm{Cov}(X_n, X_m) = \begin{cases} \frac{\sigma^2}{(q+1)^2}(q+1-|m-n|) & \text{if } |m-n| \leq q \\ 0 & \text{if } |m-n| > q \end{cases}$$

Consider the function $\varphi(e^{i\theta})$ whose sequence of Fourier coefficients is given by

$$c_n = \begin{cases} \frac{\sigma^2}{(q+1)^2}(q+1-|n|) & \text{if } |n| \leq q \\ 0 & \text{if } |n| > q \end{cases}$$

Clearly

$$\varphi(e^{i\theta}) = \frac{\sigma^2}{(q+1)^2}\sum_{k=-q}^{q}(q+1-|k|)e^{ik\theta}.$$

Straightforward computations lead to the following simpler representation of φ

$$\varphi(e^{i\theta}) = \frac{\sigma^2}{(q+1)^2}\left[1 + \frac{\sin\frac{(q+2)\theta}{2}\sin\frac{q\theta}{2}}{\sin^2\frac{\theta}{2}}\right].$$

Applying Theorem 2 to $G = \mathbb{Z}$, $\Gamma = \mathbb{T}$, $d\mu = \varphi(e^{i\theta})\frac{d\theta}{2\pi}$ one gets that $\mathcal{H}_\mu$ consists of those absolutely summable sequences $(z_n)_{n\in\mathbb{Z}}$ of complex numbers for which conditions (18) and (19) hold. $\square$

References

1. Aronszajn, N. (1950) Theory of Reproducing Kernels, Trans. Amer. Math. Soc. **68**, 337–404.
2. Chevet, S. (1981) Kernel Associated with a Cylindrical Measure, Lecture notes in Mathematics **860**, 51–84, Springer-Verlag, New York.
3. Duncan, T. E. (2000) Some Applications of Fractional Brownian Motion to Linear Systems, Kluwer Internat. Ser. Engr. and Comput. Sci., **518**, 97–105, Kluwer Acad. Publ., Boston, MA, 2000.
4. Duttweiler, D. L. and Kailath, T. (1972) RKHS Approach to Detection and Estimation Problems III: Generalized Innovations Representations and a Likelihood-Ratio Formula, IEEE Trans. Information Theory **18**, 730–745.
5. Duttweiler, D. L. and Kailath, T. (1972) RKHS Approach to Detection and Estimation Problems IV: Non-Gaussian Detection, IEEE Trans. Information Theory **19**, 19–28.
6. Duttweiler, D. L. and Kailath, T. (1972) RKHS Approach to Detection and Estimation Problems V: Parameter Estimation, IEEE Trans. Information Theory **19**, 29–36.
7. Halmos, P. R. (1974) Measure Theory, Springer-Verlag, New York, Heidelberg, Berlin.
8. Halmos, P. R. (1980) A Hilbert Space Problem Book, 2-d edition, Springer-Verlag, New York, Heidelberg, Berlin.
9. Hewitt, E. and Ross, K. A. (1963) Abstract Harmonic Analysis, Springer-Verlag, Berlin-New York.
10. Hida, T. (1970) Stationary Stochastic Processes, Princeton University Press, Princeton, NJ.
11. Jain, N. C. and Kallianpur, G. (1970) A Note on Uniform Convergence of Stochastic Processes, Ann. Math. Statist. **41**, 1360–1362.
12. Jain, N. C. and Monrad, D. (1981) Gaussian Measures in Certain Function Spaces, Lecture Notes in Math. **860**, 246–256, Springer-Verlag, New York.
13. Kailath, T. (1971) RKHS Approach to Detection and Estimation Problems I: Deterministic Signals in Gaussian Noise, IEEE Trans. Information Theory **17**, 530–549.
14. Kailath, T. and Wienert, H. L. (1975) RKHS Approach to Detection and Estimation Problems II: Gaussian Signal Detection, IEEE Trans. Information Theory **21**, 15–23.
15. Karlin, S. and Taylor, H. M. (1998) An Introduction to Stochastic Modelling, 3-d edition, Academic Press, New York.
16. Koski, T. and Sundar, P. (2001) Two Applications of Reproducing Kernel Hilbert Spaces in Stochastic Analysis, Stochastic in Finite and Infinite Dimensions, 195–206, Trends Math., Birkhäuser Boston, Boston, MA.
17. Le Page, R. (1973) Subgroups of Paths and Reproducing Kernels, Ann. of Probability, **1**, 345–347.
18. Loomis, L. H. (1953) An Introduction to Abstract Harmonic Analysis, D. Van Nostrand, Princeton.
19. Neveu, J. (1965) Mathematical Foundations of the Calculus of Probability, Holden-Day, San Francisco.
20. Parzen, E. (1962) Stochastic Processes, Holden-Day, San Francisco.

21. Parzen, E. (1963) Probability Density Functionals and Reproducing Kernel Hilbert Spaces, in Proceedings of the Symposium on Time Series Analysis, 196, Brown University, M. Rosenblatt, Editor, 155–169, Wiley, New York.
22. Patarasarathy, K. R. (1967) Probability Measures on Metric Spaces, Academic Press, New York.
23. Rudin, W. (1962) Fourier Analysis on Groups, Wiley, New York.
24. Rudin, W. (1966) Real and Complex Analysis, McGraw-Hill, New York.
25. Stein, E. M. and Weiss, G. (1971) Introduction to Fourier Analysis on Euclidean Spaces, Princeton University Press, Princeton.

Error Analysis of a Max-Plus Algorithm for a First-Order HJB Equation*

This paper is dedicated to Prof. Tyrone E. Duncan

William M. McEneaney[1]

Dept. of Mathematics/Dept. of Mechanical and Aerospace Engineering, UC San Diego, La Jolla, CA 92093-0112, USA. http://www4.ncsu.edu/~wmm/, wmm@eos.ncsu.edu

Abstract. The H_∞ problem for a nonlinear system is considered. The corresponding dynamic programming equation is a fully nonlinear, first-order, steady-state partial differential equation (PDE). The computation of the solution of a nonlinear, steady-state, first-order PDE is typically quite difficult. We consider an entirely new class of methods for the obtaining the solution of such PDEs. These methods are based on the linearity of the associated semi-group over the max-plus algebra. In particular, solution of the PDE is reduced to solution of a max-plus eigenvector problem for known unique eigenvalue 0. We consider the error analysis for such an algorithm. The errors are due to both the truncation of the basis expansion and computation of the matrix whose eigenvector one computes.

Keywords: Nonlinear control, H_∞, dynamic programming, numerical methods, partial differential equations, max-plus algebra.

1 Introduction

We consider the H_∞ problem for a nonlinear system. The corresponding dynamic programming equation (DPE) is a fully nonlinear, first-order, steady-state PDE, possessing a term which is quadratic in the gradient (cf. [1], [2], [7]). The solutions are typically nonsmooth, and further, there are multiple viscosity solutions. The computation of the solution of a nonlinear, steady-state, first-order PDE is typically quite difficult, and possibly even more so in the presence of the non-uniqueness mentioned above. In [9], the mathematical background and basic algorithm for a class of numerical methods for such PDEs was discussed. This class of methods employs the max–plus linearity of the associated semi-group. It is a completely new class of methods. The approach is appropriate for PDEs associated with nonlinear (infinite time-horizon) H_∞ conrtrol problems where the Hamiltonian is convex (or concave) over the gradient variable.

In [4], the max–plus linearity of the semi-group associated with a robust filter ([13], [5]) was noted, and provided a key ingredient in the development

* Research partially supported by NSF grant DMS-9971546.

B. Pasik-Duncan: Stochastic Theory and Control, LNCIS 280, pp. 335–351, 2002.

of a numerical algorithm. This linearity had previously been noted in [8]. A second key ingredient (first noted to our knowledge in [4]) was the development of an appropriate basis for the solution space over the max-plus algebra (i.e. with the max–plus algebra replacing the standard underlying field). This reduced the problem of propagation of the solution of the PDE forward in time to max–plus matrix-vector multiplication – with dimension being that of the number of basis functions being used.

Returning to the (design case) H_∞ problem, the associated steady–state PDE is solved to determine whether this is indeed an H_∞ controller with that disturbance attenuation level. (See for instance [1], [15].) The Hamiltonian is concave in the gradient variable. An example of an HJB PDE associated with H_∞ control is (6). There are typically multiple solutions of such PDEs. In [14], for the class of problems considered here, a specific quadratic growth bound was given which isolated this correct solution as the unique, non-negative solution satisfying $0 \leq W(x) \leq C|x|^2$ for a specific C depending on the problem data. The max–plus based methods make use of the fact that the solutions are actually fixed points of the associated semi-group, that is

$$W = S_\tau[W] \tag{1}$$

where S_τ is the semi-group with time-step τ. In this case, one does not actually use the infinitesimal version of the semi-group (the PDE).

The max–plus algebra is a commutative semi-field over $\mathbf{R} \cup \{-\infty\}$ with addition and multiplication given by

$$a \oplus b = \max\{a, b\}, \qquad a \otimes b = a + b. \tag{2}$$

Note that since 0 is the multiplicative identity, we can rewrite the above fixed point equation as

$$0 \otimes W = S_\tau[W]. \tag{3}$$

Since S_τ is linear over the max–plus algebra, one then thinks of W as an infinite-dimensional eigenvector for S_τ corresponding to eigenvalue 0. If one approximates W by some finite-dimensional vector of coefficients in a max–plus basis expansion, then (3) can be re-cast as a finite-dimensional max–plus eigenvector equation (approximating the true solution). Thus, the nonlinear PDE problem is reduced to the solution of a (max-plus) linear eigenvector problem.

Since, in reality, the value function would not have a *finite* max–plus expansion in any but the most unusual cases, we must consider the errors introduced by truncation of the expansion. In [6], the question was addressed in a broad sense. In [11], it was shown that as the number of basis functions increased, the approximation obtained by the algorithm converged to the true value function (assuming perfect computation of the matrix whose eigenvector one wants). We will now obtain some error estimates for the size of the errors introduced by this basis truncation. We also consider errors introduced by the approximation of the elements of the matrix corresponding to the H_∞ problem.

First we need to review some results from [9] and other earlier papers. This is done in Section 2. In Section 3, we obtain a bound on the size of the errors in the computation of the finite–dimensional matrix beyond which, one cannot guarantee that the method will produce an approximate solution. Then in Section 4, we consider the errors in the solution introduced by truncation of the basis functions. In Section 5, we briefly discuss how these error sources combine to obtain overall convergence rates. We also indicate some behavior that is different than that of finite-element methods.

2 Review of the Max–Plus Based Algorithm

We note again that the purpose of this section is to review background needed in later sections; proofs of the results in this section can be found in the references.

We will consider the infinite time–horizon H_∞ problem in the fixed feedback case where the control is built into the choice of dynamics. Recall that the case of active control computation (i.e. the game case) is discussed in [9] and [12]. Consider the system

$$\dot{X} = f(X) + \sigma(X)w, \qquad X(0) = x \tag{4}$$

where X is the state taking values in $\mathbf{R}^m$, f represents the nominal dynamics, the disturbance w lies in $\mathcal{W} \doteq \{w : [0,\infty) \to \mathbf{R}^\kappa : w \in L_2[0,T]\ \forall T < \infty\}$, and σ is an $m \times \kappa$ matrix–valued multiplier on the disturbance.

We will assume that the functions f, σ and l are smooth. We will also assume that $f(0) = 0$ and that there exist $K, c, M, \alpha, \beta \in (0,\infty)$ such that the following hold:

$$(x-y)^T(f(x)-f(y)) \le -c|x-y|^2 \quad \forall x,y \in \mathbf{R}^m, \qquad |f_x(x)| \le K \quad \forall x \in \mathbf{R}^m \tag{A1}$$

$$|\sigma(x)| \le M, \qquad |\sigma^{-1}(x)| \le M, \qquad |\sigma_x(x)| \le K \qquad \forall x \in \mathbf{R}^m \tag{A2}$$

$$l_{xx}(x) \le \beta, \qquad 0 \le l(x) \le \alpha|x|^2 \qquad \forall x \in \mathbf{R}^m \tag{A3}$$

where σ^{-1} is the Moore-Pensose pseudoinverse. The H_∞ available storage function for an attenuation bound γ (alternatively, the value function) is

$$W(x) = \sup_{w \in \mathcal{W}} \sup_{T<\infty} \int_0^T l(X(t)) - \frac{\gamma^2}{2}|w(t)|^2\, dt. \tag{5}$$

The corresponding DPE is

$$0 = -\sup_{w\in\mathbf{R}^\kappa}\left\{[f(x)+\sigma(x)w]^T\nabla W + l(x) - \tfrac{\gamma^2}{2}|w|^2\right\}$$
$$= -\left[\tfrac{1}{2\gamma^2}(\nabla W)^T\sigma(x)\sigma^T(x)\nabla W + f^T(x)\nabla W + l(x)\right]. \tag{6}$$

Since W itself does not appear in (6), one can always scale by an additive constant. It will be assumed throughout that we are looking for a solution satisfying $W(0)=0$. We will also suppose that the above constants satisfy

$$\frac{\gamma^2}{2M^2} > \frac{\alpha}{c^2}. \tag{A4}$$

Alternatively, define the semi–group (with X satisfying (4))

$$S_\tau[W(\cdot)](x) = \sup_{w\in\mathcal{W}}\{\int_0^\tau l(X(t)) - \frac{\gamma^2}{2}|w(t)|^2\,dt + W(X(\tau))\}.$$

Theorem 1. *There exists a unique continuous viscosity solution of (6) in the class*

$$0 \le W(x) \le c\frac{(\gamma-\delta)^2}{2M^2}|x|^2 \tag{7}$$

for sufficiently small $\delta>0$. This solution is given by (5). Further, for any $\tau\in[0,\infty)$, W given in (5) is also the unique solution of $S_\tau[W]=W$, in the class (5).

To the author's best knowledge, the first statement of the following result is due to Maslov [8].

Theorem 2. *The solution operator, S_τ, is linear in the max–plus algebra.*

As noted in the introduction, the above linearity is a key to the development of the algorithms. A second key is the use of the space of semiconvex functions and a max–plus basis for the space. A function ϕ is called **semiconvex** if for every $R<\infty$, there exists C_R such that $\widehat{\phi}(x) \doteq \phi(x)+(C_R/2)|x|^2$ is convex on the ball $B_R \doteq \{x\in\mathbf{R}^n : |x|<R\}$. The infimum over such C_R will be known as the semiconvexity constant for ϕ over B_R. We denote the space of semiconvex functions by $\mathcal{S}$, and the space of semiconvex functions which have semiconvexity constant C_R over B_R by $\mathcal{S}_{R,C_R}$. (The second subscript may sometimes be a positive definite matrix where the condition becomes $\phi(x)+(1/2)x^TC_Rx$ being convex; the case will be clear from context.) We note that given a radius, R, and a semiconvexity constant, C_R, there exists a corresponding finite, minimal (scalar) Lipschitz constant, L_R, such that if $\phi\in\mathcal{S}_{R,C_R}$ then ϕ is Lipschitz over B_R with constant L_R. It is essential that the value, W, lie in this space, and that is given by the next result.

Theorem 3. *W lies in $\mathcal{S}$; for any $R < \infty$, there exists $C_R < \infty$ such that $W \in \mathcal{S}_{R,C_R}$.*

We now turn to the max–plus basis over $\mathcal{S}_{R,C_R}$. Let $\phi \in \mathcal{S}$. Let $\{x_i\}$ be a countable, dense set over $B_{L_R/C_R}(0)$, and let symmetric $C - C_R I > 0$ where (again) $C_R > 0$ is a semiconvexity constant for ϕ over $B_R(0)$. Define $\psi_i(x) \doteq -\frac{1}{2}(x - x_i)^T C(x - x_i)$ for each i. Then one finds

$$\phi(x) = \bigoplus_{i=1}^{\infty} [a_i \otimes \psi_i(x)] \qquad \forall x \in B_R \tag{8}$$

where

$$a_i \doteq -\max_{x \in B_R} [\psi_i(x) - \phi(x)]. \tag{9}$$

This is a countable max–plus basis expansion for ϕ. More generally, the set $\{\psi_i\}$ forms a max–plus basis for the space of semiconvex functions over $B_R(0)$ with semiconvexity constant, C_R, i.e. $\mathcal{S}_{R,C_R}$. We now have the following.

Theorem 4. *Given $R < \infty$, there exists semiconvexity constant $C_R < \infty$ for W over $B_R(0)$, and a corresponding Lipschitz constant, L_R. Let $C - C_R I > 0$ and $\{x_i\}$ be dense over $B_{L_R/C_R}(0)$, and define the basis $\{\psi_i\}$ as above. Then*

$$W(x) = \bigoplus_{i=1}^{\infty} [a_i \otimes \psi_i(x)] \qquad \forall x \in B_R \tag{10}$$

where

$$a_i \doteq -\max_{x \in B_R} [\psi_i(x) - W(x)]. \tag{11}$$

For the remainder of the section, fix any $\tau \in (0, \infty)$. We assume throughout this section that one may choose C such that $C - C_R I > 0$ and such that

$$S_\tau[\psi_i] \in \mathcal{S}_{R,C} \text{ for all } i. \tag{A5}$$

We will not discuss this assumption here, but simply note that we have verified that this assumption holds for several problems where candidate values of C were obtained by requiring solution of a certain Riccati inequality everywhere in B_R. We also note that this assumption will need to be replaced by a slightly stricter assumption ($A5'$) in Section 4 for the results there and beyond.

Previously, rather than proving convergence results for the algorithms, drastic assumptions were made so that the basic concept could be presented, while still keeping the paper to a reasonable length. In particular, it was simply assumed that there was a finite set of basis functions, $\{\psi_i\}_{i=1}^n$, such that W had a finite max–plus basis expansion over B_R in those functions, that is,

$$W(x) = \bigoplus_{i=1}^{n} e_i \otimes \psi_i, \tag{12}$$

and we let $e^T \doteq (e_1\ e_2 \cdots e_n)$, and $B_{j,i} = -\max_{x\in B_R}(\psi_j(x) - S_\tau(\psi_i(x)))$. Note that B actually depends on τ, but we suppress the dependence in the notation. We made the further drastic assumption that for each $j \in \{1, 2, \ldots, n\}$, $S_\tau[\psi_j]$ also had a finite basis expansion in the same set of basis functions, $\{\psi_i\}_{i=1}^n$, so that

$$S_\tau[\psi_j(x)] = \bigoplus_{i=1}^{n} B_{j,i} \otimes \psi_i(x) \tag{13}$$

for all $x \in B_R(0)$. Although these assumptions are obviously unrealistic, they allow one to easily see the structure of the approach. Specifically, under (12), (13) one has

Theorem 5. *$S_\tau[W] = W$ if and only if $e = B \otimes e$ where $B \otimes e$ represents max–plus matrix multiplication.*

Suppose that one has computed B exactly (errors in B will also be considered below). We should note that B has a unique eigenvalue, although possibly many eigenvectors corresponding to that eigenvalue [3]. By the above results, this eigenvalue must be zero. One can compute the eigenvector via the power method. In the power method, one computes an eigenvector, e by

$$e = \lim_{N\to\infty} B^N \otimes \mathbf{0}$$

where the power is to be understood in the max–plus sense and $\mathbf{0}$ is the zero vector. Throughout, the paper, we let the $\{x_j\}$ be such that $x_1 = 0$. Since this is simply an approach to arrangement of the basis functions, we do not annotate it as an assumption. The fact that the power method works under (12), (13) is encapsulated in the following sequence of results.

Theorem 6. *There exists $\delta > 0$ such that the following holds. Let $N \in \{1, 2, \ldots, n\}$, $\{k_i\}_{i=1}^{i=N+1}$ such that $1 \le k_i \le n$ for all i and $k_{N+1} = k_1$. Suppose we are not in the case $k_i = 1$ for all i. Then*

$$\sum_{i=1}^{N} B_{k_i,k_{i+1}} \le -\delta.$$

The structure given in Theorem 6 (combined with the fact that $B_{1,1} = 0$) leads to the following.

Theorem 7. *$\lim_{N\to\infty} B^N \otimes 0$ exists, converges in a finite number of steps, and satisfies $e = B \otimes e$.*

Corollary 1. *There is a unique max–plus eigenvector, e up to a max–plus multiplicative constant, and of course, this is the output of the above power method.*

Thus, under the drastic assumptions above, one finds that the power method converges to the unique solution of the eigenvector problem (in a finite number of steps), and that this eigenvector is the finite set of coefficients in the max–plus basis expansion of the value function, W. The next sections will deal with the facts that we actually need to truncate infinite basis expansions, and that the computations of the elements of B are only approximate. Error analysis will be performed.

3 Allowable Errors in Computation of B

The guaranteed convergence of the power method relies on Theorem 6 since this implies a certain structure to a directed graph associated with B (see [9], [12]). If there was a sequence $\{k_i\}_{i=1}^{i=N+1}$ such that $1 \le k_i \le n$ for all i and $k_{N+1} = k_1$ such that one does not have $k_i = 1$ for all i, and such that $\sum_{i=1}^{N} B_{k_i,k_{i+1}} \ge 0$ then there would be no guarantee of convergence of the power method (nor the ensuing uniqueness result for that matter). In order to determine more exactly, the allowable errors in the computation of the elements of B, we need to obtain a more exact expression for the δ that appears in Theorem 6. We now proceed to do this.

Lemma 1. *Let X satisfy (4) with initial state $X(0) = x \in \mathbf{R}^n$. Let $K, \tau \in (0, \infty)$ and any $w \in L_2[0, \tau]$. Suppose $\delta > 0$ sufficiently small so that*

$$\delta \le KM^2/[c(1 - e^{-c\tau})]. \tag{14}$$

Then

$$K|X(\tau) - x|^2 + \delta \|w\|^2_{L_2[0,\tau]} \ge \frac{\delta c}{8M^2} |x|^2 \left(1 - e^{-c\tau}\right)^4.$$

Remark 1. It may be of interest to note that the assumption on the size of δ does not seem necessary. At one point in the proof to follow, this assumption is used in order to eliminate a case which would lead to a more complex expression on the right-hand side in the result in the lemma statement. If some later technique benefited from not having such an assumption, the lemma proof could be revisited in order to eliminate it. However, at this point, that would seem to be a needless technicality.

PROOF. Note that by (4) and Assumptions $(A1)$ and $(A2)$,

$$\frac{d}{dt}|X|^2 \le -2c|X|^2 + 2M|X||w| \le -c|X|^2 + \tfrac{M^2}{c}|w|^2. \tag{15}$$

Consequently, for any $t \in [0, \tau]$,

$$|X(t)|^2 \le e^{-ct}|x|^2 + \tfrac{M^2}{c} \int_0^t |w(r)|^2 \, dr$$

and so

$$\|w\|^2_{L_2(0,t)} \geq \tfrac{c}{M^2}\Big[|X(t)|^2 - |x|^2\Big] \qquad \forall t \in [0,\tau]. \tag{16}$$

We may suppose

$$|X(t)| \leq \sqrt{1+(1-e^{-ct})^4/2}\,|x| \qquad \forall t \in [0,\tau]. \tag{17}$$

Otherwise by (16) and the reverse of (17), there exists $t \in [0,\tau]$ such that

$$K|X(t)-x|^2 + \delta\|w\|^2_{L_2[0,\tau]} \geq \delta\|w\|^2_{L_2[0,t]} \geq \tfrac{\delta c}{2M^2}(1-e^{-c\tau})^4|x|^2 \tag{18}$$

in which case one already has the result. Define $\overline{K} \doteq \sqrt{1+(1-e^{-c\tau})^4/2}$.

Recalling (15), and applying (17), one has

$$\frac{d}{dt}|X(t)|^2 \leq -2c|X(t)|^2 + 2M\overline{K}|x||w(t)|.$$

Solving this ODI for $|X(t)|^2$, and using the Hölder inequality, yields the bound

$$|X(\tau)|^2 \leq |x|^2 e^{-2c\tau} + \frac{M\overline{K}|x|\|w\|}{\sqrt{c}}\big(1-e^{-4c\tau}\big)^{1/2}. \tag{19}$$

This implies

$$|X(\tau)| \leq |x|e^{-c\tau} + \frac{1}{c^{1/4}}\sqrt{M\overline{K}|x|\|w\|}\big(1-e^{-4c\tau}\big)^{1/4}. \tag{20}$$

We will consider two cases separately. First we consider the case where $|X(\tau)| \leq |x|$. Then, by (20)

$$|X(\tau)-x| \geq |x| - |X(\tau)| \geq |x|(1-e^{-c\tau}) - \frac{1}{c^{1/4}}\sqrt{M\overline{K}|x|\|w\|}\big(1-e^{-4c\tau}\big)^{1/4}. \tag{21}$$

Now note that for general $a, b, c \in \mathbf{R}$, $a + c \geq b$ implies

$$a^2 \geq \frac{b^2}{2} - c^2. \tag{22}$$

By (21) and (22) (and noting the non-negativity of the norm),

$$|X(\tau)-x|^2 \geq \max\left\{\tfrac{1}{2}|x|^2(1-e^{-c\tau})^2 - \frac{M\overline{K}}{\sqrt{c}}|x|\|w\|\big(1-e^{-4c\tau}\big)^{1/2}, 0\right\}$$

which implies

$$K|X(\tau)-x|^2 + \delta\|w\|^2 \geq \max\left\{\frac{K}{2}|x|^2(1-e^{-c\tau})^2 - \frac{KM\overline{K}}{\sqrt{c}}|x|\|w\|\big(1-e^{-4c\tau}\big)^{1/2} + \delta\|w\|^2, \delta\|w\|^2\right\}. \tag{23}$$

The right hand side of (23) is a maximum of two convex quadratic functions of $\|w\|$. The second is monotonically increasing, while the first is positive at $\|w\| = 0$ and initially decreasing. This implies that there are two possibilities for the location of the minimum of the maximum of the two functions. If the minimum of the first function is to the left of the point where the two functions intersect, then the minimum occurs at the minimum of the first function; alternatively it occurs where the two functions intersect. The minimum of the first function occurs at $\|w\|_{\min}$ (where we are abusing notation here, using the *min* subscript on the norm to indicate the value of $\|w\|$ at which the minimum occurs), and this is given by

$$\|w\|_{\min} = \frac{KM\overline{K}|x|(1-e^{-4c\tau})^{1/2}}{2\sqrt{c}\delta}. \tag{24}$$

The point of intersection of the two functions occurs at

$$\|w\|_{\rm int} = \frac{\sqrt{c}|x|(1-e^{-c\tau})^2}{2M\overline{K}(1-e^{-4c\tau})^{1/2}}. \tag{25}$$

The two points coincide when

$$\delta = \frac{KM^2\overline{K}^2(1-e^{-4c\tau})}{c(1-e^{-c\tau})^2} = \frac{KM^2[1+(1-e^{-c\tau})^4/2](1-e^{-4c\tau})}{c(1-e^{-c\tau})^2},$$

and $\|w\|_{\rm int}$ occurs to the left of $\|w\|_{\min}$ for δ less than this. It is easy to see that assumption (14) implies that δ is less than the value at which the points coincide, and consequently, the minimum of the right hand side of (23) occurs at $\|w\|_{\rm int}$.

Using the value of the right hand side of (23) corresponding to $\|w\|_{\rm int}$, we find that for any disturbance, w,

$$K|X(\tau)-x|^2 + \delta\|w\|^2 \geq \frac{\delta c|x|^2}{4M^2\overline{K}^2}\frac{(1-e^{-c\tau})^4}{(1-e^{-4c\tau})}$$

which, using definition of $\overline{K}$

$$= \frac{\delta c|x|^2}{4M^2}\frac{(1-e^{-c\tau})^4}{(1-e^{-4c\tau})[1+(1-e^{-c\tau})^4/2]} \geq \frac{\delta c|x|^2}{8M^2}(1-e^{-c\tau})^4. \tag{26}$$

Now we turn to the second case,

$$|X(\tau)| > |x|. \tag{27}$$

In this case, (27) and (20) yield

$$|x|e^{-c\tau} + \frac{1}{c^{1/4}}\sqrt{M\overline{K}|x|\|w\|}\left(1-e^{-4c\tau}\right)^{1/4} > |x|. \tag{28}$$

Upon rearrangement, (28) yields

$$\|w\| > \frac{\sqrt{c}|x|}{M\overline{K}}\frac{(1-e^{-c\tau})^2}{(1-e^{-4c\tau})^{1/2}}.$$

Consequently, using the definition of $\overline{K}$ and some simple manipulations,

$$K|X(\tau)-x|^2+\delta\|w\|^2 \geq \frac{\delta c|x|^2(1-e^{-c\tau})^4}{M^2(1-e^{-4c\tau})[1+(1-e^{-c\tau})^4/2]}$$
$$\geq \frac{\delta c|x|^2}{2M^2}(1-e^{-c\tau})^4. \tag{29}$$

Combining (26) and (29) completes the proof. □

Now we turn to how Lemma 1 can be used to obtain a more detailed replacement for the δ that appears in Theorem 6. Fix $\tau > 0$. Let

$$\hat{\gamma}_0^2 \in \left(\frac{2M^2\alpha}{c^2}, \gamma^2\right), \tag{30}$$

and in particular, let $\hat{\gamma}_0^2 = \gamma^2 - \delta$ where δ is sufficiently small so that

$$\delta < \gamma^2 - \frac{2M^2\alpha}{c^2}. \tag{31}$$

Then all results of Section 2 for W hold with γ^2 replaced by $\hat{\gamma}_0^2$, and we denote the corresponding value by $W^{\hat{\gamma}_0}$. In particular, by Theorem 4, for any $R < \infty$ there exists semiconvexity constant $C_R^0 < \infty$ for $W^{\hat{\gamma}_0}$ over $B_R(0)$, and a corresponding Lipschitz constant, L_R^0. Note that the required constants satisfy $C_R^0 < C_R$ (see proof of Theorem 4 as given in [9]). We will be solving for W over $B_R(0)$ for some $R < \infty$, and we now modify our basis functions by choosing $\{x_i\}$ to be dense over $B_{L_R^0/C_R^0}(0)$. For $C - C_R I > 0$ (which implies $C - C_R^0 I > 0$), let the basis functions be

$$\psi_i(x) \doteq -\frac{1}{2}(x-x_i)^T C(x-x_i)$$

where again we require $x_1 = 0$, the origin. Then, as before, the set $\{\psi_i\}$ forms a max–plus basis for the space of semiconvex functions over $B_R(0)$ with semiconvexity constant, C_R^0, i.e. $\mathcal{S}_{R,C_R^0}$.

For any j, let

$$\overline{x}_j \in \operatorname{argmax}\{\psi_j(x) - W^{\hat{\gamma}_0}(x)\}. \tag{32}$$

Then for any x,

$$\psi_j(x) - \psi_j(\overline{x}_j) \leq W^{\hat{\gamma}_0}(x) - W^{\hat{\gamma}_0}(\overline{x}_j) - K_0|x-\overline{x}_j|^2 \tag{33}$$

where $K_0 > 0$ is the minimum eigenvalue of $C - C_R^0 I > 0$. Note that K_0 depends on $\hat{\gamma}_0$.

Theorem 8. *Let $\delta \doteq \frac{\gamma^2}{2} - \frac{\hat{\gamma}_0^2}{2} > 0$ with $\hat{\gamma}_0$ satisfying (30) and such that (14) holds for δ. Let K_0 be as in (33). Then, for any $j \neq 1$,*

$$B_{j,j} \leq \frac{-\delta c|\overline{x}_j|^2}{8M^2}(1-e^{-c\tau})^4.$$

(Recall that by the choice of ψ_1 as the basis function centered at the origin, $B_{1,1} = 0$; see Theorem 6.)

PROOF. Let K_0, τ, δ satisfy the assumptions (i.e. (14), (31), (33)). Then

$$S_\tau[\psi_j](\overline{x}_j) - \psi_j(\overline{x}_j) = \sup_{w\in L_2} \left\{ \int_0^\tau l(X(t)) - \frac{\gamma^2}{2}|w(t)|^2\,dt + \psi_j(X(\tau)) - \psi_j(\overline{x}_j) \right\} \tag{34}$$

where X satisfies (4) with $X(0) = \overline{x}_j$. Let $\varepsilon > 0$ and w^ε be ε–optimal . Then this implies

$$S_\tau[\psi_j](\overline{x}_j) - \psi_j(\overline{x}_j) \le \int_0^\tau l(X^\varepsilon(t)) - \frac{\gamma^2}{2}|w^\varepsilon(t)|^2\,dt + \psi_j(X^\varepsilon(\tau)) - \psi_j(\overline{x}_j) + \varepsilon,$$

and by (33) and the definition of $\hat{\gamma}_0$

$$\le \int_0^\tau l(X^\varepsilon(t)) - \frac{\hat{\gamma}_0^2}{2}|w^\varepsilon(t)|^2 - \delta|w^\varepsilon(t)|^2\,dt + W^{\hat{\gamma}_0}(X^\varepsilon(\tau)) - W^{\hat{\gamma}_0}(\overline{x}_j)$$
$$-K_0|X^\varepsilon(\tau) - \overline{x}_j|^2 + \varepsilon$$

and by Theorem 1 (for $W^{\hat{\gamma}_0}$),

$$\le -\delta\|w^\varepsilon\|^2_{L_2[0,\tau]} - K_0|X^\varepsilon(\tau) - \overline{x}_j|^2 + \varepsilon.$$

Combining this with Lemma 1 yields

$$S_\tau[\psi_j](\overline{x}_j) - \psi_j(\overline{x}_j) \le \frac{-\delta c|\overline{x}_j|^2}{8M^2}\left(1 - e^{-c\tau}\right)^4 + \varepsilon.$$

Since this is true for all $\varepsilon > 0$, one has

$$S_\tau[\psi_j](\overline{x}_j) - \psi_j(\overline{x}_j) \le \frac{-\delta c|\overline{x}_j|^2}{8M^2}\left(1 - e^{-c\tau}\right)^4. \tag{35}$$

But, using (35),

$$B_{j,j} = \min_x \{S_\tau[\psi_j](x) - \psi_j(x)\} \le \frac{-\delta c|\overline{x}_j|^2}{8M^2}\left(1 - e^{-c\tau}\right)^4. \quad \square \tag{36}$$

Theorem 9. *Let K_0 be as in (33), and let $\delta > 0$ be given by*

$$\delta = \min\left\{ \frac{K_0 M^2}{c}, \frac{\gamma^2}{2} - \frac{\hat{\gamma}_0^2}{2} \right\} \tag{37}$$

(which is somewhat tighter than the requirement in the previous theorem). Let $N \in \mathcal{N}$, $\{k_i\}_{i=1}^{i=N+1}$ such that $1 \le k_i \le n$ for all i and $k_{N+1} = k_1$. Suppose we are not in the case $k_i = 1$ for all i. Then

$$\sum_{i=1}^N B_{k_i,k_{i+1}} \le -max_{k_i}|\overline{x}_{k_i}|^2 \frac{\delta c}{8M^2}\left(1 - e^{-cN\tau}\right)^4.$$

We only sketch the proof. By Theorem 8, this is true for $N = 1$. For $N > 1$, the proof relies on repeated application of S_τ and the monotonicity of the semi–group in the sense that if $g_1(x) \le g_2(x)$ for all x, then $S_\tau[g_1](x) \le S_\tau[g_2](x)$ for all x.

The convergence of the power method (described in the previous section) relied on a certain structure of B ($B_{1,1}=0$ and strictly negative loop sums as described in the assumptions of Theorem 6). Combining this with the above result on the size of loop sums, one can obtain a condition which guarantees convergence of the power method to a unique eigenvector corresponding to eigenvalue zero. This is given in the next theorem.

Theorem 10. *Let B be given by $B_{j,i} = -\max_{x\in B_R}\big(\psi_j(x) - S_\tau[\psi_i](x)\big)$ for all $i,j \le n$, and let $\widetilde{B}$ be an approximation of B with $\widetilde{B}_{1,1}=0$ and such that there exists $\varepsilon > 0$ such that*

$$|\widetilde{B}_{i,j} - B_{i,j}| \le \max\{|\overline{x}_i|^2, |\overline{x}_j|^2\} \left(\frac{\delta c}{8M^2}\right) \frac{\left(1-e^{-c\tau}\right)^4}{n^2} - \varepsilon \tag{38}$$

$\forall i,j$ such that $(i,j)\neq(1,1)$ where

$$\delta = \min\left\{\frac{K_0 M^2}{c}, \frac{\gamma^2}{2} - \frac{\hat{\gamma}_0^2}{2}\right\}. \tag{39}$$

Then the power method applied to $\widetilde{B}$ converges in a finite number of steps to the unique eigenvector $\widetilde{e}$ corresponding to eigenvalue zero, that is

$$\widetilde{e} = \widetilde{B} \otimes \widetilde{e}.$$

If the conditions of Theorem 10 are met, then one can ask what the size of the errors in the corresponding eigenvector are. Specifically, if eigenvector $\tilde{e}$ is computed using approximation $\widetilde{B}$, what is a bound on the size of the difference between e (the eigenvector of B) and $\tilde{e}$? The following theorem gives a rough, but easily obtained, bound. In particular, the proof follows by noting that the sum around any loop of the matrix other than the trivial loop, $B_{1,1}$, is strictly negative.

Theorem 11. *Let B be given by $B_{i,j} = -\max_{x\in B_R}\big(\psi_j(x) - S_\tau[\psi_i](x)\big)$ for all $i,j \le n$, and let $\widetilde{B}$ be an approximation of B with $\widetilde{B}_{1,1}=0$ and such that there exists $\varepsilon > 0$ such that*

$$|\widetilde{B}_{i,j} - B_{i,j}| \le \max\{|\overline{x}_i|^2, |\overline{x}_j|^2\} \left(\frac{\delta c}{8M^2}\right) \frac{\left(1-e^{-c\tau}\right)^4}{n^\mu} - \varepsilon \qquad \forall i,j \tag{40}$$

where $\mu \in \{2,3,4,...\}$ and δ is given by (39). Let e and $\tilde{e}$ be the eigenvectors of B and $\widetilde{B}$ respectively. Then,

$$\|e-\tilde{e}\| \doteq \max_i |e_i - \tilde{e}_i| \le \left(L_R^0/C_R^0\right)^2 \left(\frac{\delta c}{8M^2}\right) \frac{\left(1-e^{-c\tau}\right)^4}{n^{\mu-2}} - \varepsilon.$$

We remark that by taking ε sufficiently small, and noting that $1-e^{-c\tau} \le c\tau$ for nonnegative τ, Theorem 11 implies (under its assumptions)

$$\|e-\tilde{e}\| = \max_i |e_i - \tilde{e}_i| \le \left(L_R^0/C_R^0\right)^2 \left(\frac{\delta c^5}{16M^2}\right) \frac{\tau^4}{n^{\mu-2}}. \tag{41}$$

4 Convergence and Truncation Errors

In this section we consider the approximation due to using only a finite number of functions in the max–plus basis expansion. It will be shown that as the number of functions increases (in a reasonable way), the approximate solution obtained by the eigenvector computation of Section 2 converges from below to the value function, W. Specific truncation error bounds are obtained. The estimates may be rather conservative due to the form of the truncation error bound used; this question will become more clear below. Note that these are only the errors due to truncation to a finite number of basis functions; as noted above, analysis of the errors due to approximation of the entries in the B matrix is discussed further below.

Recall that we choose the basis functions throughout such that $x_1^{(n)} = 0$, or in other words, $\psi_1^{(n)}(x) = \frac{-1}{2} x^T C x$ for all n. We will use the notation $S_\tau^{(n)}$ to indicate the projection of the semi-group down into the finite dimensional space spanned by the basis functions (so $S_\tau^{(n)}$ is equivalent to B), and let $\phi_0^{(n)}$ be the finite dimensional approximation of ϕ_0 which is the initial starting function for the algorithm (one may take $\phi_0(x) \equiv 0$). Then define

$$W_{N,\tau}^{(n)}(x) \doteq {S_\tau^{(n)}}^N [\phi_0^{(n)}](x)$$

where the N superscript indicates repeated application of the operator N times. Note that since the power method converges in a finite number of steps, $W_{N,\tau}^{(n)}$ converges to some $W^{(n)^\infty}$ in a finite number of steps (where $W^{(n)^\infty}(x) = \bigoplus_i e_i \otimes \psi_i(x)$).

To specifically set C, we will replace Assumption $(A5)$ of Section 2 with the following. We assume throughout the remainder of the paper that one may choose matrix $C > 0$ and $\delta' \in (0,1)$ such that with $C' \doteq (1-\delta')C$

$$S_\tau[\psi_i] \in \mathcal{S}_{R,C'} \text{ for all } i. \tag{$A5'$}$$

Again, we simply note that we have verified that this assumption holds for some problems where candidate values of C were obtained by requiring solution of a certain Riccati inequality everywhere in B_R. Also note that one could be more general, allowing C' to be a more general positive definite symmetric matrix such that $C - C' > 0$, but we will not include that here. Finally, it should be noted that δ' would depend on τ; as $\tau \downarrow 0$, one would need to take $\delta' \downarrow 0$. Since δ' will appear in the denominator of the error bound of the next lemma (as well as implicitly in the denominator of the fraction on the right–hand side of the error bound in Theorem 12), this implies that one does not want to take $\tau \downarrow 0$ as the means for reducing the errors. This will be discussed further in the next section.

Recall that given any set of semiconvex functions $\mathcal{S}_{R,C'}$ over $B_R(0)$ for any $R < \infty$, there exists a corresponding Lipschitz constant, which we denote by

$L' = L'(C', R)$ such that all $\phi \in \mathcal{S}_{R,C'}$ are Lipschitz over $B_R(0)$ with constant L'.

The following lemma is a general result about the errors due to truncation when using the above max–plus basis expansion. The proof is quite tedious, and so we do not include it here.

Lemma 2. *Let δ', C' be as in Assumption (A5′), and let $\phi \in \mathcal{S}_{R,C'}$ with $\phi(0) = 0$, ϕ differentiable at zero with $\nabla_x \phi(0) = 0$, and $-\frac{1}{2} x^T C' x \leq \phi(x) \leq \frac{1}{2}\widehat{\mathcal{M}}|x|^2$ for all x for some $\widehat{\mathcal{M}} < \infty$. Let $\{\psi_i\}_{i=1}^n$ consist of basis functions with matrix C, centers $\{x_i\} \subseteq \overline{B}_{|L'(C')^{-1}|}(0)$ such that $C - C'I > 0$, and let $\Delta \doteq \max_{x \in \overline{B}_{|L'(C')^{-1}|}(0)} \min_i |x - x_i|$ Let*

$$\phi^\Delta(x) = \max_i [a_i + \psi_i(x)] \qquad \forall x \in B_R(0)$$

where

$$a_i = -\max_{x \in \overline{B}_R(0)} [\psi_i(x) - \phi(x)] \qquad \forall i.$$

Then

$$0 \leq \phi(x) - \phi^\Delta(x) \leq \begin{cases} |C|\left[2\widehat{\beta} + 1 + |C|/(\delta' C_R)\right]|x|\Delta & if |x| \geq \Delta \\ \frac{1}{2}\left[\widehat{\mathcal{M}} + |C|\right]|x|\Delta & otherwise \end{cases}$$

where $\widehat{\beta}$ is specified in the proof.

In order to apply this general lemma concerning max–plus basis expansions to the problem at hand, one must consider the effect of repeated application of the truncated operator $S_\tau^{(n)}$. Note that $S_\tau^{(n)}$ may be written as the composition of S_τ and a truncation operator, $\mathcal{T}^{(n)}$ where we have $\mathcal{T}^{(n)}[\phi] = \phi^\Delta$ in the notation of the previous lemma. In other words, one has the following equivalence of notation

$$S_\tau^{(n)}[\phi] = \{\mathcal{T}^{(n)} \circ S_\tau\}[\phi] = \{S_\tau[\phi]\}^\Delta. \tag{42}$$

We now proceed to consider how truncation errors accumulate. Let

$$\mathcal{M}_{C'} \doteq \max\left\{|C|\left[2\widehat{\beta} + 1 + |C|/(\delta' C_R)\right],\ \tfrac{1}{2}\left[\widehat{\mathcal{M}} + |C|\right]\right\}.$$

Fix Δ. We suppose that we have n sufficiently large (with properly distributed basis function centers) so that

$$\max_{x \in \overline{B}_{|L'(C')^{-1}|}(0)} \min_i |x - x_i| \leq \Delta.$$

Let ϕ_0 satisfy the conditions on ϕ in Lemma 2. (One can simply take $\phi_0 \equiv 0$.) Then, by Lemma 2,

$$\phi_0(x) - \mathcal{M}_{C'}|x|\Delta \leq \phi_0^{(n)}(x) \leq \phi_0(x) \qquad \forall x \in \overline{B}_R(0). \tag{43}$$

Now, for any $x \in \overline{B}_R(0)$, let $w_x^{1,\overline{\varepsilon}}$ be $\overline{\varepsilon}/2$–optimal for $S_\tau[\phi_0](x)$, and let $X_x^{1,\overline{\varepsilon}}$ be the corresponding trajectory. Then, using (43),

$$\begin{aligned}0 \leq\ & S_\tau[\phi_0](x) - S_\tau[\phi_0^{(n)}](x) \\ \leq\ & \phi_0(X_x^{1,\overline{\varepsilon}}(\tau)) - \phi_0^{(n)}(X_x^{1,\overline{\varepsilon}}(\tau)) + \frac{\overline{\varepsilon}}{2} \leq \mathcal{M}_{C'}|X_x^{1,\overline{\varepsilon}}(\tau)|\Delta + \frac{\overline{\varepsilon}}{2}.\end{aligned} \tag{44}$$

Proceeding along, one then finds

$$\begin{aligned}0 \leq\ & S_\tau[\phi_0](x) - S_\tau^{(n)}[\phi_0^{(n)}](x) \\ =\ & S_\tau[\phi_0](x) - S_\tau[\phi_0^{(n)}](x) + S_\tau[\phi_0^{(n)}](x) - S_\tau^{(n)}[\phi_0^{(n)}](x)\end{aligned}$$

which by Lemma 2, Assumption $(A5')$, and (44)

$$\leq \mathcal{M}_{C'}|X_x^{1,\overline{\varepsilon}}(\tau)|\Delta + \mathcal{M}_{C'}|x|\Delta + \frac{\overline{\varepsilon}}{2}. \tag{45}$$

One can continue this process using $\overline{\varepsilon}/4$–optimal w over the next time–step, $\overline{\varepsilon}/8$–optimal w over the time–step following that, and so on. An induction argument yields the following.

Lemma 3. *One has*

$$0 \leq S_{N\tau}[\phi_0](x) - {S_\tau^{(n)}}^N[\phi_0](x) \leq \mathcal{M}_{C'}\Delta \sum_{i=0}^{N} |\overline{X}_x^{N,\overline{\varepsilon}}(i\tau)| + \sum_{i=1}^{N} \frac{\overline{\varepsilon}}{2^i}. \tag{46}$$

where the construction of $\overline{\varepsilon}$–optimal $\overline{X}_x^{N,\overline{\varepsilon}}(\cdot)$ by induction follows in the obvious way as above.

Using this lemma, one may obtain the following theorem in a straight–forward manner.

Theorem 12. *Let $\{\psi_i\}_{i=1}^n$, C' and Δ be as in Lemma 2. Then, there exists $\overline{m}, \overline{\lambda} \in (0,\infty)$ such that*

$$0 \leq W(x) - {W^{(n)}}^\infty(x) \leq \mathcal{M}_{C'}\left(\frac{e^{\overline{m}}}{1-e^{-\overline{\lambda}\tau}}\right)|x|\Delta \qquad \forall x \in \overline{B}_R(0). \tag{47}$$

5 Error Summary

We now combine the elements of the error analyses of the previous sections.

Note that [11] contains a short discussion of computing an approximation $\widehat{B}(\tau)$ to $B(\tau)$ where we are now explicitly indicating the dependence of B on τ. From Section 3, a sufficient condition on the size of the errors in this approximation which guarantees convergence of the power method (and accompanying results) with approximation $\widehat{B}(\tau)$ to $B(\tau)$, is (using a particular value for ε)

$$|\widehat{B}_{i,j}(\tau) - B_{i,j}(\tau)| \leq \max\{|\overline{x}_i|^2, |\overline{x}_j|^2\}\left(\frac{\delta c}{9M^2}\right)\frac{\left(1-e^{-c\tau}\right)^4}{n^\mu} \qquad \forall i,j \tag{48}$$

for some $\mu \in \{2,3,\cdots\}$. Note that (48) is a bound on the allowable error sizes—not the errors in the computation/approximation of $B(\tau)$. A Runge–Kutta technique may be applied to actually compute the components of $\widehat{B}(\tau)$ [11]. Then, as in Theorem 11, one obtains a corresponding error in the eigenvector.

As discussed in Section 4, there are further errors in the eigenvector due to the truncation of the basis expansion. Note that these errors go down linearly with $1/N_D$ where N_D is the number of basis functions per space dimension. In order for this to be useful, the errors in $\widehat{B}$ must also be reduced at the correct rate. However, we note that one does not want to reduce τ as one increases N_D; this is in contradistinction to finite-element methods. This is easily seen by considering (48) where the errors grow like $1/\tau$ for small τ. Thus, although the errors in $\widehat{B}(\tau) - B(\tau)$ can be reduced by reduction of τ, this is not desirable since that would have the reverse effect on the truncation errors (48). An obvious solution is to fix τ while using an increasingly fine mesh in the Runge–Kutta (or analogous) technique used in computation of $\widehat{B}(\tau)$.

Lastly, it is conjectured that if the solution is C^1 rather than merely a continuous viscosity solution, then the errors would drop like $(1/N_D)^2$ rather than $1/N_D$.

Thanks

The author would like to thank Professors Wendell H. Fleming and Matthew R. James for helpful discussions at Australian National University.

References

1. T. Basar and P. Bernhard, H_∞–Optimal Control and Related Minimax Design Problems, Birkhäuser (1991).
2. J. A. Ball and J. W. Helton, H_∞ control for nonlinear plants: connections with differential games, Proc. 28th IEEE Conf. Dec. Control, Tampa FL (1989), 956–962.
3. F. L. Baccelli, G. Cohen, G.J. Olsder and J.-P. Quadrat, Synchronization and Linearity, John Wiley (1992).
4. W. H. Fleming and W. M. McEneaney, A max–plus based algorithm for an HJB equation of nonlinear filtering, SIAM J. Control and Optim., 38 (2000), pp. 683–710.
5. W. H. Fleming, Deterministic nonlinear filtering, Annali Scuola Normale Superiore Pisa, Cl. Scienze Fisiche e Matematiche, Ser. IV, **25** (1997), 435–454.
6. E. Gallestey, M. R. James and W. M. McEneaney, Max–plus approximation methods in partially observed H_∞ control, 38th IEEE Conf. on Decision and Control, 3011–3016.
7. M. R. James, A partial differential inequality for dissipative nonlinear systems, Systems and Control Letters, **21** (1993) 315–320.

8. V. P. Maslov, On a new principle of superposition for optimization problems, Russian Math. Surveys, **42** (1987) 43–54.
9. W. M. McEneaney, Max-Plus Eigenvector Representations for Solution of Nonlinear H_∞ Problems: Basic Concepts, Submitted to IEEE Trans. Auto. Control.
10. W. M. McEneaney, Error Analysis of a Max-Plus Algorithm for a First-Order HJB Equation, Proc. Workshop On Max-Plus Algebras and Their Applications to Discrete-event Systems, Theoretical Computer Science, and Optimization, Prague 27-29 August 2001.
11. W. M. McEneaney, Convergence and error analysis for a max–plus algorithm, Proc. 39th IEEE Conf. on Decision and Control (2000), 1194–1199.
12. W. M. McEneaney and M. Horton, Max–plus eigenvector representations for nonlinear H_∞ value functions, 37th IEEE Conf. on Decision and Control (1998), 3506–3511.
13. W. M. McEneaney, Robust/H_∞ filtering for nonlinear systems, Systems and Control Letters, Vol. 33 (1998), 315–325.
14. W. M. McEneaney, A Uniqueness result for the Isaacs equation corresponding to nonlinear H_∞ control, Math. Controls, Signals and Systems, Vol. 11 (1998), 303–334..
15. P. Soravia, H_∞ control of nonlinear systems: differential games and viscosity solutions, SIAM J. Control and Optim., **34** (1996), 1071–1097.

Optimal Strategies for Ergodic Control Problems Arising from Portfolio Optimization

Hideo Nagai

Department of Mathematical Science, Graduate School of Engineering Science, Osaka University, Toyonaka, 560-8531, Japan. nagai@sigmath.es.osaka-u.ac.jp

Abstract. We consider constructing optimal strategies for risk-sensitive portfolio optimization problems on an infinite time horizon for general factor models, where the mean returns and the volatilities of individual securities or asset categories are explicitly affected by economic factors. The factors are assumed to be general diffusion processes. In studying the ergodic type Bellman equations of the risk-sensitive portfolio optimization problems we introduce some auxiliary classical stochastic control problems with the same Bellman equations as the original ones. We show that the optimal diffusion processes of the problem are ergodic and that under some condition related to integrability by the invariant measures of the diffusion processes we can construct optimal strategies for the original problems by using the solution of the Bellman equations.

1 Introduction

Let us consider the following ergodic type Bellman equation of risk-sensitive control:

$$\chi = \frac{1}{2}\mathrm{tr}(a(x)D^2 v) + \frac{\theta}{2}(\nabla v)^* a(x)\nabla v + \inf_{z.}\{b(x,z)^*\nabla v + \Phi(x,z)\}, \qquad (1.1)$$

where the pair (v, χ) of a function v and a constant χ is considered to be a solution and the constant χ is considered to characterize the minimum of the risk-sensitive long-run criterion:

$$\liminf_{T\to\infty} \frac{1}{\theta T} \log E_x[e^{\theta\int_0^T \Phi(X_s,z_s)ds}]$$

subject to

$$dX_t = b(X_t, z_t)dt + \sigma(X_t)dW_t, \quad X_0 = x,$$

where W_t is a standard Brownian motion process on a filtered probability space, $\sigma(x)$ is a matrix valued function such that $\sigma\sigma^*(x)$ and z_t is a control process taking its value on a control region Z. Suppose that (1.1) has a smooth solution, the infimum in (1.1) is attained by a function $z(x)$:

$$b(x, z(x))^*\nabla v(x) + \Phi(x, z(x)) = \inf_{z\in Z}\{b(x,z)^*\nabla v(x) + \Phi(x,z)\},$$

B. Pasik-Duncan: Stochastic Theory and Control, LNCIS 280, pp. 353–368, 2002.

and the stochastic differential equation

$$d\hat{X}_t = b(\hat{X}_t, z(\hat{X}_t))dt + \sigma(\hat{X}_t)dW_t, \quad \hat{X}_0 = x$$

has a solution, then $z(\hat{X}_t)$ may be considered an optimal strategy. However it is not always the case. Indeed Fleming and Sheu has noticed, by taking up one dimensional problems concerning risk-sensitive portofolio optimizaion, that the solution of the corresponding Bellman equation does not always construct an optimal strategy (cf. Fleming and Sheu [8]). We have also noticed such a situation occurs in discussing risk sensitive control problems related to eigenvalue problems for certain Schrödinger operators ([14]). So, there arises a problem to know the conditions under which the solutions define optimal strategies. In the case of linear Gaussian factor models, several authors have tried to find the conditions for the solution of the ergodic type Bellman equation to construct optimal strategies for risk-sensitive portfolio optimization (cf. [14],[16],[9],[10],[15],[17]). Bielecki and Pliska [16] and Fleming and Sheu [9],[10] have shown that, if the risk-sensitive parameter θ is small, then the construction of an optimal strategy by the solution is done well. For a general size of θ we have constructed an optimal strategy under some condition in a full information case in [15] as well as a partial information case in [17].

In the present paper we shall consider the problems of constructing optimal strategies for portfolio optimization on infinite time horizon for general factor models, by using the solutions of corresponding ergodic type Bellman equations, and present the results such that the solutions define optimal strategies under the condition which amounts to an integrability condition by the invariant measures of underlying ergodic diffusion processes. The ergodic diffusion processes are the optimal ones of other classical ergodic control problems with the same Bellman equations of ergodic type as the original ones. Furthemore the integrabiblity condition is checked precisely in the case of linear Gaussian models in section 5. The full paper will be seen elsewhere.

2 Finite Time Horizon Case

We consider a market with $m+1 \geq 2$ securities and $n \geq 1$ factors. We assume that the set of securities includes one bond, whose price is defined by ordinary differential equation:

$$dS^0(t) = r(X_t)S^0(t)dt, \quad S^0(0) = s^0, \tag{2.1}$$

where $r(x)$ is a nonnegative bounded function. The other security prices S_t^i, $i = 1, 2, \ldots, m$ and factors X_t are assumed to satisfy the following stochastic differential equations:

$$\begin{aligned} dS^i(t) &= S^i(t)\{g^i(X_t)dt + \textstyle\sum_{k=1}^{n+m} \sigma_k^i(X_t)dW_t^k\}, \\ S^i(0) &= s^i, \quad i = 1, \ldots, m \end{aligned} \tag{2.2}$$

and

$$dX_t = b(X_t)dt + \lambda(X_t)dW_t,$$
$$X(0) = x \in R^n, \tag{2.3}$$

where $W_t = (W_t^k)_{k=1,\dots,(n+m)}$ is a $m+n$ dimensional standard Brownian motion process defined on a filtered probability space $(\Omega, \mathcal{F}, P, \mathcal{F}_t)$. Here σ and λ are respectively $m \times (m+n)$,$n \times (m+n)$ matrix valued functions. We assume that

$$g,\ \sigma,\ b,\ \lambda \text{ are locally Lipshitz}$$

$$c_1|\xi|^2 \le \xi^* \sigma\sigma^*(x)\xi \le c_2|\xi|^2, \quad c_1, c_2 > 0 \tag{2.4}$$

$$x^* b(x) + \frac{1}{2} \parallel \lambda\lambda^*(x) \parallel \le K(1+|x|^2)$$

where σ^* stands for the transposed matrix of σ.

Let us denote investment strategy to i-th security $S^i(t)$ by $h^i(t)$, $i = 0, 1, \dots,$ m and set

$$S(t) = (S^1(t), S^2(t), \dots, S^m(t))^*,$$

$$h(t) = (h^1(t), h^2(t), \dots,, h^m(t))^*$$

and

$$\mathcal{G}_t = \sigma(S(u), X(u); u \le t).$$

Here S^* stands for transposed matrix of S.

Definition 1 *$(h^0(t), h(t)^*)_{0 \le t \le T}$ is said an investment strategy if the following conditions are satisfied*

i) $h(t)$ is a R^m valued $\mathcal{G}_t$ progressively measurable stochastic process such that

$$\sum_{i=1}^m h^i(t) + h^0(t) = 1 \tag{2.5}$$

ii)

$$P\left(\int_0^T |h(s)|^2 ds < \infty\right) = 1.$$

The set of all investment strategies will be denoted by $\mathcal{H}(T)$. When $(h^0(t), h(t)^*)_{0\le t\le T} \in \mathcal{H}(T)$ we will often write $h \in \mathcal{H}(T)$ for simplicity since h^0 is determined by (2.5).

For given $h \in \mathcal{H}(T)$ the process $V_t = V_t(h)$ representing the investor's capital at time t is determined by the stochastic differential equation:

$$\frac{dV_t}{V_t} = \sum_{i=0}^m h^i(t)\frac{dS^i(t)}{S^i(t)}$$

$$= h^0(t)r(X_t)dt + \sum_{i=1}^m h^i(t)\{g^i(X_t)dt + \sum_{k=1}^{m+n} \sigma^i_k(X_t)dW^k_t\}$$

$$V_0 = v.$$

Then, taking (2.5) into account it turns out to be a solution of

$$\frac{dV_t}{V_t} = r(X_t)dt + h(t)^*(g(X_t) - r(X_t)\mathbf{1})dt + h(t)^*\sigma(X_t)dW_t, \qquad (2.6)$$

$$V_0 = v,$$

where $\mathbf{1} = (1, 1, \ldots, 1)^*$.

We first consider the following problem. For a given constant $\theta > -2$, $\theta \neq 0$ maximize the following risk-sensitized expected growth rate up to time horizon T:

$$J(v, x; h; T) = -\frac{2}{\theta}\log E[e^{-\frac{\theta}{2}\log V_T(h)}], \qquad (2.7)$$

where h ranges over the set $\mathcal{A}(T)$ of all admissible strategies defined later. Then we consider the problem of maximizing the risk-sensitized expected growth rate per unit time

$$J(v, x; h) = \limsup_{T\to\infty}\left(\frac{-2}{\theta T}\right)\log E[e^{-\frac{\theta}{2}\log V_T(h)}], \qquad (2.8)$$

where h ranges over the set of all investment straregies such that $h \in \mathcal{A}(T)$ for each T.

Since V_t satisfies (2.6) we have

$$V_t^{-\theta/2} = v^{-\theta/2}\exp\{\tfrac{\theta}{2}\int_0^t \eta(X_s, h_s))ds$$

$$- \tfrac{\theta}{2}\int_0^t h_s^*\sigma(X_s)dW_s - \tfrac{\theta^2}{8}\int_0^t h_s^*\sigma\sigma^*(X_s)h_s ds\},$$

where

$$\eta(x, h) = \left(\frac{\theta+2}{4}\right)h^*\sigma\sigma^*(x)h - r(x) - h^*(g(x) - r(x)\mathbf{1}).$$

If a given investment strategy h satisfies

$$E[e^{-\frac{\theta}{2}\int_0^T h(s)^*\sigma^*(X_s)dW_s-\frac{\theta^2}{8}\int_0^T h(s)^*\sigma\sigma^*(X_s)h(s)ds}]=1, \tag{2.9}$$

then we can introduce a probability measure P^h given by

$$P^h(A)=E[e^{-\frac{\theta}{2}\int_0^T h^*(s)\sigma(X_s)dW_s-\frac{\theta^2}{8}\int_0^T h^*(s)\sigma\sigma^*(X_s)h(s)ds};A]$$

for $A\in\mathcal{F}_T$, $T>0$. By the probability measure P^h our criterion $J(v,x;h;T)$ and $J(v,x;h)$ can be written as follows:

$$J(v,x;h,T)=\log v-\frac{2}{\theta}\log E^h[e^{\frac{\theta}{2}\int_0^T \eta(X_s,h(s))ds}] \tag*{(2.7)$'$}$$

and

$$J(v,x;h)=\liminf_{T\to\infty}-\frac{2}{\theta T}\log E^h[e^{\frac{\theta}{2}\int_0^T \eta(X_s,h(s))ds}]. \tag*{(2.8)$'$}$$

On the other hand, under the probability measure,

$$W_t^h=W_t-\langle W_\cdot,-\tfrac{\theta}{2}\int_0^\cdot h^*(s)\sigma(X_s)dW_s\rangle_t$$

$$=W_t+\tfrac{\theta}{2}\int_0^t\sigma^*(X_s)h(s)ds$$

is a standard Brownian motion process, and therefore, the factor process X_t satisfies the following stochastic differential equation

$$dX_s=(b(X_s)-\frac{\theta}{2}\lambda\sigma^*(X_s)h(s))ds+\lambda(X_s)dW_s^h. \tag{2.10}$$

We regard (2.10) as a stochastic differential equation controlled by h and the criterion function is written by P^h as follows:

$$J(v,x;h;T-t)=\log v-\frac{2}{\theta}\log E^h[e^{\frac{\theta}{2}\int_0^{T-t}\eta(X_s,h(s))ds}] \tag{2.11}$$

and the value function

$$u(t,x)=\sup_{h\in\mathcal{H}(T-t)}J(v,x;h;T-t),\ 0\le t\le T. \tag{2.12}$$

Then, according to Bellman's dynamic programming principle, it should satisfy the following Bellman equation

$$\frac{\partial u}{\partial t}+\sup_{h\in R^m}L^h u=0, \tag{2.13}$$

$$u(T,x)=\log v,$$

where L^h is defined by

$$L^h u(t,x) = \frac{1}{2}\mathrm{tr}(\lambda\lambda^*(x)D^2u) + \left(b(x) - \frac{\theta}{2}\lambda\sigma^*(x)h\right)^* Du$$
$$- \frac{\theta}{4}(Du)^*\lambda\lambda^*(x)Du - \eta(x,h).$$

Note that $\sup_{h\in R^m} L^h u$ can be written as

$$\sup_{h\in R^m} L^h u(t,x) = \tfrac{1}{2}\mathrm{tr}(\lambda\lambda^*(x)D^2u) + (b - \tfrac{\theta}{\theta+2}\lambda\sigma^*(\sigma\sigma^*)^{-1}(g - r\mathbf{1}))^* Du$$
$$-\tfrac{\theta}{4}(Du)^*\lambda(I - \tfrac{\theta}{\theta+2}\sigma^*(\sigma\sigma^*)^{-1}\sigma)\lambda^* Du$$
$$+\tfrac{1}{\theta+2}(g - r\mathbf{1})^*(\sigma\sigma^*)^{-1}(g - r\mathbf{1})$$

Therefore our Bellman equation (2.13) is written as follows:

$$\begin{aligned} &\tfrac{\partial u}{\partial t} + \tfrac{1}{2}\mathrm{tr}(\lambda\lambda^* D^2u) + B(x)^* Du - (Du)^*\lambda N^{-1}\lambda^* Du + U(x) = 0, \\ &u(T,x) = \log v, \end{aligned} \tag{2.14}$$

where

$$\begin{aligned} B(x) &= b(x) - \tfrac{\theta}{\theta+2}\lambda\sigma^*(\sigma\sigma^*)^{-1}(g(x) - r(x)\mathbf{1}) \\ N^{-1}(x) &= \tfrac{\theta}{4}(I - \tfrac{\theta}{\theta+2}\sigma^*(\sigma\sigma^*)^{-1}\sigma(x)) \\ U(x) &= \tfrac{1}{\theta+2}(g - r\mathbf{1})^*(\sigma\sigma^*)^{-1}(g - r\mathbf{1}). \end{aligned} \tag{2.15}$$

As for (2.14) we note that if $\theta > 0$, then

$$\frac{\theta}{2(\theta+2)}I \le N^{-1} \le \frac{\theta}{4}I$$

and therefore we have

$$-\frac{\theta}{4}\lambda\lambda^* \le -\lambda N^{-1}\lambda^* \le -\frac{\theta}{2(\theta+2)}\lambda\lambda^*.$$

Such kinds of equations have been studied in Nagai [16], or Bensoussan, Frehse and Nagai [3]. Here we can obtain the following result along the line of [3], Theorem 5.1 with refinement on estimate (2.17).

Theorem 12 *i) If, in addition to (2.4), $\theta > 0$ and*

$$\nu_r|\xi|^2 \leq \xi^*\lambda\lambda^*(x)\xi \leq \mu_r|\xi|^2, \quad r = |x|, \quad \nu_r, \ \mu_r > 0, \tag{2.16}$$

then we have a solution of (2.14) such that

$$u, \ \frac{\partial u}{\partial t}, \ D_k u, \ D_{kj} u \in L^p(0,T; L^p_{loc}(R^n)), \quad 1 < \forall p < \infty$$

$$\frac{\partial^2 u}{\partial t^2}, \ \frac{\partial D_k u}{\partial t}, \ \frac{\partial D_{kj} u}{\partial t}, D_{kjl} u \in L^p(0,T; L^p_{loc}(R^n)), \quad 1 < \forall p < \infty$$

$$u \geq \log v, \quad \frac{\partial u}{\partial t} \leq 0.$$

Furthermore we have the estimate

$$\begin{aligned} |\nabla u|^2(t,x) - \frac{c_0}{\nu_r}\frac{\partial u}{\partial t}(t,x) &\leq c_r(|\nabla Q|^2_{2r} + |Q|^2_{2r} + |\nabla(\lambda\lambda^*)|^2_{2r} \\ &+ |\nabla B|_{2r} + |B|^2_{2r} + |U|_{2r} + |\nabla U|^2_{2r} + 1), \qquad x \in B_r, \ t \in [0,T) \end{aligned} \tag{2.17}$$

where

$$Q = \lambda N^{-1}\lambda^*, \quad c_0 = \frac{4(1+c)(\theta+2)}{\theta}, \quad c > 0$$

$$|\cdot|_{2r} = \|\cdot\|_{L^\infty(B_{2r})}$$

and c_r is a positive constant depending on n, r, ν_r, μ_r and c.

ii) If, in addition to the above conditions,

$$\inf_{|x|\geq r} U(x) \to \infty, \quad \text{as } r \to \infty,$$

then the above solution u satisfies

$$\inf_{|x|\geq r, t\in(0,T)} u(x,t) \to \infty, \quad \text{as } r \to \infty.$$

Moreover, there exists at most one such solution in $L^\infty(0,T; W^{1,\infty}_{loc}(R^n))$

Remark. If

$$\frac{1}{\nu_r}, \ \mu_r \leq M(1 + r^m), \quad \exists m > 0, \tag{2.21}$$

then we have

$$c_r \leq M'(1 + r^{m'}), \ \exists m'$$

in estimate (2.17). In particular, if $m = 0$, then c_r can be taken independent of r.

Let us define a class of admissible investment strategy $\mathcal{A}_T$ as the set of investment strategies satisfying (2.9). Then, thanks to the above theorem and remark we have the following proposition.

Proposition 13 *i) We assume the assumptions of the above theorem and let u be a solution of (2.14). Define*

$$\hat{h}_t = \hat{h}(t, X_t)$$

$$\hat{h}(t,x) = \tfrac{2}{\theta+2}(\sigma\sigma^*)^{-1}(g - r\mathbf{1} - \tfrac{\theta}{2}\sigma\lambda^* Du)(t,x),$$

where X_t is the solution of (2.3), then, under the assumption that

$$E[e^{-\int_0^T (2N^{-1}\lambda^* Du+\theta K)^*(x_s)dW_s - \frac{1}{2}\int_0^T (2N^{-1}\lambda^* Du+\theta K)^*(2N^{-1}\lambda^* Du+\theta K)(x_s)ds}]$$
$$= 1, \tag{2.22}$$

where

$$K = \frac{1}{\theta+2}\sigma^*(\sigma\sigma^*)^{-1}(g - r\mathbf{1}),$$

$\hat{h}_t \in \mathcal{A}_T$ is an optimal strategy for the portfolio optimization problem of maximizing the criterion (2.7).

ii) if

$$c_1|\xi|^2 \leq \xi^*\lambda\lambda^*(x)\xi \leq c_2|\xi|^2, \quad c_1, c_2 > 0 \tag{2.23}$$

$$g,\ b,\ \lambda,\ \sigma \ \text{ are globally Lipshitz,}$$

then (2.22) is valid.

To disccuss the problem on infinite time horizon we introduce another stochastic control problem on a finite time horizon with the same Bellman equation as (2.14) and then consider its ergodic counter part. For that let us set

$$G = b - \lambda\sigma^*(\sigma\sigma^*)^{-1}(g - r\mathbf{1})$$

and rewrite equation (2.14) as

$$\frac{\partial u}{\partial t} + \frac{1}{2}\mathrm{tr}(\lambda\lambda^*(x)D^2u) + G(x)^*Du$$
$$-(-\lambda^*Du + NK)^*N^{-1}(-\lambda^*Du + NK)(x) + \tfrac{\theta+2}{2}K^*NK(x) = 0,$$
$$u(T,x) = \log v. \tag{2.24}$$

Since

$$-(-\lambda^* Du+NK)^* N^{-1}(-\lambda^* Du+NK) = \inf_{z\in R^{n+m}} \{z^* Nz+2z^* NK-2(\lambda z)^* Du\},$$

we can regard (2.21) as the Bellman equation of the following stochastic control problem. Set

$$u(t,x) = \inf_{Z_\cdot} E_x\Big[\int_0^{T-t} \{Z_s^* N(Y_s)Z_s + 2Z_s^* NK(Y_s) + \tfrac{\theta+2}{2} K^* NK(Y_s)\}\, ds + \log v\Big], \tag{2.25}$$

where Y_t is a controlled process governed by the stochastic differential equatinion

$$dY_t = \lambda(Y_t)dW_t + (G(Y_t) - 2\lambda(Y_t)Z_t)dt, \tag{2.26}$$

and Z_t is a control taking its value on R^{n+m}. We define the set of admissible controls Z_t as all progressively measurable processes satisfying

$$E_x[\int_0^T |Z_s|^{2q} ds] < \infty, \quad \forall q \geq 1.$$

An ergodic counterpart of the above problem is formulated as follows. Consider the problem:

$$\chi = \inf_{Z_\cdot} \liminf_{T\to\infty} \frac{1}{T} E_x[\int_0^T \{Z_s^* N(Y_s)Z_s + 2Z_s^* NK(Y_s) + \frac{\theta}{2} K^* NK(Y_s)\}ds] \tag{2.27}$$

with controlled process Y_t governed by (2.26). Then, corresponding Bellman equation is written as

$$\chi = \frac{1}{2}\mathrm{tr}(\lambda\lambda^*(x)D^2 w) + G(x)^* Dw - (-\lambda^* Dw + NK)^* N^{-1}(-\lambda^* Dw + NK)(x) + \tfrac{\theta+2}{2} K^* NK(x), \tag{2.28}$$

whose original one is

$$\chi = \frac{1}{2}\mathrm{tr}(\lambda\lambda^*(x)D^2 w) + B(x)^* Dw - (Dw)^* \lambda N^{-1}\lambda^*(x)Dw + U(x) = 0, \tag{2.29}$$

namely,

$$\chi = \frac{1}{2}\mathrm{tr}(\lambda\lambda^*(x)D^2 w) + (b - \frac{\theta}{\theta+2}\lambda\sigma^*(\sigma\sigma^*)^{-1}(g - r\mathbf{1}))^* Dw$$
$$- \tfrac{\theta}{4}(Dw)^*\lambda(I - \tfrac{\theta}{\theta+2}\sigma^*(\sigma\sigma^*)^{-1}\sigma)\lambda^* Dw + \tfrac{1}{\theta+2}(g - r\mathbf{1})^*(\sigma\sigma^*)^{-1}(g - r\mathbf{1}).$$

In the following section we shall analyze the Bellman equation of ergodic type (2.28). Indeed we shall deduce equation (2.28), accordingly (2.29), as the limit of parabolic type equation (2.24) as $T \to \infty$ under suitable conditions.

Remark. To regard our Bellman equation as (2.24) has a meaning from financial view points. Indeed, under the minimal martingale measure $\tilde{P}$ (cf. [6] Proposition 1.8.2 as for minimal martingale measures), which is defined by

$$\left.\frac{d\tilde{P}}{dP}\right|_{\mathcal{F}_T} = e^{-\int_0^T \zeta(X_s)^* dW_s - \frac{1}{2}\int_0^T |\zeta(X_s)|^2 ds},$$

$\zeta(x) = \sigma^*(\sigma\sigma^*)^{-1}(x)(g(x) - r(x)\mathbf{1})$ factor process X_t is the diffusion process with the generator

$$L = \frac{1}{2}\mathrm{tr}(\lambda\lambda^*(x)D^2) + G(x)^* D,$$

namely, it is governed by the SDE

$$dX_t = \lambda(X_t)d\tilde{W}_t + G(X_t)dt$$

Here $\tilde{W}_t = W_t + \int_0^t \zeta(X_s)ds$ and it is a Brownian motion under the probability measure $\tilde{P}$.

3 Ergodic Type Bellman Equation

In what follows we assume that

$$\frac{1}{2}\mathrm{tr}(\lambda\lambda^*(x)) + x^*G(x) + \frac{\kappa}{2}\frac{x^*\lambda\lambda^*(x)x}{\sqrt{1+|x|^2}} \le 0, \qquad |x| \ge \exists r > 0,\ \ \kappa > 0 \quad (3.1)$$

and set

$$L = \frac{1}{2}\mathrm{tr}(\lambda\lambda^*(x)D^2) + G^*(x)D.$$

Proposition 14 *We assume (2.4), (3.1) and (2.16) with*

$$\nu_r \ge e^{-\frac{\kappa - c}{8}r}, \qquad r >> 1,\ c > 0, \quad (3.2)$$

then L diffusion process $(\tilde{P}_x, X_t)$ is ergodic and satisfies

$$\tilde{E}_x[e^{\kappa\sqrt{1+|X_t|^2}}] \le e^{\kappa\sqrt{1+|x|^2}} \quad (3.3)$$

Theorem 15 *Assume the assumptions of Theorem 2.1, (3.1) and that*

$$\frac{1}{\nu_r},\ \mu_r \le K(1+r^m)$$

$$|Q|,\ |\nabla Q|,\ |B|,\ |\nabla B|,\ U,\ |\nabla U|,\ |\nabla(\lambda\lambda^*)| \le K(1+|x|^m),$$

then, as $T \to \infty$,

$$u(0,x;T) - u(0,0;T) \to w(x),$$

$$\tfrac{1}{T}u(0,x;T) \to \chi,$$

uniformly on each compact set, where (w,χ) *is the solution of (2.28) such that* $w \in C^2(R^n)$.

Our Bellman equation of ergodic type (2.28) is rewritten as

$$\chi = \frac{1}{2}\mathrm{tr}(\lambda\lambda^*(x)D^2w) + G(x)^*Dw$$
$$\inf_{z\in R^{n+m}}\{z^*Nz + 2z^*NK - 2(\lambda z)^*Dw\} + \tfrac{\theta+2}{2}K^*NK(x), \tag{3.10}$$

and the infimum is attained by

$$\hat{z}(x) = N^{-1}\lambda^*(x)Dw(x) - K(x),$$

which define the following elliptic operator considered as the generator of the optimal diffusion for (2.27)

$$\hat{L} = \frac{1}{2}\mathrm{tr}(\lambda\lambda^*(x)D^2) + G^*(x)D - 2(\lambda N^{-1}\lambda^*(x)Dw(x) - \lambda K(x))^*D.$$

Then we have the following proposition.

Proposition 16 *Under the assumption of Theorem 3.1* $\hat{L}$ *diffusion process is ergodic.*

4 Optimal Strategy for Portfolio Optimization on Infinite Time Horizon

Define the set of admissible strategies $\mathcal{A}$ by

$$\mathcal{A} = \{h : h \in \mathcal{A}(T), \forall T\}$$

and set

$$\hat{H}_t = \hat{H}(X_t)$$

$$\hat{H}(x) = \tfrac{2}{\theta+2}(\sigma\sigma^*)^{-1}(g - r\mathbf{1} - \tfrac{\theta}{2}\sigma\lambda^*Dw)(x),$$

where X_t is the solution of SDE (2.3), then we have the following theorem.

Theorem 17 *In addition to the assumptions of Theorem 3.1 we assume (2.23) and that*

$$\frac{4}{\theta^2}(g-r\mathbf{1})^*(\sigma\sigma)^{-1}(g-r\mathbf{1})-(Dw)^*\lambda\sigma^*(\sigma\sigma^*)^{-1}\sigma\lambda^*Dw\to\infty,\ \ |x|\to\infty, \tag{4.1}$$

then $\hat{H}_t$ is an optimal strategy for portfolio optimization maximizing long run criterion (2.8)

$$J(v,x;\hat{H})=\sup_{h\in\mathcal{A}}J(v,x;h).$$

Remark. Under the probability measure $\hat{P}_x$ the factor process is an ergodic diffusion process with the generator $\hat{L}$. In fact, by calculation, we can see that

$$\tfrac{1}{2}\mathrm{tr}(\lambda\lambda^*D^2)+(b-\tfrac{\theta}{2}\lambda\sigma^*\hat{H}-\tfrac{\theta}{2}\lambda\lambda^*Dw)^*D$$

$$=\tfrac{1}{2}\mathrm{tr}(\lambda\lambda^*(x)D^2)+G^*(x)D-2(\lambda N^{-1}\lambda^*(x)Dw(x)-\lambda K(x))^*D.$$

Then, under assumption (4.1), $\hat{L}$ diffusion process $(\hat{P}_x,X_t)$ satisfies

$$\hat{E}_x[e^{\frac{\theta}{2}w(X_T)}]\to\int e^{\frac{\theta}{2}w(x)}\mu(dx)<\infty,\quad \text{as } T\to\infty$$

where μ is the invariant measure of $(\hat{P}_x,X_t)$.

5 Example

Example (Linear Gaussian case) Let us consider the case where

$$g(x)=a+Ax,\ \sigma(x)=\Sigma,\quad b(x)=b+Bx,\ \lambda(x)=\Lambda,\quad r(x)=r,$$

where $A,\ B,\ \Sigma,\ \Lambda$ are all constant matrices and a and b are constant vectors. Such a case has been considered by Bielecki and Pliska [14], [16],Fleming and Sheu [9],[10] and Kuroda and Nagai [15].

In this case the solution $u(t,x)$ of (2.14) has the following explicit form

$$u(t,x)=\frac{1}{2}x^*P(t)x+q(t)^*x+k(t)$$

where $P(t)$ is a solution of the Riccati differential equation:

$$\dot{P}(t)-P(t)K_0P(t)+K_1^*P(t)+P(t)K_1+\tfrac{2}{\theta+2}A^*(\Sigma\Sigma^*)^{-1}A=0, \tag{5.1}$$

$$P(T)=0,$$

and

$$K_0 = \frac{\theta}{2}\Lambda(I - \frac{\theta}{\theta+2}\Sigma^*(\Sigma\Sigma^*)^{-1}\Sigma)\Lambda^*, \quad K_1 = B - \frac{\theta}{\theta+2}\Lambda\Sigma^*(\Sigma\Sigma^*)^{-1}A.$$

The term $q(t)$ is a solution of linear differential equation:

$$\begin{aligned} &\dot{q}(t) + (K_1^* - P(t)K_0)q(t) + P(t)b \\ &\quad + \left(\frac{2}{\theta+2}A^* - \frac{\theta}{\theta+2}P(t)\Lambda\Sigma^*\right)(\Sigma\Sigma^*)^{-1}(a - r\mathbf{1}) = 0 \\ &q(T) = 0 \end{aligned}$$

and $k(t)$ a solution of

$$\begin{aligned} &\dot{k}(t) + \frac{1}{2}\mathrm{tr}(\Lambda\Lambda^* P(t)) - \frac{\theta}{4}q(t)^*\Lambda\Lambda^* q(t) + b^* q(t) + r \\ &\quad + \frac{1}{\theta+2}(a - r\mathbf{1})^*(\Sigma\Sigma^*)^{-1}(a - r\mathbf{1}) \\ &\quad + \frac{\theta^2}{4(\theta+2)}q(t)^*\Lambda\Sigma^*(\Sigma\Sigma^*)^{-1}\Sigma\Lambda^* q(t) \\ &\quad - \frac{\theta}{\theta+2}(a - r\mathbf{1})^*(\Sigma\Sigma^*)^{-1}\Sigma\Lambda^* q(t) = 0 \\ &k(T) = \log v. \end{aligned}$$

If

$$G \equiv B - \Lambda\Sigma^*(\Sigma\Sigma^*)^{-1}A \quad \text{is stable,}$$

then
i) $P(0) = P(0;T)$ converges, as $T \to \infty$, to a nonnegative definite matrix $\tilde{P}$, which is a solution of algebraic Riccati equation:

$$K_1^*\tilde{P} + \tilde{P}K_1 - \tilde{P}K_0\tilde{P} + \frac{2}{\theta+2}A^*(\Sigma\Sigma^*)^{-1}A = 0.$$

Moreover $\tilde{P}$ satisfies the estimate

$$0 \le \tilde{P} \le \frac{2}{\theta}\int_0^\infty e^{sG^*}A^*(\Sigma\Sigma^*)^{-1}Ae^{sG}ds. \tag{5.2}$$

ii) $q(0) = q(0;T)$ converges, as $T \to \infty$, to a constant vector $\tilde{q}$, which satisfies

$$(K_1^* - \tilde{P}K_0)\tilde{q} + \tilde{P}b + (\frac{2}{\theta+2}A^* - \frac{\theta}{\theta+2}\tilde{P}\Lambda\Sigma^*)(\Sigma\Sigma^*)^{-1}(a - r\mathbf{1}) = 0$$

iii) $k(0;T)/T$ converges to a constant $\rho(\theta)$ defined by

$$\rho(\theta) = \frac{1}{2}\mathrm{tr}(\tilde{P}\Lambda\Lambda^*) - \frac{\theta}{4}\tilde{q}^*\Lambda\Lambda^*\tilde{q} + b^*\tilde{q} + r$$

$$
\begin{aligned}
&+ \frac{1}{\theta+2}(a - r\mathbf{1})^*(\Sigma\Sigma^*)^{-1}(a - r\mathbf{1}) \\
&+ \frac{\theta^2}{4\theta+8}\tilde{q}^* \Lambda\Sigma^*(\Sigma\Sigma^*)^{-1}\Sigma\Lambda^*\tilde{q} \\
&- \frac{\theta}{\theta+2}(a - r\mathbf{1})^*(\Sigma\Sigma^*)^{-1}\Sigma\Lambda^*\tilde{q}
\end{aligned}
$$

if, moreover,

$$(B^*, A^*(\Sigma\Sigma^*)^{-1}\Sigma) \text{ is controllable,} \tag{5.3}$$

then
iv) the solution $\tilde{P}$ of the above algebraic Riccati equation is strictly positive definite.

Finally, if, in addition to the above conditions,

$$(B, \Lambda) \text{ is controllable,} \tag{5.4}$$

then
v) the investment strategy $\tilde{h}_t$ defined by

$$\tilde{h}_t = \frac{2}{\theta+2}(\Sigma\Sigma^*)^{-1}[a - r\mathbf{1} - \frac{\theta}{2}\Sigma\Lambda^*\tilde{q} + (A - \frac{\theta}{2}\Sigma\Lambda^*\tilde{P})X_t]$$

is optimal for the portofolio optimization on infinite time horizon maximizing the criterion (2.8):

$$\sup_{h \in \mathcal{A}} J(v, x; h) = J(v, x; \tilde{h}_.) = \rho(\theta)$$

if and only if

$$\hat{P}\Lambda\Sigma^*(\Sigma\Sigma^*)^{-1}\Sigma\Lambda^*\hat{P} < A^*(\Sigma\Sigma^*)^{-1}A, \tag{5.5}$$

where $\hat{P} = \frac{\theta}{2}\tilde{P}$ (cf. [15]).

Set

$$w(x) = \frac{1}{2}x^*\tilde{P}x + \tilde{q}^*x,$$

then $w(x)$ satisfies (2.28) and (5.5) is equivalent to

$$\int e^{\frac{\theta}{2}w(x)}\mu(dx) < \infty$$

under the assumptions (5.3) and (5.4), where $\mu(dx)$ is the invariant measure of $\hat{L}$ diffusion process. We consider the case where $n = m = 1$. Then $\Sigma\Sigma^*$, $\Lambda\Sigma^*$, A, B are all scalars and (5.5) is written as

$$\frac{\theta^2}{4}\tilde{P}^2(\Lambda\Sigma^*)^2 < A^2 \tag{5.5'}$$

We can find sufficient condition for (5.5') by using estimate (5.2). Indeed, If

$$A^2(\Lambda\Sigma^*)^2(\Sigma\Sigma^*)^{-2}\left(\int_0^\infty e^{2sG}ds\right)^2 < 1 \tag{5.6}$$

then (5.5') holds. (5.6) is equivalent to

$$(2B(\Sigma\Sigma^*) - 3(\Lambda\Sigma^*)A)(2B(\Sigma\Sigma^*) - (\Lambda\Sigma^*)A) > 0,$$

from which we see that

$$B < \frac{1}{2}\Lambda\Sigma^*(\Sigma\Sigma^*)^{-1}A \quad \text{if} \quad \Lambda\Sigma^* A > 0 \tag{5.7}$$

$$B < \frac{3}{2}\Lambda\Sigma^*(\Sigma\Sigma^*)^{-1}A \quad \text{if} \quad \Lambda\Sigma^* A < 0 \tag{5.8}$$

since $G = B - \Lambda\Sigma^*(\Sigma\Sigma^*)^{-1}A < 0$ by the stability assumption.

References

1. Bhattacharya, R. N. (1978) Criteria for recurrence and existence of invariant measures for multidimensional diffusions, The Annals of Probability **6**, 541–553
2. Bensoussan, A. (1992) Stochastic Control of Partially Observable Systems, Cambridge Univ. Press.
3. Bensoussan, A., Frehse, J., and Nagai, H. (1998) Some Results on Risk-sensitive with full observation, Appl. Math. and its Optimization **37**, 1–41.
4. Bielecki, T. R. and Pliska, S. R. (1999) Risk-Sensitive Dynamic Asset Management. Appl. Math. Optim. **39**, 337–360.
5. Bielecki, T. R. and Pliska, S. R. (2001) Risk-Sensitive Intertemporal CAPM, With Application to Fixed Income Management, preprint.
6. El Karoui, N. and Quenez, M.-C. (1995) Dynamic Programming pricing of contingent claims in an incomplete market, SIAM J. Cont. Optim. **33**, 29–66.
7. Fleming, W. H. and McEneaney, W. M. (1995) Risk-sensitive control on an infinite horizon, SIAM J. Control and Optimization **33**, 1881–1915.
8. Fleming, W. H. and Sheu, S. J. (1999) Optimal Long Term Growth Rate of Expected Utility of Wealth, Ann. Appl. Prob. **9**, 871–903.
9. Fleming, W. H. and Sheu, S. J. (2000) Risk-sensitive control and an optimal investment model, Mathematical Finance **10**, 197–213.
10. Fleming, W. H. and Sheu, S. J. (2001) Risk-sensitive control and an optimal investment model (II), preprint.
11. Has'minskii R. Z. (1980) Stochastic stability of differential equations, Sijthoff and Noordhoff, Alphen aan den Rijin.
12. Kaise, H. and Nagai, H. (1998) Bellman-Isaacs equations of ergodic type related to risk-sensitive control and their singular limits, Asymptotic Analysis **16**, 347–362.
13. Kaise, H. and Nagai, H. (1999) Ergodic type Bellman equations of risk-sensitive control with large parameters and their singular limits, Asymptotic Analysis **20**, 279–299.

14. Kuroda, K. and Nagai, H. (2001) Ergodic type Bellman equation of risk-sensitive control and potfolio optimization on infinite time horizon, A volume in honor of A. Benosussan, Eds. J. L. Menaldi et al., IOS Press, 530–538.
15. Kuroda, K. and Nagai, H. (2000) Risk-sensitive portfolio optimization on infinite time horizon, Preprint.
16. Nagai, H. (1996) Bellman equations of risk-sensitive control, SIAM J. Control and Optimization **34**, 74–101.
17. Nagai, H. and Peng, S. (2001) Risk-sensitive dynamic portfolio optimization with partial information on infinite time horizon, Annals of Appl. Prob., to appear.
18. Wonham, W. M. (1968) On a Matrix Riccati Equation of Stochastic Control, SIAM J. Control Optim. **6**, 681–697.

Finite Horizon Full-State Feedback kCC Control in Civil Structures Protection

Khanh D. Pham[1], Michael K. Sain[1], and Stanley R. Liberty[2]

[1] University of Notre Dame, Notre Dame, IN 46556, USA
[2] Bradley University, Peoria IL 61625, USA

Abstract. The following study presents a finite horizon full-state feedback k-Cost-Cumulant (kCC) control problem, in which the objective function representing a linear combination of k cumulant indices of a finite time integral quadratic performance measure associated to a linear stochastic system, is minimized. A dynamic programming approach is used to obtain the optimal control solution. This control algorithm is then applied to the First Generation Structure Benchmark for Seismically Excited Buildings. Simulation results indicate that the kCC control has both performance and control design flexibility advantages when compared with popular control strategies in building protection.

1 Preliminaries

Consider controlling a linear dynamical system with state $x(\cdot) \in \mathcal{R}^n$ modeled on $T = [t_0, t_f]$ by

$$\frac{d}{dt}x(t) = A(t)x(t) + B(t)u(t) + G(t)w(t)\,, \qquad x(t_0) = x_0 \tag{1}$$

where matrix-valued continuous coefficients $A(\cdot) \in \mathcal{C}([t_0,t_f];R^{n\times n})$, $B(\cdot) \in \mathcal{C}([t_0,t_f];R^{n\times m})$ and $G(\cdot) \in \mathcal{C}([t_0,t_f];R^{n\times p})$. The process $w(\cdot) \in R^p$ is zero-mean white Gaussian with covariance $E\left\{w(\tau)w^T(\xi)\right\} = W(\tau)\delta(\tau-\xi)$. Also assume the initial condition, $x(t_0)$, is known.

For a given (t_0, x_0), the state process $x(\cdot)$ is uniquely determined from $u(\cdot)$. The cost functional associated to (1) is thus considered to be dependent on $u(\cdot)$ and is denoted by the mapping $J : R^m[t_0,t_f] \mapsto R^+$ with the rule of action

$$J(u) = x^T(t_f)Q_f x(t_f) + \int_{t_0}^{t_f} \left[x^T(\tau)Q(\tau)x(\tau) + u^T(\tau)R(\tau)u(\tau)\right] d\tau \tag{2}$$

where $Q(\cdot) \in \mathcal{C}([t_0,t_f];R^{n\times n})$ and $R(\cdot) \in \mathcal{C}([t_0,t_f];R^{m\times m})$ are symmetric with $Q(\cdot)$ positive semidefinite and $R(\cdot)$ positive definite. Also $Q_f \in R^{n\times n}$ is symmetric and positive semidefinite.

Suppose further that there is no observation noise. The state at any instant of time is what it takes for describing completely the future behavior of the system process with different control functions applied. Thus, the optimal

B. Pasik-Duncan: Stochastic Theory and Control, LNCIS 280, pp. 369–383, 2002.

control at time t must be a function of state at time t and of the time t, i.e. $\psi : [t_0, t_f] \times R^n \mapsto R^m$ with control action given by

$$u(t) = \psi(t, x(t))\,, \qquad t \in [t_0, t_f]$$

in such a way that $u(\cdot) \in \mathcal{C}([t_0, t_f]; R^m)$. It is well known that when there is no actuator noise present in (1) the control that minimizes the cost functional, J, is generated via a linear-memoryless feedback law. Also, when the noise is present, a linear-memoryless feedback generates the control that minimizes the expected value of J, which is the first cumulant of J. Furthermore, as shown in [3], under linear, state-feedback, control all cumulants of J have the same quadratic-affine functional form. This common form of the cumulants facilitates the definition of a cumulant based performance index and the formulation of an optimization problem involving a finite number of the cumulants of J. Thus we restrict our search for optimal control solutions to kCC control problems to linear time-varying feedback laws generated from the state $x(\cdot)$ by

$$u(t) = K(t)x(t)\,, \qquad t \in [t_0, t_f] \tag{3}$$

where $K(\cdot) \in \mathcal{C}([t_0, t_f]; R^{m\times n})$ is in the class of admissible feedback gains defined in the following Definition 3.

As shown in [3] with an admissible gain $K(\cdot)$ and $(t_0, x_0) \in \{t_0\} \times R^n$ being given, the kth cost cumulant of J with fixed $k \in Z^+$ has the form

$$\kappa_k(t_0, x_0; K) = x_0^T H(t_0, k)x_0 + D(t_0, k)\,, \tag{4}$$

in which cost cumulant building entities $H(\alpha, k)$ and $D(\alpha, k)$ evaluated at $\alpha = t_0$ satisfy the matrix differential equations

$$\begin{aligned}\frac{d}{d\alpha}H(\alpha,1) &= -[A(\alpha)+B(\alpha)K(\alpha)]^T H(\alpha,1) - K^T(\alpha)R(\alpha)K(\alpha)\\ &\quad -H(\alpha,1)[A(\alpha)+B(\alpha)K(\alpha)] - Q(\alpha)\,, \end{aligned} \tag{5}$$

$$\begin{aligned}\frac{d}{d\alpha}H(\alpha,i) &= -[A(\alpha)+B(\alpha)K(\alpha)]^T H(\alpha,i)\\ &\quad -H(\alpha,i)[A(\alpha)+B(\alpha)K(\alpha)]\\ &\quad -\sum_{j=1}^{i-1}\frac{2i!}{j!(i-j)!}H(\alpha,j)G(\alpha)W(\alpha)G^T(\alpha)H(\alpha,i-j)\end{aligned} \tag{6}$$

and

$$\frac{d}{d\alpha}D(\alpha,\, i) = -Tr\big\{H(\alpha,i)G(\alpha)W(\alpha)G^T(\alpha)\big\}\,, \quad 1 \le i \le k \tag{7}$$

with terminal conditions $H(t_f, 1) = Q_f$, $H(t_f, i) = 0$ for $2 \le i \le k$, and $D(t_f, i) = 0$ for $1 \le i \le k$.

2 Finite Horizon Full-State Feedback kCC Control Problem Statements

Throughout the following development, the subset of symmetric matrices of the vector space of all $n \times n$ matrices with real elements is denoted by S^n. Let k-tuple variables $\mathcal{H}$ and $\mathcal{D}$ be defined as follows

$$\mathcal{H}(\cdot) = (\mathcal{H}_1(\cdot), \ldots, \mathcal{H}_k(\cdot))$$
$$\mathcal{D}(\cdot) = (\mathcal{D}_1(\cdot), \ldots, \mathcal{D}_k(\cdot))$$

with each element $\mathcal{H}_i(\cdot) \in C^1([t_0, t_f]; S^n)$ of $\mathcal{H}$ and $\mathcal{D}_i(\cdot) \in C^1([t_0, t_f]; R)$ of $\mathcal{D}$ having the representations

$$\mathcal{H}_i(\cdot) = H(\cdot, i)$$
$$\mathcal{D}_i(\cdot) = D(\cdot, i)$$

with the right members satisfying the dynamic equations (5)-(7) on $[t_0, t_f]$. Then, for notational simplicity it is convenient to introduce the mappings

$$\mathcal{F}_i : [t_0, t_f] \times \underbrace{S^n \times \cdots \times S^n}_{\text{k-times}} \times R^{m \times n} \mapsto S^n$$
$$\mathcal{G}_i : [t_0, t_f] \times \underbrace{S^n \times \cdots \times S^n}_{\text{k-times}} \mapsto R$$

with the actions

$$\begin{aligned}
\mathcal{F}_1(\alpha, \mathcal{H}, K) &= -[A(\alpha) + B(\alpha)K(\alpha)]^T \mathcal{H}_1(\alpha) - \mathcal{H}_1(\alpha)[A(\alpha) + B(\alpha)K(\alpha)] \\
&\quad - K^T(\alpha)R(\alpha)K(\alpha) - Q(\alpha)\,, \\
\mathcal{F}_i(\alpha, \mathcal{H}, K) &= -[A(\alpha) + B(\alpha)K(\alpha)]^T \mathcal{H}_i(\alpha) - \mathcal{H}_i(\alpha)[A(\alpha) + B(\alpha)K(\alpha)] \\
&\quad - \sum_{j=1}^{i-1} \frac{2i!}{j!(i-j)!} \mathcal{H}_j(\alpha)G(\alpha)W(\alpha)G^T(\alpha)\mathcal{H}_{i-j}(\alpha),\ 2 \le i \le k \\
\mathcal{G}_i(\alpha, \mathcal{H}) &= -Tr\{\mathcal{H}_i(\alpha)G(\alpha)W(\alpha)G^T(\alpha)\}\,, \quad 1 \le i \le k.
\end{aligned}$$

Now it is straightforward to establish the product mappings

$$\mathcal{F}_1 \times \cdots \times \mathcal{F}_k : [t_0, t_f] \times \underbrace{S^n \times \cdots \times S^n}_{\text{k-times}} \times R^{m \times n} \mapsto \underbrace{S^n \times \cdots \times S^n}_{\text{k-times}}$$
$$\mathcal{G}_1 \times \cdots \times \mathcal{G}_k : [t_0, t_f] \times \underbrace{S^n \times \cdots \times S^n}_{\text{k-times}} \mapsto \underbrace{R \times \cdots \times R}_{\text{k-times}}$$

along with the corresponding notations

$$\mathcal{F}(\cdot,\cdot,\cdot) = (\mathcal{F}_1(\cdot,\cdot,\cdot), \ldots, \mathcal{F}_k(\cdot,\cdot,\cdot))$$
$$\mathcal{G}(\cdot,\cdot) = (\mathcal{G}_1(\cdot,\cdot), \ldots, \mathcal{G}_k(\cdot,\cdot))$$

Thus, the dynamic equations of motion (5)-(7) can be rewritten as

$$\frac{d}{d\alpha}\mathcal{H}(\alpha) = \mathcal{F}(\alpha, \mathcal{H}(\alpha), K(\alpha)), \qquad \mathcal{H}(t_f) = \mathcal{H}_f$$
$$\frac{d}{d\alpha}\mathcal{D}(\alpha) = \mathcal{G}(\alpha, \mathcal{H}(\alpha)), \qquad \mathcal{D}(t_f) = \mathcal{D}_f$$

where terminal value conditions $\mathcal{H}_f = (Q_f, 0, \ldots, 0)$ and $\mathcal{D}_f = (0, \ldots, 0)$.

Note that the product system uniquely determines $\mathcal{H}$ and $\mathcal{D}$ once an admissible control gain K is specified. Hence, $\mathcal{H}$ and $\mathcal{D}$ are considered as $\mathcal{H}(\cdot, K)$ and $\mathcal{D}(\cdot, K)$, respectively. The performance index in kCC control problems can now be formulated in terms of K.

Definition 1. (Performance Index)
Fix a $k \in Z^+$ and a sequence $\mu = \{\mu_i \geq 0\}_{i=1}^{k}$ with $\mu_1 > 0$. Then for a given initial condition x_0, the performance index

$$\phi_0 : \{t_0\} \times \underbrace{S^n \times \cdots \times S^n}_{\text{k-times}} \times \underbrace{R \times \cdots \times R}_{\text{k-times}} \mapsto R^+$$

of the finite horizon full-state feedback kCC control problem is defined by

$$\phi_0\left(t_0, \mathcal{H}(t_0, K), \mathcal{D}(t_0, K)\right) = \sum_{i=1}^{k} \mu_i \kappa_i(t_0, x_0; K)$$
$$= x_0^T \sum_{i=1}^{k} \mu_i \mathcal{H}_i(t_0, K) x_0 + \sum_{i=1}^{k} \mu_i \mathcal{D}_i(t_0, K) \quad (8)$$

where the scalar, real constants μ_i represent parametric control freedom and unique symmetric solutions $\{\mathcal{H}_i(t_0, K) \geq 0\}_{i=1}^{k}$ and $\{\mathcal{D}_i(t_0, K) \geq 0\}_{i=1}^{k}$ evaluated at $\alpha = t_0$ satisfy

$$\frac{d}{d\alpha}\mathcal{H}(\alpha) = \mathcal{F}(\alpha, \mathcal{H}(\alpha), K(\alpha)), \qquad \mathcal{H}(t_f) = \mathcal{H}_f$$
$$\frac{d}{d\alpha}\mathcal{D}(\alpha) = \mathcal{G}(\alpha, \mathcal{H}(\alpha)), \qquad \mathcal{D}(t_f) = \mathcal{D}_f$$

in which $\mathcal{H}_f = (Q_f, 0, \ldots, 0)$ and $\mathcal{D}_f = (0, \ldots, 0)$.

From Definition 1, it is clear that the kCC problem is an initial cost problem, in contrast with the more traditional terminal cost class of investigations. One may address an initial cost problem by introducing changes of variables which convert it to a terminal cost problem. However, this modifies the natural context of kCC, which it is preferable to retain. Instead, one may take a more direct dynamic programming approach to the initial cost problem. Such an approach is illustrative of the more general concept of the principle of optimality, an idea tracing its roots back to the 17th century. The development in the sequel is motivated by the excellent treatment in [2], and is intended to follow it closely. Because [2] embodies the traditional endpoint problem and corresponding use of dynamic programming, however, it is necessary to make appropriate modifications in the sequence of results, as well as to introduce the terminology of kCC.

If the terminal time t_f and the pair $(\mathcal{H}_f, \mathcal{D}_f)$ are given, then the other end condition involving the initial time t_0 and state pair $(\mathcal{H}_0, \mathcal{D}_0)$ can be specified by a target set requirement.

Definition 2. (Target Set)

$$(t_0, \mathcal{H}_0, \mathcal{D}_0) \in \mathcal{M}$$

where the target set $\mathcal{M}$ is a closed subset of

$$[t_0, t_f] \times \underbrace{S^n \times \cdots \times S^n}_{\text{k-times}} \times \underbrace{R \times \cdots \times R}_{\text{k-times}}$$

Hence, for given terminal data $(t_f, \mathcal{H}_f, \mathcal{D}_f)$, the class $\mathcal{K}_{t_f, \mathcal{H}_f, \mathcal{D}_f; \mu}$ of admissible feedback gains may be defined as follows.

Definition 3. (Admissible Feedback Gains)
Let the compact subset $\overline{K} \subset R^{m \times n}$ be the allowable set of gain values. For a given $k \in Z^+$ and a sequence $\mu = \{\mu_i \geq 0\}_{i=1}^k$ with $\mu_1 > 0$, let $\mathcal{K}_{t_f, \mathcal{H}_f, \mathcal{D}_f; \mu}$ be the class of $\mathcal{C}([t_0, t_f]; R^{m \times n})$ with values $K(\cdot) \in \overline{K}$ for which solutions to the dynamic equations of motion

$$\frac{d}{d\alpha}\mathcal{H}(\alpha) = \mathcal{F}(\alpha, \mathcal{H}(\alpha), K(\alpha)), \qquad \mathcal{H}(t_f) = \mathcal{H}_f \tag{9}$$

$$\frac{d}{d\alpha}\mathcal{D}(\alpha) = \mathcal{G}(\alpha, \mathcal{H}(\alpha)), \qquad \mathcal{D}(t_f) = \mathcal{D}_f \tag{10}$$

exist and meet both the terminal value conditions and the target constraint $(t_0, \mathcal{H}_0, \mathcal{D}_0) \in \mathcal{M}$.

Then one may state the finite horizon full-state feedback kCC optimization problem.

Definition 4. (kCC)
Suppose $k \in Z^+$ and a sequence $\mu = \{\mu_i \geq 0\}_{i=1}^k$ with $\mu_1 > 0$ are fixed. Then the finite horizon full-state feedback kCC control optimization problem is given by

$$\min_{K(\cdot) \in \mathcal{K}_{t_f, \mathcal{H}_f, \mathcal{D}_f; \mu}} \phi_0\left(t_0, \mathcal{H}(t_0, K), \mathcal{D}(t_0, K)\right) \tag{11}$$

subject to the dynamic equations of motion, for $\alpha \in [t_0, t_f]$

$$\frac{d}{d\alpha}\mathcal{H}(\alpha) = \mathcal{F}(\alpha, \mathcal{H}(\alpha), K(\alpha)), \qquad \mathcal{H}(t_f) = \mathcal{H}_f \tag{12}$$

$$\frac{d}{d\alpha}\mathcal{D}(\alpha) = \mathcal{G}(\alpha, \mathcal{H}(\alpha)), \qquad \mathcal{D}(t_f) = \mathcal{D}_f. \tag{13}$$

In the dynamic programming framework of [2], the terminal time and states of finite horizon full-state feedback kCC optimization problems should be denoted by $(\varepsilon, \mathcal{Y}, \mathcal{Z})$ rather than $(t_f, \mathcal{H}_f, \mathcal{D}_f)$. Thus, the solution of these optimization problems depends on their terminal conditions.

Definition 5. (Value Function)
Suppose that

$$(\varepsilon, \mathcal{Y}, \mathcal{Z}) \in [t_0, t_f] \times \underbrace{S^n \times \cdots \times S^n}_{\text{k-times}} \times \underbrace{R \times \cdots \times R}_{\text{k-times}}$$

is given and fixed. Then the value function $\mathcal{V}(\varepsilon, \mathcal{Y}, \mathcal{Z})$ is defined by

$$\mathcal{V}(\varepsilon, \mathcal{Y}, \mathcal{Z}) = \inf_{K(\cdot) \in \mathcal{K}_{\varepsilon,\mathcal{Y},\mathcal{Z};\mu}} \phi_0\left(t_0, \mathcal{H}(t_0, K), \mathcal{D}(t_0, K)\right)$$

Conventionally, set $\mathcal{V}(\varepsilon, \mathcal{Y}, \mathcal{Z}) = \infty$ when $\mathcal{K}_{\varepsilon,\mathcal{Y},\mathcal{Z};\mu}$ is empty. Next, some properties of the value function may be mentioned.

Theorem 1. *(Property 1: Necessary Condition)*
The value function evaluated along any trajectory corresponding to a control gain feasible for its terminal states is a non-increasing function of time.

Theorem 2. *(Property 2: Necessary Condition)*
The value function evaluated along any optimal trajectory is constant.

It is important to note that these properties are necessary conditions for optimality. The following theorem shows that these conditions are also sufficient. Unless otherwise specified, the dependence of trajectory solutions $\mathcal{H}$ and $\mathcal{D}$ on K are now omitted for notational simplicity.

Theorem 3. *(Sufficient Condition)*
Let $\mathcal{W}(\varepsilon, \mathcal{Y}, \mathcal{Z})$ be an extended real-valued function defined on

$$[t_0, t_f] \times \underbrace{S^n \times \cdots \times S^n}_{\text{k-times}} \times \underbrace{R \times \cdots \times R}_{\text{k-times}}$$

such that $\mathcal{W}(\varepsilon, \mathcal{Y}, \mathcal{Z}) = \phi_0\big(\varepsilon, \mathcal{Y}, \mathcal{Z}\big)$ if $(\varepsilon, \mathcal{Y}, \mathcal{Z}) \in \mathcal{M}$.

Let t_f, $\mathcal{H}_f$, $\mathcal{D}_f$ be given terminal conditions, and suppose, for each trajectory pair $(\mathcal{H}, \mathcal{D})$ corresponding to a control gain K in $\mathcal{K}_{t_f,\mathcal{H}_f,\mathcal{D}_f;\mu}$, that $\mathcal{W}(\alpha, \mathcal{H}(\alpha), \mathcal{D}(\alpha))$ is finite and non-increasing on $[t_0, t_f]$.

If K^* is a control gain in $\mathcal{K}_{t_f,\mathcal{H}_f,\mathcal{D}_f;\mu}$ such that for the corresponding trajectory pair $(\mathcal{H}^*, \mathcal{D}^*)$, $\mathcal{W}(\alpha, \mathcal{H}^*(\alpha), \mathcal{D}^*(\alpha))$ is constant then K^* is an optimal control gain and $\mathcal{W}(t_f, \mathcal{H}_f, \mathcal{D}_f) = \mathcal{V}(t_f, \mathcal{H}_f, \mathcal{D}_f)$.

Corollary 1. *(Restriction of K)*
Let K^* be an optimal control gain in $\mathcal{K}_{t_f,\mathcal{H}_f,\mathcal{D}_f;\mu}$ and $(\mathcal{H}^*,\mathcal{D}^*)$ the corresponding trajectory pair of the dynamic equations

$$\frac{d}{d\alpha}\mathcal{H}(\alpha) = \mathcal{F}(\alpha,\mathcal{H}(\alpha),K(\alpha)), \qquad \mathcal{H}(t_f) = \mathcal{H}_f$$

$$\frac{d}{d\alpha}\mathcal{D}(\alpha) = \mathcal{G}(\alpha,\mathcal{H}(\alpha)), \qquad \mathcal{D}(t_f) = \mathcal{D}_f\,;$$

then the restriction of K^* to $[t_0,\alpha]$ is an optimal control gain for each control problem with terminal conditions $(\alpha,\mathcal{H}^*(\alpha),\mathcal{D}^*(\alpha))$ when $t_0 \le \alpha \le t_f$.

Remarks. Both necessary and sufficient conditions implied by these properties for a control gain to be optimal give hints that one may find a function $\mathcal{W}(\varepsilon,\mathcal{Y},\mathcal{Z})$: $[t_0,t_f]\times\underbrace{S^n\times\cdots\times S^n}_{\text{k-times}}\times\underbrace{R\times\cdots\times R}_{\text{k-times}}\mapsto R^+$such that $\mathcal{W}(\varepsilon,\mathcal{Y},\mathcal{Z}) = \phi_0\big(\varepsilon,\mathcal{Y},\mathcal{Z}\big)$ on $\mathcal{M}$; $\mathcal{W}(\varepsilon,\mathcal{Y},\mathcal{Z})$ is constant on the corresponding trajectory pair; and $\mathcal{W}(\varepsilon,\mathcal{Y},\mathcal{Z})$ is non-increasing on other trajectories.

Section 3 will discuss the construction of scalar-valued functions $\mathcal{W}(\varepsilon,\mathcal{Y},\mathcal{Z})$ and Section 4 carries out such tasks for an important modern application: protection of civil structures.

Definition 6. (Reachable Set)
Define the reachable set $\mathcal{Q}_1$ by

$$\mathcal{Q}_1 = \left\{(\varepsilon,\mathcal{Y},\mathcal{Z}) \in [t_0,t_f]\times\underbrace{S^n\times\cdots\times S^n}_{\text{k-times}}\times\underbrace{R\times\cdots\times R}_{\text{k-times}} : \mathcal{K}_{\varepsilon,\mathcal{Y},\mathcal{Z};\mu}\neq 0\right\}.$$

Then, the next theorem says that the value function must satisfy both a partial differential inequality and an equation at each interior point of the reachable set at which it is differentiable.

Theorem 4. *(HJB for kCC)*
Let $(\varepsilon,\mathcal{Y},\mathcal{Z})$ be any interior point of the reachable set $\mathcal{Q}_1$ *at which the scalar-valued function $\mathcal{V}(\varepsilon,\mathcal{Y},\mathcal{Z})$ is differentiable. Then $\mathcal{V}(\varepsilon,\mathcal{Y},\mathcal{Z})$ satisfies the partial differential inequality*

$$\frac{\partial}{\partial\varepsilon}\mathcal{V}(\varepsilon,\mathcal{Y},\mathcal{Z}) + \frac{\partial}{\partial vec(\mathcal{Y})}\mathcal{V}(\varepsilon,\mathcal{Y},\mathcal{Z})\cdot vec(\mathcal{F}(\varepsilon,\mathcal{Y},K))$$
$$+\frac{\partial}{\partial vec(\mathcal{Z})}\mathcal{V}(\varepsilon,\mathcal{Y},\mathcal{Z})\cdot vec(\mathcal{G}(\varepsilon,\mathcal{Y})) \le 0$$

for all $K\in\overline{K}$ and $vec(\cdot)$ the vectorizing operator of enclosed entities.

If there is an optimal control gain K^* in $\mathcal{K}_{\varepsilon,\mathcal{Y},\mathcal{Z};\mu}$, then the partial differential equation of dynamic programming

$$\min_{K\in\overline{K}}\left\{\frac{\partial}{\partial\varepsilon}\mathcal{V}(\varepsilon,\mathcal{Y},\mathcal{Z}) + \frac{\partial}{\partial vec(\mathcal{Y})}\mathcal{V}(\varepsilon,\mathcal{Y},\mathcal{Z})\cdot vec(\mathcal{F}(\varepsilon,\mathcal{Y},K))\right.$$

$$+\frac{\partial}{\partial vec(\mathcal{Z})}\mathcal{V}(\varepsilon,\mathcal{Y},\mathcal{Z})\cdot vec(\mathcal{G}(\varepsilon,\mathcal{Y}))\Big\} = 0 \tag{14}$$

is satisfied.

One may now refine the definition of admissible feedback control gains.

Definition 7. (Refined Reachable Set)
Let the admissible control gain K be a function

$$K = K(\alpha,\mathcal{H},\mathcal{D})$$

from a subset $\mathcal{Q}_{11}$ of

$$[t_0,t_f]\times\underbrace{S^n\times\cdots\times S^n}_{\text{k-times}}\times\underbrace{R\times\cdots\times R}_{\text{k-times}}$$

into $\overline{K}$ such that for each $(\varepsilon,\mathcal{Y},\mathcal{Z})$ in $\mathcal{Q}_{11}$, there is a unique solution pair $(\mathcal{H}(\alpha;\varepsilon,\mathcal{Y}),\mathcal{D}(\alpha;\varepsilon,\mathcal{Z}))$ to the dynamic equations of motion

$$\frac{d}{d\alpha}\mathcal{H}(\alpha) = \mathcal{F}(\alpha,\mathcal{H},K(\alpha,\mathcal{H},\mathcal{D}))\,,$$
$$\frac{d}{d\alpha}\mathcal{D}(\alpha) = \mathcal{G}(\alpha,\mathcal{H})$$

on an interval $t_0(\varepsilon,\mathcal{Y},\mathcal{Z})\le\alpha\le\varepsilon$ with $\mathcal{H}(\varepsilon;\varepsilon,\mathcal{Y})=\mathcal{Y}$ and $\mathcal{D}(\varepsilon;\varepsilon,\mathcal{Z})=\mathcal{Z}$, such that $(\alpha,\mathcal{H}(\alpha;\varepsilon,\mathcal{Y}),\mathcal{D}(\alpha;\varepsilon,\mathcal{Z}))\in\mathcal{Q}_{11}$ for $t_0(\varepsilon,\mathcal{Y},\mathcal{Z})\le\alpha\le\varepsilon$ and

$$(t_0(\varepsilon,\mathcal{Y},\mathcal{Z}),\mathcal{H}(t_0(\varepsilon,\mathcal{Y},\mathcal{Z});\varepsilon,\mathcal{Y}),\mathcal{D}(t_0(\varepsilon,\mathcal{Y},\mathcal{Z});\varepsilon,\mathcal{Z}))\in\mathcal{M}.$$

Obviously, $\mathcal{Q}_{11}$ is a subset of the reachable set $\mathcal{Q}_1$. Whenever the finite horizon kCC control problem formulated for any set of terminal conditions $(\varepsilon,\mathcal{Y},\mathcal{Z})$ in the reachable set $\mathcal{Q}_{11}$ admits an optimal feedback control gain, the equality

$$\mathcal{V}(\varepsilon,\mathcal{Y},\mathcal{Z}) = \phi_0\left(t_0(\varepsilon,\mathcal{Y},\mathcal{Z}),\mathcal{H}(t_0(\varepsilon,\mathcal{Y},\mathcal{Z});\varepsilon,\mathcal{Y}),\mathcal{D}(t_0(\varepsilon,\mathcal{Y},\mathcal{Z});\varepsilon,\mathcal{Z})\right)$$

thus holds between the performance index of Definition 1 and the value function.

Theorem 5. *(Differentiability of Value Function)*
Suppose that $K^*(\alpha,\mathcal{H},\mathcal{D})$ is an optimal feedback control gain and $t_0(\varepsilon,\mathcal{Y},\mathcal{Z})$ and $(\mathcal{H}(t_0(\varepsilon,\mathcal{Y},\mathcal{Z});\varepsilon,\mathcal{Y}),\mathcal{D}(t_0(\varepsilon,\mathcal{Y},\mathcal{Z});\varepsilon,\mathcal{Z}))$ are the initial time and initial states for the trajectories of

$$\frac{d}{d\alpha}\mathcal{H}(\alpha) = \mathcal{F}(\alpha,\mathcal{H},K^*(\alpha,\mathcal{H},\mathcal{D}))\,,$$
$$\frac{d}{d\alpha}\mathcal{D}(\alpha) = \mathcal{G}(\alpha,\mathcal{H})$$

with terminal condition $(\varepsilon,\mathcal{Y},\mathcal{Z})$. Then the value function $\mathcal{V}(\varepsilon,\mathcal{Y},\mathcal{Z})$ is differentiable at each point at which $t_0(\varepsilon,\mathcal{Y},\mathcal{Z})$ and $\mathcal{H}(t_0(\varepsilon,\mathcal{Y},\mathcal{Z});\varepsilon,\mathcal{Y})$ and $\mathcal{D}(t_0(\varepsilon,\mathcal{Y},\mathcal{Z});\varepsilon,\mathcal{Z})$ are differentiable with respect to $(\varepsilon,\mathcal{Y},\mathcal{Z})$.

The verification theorem is next stated in the notation of this paper.

Theorem 6. *(Verification Theorem)*
Fix $k \in Z^+$ and let $\mathcal{W}(\varepsilon, \mathcal{Y}, \mathcal{Z})$ be a continuously differentiable solution of the equation (14) which satisfies the boundary condition

$$\mathcal{W}(\varepsilon, \mathcal{Y}, \mathcal{Z}) = \phi_0(\varepsilon, \mathcal{Y}, \mathcal{Z}), \qquad \text{for } (\varepsilon, \mathcal{Y}, \mathcal{Z}) \in \mathcal{M}. \tag{15}$$

Let $(t_f, \mathcal{H}_f, \mathcal{D}_f)$ be a point of $\mathcal{Q}_{11}$, K a control gain in $\mathcal{K}_{t_f,\mathcal{H}_f,\mathcal{D}_f;\mu}$ and $\mathcal{H}$ and $\mathcal{D}$ the corresponding solutions of the equations of motion (12)-(13). Then $\mathcal{W}(\alpha, \mathcal{H}(\alpha), \mathcal{D}(\alpha))$ is a non-increasing function of α. If K^* is a control gain in $\mathcal{K}_{t_f,\mathcal{H}_f,\mathcal{D}_f;\mu}$ defined on $[t_0^*, t_f]$ with corresponding solution, $\mathcal{H}^*$ and $\mathcal{D}^*$ of the equations (12)-(13) such that for $\alpha \in [t_0^*, t_f]$

$$\begin{aligned} &\frac{\partial}{\partial \varepsilon}\mathcal{W}(\alpha, \mathcal{H}^*(\alpha), \mathcal{D}^*(\alpha)) \\ &+\frac{\partial}{\partial\, vec(\mathcal{Y})}\mathcal{W}(\alpha, \mathcal{H}^*(\alpha), \mathcal{D}^*(\alpha)) \cdot vec(\mathcal{F}(\alpha, \mathcal{H}^*(\alpha), K^*(\alpha))) \\ &+\frac{\partial}{\partial vec(\mathcal{Z})}\mathcal{W}(\alpha, \mathcal{H}^*(\alpha), \mathcal{D}^*(\alpha)) \cdot vec(\mathcal{G}(\alpha, \mathcal{H}^*(\alpha))) = 0 \end{aligned} \tag{16}$$

then K^* is an optimal control gain in $\mathcal{K}_{t_f,\mathcal{H}_f,\mathcal{D}_f;\mu}$ and

$$\mathcal{W}(\varepsilon, \mathcal{Y}, \mathcal{Z}) = \mathcal{V}(\varepsilon, \mathcal{Y}, \mathcal{Z}) \tag{17}$$

where $\mathcal{V}(\varepsilon, \mathcal{Y}, \mathcal{Z})$ is the value function.

In the following section, $\mathcal{M} = \{t_0\} \times \underbrace{S^n \times \cdots \times S^n}_{\text{k-times}} \times \underbrace{R \times \cdots \times R}_{\text{k-times}}.$

3 Finite Horizon Full-State Feedback *k*CC Control Solution

The approach to obtaining a control solution to the finite horizon full-state feedback kCC problem using the dynamic programming methodology requires one to parameterize the terminal time and states of the optimization problem which we choose to denote by $(\varepsilon, \mathcal{Y}, \mathcal{Z})$ rather than $(t_f, \mathcal{H}_f, \mathcal{D}_f)$. That is, for $\varepsilon \in [t_0, t_f]$ and $1 \leq i \leq k$, the states of the system (12)-(13) defined on the interval $[t_0, \varepsilon]$ have terminal values denoted by $\mathcal{H}(\varepsilon) = \mathcal{Y}$ and $\mathcal{D}(\varepsilon) = \mathcal{Z}$. Furthermore, the performance index (8) is quadratic affine in terms of arbitrarily fixed x_0. Along lines suggested by [4] solution to the HJB equation (14) may be sought in the form

$$\mathcal{W}(\varepsilon, \mathcal{Y}, \mathcal{Z}) = x_0^T \sum_{i=1}^{k} \mu_i(\mathcal{Y}_i + \mathcal{E}_i(\varepsilon))\, x_0 + \sum_{i=1}^{k} \mu_i(\mathcal{Z}_i + \mathcal{T}_i(\varepsilon)) \tag{18}$$

where parametric functions of time $\mathcal{E}_i(\varepsilon) \in \mathcal{C}^1([t_0, t_f]; S^n)$ and $\mathcal{T}_i(\varepsilon) \in \mathcal{C}^1([t_0, t_f]; R)$ are to be determined. The next lemma shows how the time derivative of $\mathcal{W}(\varepsilon, \mathcal{Y}, \mathcal{Z})$ looks like using inverse vectorizing transformation

Lemma 1. *Fix a $k \in Z^+$ and let $(\varepsilon, \mathcal{Y}, \mathcal{Z})$ be any interior point of the reachable set $\mathcal{Q}_1$ at which the real-valued function $\mathcal{W}(\varepsilon, \mathcal{Y}, \mathcal{Z})$ (18) is differentiable. Then*

$$\frac{d}{d\varepsilon}\mathcal{W}(\varepsilon, \mathcal{Y}, \mathcal{Z}) = \frac{\partial}{\partial \varepsilon}\mathcal{W}(\varepsilon, \mathcal{Y}, \mathcal{Z}) + \sum_{i=1}^{k} \frac{\partial}{\partial \mathcal{Y}_i}\mathcal{W}(\varepsilon, \mathcal{Y}, \mathcal{Z})\mathcal{F}_i(\varepsilon, \mathcal{Y}, K) + \sum_{i=1}^{k} \frac{\partial}{\partial \mathcal{Z}_i}\mathcal{W}(\varepsilon, \mathcal{Y}, \mathcal{Z})\mathcal{G}_i(\varepsilon, \mathcal{Y})$$

for all $K \in \overline{K}$.

Finally, it is now appropriate to state the optimal control solution for finite horizon full-state feedback kCC control problems. Because of space limitations, the proof is omitted.

Theorem 7. *(Finite Horizon Full-State Feedback kCC Control Solution) Let $A(\cdot) \in \mathcal{C}([t_0, t_f]; R^{n \times n})$; $B(\cdot) \in \mathcal{C}([t_0, t_f]; R^{n \times m})$; $Q(\cdot) \in \mathcal{C}([t_0, t_f]; R^{n \times n})$ symmetric positive semidefinite; and $R(\cdot) \in \mathcal{C}([t_0, t_f]; R^{m \times m})$ symmetric positive definite. Suppose further both $k \in Z^+$ and the sequence $\mu = \{\mu_i \geq 0\}_{i=1}^{k}$ where $\mu_1 > 0$ are fixed. Then optimal full-state feedback is achieved by the gain $K^*(\cdot)$ given by*

$$K^*(\alpha) = -R^{-1}(\alpha)B^T(\alpha)\left[H^*(\alpha, 1) + \sum_{i=2}^{k} \mu_i H^*(\alpha, i)\right], \tag{19}$$

whenever $\{H^*(\alpha, i)\}_{i=1}^{k}$ are solutions of the backward differential equations

$$\begin{aligned}\frac{d}{d\alpha}H^*(\alpha, 1) = &-[A(\alpha) + B(\alpha)K^*(\alpha)]^T H^*(\alpha, 1) - Q(\alpha) \\ &-H^*(\alpha, 1)[A(\alpha) + B(\alpha)K^*(\alpha)] - K^{*T}(\alpha)R(\alpha)K^*(\alpha),\end{aligned} \tag{20}$$

$$\begin{aligned}\frac{d}{d\alpha}H^*(\alpha, i) = &-[A(\alpha) + B(\alpha)K^*(\alpha)]^T H^*(\alpha, i) \\ &-H^*(\alpha, i)[A(\alpha) + B(\alpha)K^*(\alpha)] \\ &-\sum_{j=1}^{i-1} \frac{2i!}{j!(i-j)!} H^*(\alpha, j)G(\alpha)W(\alpha)G^T(\alpha)H^*(\alpha, i-j)\end{aligned} \tag{21}$$

with terminal conditions $H^*(t_f, 1) = Q_f$, and $H^*(t_f, i) = 0$ when $2 \leq i \leq k$.

Remarks. Observe that the structure of the optimal control depends on the structure of the controlled system, the weighting matrices and the second order statistics of the disturbance process. Once all of these are specified, the optimal control can be implemented by a linear feedback gain described by (19), which reduces to well known results when $k = 1$. Further, it should be noted that equations (20)-(21) are coupled. Thus, in solving this finite horizon full-state feedback kCC problem, the equations (20)-(21) must be simultaneously integrated over a finite time interval of interest.

4 Earthquake-Protection Structure Benchmark

Reference [10] presents the First Generation Structure Benchmark for protection of civil structures from earthquakes. The benchmark defines a model for the structural behavior of a realistic, actively controlled, three-story, single-bay, experimental building with an active mass driver (AMD) installed on the top floor. See Fig. 1 for details. The test structure is designed to be a scale model of the prototype building discussed in [1] and is subject to one-dimensional ground motion. The input excitations come from 1940 El Centro and 1968 Hachinohe historical earthquake records. The performance of a candidate controller is then judged by ten performance criteria along with several implementation constraints as described in the Benchmark Problem Statement. Criteria 1-5 and 6-10 attach to rms and peak performance, respectively. In each group, criteria 1-2 and 6-7 refer to floor displacements and accelerations, whereas 3-5 and 8-10 measure actuator displacement, velocity and acceleration. The formulae for $\{J_i\}$ involve normalization and consideration of all three floors. The reader may consult [10] for additional details.

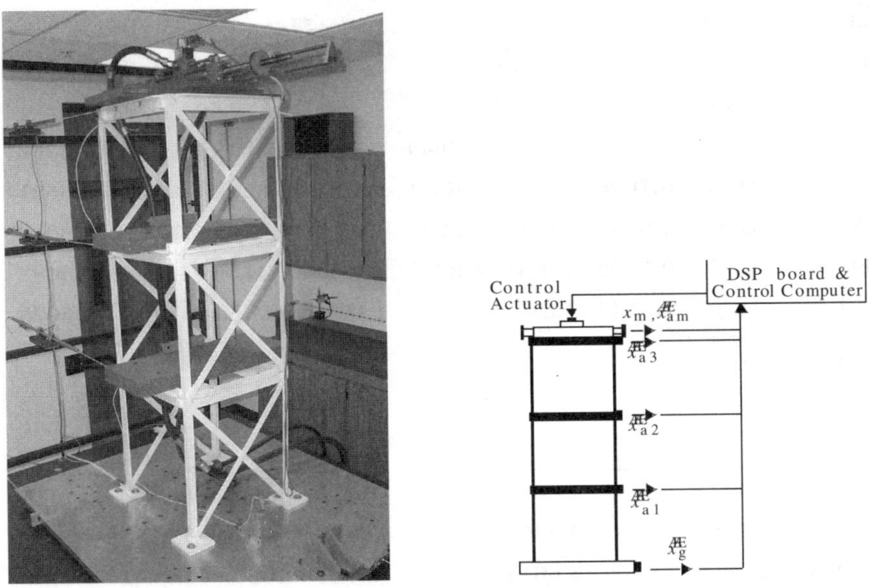

Fig. 1. First Generation Structure Benchmark of Seismically Excited 3DOF Building. (**a**) AMD-Controlled Structure. (**b**) Schematic Diagram of Experiment Setup

Simulation Results

Numerical simulations have been performed to assess the effects produced by the cost cumulants, as their features are incorporated into the controller. Table 1 summarizes the trends in structural responses and control effort for the test structure subject to El Centro and Hachinohe earthquake excitations when controllers use the 1st and ith cost cumulants where $i = 2, 3, 4$. As can be seen from Table 1, the rms and peak structural responses can both be made smaller than in the case of baseline LQG control for different values of the control weightings μ_i. As expected, there is some trade-off behavior of control effectiveness versus control resource. That is, more actuator control input is required for the better control performance. Moreover, all CC controllers remain within implementation constraints. Note that bold entries denote which quake produced the bigger effect.

Table 1. Effects of Second, Third, and Fourth Cumulants vs. Baseline LQG Control

Eval. Criteria	Baseline LQG $\mu_1 = 1.0$		2nd Cumulant $\mu_2 = 1.2 \times 10^{-5}$		3rd Cumulant $\mu_3 = 5.0 \times 10^{-11}$		4th Cumulant $\mu_4 = 1.4 \times 10^{-16}$	
J1	0.2897		0.2071		0.1942		0.1896	
J2	0.4439		0.3127		0.2946		0.2883	
J3	0.4840		0.7258		0.7784		0.7988	
J4	0.4853		0.7195		0.7698		0.7893	
J5	0.5976		0.7452		0.7863		0.8027	
J6	0.4021	**0.4556**	0.3321	**0.3844**	0.3159	**0.3788**	0.3096	**0.3766**
J7	0.6361	**0.7096**	0.5095	**0.6681**	0.4787	**0.6562**	0.4671	**0.6521**
J8	0.5953	**0.6691**	1.0686	**1.3287**	1.1787	**1.5225**	1.2196	**1.6013**
J9	0.6058	**0.7802**	1.0834	**1.3237**	1.1862	**1.4771**	1.2254	**1.5383**
J10	0.9397	**1.3142**	1.0971	**1.5189**	1.1511	**1.5852**	1.1641	**1.6073**
uRMS	0.1441		0.2345		0.2549		0.2628	
amRMS	1.0696		1.3338		1.4074		1.4368	
xmRMS	0.6341		0.9508		1.0197		1.0464	
uMAX	**0.5248**	0.3065	**1.0079**	0.6193	**1.1113**	0.7096	**1.1498**	0.7461
amMAX	**4.7454**	3.3907	**5.5402**	3.9188	**5.8129**	4.0899	**5.8785**	4.1467
xmMAX	**2.0059**	1.1106	**3.6012**	2.2056	**3.9721**	2.5273	**4.1101**	2.6582
	Centro	Hach	Centro	Hach	Centro	Hach	Centro	Hach

Benchmark Evaluation Comparisons

Tables 2 and 3 contain performance comparisons for the baseline LQG, the 4CC extension of baseline LQG, and various control designs in [10]. As can be observed, with chosen $\mu_2 = 1.0 \times 10^{-5}$, $\mu_3 = 9.0 \times 10^{-12}$ and $\mu_4 = 2.0 \times 10^{-20}$ the performance of the 4CC controller exceeds those of baseline LQG, Multi-Objective, Covariance, Probabilistic, and Fuzzy controllers and is comparable to those produced by the Sliding Mode and μ-Synthesis controllers. Herein by adopting the design weighting parameters from the baseline LQG, the 4CC control design was done in a short amount of time. Yet, its control performance was still able to compete well with other, more established, control algorithms not so constrained. Not surprisingly, the 4CC design employs more actuation effort to further decrease structural displacement and acceleration both in rms and peak quantities. The 4CC control design simply exploits the available actuator constraint. Finally, Fig. 2 indicates that the 4CC design maintains a comparable stability margin as compared to that of the baseline

Table 2. Benchmark Evaluation Comparisons

	Sample LQG		Multi-Objective		Sliding Mode		4CC	
	Centro	Hach	Centro	Hach	Centro	Hach	Centro	Hach
J1	0.2896		0.3130		0.1878		0.1869	
J2	0.4439		0.4795		0.2846		0.2845	
J3	0.4840		0.4333		0.8444		0.8113	
J4	0.4853		0.4383		0.8259		0.8012	
J5	0.5976		0.3693		0.8327		0.8136	
J6	0.4021	**0.4556**	0.4144	**0.4487**	0.3083	**0.3779**	0.3058	**0.3752**
J7	0.6361	**0.7096**	0.6300	**0.7492**	0.4725	**0.6616**	0.4594	**0.6491**
J8	0.5953	**0.6691**	0.6182	**0.6878**	1.2479	**1.6240**	1.2439	**1.6494**
J9	0.6058	**0.7802**	0.6067	**0.9540**	1.2098	**1.4811**	1.2481	**1.5758**
J10	0.9397	**1.3142**	0.9199	**1.1568**	1.1078	**1.6491**	1.1748	**1.6333**
uRMS	0.1441		0.1469		0.2710		0.2676	
amRMS	1.0696		0.6611		1.4905		1.4563	
xmRMS	0.6341		0.5677		1.1062		1.0629	
uMAX	**0.5248**	0.3065	**0.6030**	0.3501	**1.1601**	0.7539	**1.1724**	0.7686
amMAX	**4.7454**	3.3907	**4.6455**	2.9846	**5.5942**	4.2546	**5.9327**	4.2139
xmMAX	**2.0059**	1.1106	**2.0834**	1.1417	**4.2056**	2.6958	**4.1921**	2.7379

Table 3. Benchmark Evaluation Comparisons

	μ-Synthesis		Covariance		Probabilistic		Fuzzy Control	
J1	0.1915		0.2762		0.2070		0.3232	
J2	0.2934		0.4205		0.3453		0.5087	
J3	0.8410		0.5161		0.8507		0.4894	
J4	0.8363		0.5200		0.8318		0.4137	
J5	0.8131		0.5001		0.6834		0.5891	
J6	0.3122	**0.3780**	0.3966	**0.4369**	0.3451	**0.3796**	0.4158	**0.4748**
J7	0.4790	**0.6791**	0.6267	**0.6908**	0.5347	**0.6843**	0.8145	**0.8666**
J8	1.2670	**1.4289**	0.6322	**0.7197**	1.3408	**1.6435**	0.5428	**0.6249**
J9	1.2141	**1.5374**	0.6779	**0.9257**	1.2003	**1.5579**	0.5847	**0.6474**
J10	1.0951	**1.2368**	0.9566	**1.0589**	0.8594	**0.9358**	1.1000	**1.2994**
uRMS	0.2792		0.1585		0.2956		0.1580	
amRMS	1.4555		0.8952		1.2233		1.0545	
xmRMS	1.1016		0.6761		1.1144		0.6411	
uMAX	**1.1889**	0.6614	**0.5957**	0.3387	**1.2605**	0.7576	**0.5506**	0.2824
amMAX	**5.5301**	3.1909	**4.8307**	2.7320	**4.3401**	2.4145	**5.5550**	3.3525
xmMAX	**4.2699**	2.3719	**2.1306**	1.1947	**4.5185**	2.7282	**1.8291**	1.0374
	Centro	Hach	Centro	Hach	Centro	Hach	Centro	Hach

LQG in addition to its better structural performance, where the distance from the origin to a Nyquist plot is a measure of the stability margin.

5 Conclusions

Finite linear combinations of cost cumulants as discussed herein are an extension of the risk sensitive idea, with one design parameter available for each additional cost cumulant. When $k > 2$, they represent an extension [5]-[6] of the ideas [7]-[9] introduced by Sain in the mid 1960s. The optimum multiple cost cumulants offer the designer substantial parametric freedom to be utilized in attempting to achieve better structural control performance. Moreover, the present results involving the first four cost cumulants appear to be the first of their type to appear in the literature. These performance results also indicate in a most encouraging way that both third and fourth cost cumulants may be able to support second and first cost cumulants in

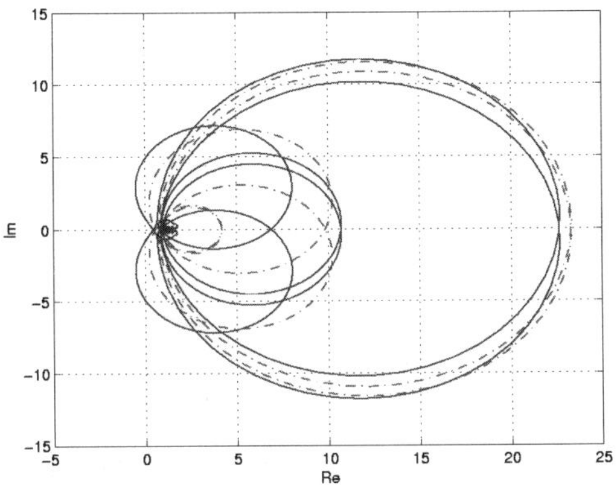

Fig. 2. Nyquist Plots of Determinant of Return Difference Matrices for baseline LQG and 4CC control designs. The return difference as depicted by (*dashed-dotted curve*) is that of the baseline LQG in the Benchmark. The other plot as described by the (*solid curve*) corresponds to the 4CC design.

achieving desirable control system behaviors, though of course not all with the same level of influence.

References

1. Chung L.L., Lin R.C., Soong T.T., Reinhorn A.M. (1989) Experiments on Active Control for MDOF Seismic Structures. Journal of Engineering Mechanics, Vol. 115, No. 8: 1609-1627
2. Fleming W.H., Rishel R.W. (1975) Deterministic and Stochastic Optimal Control. Springer, New York
3. Liberty S.R., Hartwig R.C. (1976) On the Essential Quadratic Nature of LQG Control-Performance Measure Cumulants. Information and Control, Vol. 32, No. 3:276-305
4. Mou L., Liberty S.R., Pham K.D., Sain M.K. (2000) Linear Cumulant Control and Its Relationship to Risk-Sensitive Control. Proceedings Allerton Conference on Communication, Control, and Computing, 422-430
5. Pham K.D. (1998) Design of kth-Order Linear Optimal Cost Cumulant Controller. MS Thesis, University of Nebraska at Lincoln, Nebraska
6. Pham K.D., Liberty S.R., Sain M.K., Spencer B.F. (1999) Generalized Risk Sensitive Building Control: Protecting Civil Structures with Multiple Cost Cumulants. Proceedings American Control Conference. San Diego, California, 500-504
7. Sain M.K. (1965) On Minimal-Variance Control of Linear Systems with Quadratic Loss. PhD Dissertation, University of Illinois at Urbana, Illinois
8. Sain M.K. (1965) Relative Costs of Mean and Variance Control for a Class of Linear, Noisy Systems. Proceedings Allerton Conference on Circuit and System Theory, 121-129
9. Sain M.K. (1966) Control of Linear Systems According to the Minimal Variance Criterion-A New Approach to the Disturbance Problem. IEEE Transactions on Automatic Control, AC-11, No. 1: 118-122
10. Spencer B.F., Dyke S.J., Deoskar H.S. (1998) First Generation Benchmark. Special Issue of Earthquake Engineering and Structural Dynamics, Vol. 27, No. 11: 1127-1139

Robust Stochastic Maximum Principle: A Measured Space as Uncertainty Set

Alex S. Poznyak

CINVESTAV-IPN, Departmento de Control Automatico, AP 14-740, CP 07300, Mexico D.F., Mexico. apoznyak@ctrl.cinvestav.mx

Abstract. This paper develops a version of Robust Stochastic Maximum Principle (*RSMP*) applied to the Minimax Mayer Problem formulated for stochastic differential equations with the control-dependent diffusion term. The parametric families of first and second order adjoint stochastic processes are introduced to construct the corresponding Hamiltonian formalism. The Hamiltonian function used for the construction of the robust optimal control is shown to be equal to the Lebesque integral over a parametric set of the standard stochastic Hamiltonians corresponding to a fixed value of the uncertain parameter. The paper deals with a cost function given at finite horizon and containing the mathematical expectation of a terminal term. A terminal condition, covered by a vector function, is also considered. The optimal control strategies, adapted for available information, for the wide class of uncertain systems given by an stochastic differential equation with unknown parameters from a given compact set, are constructed. This problem belongs to the class of minimax stochastic optimization problems. The proof is based on the recent results obtained for Minimax Mayer Problem with a finite uncertainty set [14], [43], [44] and [45] as well as on the variation results of [53] derived for Stochastic Maximum Principle for nonlinear stochastic systems under complete information. The corresponding discussion of the obtain results concludes this study.

Keywods: Robust Control, Maximum Principle, Minimax Problem, Stochastic Processes.

AMS (MOS) Subject Classification: 05C38, 15A15, 05A15, 15A18.

1 Introduction

During the last decades, the minimax control problem, dealing with different classes of nonlinear systems, has received much attention from many researchers because of its theoretical and practical importance. Basically, the results of this area are based on two classical approaches: *Maximum Principle* (MP) [41] and *Dynamic Programming method* (DP) [3]. In the case of a complete model description, both of them can be directly applied to construct the optimal control. Various forms of the *Stochastic Maximum Principle* have been published in the literature [37], [28], [30], [8], [9], [4], [5], [31], [32] and [25], [57], [56], [53]. The DP approach has been studied in [20], [40], [35], [18], [19] and [51].

Faced with some *uncertainties* (parametric type, unmodelled dynamics, external perturbations etc.) these results cannot be applied. There are two

B. Pasik-Duncan: Stochastic Theory and Control, LNCIS 280, pp. 385–397, 2002.

ways to overcome the uncertainty problems: the adaptive approach [26], [27] and *minimax or robust control*, where the maximization is taken over a set of possible uncertainties and the minimization is taken over all of the control strategies within a given set [1], [23], [34], [36], [55], [24], [29], [39], [33], [38], [46], [22]and [47].

Recently, the robust MP was derived in [14] for a deterministic Mayer problem and generalized in [15] for Bolza and Lagrange problems. For *stochastic uncertain systems*, a minimax control of a class of dynamic systems with mixed uncertainties was investigated in [2], [42], [48], [52], [21], [7], [17], [6] and [50].

The purpose of this paper is to explore the possibilities of the MP approach for a class of minimax control problems for uncertain systems given by a system of stochastic differential equations with a controlled diffusion term and unknown parameters within a given measured compact set. The proof is based on the recent results obtained for Minimax Mayer Problem with a finite uncertainty set [14], [43], [44] and [45] as well as on the results of [53] derived for Stochastic Maximum Principle for nonlinear stochastic systems under complete information. The *Tent Method* [10], [11], [12] and [13] is used to formulate the necessary conditions of optimality in Hamiltonian form.

2 Problem Setting

2.1 Stochastic Uncertain System

Let $(\Omega, \mathcal{F}, \{\mathcal{F}_t\}_{t\geq 0}, \mathbf{P})$ be a given filtered probability space, that is, the probability space $(\Omega, \mathcal{F}, \mathbf{P})$ is complete, the sigma-algebra $\mathcal{F}_0$ contains all the $\mathbf{P}$ - null sets in $\mathcal{F}$ and the filtration $\{\mathcal{F}_t\}_{t\geq 0}$ is right continuous ($\mathcal{F}_{t+} := \cap_{s>t}\mathcal{F}_s = \mathcal{F}_t$). On this probability space an m- dimensional standard Brownian motion is defined. Consider the stochastic nonlinear controlled continuous-time system with the dynamics $x(t)$ given by

$$\begin{cases} dx(t) = b^{\alpha}(t, x(t), u(t))dt + \sigma^{\alpha}(t, x(t), u(t))dW(t) \\ x(0) \; = x_0, \quad t \in [0, T]\,(T > 0) \end{cases} \tag{1}$$

In the above $u(t) \in U$ is a control at time t and $b^{\alpha} : [0, T] \times \mathbb{R}^n \times U \to \mathbb{R}^n$, $\sigma^{\alpha} : [0, T] \times \mathbb{R}^n \times U \to \mathbb{R}^{n\times m}$. The parameter α is supposed to be a priory unknown and running a given parametric set $\mathcal{A}$ from a space with a *countable additive measure* m. For any $\alpha \in \mathcal{A}$ denote

$$\begin{aligned} b^{\alpha}(t, x, u) &:= (b_1^{\alpha}(t, x, u), \dots, b_n^{\alpha}(t, x, u))^{\mathsf{T}} \\ \sigma^{\alpha}(t, x, u) &:= \left(\sigma^{1\alpha}(t, x, u), \dots, \sigma^{n\alpha}(t, x, u)\right) \\ \sigma^{j\alpha}(t, x, u) &:= \left(\sigma_1^{j\alpha}(t, x, u), \dots, \sigma_m^{j\alpha}(t, x, u)\right)^{\mathsf{T}} \end{aligned}$$

It is assumed that

A1: $\{\mathcal{F}_t\}_{t\geq 0}$ is the natural filtration generated by $(W(t), t \geq 0)$ and augmented by the **P**- null sets from $\mathcal{F}$ and

A2: (U, d) is a separable metric space with a metric d.

Definition 1. The function $f : [0,T] \times \mathbb{R}^n \times U \to \mathbb{R}^{n\times m}$ is said to be an $L_\phi(C^2)$ -mapping if it is Borel measurable, it is C^2 in x for any $t \in [0,T]$ and any $u \in U$, there exist a constant L and a modulus of continuity $\phi : [0,\infty) \to [0,\infty)$ such that for any $t \in [0,T]$ and for any $x, u, \hat{x}, u \in \mathbb{R}^n \times U \times \mathbb{R}^n \times U$

$$\begin{aligned}
\|f(t,x,u) - f(t,\hat{x},\hat{u})\| &\leq L\|x-\hat{x}\| + \phi(d(u,\hat{u})), \quad \|f(t,0,u)\| \leq L \\
\|f_x(t,x,u) - f_x(t,\hat{x},\hat{u})\| &\leq L\|x-\hat{x}\| + \phi(d(u,\hat{u})) \\
\|f_{xx}(t,x,u) - f_{xx}(t,\hat{x},\hat{u})\| &\leq \phi(\|x-\hat{x}\| + d(u,\hat{u}))
\end{aligned}$$

(here $f_x(\cdot,x,\cdot)$ and $f_{xx}(\cdot,x,\cdot)$ are the partial derivatives of the first and the second order).

A3: for any $\alpha \in \mathcal{A}$ both $b^\alpha(t,x,u)$ and $\sigma^\alpha(t,x,u)$ are supposed to be $L_\phi(C^2)$-mappings.

Let $\mathcal{A}_0 \subset \mathcal{A}$ be measurable subsets with a finite measure, that is, $m(\mathcal{A}_0) < \infty$.

A4: All components $b^\alpha(t,x,u)$, $\sigma^\alpha(t,x,u)$ are measurable with respect to α, that is, for any $i = 1,\dots,n$, $j = 1,\dots,m$, $c \in R^1$, $x \in R^n$, $u \in U$ and $t \in [0,T]$

$$\{\alpha : b_i^\alpha(t,x,u) \leq c\} \in \mathcal{A}, \quad \{\alpha : \sigma_j^{i\alpha}(t,x,u) \leq c\} \in \mathcal{A}$$

Moreover, every considered function of α is assume to be measurable with respect to α. It is assumed that *the past information is available* for the controller. To emphasize the dependence of the random trajectories on the parameter $\alpha \in \mathcal{A}$ the equation (1) is rewritten as

$$\begin{cases}
dx^\alpha(t) = b^\alpha(t, x^\alpha(t), u(t))dt + \sigma^\alpha(t, x^\alpha(t), u(t))dW(t) \\
x^\alpha(0) \;\; = x_0, \quad t \in [0,T] \, (T > 0)
\end{cases} \tag{2}$$

2.2 A Terminal Condition, A Feasible and Admissible Control

Definition 2. A stochastic control $u(\cdot)$ is called a **feasible** in the stochastic sense (or, s- feasible) for the system (2) if

$$u(\cdot) \in \mathcal{U}[0,T] := \{u : [0,T] \times \Omega \to U \mid u(\cdot) \text{ is } \{\mathcal{F}_t\}_{t\geq 0}\text{-adapted}\}$$

$x^\alpha(t)$ is the unique solution of (2) in the sense that for any $x^\alpha(t)$ and $\hat{x}^\alpha(t)$, satisfying (2),

$$\mathbf{P}\{\omega \in \Omega : x^\alpha(t) = \hat{x}^\alpha(t)\} = 1$$

The set of all s- feasible controls is denoted by $\mathcal{U}^s_{\text{feas}}[0,T]$. The pair $(x^\alpha(t); u(\cdot))$, where $x^\alpha(t)$ is the solution of (2) corresponding to this $u(\cdot)$, is called an *s-feasible pair*.

The assumptions A1–A4 guarantee that any $u(\cdot)$ from $\mathcal{U}[0,T]$ is s-feasible.
In addition, it is required that the following terminal state constraints are satisfied:

$$E\{h^j(x^\alpha(T))\} \geq 0 \qquad j = 1, \ldots, l \tag{3}$$

where $h^j : \mathbb{R}^n \to \mathbb{R}$ are given functions.

A5: For $j = 1, \ldots, l$ the functions h^j are $L_\phi(C^2)$-mappings.

Definition 3. The control $u(\cdot)$ and the pair $(x^\alpha(t); u(\cdot))$ are called an *s-admissible* control or *realizing the terminal condition* (3) and an s-admissible pair, respectively, if $u(\cdot) \in \mathcal{U}^s_{\text{feas}}[0,T]$, $x^\alpha(t)$ is the solution of (2), corresponding to this $u(\cdot)$, such that the inequalities (3) are satisfied. The set of all s-admissible controls is denoted by $\mathcal{U}^s_{\text{adm}}[0,T]$.

2.3 Highest Cost Function and Robust Optimal Control

For any scalar-valued function $\varphi(\alpha)$ bounded on $\mathcal{A}$ define *the m-true (or m-essential) maximum* of $\varphi(\alpha)$ on $\mathcal{A}$ as follows:

$$\underset{\alpha\in\mathcal{A}}{\text{m-vraimax}}\, \varphi(\alpha) := \max \varphi^+$$

such that

$$m\left\{\alpha \in \mathcal{A} : \varphi(\alpha) > \varphi^+\right\} = 0$$

It can be easily shown (see, for example, [54]) that the following *integral presentation* for the true maximum holds:

$$\underset{\alpha\in\mathcal{A}}{\text{m-vraimax}}\, \varphi(\alpha) = \sup_{\mathcal{A}_0 \subset \mathcal{A}: m(\mathcal{A}_0)>0} \frac{1}{m(\mathcal{A}_0)} \int_{\mathcal{A}_0} \varphi(\alpha)\, dm \tag{4}$$

where the Lebesgue-Stieltjes integral is taken over all subsets $\mathcal{A}_0 \subset \mathcal{A}$ with positive measure $m(\mathcal{A}_0)$. Consider the cost function h^α containing a terminal term, that is,

$$h^\alpha := E\left\{h^0(x^\alpha(T))\right\} \tag{5}$$

Here $h_0(x)$ is a positive, bounded and smooth *cost function* defined on $\mathbb{R}^n$. The end time-point T is assumed to be finite and $x^\alpha(t) \in \mathbb{R}^n$. If an admissible control is applied, for every $\alpha \in \mathcal{A}$ we deal with the cost value $h^\alpha =$

$E\{h_0(x^\alpha(T))\}$ calculated at the terminal point $x^\alpha(T) \in \mathbb{R}^n$. Since the realized value of α is a priory unknown, define the *worst (highest) cost*

$$F = \sup_{\mathcal{A}_0 \subset \mathcal{A}: m(\mathcal{A}_0) > 0} \frac{1}{m(\mathcal{A}_0)} \int_{\mathcal{A}_0} E\left\{h^0(x^\alpha(T))\right\} dm = \underset{\alpha \in \mathcal{A}}{\text{m-vraimax}}\, h^\alpha \tag{6}$$

The function F depends only on the considered admissible control $u(t)$, $t_0 \leq t \leq t_1$.

Definition 4. The control $\bar{u}(t)$, $0 \leq t \leq T$ is said to be *robust optimal* if (i) it satisfies the terminal condition, that is, it is admissible; (ii) it achieves the *minimal* worst (highest) cost F^0 (among all admissible controls satisfying the terminal condition).

Thus the *Robust Optimization Problem* consists of finding an admissible control action $u(t)$, $0 \leq t \leq T$, which provides

$$F^0 := F = \min_{u(t)} \underset{\alpha \in \mathcal{A}}{\text{m-vraimax}}\, h^\alpha = \min_{u(t)} \max_{\lambda \in \Lambda} \int_{\lambda \in \Lambda} \lambda(\alpha) E\left\{h^0(x^\alpha(T))\right\} dm(\alpha) \tag{7}$$

where the minimum is taken over all admissible control strategies and the maximum over all functions $\lambda(\alpha)$ within, so-called, the set of "*distribution densities*" Λ defined by

$$\Lambda := \left\{ \lambda = \lambda(\alpha) = \mu(\alpha) \left(\int_{\alpha \in \mathcal{A}} \mu(\alpha) dm(\alpha) \right)^{-1} \geq 0, \int_{\alpha \in \mathcal{A}} \lambda(\alpha) dm(\alpha) = 1 \right\} \tag{8}$$

This is the *Stochastic Minimax Bolza Problem.*

3 Robust Stochastic Maximum Principle

3.1 First and Second Order Adjoint Processes

The adjoint equations and the associated Hamiltonian function are introduced in this section to present the *necessary conditions* of the robust optimality for the considered class of partially unknown stochastic systems which is called the *Robust Stochastic Maximum Principle* (RSMP). Following [56], for any $\alpha \in \mathcal{A}$ and any admissible control $u(\cdot) \in \mathcal{U}^s_{\text{adm}}[0, T]$ consider

— the 1-st order vector adjoint equations:

$$d\psi^\alpha(t) = \left[-b_x^\alpha\left(t, x^\alpha(t), u(t)\right)^\top \psi^\alpha(t) \right.$$

$$
+\sum_{j=1}^{m}\sigma_x^{\alpha j}\left(t,x^\alpha(t),u(t)\right)^\intercal q_j^\alpha(t)\Bigg]dt+q^\alpha(t)dW(t)
$$

$$
\psi^\alpha(T)=c^\alpha,\quad t\in[0,T] \tag{9}
$$

— the 2-st order matrix adjoint equations:

$$
\begin{aligned}
d\Psi^\alpha(t)=-\Bigg[&b_x^\alpha(t,x^\alpha(t),u(t))^\intercal\Psi^\alpha(t)+\Psi^\alpha(t)b_x^\alpha(t,x^\alpha(t),u(t))\\
&+\sum_{j=1}^{m}\sigma_x^{\alpha j}(t,x^\alpha(t),u(t))^\intercal\Psi^\alpha(t)\sigma_x^{\alpha j}(t,x^\alpha(t),u(t))\\
&+\sum_{j=1}^{m}(\sigma_x^{\alpha j}(t,x^\alpha(t),u(t))^\intercal Q_j^\alpha(t)+Q_j^\alpha(t)\sigma_x^{\alpha j}(t,x^\alpha(t),u(t))\\
&+H_{xx}^\alpha(t,x^\alpha(t),u(t),\psi^\alpha(t),q^\alpha(t))\Bigg]dt+\sum_{j=1}^{m}Q_j^\alpha(t)dW^j(t)
\end{aligned}
$$

$$
\Psi^\alpha(T)=C^\alpha,\quad t\in[0,T] \tag{10}
$$

Here $c^\alpha\in L^2_{\mathcal{F}_T}(\Omega,\mathbb{R}^n)$ is a square integrable $\mathcal{F}_T$-measurable $\mathbb{R}^n$-valued random vector, $\psi^\alpha(t)\in L^2_{\mathcal{F}_t}(\Omega,\mathbb{R}^n)$ is a square integrable $\{\mathcal{F}_t\}_{t\geq 0}$-adapted $\mathbb{R}^n$-valued vector random process and $q^\alpha(t)\in L^2_{\mathcal{F}_t}(\Omega,\mathbb{R}^{n\times m})$ is a square integrable $\{\mathcal{F}_t\}_{t\geq 0}$-adapted $\mathbb{R}^{n\times m}$-valued matrix random process. Similarly, $C^\alpha\in L^2_{\mathcal{F}_T}(\Omega,\mathbb{R}^{n\times n})$ is a square integrable $\mathcal{F}_T$-measurable $\mathbb{R}^{n\times n}$-valued random matrix, $\Psi^\alpha(t)\in L^2_{\mathcal{F}_t}(\Omega,\mathbb{R}^{n\times n})$ is a square integrable $\{\mathcal{F}_t\}_{t\geq 0}$-adapted $\mathbb{R}^{n\times n}$-valued matrix random process $Q_j^\alpha(t)\in L^2_{\mathcal{F}_t}(\Omega,\mathbb{R}^{n\times m})$ is a square integrable $\{\mathcal{F}_t\}_{t\geq 0}$-adapted $\mathbb{R}^{n\times n}$-valued matrix random process. $b_x^\alpha(t,x^\alpha,u)$ and $H_{xx}^\alpha(t,x^\alpha,u,\psi^\alpha,q^\alpha)$ is the first and, correspondingly, the second derivatives of these functions by x^α. The function $H^\alpha(t,x,u,\psi,q)$ is defined as

$$
H^\alpha(t,x,u,\psi,q):=b^\alpha(t,x,u)^\intercal\psi+tr[q^\intercal\sigma^\alpha] \tag{11}
$$

As it is seen from (10), if $C^\alpha=C^{\alpha\intercal}$ then for any $t\in[0,T]$ the random matrix $\Psi^\alpha(t)$ is symmetric (but not necessarily positive or negative definite). In (9) and (10), which are the backward stochastic differential equations with the $\{\mathcal{F}_t\}_{t\geq 0}$-adapted solutions, the unknown variables to be selected are the pair of terminal conditions c^α,C^α and the collection $(q^\alpha,Q_j^\alpha,j=1,\ldots,l)$ of $\{\mathcal{F}_t\}_{t\geq 0}$-adapted stochastic matrices. Note that the equations (2) and (9) can be rewritten in Hamiltonian form as

$$
\begin{cases}
dx^\alpha(t)=H_\psi^\alpha(t,x^\alpha(t),u(t))^\intercal\psi^\alpha(t),q^\alpha(t))dt+\sigma^\alpha(t,x^\alpha(t),u(t))dW(t)\\
x^\alpha(0)\ =x_0,\quad t\in[0,T]
\end{cases} \tag{12}
$$

$$\begin{cases} d\psi^{\alpha}(t) = -H_x^{\alpha}(t, x^{\alpha}(t), u(t))^{\intercal}\psi^{\alpha}(t), q^{\alpha}(t))dt + q^{\alpha}(t)dW(t) \\ \psi^{\alpha}(T) = c^{\alpha}, \quad t \in [0, T] \end{cases} \tag{13}$$

3.2 Main Result

Theorem 1 (Robust Stochastic Maximum Principle). *Let A1–A5 be fulfilled and $(\bar{x}^{\alpha}(\cdot), \bar{u}(\cdot))$ be the α-robust optimal pairs ($\alpha \in \mathcal{A}$). The parametric uncertainty set $\mathcal{A}$ is a space with countable additive measure $m(\alpha)$ which assumed to be given. Then for every $\varepsilon > 0$ there exist collections of terminal conditions $c^{\alpha,(\varepsilon)}$, $C^{\alpha,(\varepsilon)}$, $\{\mathcal{F}_t\}_{t\geq 0}$-adapted stochastic matrices $(q^{\alpha,(\varepsilon)}, Q_j^{\alpha,(\varepsilon)}$, $j = 1, \ldots, l)$ in (9) and (10), and nonnegative constants $\mu_{\alpha}^{(\varepsilon)}$ and $\nu_{\alpha j}^{(\varepsilon)}$, $j = 1, \ldots, l$ such that the following conditions are fulfilled:*

1. (Complementary slackness condition): *For any $\alpha \in \mathcal{A}$*
   (i) *the inequality $\|E\{h^0(\bar{x}^{\alpha}(T))\} - \max_{\alpha\in\mathcal{A}} E\{h^0(\bar{x}^{\alpha}(T))\}| < \varepsilon$ holds or $\mu_{\alpha}^{(\varepsilon)} = 0$;*
   (ii) *moreover, either the inequality $|E\{h^j(\bar{x}^{\alpha}(T))\}| < \varepsilon$ holds or $\nu_{\alpha j}^{(\varepsilon)} = 0$, $j = 1, \ldots, l$;* (14)
2. (Transversality condition): *For any $\alpha \in \mathcal{A}$ the inequality*

$$\left\| c^{\alpha,(\varepsilon)} + \mu_{\alpha}^{(\varepsilon)} h_x^0(\bar{x}^{\alpha}(T)) + \sum_{j=1}^{l} \nu_{\alpha j}^{(\varepsilon)} h_x^j(\bar{x}^{\alpha}(T)) \right\| < \varepsilon \qquad \mathbf{P}\text{-}a.s. \tag{15}$$

$$\left\| C^{\alpha,(\varepsilon)} + \mu_{\alpha}^{(\varepsilon)} h_{xx}^0(\bar{x}^{\alpha}(T)) + \sum_{j=1}^{l} \nu_{\alpha j}^{(\varepsilon)} h_{xx}^j(\bar{x}^{\alpha}(T)) \right\| < \varepsilon \qquad \mathbf{P}\text{-}a.s. \tag{16}$$

   hold;
3. (Nontriviality condition): *There exists a set $\mathcal{A}_0 \subset \mathcal{A}$ with positive measure $m(\mathcal{A}_0) > 0$ such that for every $\alpha \in \mathcal{A}_0$ either $c^{\alpha,(\varepsilon)} \overset{a.s.}{\neq} 0$ or, at least, one of the numbers $\mu_{\alpha}^{(\varepsilon)}, \nu_{\alpha j}^{(\varepsilon)}$, $j = 1, \ldots, l$, is distinct from 0, that is, with probability one*

$$\forall \alpha \in \mathcal{A}_0 \in \mathcal{A}: \quad \left|c^{\alpha,(\varepsilon)}\right| + \mu_{\alpha}^{(\varepsilon)} + \sum_{j=1}^{l} \nu_{\alpha j}^{(\varepsilon)} > 0 \tag{17}$$

4. (Maximality condition): *the robust optimal control $\bar{u}(\cdot)$ for almost all $t \in [0, T]$ maximizes the generalized Hamiltonian function*

$$\mathcal{H}\left(t, \bar{x}^{\diamond}(t), u, \psi^{\diamond,(\varepsilon)}(t), \Psi^{\diamond,(\varepsilon)}(t), q^{\diamond,(\varepsilon)}(t)\right)$$

$$:= \int_{\mathcal{A}} \mathcal{H}^{\alpha}\left(t, \bar{x}^{\alpha}(t), u, \psi^{\alpha,(\varepsilon)}(t), \Psi^{\alpha,(\varepsilon)}(t), q^{\alpha,(\varepsilon)}(t)\right) dm(\alpha) \quad (18)$$

where

$$\mathcal{H}^{\alpha}\left(t, \bar{x}^{\alpha}(t), u, \psi^{\alpha,(\varepsilon)}(t), \Psi^{\alpha,(\varepsilon)}(t), q^{\alpha,(\varepsilon)}(t)\right)$$
$$:= H^{\alpha}\left(t, \bar{x}^{\alpha}(t), u, \psi^{\alpha,(\varepsilon)}(t), q^{\alpha,(\varepsilon)}(t)\right) - \frac{1}{2}\mathrm{tr}\left[\bar{\sigma}^{\alpha\mathsf{T}}\Psi^{\alpha,(\varepsilon)}(t)\bar{\sigma}^{\alpha}\right]$$
$$+ \frac{1}{2}\mathrm{tr}\left[(\sigma^{\alpha}(t, \bar{x}^{\alpha}(t), u) - \bar{\sigma}^{\alpha})^{\mathsf{T}}\Psi^{\alpha,(\varepsilon)}(t)(\sigma^{\alpha}(t, \bar{x}^{\alpha}(t), u) - \bar{\sigma}^{\alpha})\right] \quad (19)$$

the function $H^{\alpha}(t, \bar{x}^{\alpha}(t), u, \psi^{\alpha,(\varepsilon)}(t), q^{\alpha,(\varepsilon)}(t)$ *is given by (11),*

$$\bar{\sigma}^{\alpha} := \sigma^{\alpha}\left(t, \bar{x}^{\alpha}(t), \bar{u}(t)\right) \quad (20)$$

$$\bar{x}^{\diamond}(t) := \left(\bar{x}^{1\mathsf{T}}(t), \ldots, \bar{x}^{N\mathsf{T}}(t)\right)^{\mathsf{T}},$$
$$\psi^{\diamond,(\varepsilon)}(t) := \left(\psi^{1,(\varepsilon)\mathsf{T}}(t), \ldots, \psi^{N,(\varepsilon)\mathsf{T}}(t)\right)^{\mathsf{T}}$$
$$q^{\diamond,(\varepsilon)}(t) := \left(q^{1,(\varepsilon)}(t), \ldots, q^{N,(\varepsilon)}(t)\right),$$
$$\Psi^{\diamond,(\varepsilon)}(t) := \left(\Psi^{1,(\varepsilon)}(t), \ldots, \Psi^{N,(\varepsilon)}(t)\right)$$

and $\psi^{i,(\varepsilon)\mathsf{T}}(t)$, $\Psi^{i,(\varepsilon)}(t)$ *verify (9) and (10) with the terminal conditions* $c^{\alpha,(\varepsilon)}$ *and* $C^{\alpha,(\varepsilon)}$, *respectively, i.e., for almost all* $t \in [0, T]$

$$\bar{u}(t) = \arg\max_{u \in U} \mathcal{H}\left(t, \bar{x}^{\diamond}(t), u, \psi^{\diamond,(\varepsilon)}(t), \Psi^{\diamond,(\varepsilon)}(t), q^{\diamond,(\varepsilon)}(t)\right) \quad (21)$$

The proof follows [44] with some slight modifications corresponding to the considered case of a measured space as uncertainty set.

4 Discussions

The Hamiltonian function $\mathcal{H}$ used for the construction of the robust optimal control $\bar{u}(t)$ is equal (see (18)) to the Lebesque integral over the uncertainty set of the standard stochastic Hamiltonians $\mathcal{H}^{\alpha}$ corresponding to each fixed value of the uncertain parameter.

From the Hamiltonian structure (19) it follows that if $\sigma^{\alpha j}\left(t, \bar{x}^{\alpha}(t), u(t)\right)$ does not depend on $u(t)$, the *2-nd order adjoint process* does not participate in the robust optimal constructions.

If the stochastic plant is completely known , that is, there is no parametric uncertainty ($\mathcal{A} = \alpha_0, dm(\alpha) = \delta\left(\alpha - \alpha_0\right) d\alpha$), then if $\varepsilon \to 0$, it follows that, in this case, RSMP converts to *Stochastic Maximum Principle* (see [28], [56] and [53]).

In the deterministic case, when there is no any stochastics ($\sigma^{\alpha}(t, \bar{x}^{\alpha}(t), u(t)) \equiv 0$), the Robust Maximum Principle for minimax problems (in Mayer form) stated in [14] and [16] is obtained directly.

When, in addition, there is no parametric uncertainties, that is,

$$\mathcal{A} = \alpha_0, dm(\alpha) = \delta(\alpha - \alpha_0)\, d\alpha$$

the *Classical Maximum Principle* for the optimal control problems (in Mayer form), is obtained [41].

The Theorem above gives the Robust Maximum Principle only for the problem with a fixed horizon. The non-fixed horizon case demands a special construction and implies another formulation of RMP.

If no a priory information on some or others parameter values and the distance on a compact $\mathcal{A} \subset \mathbb{R}^s$ is defined by the natural way as $\|\alpha_1 - \alpha_2\|$, then the *Maximum Condition* can be formulated (and proved) as follows:

$$u(t) \in Arg \max_{u \in U} \int_{\mathcal{A}} \mathcal{H}^{\alpha}\left(t, \bar{x}^{\alpha}(t), u, \psi^{\alpha,(\varepsilon)}(t), \Psi^{\alpha,(\varepsilon)}(t), q^{\alpha,(\varepsilon)}(t)\right) d\alpha$$

almost everywhere on $[t_0, t_1]$, that represents, evidently, a partial case of the general condition (21) with an uniform absolutely continuous measure, that is, when

$$dm(\alpha) = p(\alpha) d\alpha = \frac{1}{m(\mathcal{A})} d\alpha$$

with $p(\alpha) = m^{-1}(\mathcal{A})$.

If the uncertainty set $\mathcal{A}$ is *finite*, the Robust Maximum Principle, proved above, gives the result contained in [15], [43] and [44] and in the complementary slackness condition we have the equalities.

It is naturally to ask: is it possible, in general case, to replace the inequalities by the equalities as it was done above or not? The answer is negative: the inequalities in the main theorem **cannot be replaced by equalities**.

5 Conclusion

In this paper the Robust Stochastic Maximum Principle (in Mayer form) is presented for a class of nonlinear continuous-time stochastic systems containing an unknown parameter from a given measured set and subject to terminal constraints. Its proof is based on the use of the Tent Method with the special technique specific for stochastic calculus. *The Hamiltonian function used for these constructions is equal to the Lebesque integral over the given uncertainty set of the standard stochastic Hamiltonians corresponding to a fixed value of the uncertain parameter.*

The future investigation may be focused on the Linear Quadratic Stochastic Problem which seems to us to be solvable by this technique also. However, it will be considered in a subsequent paper. Furthermore, it makes sense to continue this study in the following directions:

- formulate RSMP for minimax problem in the general Bolza form,
- consider the terminal constraints with the additional integral terms,
- consider the terminal constraints including the probability of some events,
- formulate RSMP for minimax problem not for a fixed horizon but for a random stopping time.

References

1. Basar, T. and Bernhard, P. (1991) H^{∞} Optimal Control and Related Minimax Design Problems: A Dynamic Game Approach. Boston, MA: Birkhäuser.
2. Basar, T. (1994) Minimax Control of Switching Systems Under Sampling, in Proceedings of the 33rd Conference on Decision and Control, Lake Buena Vista, Florida, 716–721.
3. Bellman, R. (1957) Dynamic Programming, Princeton University Press.
4. Bensoussan, A. (1983) Lecture on Stochastic Control, Part 1, Lecture Notes in Math. **972**, 1–39.
5. Bensoussan, A. (1992) Stochstic Control of Partially Observable Systems, Cambridge University Press.
6. Bernhard, P. (1994) Minimax Versus Stochastic Partial Information Control. Proceedings of the 33rd Conference on Decision and Control, Lake Buena Vista, Florida, 2572–2577.
7. Blom, H. and Everdij, M. (1993) Embedding Adaptive JLQG into LQ Martingale with a Complete Observable Stochastic Control Matrix, in Proceedings of the 32nd Conference on Decision and Control, San Antonio, Texas, 849–854.
8. Bismut, J. M. (1977) On Optimal Control of Linear Stochastic Equations with a Linear Quadratic Criteria, SIAM J. Control, **15**.
9. Bismut, J. M. (1978) An Introductory Approach to Duality in Stochastic Control, SIAM Review **20**, 62–78.
10. Boltyansky, V. G. (1975) The Method of Tents in the Theory of Extremal Problems, Uspehi Mat. Nauk **30**, 3–65.
11. Boltyansky, V. G. (1978) Optimal Control of Discrete Systems. A Halsted Press Book. Jon Willey & Sons, New-York, Toronto, Ont.
12. Boltyansky, V. G. (1987) The Tent Method in Topological Vector Spaces, Soviet. Math. Dokl. **34**, 176–179.
13. Boltyansky, V. G. (2002) The Tent Method in Banach Spaces, to be published.
14. Boltyansky, V. G. and Poznyak, A. S. (1999) Robust Maximum Principle in Minimax Control, Int. J. Control **72**, 305–314.
15. Boltyansky, V. G. and Poznyak, A. S. (1999) Robust Maximum Principle for Minimax Bolza Problem with Terminal Set, in Proceedings of IFAC-99, Beijing, 263–268.

16. Boltyansky, V. G. and Poznyak, A. S. (2001) Robust Maximum Principle for a Measured Space as Uncertainty Set, Dynamic Systems and Applications, to appear.
17. Boukas, E. K., Shi, P., and Andijani, A. (1999) Robust Inventory-Production Control Problem with Stochastic Demand, Optimal Control Applications and Methods **20**, 1–20.
18. Clarke, F. H. and Vinter, R. B. (1983) Local Optimality Conditions and Lipshitzian Solutions to the Hamilton-Jacobi Equations, SIAM J.Control **21**, 856–870.
19. Clarke, F. H. (1983) Optimization and Nonsmooth Analysis. John Wiley, NY.
20. Crandall, M. G. and Lions, P. L. (1983) Viscosity Solution of Hamilton-Jacobi Equations, Trans. Amer. Math. Soc. **277**, 1–42.
21. Didinsky, G. and Basar, T. (1991) Minimax controller and Filter designs for Discrete-Time Linear Systems Under Soft-Constrained Quadratic Performance Indices, in Proceedings of the 30th Conference on Decision and Control, Brighton, England, 585–590.
22. Didinsky, G. and Basar, T. (1994) Minimax Adaptive Control of Uncertain Plants, in Proceedings of the 33rd Conference on Decision and Control, Lake Buena Vista, Florida, 2839–2844.
23. Dorato, P. and Drenick, R. F. (1966) Optimality, Insensitivity and Game Theory, in Sensitivity Methods in Control Theory, L. Radanovich, ed., New York, NY: Pergamon Press, 78–102.
24. Doyle, J., Glover, K., Khargonekar, P., and Francis, B. (1989) State-Space Solutions to Standard H_2 and H^∞ Control Problems, IEEE Transactions on Automatic Control **34**, 831–847.
25. Duncan, T. E. (1967) Doctoral Dissertation, Dept. of El. Eng., Stanford University.
26. Duncan, T. E., Guo, L., and Pasik-Duncan, B. (1999) Adaptive Continuos-Time Linear Quadratic Gaussian Control, IEEE Transation on AC **44**, 1653–1662.
27. Duncan, T. E. and Varaiya, P. P. (1971) On the Solution of a Stochastic Control System, SIAM J. Control **9**, 354–371.
28. Fleming, W. H. and Rishel, R. W. (1975) Optimal Deterministic and Stochastic Control. Springer-Verlag, Berlin.
29. Glover, K. and Doyle, J. (1988) State-Space Formulae for all stabilizing Controllers that Satisfy an H^∞-norm Bound and Relations to Risk Sensitivity, Systems and Control Letters **11**, 167–172.
30. Haussman, U. G. (1981) Some Examples of Optimal Control, or: The Stochastic maximum Principle at Work, SIAM Review **23**, 292–307.
31. Haussman, U. G. (1982) On the Existence of Optimal Controls for Partially Observed Diffusions, SIAM J. Control **20**, 385–407.
32. Kallianpur, G. (1980) Stochastic Filtering Theory, Springer-Verlag, NY.
33. Khargonekar, P. P. (1991) State-Space H^∞-Control Theory and the LQG Control Problem, in Mathematical System Theory: The Influence of R. E. Kalman, A. C. Antoulas, ed., Springer-Verlag.
34. Krasovskii, N. N. (1969) Game Problem on Movements Correction, Applied Mathematic and Mechanics **33**, (in Russian).
35. Krylov, N. V. (1980) Controlled Diffusion Processes, Springer-Verlag, NY.
36. Kurjanskii, A. B. (1977) Control and Observations in Uncertain Conditions. Nauka, Moscow (in Russian).

37. Kushner, H. (1972) Necessary conditions for continuous Parameter Stochastic Optimization Problems, SIAM J. Control **10**, 550–565.
38. Ming, K., Lau, B. S., Kosut, R. L., and Franklin, G. F. (1991) Robust Control Desigh for Ellipsoidal Plant Set, in Proceedings of the 30th Conference on Decision and Control, Brighton, England, 291–296.
39. Limebeer, D., Anderson, B. D. O., Khargonekar, P., and Green, M. (1989) A game theoretic Approach to H^{∞}-Control for Time Varying Systems, in Proceedings of Int. Symposium on the Mathematical Theory of Networks and Systems, Amsterdam.
40. Lions, P. L. (1983) Optimal Control of Diffusion Processes and Hamilton-Jacobi-Bellman Equations, Part 1: Viscosity Solutions and Uniqueness, Comm. Partial Diff. Equ. **11**, 1229–1276.
41. Pontryagin, L. S., Boltyansky, V. G., Gamkrelidze, R. V., and Mishenko, E. F. (1962) The Mathematical Theory of Optimal Processes, Interscience, New York (translated from Russian, 1969).
42. Poznyak, A. and Taksar, M. (1996) Robust Control of Linear Stochastic Systems with Fully Observable State, Applicationes Mathematicae, **24**, 35–46.
43. Poznyak, A. S., Duncan, T. E., Pasik-Duncan, B., and Boltyansky, V. G. Robust Stochastic Maximum Principle. Proceedings of the ....
44. Poznyak, A. S., Duncan, T. E., Pasik-Duncan, B., and Boltyansky, V. G. Robust Stochastic Maximum Principle,International Journal of Control, submitted.
45. Poznyak, A. S., Duncan, T. E., Pasik-Duncan, B., and Boltyansky, V. G. Robust Optimal Control for Minimax Stochastic Linear Quadratic Problem, International Journal of Control, submitted.
46. Pytlak, R. (1990) Strong. Variation Algoritm for Minimax Control: Parallel Approach, in Proceedings of the 29th Conference on Decision and Control, Honolulu, Hawall, 2469–2474.
47. Siljak, D. D. (1989) Parameter Space Methods for Robust Control Design: A Guided Tour, IEEE Transactions on Automatic Control 34 (7), 674- 688.
48. Taksar M.I., Poznyak, A. S., and Iparraguirre, A. (1998) Robust output Feedback Control for Linear Stochastic Systems in Continuous Time with Time-Varying Parameters, IEEE Transactions on AC **43**, 1133–1136.
49. Taksar, M.I. and Zhou, X.Y. (1998) Optimal risk and divident control for a company with a debt liability, Insurance: Math and Econom. **22**, 105–122.
50. Ugrinovskii, V. A. and Petersen, I. R. (1997) Finite Horizon Minimax Optimal Control of Nonlinear Continuous Time Systems with Stochastic Uncertainties, in Proceedings of 36th IEEE CDC, San Diego, 2265–2270.
51. Vinter, R. B. (1988) New Results on the Relationship Between Dynamic Programming and Their Connection in Deterministic Control, Math. Control. Sign. Syst. **1**, 97–105.
52. Yaz, G. (1991) Minimax Control of Discrete Nonlinear Stochastic Systems with White Noise Uncertainty, in Proceedings of the 30th Conference on Decision and Control, Brighton, England, 1815–1816.
53. Yong, J. and Zhou, X. Y. (1999) Stochastic controls: Hamiltonian Systems and HJB Equations, Springer-Verlag.
54. Yoshida, K. (1979) Functional Analysis, Narosa Publishing House, New Delhi.
55. Zames, J. (1981) Feedback and Optimality sensitivity: Model Reference Transformation, Multiplicative Seminorms and Approximate Inverses, IEEE Transactions on Automatic Control **26**, 301–320.

56. Zhou, X. Y. (1991) A Unified Treatment of Maximum Principle and Dynamic Programming in Stochastic Controls, Statistics and Stochastic Reports **36**, 137–161.
57. Zakai, M. (1969) On the Optimal Filtering of Diffusion Processes, Z. Wahrsch. Geb. **11**, 230–243.

On Optimality of Stochastic N-Machine Flowshop with Long-Run Average Cost

Ernst Presman[1], Suresh P. Sethi[2], Hanqin Zhang[3], and Qing Zhang[4]

[1] Central Economics and Mathematics Institute, The Russian Academy of Sciences, Moscow, Russia
[2] School of Management, The University of Texas at Dallas, Mail Station JO4.7, Box 830688, Richardson, TX 75083-0688, USA
[3] Institute of Applied Mathematics, Academia Sinica, Beijing, 100080, P. R. China
[4] Department of Mathematics, University of Georgia, Athens, GA 30602, USA

Abstract. This paper is concerned with the problem of production planning in a stochastic manufacturing system with serial machines that are subject to breakdown and repair. The machine capacities are modeled by a Markov chain. The objective is to choose the input rates at the various machines over time in order to meet the demand for the system's production at the minimum long-run average cost of production and surplus, while ensuring that the inventories in internal buffers between adjacent machines remain nonnegative. The problem is formulated as a stochastic dynamic program. We prove a verification theorem and derive the optimal feedback control policy in terms of the directional derivatives of the potential function.

1 Introduction

Beginning with Bielecki and Kumar (1988), there has been a considerable interest in studying the problem of production planning in stochastic manufacturing systems with the objective of minimizing long-run average cost. Bielecki and Kumar (1988) deal with a single machine with two states: up and down), single product problem with linear holding and backlog costs. Because of the simple structure of their problem, they are able to obtain an explicit solution for the problem, and thus verify the optimality of the resulting policy (hedging point policy). Sharifinia (1988) deal with an extension of the Bielecki-Kumar model with more than two machine states. Liberopoulous and Caramanis (1998) show that Sharifinia's method for evaluating hedging point policies applies even when the transition rates of the machine states depend on the production rate. Liberopoulous and Hu (1995) obtain monotonicity of the threshold levels corresponding to different machine states. At the same time, there has been a number of heuristic analyses of the multi-product problem. Srisvatsan (1993) and Srisvatsan and Dallery (1998) consider a two-product problem. Caramanis and Sharifinia (1991) decompose a multi-product problem into an analytically tractable single-product problem in order to obtain near-optimal hedging points for the problem. All of

B. Pasik-Duncan: Stochastic Theory and Control, LNCIS 280, pp. 399–417, 2002.

these papers, however, are heuristic in nature, since they do not rigorously prove the optimality of the policies for their extensions of the Bielecki-Kumar model.

The difficulty in proving the optimality lies in the fact that when the problem is generalized to include convex costs and multiple machine capacity levels, explicit solutions are no longer possible, whereas the Bielecki-Kumar proof of optimality depends on being able to explicitly obtain the value function of the problem. One needs, therefore, to develop appropriate dynamic programming equations, existence of their solutions, and verification theorems for optimality. This is accomplished by Sethi et al. (1997) and Sethi et al. (1998) for single and multi-product problems. They use the vanishing discount approach to prove the optimality of hedging point policies for convex surplus costs and linear/convex production costs. These models are also considered by Duncan *et al.* (2001) with additional variables such Markov product demand. Their results make precise the heuristic treatments of several manufacturing system problems carried out by Sharifinia (1988), Srisvatsan and Dallery (1998), and others.

Presman *et al.* (1995, 1997) considered flowshops and obtained optimal control policies for such problems in the context of the discounted cost criterion. A characteristic difficulty of the flowshop is the presence of the state constraints arising from the requirement that the number of parts in the internal buffers between any two adjacent machines must remain non-negative. Our objective in this paper is to treat their problem with a view to minimizing the long-run average of expected production and surplus costs. We write the Hamilton-Jacobi-Bellman equation in terms of directional derivatives and prove a verification theorem. Using the vanishing discount approach for the average-cost problem, we obtain a solution of the Hamilton-Jacobi-Bellman (HJB) equation.

Two major contributions are made in order to implement the vanishing discount approach. One is in constructing a control policy which takes any given system state to any other state in a time whose rth moment has a finite expectation. The other is in obtaining a solution of the HJB equation for the problem in terms of directional derivatives by a limit procedure for the discounted cost problem as the discount rate tends to zero.

The plan of this paper is as follows. In Section 2, we formulate an N -machine flowshop as a continuous-time stochastic optimal control problem and specify required assumptions. In Section 3, the HJB equation for the problem in terms of directional derivatives is specified, and a verification theorem for optimality over a class of admissible controls is given. Moreover, by using the vanishing discount approach, it is shown that the HJB equation for the problem has a solution. Section 5 concludes the paper.

2 Problem Formulation

We consider a manufacturing system producing a single finished product using N machines in tandem that are subject to breakdown and repair. We are given a stochastic process $\boldsymbol{m}(t) = (m_1(t), \ldots, m_N(t))$ on the standard probability space $(\Omega, \mathcal{F}, P)$, where $m_k(t), k = 1, \ldots, N$, is the capacity of the kth machine at time t. We use $u_k(t)$ to denote the input rate to the kth machine, $k = 1, \ldots, N$, and $x_k(t) \geq 0$ to denote the number of parts in the buffer between the kth and $(k+1)$th machines, $k = 1, \ldots, N-1$. We assume a constant demand rate d. The difference between cumulative production and cumulative demand, called surplus, is denoted by $x_N(t)$. If $x_N(t) > 0$, we have finished good inventories, and if $x_N(t) < 0$, we have backlogs.

The dynamics of the system can be written as follows:

$$\dot{x}_k(t) = u_k(t) - u_{k+1}(t), \; x_k(0) = x_k^0, \; k = 1, \ldots, N, \tag{1}$$

where $u_{N+1}(t) = d$. This relation can be written in the following vector form:

$$\dot{\boldsymbol{x}}(t) = A\boldsymbol{u}(t), \; \boldsymbol{x}(0) = \boldsymbol{x}^0, \tag{2}$$

where $A : R^{N+1} \to R^N$ is the corresponding linear operator. Here and elsewhere we use boldface letters to stand for vectors. Since the number of parts in the internal buffers cannot be negative, we impose the state constraints $x_k(t) \geq 0$, $k = 1, \ldots, N-1$. To formulate the problem precisely, let $S = [0, \infty)^{N-1} \times (-\infty, \infty) \in R^N$ denote the state constraint domain, $b(S)$ denote the boundary of S, and $S^o = S \setminus b(S)$. For $\boldsymbol{m} = (m_1, \ldots, m_N), m_k \geq 0, k = 1, \ldots, N$, let

$$U(\boldsymbol{m}) = \{\boldsymbol{u} = (u_1, \ldots, u_N, d) : 0 \leq u_k \leq m_k, \quad k = 1, \ldots, N\}, \tag{3}$$

and for $\boldsymbol{x} \in S$, let

$$U(\boldsymbol{x}, \boldsymbol{m}) = \{\boldsymbol{u} \in U(\boldsymbol{m}) : \; x_k = 0 \Rightarrow u_k - u_{k+1} \geq 0, \; k = 1, \ldots, N-1\}. \tag{4}$$

Let the sigma algebra $\mathcal{F}_t = \sigma\{\boldsymbol{m}(s) : 0 \leq s \leq t\}$. We now define the concept of admissible controls.

Definition 2.1. We say that a control $\boldsymbol{u}(\cdot) = (u_1(\cdot), \ldots, u_N(\cdot), d)$ is admissible with respect to the initial state vector $\boldsymbol{x}^0 = (x_1^0, \ldots, x_N^0) \in S$, if

(i) $\boldsymbol{u}(\cdot)$ is an $\mathcal{F}_t$-adapted process;

(ii) $\boldsymbol{u}(t)$ is Borel measurable in t a.s. and $\boldsymbol{u}(t) \in U(\boldsymbol{m}(t))$ for all $t \geq 0$;

(iii) the corresponding state process $\boldsymbol{x}(t) = (x_1(t), \ldots, x_N(t)) \in S$ for all $t \geq 0$.

The problem is to find an admissible control $\boldsymbol{u}(\cdot)$ that minimizes the long-run average cost function

$$J(\boldsymbol{x}^0, \boldsymbol{m}^0, \boldsymbol{u}(\cdot)) = \limsup_{T \to \infty} \frac{1}{T} E \int_0^T h(\boldsymbol{x}(t), \boldsymbol{u}(t)) dt, \tag{5}$$

where $h(\cdot,\cdot)$ defines the cost of surplus and production and $\boldsymbol{m}^0$ is the initial value of $\boldsymbol{m}(t)$. We impose the following assumptions on the process $\boldsymbol{m}(t) = (m_1(t),\dots,m_N(t))$ and the cost function $h(\cdot,\cdot)$ throughout this paper.

(A.1) Let $\mathcal{M} = \{\boldsymbol{m}^1,\dots,\boldsymbol{m}^p\}$ for some integer $p \geq 1$, where $\boldsymbol{m}^j = (m_1^j,\dots,m_N^j)$, with $m_k^j, k = 1,\dots,N$, denote the capacity of the kth machine, $j = 1,\dots,p$. The capacity process $\boldsymbol{m}(t) \in \mathcal{M}$ is a finite state Markov chain with the infinitesimal generator

$$Qf(\boldsymbol{m}) = \sum_{\boldsymbol{m}' \neq \boldsymbol{m}} q_{\boldsymbol{m}'\boldsymbol{m}}[f(\boldsymbol{m}') - f(\boldsymbol{m})]$$

for some $q_{\boldsymbol{m}'\boldsymbol{m}} \geq 0$ and any function $f(\cdot)$ on $\mathcal{M}$. Moreover, the Markov process is strongly irreducible and has the stationary distribution $p_{\boldsymbol{m}^j}$, $j = 1,\dots,p$.

(A.2) Let $p_k = \sum_{j=1}^p m_k^j p_{\boldsymbol{m}^j}$. Assume that $\min_{1\leq k\leq N} p_k > d$.

(A.3) $h(\cdot,\cdot)$ is a non-negative, jointly convex function that is strictly convex in either $\boldsymbol{x}$ or $\boldsymbol{u}$ or both. For all $\boldsymbol{x},\boldsymbol{x}' \in S$ and $\boldsymbol{u},\boldsymbol{u}' \in U(\boldsymbol{m}^j), j = 1,\dots,p$, there exist constants C_0 and $K_h \geq 1$ such that

$$|h(\boldsymbol{x},\boldsymbol{u}) - h(\boldsymbol{x}',\boldsymbol{u}')| \leq C_0(1 + |\boldsymbol{x}|^{K_h} + |\boldsymbol{x}'|^{K_h})(|\boldsymbol{x} - \boldsymbol{x}'| + |\boldsymbol{u} - \boldsymbol{u}'|).$$

We use $\mathcal{A}(\boldsymbol{x}^0,\boldsymbol{m}^0)$ to denote the set of all admissible controls with respect to $\boldsymbol{x}^0 \in S$ and $\boldsymbol{m}(0) = \boldsymbol{m}^0$. Let $\lambda(\boldsymbol{x}^0,\boldsymbol{m}^0)$ denote the minimal expected cost, i.e.,

$$\lambda(\boldsymbol{x}^0,\boldsymbol{m}^0) = \inf_{\boldsymbol{u}(\cdot)\in\mathcal{A}(\boldsymbol{x}^0,\boldsymbol{m}^0)} J(\boldsymbol{x}^0,\boldsymbol{m}^0). \tag{6}$$

For writing the HJB equation for our problem, we first introduce some notation. Let $\mathcal{G}$ denote the family of real-valued functions $f(\cdot,\cdot)$ defined on $S \times \mathcal{M}$ such that

(i) $f(\cdot,\boldsymbol{m})$ is convex for any $\boldsymbol{m} \in\mathcal{M}$;

(ii) there exists a function $C(\boldsymbol{x},\boldsymbol{x}')$ such that for any $\boldsymbol{m} \in\mathcal{M}$ and any $\boldsymbol{x},\boldsymbol{x}'\in S$,

$$|f(\boldsymbol{x},\boldsymbol{m}) - f(\boldsymbol{x}',\boldsymbol{m})| \leq C(\boldsymbol{x},\boldsymbol{x}')|\boldsymbol{x} - \boldsymbol{x}'|.$$

Remark. By Theorem 10.4 on Page 86 of Rockafellar (1972), (i) and (ii) imply that $f(\cdot,\boldsymbol{m})$ is Lipschitzian on any closed bounded subset of S for any $\boldsymbol{m} \in \mathcal{M}$.

Formally, we write the HJB equation in terms of directional derivatives for our problem as

$$\lambda = \inf_{\boldsymbol{u}\in U(\boldsymbol{x},\boldsymbol{m})} \{\partial_{A\boldsymbol{u}} f(\boldsymbol{x},\boldsymbol{m}) + h(\boldsymbol{x},\boldsymbol{u})\} + Qf(\boldsymbol{x},\cdot)(\boldsymbol{m}), \tag{7}$$

where λ is a constant, $f(\cdot,\cdot) \in \mathcal{G}$, and $\partial_{A\boldsymbol{u}} f(\boldsymbol{x},\boldsymbol{m})$ denotes the directional derivative of function $f(\boldsymbol{x},\boldsymbol{m})$ along the direction $A\boldsymbol{u} \in R^N$.

3 Main Results

First we have the following verification theorem.

Theorem 3.1. *Assume that* $(\lambda, f(\cdot,\cdot))$ *with* $f(\cdot,\cdot)$ *convex on* $S \times \mathcal{M}$ *satisfies* (7), *there exists a constant function* $\boldsymbol{u}^*(\boldsymbol{x},\boldsymbol{m})$ *for which*

$$\begin{aligned}\inf_{\boldsymbol{u}\in U(\boldsymbol{x},\boldsymbol{m})} & \{\partial_{A\boldsymbol{u}} f(\boldsymbol{x},\boldsymbol{m}) + h(\boldsymbol{x},\boldsymbol{u})\} \\ &= \partial_{A\boldsymbol{u}^*(\boldsymbol{x},\boldsymbol{m})} f(\boldsymbol{x},\boldsymbol{m}) + h(\boldsymbol{x},\boldsymbol{u}^*(\boldsymbol{x},\boldsymbol{m})),\end{aligned} \tag{8}$$

and the equation $\dot{\boldsymbol{x}}(t) = A\boldsymbol{u}^*(\boldsymbol{x}(t),\boldsymbol{m}(t))$, *has for any initial condition* $(\boldsymbol{x}^*(0), \boldsymbol{m}(0)) = (\boldsymbol{x}^0,\boldsymbol{m}^0)$, *a solution* $\boldsymbol{x}^*(t)$ *such that*

$$\lim_{T\to\infty} \frac{Ef(\boldsymbol{x}^*(T),\boldsymbol{m}(T))}{T} = 0. \tag{9}$$

Then $\boldsymbol{u}^*(t) = \boldsymbol{u}^*(\boldsymbol{x}^*(t),\boldsymbol{m}(t))$ *is an optimal control. Furthermore,* $\lambda(\boldsymbol{x}^0,\boldsymbol{m}^0)$ *does not depend on* $\boldsymbol{x}^0$ *and* $\boldsymbol{m}^0$, *and it coincides with* λ. *Moreover, for any* $T > 0$,

$$\begin{aligned}& f(\boldsymbol{x}^0,\boldsymbol{m}^0) \\ &\quad = \inf_{\boldsymbol{u}(\cdot)\in\mathcal{A}(\boldsymbol{x}^0,\boldsymbol{m}^0)} E\left[\int_0^T (h(\boldsymbol{x}(t),\boldsymbol{u}(t)) - \lambda)\, dt + f(\boldsymbol{x}(T),\boldsymbol{m}(T))\right] \\ &\quad = E\left[\int_0^T (h(\boldsymbol{x}^*(t),\boldsymbol{u}^*(t)) - \lambda)\, dt + f(\boldsymbol{x}^*(T),\boldsymbol{m}(T))\right].\end{aligned} \tag{10}$$

Proof. Since $(\lambda, f(\cdot,\cdot))$ is a solution to (7) and $(\boldsymbol{x}^*(t),\boldsymbol{u}^*(t))$ satisfy condition (8), we have

$$\partial_{A\boldsymbol{u}(t)} f(\boldsymbol{x}^*(t),\boldsymbol{m}(t)) + Qf(\boldsymbol{x}^*(t),\cdot)(\boldsymbol{m}(t)) = \lambda - h(\boldsymbol{x}^*(t),\boldsymbol{u}^*(t)). \tag{11}$$

Since $f(\cdot,\cdot) \in \mathcal{G}$, we apply Dynkin's formula and use (11) to get

$$\begin{aligned}& Ef(\boldsymbol{x}^*(T),\boldsymbol{m}(T)) \\ &= f(\boldsymbol{x}^0,\boldsymbol{m}^0) + E\int_0^T \left[\partial_{A\boldsymbol{u}(t)} f(\boldsymbol{x}^*(t),\boldsymbol{m}(t)) + Qf(\boldsymbol{x}^*(t),\cdot)(\boldsymbol{m}(t))\right] dt \\ &= f(\boldsymbol{x}^0,\boldsymbol{m}^0) + E\int_0^T [\lambda - h(\boldsymbol{x}^*(t),\boldsymbol{u}^*(t)]dt \\ &= f(\boldsymbol{x}^0,\boldsymbol{m}^0) + \lambda T - E\int_0^T h(\boldsymbol{x}^*(t),\boldsymbol{u}^*(t))dt.\end{aligned} \tag{12}$$

We can rewrite (12) as

$$\lambda = \frac{1}{T}\left[Ef(\boldsymbol{x}^*(T),\boldsymbol{m}(T)) - f(\boldsymbol{x}^0,\boldsymbol{m}^0)\right] + \frac{1}{T}E\int_0^T h(\boldsymbol{x}^*(t),\boldsymbol{u}^*(t))dt. \tag{13}$$

Using (9) and taking the limit as $T \to \infty$, we get

$$\lambda \geq \limsup_{T\to\infty} \frac{1}{T} E \int_0^T h(\boldsymbol{x}^*(t), \boldsymbol{u}^*(t))dt. \tag{14}$$

Moreover, for any $\boldsymbol{u}(\cdot) \in \mathcal{A}(\boldsymbol{x}^0, \boldsymbol{m}^0)$, we have from (7) that

$$\partial_{A\boldsymbol{u}(t)} f(\boldsymbol{x}^*(t), \boldsymbol{m}(t)) + Qf(\boldsymbol{x}^*(t), \cdot)(\boldsymbol{m}(t)) \geq \lambda - h(\boldsymbol{x}^*(t), \boldsymbol{u}(t)).$$

Similar to (12), we get

$$\lambda \leq \limsup_{T\to\infty} \frac{1}{T} E \int_0^T h(\boldsymbol{x}^*(t), \boldsymbol{u}(t))dt. \tag{15}$$

By (14) and (15) we obtain that $\boldsymbol{u}^*(t)$ is an optimal control and $\lambda(\boldsymbol{x}^0, \boldsymbol{m}^0) = \lambda$.

For proving (10), consider for a finite horizon T, the problem of minimization of functional

$$E\left[\int_0^T (h(\mathrm{x}(t), \mathrm{u}(t)) - \lambda)\, dt + f(\mathrm{x}(T), \boldsymbol{m}(T))\right].$$

Functions $f(\boldsymbol{x}, \boldsymbol{m})$ and $\boldsymbol{u}(\boldsymbol{x}, \boldsymbol{m})$ satisfy the HJB equation on the time interval $[0, T]$. According to the verification theorem for a finite time interval, $f(\boldsymbol{x}, \boldsymbol{m})$ coincides with the optimal value of the functional. This completes the proof of the theorem. □

Our goal in the remainder of the paper is to construct a pair $(\lambda, W(\cdot,\cdot))$ which satisfies (7). To get this pair, we use the vanishing discount approach. Consider the corresponding control problem with the cost discounted at the rate ρ. For $\boldsymbol{u}(\cdot) \in \mathcal{A}(\boldsymbol{x}^0, \boldsymbol{m}^0)$, we define the expected discounted cost as

$$J^\rho(\boldsymbol{x}^0, \boldsymbol{m}^0, \boldsymbol{u}(\cdot)) = E\int_0^\infty e^{-\rho t} h(\boldsymbol{x}(t), \boldsymbol{u}(t))dt.$$

Define the value function of the discounted cost problem as

$$V^\rho(\boldsymbol{x}^0, \boldsymbol{m}^0) = \inf_{\boldsymbol{u}(\cdot)\in\mathcal{A}(\boldsymbol{x}^0,\boldsymbol{m}^0)} J^\rho(\boldsymbol{x}^0, \boldsymbol{m}^0, \boldsymbol{u}(\cdot)).$$

In order to get the solution of (7), we need the following result, which is also of independent interest.

Theorem 3.2. *For any $(\boldsymbol{x}^0, \boldsymbol{m}^0) \in S \times \mathcal{M}$ and $(\boldsymbol{y}, \boldsymbol{m}') \in S \times \mathcal{M}$, there exists a control policy $\boldsymbol{u}(t)$, $t \geq 0$, such that for any $r \geq 1$,*

$$E\eta^r \leq C_1(r)\left(1 + \sum_{k=1}^{N-1} |x_k^0 - y_k|^r\right), \tag{16}$$

where

$$\eta = \inf\{t \geq 0 : \boldsymbol{x}(t) = \boldsymbol{y},\, \boldsymbol{m}(t) = \boldsymbol{m}'\},$$

and $\boldsymbol{x}(t), t \geq 0$, *is the surplus process corresponding to the control policy* $\boldsymbol{u}(t)$ *and the initial condition* $(\boldsymbol{x}(0), \boldsymbol{m}(0)) = (\boldsymbol{x}^0, \boldsymbol{m}^0)$.

To prove the theorem, we first establish the following lemma concerning the difference between a Markov process and its mean.

Lemma 3.1. *Let* $\tilde{\tau}$ *be a Markov time with respect to an ergodic Markov process* $\tilde{\boldsymbol{m}}(t)$ *with a finite state space* $\tilde{\mathcal{M}} = \{\tilde{\boldsymbol{m}}^1, \ldots, \tilde{\boldsymbol{m}}^p\}$ *with* $\tilde{\boldsymbol{m}}^j = (\tilde{m}_1^j, \ldots, \tilde{m}_N^j)$ $(j = 1, \ldots, p)$. *Let* $\tilde{p}_{\tilde{\boldsymbol{m}}^j}$ *be its stationary distribution, and* $\tilde{\boldsymbol{p}} = (\tilde{p}_1, \ldots, \tilde{p}_N)$ *be its stationary expectation, i.e.* $\tilde{p}_k = \sum_{j=1}^p \tilde{m}_k^j \tilde{p}_{\tilde{\boldsymbol{m}}^j}$. *Then for any linear function* $l(\tilde{\boldsymbol{m}})$, *there exists a constant* C_2 *such that for any* $T > 0$,

$$E \exp\left\{\frac{1}{\sqrt{T}} \sup_{0 \leq t \leq T} \left| \int_{\tau}^{\tau+t} (l(\tilde{\boldsymbol{m}}(s)) - l(\tilde{\boldsymbol{p}}))\, ds \right| \right\} \leq C_2.$$

Proof. Similar to Corollary C.2 in Sethi and Zhang (1994), we can prove that for any Markov time $\tilde{\tau}$ with respect to $\tilde{\boldsymbol{m}}(t)$, there exists a $C_2(A)$ for any $T > 0$ and $A > 0$ such that

$$E \exp\left\{\frac{A}{\sqrt{T}} \sup_{0 \leq t \leq T} \left| \int_{\tilde{\tau}}^{\tilde{\tau}+t} \left(I_{\{\tilde{\boldsymbol{m}}(s) = \tilde{\boldsymbol{m}}^j\}} - \tilde{p}_{\tilde{\boldsymbol{m}}^j} \right) ds \right| \right\} \leq C_2(A). \tag{17}$$

First we show that there exists a constant C_3 such that for any $T > 0$,

$$E \exp\left\{\frac{1}{\sqrt{T}} \sup_{0 \leq t \leq T} \left| \int_{\tilde{\tau}}^{\tilde{\tau}+t} (\tilde{m}_i(s) - \tilde{p}_i)\, ds \right| \right\} \leq C_3. \tag{18}$$

To do this, we note that

$$\begin{aligned}
&E \exp\left\{\frac{1}{\sqrt{T}} \sup_{0 \leq t \leq T} \left| \int_{\tilde{\tau}}^{\tilde{\tau}+t} (\tilde{m}_i(s) - \tilde{p}_i)\, ds \right| \right\} \\
&\quad = E \exp\left\{\frac{1}{\sqrt{T}} \sup_{0 \leq t \leq T} \left| \int_{\tilde{\tau}}^{\tilde{\tau}+t} \sum_{j=1}^p \tilde{m}_i^j \left(I_{\{\tilde{\boldsymbol{m}}(s) = \tilde{\boldsymbol{m}}^j\}} - \tilde{p}_{\tilde{\boldsymbol{m}}^j} \right) ds \right| \right\} \\
&\quad \leq E \exp\left\{\frac{1}{\sqrt{T}} \sum_{j=1}^p \tilde{m}_i^j \sup_{0 \leq t \leq T} \left| \int_{\tilde{\tau}}^{\tilde{\tau}+t} \left(I_{\{\tilde{\boldsymbol{m}}(s) = \tilde{\boldsymbol{m}}^j\}} - \tilde{p}_{\tilde{\boldsymbol{m}}^j} \right) ds \right| \right\}.
\end{aligned}$$

Using the Schwarz inequality we get

$$E \exp\left\{\frac{1}{\sqrt{T}} \sum_{j=1}^p \tilde{m}_i^j \sup_{0 \leq t \leq T} \left| \int_{\tilde{\tau}}^{\tilde{\tau}+t} \left(I_{\{\tilde{\boldsymbol{m}}(s) = \tilde{\boldsymbol{m}}^j\}} - \tilde{p}_{\tilde{\boldsymbol{m}}^j} \right) ds \right| \right\}$$

$$\leq \prod_{j=1}^{p} \left(E \exp\left\{ \frac{2}{\sqrt{T}} \tilde{m}_i^j \sup_{0 \leq t \leq T} \left| \int_{\tilde{\tau}}^{\tilde{\tau}+t} \left(I_{\{\tilde{\boldsymbol{m}}(s)=\tilde{\boldsymbol{m}}^j\}} - \tilde{p}_{\tilde{\boldsymbol{m}}^j} \right) ds \right| \right\} \right)^{\frac{m_i^j}{\sum_{k=1}^p m_i^k}} .$$

From here and (17), we get (18). From (18), by the Schwarz inequality, we get the statement of the lemma. □

Proof of Theorem 3.2. The proof is divided into six steps.

Step 1. We construct an auxiliary process $\hat{\boldsymbol{m}}(t)$. It follows from (A.2) that we can select vectors $\hat{\boldsymbol{m}}^j = (\hat{m}_1^j, \ldots, \hat{m}_N^j)$, $j = 1, \ldots, p$, such that $\hat{m}_1^j = m_1^j$, $\hat{m}_k^j \leq m_k^j$, $j = 1, \ldots, p$, $k = 2, \ldots, N$, and

$$\hat{p}_k := \sum_{j=1}^{p} \hat{m}_k^j p_{\boldsymbol{m}^j} > \hat{p}_{k+1} := \sum_{j=1}^{p} \hat{m}_{k+1}^j p_{\boldsymbol{m}^j} > d, \quad k = 1, \ldots, N-1. \tag{19}$$

Let us define the process $\hat{\boldsymbol{m}}(t)$ as follows:

$$\hat{\boldsymbol{m}}(t) = \hat{\boldsymbol{m}}^j \text{ whenever } \boldsymbol{m}(t) = \boldsymbol{m}^j.$$

We know that $\hat{\boldsymbol{m}}(t) \in \mathcal{M}$ is strongly irreducible and has the stationary distribution $p_{\hat{\boldsymbol{m}}^j}, j = 1, \ldots, p$, where $p_{\hat{\boldsymbol{m}}^j} = p_{\boldsymbol{m}^j}$. Thus, $\hat{\boldsymbol{p}} = (\hat{p}_1, \ldots, \hat{p}_N)$ corresponds to its stationary expectation, and (19) gives

$$p_1 = \hat{p}_1 > \hat{p}_2 > \cdots > \hat{p}_N > d. \tag{20}$$

Step 2. We construct a family of auxiliary processes $\boldsymbol{x}^0(t|s, \boldsymbol{x})$, $t \geq s \geq 0$ and $\boldsymbol{x} \in S$. Consider the following function $\boldsymbol{u}^0(\boldsymbol{x}, \boldsymbol{m}) = (u_1^0(\boldsymbol{x}, \boldsymbol{m}), \ldots, u_N^0(\boldsymbol{x}, \boldsymbol{m}))$:

$$\begin{aligned} u_1^0(\boldsymbol{x}, \boldsymbol{m}) &= m_1, \\ u_k^0(\boldsymbol{x}, \boldsymbol{m}) &= \begin{cases} m_k, & \text{if } x_{k-1} > 0, \\ m_k \wedge u_{k-1}^0(\boldsymbol{x}, \boldsymbol{m}), & \text{if } x_{k-1} = 0, \end{cases} \end{aligned} \tag{21}$$

$k = 2, \ldots, N$. We define $\boldsymbol{x}^0(t|s, \boldsymbol{x})$ as the process which satisfies the following equation (see (2)):

$$\dot{\boldsymbol{x}}^0(t|s, \boldsymbol{x}) = A\boldsymbol{u}^0(\boldsymbol{x}^0(t|s, \boldsymbol{x}), \hat{\boldsymbol{m}}(t)), \quad \boldsymbol{x}^0(s|s, \boldsymbol{x})) = \boldsymbol{x}.$$

Clearly $\boldsymbol{x}^0(t|s, \boldsymbol{x}) \in S$ for all $t \geq s$. For a fixed s, $\boldsymbol{x}^0(t|s, \boldsymbol{x})$ is the state of the system with the production rate which is obtained by using the maximum admissible modified capacity at each machine.

Define now the Markov time

$$\tau(s, \boldsymbol{x}) = \inf\{t \geq s : x_1^0(t|s, \boldsymbol{x}) \geq y_1, \; x_k^0(t|s, \boldsymbol{x}) \geq a + y_k, \; k = 2, \ldots, N\}, \tag{22}$$

where $a > 0$ is a constant specified later. It follows from this definition that $\tau(s, \boldsymbol{x})$ is the first time when the state process $\boldsymbol{x}^0(t|s, \boldsymbol{x})$ exceeds $(y_1, a + y_2, \ldots, a + y_N)$ under the production rate $\boldsymbol{u}^0(\boldsymbol{x}^0(t|s, \boldsymbol{x}), \hat{\boldsymbol{m}}(t))$. Since each machine's modified average capacity is larger than the modified average capacity of the machine that follows it, and since the last machine's modified average capacity is larger than the required rate d (see (10)), we establish the following result.

Step 3. We prove that there exists a constant $C_4 = C_4(r)$ such that

$$E\left(\tau(s, \boldsymbol{x}) - s\right)^{2r} < C_4 \left(1 + \sum_{k=1}^{N-1} \left((y_k - x_k)^+\right)^r\right)^2. \tag{23}$$

For simplicity in exposition, we will write $\tau, \theta, \boldsymbol{x}^r(t)$, and $\boldsymbol{u}^r(t)(r = 0, 1)$ instead of $\tau(s, \boldsymbol{x}), \theta(s, \boldsymbol{x})$, $\boldsymbol{x}^r(t|s, \boldsymbol{x})$, and $\boldsymbol{u}^r(\boldsymbol{x}^r(t|s, \boldsymbol{x}), \hat{\boldsymbol{m}}(t))$, respectively. Let $\overline{m}_k = max_{1 \le j \le p} \hat{m}_k^j$, $k = 1, \ldots, n$, $\overline{m}_{N+1} = \hat{p}_{N+1} = d$. We can choose $\varepsilon > 0$ such that

$$(\hat{p}_k - \hat{p}_{k+1})(1 - \varepsilon) - \varepsilon \overline{m}_{k+1} =: b_k > 0 \text{ for all } 1 \le k \le N.$$

Let $a_1 = 0$ and $a_k = a$ for $2 \le k \le N$. By the definition of τ,

$$\begin{aligned}
&P(\tau - s > t) \\
&\quad \le \sum_{k=1}^{N} P(x_k^0(s+t) < a_k + y_k) \le P\left(\inf_{s+\varepsilon t \le v \le s+t} x_1^0(v) = 0\right) \\
&\qquad + \sum_{k=1}^{N-2} P\left(\bigcap_{i=1}^{k} \left\{\inf_{s+\varepsilon t \le v \le s+t} x_i^0(v) > 0\right\}\right. \\
&\qquad\qquad \left.\cap \left\{\inf_{s+\varepsilon t \le v \le s+t} x_{k+1}^0(v) = 0\right\}\right) \\
&\qquad + \sum_{k=1}^{N} P\left(\bigcap_{i=1}^{N-1} \left\{\inf_{s+\varepsilon t \le v \le s+t} x_i^0(v) > 0\right\}\right. \\
&\qquad\qquad \left.\cap \left\{x_k^0(s+t) < a_k + y_k\right\}\right).
\end{aligned} \tag{24}$$

First we estimate the first term on the right-hand side of (24). Note that $u_2^0(v) \le \hat{m}_2(v)$. Thus using Lemma 3.1 we get

$$\begin{aligned}
&P\left(\inf_{s+\varepsilon t \le v \le s+t} x_1^0(v) = 0\right) \\
&\quad \le P\left(\inf_{\varepsilon t \le v \le t} \int_s^{s+v} (\hat{m}_1(r) - \hat{m}_2(r))dr \le 0\right)
\end{aligned}$$

$$\leq P\left(\inf_{\varepsilon t\leq v\leq t}\int_s^{s+v}(\hat{m}_1(r)-\hat{m}_2(r)-(\hat{p}_1-\hat{p}_2))dr\leq-\varepsilon(\hat{p}_1-\hat{p}_2)t\right)$$
$$\leq P\left(\sup_{0\leq v\leq t}\left|\int_s^{s+v}[(\hat{m}_1(r)-\hat{m}_2(r))-(\hat{p}_1-\hat{p}_2)]dr\right|\geq\varepsilon(\hat{p}_1-\hat{p}_2)t\right)$$
$$\leq C_2\exp\left\{-\varepsilon(\hat{p}_1-\hat{p}_2)\sqrt{t}\right\}. \tag{25}$$

If $\inf_{s+\varepsilon t\leq v\leq s+t} x_i(v)>0$ for all $1\leq i\leq k$, $k\leq N-2$, then $u^0_{k+1}(v)=\hat{m}_{k+1}(v)$ and $u^0_{k+2}(v)\leq\hat{m}_{k+2}(v)$ for $v\in(s+\varepsilon t,s+t)$. So, just as in the proof of (25), we can show that for $k=1,\ldots,N-2$,

$$P\left(\bigcap_{i=1}^k\left\{\inf_{s+\varepsilon t\leq v\leq s+t}x^0_i(v)>0\right\}\cap\left\{\inf_{s+\varepsilon t\leq v\leq s+t}x^0_{k+1}(v)=0\right\}\right)$$
$$\leq C_2\exp\left\{-\varepsilon(\hat{p}_{k+1}-\hat{p}_{k+2})\sqrt{t}\right\}. \tag{26}$$

Now we consider the members of the last sum on the right-hand side of (24). According to the definition of $u^0_k(\cdot)$,

$$P\left(\bigcap_{k=1}^{N-1}\left\{\inf_{s+\varepsilon t\leq v\leq s+t}x^0_k(v)>0\right\}\cap\left\{x^0_k(s+t)<a_k+y_k\right\}\right)$$
$$\leq P\left(x_k-\varepsilon t\overline{m}_{k+1}+\int_{s+\varepsilon t}^{s+t}(\hat{m}_k(r)-\hat{m}_{k+1}(r))dr<a_k+y_k\right)$$
$$\leq P\Bigg(\int_{s+\varepsilon t}^{s+t}(\hat{m}_k(r)-\hat{m}_{k+1}(r)-(\hat{p}_k-\hat{p}_{k+1})dr$$
$$<(a_k+y_k-x_k)^+-b_kt\Bigg). \tag{27}$$

Applying Lemma 3.1 we have from (27):

$$P\left(\bigcap_{k=1}^{N-1}\left\{\inf_{s+\varepsilon t\leq v\leq s+t}x^0_k(v)>0\right\}\cap\left\{x^0_k(s+t)<a_k+y_k\right\}\right)$$
$$\leq\begin{cases}1 & \text{for } 0\leq t\leq(a_k+y_k-x_k)^+/b_k,\\ C_2\exp\left\{-\frac{b_kt-(a_k+y_k-x_k)^+}{\sqrt{t}}\right\} & \text{for } t\geq(a_k+y_k-x_k)^+/b_k.\end{cases} \tag{28}$$

Note that $E(\tau-s)^{2r}=\int_0^\infty t^{2r-1}P(\tau-s>t)dt$. By substituting from (24), (25), (26), and (28) into this relation, we get (23).

Step 4. We construct a family of auxiliary processes $\boldsymbol{x}^1(t|s,\boldsymbol{x})$, $t\geq s\geq 0$ and $\boldsymbol{x}\in S$. Consider the following function $\boldsymbol{u}^1(\boldsymbol{x},\boldsymbol{m})=(u^1_1(\boldsymbol{x},\boldsymbol{m}),\ldots,u^1_N(\boldsymbol{x},\boldsymbol{m}))$, which is defined only for $\boldsymbol{x}$ such that $x_i \geq y_i$ with $1 \leq i \leq N-1$:

$$u_1^1(\boldsymbol{x}, \boldsymbol{m}) = 0,$$

$$u_k^1(\boldsymbol{x}, \boldsymbol{m}) = \begin{cases} m_k, & \text{if } x_{k-1} > y_{k-1}, \\ m_k \wedge u_{k-1}^1(\boldsymbol{x}, \boldsymbol{m}), & \text{if } x_{k-1} = y_{k-1} \end{cases} \qquad k = 2, \ldots, N. \tag{29}$$

We define $\boldsymbol{x}^1(t|s, \boldsymbol{x})$ as a continuous process which coincides with $\boldsymbol{x}^0(t|s, \boldsymbol{x})$ for $s \leq t \leq \tau(s, \boldsymbol{x})$, which satisfies the following equation (see (2)):

$$\dot{\boldsymbol{x}}^1(t|s, \boldsymbol{x}) = A\boldsymbol{u}^1(\boldsymbol{x}^1(t|s, \boldsymbol{x}), \hat{\boldsymbol{m}}(t)), \; t \geq \tau(s, \boldsymbol{x}).$$

Clearly $\boldsymbol{x}^1(t|s, \boldsymbol{x}) \in S$ for all $t \geq s$, and $x_i^1(t|s, \boldsymbol{x}) \geq y_i$ $(1 \leq i \leq N-1)$ for $t \geq \tau(s, \boldsymbol{x})$. This process corresponds to a policy in which after $\tau(s, \boldsymbol{x})$, we stop production at the first machine and have the maximum possible production rate at other machines under the restriction that the content of each buffer k, $1 \leq k \leq N-1$, is not less than y_k.

We define now a Markov time

$$\theta(s, \boldsymbol{x}) = \inf\{t \geq \tau(s, \boldsymbol{x}) : x_N^1(t|s, \boldsymbol{x}) = y_N\}. \tag{30}$$

Step 5. We establish that (i) a constant a given by (22) can be chosen in such a way that for all $s, \boldsymbol{x}$,

$$P\left(\boldsymbol{x}^1(\theta(s, \boldsymbol{x})|s, \boldsymbol{x}) = \boldsymbol{y}, \; \boldsymbol{m}(\theta(s, \boldsymbol{x}) = \boldsymbol{m}'\right) \geq 1 - q > 0, \text{ and} \tag{31}$$

(ii) there exists a constant C_5 such that

$$\frac{a}{d} \leq \theta(s, \boldsymbol{x}) - s \leq \frac{1}{d}\left(\sum_{k=1}^N x_k - \sum_{k=1}^N y_k + C_5[\tau(s, \boldsymbol{x}) - s]\right), \tag{32}$$

$$\sum_{k=1}^N x_k^1(\theta(s, \boldsymbol{x})|s, \boldsymbol{x}) \leq \sum_{k=1}^N x_k + C_5[\tau(s, \boldsymbol{x}) - s]. \tag{33}$$

First taking the sum of all the equations in (1), we have $\sum_{k=1}^N x_k^1(t) = \sum_{k=1}^N x_k + \int_s^t (u_1^0(v) - d)dv$ for $s \leq t \leq \tau$. Consequently,

$$\sum_{k=1}^N x_k^1(\tau) \leq \sum_{k=1}^N x_k + (\bar{m}_1 - d)(\tau - s). \tag{34}$$

Since $u_1^1(t) = 0$ for $t > \tau$, we have as before that $\sum_{k=1}^N x_k^1(\theta) = \sum_{k=1}^N x_k^1(\tau) - d(\theta - \tau)$. Since $\theta > \tau$ and $x_k^1(\theta) \geq y_k$, we have

$$\theta - \tau \leq \frac{1}{d}\left(\sum_{k=1}^N x_k^1(\tau) - \sum_{k=1}^N y_k\right), \quad \sum_{k=1}^N x_k^1(\theta) \leq \sum_{k=1}^N x_k^1(\tau). \tag{35}$$

From the definitions of θ and τ, and (1) with $k = N$, we have that $y_N = x_N^1(\theta) = x_N^1(\tau) + \int_\tau^\theta (u_N^1(v) - d)ds \geq y_N + a - d(\theta - \tau)$, i.e., $\theta - \tau \geq a/d$. This relation (34) and (35) prove (32)-(33). To prove (31), we introduce the following notations:

$$\theta(k) = \inf\{t \geq \tau : \ x_k^1(t) = y_k\}, \quad k = 1, \dots, N, \quad \tilde{\mathcal{S}} = \{\omega : \ \boldsymbol{x}^1(\theta) = \boldsymbol{y}\},$$

$$\mathcal{S}(k) = \left\{\omega : \inf_{0 \leq t < \infty} \int_\tau^{\tau+t} [\hat{m}_k(v) - \hat{m}_{k+1}(v)\} dv > -a/2 \right\}, \qquad k = 2, \dots, N,$$

$$\bar{\mathcal{S}} = \bigcap_{k=2}^{N} \mathcal{S}(k), \quad \mathcal{S}(0) = \{\omega : \ \boldsymbol{m}(\theta) = \boldsymbol{m}'\}, \quad \mathcal{S} = \tilde{\mathcal{S}} \cap \mathcal{S}(0).$$

Note that $\tilde{\mathcal{S}} = \{\omega : \theta(N) \geq \max_{1 \leq k \leq N-1} \theta(k)\}$. From the definition of $\boldsymbol{u}^1(\boldsymbol{x}, \boldsymbol{m})$ and $\boldsymbol{x}^1(t)$, it follows that if $\omega \in \bar{\mathcal{S}}$, then

$$u_{k+1}^1(t) = \begin{cases} \hat{m}_{k+1}(v) \text{ for } \tau < t \leq \theta(k) \ 1 \leq k \leq N-1, \\ 0 \quad \text{ for } \quad t > \theta(k) \quad 1 \leq k < N-1, \end{cases}$$

and

$$\theta(k) - \theta(k-1) \geq \frac{x_k^1(\theta(k-1)) - y_k}{\bar{m}_k} \geq \frac{a}{\bar{m}_k}, \quad k = 2, \dots, N. \tag{36}$$

Therefore, $\bar{\mathcal{S}} \subseteq \tilde{\mathcal{S}}$ and

$$P[\mathcal{S}^c] \leq \sum_{k=2}^{N} P(\mathcal{S}^c(k)) + P(\bar{\mathcal{S}} \cap \mathcal{S}^c(0)). \tag{37}$$

Note that if η_1 and η_2 are Markov times and $\eta_2 - \eta_1 > 1$, then there exists $q_1 < 1$ such that

$$\max_{1 \leq j \leq p} P(\boldsymbol{m}(\eta_2) \neq \boldsymbol{m}' | \boldsymbol{m}(\eta_1) = \boldsymbol{m}^j) < q_1 < 1. \tag{38}$$

Taking the conditional probability with respect to $\theta(N-1)$ and using (36) with $k = N$, $a > 2/\sum_{k=2}^{N}(1/\bar{m}_k)$, and (38), we have

$$P(\bar{\mathcal{S}} \cap \mathcal{S}^c(0)) < q_1 < 1. \tag{39}$$

Applying Lemma 3.1 we have

$$
\begin{aligned}
P(S^c(k)) &\le \textstyle\sum_{n=1}^{\infty} P\left(\int_{\tau}^{\tau+n}(\hat{m}_k(v)-\hat{m}_{k+1}(v))dv < -a+\bar{m}_{k+1}\right)\\
&\le \textstyle\sum_{n=1}^{\infty} P\left(\left|\int_{\tau}^{\tau+n}\left[(\hat{m}_k(v)-\hat{m}_{k+1}(v))-(\hat{p}_k-\hat{p}_{k+1})\right]ds\right|\right.\\
&\qquad\qquad \left. > a+n(\hat{p}_k-\hat{p}_{k+1})-\bar{m}_{k+1}\right)\\
&\le C_5\textstyle\sum_{n=1}^{\infty}\exp\left\{-\frac{a+n(\hat{p}_k-\hat{p}_{k+1})-\bar{m}_{k+1}}{\sqrt{n}}\right\} \le C_6 e^{-C_7\sqrt{a}}.
\end{aligned}
\tag{40}
$$

It follows from (37), (39), and (40) that we can choose a and q such that $P(S^c)\le q<1$. This proves (31).

Step 6. We construct a process $\boldsymbol{x}(t)$ $(t\ge 0)$ and the corresponding control policy $\boldsymbol{u}(t)$, which satisfies the statement of Theorem 2.4.

Define a sequence of Markov times $(\theta_i)_{i=0}^{\infty}$ and the process $\boldsymbol{x}(t)$ for $\theta_i\le t<\theta_{i+1}$ $(i=1,2,\cdots)$ as follows: $\theta_0=0$, $\theta_1=\theta(0,\boldsymbol{x}^0)$ and $\boldsymbol{x}(t)=\boldsymbol{x}^1(t|0,\boldsymbol{x}^0)$ with $0\le t\le\theta_1$. If θ_i is defined for $i\ge 1$ and $\boldsymbol{x}(t)$ is defined for $0\le t\le\theta_i$, then we let $\theta_{i+1}=\theta(\theta_i,\boldsymbol{x}(\theta_i))$ and $\boldsymbol{x}(t)=\boldsymbol{x}^1(t|\theta_i,\boldsymbol{x}(\theta_i))$ with $\theta_i\le t\le\theta_{i+1}$. According to the left inequality in (32), the process $\boldsymbol{x}(t)$ is defined for all $t\ge 0$. Let $\tau_i=\tau(\theta_i,\boldsymbol{x}(\theta_i))$. The control policy corresponding to the process $\boldsymbol{x}(t)$ is given by

$$
\boldsymbol{u}(t)=\begin{cases}\boldsymbol{u}^0(\boldsymbol{x}(t),\hat{\boldsymbol{m}}(t)), & \text{if } \theta_{i-1}\le t<\tau_i,\\ \boldsymbol{u}^1(\boldsymbol{x}(t),\hat{\boldsymbol{m}}(t)), & \text{if } \tau_i\le t<\theta_i,\end{cases}\qquad i=1,2\ldots, \tag{41}
$$

It is clear that $\boldsymbol{u}(t)\in\mathcal{A}(\boldsymbol{x}^0,\boldsymbol{m}^0)$. For the process $\boldsymbol{x}(t)$, a Markov time is defined as

$$
\eta=\inf\{t\ge 0:\boldsymbol{x}(t)=\boldsymbol{y},\ \boldsymbol{m}(t)=\boldsymbol{m}'\}.
$$

Let $\mathcal{S}_i=\{\omega:\boldsymbol{x}(\theta_i)=\boldsymbol{y},\boldsymbol{m}(\theta_i)=\boldsymbol{m}\}$. Using conditional probabilities, we have from (31) that

$$
P\left(\cap_{l=1}^{i}\mathcal{S}_l^c\right)\le q^i,\quad i=1,2,\ldots. \tag{42}
$$

Using (42) and the definition of $\boldsymbol{x}(t)$ we get:

$$
\eta^r=\sum_{i=1}^{\infty}\theta_i^r I_{\{\cap_{l=0}^{i-1}\mathcal{S}_l^c\cap\mathcal{S}_i\}},\ a.s., \tag{43}
$$

where $\mathcal{S}_0^c=\Omega$. Using (32) and (33) we have for $n=1,2,\ldots$,

$$\theta_n - \theta_{n-1} \le \frac{1}{d}\left(\sum_{k=1}^{N} x_k(\theta_{n-1}) - \sum_{k=1}^{N} y_k + C_3(\tau_n - \theta_{n-1})\right), \tag{44}$$

$$\sum_{k=1}^{N} x_k(\theta_n) \le \sum_{k=1}^{N} x_k(\theta_{n-1}) + C_3(\tau_n - \theta_{n-1}). \tag{45}$$

Using (44) and (45) we have for $i = 1, 2, \ldots,$

$$\theta_i \le \frac{1}{d}\left(\sum_{k=1}^{N}(x_k^0 - y_k)^+ + C_3\sum_{n=1}^{i}(\tau_n - \theta_{n-1})\right),$$

or

$$\theta_i^r \le C_4 i^r\left(\sum_{k=1}^{N}\left((x_k^0 - y_k)^+\right)^r + \sum_{n=1}^{i}(\tau_n - \theta_{n-1})^r\right). \tag{46}$$

Note now that $\boldsymbol{x}(\theta_n) \ge \boldsymbol{y}$ for $n = 1, 2, \cdots$. Using the Schwarz inequality (Corollary 3 in page 104 of Chow and Teicher, 1988), we get from (42) and Lemma 3.1:

$$\begin{aligned} &E\left(\tau_1^r I_{\{\cap_{l=1}^{i-1}\mathcal{S}_l^c \cap \mathcal{S}_i\}}\right) \\ &\quad \le q^{(i-1)/2}\left(E(\tau_1^{2r})\right)^{1/2} \\ &\quad \le C_2(r) q^{(i-1)/2}\left(1 + \sum_{k=1}^{N-1}\left((y_k - x_k^0)^+\right)^r\right), \qquad i = 1, 2, \ldots, \end{aligned} \tag{47}$$

and

$$\begin{aligned} &E\left((\tau_n - \theta_{n-1})^r I_{\{\cap_{l=1}^{i-1}\mathcal{S}_l^c \cap \mathcal{S}_i\}}\right) \\ &\quad \le q^{(i-1)/2}\left(E((\tau_n - \theta_{n-1})^{2r}\right)^{1/2} \\ &\quad \le C_2(r) q^{(i-1)/2}, \qquad 2 \le n \le i = 2, 3, \ldots. \end{aligned} \tag{48}$$

Substituting (46) into (43), taking expectation, and using (47) and (48), we get (16). □

The next two theorems are concerned with the solution of (7).

Theorem 3.3. *There exists a sequence $\{\rho_k : k \ge 1\}$ with $\rho_k \to 0$ as $k \to \infty$ such that for $(\boldsymbol{x}, \boldsymbol{m}) \in S \times \mathcal{M}$:*

$$\lim_{k\to\infty} \rho_k V^{\rho_k}(\boldsymbol{x}, \boldsymbol{m}) = \lambda,$$

$$\lim_{k\to\infty} [V^{\rho_k}(\boldsymbol{x},\boldsymbol{m}) - V^{\rho_k}(0,\boldsymbol{m}^0)] = W(\boldsymbol{x},\boldsymbol{m}),$$

where $W(\boldsymbol{x},\boldsymbol{m})$ *is convex in* $\boldsymbol{x}$ *for any given* $\boldsymbol{m}$.

Proof. For the value function $V^\rho(\boldsymbol{x},\boldsymbol{m})$ of the discounted cost problem, we define the differential discounted value function, known also as the potential function,

$$W^\rho(\boldsymbol{x},\boldsymbol{m}) = V^\rho(\boldsymbol{x},\boldsymbol{m}) - V^\rho(0,\boldsymbol{m}).$$

Thus, the function $W^\rho(\boldsymbol{x},\boldsymbol{m})$ is convex in $\boldsymbol{x}$. Following the line of the proof of Theorem 3.2 in Sethi *et al.* (1997), we see that there exist constants ρ_0 and $C_7 > 0$ such that for $0 < \rho \le \rho_0$,

$$\rho V^\rho(0,\boldsymbol{m}) \le C_7.$$

Thus, there exists a sequence $\{\rho_k : k \ge 1\}$ with $\rho_k \to 0$ as $k \to \infty$ such that for $(\boldsymbol{x},\boldsymbol{m}) \in S \times \mathcal{M}$,

$$\lim_{k\to\infty} \rho_k V^{\rho_k}(0,\boldsymbol{m}) = \lambda. \tag{49}$$

Note that the first statement of Theorem 3.3 follows from (49) and the last statement of Theorem 3.2. So, it remains to prove the last statement of Theorem 3.3.

To do this we first show that there is a constant $C_8 > 0$ such that

$$|W^\rho(\boldsymbol{x},\boldsymbol{m})| \le C_8\left(1 + |\boldsymbol{x}|^{K_h+2}\right), \tag{50}$$

for all $(\boldsymbol{x},\boldsymbol{m}) \in S \times \mathcal{M}$ and $\rho > 0$. Without loss of generality we suppose that $V^\rho(\boldsymbol{x},\boldsymbol{m}) \ge V^\rho(0,\boldsymbol{m})$ (the case $V^\rho(\boldsymbol{x},\boldsymbol{m}) \le V^\rho(0,\boldsymbol{m})$ is treated in the same way).

By Theorem 3.2 there exists a control policy $\boldsymbol{u}(\cdot)$ such that

$$E\eta^r \le C_1(r)\left(1 + \sum_{k=1}^N |x_k|^r\right), \tag{51}$$

where $\eta = \inf\{t > 0 : (\boldsymbol{x}(t),\boldsymbol{m}(t)) = (0,\boldsymbol{m})\}$, and $\boldsymbol{x}(t)$ is the state process corresponding to $\boldsymbol{u}(t)$ with the initial condition $(\boldsymbol{x}(0),\boldsymbol{m}(0)) = (\boldsymbol{x},\boldsymbol{m})$. From the dynamic programming principle we have

$$\begin{aligned} V^\rho(\boldsymbol{x},\boldsymbol{m}) &\le E\left(\int_0^\eta \exp(-\rho t) h(\boldsymbol{x}(t),\boldsymbol{u}(t))dt + \exp(-\rho\eta) V^\rho(\boldsymbol{x}(\eta),\boldsymbol{m}(\eta))\right) \\ &= E\left(\int_0^\eta \exp(-\rho t) h(\boldsymbol{x}(t),\boldsymbol{u}(t))dt + \exp(-\rho\eta) V^\rho(0,\boldsymbol{m})\right) \\ &\le E\left(\int_0^\eta \exp(-\rho t) h(\boldsymbol{x}(t),\boldsymbol{u}(t))dt\right) + V^\rho(0,\boldsymbol{m}). \end{aligned}$$

Therefore,

$$
\begin{aligned}
|W^\rho(\boldsymbol{x},\boldsymbol{m})| &= V^\rho(\boldsymbol{x},\boldsymbol{m}) - V^\rho(0,\boldsymbol{m}) \\
&\le E\left(\int_0^\eta \exp(-\rho t)h(\boldsymbol{x}(t),\boldsymbol{u}(t))dt\right). \qquad (52)
\end{aligned}
$$

By Assumption (A.3), there exists a $\tilde{C}_0 > 0$ such that

$$h(\boldsymbol{x}(t),\boldsymbol{u}(t)) \le \tilde{C}_0(1 + |\boldsymbol{x}|^{K_h+1} + t^{K_h+1}), \qquad (53)$$

where we use the fact that $\boldsymbol{u}(\cdot)$ is bounded. Therefore, (51) implies that

$$
\begin{aligned}
&E\int_0^\eta \exp(-\rho t)h(\boldsymbol{x}(t),\boldsymbol{u}(t))dt \le E\int_0^\eta \tilde{C}_0(1 + |\boldsymbol{x}|^{K_h+1} + t^{K_h+1})dt \\
&= \tilde{C}_0(E\eta + |\boldsymbol{x}|E\eta + E(\eta)^{K_h+2}) \le C_9(1 + \sum_{k=1}^N |x_k|^{K_h+2}),
\end{aligned}
$$

for some $C_9 > 0$. Thus (52) gives (50).

For $\delta \in (0,1)$, let $B^\delta = [\delta, 1/\delta]^{N-1} \times [-1/\delta, 1/\delta]$. Based on (50) it follows from Theorem 10.6 on page 88 of Rockafellar(1972) that there is a $C(\delta)$ such that for $\boldsymbol{x},\boldsymbol{x}' \in B^\delta$,

$$|W^\rho(\boldsymbol{x},\boldsymbol{m}) - W^\rho(\boldsymbol{x}',\boldsymbol{m})| \le C(\delta)|\boldsymbol{x} - \boldsymbol{x}'|. \qquad (54)$$

Without loss of generality we assume that $C(\delta)$ is a decreasing function in δ. For $1 \le n \le N-1$ and $1 \le i_1 < \dots < i_n \le N-1$, let $S_{i_1\dots i_n} = \{\boldsymbol{x} \in b(S) : x_{i_\ell} = 0 \text{ for } \ell = 1,\dots,n\}$ and $S^o_{i_1\dots i_n} = \{\boldsymbol{x} \in S_{i_1\dots i_n} : x_j > 0, 1 \le j \le N-1,\ j \notin \{i_1,\dots,i_n\}\}$. That is, $S^o_{i_1\dots i_n}$ is the interior of $S_{i_1\dots i_n}$ relative to $[0,\infty)^{N-n-1} \times (-\infty,+\infty)$. Note that the function $V^\rho(\boldsymbol{x},\boldsymbol{m})$ is still convex on $S_{i_1\dots i_n}$. Let

$$B^\delta_{i_1\dots i_n} = \Pi_{\ell=1}^{N-1}\Upsilon^\delta_\ell \times [-\frac{1}{\delta}, \frac{1}{\delta}]$$

with

$$\Upsilon^\delta_\ell = \begin{cases} \{0\}, & \text{if } \ell \in \{i_1,\dots,i_n\} \\ [\delta, \frac{1}{\delta}], & \text{if } \ell \notin \{i_1,\dots,i_n\}. \end{cases}$$

Using again Theorem 10.6 on Page 88 of Rockafellar (1972), in view of (50), there is a $C_{i_1\dots i_n}(\delta) > 0$ such that for $\boldsymbol{x},\boldsymbol{x}' \in B^\delta_{i_1\dots i_n}$

$$|W^\rho(\boldsymbol{x},\boldsymbol{m}) - W^\rho(\boldsymbol{x}',\boldsymbol{m})| \le C_{i_1\dots i_n}(\delta)|\boldsymbol{x} - \boldsymbol{x}'|. \qquad (55)$$

Also we assume that $C_{i_1\dots i_n}(\delta)$ is a decreasing function in δ. From the arbitrariness of δ and (54)-(55), there exist $W(\boldsymbol{x},\boldsymbol{m})$ and a sequence of $\{\rho_k : k \ge 1\}$ with $\rho_k \to 0$ as $k \to \infty$ such that for $(\boldsymbol{x},\boldsymbol{m}) \in S \times \mathcal{M}$,

$$\lim_{k\to\infty} [V^{\rho_k}(\boldsymbol{x},\boldsymbol{m}) - V^{\rho_k}(0,\boldsymbol{m})] = W(\boldsymbol{x},\boldsymbol{m}). \qquad (56)$$

It follows from the convexity of $W^{\rho_k}(\boldsymbol{x},\boldsymbol{m})$ that the limit function $W(\boldsymbol{x},\boldsymbol{m})$ is also convex on $S \times \mathcal{M}$. □

Let $\bigtriangledown(V^{\rho_k}(\boldsymbol{x},\boldsymbol{m}) - V^{\rho_k}(0,\boldsymbol{m}^0))$ be the derivative of $V^{\rho_k}(\boldsymbol{x},\boldsymbol{m}) - V^{\rho_k}(0,\boldsymbol{m}^0))$ at the point $\boldsymbol{x}$ when the derivative exists.

Theorem 3.4. (i) *$\lambda(\boldsymbol{x}^0,\boldsymbol{m}^0)$ does not depend on $(\boldsymbol{x}^0,\boldsymbol{m}^0)$.*

(ii) *The pair $(\lambda, W(\cdot,\cdot))$ defined in Theorem* 3.3 *satisfies* (7) on S^0.

(iii) *If there exists an open subset $\hat{S}$ of S such that* (a) $\{\bigtriangledown(V^{\rho_k}(\boldsymbol{x},\boldsymbol{m}) - V^{\rho_k}(0,\boldsymbol{m}^0)) : k \geq 1\}$ *is uniformly equi-Lipschitzian on $\hat{S}$, and* (b) $b(\hat{S}) \subseteq b(S)$, *where $b(S)$ and $b(\hat{S})$ are the boundaries of S and $\hat{S}$, respectively. Then the pair $(\lambda, W(\cdot,\cdot))$ defined in Theorem* 3.3 *satisfies* (7) *on S, i.e., it is a solution to* (7).

Proof. Let $\mathcal{O}$ be the set of all points in the interior of $\mathcal{S}$ on which $W(\boldsymbol{x},\boldsymbol{m})$ is differentiable. From the convexity of $W(\boldsymbol{x},\boldsymbol{m})$ we know that $\mathcal{O}$ is dense in $\mathcal{S}$. It follows from the properties of convex functions that for $\boldsymbol{x} \in \mathcal{O}$ and any $\boldsymbol{r}$,

$$\lim_{k\to\infty} \partial_{\boldsymbol{r}} W^{\rho_k}(\boldsymbol{x},\boldsymbol{m}) = \partial_{\boldsymbol{r}} W(\boldsymbol{x},\boldsymbol{m}). \tag{57}$$

Presman *et al.* (1995) proved that for any $\boldsymbol{x} \in S$, the value function $V^{\rho_k}(\boldsymbol{x},\boldsymbol{m})$ of the discounted cost problem satisfies

$$\rho_k V^{\rho_k}(\boldsymbol{x},\boldsymbol{m}) = \inf_{\boldsymbol{u}\in U(\boldsymbol{x},\boldsymbol{m})} \{\partial_{A\boldsymbol{u}} V^{\rho_k}(\boldsymbol{x},\boldsymbol{m}) + h(\boldsymbol{x},\boldsymbol{m})\} + QV^{\rho_k}(\boldsymbol{x},\boldsymbol{m}).$$

This implies that

$$\rho_k V^{\rho_k}(\boldsymbol{x},\boldsymbol{m}) = \inf_{\boldsymbol{u}\in U(\boldsymbol{x},\boldsymbol{m})} \{\partial_{A\boldsymbol{u}} W^{\rho_k}(\boldsymbol{x},\boldsymbol{m}) + h(\boldsymbol{x},\boldsymbol{m})\} + QW^{\rho_k}(\boldsymbol{x},\boldsymbol{m}). \tag{58}$$

Taking the limit on both sides, we have that for $\boldsymbol{x} \in \mathcal{O}$,

$$\lambda = \inf_{\boldsymbol{u}\in U(\boldsymbol{x},\boldsymbol{m})} \{\partial_{A\boldsymbol{u}} W(\boldsymbol{x},\boldsymbol{m}) + h(\boldsymbol{x},\boldsymbol{m})\} + QW(\boldsymbol{x},\boldsymbol{m}). \tag{59}$$

If $\boldsymbol{x} \notin \mathcal{O}$, $\boldsymbol{x} \in S^0$ then for any direction $\boldsymbol{r}$, there exist a sequence $\{\boldsymbol{x}_n\}_{n=1}^{\infty}$ such that $\boldsymbol{x}_n \in \mathcal{O}$ and $\partial_{\boldsymbol{r}} W(\boldsymbol{x}_n,\boldsymbol{m}) \to \partial_{\boldsymbol{r}} W(\boldsymbol{x},\boldsymbol{m})$. From this fact and from continuity of $W(\boldsymbol{x},\boldsymbol{m})$, it follows that (59) holds for all $\boldsymbol{x}$ in the interior of S.

Consider now the boundary $b(S)$ of S. From the uniformly equi-Lipschitzian property of $\{\bigtriangledown W^{\rho_k}(\boldsymbol{x},\boldsymbol{m}) : k \geq 1\}$ on $\hat{S}$, we know that (57) holds for all $\boldsymbol{x} \in b(S)$. Therefore, we have (59) in $b(S)$. □

4 Concluding Remarks

In this paper, we have developed a theory of dynamic programming in terms of directional derivative for an N-machine flowshop with convex costs and the long-run average cost minimization criterion. Further research should focus on extending this analysis to N-machine flowshops with limited buffers. For such systems with two machines, see Presman *et al.* (2000).

References

1. Bielecki, T. and Kumar, P. R. (1988) Optimality of zero-inventory policies for unreliable manufacturing systems, Operations Research **36**, 532–546.
2. Caramanis, M. and Sharifinia, A. (1991) Optimal manufacturing flow control design, International J. Flexible Manufacturing Systems **3**, 321–336.
3. Chow, Y. S. and Teicher, H. (1988) Probability Theory, Springer Verlag, New York.
4. Clarke, F. (1983) Optimization and Nonsmooth Analysis, Wiley-Intersciences, New York.
5. Duncan, T. E., Pasik-Duncan, B., and Stettner, L. (2001) Average cost per unit time control of stochastic manufacturing systems: Revisited, Math Meth Oper Res **54**, 259–278.
6. Fleming, W. and Soner, H. (1992) Controlled Markov Processes and Viscosity Solutions. Springer Verlag, New York.
7. Liberopoulos, G. and Caramanis, M. (1993) Production control of manufacturing systems with production rate dependent failure rate, IEEE Trans. Auto. Control **38**, 889–895.
8. Liberopoulos, G. and Hu, J. (1995) On the ordering of optimal hedging points in a class of manufacturing flow control models, IEEE Trans. Auto. Control **40**, 282–286.
9. Presman, E., Sethi, S., and Suo, W. (1997) Existence of optimal feedback production plans in stochastic flowshops with limited buffers, Automatica **33**, 1899–1903.
10. Presman, E., Sethi, S., Zhang, H., and Bisi, A. (2000) Average cost optimal policies for an unreliable two-machine flowshops with limited internal buffer, Annals of Operations Research **98**, 333–351.
11. Presman, E., Sethi, S., and Zhang, Q. (1995) Optimal feedback production planning in a stochastic N-machine flowshop, Automatica, **31**, 1325–1332.
12. Rockafellar, R. T. (1996) Convex Analysis, Princeton University Press, Princeton, NJ, Reprint Edition.
13. Sethi, S. P., Suo, W., Taksar, M. I., and Zhang, Q. (1997) Optimal production planning in a stochastic manufacturing system with long-run average cost, J. of Optimization Theory and Applications **92**, 161–188.
14. Sethi, S. P., Suo, W., Taksar, M. I., and Yan, H. (1998) Optimal production planning in a multi-product stochastic manufacturing system with long-run average cost, Discrete Event Dynamic Systems: Theory and Applications **8**, 37–54.
15. Sharifinia, S. (1988) Production control of a manufacturing system with multiple machine states, IEEE Trans. Auto. Control, **33**, 620–625.
16. Soner, H. (1986) Optimal stochastic control with state-space constraints II, SIAM J. on Control and Optimization **24**, 1110–1123.
17. Srisvatsan, N. Synthesis of Optimal Policies in Stochastic Manufacturing Systems, Ph.D Thesis, OR Center, MIT.

18. Srisvatsan, N. and Dallery, Y. (1997) Partial characterization of optimal hedging point policies in unreliable two-part-type manufacturing systems, Operations Research **46**, 36–45.

A Risk-Sensitive Generalization of Maximum A Posterior Probability (MAP) Estimation

Vahid Reza Ramezani and Steven I. Marcus

Department of Electrical and Computer Engineering and Institute for Systems Research, University of Maryland, College Park, MD 20742, USA

Abstract. A sequential filtering scheme for the risk-sensitive state estimation of partially observed Markov chains is presented. The previously introduced risk-sensitive filters are unified in the context of risk-sensitive Maximum A Posterior Probability (MAP) estimation. Structural results for the filter banks are given. The influence of the availability of information and the transition probabilities on the decision regions and the behavior of risk-sensitive estimators are studied.

1 Introduction

The exponential (risk-sensitive) criterion of a quadratic function of the state and control for full state control was first proposed by Jacobson [2]. Whittle [16] produced the controller for the linear/quadratic partial observation case which required a state estimator *separated* from the control policy in a fashion analogous to the policy separation for the partial observation in the risk-neutral case. Speyer [6] treated the estimation problem for the linear/quadratic partial observation case and showed that a linear estimator is optimal among all non-linear and linear estimators. The non-linear discrete-time stochastic problem for the partially observed control problem was solved by a change of measure technique [4]. This so called *reference probability* approach was later used by Dey and Moore [7] to solve an estimation problem for a *partially observed Markov chain* or, as it is commonly referred to in signal processing literature, a *Hidden Markov Model* (HMM).

Our work can be traced back to Speyer's paper [6] nearly a decade ago and is related to the Dey-Moore filter [7]. We are also interested in the estimation problem with the exponential criterion for HMM's. However, we have a different perspective; we view the estimator as a dynamical system whose dynamics are inherited from the Markov chain through the partial observation and an optimization criterion. We are not only interested in the computation of the optimal estimator for an HMM and its properties for exponential criterion, but also in the *qualitative analysis* of its sample paths. Under perfect observation the dynamics of the estimator revert back to the Markov chain itself but partial observation produces non-trivial results.

It is often said that risk-sensitive filters take into account the "higher order" moments of the estimation error. Roughly speaking, this follows from

B. Pasik-Duncan: Stochastic Theory and Control, LNCIS 280, pp. 419–433, 2002.

the analytic property of the exponential $e^x = \sum_{k=0}^{\infty} x^k/k!$ so that if Ψ stands for the sum of the error functions over some interval of time then

$$E[\exp(\gamma\Psi)] = E[1 + \gamma\Psi + (\gamma)^2(\Psi)^2/2 + \cdots].$$

Thus, at the expense of the mean error cost, the higher order moments are included in the minimization of the expected cost, reducing the "risk" of large deviations and increasing our "confidence" in the estimator. The parameter $\gamma > 0$ controls the extent to which the higher order moments are included. In particular, the first order approximation, $\gamma \to 0$, $E[\exp(\gamma\Psi] \cong 1 + \gamma E\Psi$, indicates that the original minimization of the sum criterion or the risk-neutral problem is recovered as the small risk limit of the exponential criterion.

In this paper, we introduce a filtering scheme for the risk sensitive Maximum A Posterior Probability (MAP) estimation for HMM's and analyze its structural properties. The previously introduced risk sensitive filters will be considered as special cases.

In section 2, the problem set up, our basic notation and the risk-sensitive filter banks are introduced. In section 3, structural results for the filters are given and in section 4, the behavior of the sample paths is analyzed.

2 Risk Sensitive Filter Banks

2.1 The Estimation of Hidden Markov Models

Define a Hidden Markov Model as a five tuple $< \mathbf{X}, \mathbf{Y}, \mathbf{X}, \mathcal{A}, Q >$; here $\mathcal{A}$ is the transition matrix, $\mathbf{Y} = \{1, 2, \ldots, N_{\mathbf{Y}}\}$ is the set of observations and $\mathbf{X} = \{1, 2, \ldots, N_{\mathbf{X}}\}$ is the finite set of (internal) states as well as the set of estimates or decisions. In addition, we have that $Q := [q_{x,y}]$ is the $N_{\mathbf{X}} \times N_{\mathbf{Y}}$ state/observation matrix, i.e., $q_{x,y}$ is the probability of observing y when the state is x. We consider the following information pattern. At decision epoch t, the system is in the (unobservable) state $X_t = i$ and the corresponding observation Y_t is gathered, such that

$$P(Y_t = j | X_t = i) = q_{i,j}. \tag{1}$$

The estimators V_t are functions of observations $(Y_0, \ldots, Y_t)$ and are chosen according to some specified criterion. Throughout this paper, we use upper case letters to denote estimators and script upper case letters to denote "estimation maps " from observations to the set $\mathbf{X}$. If Y_t is an observation and V_t an estimator: $V_t = \mathcal{V}_t \circ Y_t$. When it causes no confusion, we may use upper case letters for both.

2.2 Maximum A Posterior Probability Estimator (MAP) for HMM's

Consider a sequence of finite dimensional random variables X_t and the corresponding observations Y_t defined on the common probability space $(\Omega, F, \mathbf{P})$.

The Maximum A Posteriori Probability (MAP) estimator $\hat{X}_t$ is a Borel measurable function of the filtration generated by observations up to Y_t denoted by $\mathcal{Y}_t$ which satisfies for $\omega \in \Omega$:

$$\hat{X}_t(\omega) = \text{argmin}_{\zeta \in \mathbf{X}} E[\rho(X_t, \zeta) \mid \mathcal{Y}_t = \mathcal{Y}_t(\omega)] \qquad t = 0, 1, \dots \tag{2}$$

where

$$\rho(X_t, \zeta) = \begin{cases} 0 & \text{if } X_t = \zeta; \\ 1 & \text{otherwise.} \end{cases}$$

The usual definition of MAP as the argument with the greatest probability given the observation follows from the above [10]. We will need the following simple Lemma:

Lemma 2.2.1: MAP also results from the additive cost minimization:

$$(\hat{X}_0, \dots, \hat{X}_N)(\omega) = \text{argmin}_{\zeta_0 \cdots \zeta_N \in \mathbf{X}^N} E\left[\sum_{i=0}^{N} \rho(X_i, \zeta_i) | \, \mathcal{Y}_i = \mathcal{Y}_i(\omega)\right] \tag{3}$$

where $\mathbf{X}^N$ is the product space and each $\hat{X}_i$ is $\mathcal{Y}_i$ measurable.

Proof: The proof follows easily from the linearity of the conditional expectation and term by term minimization of the resulting sum.

2.3 Change of Measure

To carry out the computations, we will use a change of measure technique introduced in [3] and [4]. Let $(\Omega, F, \mathbf{P})$ be the canonical probability space on which all of our time series are defined. Let $\mathcal{Y}_t$ be the filtration generated by the available observations up to decision epoch t, and let $\mathcal{G}_t$ be the filtration generated by the sequence of states and observations up to that time. Then a new probability measure $\mathcal{P}^\dagger$ is defined by the restriction of the Radon-Nikodym derivative on $\mathcal{G}_t$ to

$$\frac{d\mathcal{P}}{d\mathcal{P}^\dagger}|_{\mathcal{G}_t} = \lambda_t := N_{\mathbf{Y}}^t \cdot \Pi_{k=1}^t q_{X_k, Y_k} \tag{4}$$

under which $\{Y_t\}$ is independently and identically distributed (i.i.d). Each distribution is uniform over the set $\mathbf{Y}$ and $\{Y_t\}$ is independent of $\{X_t\}$. That such a measure exists follows directly from Kolmogorov extension theorem.

Before we can introduce our filter, we need an optimization result. Let V_t be measurable functions of observations up to t taking values in $\{X_t\}$ and $\rho(\cdot, \cdot)$ as above. Fix $V_0, \dots, V_{k-1}$. We would like to find $\hat{V}_k, \dots, \hat{V}_{H-1}$ such that the following criterion is minimized:

$$S^\gamma(V_0, \dots, V_{H-1}) := E\big[\exp(\gamma \cdot \mathcal{C}_H)\big], \tag{5}$$

where

$$\mathcal{C}_H := \sum_{t=0}^{H-1} \rho(X_t, V_t) \tag{6}$$

and γ is a strictly positive parameter. The minimum value will be denoted by $\bar{S}^\gamma(\hat{V}_k, \ldots, \hat{V}_{H-1})$.

This optimization problem can be solved via dynamic programming. We need to define recursively an *information state*

$$\sigma_{t+1}^\gamma = N_{\mathbf{Y}} \cdot \overline{Q}(Y_{t+1}) \mathcal{D}^T(V_t) \cdot \sigma_t^\gamma, \tag{7}$$

where $\overline{Q}(y) := \mathrm{diag}(q_{i,y})$, $\mathcal{A}^T$ denotes the transpose of the matrix $\mathcal{A}$ and the matrix $\mathcal{D}$ is defined by

$$[\mathcal{D}(v)]_{i,j} := a_{i,j} \cdot \exp(\gamma \rho(i, v)). \tag{8}$$

σ_0^γ is set equal to $N_{\mathbf{Y}} \cdot \overline{Q}(Y_0) p_0$, where p_0 is the initial distribution of the state and is assumed to be known. In the context of the risk-sensitive estimation of Markov chains, the meaning of σ_t^γ will become clear.

We define the matrix

$$L(v, y) := N_{\mathbf{Y}} \cdot \overline{Q}(y) \mathcal{D}^T(v). \tag{9}$$

It can be shown that [1]

$$S^\gamma(V_0, \ldots, V_{H-1}) = E\big[\exp(\gamma \cdot \mathcal{C}_H)\big] = E^\dagger \left[\sum_{i=1}^{N_X} \sigma_H^\gamma(i) \right], \tag{10}$$

where $E^\dagger$ is the expectation with respect to the new measure. Define the value functions $J^\gamma(\cdot, H-j) : R_+^{N_{\mathbf{X}}} \to R$, $j = H, \ldots, H-k$, as

$$J^\gamma(\sigma, H-j) := \min_{V_{H-j} \ldots V_{H-1}} \left\{ E^\dagger \left\{ \sum_{i=1}^{N_X} \sigma_H^\gamma(i) \mid \sigma_{H-j}^\gamma = \sigma \right\} \right\}. \tag{11}$$

Lemma 2.3.1: Let $V_0, \ldots, V_{k-1}$ be given. Then the value functions defined above are obtained from the following dynamic programming equations:

$$\begin{cases} \bar{J}^\gamma(\sigma, H) & = \sum_{i=1}^{N_{\mathbf{X}}} \sigma(i); \\ \\ \bar{J}^\gamma(\sigma, H-j) & = \min_{v \in \mathbf{X}} \left\{ E^\dagger \left[\bar{J}^\gamma(L(v, Y_{H-j+1}) \cdot \sigma, H-j+1) \right] \right\}, \end{cases} \tag{12}$$

$$j = 1, 2, \ldots\ldots, H-k$$

The estimation maps $\hat{\mathcal{V}}_k, \ldots, \hat{\mathcal{V}}_{H-1}$ obtained from (12) are risk optimal, i.e., $\hat{\mathcal{V}}_k(\sigma_k^\gamma), \ldots, \hat{\mathcal{V}}_{H-1}(\sigma_{H-1}^\gamma)$ achieve the minimum in (5).

Proof: See [1].

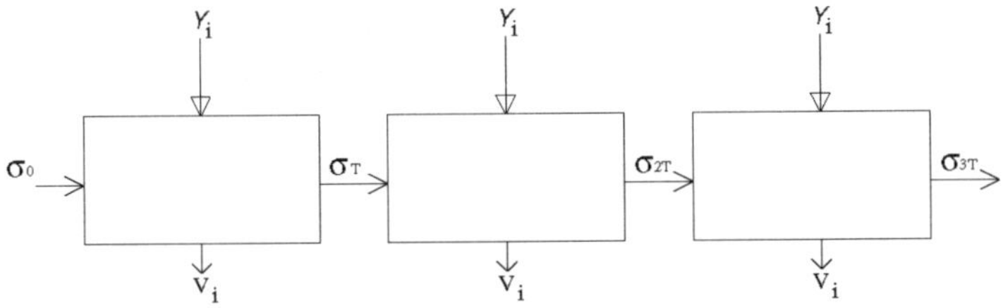

Fig. 1. The T-step risk sensitive filter banks.

2.4 The T-Step MAP Risk-Sensitive Estimator (TMAP)

The T-step MAP risk-sensitive estimator (TMAP) is defined by the following criterion:

$$\hat{V}_{NT}, \dots, \hat{V}_{(N+1)T-1} = \underset{\substack{V_t \in X, \\ t=NT,\dots,(N+1)T-1}}{\operatorname{argmin}} E\left\{\exp\left(\gamma \cdot \left[\sum_{t=0}^{NT-1} \rho(X_t, \hat{V}_t) + \sum_{t=NT}^{(N+1)T-1} \rho(X_t, V_t)\right]\right)\right\}, \tag{13}$$

where V_t is $\mathcal{Y}_t$ measurable, T is the size of the filter and $N = 0, 1, \dots,$ is the index of filtering segments. This exponential criterion is a generalization of the risk-sensitive filtering idea introduced in [7] for the quadratic cost with the filtering performed in single steps, i.e., for $T = 1$; we will look at this special case for TMAP in sections 3 and 4 and show that it is essentially a "greedy algorithm".

Theorem 2.4.1: The TMAP can be computed recursively by the following procedure:
1) Set $\sigma_0 = N_{\mathbf{Y}} \cdot \overline{Q}(Y_0) p_0$.
2) Given σ_{NT}, use the minimizing sequence of the value functions obtained from the following dynamic programming equations

$$\begin{cases} \bar{J}^{\gamma}(\sigma, T) & = \sum_{i=1}^{N_{\mathbf{X}}} \sigma(i); \\ \bar{J}^{\gamma}(\sigma, T-j) & = \min_{v \in \mathbf{X}} \left\{ E^{\dagger}\left[\bar{J}^{\gamma}(L(v, Y_{NT+T-j+1}) \cdot \sigma, T-j+1)\right]\right\} \end{cases} \tag{14}$$

$$j = 1, 2, \dots, T$$

to determine the value of the optimum estimates $\hat{V}_{NT}, \dots, \hat{V}_{(N+1)T-1}$ as a function of the information state $\sigma_{NT}, \dots, \sigma_{(N+1)T-1}$ obtained by (7).
3) Apply (7) once more to obtain $\sigma_{(N+1)T}$ and repeat steps (2) and (3) starting at (N+1)T.
Furthermore, for any given N as $\gamma \to 0$, TMAP (i.e. the above algorithm) reduces to MAP.

Proof: The proof follows from repeated applications of Lemma 2.3.1. We will skip the details. The limiting result follows from the first order approximation of the exponential function and the observation that as $\gamma \to 0$, the matrix $\mathcal{D}(v) \to \mathcal{A}$ element wise. This implies that in the limit the input to each filtering step is the unnormalized conditional distribution and thus by Lemma 2.2.1 the filtering process reduces to the well known MAP estimation of HMM's.
Note that although the size of the sum $\sum_{i=0}^{NT-1} \rho(X_t, \hat{V}_t)$ increases with N, all we need to track is the information state, computed recursively. The optimal estimates $\hat{\mathcal{V}}_{NT}(\sigma_{NT}), \dots , \hat{\mathcal{V}}_{(N+1)T-1}(\sigma_{(N+1)T-1})$ are measurable functions of the information state alone. Since our Markov chain is homogeneous and under the new measure the observations are i.i.d, (14) depends only on T and not on N. This justifies the cascade filter banks representation of Figure 1.
We point out that theorem 2.4.1 has no control counterpart. In this case, the estimators $\{V_t\}$ have no influence on the dynamics and thus estimation can be broken down into separate segments with the information state reinitialized. The same cannot be said for a controlled Markov chain due to the influence of the controller on the dynamics; the separate segments cannot be joined to represent the entire process in the above limiting sense as the "decoupling" Lemma 2.2.1 no longer holds. Note also that controlled Markov chains are not homogeneous.

3 Structural Results: The Filter Banks and the Information State

It is clear from the above that to describe the behavior of TMAP we must understand the operation of each filtering segment and understand the meaning of the information state. The key in understanding the filter's operation is the analysis of the value functions which are obtained via dynamic programming.

Lemma 3.1.1: The value functions are continuous and concave functions of the information state $\sigma \in R_+^{N_{\mathbf{X}}}$.

Proof: See [1].

Next, for P a finite set of vectors in $R_+^{N_{\mathbf{X}}}$, denote by $\mathcal{O}(P)$ the set

$$\mathcal{O}(P) := \Big\{ \frac{1}{N_{\mathbf{Y}}} \sum_{y=1}^{N_{\mathbf{Y}}} \alpha_y \cdot L(v, y) \mid \alpha_y \in P, v \in \mathbf{X} \Big\}. \tag{15}$$

Note that if P is finite so is $\mathcal{O}(P)$, since $|\mathcal{O}(P)| \le |P|^{N_{\mathbf{Y}}} \cdot N_{\mathbf{X}}$.

Lemma 3.1.2: The value functions given by (14) are piecewise linear functions (hyper-planes through the origin) of $\sigma \in R_+^{N_{\mathbf{X}}}$, such that if P_{j-1} indicates the vectors in $R_+^{N_{\mathbf{X}}}$ which specify the set of hyper planes for $\bar{J}^\gamma(\sigma, T -$

$j+1$) then

$$\bar{J}^{\gamma}(\sigma, T-j+1) = \min_{\alpha \in P_{j-1}} \{\alpha \cdot \sigma\} \quad \bar{J}^{\gamma}(\sigma, T-j) = \min_{\alpha \in \mathcal{O}(P_{j-1})} \{\alpha \cdot \sigma\}, \quad (16)$$

where $P_0 = \bar{\mathbf{1}} := (\sum_{k=1}^{N_{\mathbf{X}}} e_k)^{\mathbf{T}}$ and $\{e_k\}$ are the unit vectors in $R^{\mathbf{N_x}}$.

Proof: See [1].

Lemma 3.1.3: The optimal estimates $\{\hat{V}_t\}$ are constant along rays through the origin, i.e., let $\sigma \in R_+^{N_{\mathbf{X}}}$ then $\hat{\mathcal{V}}_t(\sigma') = \hat{\mathcal{V}}_t(\sigma)$, for all $\sigma' = \lambda\sigma$, $\lambda > 0$.

Proof: From Lemma 3.1.2, we see that $\bar{J}^{\gamma}(\sigma', T-j) = \lambda \bar{J}^{\gamma}(\sigma, T-j)$. The result follows from Theorem 2.4.1.

Definition: A cone in $R_+^{N_{\mathbf{X}}}$ is a set defined by $C_S := \{\sigma | \sigma = \lambda x, x \in S \subset R_+^{N_{\mathbf{X}}}, \lambda > 0\}$.

Definition: For j=1,2, ... , T and $v \in X$, let

$$\begin{aligned}\bar{J}_v^{\gamma}(\sigma, T-j) &:= E^{\dagger}\big[\bar{J}^{\gamma}(L(v, Y_{T-j+1}) \cdot \sigma, T-j+1)\big] \\ &= \frac{1}{N_{\mathbf{Y}}} \sum_{y=1}^{N_{\mathbf{Y}}} \big[\bar{J}^{\gamma}(L(v, y) \cdot \sigma, T-j+1)\big]. \quad (17)\end{aligned}$$

Definition: The decision region $DR_v^j \subset R_+^{N_{\mathbf{X}}}$ for the estimate $v \in \mathbf{X}$, at the $T-j$ decision epoch, is defined as

$$DR_v^j := \big\{\sigma \mid \sigma \in R_+^{N_{\mathbf{X}}}, \bar{J}^{\gamma}(\sigma, T-j) = \bar{J}_v^{\gamma}(\sigma, T-j)\big\}. \quad (18)$$

It follows from the definition of $\hat{\mathcal{V}}_{NT+T-j}(\sigma)$ that

$$DR_v^j := \big\{\sigma \mid \sigma \in R_+^{N_{\mathbf{X}}}, \quad \hat{\mathcal{V}}_{NT+T-j}(\sigma) = v\big\}. \quad (19)$$

We say a decision is made "strictly", if it is the only possible decision.

Theorem 3.1.1: For each $v = i \in \{1, 2, \ldots, \mathbf{N_X}\}$ and for every $j = 1, 2, \ldots$, T, the decision region DR_i^j is always non-empty and includes a cone about the σ_i axis within which the decision made is (strictly) $\hat{\mathcal{V}}_{NT+T-j}(\sigma) = i$.

Proof: We state the proof for $N_X = 2$, from which the proof for the general case will become evident for the reader. On the σ_1 axis, we have by definition

$$\bar{J}_1^{\gamma}(\sigma, T-j) = \frac{1}{N_{\mathbf{Y}}} \sum_{y=1}^{N_{\mathbf{Y}}} \big[\bar{J}^{\gamma}(N_{\mathbf{Y}} \cdot \overline{Q}(y) \mathcal{A}^T \begin{bmatrix} 1 & 0 \\ 0 & e^{\gamma} \end{bmatrix} \begin{bmatrix} \sigma_1 \\ 0 \end{bmatrix}, T-j+1)\big].$$

$$\bar{J}_2^{\gamma}(\sigma, T-j) = \frac{1}{N_{\mathbf{Y}}} \sum_{y=1}^{N_{\mathbf{Y}}} \big[\bar{J}^{\gamma}(N_{\mathbf{Y}} \cdot \overline{Q}(y) \mathcal{A}^T \begin{bmatrix} e^{\gamma} & 0 \\ 0 & 1 \end{bmatrix} \begin{bmatrix} \sigma_1 \\ 0 \end{bmatrix}, T-j+1)\big].$$

$$\bar{J}_1^{\gamma}(\sigma, T-j) = \frac{1}{N_{\mathbf{Y}}} \sum_{y=1}^{N_{\mathbf{Y}}} [\bar{J}^{\gamma}(N_{\mathbf{Y}} \cdot \overline{Q}(y) \mathcal{A}^T \begin{bmatrix} \sigma_1 \\ 0 \end{bmatrix}, T-j+1)].$$

$$\bar{J}_2^{\gamma}(\sigma, T-j) = \frac{1}{N_{\mathbf{Y}}} \sum_{y=1}^{N_{\mathbf{Y}}} [\bar{J}^{\gamma}(N_{\mathbf{Y}} \cdot \overline{Q}(y) \mathcal{A}^T \begin{bmatrix} e^{\gamma}\sigma_1 \\ 0 \end{bmatrix}, T-j+1)].$$

Applying Lemma 3.1.2 to each term of the summation on the right-hand side of $\bar{J}_2^{\gamma}(\sigma, T-j)$, we get

$$\bar{J}_2^{\gamma}(\sigma, T-j) = \frac{e^{\gamma}}{N_{\mathbf{Y}}} \sum_{y=1}^{N_{\mathbf{Y}}} [\bar{J}^{\gamma}(N_{\mathbf{Y}} \cdot \overline{Q}(y) \mathcal{A}^T \begin{bmatrix} \sigma_1 \\ 0 \end{bmatrix}, T-j+1)].$$

Therefore, we can write

$$\bar{J}_2^{\gamma}(\sigma, T-j) - \bar{J}_1^{\gamma}(\sigma, T-j)$$
$$= (e^{\gamma} - 1)\{\frac{1}{N_{\mathbf{Y}}} \sum_{y=1}^{N_{\mathbf{Y}}} [\bar{J}^{\gamma}(N_{\mathbf{Y}} \cdot \overline{Q}(y) \mathcal{A}^T \begin{bmatrix} \sigma_1 \\ 0 \end{bmatrix}, T-j+1)]\}.$$

But $e^{\gamma} > 1$ since $\gamma > 0$ and for every j, the value functions are strictly positive being integrals of the exponential functions. Thus from the above on the σ_1 axis we have the strict inequality

$$\bar{J}_1^{\gamma}(\sigma, T-j) < \bar{J}_2^{\gamma}(\sigma, T-j)$$

which implies DR_1^j includes the σ_1 axis; fix σ on that axis, then by Lemma 3.1.1 and because sums and compositions of continuous functions are continuous, there exists a disk of positive radius in the R_+^2, i.e., $d(r,\sigma) \bigcap R_+^2, r > 0$ such that $\bar{J}_1^{\gamma}(\sigma, T-j) < \bar{J}_2^{\gamma}(\sigma, T-j)$ for every $x \in d(r,\sigma) \bigcap R_+^2$. Therefore, Lemma 3.1.3 implies that on the cone $\mathbf{C}_{d(r,\sigma)\cap R_+^2} \subset DR_1^j$ the decision is strictly $v = 1$.

The same proof works in higher dimensions by fixing an axis σ_l and making pairwise comparisons between $\bar{J}_l^{\gamma}(\sigma, T-j)$ and $\bar{J}_k^{\gamma}(\sigma, T-j)$, $k \neq l$ along the σ_l axis. The "strict" cone around the σ_l axis will be the intersection of all cones obtained from pairwise comparisons.

In general, the boundaries among the decision regions are not of "threshold type" (unlike MAP). We give a two dimensional example. We consider the TMAP with $N_{\mathbf{X}} = 2$.

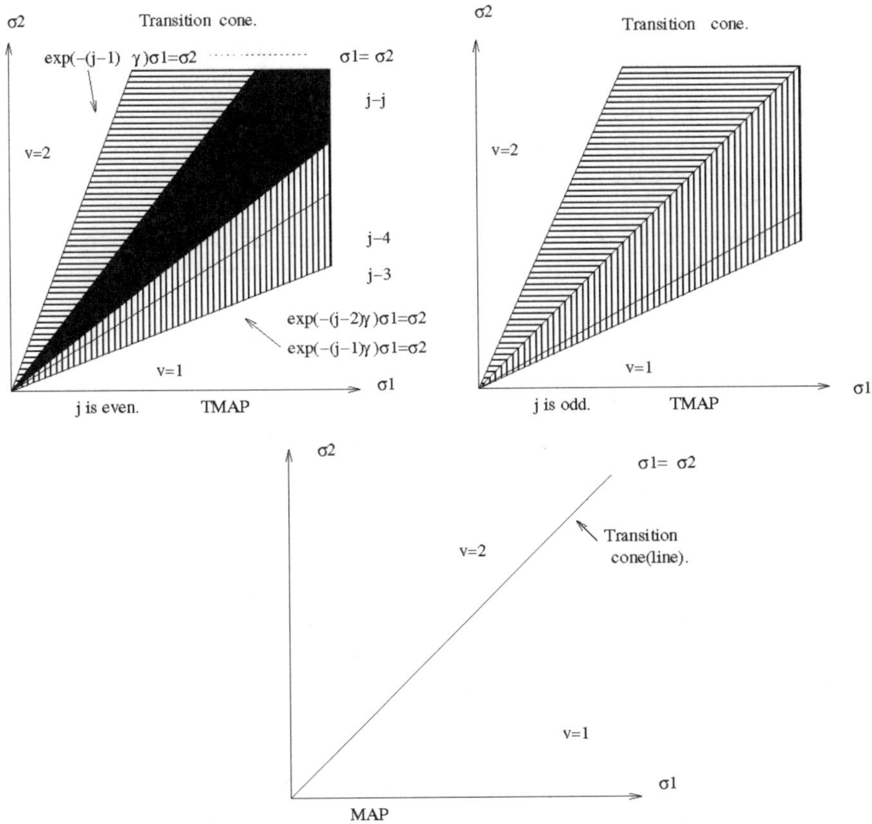

Fig. 2. Decision cones for $a_{11} = a_{22} = 1, q_{xy} = 1/2$.

Remark: The transition cone $DR_1^j \bigcap DR_2^j$ is not, in general, of threshold type, i.e., the cone: $DR_1^j \bigcap DR_2^j$ does not degenerate to a line; we give a simple counter example.

Let $a_{11} = a_{22} = 1$ and $q_{xy} = 1/2$; then it can be shown (by straightforward induction) that the transition cones are not degenerate.

Let's look at the counter example more closely (Figure 2). The cone where the decision is strictly v=1, i.e., $R_+^2 \cap (DR_2^j)^c$ is given by $\sigma_1 > \sigma_2 \cdot \exp(\gamma(j-1))$ and by symmetry $R_+^2 \cap (DR_1^j)^c$ is given by $\sigma_2 > \sigma_1 \cdot \exp(\gamma(j-1))$. The transition cone (where either decision is acceptable) is given by the complement of the union of these two regions (the colored area). The value functions are given by $j+1$ hyper-planes: $\sigma_1 + \exp(\gamma(j))\sigma_2$, $\sigma_1 \exp(\gamma(1)) + \exp(\gamma(j-1))\sigma_2$, $\sigma_1 \exp(\gamma(2)) + \exp(\gamma(j-2))\sigma_2$ ..., $\sigma_2 + \exp(\gamma(j))\sigma_1$ on the $j+1$ cones beginning with $\sigma_1 > \sigma_2 \cdot \exp(\gamma(j-1))$ and ending with $\sigma_2 > \sigma_1 \cdot \exp(\gamma(j-1))$. The transition cone between them is the union of $j-1$ cones whose boundaries are lines $\exp(-(j-1)\gamma)\sigma_1 = \sigma_2$, $\exp(-(j-2)\gamma)\sigma_1 = \sigma_2$,

$\ldots, \sigma_1 = \sigma_2, \ldots, \exp(-(j-2)\gamma)\sigma_2 = \sigma_1, \ldots, \exp(-(j-1)\gamma)\sigma_2 = \sigma_1$. When j is odd, the line $\sigma_1 = \sigma_2$ is a boundary (boundary in the sense of slope change in the cost and not decision); when j is even, it is not. (The solid cone which includes $\sigma_1 = \sigma_2$ for even values of j is meant to emphasize this). On the transition cone either decision is allowed. We can interpret this region as the zone of uncertainty. For MAP and TMAP with T=1 (we will show this later) this region is the threshold $\sigma_1 = \sigma_2$, but as the above example shows, for TMAP with $T > 1$, it may be a non-degenerate cone. We could interpret this as reflecting the "conservative" nature of the risk-sensitive estimation. We are expanding the zone of "uncertainty" at the expense of the region of "certainty". We will show in the subsequent sections that this is not always the manner in which risk-sensitivity manifests itself in the structure of decision regions. It is possible that the the transition cone remains degenerate and either of the two other cones expands at the expense of the other or the decision regions are not affected by risk sensitivity at all, i.e., they remain identical to that of MAP. In two dimensions for example, $DR_1^j = \{\sigma | \sigma_1 > \sigma_2\}$ and $DR_2^j = \{\sigma | \sigma_2 > \sigma_1\}$.

The above theorem only guarantees the existence of of non-degenerate cones around the σ_l axis but says nothing about their size. In fact, observe that in the above example the size of these cones becomes arbitrarily small as $\gamma \to \infty$ since the slopes of the lines $\exp(-(j-1)\gamma)\sigma_1 = \sigma_2$ and $\exp(-(j-1)\gamma)\sigma_2 = \sigma_1$, for every $j > 1$, converge to zero and infinity respectively.

Two special cases ($N = 0$, $T = M$) and ($T = 1$, $N = 0, \ldots, M-1$) are of interest. In both cases, the index t ranges from t=0 to t=M-1. In the first case, TMAP reduces to the exponential/sum criterion for HMM's which is the discrete and finite dimensional version of the risk-sensitive L^2 filter introduced by Speyer and others. The second case would be the MAP version of the quadratic cost risk-sensitive filtering introduced first to our best knowledge by Dey and Moore in [7]. Obviously, the structural results obtained so far apply to these special cases.

Theorem 3.1.2: Let $\mathcal{EX}(v)$ be the diagonal matrix $\mathrm{diag}\big[\exp(\gamma\rho(i,v))\big]$, $i = 1, \ldots, N_{\mathbf{X}}$. Then the one step TMAP decision regions are given by $\hat{\mathcal{V}}_t(\sigma) = i$ if $\sigma_i \geq \sigma_j$, $\forall j \neq i$.

Proof: From the definition, we have: $\mathcal{D}^T(v) = \mathcal{A}^T \mathcal{EX}(v)$ and that $\bar{J}^\gamma(\sigma, T) = <\sigma, \bar{\mathbf{1}}>$ and thus

$$\begin{aligned}
\bar{J}_v^\gamma(\sigma, T-1) &= \frac{1}{N_{\mathbf{Y}}} \sum_{y=1}^{N_{\mathbf{Y}}} \big[\bar{J}^\gamma(L(v,y)\cdot\sigma, \ \ T)\big] \\
&= \frac{1}{N_{\mathbf{Y}}} \sum_{y=1}^{N_{\mathbf{Y}}} \big[\bar{J}^\gamma(N_{\mathbf{Y}} \cdot \overline{Q}(y)\mathcal{A}^T\mathcal{EX}(v)\cdot\sigma, T)\big] \\
&= \sum_{y=1}^{N_{\mathbf{Y}}} \langle \overline{Q}(y)\big(\mathcal{A}^T\mathcal{EX}(v)\cdot\sigma\big), \bar{\mathbf{1}}\rangle
\end{aligned}$$

$$= \left\langle \left(\sum_{y=1}^{N_{\mathbf{Y}}} \overline{Q}(y) \right) (\mathcal{A}^T \mathcal{E}\mathcal{X}(v) \cdot \sigma), \bar{\mathbf{1}} \right\rangle$$
$$= \langle I \mathcal{A}^T \mathcal{E}\mathcal{X}(v) \cdot \sigma, \bar{\mathbf{1}} \rangle$$
$$= \langle \mathcal{E}\mathcal{X}(v) \cdot \sigma, \mathcal{A}\bar{\mathbf{1}} \rangle \;=\; \langle \mathcal{E}\mathcal{X}(v) \cdot \sigma, \bar{\mathbf{1}} \rangle. \tag{20}$$

A little calculation shows that given σ the above is minimized, if we set v equal to the index of the largest component of σ, i.e., if σ_l is the largest component of σ then $v = l$. This is precisely how the decision regions for MAP are defined.

Note that TMAP for T=1 is not reduced to MAP; although the decision regions are the same, the information states are different. In the case of MAP, the information state is the conditional distribution, while in the case of TMAP the information state is given by (7).

Theorem 3.1.3: The value function $\bar{J}^{\gamma}(\sigma, T-j)$ when $\forall(x,y),\ q_{yx} = 1/N_{\mathbf{Y}}$ is given by

$$\bar{J}^{\gamma}(\sigma, T-j)$$
$$= \min_{v_{T-j},\cdots,v_{T-1}} \langle \sigma,\ \mathcal{E}\mathcal{X}(v_{T-j}) \cdot \mathcal{A} \cdot \mathcal{E}\mathcal{X}(v_{T-j+1}) \cdots \mathcal{A} \cdot \mathcal{E}\mathcal{X}(v_{T-1}) \cdot \bar{\mathbf{1}} \rangle.$$

Proof: The proof is based on the same technique used in the previous theorem and will be omitted.

In the above counter example when $\mathcal{A} = I_{2\times 2}$ and $q_{yx} = 1/2$, by the above theorem, we have

$$\bar{J}^{\gamma}(\sigma, T-j) = \min_{v_{T-j},\cdots,v_{T-1}} < \sigma, \mathcal{E}\mathcal{X}(v_{T-j}) \cdot ... \mathcal{E}\mathcal{X}(v_{T-j}) \cdot \bar{\mathbf{1}} > .$$

If we let the number of times we choose $x = 2$ be n_2, and likewise for $x = 1$, $n_1 = T - n_2$, a little algebra will show that the total cost $\bar{J}^{\gamma}(\sigma, 0)$ is given by

$$\sigma_1 \exp\{\gamma n_2\} + \sigma_2 \exp\{\gamma(T - n_2)\}.$$

By differentiation with respect to n_2, a few rearrangements and taking logarithms, the minimum cost is obtained when (modulo the integer parts)

$$T/2 - \frac{1}{2\gamma} \log(\sigma_1/\sigma_2) = n_2; \qquad 0 \le n_2 \le T. \tag{21}$$

This suggests that for large values of γ regardless of σ, we choose the two states an approximately equal number of times. Note that if we let $\gamma \to 0$, assuming $\sigma_1 > \sigma_2$, $n_2 \to 0$ as expected.

4 Risk-Sensitivity and the Sample Path Perspective

Through the sample path perspective, we can explain the behavior of the risk-sensitive estimators. In HMM's all sample paths pass through a finite number of states. We can think of the transition probabilities as determining a flow in the system. So far the only example we have considered was a non-mixing dynamical system. The transition probabilities were set to be zero and there was no flow between the states. This is important from the error smoothing point of view, for as the flow passes through the states, so does the history of the errors. If sample paths that have accumulated estimation error remain in a particular state, the estimator will be "attracted" to the state in order to relieve the error accumulated there. This explains the oscillatory behavior of our two state example in which no flow between the states was allowed. On the other hand, if the transition probabilities are non-zero this "attraction" is somewhat relieved; we have verified through simulations that mixing indeed inhibits the oscillatory behavior. This will also have an effect on the decision regions as we will see shortly. But if we go through a state "too quickly", we cannot use that state to smoothen the error accumulated in the path effectively. Both these cases lead to certain type of singularities in the decision regions.

The second issue is the role of information. If we expect to receive good information about the system in the future, which will in turn reduce the error accumulation, we are likely to be less conservative at the present about our decisions. This means that we expect TMAP's decision regions to become less conservative and look more like MAP's under increased availability of information. This too will be shown in the following example.

We will study the decision regions for T=2 TMAP for an HMM with $N_{\mathbf{X}} = N_{\mathbf{Y}} = 2$ and

$$\mathcal{A} = \begin{bmatrix} 1/2 & 1/2 \\ \delta & 1-\delta \end{bmatrix}; \quad Q = \begin{bmatrix} 1/2+I & 1/2-I \\ 1/2-I & 1/2+I \end{bmatrix}. \tag{22}$$

The parameter I controls the availability of information. When $I = 0$, no information is available (the case of pure prediction) and as $I \to 1/2$, the HMM becomes perfectly observable. The parameter δ determines the transition probabilities of the second state and in particular $\delta = 0$ will make the Markov chain non-mixing.

As shown before for T=1, the decision regions are identical to those of MAP. First let $I = 0$, then it can be shown that for $T = 2$, the decision regions of j=2 (the first stage of the two step filter) are of the threshold type determined by a transition line L with the slope $m(\delta)$ followed by the equi-partition decision regions (identical to the decisions regions of MAP). The decision regions of the first stage are given by $\sigma_2 < m(\delta)\sigma_1$ choose 1 and if $\sigma_2 > m(\delta)\sigma_1$ choose 2. The slope $m(\delta)$ is given by

$$m(\delta) = \begin{cases} \frac{e^{\gamma}+1}{2} \cdot \frac{1}{\delta e^{\gamma}+1-\delta} & \delta < 1/2; \\ \\ \frac{e^{\gamma}+1}{2} \cdot \frac{1}{(1-\delta)e^{\gamma}+\delta} & \delta > 1/2. \end{cases}$$

Simple calculations show that the slope is always greater than or equal to 1 (only when $\delta = 1/2$), so that the decision region of the first state is enlarged at the expense of the second. As expected when $\gamma \to 0$, the decision regions equalize. When $\gamma \to \infty$, the slope is given by

$$m(\delta) = \begin{cases} \frac{1}{2\delta} & \delta < 1/2; \\ \\ \frac{1}{2(1-\delta)} & \delta > 1/2. \end{cases}$$

When either $\delta = 0$ or $\delta = 1$, the slope becomes infinite. These are the singularities that we mentioned earlier. The equalization of the two regions at $\delta = 1/2$ is a general property which holds true even when no constraint is put on the available information as the following theorem demonstrates:

Theorem 4.1.1: Consider the HMM described by (1). The risk-sensitive decision regions are equalized under uniform flow: $a_{ij} = 1/N_{\mathbf{X}} \;\; \forall(i,j)$. Furthermore, TMAP reduces to MAP for every choice of T and γ.

Proof: See [1].

In the above result, the observation matrix $\overline{Q}(y)$ plays no role under the assumption of uniform flow in the determination of the decision regions. But this is the exception. In general, the availability of information appears to have an elegant relation to the structure of the decision regions as the following shows.

Proposition 4.1.2: In (22), let $\delta = 0$ and $I \geq \frac{1}{2(1+e^{-\gamma})}$; then the decision regions for TMAP, $T = 2$ are equalized.

Proof: The proof follows from the solution of a system of simultaneous inequalities defined by (14) with the constraints $\delta = 0$ and $I \geq \frac{1}{2(1+e^{-\gamma})}$. We skip the tedious algebra.

As we mentioned earlier, this does not imply that TMAP for $T = 2$ reduces to MAP because the recursions governing the evolution of the information states for the two cases are different. But if for some T the decision regions are equalized then TMAP with filter size T does reduce to the TMAP with filter size $T = 1$. This would be significant both conceptually and computationally if we could determine conditions under which the decision regions are equalized. Note in the above for computational reasons, we had constrained the observation matrix to be symmetric. This produces only a sufficient condition as stated in Proposition 4.1.2. The right measure of the minimum quantity of information needed must be free of such constraints (for example, Shannon's mutual information among the states and the observations). Observe that in the above example, the amount of needed information grows with increasing γ. Clearly $\frac{1}{2(1+e^{-\gamma})} \to 1/2$ as $\gamma \to \infty$ which implies under infinite risk, we need perfect observation to equalize the decision regions.

We saw that under the condition of uniform flow TMAP reduces to MAP, i.e., the estimated process based on the discrete metric with the assumption of uniform flow is invariant under risk-sensitivity. We may be led to believe that perhaps for large values of γ, risk-sensitivity tends to move the estimator toward this stable invariance, making the estimator look more and more "uniform". One has to be careful about what this means. In fact, risk-sensitivity tends to increase oscillatory behavior and not relieve it. A more conservative estimator tends to move around the state space more rapidly from sample path to sample path and not for too long "trust" the correctness of the sample path it may be following. It is in this sense that the estimates are made more "uniform".

Finally, we point out that many of the results of this paper depend on the properties of the discrete metric (used to define MAP) which is not the natural metric for R^n. Therefore, our structural results do not directly illuminate the linear-Gaussian risk-sensitive estimation case. However, the intuition gained in the discrete finite dimensional setting about the behavior of the sample paths may lead to a better understanding of the linear-Gaussian risk-sensitive estimator as well.

Acknowledgement

This work was supported in part by the National Science Foundation under Grant DMI-9988867, by the Air Force Office of Scientific Research under Grant F496200110161, and by ONR contract 01-5-28834 under the MURI Center for Auditory and Acoustics Research.

References

1. Vahid R. Ramezani. PhD dissertation (2001). Dept. of Electrical and Computer Engineering, University of Maryland, College Park. Institute for Systems Research Technical Reports PhD 2001-7. http://www.isr.umd.edu/TechReports/ISR/2001/PhD_2001-7
2. D. H. Jacobson. Optimal stochastic linear systems with exponential performance criteria and their relation to deterministic games, IEEE Trans. Aut. Control, vol. AC-18, no. 2, April 1973.
3. R. J. Elliot, L. Aggoun, and J. B. Moore. Hidden Markov Models: Estimation and control. Springer, Berlin, Heidelberg, New York, 1994.

4. M. R. James, J. S. Baras, and R. J. Elliot. Risk sensitive Control and dynamic games for partially Observed Discrete-Time Nonlinear Systems. IEEE Trans. Aut. Control, vol. 39 (4), pp. 780-792, 1994.
5. J. S. Baras and M. R. James. Robust and Risk-sensitive Output Feedback Control for Finite State Machines and Hidden Markov Models. Institute for Systems Research, University of Maryland at college Park, Technical Research Reports T. R. 94-63.
6. J. L. Spyer, C. Fan and R. N. Banava. Optimal Stochastic Estimation with Exponential Cost Criteria. Proceedings of 31th conference on decision and control. pp. 2293-2298, 1992.
7. S. Dey and J. Moore. Risk sensitive filtering and smoothing for hidden Markov models. Systems and control letters. vol. 25, pp. 361-366, 1995.
8. S. Dey and J. Moore. Risk sensitive filtering and smoothing via reference probability methods. IEEE Trans. Aut. Control. vol. 42, no. 11 pp. 1587-91, 1997.
9. R. K. Boel, M. R. James, and I. R. Petersen. Robustness and risk-sensitive filtering. Proceedings of the 36th IEEE Conference on Decision and Control. Cat. No. 97CH36124, 1997.
10. H. Vincent Poor. An introduction to signal detection and estimation. Springer Verlag, 1994.
11. P. Whittle. Risk-sensitive optimal control. John Wiley & sons, 1990.
12. E. Fernandez-Gaucherand and S. I. Marcus. Risk-sensitive optimal control of Hidden Markov Models: a case study. IEEE Transactions on Automatic Control, vol. 42, no. 10, pp. 1418-1422, 1997.
13. R. J. Elliot. Stochastic Calculus and Applications. New York: Springer-Verlag, 1982.
14. A. N. Shiryayev. probability. New York: Springer-Verlag, 1984.
15. P. Whittle. Risk-sensitive optimal control. John Wiley & sons, 1990.
16. P. Whittle. Risk-sensitive linear/quadratic/Gaussian control. Adv. Applied probability. 13, 746-777, 1981, applied probability trust.

Bayesian Adaptive Control of Discrete Time Partially Observed Markov Processes*

Dedicated to Prof. Tyrone Duncan on his 60th birthday

L. Stettner

Institute of Mathematics, Polish Academy of Sciences, Sniadeckich 8, 00-950 Warsaw, Poland

Abstract. In this paper adaptive control of partially observed discrete time Markov processes with transition probability depending on an unknown random variable is studied. Although the techniques and methods used in the paper can be extended in various directions, the author concentrated himself on an analysis of three models called respectively: mixed observation model, model with observed regeneration, and rich observation model for which partially observed control problem with known transition probability and average cost per unit time functional can be solved.

Keywords: Bayesian adaptive control, average cost per unit time functional, filtering process, invariant measure, Markov process, partial observation
AMS subject classification: 93E20, 93E11, 93C40

1 Introduction

On a probability space $(\Omega, \mathcal{F}, \mathcal{P})$ consider a controlled Markov process (x_n) taking values in a complete separable metric space E endowed with the Borel σ-algebra $\mathcal{E}$. Assume that x_n has a controlled transition operator $P^{v_n,\alpha^0}(x_n, \cdot)$, where v_n is the control at time n taking values in a compact metric space U and adapted to the observation σ-field $Y^n = \sigma\{y_1, \ldots, y_n\}$ and α^0 is an unknown random variable taking values in a compact metric space A with known distribution η.

We assume that the observation process (y_n) is statistically dependent on $X^n = \sigma\{x_0, x_1, \ldots, x_n\}$ via

$$P\{y_{n+1} \in B \mid X^{n+1}, Y^n\} = \int_B r(x_{n+1}, y)dy \tag{1}$$

for a Borel measurable function $r : E \times R^d \to R^+$, $B \in \mathcal{B}(R^d)$.

Let $c : E \times U \to R^+$ be a continuous bounded function. Our aim is to minimize the following average cost per unit time functional

$$\limsup_{n\to\infty} \frac{1}{n} \left\{ \sum_{i=0}^{n-1} c(x_i, v_i) \right\} \tag{2}$$

* Supported by the Center of Excellence IMPAN-BC

B. Pasik-Duncan: Stochastic Theory and Control, LNCIS 280, pp. 435–446, 2002.

which often leads to the expected average per unit time cost functional

$$J_\mu^{\alpha^0}((v_n)) = \limsup_{n\to\infty} \frac{1}{n} E_\mu^{\alpha^0} \left\{ \sum_{i=0}^{n-1} c(x_i, v_i) \right\} \tag{3}$$

over all Y^n - adapted controls v_n, where $\mu \in \mathcal{P}(E)$ is the initial law of (x_n). Denote by λ_{α^0} the optimal value of the cost functional (3). One would like to construct a strategy (v_n) for which we obtain in (2) the value P a.e. close to λ_{α^0}.

There is a number of difficulties we encounter in the above adaptive control problem. First of all average cost per unit time problem for partially observed Markov processes is only solved under certain restrictive assumptions in [12] (see also [11]) and [2]. Ergodicity of controlled filtering processes is in known in particular cases only, namely in the case of mixed observation or the case with observed regeneration studied in [11] and in the rich observation model considered in [4]. Adaptive control with partial observation using parametrical approach was investigated in [8]. The approach used in [8] required uniform in α law of large numbers which is difficult to show. Bayesian approach to adaptive completely observed control problems was considered in a series of papers [4], [5], [6], [7] and [1]. To the best of the author's knowledge there are no papers on Bayesian approach to partially observed control of Markov processes.

Given a control $v = (v_n)$ and $\alpha \in A$ define a so called controlled filtering process $\pi_n^{v\alpha}$ by the following recursive formula

$$\pi_{n+1}^{v\alpha}(B) = \frac{\int_B r(z, y_{n+1}) P^{v_n,\alpha}(\pi_n^{v\alpha}, dz)}{\int_E r(z, y_{n+1}) P^{v_n,\alpha}(\pi_n^{v\alpha}, dz)} := M^{v_n,\alpha}(y_{n+1}, \pi_n^{v\alpha})(B), \tag{4}$$

for $B \in \mathcal{E}$ with $\pi_0 = \mu$ the initial law of (x_n). Morevover define another measure valued process

$$\begin{aligned} \hat{\alpha}_{n+1}(D) &= \frac{\int_D \int_E r(z, y_{n+1}) P^{v_n,\alpha}(\pi_n^{v\alpha}, dz) \hat{\alpha}_n(d\alpha)}{\int_A \int_E r(z, y_{n+1}) P^{v_n,\alpha}(\pi_n^{v\alpha}, dz) \hat{\alpha}_n(d\alpha)} \\ &:= G^{v_n}(y_{n+1}, \hat{\alpha}_n, (\pi_n^{v\alpha})_{\alpha\in A})(D) \end{aligned} \tag{5}$$

for $D \in \mathcal{B}(A)$ - the set of Borel subsets of A with $\hat{\alpha}_0(\alpha^0) = \eta$. Following the proof of Lemma 1.1 of [11] one can show that

$$P\{x_n \in B | Y^n, \alpha^0\} = \pi_n^{v\alpha^0}(B) \tag{6}$$

and

$$P\{\alpha^0 \in D | Y^n\} = \hat{\alpha}_n(D) \tag{7}$$

P a.e., for $B \in \mathcal{E}$ and $D \in \mathcal{B}(A)$.

2 Identification

Using Bayesian approach we easily obtain the convergence of the estimation since by martingale convergence theorem we immediately have that $\hat{\alpha}_n$ converges weakly to $\hat{\alpha}_\infty$ as $n \to \infty$. Another problem is to identify the limit $\hat{\alpha}_\infty$. When $\hat{\alpha}_\infty = \delta_{\alpha^0}$, where δ_α is the Dirac measure concentrated in α, we say we have *identifiability*, which is one of the purposes of our studies. The disadvantage of the Bayesian approach is the dimension of the problem is increasing since that by (5) we have to observe the vector $\begin{bmatrix} \hat{\alpha}_n \\ (\pi_n^{v\alpha})_{\alpha \in A} \end{bmatrix}$.

For $v \in U$, $\alpha \in A$, $y \in R^d$ and $\mu \in P(E)$ - the set of probability measures on E define

$$R^{v\alpha}\mu(y) := \int_E r(z,y) \int_E P^{v,\alpha}(z_1, dz)\mu(dz_1). \tag{8}$$

Given a sequence of random variables (ξ_n) we say that z is a *strictly frequent point* of (ξ_n) if there is a subsequence n_k, $n_k \to \infty$ such that $\xi_{n_k} = z$. If for every $\epsilon > 0$ the ball $B(z,\epsilon)$ with the center in z and radius ϵ contains infinitely many elements of (ξ_n) with probability 1, we say that z is a *frequent point* of (ξ_n). In what follows we shall consider $(\mu_\alpha)_{\alpha \in A}$, as an element of the space $C(A, P(E))$ of continuous mappings from A into $P(E)$ with supremum norm, and a norm in $P(E)$ generating the weak convergence topology. We have

Proposition 18 *If* $\begin{bmatrix} v \\ y \\ (\mu_\alpha)_{\alpha \in A} \end{bmatrix}$ *is a frequent point of the sequence* $\begin{bmatrix} v_n \\ y_{n+1} \\ (\pi_n^{v\alpha})_{\alpha \in A} \end{bmatrix}$ *and the mappings* $(v,\alpha,y,\mu) \mapsto R^{v\alpha}\mu(y)$ *as well as* $\alpha \mapsto \mu_\alpha$ *are continuous, then for* α, $\alpha' \in$ *supp* $\hat{\alpha}_\infty$, *the support of the measure* $\hat{\alpha}_\infty$, *we have* $R^{v\alpha}\mu_\alpha(y) = R^{v\alpha'}\mu_{\alpha'}(y)$.

Proof. Notice that by (5) for $\phi \in C(A)$ - the space of continuous bounded functions on A we have

$$\hat{\alpha}_{n+1}(\phi) \int_A \int_E r(z, y_{n+1}) P^{v_n,\alpha}(\pi_n^{v\alpha}, dz)\hat{\alpha}_n(d\alpha) =$$
$$= \int_A \phi(\alpha) \int_E r(z, y_{n+1}) P^{v_n,\alpha}(\pi_n^{v\alpha}, dz)\hat{\alpha}_n(d\alpha)$$

Letting $n \to \infty$ along a subsequence for which the vector $\begin{bmatrix} v \\ y \\ (\mu_\alpha)_{\alpha \in A} \end{bmatrix}$ is strictly frequent we obtain that

$$\int_A \phi(\alpha)(R^{v\alpha}\mu_\alpha(y) - \int_A R^{v\alpha'}\mu_{\alpha'}(y)\hat{\alpha}_\infty(d\alpha'))\hat{\alpha}_\infty(d\alpha) = 0 \tag{9}$$

Consequently for α, $\alpha' \in supp\ \hat{\alpha}_\infty$ we obtain $R^{v\alpha}\mu_\alpha(y) = R^{v\alpha'}\mu_{\alpha'}(y)$. The case with frequent vector follows similarly using suitable continuity arguments (see the proof of Proposition 2 in [5]). □

Notice that when $R^{v\alpha}\mu_\alpha(y) = R^{v\alpha'}\mu_{\alpha'}(y)$ for $y \in R^d$ or in a certain recurrent set implies coincidence of α and α', by Proposition 1 we have the identifiability of α^0. In general it is difficult to find a frequent sequence of measures $(\mu_\alpha)_{\alpha \in A}$. This is satisfied however in the two special cases considered in the next sections.

2.1 Mixed Observation Model

Assume that $E = R^d$ and the state process (x_n) is completely observed in a recurrent closed set $\Gamma \subset R^d$ while the observation y_n outside of Γ is described as follows

$$P\{y_{n+1} \in B \cap \Gamma^c \mid X^{n+1}, Y^n\} = \int_{B \cap \Gamma^c} r(x_{n+1}, y)dy \tag{10}$$

for $B \in \mathcal{B}(R^d)$. Assume moreover that $P^{v,\alpha}(x, B) = \int_B p(x, z, v, \alpha)dz$, where the density is a continuous function of its arguments. Furthermore, assume that there is a compact set $\Gamma_1 \subset \Gamma$ such that for any initial law of μ and control $v = (v_n)$

$$E_\mu^{v\alpha} T_{\Gamma_1} < \infty \tag{11}$$

and

$$\sup_{\alpha \in A} \sup_{x \in \Gamma_1} \sup_{v} E_x^{v\alpha} \tau^2 < \infty \tag{12}$$

where $T_{\Gamma_1} = \inf\{j \geq 0 : x_j \in \Gamma_1\}$ or $+\infty$ whenever the set $\{j \geq 0 : x_j \in \Gamma_1\}$ is empty and $\tau = T_{\Gamma^c} + T_{\Gamma_1}\Theta_{T_{\Gamma^c}}$ with Θ being the time shift operator, is the first entry time to Γ_1 after leaving the set Γ. In the above case we have that

$$\pi_{n+1}^{v\alpha}(B) = 1_{B\cap\Gamma}(y_{n+1}) + 1_{\Gamma^c}(y_{n+1}) \frac{\int_{B\cap\Gamma^c} r(z, y_{n+1}) P^{v_n,\alpha}(\pi_n^{v\alpha}, dz)}{\int_{\Gamma^c} r(z, y_{n+1}) P^{v_n,\alpha}(\pi_n^{v\alpha}, dz)} \tag{13}$$

and

$$\hat{\alpha}_{n+1}(D) = 1_{\Gamma}(y_{n+1})\frac{\int_D \int_E p(z,x_{n+1},v_n,\alpha)\pi_n^{v\alpha}(dz)\hat{\alpha}_n(d\alpha)}{\int_A \int_E p(z,x_{n+1},v_n,\alpha)\pi_n^{v\alpha}(dz)\hat{\alpha}_n(d\alpha)} +$$
$$1_{\Gamma^c}(y_{n+1})\frac{\int_D \int_E r(z,y_{n+1})P^{v_n,\alpha}(\pi_n^{v\alpha},dz)\hat{\alpha}_n(d\alpha)}{\int_A \int_E r(z,y_{n+1})P^{v_n,\alpha}(\pi_n^{v\alpha},dz)\hat{\alpha}_n(d\alpha)} \tag{14}$$

Consequently when the density $p(x,z,v,\alpha)$ is positive for $x,z \in \Gamma_1$, measures δ_z, $z \in \Gamma_1$ are frequent for the sequence $(\pi_n^{v\alpha})_{\alpha\in A}$. In particular, by an analog of Proposition 1 for α, $\alpha' \in supp\ \hat{\alpha}_\infty$, $x \in \Gamma_1$, $y \in \Gamma$ we obtain

$$p(x,y,v,\alpha) = p(x,y,v,\alpha'), \tag{15}$$

while for $y \in \Gamma^c$

$$\int_{\Gamma^c} r(z,y)\int_E P^{v,\alpha}(x,dz) = \int_{\Gamma^c} r(z,y)\int_E P^{v,\alpha'}(x,dz) \tag{16}$$

provided $\begin{bmatrix} v \\ y \end{bmatrix}$ is frequent in the sequence $\begin{bmatrix} v_n \\ y_{n+1} \end{bmatrix}$.

2.2 Model With Observed Regeneration

Assume that there is a sequence (τ_n) of Y^n adapted stopping times such that x_{τ_n+1} are i.i.d. with law ζ_{α^0}. This is the case when there are sets $K \subset R^d$ and $\Lambda \in \mathcal{E}$ such that whenever $y_n \in K$ we know that $x_n \in \Lambda$ (which means that $r(x,y)=0$ for $y \in K$ and $x \notin \Lambda$), $P^{v,\alpha}(x,\cdot) = \zeta_\alpha(\cdot)$ for $x \in \Lambda$, and

$$\inf_{x\in\Lambda}\int_K r(x,y)dy > 0,$$

$$\inf_{v\in U}\inf_{\alpha\in A}\inf_{x\in E\setminus\Lambda} P^{v,\alpha}(x,\Lambda) > 0.$$

Then we clearly have that

$$\pi_{\tau_n+1}^{v\alpha}(B) = \frac{\int_B r(z,y_{\tau_n+1})\zeta_\alpha(dz)}{\int_E r(z,y_{\tau_n+1})\zeta_\alpha(dz)}, \tag{17}$$

and consequently one can expect to find frequent points of the sequence $\begin{bmatrix} v_{\tau_{n+1}} \\ y_{\tau_n+2} \\ (\pi_{\tau_{n+1}}^{v\alpha})_{\alpha\in A} \end{bmatrix}$.

3 Control Approximations

There are various possibilities to control partially observed adaptive models. However taking into account applicability it seems reasonable to consider relaxed control choosing at time n the control $u_\alpha(\pi_n^{v\alpha})$ with measure $\hat{\alpha}_n$, where $u_\alpha(\pi_n^{v\alpha})$ is a control which is nearly optimal for the value of unknown parameter equal to α. Then one step cost is of the form

$$E^{\alpha^0}\left\{\int_A c(x_n, u_\alpha(\pi_n^{v\alpha}))\hat{\alpha}_n(d\alpha)\right\} =$$
$$E^{\alpha^0}\left\{\int_A \int_E \int_A c(z, u_\alpha(\pi_n^{v\alpha}))\hat{\alpha}_n(d\alpha)\pi_n^{\alpha'}(dz)\hat{\alpha}_n(d\alpha')\right\} \tag{18}$$

The above relaxed control is still infinite dimensional and therefore it is important to decrease its dimensionality by an approximation the measure $\hat{\alpha}_n$. For this purpose we shall need additional assumptions. From now on we shall assume that the set A is countable and the observation process y_n is of the form

$$y_n = h(x_n, w_n), \tag{19}$$

where (w_n) is a sequence of i.i.d. R^d valued respectively random variables with strictly positive density (distribution) g. We assume furthermore that for each $x \in E$, the function $h(x, \cdot)$ is a C^1 diffeomorphism of R^d.

Consider a new probability measure P^o on Ω such that the restrictions P_n^o, P_n of P^o and P respectively to the σ-field $F_n = \sigma\{X^n, Y^n\}$ satisfy

$$P_n(d\omega) = \prod_{i=1}^n \frac{r(x_i, y_i)}{g(y_i)} P_n^o(d\omega). \tag{20}$$

Recalling the proof of Lemma 1.8 of [12] we have

Lemma 1. *Under the measure P^o, the observations y_n are i.i.d. with common density g, independent on α^0 and independent of x_j for $j \leq n$, while (x_n) is a controlled Markov process having for n and Y^n - adapted control v_n the same transition probability $P^{v_n,\alpha^0}(x_n, dy)$ as under P.*

Given $\epsilon > 0$ let A^ϵ be a finite subset of A such that

$$P\left\{\alpha^0 \in A^\epsilon\right\} \geq 1 - \epsilon \tag{21}$$

Let for $D \subset A^\epsilon$

$$\bar{\alpha}_n^\epsilon(D) = \frac{\hat{\alpha}_n(D)}{\hat{\alpha}_n(A^\epsilon)}. \tag{22}$$

By direct calculation we obtain the following recursion

$$\bar{\alpha}_{n+1}^\epsilon(D) = \frac{\sum_{\alpha\in D}\int_E r(z, y_{n+1})P^{v_n,\alpha}(\pi_n^{v\alpha}, dz)\bar{\alpha}_n^\epsilon(\alpha)}{\sum_{\alpha\in A^\epsilon}\int_E r(z, y_{n+1})P^{v_n,\alpha}(\pi_n^{v\alpha}, dz)\bar{\alpha}_n^\epsilon(\alpha)}, \tag{23}$$

with $\bar{\alpha}_0^\epsilon(D) = \eta_\epsilon(D) := \frac{\eta(D)}{\eta(A^\epsilon)}$.

Moreover

$$
\begin{aligned}
&|E^{\alpha^0}\left\{\int_A c(x_n, u_\alpha(\pi_n^{v\alpha}))\hat{\alpha}_n(d\alpha)\right\} \\
&-E^{\alpha^0}\left\{\sum_{\alpha\in A^\epsilon} c(x_n, u_\alpha(\pi_n^{v\alpha}))\bar{\alpha}_n^\epsilon(\alpha)\right\}| \\
&\leq 2\|c\|\epsilon, \qquad (24)
\end{aligned}
$$

with $\|c\|$ standing for the supremum norm of c.

The following result appears to be crucial for the approximation

Lemma 2. *We have that*

$$
\begin{aligned}
&|E^{\alpha^0}\left\{\sum_{\alpha\in A^\epsilon} c(x_n, u_\alpha(\pi_n^{v\alpha}))\bar{\alpha}_n^\epsilon(\alpha)\right\} \\
&-E^{\alpha^{0\epsilon}}\left\{\sum_{\alpha\in A^\epsilon} c(x_n, u_\alpha(\pi_n^{v\alpha}))\bar{\alpha}_n^\epsilon(\alpha)\right\}| \leq 2\epsilon\|c\| \qquad (25)
\end{aligned}
$$

where $\alpha^{0\epsilon}$ corresponds to α^0 restricted to A^ϵ and has the law η_ϵ.

Proof. Notice that by Lemma 1 we have that

$$
\begin{aligned}
&E^{\alpha^0}\left\{\sum_{\alpha\in A^\epsilon} c(x_n, u_\alpha(\pi_n^{v\alpha}))\bar{\alpha}_n^\epsilon(\alpha)\right\} = \\
&\sum_{\alpha'\in A} E^{o\alpha'}\left\{\prod_{i=1}^n \frac{r(x_i, y_i)}{g(y_i)} \sum_{\alpha\in A^\epsilon} c(x_n, u_\alpha(\pi_n^{v\alpha}))\bar{\alpha}_n^\epsilon(\alpha)\right\}\eta(\alpha') \\
&= \sum_{\alpha'\in A} E^{\alpha'}\left\{\sum_{\alpha\in A^\epsilon} c(x_n, u_\alpha(\pi_n^{v\alpha}))\bar{\alpha}_n^\epsilon(\alpha)\right\}\eta(\alpha'). \qquad (26)
\end{aligned}
$$

Now

$$
\begin{aligned}
&|\sum_{\alpha'\in A} E^{\alpha'}\left\{\sum_{\alpha\in A^\epsilon} c(x_n, u_\alpha(\pi_n^{v\alpha}))\bar{\alpha}_n^\epsilon(\alpha)\right\}\eta(\alpha') \\
&- \sum_{\alpha'\in A^\epsilon} E^{\alpha'}\left\{\sum_{\alpha\in A^\epsilon} c(x_n, u_\alpha(\pi_n^{v\alpha}))\bar{\alpha}_n^\epsilon(\alpha)\right\}\eta_\epsilon(\alpha')| \leq 2\epsilon\|c\|. \qquad (27)
\end{aligned}
$$

Finally,

$$
\begin{aligned}
&\sum_{\alpha'\in A^\epsilon} E^{\alpha'}\left\{\sum_{\alpha\in A^\epsilon} c(x_n, u_\alpha(\pi_n^{v\alpha}))\bar{\alpha}_n^\epsilon(\alpha)\right\}\eta_\epsilon(\alpha') \\
&= E^{\alpha^{0\epsilon}}\left\{\sum_{\alpha\in A^\epsilon} c(x_n, u_\alpha(\pi_n^{v\alpha}))\bar{\alpha}_n^\epsilon(\alpha)\right\} \qquad (28)
\end{aligned}
$$

and by (26), (27), (28) we obtain (25). □

By (24) and (25) we see that from approximation point of view instead of evaluation the cost functional (3) with unknown parameter α^0 it is sufficient to study the cost functional (3) with finite valued parameter $\alpha^{0\epsilon}$. Starting from this point we shall need the following assumptions:

(A1) *for* $\phi \in C(E)$ *if* $E \ni x_n \to x$, $U \ni v_n \to v$ *and* $A \ni \alpha_n \to \alpha$ *we have*

$$P^{v_n,\alpha_n}\phi(x_n) \to P^{v,\alpha}\phi(x)$$

as $n \to \infty$.
(A2) $r \in C(E \times R^d)$
(A3) *for* $\psi \in C(R^d)$ *and* $E \ni x_n \to x$ *we have*

$$\int_{R^d} \psi(y) r(x_n, y) dy \to \int_{R^d} \psi(y) r(x, y) dy$$

as $n \to \infty$.

With the use of Lemma 2 we can show the following

Theorem 1. *Assume* A *is countable, we use a randomized control* $u_\alpha(\pi_n^{v\alpha})$ *with measure* $\hat{\alpha}_n$, *where* $u_\alpha : P(E) \mapsto U$ *are continuous and we have identifiability. Furthermore, assume that there is a unique invariant measure* Φ_α *for the filtering process* π_n^α *under control* $u_\alpha(\pi_n^{v\alpha})$ *for* α *being the true value of the parameter and assumptions (A1)-(A3) are satisfied. Then the cost functional (3) is equal to*

$$\int_A \int_{P(A)} \int_E c(z, u_\alpha(\nu)) \nu(dz) \Phi_\alpha(d\nu) \eta(d\alpha) \tag{29}$$

and is the same as under control $u_{\alpha^0}(\pi_n^{\alpha^0})$.

Proof. Notice first that by (24) and (25), taking into account that weak convergence of $\hat{\alpha}_n$ to δ_{α^0} implies the weak convergence of $\bar{\alpha}_n^\epsilon$ to $\delta_{\alpha^{0\epsilon}}$ it suffices to prove theorem for finite A^ϵ with a randomized control $u_\alpha(\pi_n^{v\alpha})$ with measure $\bar{\alpha}_n^\epsilon$. Moreover when $\alpha^{0\epsilon}$ is a true parameter then the pair $\begin{bmatrix} \bar{\alpha}_n^\epsilon \\ (\pi_n^{v\alpha})_{\alpha \in A^\epsilon} \end{bmatrix}$ form a Markov process with transition operator $\Pi^v(\zeta, (\mu_\alpha)_{\alpha \in A^\epsilon}, \cdot)$ defined as follows

$$\Pi^v F(\zeta, (\mu_\alpha)_{\alpha \in A^\epsilon}) = \sum_{\alpha' \in A^\epsilon} \int_E \int_{R^d} F(G^v(y, \zeta, (\mu_\alpha)_{\alpha \in A^\epsilon}),$$

$$(M^{v,\alpha}(y, \mu_\alpha))_{\alpha \in A^\epsilon})) r(z, y) dy P^{v,\alpha'}(\mu_{\alpha'}, dz) \zeta(\alpha') \tag{30}$$

for any bounded Borel measurable function $F : P(A^\epsilon) \times \prod_{\alpha \in A^\epsilon} P(E) \mapsto R$ with v equal to $u_\alpha(\nu_\alpha)$ with measure ζ.

By identifiability for $F : P(A^\epsilon) \mapsto R$ we have

$$E^{\alpha^{0\epsilon}}\{F(\bar{\alpha}_n^\epsilon)\} \to \sum_{\alpha\in A^\epsilon} F(\delta_\alpha)\eta_\epsilon(\alpha) \tag{31}$$

as $n \to \infty$. By (A1)-(A3) and Corollary 1.5 of [11] the operator Π^v is Feller and therefore Cesaro averages of the transition operator Π^v converge to an invariant measure Φ^ϵ for the pair $\begin{bmatrix} \bar{\alpha}_n^\epsilon \\ (\pi_n^{v\alpha})_{\alpha\in A^\epsilon} \end{bmatrix}$. Consequently using the definition of the invariant measure and (31) we obtain

$$\Phi^\epsilon(F, \prod_{\alpha\in A^\epsilon} P(E)) =$$
$$\int_{P(A^\epsilon)\times\prod_{\alpha\in A^\epsilon} P(E)} E_{\zeta(\mu_\alpha)_{\alpha\in A^\epsilon}}\{F(\bar{\alpha}_n^\epsilon)\}\,\Phi^\epsilon(d\zeta, (d\mu_\alpha)_{\alpha\in A^\epsilon})$$
$$\to \int_{P(A^\epsilon)} \sum_{\alpha\in A^\epsilon} F(\delta_\alpha)\zeta(\alpha)\Phi^\epsilon(d\zeta, \prod_{\alpha\in A^\epsilon} P(E)) \tag{32}$$

as $n \to \infty$.

Therefore since by (31) the barycenter of $\Phi^\epsilon(\cdot, \prod_{\alpha\in A^\epsilon} P(E))$ is η_ϵ we have

$$\Phi^\epsilon(F, \prod_{\alpha\in A^\epsilon} P(E)) = \sum_{\alpha\in A^\epsilon} F(\delta_\alpha)\eta_\epsilon(\alpha) \tag{33}$$

Let $p(\omega_1, d\omega_2)$ be a regular conditional probability (see Theorem 3.1 of [9]) of Φ^ϵ with respect to the σ - field $\mathcal{B}(P(A)) \otimes \{\emptyset, \prod_{\alpha\in A^\epsilon} P(E)\}$. Therefore for bounded Borel measurable $F : P(A) \times \prod_{\alpha\in A^\epsilon} P(E) \mapsto R$ we have

$$\Phi^\epsilon(F) = \int_{P(A^\epsilon)} \int_{\prod_{\alpha\in A^\epsilon} P(E)} F(\omega_1, \omega_2)p(\omega_1, d\omega_2)\Phi^\epsilon(d\omega_1, \prod_{\alpha\in A^\epsilon} P(E))$$
$$= \sum_{\alpha\in A^\epsilon} \int_{\prod_{\alpha\in A^\epsilon} P(E)} F(\delta_\alpha, \omega_2)p(\delta_\alpha, d\omega_2)\eta_\epsilon(\alpha). \tag{34}$$

Since by (30)

$$\Pi^v F(\delta_\alpha, (\mu_{\alpha'})_{\alpha'\in A^\epsilon}) =$$
$$\int_E \int_{R^d} F(\delta_\alpha, (M^{u_\alpha(\mu_\alpha),\alpha'}(y, \mu_{\alpha'}))_{\alpha'\in A^\epsilon})r(z, y)dyP^{u_\alpha(\mu_\alpha),\alpha}(\mu_\alpha, dz) \tag{35}$$

by the definition of the invariant measure we therefore have

$$\Phi^\epsilon(\Pi^v F) = \sum_{\alpha\in A^\epsilon} \int_{\prod_{\alpha\in A^\epsilon} P(E)} \int_E \int_{R^d} F(\delta_\alpha, (M^{u_\alpha(\mu_\alpha),\alpha'}(y, \mu_{\alpha'}))_{\alpha'\in A^\epsilon})$$
$$r(z, y)dyP^{u_\alpha(\mu_\alpha),\alpha}(\mu_\alpha, dz)p(\delta_\alpha, (d\mu_\alpha)_{\alpha\in A^\epsilon})\eta_\epsilon(\alpha). \tag{36}$$

In particular, when $F(\zeta, (\mu_\alpha)_{\alpha \in A^\epsilon}) = f_1(\zeta, (\mu_\alpha)_{\alpha\in A^\epsilon}) f_2(\zeta)$ we have

$$\sum_{\alpha\in A^\epsilon} \int_{\prod_{\alpha\in A^\epsilon} P(E)} (\int_E \int_{R^d} f_1(\delta_\alpha, (M^{u_\alpha(\mu_\alpha),\alpha'}(y, \mu_{\alpha'}))_{\alpha'\in A^\epsilon}) r(z,y)dy$$
$$P^{u_\alpha(\mu_\alpha),\alpha}(\mu_\alpha, dz) - f_1(\delta_\alpha, (\mu_{\alpha'})_{\alpha'\in A^\epsilon})) p(\delta_\alpha, (d\mu_{\alpha'})_{\alpha'\in A^\epsilon}) f_2(\delta_\alpha)\eta_\epsilon(\alpha)$$
$$= 0. \tag{37}$$

Since (37) holds for any Borel measurable f_2 for η_ϵ almost all α we obtain

$$\int_{\prod_{\alpha\in A^\epsilon} P(E)} \int_E \int_{R^d} f_1(\delta_\alpha, (M^{u_\alpha(\mu_\alpha),\alpha'}(y, \mu_{\alpha'}))_{\alpha'\in A^\epsilon}) r(z,y)dy$$
$$P^{u_\alpha(\mu_\alpha),\alpha}(\mu_\alpha, dz) p(\delta_\alpha, (d\mu_{\alpha'})_{\alpha'\in A^\epsilon})$$
$$= \int_{\prod_{\alpha\in A^\epsilon} P(E)} f_1(\delta_\alpha, \omega_2) p(\delta_\alpha, d\omega_2). \tag{38}$$

Consequently when $f_1(\delta_\alpha, (\mu_{\alpha'})_{\alpha'\in A^\epsilon}) = f(\mu_\alpha)$ we have

$$\int_{P(E)} \int_E \int_{R^d} f(M^{u_\alpha(\mu_\alpha),\alpha}(y, \mu_\alpha)) r(z,y)dy$$
$$P^{u_\alpha(\mu_\alpha),\alpha}(\mu_\alpha, dz) p(\delta_\alpha, d\mu_\alpha \times \prod_{\alpha\in A^\epsilon\setminus\{\alpha\}} P(E))$$
$$= \int_{P(E)} f(\mu_\alpha) p(\delta_\alpha, d\mu_\alpha \times \prod_{\alpha\in A^\epsilon\setminus\{\alpha\}} P(E)) \tag{39}$$

from which we obtain that $p(\delta_\alpha, d\mu_\alpha \times \prod_{\alpha\in A^\epsilon\setminus\{\alpha\}} P(E))$ is an invariant measure for π_n^α and therefore coincides with Φ_α. For

$$F(\delta_\alpha, (\mu_{\alpha'})_{\alpha'\in A^\epsilon}) = \int_E c(z, u_\alpha(\mu_\alpha)) \mu_\alpha(dz)$$

by (34) we therefore have

$$\Phi^\epsilon(F) = \sum_{\alpha\in A^\epsilon} \int_{P(E)} \int_E c(z, u_\alpha(\nu)) \nu(dz) \Phi_\alpha(d\nu) \eta_\epsilon(\alpha). \tag{40}$$

Notice that (40) corresponds to the value of the cost functional (3) for the true parameter $\alpha^{0\epsilon}$ and control $u_\alpha(\pi_n^{v\alpha})$ with measure $\bar{\alpha}_n^\epsilon$.

Letting in (40) $\epsilon \to 0$ we finally obtain (29). Moreover in (26) we have

$$E^{\alpha^0}\left\{\int_A c(x_n, u_\alpha(\pi_n^{v\alpha}))\hat{\alpha}_n(d\alpha)\right\} =$$
$$= \int_A E^{\alpha'}\left\{\int_A c(x_n, u_\alpha(\pi_n^{v\alpha}))\hat{\alpha}_n(d\alpha)\right\} \eta(d\alpha') \tag{41}$$

from which we see that (29) is the value of the cost functional (3) for α^0 with control $u_{\alpha^0}(\pi_n^{v\alpha^0})$ that is corresponding to the known value of the parameter α^0. □

To use Theorem 1 we have to know ergodicity of the the controlled filtering processes as well as be able to show identifiability. In the next section we recall a model studied in [4] for which both ergodicity and identifiability hold.

4 Rich Observation Model

Let (x_n) be a Markov chain on $E = \{1, 2, \dots, d\}$ with controlled transition matrix $P^{v,\alpha^0}(i,j)$. The state process (x_n) is partially observed via the d-dimensional process

$$y_n = h(x_n) + w_n \tag{42}$$

where (w_n) is a sequence of d-dimensional i.i.d. random vectors with standard normal distribution and $h : E \mapsto R^d$ has components $h^i(j) = 0$ for $j \neq i$ and $h^i(i) = h_i \neq 0$. In the above model we get at each time moment a noisy observation of all states. As in Proposition 1 and Corollary 2 of [4] each filtering process (π_n^α) with control u_{α^0} is uniformly ergodic and has a unique invariant measure Φ_α. To perform identifiability following Proposition 1 we need recurrent points $(\mu_\alpha)_{\alpha \in A}$ of $\pi_n^{v\alpha}$. They can be found using so called Hilbert semimetric $\rho(\mu, \nu)$ defined on the set of nonnegative finite measures $M(E)$ as follows

$$\rho(\mu, \nu) := \ln \beta(\mu, \nu) + \ln \beta(\nu, \mu) \tag{43}$$

where $\beta(\mu, \nu) = \inf \{\lambda \geq 0 : \lambda\mu \geq \nu\}$. Let for $\mu \in M(E)$ and $B \subset E$

$$T^{v,\alpha,y}\mu(B) := \sum_{j \in B} r(z, y) P^{v,\alpha}(\mu, j) \tag{44}$$

If

$$\Delta := \sup_{x,x' \in E} \sup_{v,v' \in E} \sup_{\alpha,\alpha' \in A} \sup_{z \in E} \frac{P^{v,\alpha}(x, z)}{P^{v,\alpha}(x', z)} < \infty \tag{45}$$

then

$$\sup_{\mu,\nu \in M(E)} \sup_{(v,\alpha,y) \in U \times A \times R^d} \rho(T^{v,\alpha,y}\mu, T^{v,\alpha,y}\nu) \leq 2 \ln \Delta. \tag{46}$$

Moreover by Theorem 1.1 from [10] we have that

$$\sup_{(v,\alpha,y) \in U \times A \times R^d} \rho(T^{v,\alpha,y}\mu, T^{v,\alpha,y}\nu) \leq \tanh\left(\frac{2 \ln \Delta}{4}\right) \rho(\mu, \nu), \tag{47}$$

so that assuming (46) we have that the operator T a contraction uniformly in (v, α, y). Finally by Lemma 2.2 of [3] for $\mu, \nu \in P(E)$ we have

$$\|\mu - \nu\| \leq \frac{2}{\ln 2} \rho(\mu, \nu) \tag{48}$$

where $\|\cdot\|$ stands for the variation norm. To find recurrent points of $\pi_n^{v\alpha}$ notice that under observation of the from (42) any point of R^d is recurrent of the observation sequence. Choose a finite $\bar{\epsilon}$ net U_f in U and use Cesaro rare sequences of constant controls $v \in U_f$, which have an increasing length (see e.g. [4] section 4). Alternatively, use randomized control allowing to use any control from the ball in U with small probability (see e.g. [7]). As a result using uniform contractivity of the operator T and (48) we obtain a family of required frequent points. Notice that the procedure to construct frequent points of $\pi_n^{v\alpha}$ can be used for general models for which (46) holds.

References

1. Borkar, V. S. and Mundra, S. M. (1998) Bayesian Parameter Estimation and Adaptive Control of Markov Processes with Time-Averaged Costs, Applicationes Math. **25**, 339–358,
2. Borkar, V. S. (2000) Average Cost Dynamic Programming Equations for Controlled Markov Chains with Partial Observations, SICON 39, 673–681,
3. Borkar, V. S. (1998) Ergodic control of partially observed Markov processes, Systems & Control Letters **34**, 185–189,
4. Di Masi, G. B. and Stettner, Ł. (1994) On Adaptive Control of a Partially Observed Markov Chain, Applicationes Math. **22**, 165–180,
5. Di Masi, G. B. and Stettner, Ł. (1995) Bayesian Ergodic Control of Discrete Time Markov Processes, Stochastics and Stochastics Rep. **54**, 301–316.
6. Di Masi, G. B. and Stettner, Ł. (1997) Bayesian Ergodic Adaptive Control of Diffusion Processes, Stochastics and Stochastics Rep. **60**, 155–183.
7. Di Masi, G. B. and Stettner, Ł. (1998) Bayesian adaptive control of discrete-time Markov processes with long-run average cost, Systems and Control Letters **34**, 55–62,
8. Duncan, T. E., Pasik-Duncan, B., and Stettner, Ł. (1998) Adaptive Control of a Partially Observed Discrete Time Markov Process, JAMO 37, 269–293,
9. Ikeda, N. and Watanabe, S. (1981) Stochastic Differential Equations and Diffusion Processes, North Holland.
10. Liverani, C. (1995) Decay of correlation, Ann. Math. **142**, 239–301.
11. Runggaldier, W. J. and Stettner, Ł. (1994) Approximations of Discrete Time Partially Observed Control Problems, Giardini, Pisa.
12. Stettner, Ł. (1993) Ergodic Control of Partially Observed Markov Processes With Equivalent Transition Probabilities, Applicationes Math. **22**, 25–38.

Portfolio Optimization in Markets Having Stochastic Rates

Richard H. Stockbridge[1,2]

[1] Department of Mathematical Sciences, University of Wisconsin Milwaukee, Milwaukee, WI 53201, USA
[2] Department of Statistics, University of Kentucky, Lexington, KY 40506-0027, USA

Abstract. The Merton problem of optimizing the expected utility of consumption for a portfolio consisting of a bond and N stocks is considered when changes in the bond's interest rate, in the mean return rates and in the volatilities for the stock price processes are modelled by a finite-state Markov chain. This paper establishes an equivalent linear programming formulation of the problem. Two cases are considered. The first model assumes that these coefficients are known to the investor whereas the second model investigates a partially observed model in which the mean return rates and volatilities for the stocks are not directly observable.

1 Introduction

Consider a regime-switching market consisting of a bond B which earns interest at a rate r_t and N stocks $S = (S^1, \dots, S^N)$ which have mean return rates $\mu_t = (\mu_t^1, \dots, \mu_t^N)$ and volatility matrix $\Sigma_t = ((\sigma_t^{ij}))$ which all vary according to a finite-state Markov chain. In particular, the change in the mean return rates allow both "bull" ($\mu_t^i > r$) and "bear" ($\mu_t^i < r$) markets to occur in the model. The assets therefore satisfy

$$\begin{aligned} dB_t &= r_t B_t dt \\ dS_t^i &= S_t^i \left(\mu_t^i dt + \sum_j \sigma_t^{ij} dW_t^j \right), \qquad i = 1, \dots, N \end{aligned}$$

in which $W = (W^1, \dots, W^N)$ is an N-dimensional standard Brownian motion process and (r_t, μ_t, Σ_t) is a continuous time Markov chain having states

$$\mathbb{M} := \{(r_1, \mu_1, \Sigma_1), \dots, (r_n, \mu_n, \Sigma_n)\}, \tag{1}$$

where $\mu_i = (\mu_i^1, \dots, \mu_i^N)$, and generator Q. It is possible that only one component or two components change when the rate process changes; for example, there may be i and j such that $r_i = r_j$ but $\mu_i \neq \mu_j$ or $\Sigma_i \neq \Sigma_j$. This possibility will be important in section 4.

Let y_t^0 denote the amount that is invested in the bond and let $y_t^1, \dots, y_t^N$ denote the amounts invested in the N stocks at time t. The investor is allowed

B. Pasik-Duncan: Stochastic Theory and Control, LNCIS 280, pp. 447–458, 2002.

to consume his wealth at rate C_t at time t and this consumption is drawn from the bond. Thus the process $y = (y^0, y^1, \ldots, y^N)$ satisfies

$$dy_t^0 = (r_t y_t^0 - C_t)dt - \sum_{i=1}^{N} dM_t^i \tag{2}$$

$$dy_t^i = y_t^i \left(\mu_t^i dt + \sum_{j=1}^{N} \sigma_t^{ij} dW_t^j \right) + dM_t^i, \qquad i = 1, \ldots, N \tag{3}$$

in which the process M^i records the purchases and sales of stock i. Notice this model does not include costs for such transactions.

The objective of the investor is to maximize his expected discounted utility of consumption

$$E\left[\int_0^\infty e^{-\rho s} U(C_s)\, ds\right], \tag{4}$$

where the function U is only required to be measurable and bounded below.

The above model is a modification of a model solved by Merton [8] in which the market rates are constant and the utility function is the HARA function $U(c) = \gamma^{-1} c^\gamma$ for some $\gamma < 1$ with $\gamma \neq 0$. Using dynamic programming methods, Merton showed that the optimal solution is to maintain constant proportions $\pi = (\pi^1, \ldots, \pi^N)$ of total wealth in the stocks at all times and to fix consumption at a constant rate c of total wealth as well. The proportion of wealth invested in each stock is given by

$$\pi = \frac{1}{(1-\gamma)} (\Sigma\Sigma')^{-1} (\mu - r\underline{1})$$

and rate of consumption (as a fraction of total wealth) is

$$c = \left(\frac{1}{1-\gamma} \left[\rho - \gamma r + \frac{\gamma}{2(1-\gamma)} (\mu - r\underline{1})' (\Sigma\Sigma')^{-1} (\mu - r\underline{1}) \right] \right)^{\gamma - 1},$$

where $\mu - r\underline{1} = (\mu^1 - r, \ldots, \mu^N - r)$.

The regime-switching model has been used to evaluate European options by Di Masi, Kabanov and Runggaldier [2] and Guo [3] and to analyze American options by Buffington and Elliott [1].

This paper is organized as follows. The next section gives a careful stochastic processes formulation for the problem, including the possibility for the decisions by the investor to be determined by randomization at each time t. In section 3, we consider the completely observed problem in which it is assumed that the investor knows the interest rate r_t, the mean return rates μ_t^i and the volatilities σ_t^{ij} at each instant. The problem is formulated using the linear programming approach. In section 4 a partially observed problem is discussed where the mean return rate process μ and volatility matrix Σ are not known by the investor.

2 Formulation

Rather than deal with the model (2)-(3), we initially follow the approach of Merton and consider the total wealth process and the proportions of total wealth invested in the stocks. We then rephrase the dynamics in terms of a controlled martingale problem and introduce relaxed or randomized controls.

Let w denote the total wealth process, $w_t = \sum_{i=0}^{N} y_t^i$, and, for $i = 1, \dots, N$, let $\pi_t^i = y_t^i / w_t$ denote the proportion of wealth invested in stock i at time t and let $\pi_t^0 = 1 - \sum_{i=1}^{N} \pi_t^i$ denote the proportion of wealth invested in the bond at time t. We assume the initial wealth is fixed at $\tilde{w}_0$. In light of Merton's results that the consumption rate is a constant multiple of the wealth of the investor, we reparametrize the consumption process by setting $C_t = c_t w_t$ so that the new decision variable c_t gives the rate of consumption per unit of wealth. The wealth process then satisfies

$$\begin{aligned} dw_t &= (r_t y_t^0 - c_t w_t)dt + \sum_{i=1}^{N} y_t^i \left(\mu_t^i dt + \sum_{j=1}^{N} \sigma_t^{ij} dW_t^j \right) \\ &= \left(r_t \pi_t^0 w_t + \sum_{i=1}^{N} \mu_t^i \pi_t^i w_t - c_t w_t \right) dt + \sum_{i,j=1}^{N} \sigma_t^{ij} \pi_t^i w_t dW_t^j \\ &= \left(r_t + \sum_{i=1}^{N} (\mu_t^i - r_t) \pi_t^i - c_t \right) w_t dt + \sum_{i,j=1}^{N} \sigma_t^{ij} \pi_t^i w_t dW_t^j . \end{aligned} \tag{5}$$

Note that the differentials dM_t^i drop out of the equation since there are no costs for transactions.

The wealth process will be characterized in terms of the controlled martingale problem for its generator. We therefore begin with a definition of a solution to a controlled martingale problem and refer the reader to the paper of Kurtz [4] for more details. Let E denote the state space of a process and U denote the space of controls available to the decision maker. Let $A : \mathcal{D} \to C(E \times U)$ be an operator on functions $\mathcal{D} \subset C(E)$ and let ν_0 denote the desired initial distribution of the solution. A process (X, u) is said to be a solution of the controlled martingale problem for (A, ν_0) if there exists a filtration $\{\mathcal{F}_t\}$ such that (X, u) is $\{\mathcal{F}_t\}$-progressively measurable, $X(0)$ has distribution ν_0 and

$$f(X(t)) - \int_0^t Af(X(s), u(s))\, ds$$

is an $\{\mathcal{F}_t\}$-martingale for every $f \in \mathcal{D}$.

For our portfolio model, we take $\mathcal{D} = C_c^2(\mathbb{R}^+)$, where $\mathbb{R}^+$ denotes the possible values of the wealth process. We also need to keep track of the values of the interest rate r_t, the mean return rates μ_t and the volatilities Σ_t so we include them as state variables and therefore have functions of

(r, μ, Σ) as well as of w. However since (r_t, μ_t, Σ_t) take on only finitely many values, it suffices to restrict attention to the collection of indicator functions $\mathcal{I} = \{I_{\{(r_k,\mu_k,\Sigma_k)\}}(r, \mu, \Sigma) : k = 1, \ldots, n\}$.

Let $\mathcal{F}_t = \sigma(W_s, r_s, \mu_s, \Sigma_s : 0 \le s \le t)$. In light of the controlled martingale problem definition, we require that the decision variables c and $\pi = (\pi^1, \ldots, \pi^N)$ be progressively measurable, which essentially says that the decisions at time t are only based on the history of the solution up to time t. We say that consumption and proportion processes c and π are *admissible* if the wealth process w never becomes negative, the quadruple $(w, (r, \mu, \Sigma), \pi, c) = \{(w_t, (r_t, \mu_t, \Sigma_t), \pi_t, c_t) : t \ge 0\}$ is $\{\mathcal{F}_t\}$-progressively measurable and (5) is satisfied.

Let $(w, (r, \mu, \Sigma), \pi, c)$ be a quadruple consisting of an admissible pair (π, c), the resulting wealth process w and the market rates process (r, μ, Σ). For each $f \in \mathcal{D}$ and $I_{\{(r_k,\mu_k,\Sigma_k)\}} \in \mathcal{I}$, an application of Itô's formula gives

$$\begin{aligned}
&f(w_t)I_{\{(r_k,\mu_k,\Sigma_k)\}}(r_t, \mu_t, \Sigma_t) \\
&\quad = f(w_0)I_{\{(r_k,\mu_k,\Sigma_k)\}}(r_0, \mu_0, \Sigma_0) \\
&\qquad + \int_0^t \left(\left[\left(r_s + \sum_{i=1}^N (\mu_s^i - r_s)\pi_s^i - c_s \right) w_s f'(w_s) \right.\right. \\
&\qquad\qquad \left. + (1/2) \sum_{j=1}^N \left(\sum_{i=1}^N \sigma_s^{ij} \pi_s^i \right)^2 w_s^2 f''(w_s) \right] I_{\{(r_k,\mu_k,\Sigma_k)\}}(r_s, \mu_s, \Sigma_s) \\
&\qquad\qquad \left. + f(w_s) Q I_{\{(r_k,\mu_k,\Sigma_k)\}}(r_s, \mu_s, \Sigma_s) \right) ds \\
&\qquad + \sum_{j=1}^N \int_0^t \sum_{i=1}^N \sigma_s^{ij} \pi_s^i w_s f'(w_s) dW_s^j
\end{aligned}$$

in which the generator Q of the market rates process $\{(r_t, \mu_t, \Sigma_t) : t \ge 0\}$ is given by

$$\begin{aligned}
&Q I_{\{(r_k,\mu_k,\Sigma_k)\}}(r_i, \mu_i, \Sigma_i) \\
&\quad = \sum_{j=1}^n [I_{\{(r_k,\mu_k,\Sigma_k)\}}(r_j, \mu_j, \Sigma_j) - I_{\{(r_k,\mu_k,\Sigma_k)\}}(r_i, \mu_i, \Sigma_i)] q_{ij}.
\end{aligned}$$

Defining the generator L on functions $f I_{\{(r_k,\mu_k,\Sigma_k)\}}$ by

$$\begin{aligned}
&L(f I_{\{(r_k,\mu_k,\Sigma_k)\}})(w, (r, \mu, \Sigma), \pi, c) \qquad (6) \\
&= \left(r + \sum_{i=1}^N (\mu^i - r)\pi^i - c \right) w f'(w) + (1/2) \sum_{j=1}^N \left(\sum_{i=1}^N \sigma^{ij} \pi^i \right)^2 w^2 f''(w)
\end{aligned}$$

$$+f(w)QI_{\{(r_k,\mu_k,\Sigma_k)\}}(r,\mu,\Sigma),$$

it follows that

$$f(w_t)I_{\{(r_k,\mu_k,\Sigma_k)\}}(r_t,\mu_t,\Sigma_t) - \int_0^t L(fI_{\{(r_k,\mu_k,\Sigma_k)\}})(w_s,(r_s,\mu_s,\Sigma_s),\pi_s,c_s)\,ds$$

is an $\{\mathcal{F}_t\}$-martingale.

We incorporate one relaxation of the above requirements. Rather than require the investor to firmly decide on the values of π_t and c_t at each time t, we allow him the opportunity to select these values according to some randomization mechanism. Thus the pair (π_t, c_t) may have a non-degenerate probability distribution Λ_t on $U = \mathbb{R}^N \times \mathbb{R}^+$ (Λ_t will be a degenerate distribution when (π_t, c_t) are deterministic). Therefore the portfolio process is $(w,(r,\mu,\Sigma),\Lambda) = \{(w_t,(r_t,\mu_t,\Sigma_t),\Lambda_t) : t \geq 0\}$ which is $\{\mathcal{F}_t\}$-progressively measurable and

$$f(w_t)I_{\{(r_k,\mu_k,\Sigma_k)\}}(r_t,\mu_t,\Sigma_t) \tag{7}$$
$$-\int_0^t \int_{\mathbb{R}^N\times\mathbb{R}^+} L(fI_{\{(r_k,\mu_k,\Sigma_k)\}})(w_s,(r_s,\mu_s,\Sigma_s),\pi,c)\Lambda_s(d\pi\times dc)\,ds$$

is an $\{\mathcal{F}_t\}$-martingale for every $f \in \mathcal{D}$ and $I_{\{(r_k,\mu_k,\Sigma_k)\}} \in \mathcal{I}$. Finally recall that the initial wealth is given by $\tilde{w}_0$ and we assume the initial rates are $(r_0,\mu_0,\Sigma_0) \in \mathbb{M}$ (cf. (1)).

We summarize the formulation of the portfolio optimization problem as follows. The objective of the investor is to maximize (4) over all solutions $(w,(r,\mu,\Sigma),\Lambda)$ of the controlled martingale problem for $(L,\tilde{w}_0,(r_0,\mu_0,\Sigma_0))$; that is, such that (7) is an $\{\mathcal{F}_t\}$-martingale for every $f \in \mathcal{D}$ and $I_{\{(r_k,\mu_k,\Sigma_k)\}} \in \mathcal{I}$ and the initial values are $(\tilde{w}_0,(r_0,\mu_0,\Sigma_0))$.

3 Completely Observed Problem

The portfolio optimization problem can be reformulated as a linear program using results in Kurtz and Stockbridge [6]. We motivate the form of the linear program and quote an existence result from [6] to establish the equivalence.

For each solution $(w,(r,\mu,\Sigma),\Lambda)$ of the controlled martingale problem and $f \in \mathcal{D}$ and $I_{\{(r_k,\mu_k,\Sigma_k)\}} \in \mathcal{I}$, the fact that (7) is an $\{\mathcal{F}_t\}$-martingale implies that

$$e^{-\rho t}f(w_t)I_{\{(r_k,\mu_k,\Sigma_k)\}}(r_t,\mu_t,\Sigma_t) \tag{8}$$
$$-\int_0^t \int_{\mathbb{R}^N\times\mathbb{R}^+} e^{-\rho s}[L(fI_{\{(r_k,\mu_k,\Sigma_k)\}})(w_s,(r_s,\mu_s,\Sigma_s),\pi,c)$$
$$-\rho f(w_s)I_{\{(r_k,\mu_k,\Sigma_k)\}}(r_s,\mu_s,\Sigma_s)]\Lambda_s(d\pi\times dc)\,ds$$

is also an $\{\mathcal{F}_t\}$-martingale. As a result

$$f(\tilde{w}_0)I_{\{(r_k,\mu_k,\Sigma_k)\}}(r_0,\mu_0,\Sigma_0)$$

$$= E\Bigg[e^{-\rho t} f(w_t) I_{\{(r_k,\mu_k,\Sigma_k)\}}(r_t,\mu_t,\Sigma_t) - \int_0^t \int_{\mathbb{R}^N\times\mathbb{R}^+} e^{-\rho s}[L(fI_{\{(r_k,\mu_k,\Sigma_k)\}})(w_s,(r_s,\mu_s,\Sigma_s),\pi,c) - \rho f(w_s) I_{\{(r_k,\mu_k,\Sigma_k)\}}(r_s,\mu_s,\Sigma_s)]\Lambda_s(d\pi\times dc)\,ds\Bigg]$$

and letting $t \to \infty$,

$$\begin{aligned} &f(\tilde{w}_0) I_{\{(r_k,\mu_k,\Sigma_k)\}}(r_0,\mu_0,\Sigma_0) \\ &= -E\Bigg[\int_0^\infty \int_{\mathbb{R}^N\times\mathbb{R}^+} e^{-\rho s}[L(fI_{\{(r_k,\mu_k,\Sigma_k)\}})(w_s,(r_s,\mu_s,\Sigma_s),\pi,c) \\ &\qquad - \rho f(w_s) I_{\{(r_k,\mu_k,\Sigma_k)\}}(r_s,\mu_s,\Sigma_s)]\Lambda_s(d\pi\times dc)\,ds\Bigg]. \end{aligned} \tag{9}$$

Now define the expected discounted occupation measure m of the solution $(w,(r,\mu,\Sigma),\Lambda)$ on $\mathbb{R}^+\times\mathbb{M}\times\mathbb{R}^N\times\mathbb{R}^+$ by

$$\begin{aligned} &m(G_1\times G_2\times G_3\times G_4) \\ &= E\Bigg[\int_0^\infty \int_{\mathbb{R}^N\times\mathbb{R}^+} e^{-\rho s} I_{G_1}(w_s) I_{G_2}(r_s,\mu_s,\Sigma_s) \\ &\qquad \cdot I_{G_3}(\pi) I_{G_4}(c)\Lambda_s(d\pi\times dc)\,ds\Bigg] \end{aligned}$$

for all Borel sets $G_1\subset\mathbb{R}^+$, $G_2\subset\mathbb{M}$, $G_3\subset\mathbb{R}^N$ and $G_4\subset\mathbb{R}^+$. Notice that by setting each G_i to be the entire space $m(\mathbb{R}^+\times\mathbb{M}\times\mathbb{R}^N\times\mathbb{R}^+) = 1/\rho$. It follows from (9) that

$$\begin{aligned} &\int [L(fI_{\{(r_k,\mu_k,\Sigma_k)\}})(w,(r,\mu,\Sigma),\pi,c) \\ &-\rho f(w) I_{\{(r_k,\mu_k,\Sigma_k)\}}(r,\mu,\Sigma)]\, m(dw\times d\mu\times d\pi\times dc) \\ &= -f(\tilde{w}_0) I_{\{(r_k,\mu_k,\Sigma_k)\}}(r_0,\mu_0,\Sigma_0) \end{aligned} \tag{10}$$

for each $f\in\mathcal{D}$ and $I_{\{(r_k,\mu_k,\Sigma_k)\}}\in\mathcal{I}$. Moreover, the expected utility (4) can be written in terms of m as

$$\int U(cw)\, m(dw\times d(r,\mu,\Sigma)\times d\pi\times dc). \tag{11}$$

Thus the maximization of (4) over processes $(w,(r,\mu,\Sigma),\pi,c)$ for which (7) is a martingale corresponds to the maximization of (11) over measures m on $\mathbb{R}^+\times\mathbb{M}\times\mathbb{R}^N\times\mathbb{R}^+$ satisfying (10) and having total mass $1/\rho$.

We have seen that each solution of the controlled martingale problem has an expected discounted occupation measure which satisfies (10) for each

$f \in \mathcal{D}$ and $I_{\{(r_k,\mu_k,\Sigma_k)\}} \in \mathcal{I}$. Corollary 5.3 of Kurtz and Stockbridge [6] shows that the adjoint relation (10), in fact, characterizes the solutions of the controlled martingale problem by proving existence of a solution for each measure m satisfying (10). We phrase the existence theorem in terms of the model of this paper and refer the reader to [6] for the proof of the general result. (The application of Corollary 5.3 of [6] requires rewriting (9) in the form

$$\begin{aligned} 0 = E\Bigg[\int_0^t \int_{\mathbb{R}^N\times\mathbb{R}^+} & e^{-\rho s}[L(fI_{\{(r_k,\mu_k,\Sigma_k)\}})(w_s,(r_s,\mu_s,\Sigma_s),\pi,c) \\ & +\rho(f(\tilde{w}_0)I_{\{(r_k,\mu_k,\Sigma_k)\}}(r_0,\mu_0,\Sigma_0) \\ & -f(w_s)I_{\{(r_k,\mu_k,\Sigma_k)\}}(r_s,\mu_s,\Sigma_s))]\Lambda_s(d\pi\times dc)\,ds\Bigg] \end{aligned}$$

and renormalizing by a factor of $1/\rho$.)

THEOREM 3.1. *Let m be a measure on $\mathbb{R}^+\times\mathbb{M}\times\mathbb{R}^N\times\mathbb{R}^+$ satisfying (10) and having total mass $1/\rho$. Define m_0 to be the state marginal on $\mathbb{R}^+\times\mathbb{M}$ and η to be the regular conditional distribution of (π,c) given $(w,(r,\mu,\Sigma))$ under m, and thus m, m_0 and η are related by the relation*

$$m(dw\times d(r,\mu,\Sigma)\times d\pi\times dc) = \eta(w,(r,\mu,\Sigma),d\pi\times dc)m_0(dw\times d(r,\mu,\Sigma)).$$

Then there exists a wealth and market rates process $(w,(r,\mu,\Sigma))$ having initial distribution $(\tilde{w}_0,(r_0,\mu_0,\Sigma_0))$ such that

$$\begin{aligned} & f(w_t)I_{\{(r_k,\mu_k,\Sigma_k)\}}((r_t,\mu_t,\Sigma_t) \\ & \quad -\int_0^t\int_{\mathbb{R}^N\times\mathbb{R}^+} L(fI_{\{(r_k,\mu_k,\Sigma_k)\}})(w_s,(r_s,\mu_s,\Sigma_s),\pi,c) \\ & \qquad\qquad \cdot\eta(w_s,(r_s,\mu_s,\Sigma_s),d\pi\times dc)\,ds \end{aligned} \tag{12}$$

is a martingale with respect to $\{\mathcal{F}_t^{w,(r,\mu,\Sigma)}\}$ and

$$\begin{aligned} E\Bigg[\int_0^\infty e^{-\rho s}\int_{\mathbb{R}^N\times\mathbb{R}^+} & U(cw_s)\eta(w_s,(r_s,\mu_s,\Sigma_s),d\pi\times dc)\,ds\Bigg] \\ = \int_{\mathbb{R}^+\times\mathbb{M}\times\mathbb{R}^N\times\mathbb{R}^+} & U(cw)\,m(dw\times d(r,\mu,\Sigma)\times d\pi\times dc). \end{aligned} \tag{13}$$

Notice, in particular, that the randomized control of this solution to the controlled martingale problem is $\Lambda_s(d\pi\times dc) = \eta(w_s,(r_s,\mu_s,\Sigma_s),d\pi\times dc)$ and thus the control is specified in feedback form of the current state of the process.

As a corollary, the completely observed portfolio optimization problem is equivalent to a linear program over the space of measures.

COROLLARY 3.2. *The problem of maximizing (4) over all solutions (w, p, Λ) satisfying (7) is a martingale for all $f \in D$ and $I_{\{(r_k,\mu_k,\Sigma_k)\}} \in \mathcal{I}$ with $w_0 = \tilde{w}_0$ is equivalent to the linear program of maximizing (11) over all measures m satisfying (10) for each $f \in \mathcal{D}$ and $I_{\{(r_k,\mu_k,\Sigma_k)\}} \in \mathcal{I}$ and with total mass equal to $1/\rho$.*

4 Partially Observed Problem

A similar analysis can be applied to the portfolio management problem under the condition that the mean and volatility rates of the stocks are not observable.

Let $(w, (r, \mu, \Sigma), \Lambda)$ be a solution of the controlled martingale problem for L, having initial states $(\tilde{w}_0, (r_0, \mu_0, \Sigma_0))$. Let $\mathcal{F}_t^{w,r,\Lambda} = \sigma(w_s, r_s, \Lambda_s : 0 \leq s \leq t)$ denote the information available to the investor. Note, in particular, that the mean return rates process μ and the volatilities matrix Σ are not $\{\mathcal{F}_t^{w,r,\Lambda}\}$-adapted. (Recall, that the values $r_1, \ldots, r_n$ are not necessarily distinct so the fact that the interest rate process r is observable does not immediately imply that the processes μ and Σ are therefore indirectly observed.) Let p_t denote the conditional distribution of (r_t, μ_t, Σ_t) given $\mathcal{F}_t^{w,r,\Lambda}$ and denote the conditional probabilities by

$$p_t^i = P((r_t, \mu_t, \Sigma_t) = (r_i, \mu_i, \Sigma_i)|\mathcal{F}_t^{w,r,\Lambda}), \qquad i = 1, \ldots, n.$$

Following Kurtz and Ocone [5] and Kurtz [7], the fact that (8) is an $\{\mathcal{F}_t\}$-martingale for each $f \in \mathcal{D}$ and $I_{\{(r_k,\mu_k,\Sigma_k)\}} \in \mathcal{I}$ implies that for each $f \in \mathcal{D}$ and $I_{\{(r_k,\mu_k,\Sigma_k)\}} \in \mathcal{I}$

$$\begin{aligned}
&\langle e^{-\rho t} f(w_t) I_{\{(r_k,\mu_k,\Sigma_k)\}}(\cdot), p_t\rangle \\
&\quad - \int_0^t \int_{\mathbb{R}^N \times \mathbb{R}^+} \langle e^{-\rho s}[L(f I_{\{(r_k,\mu_k,\Sigma_k)\}}) \\
&\qquad\qquad - \rho f I_{\{(r_k,\mu_k,\Sigma_k)\}}](w_s, \cdot, \pi, c), p_s\rangle \Lambda_s(d\pi \times dc) ds \\
&= \int_{\mathbb{M}} e^{-\rho t} f(w_t) I_{\{(r_k,\mu_k,\Sigma_k)\}}(\mu) p_t(d\mu) \\
&\quad - \int_0^t \int_{\mathbb{R}^N \times \mathbb{R}^+} \int_{\mathbb{M}} e^{-\rho s}[L(f I_{\{(r_k,\mu_k,\Sigma_k)\}}) \\
&\qquad\qquad - \rho f I_{\{(r_k,\mu_k,\Sigma_k)\}}](w_s, \mu, \pi, c) p_s(d\mu) \Lambda_s(d\pi \times dc) \\
&= e^{-\rho t} f(w_t) p_t^k \\
&\quad - \int_0^t \int_{\mathbb{R}^N \times \mathbb{R}^+} \sum_{i=1}^n e^{-\rho s}[L(f I_{\{(r_k,\mu_k,\Sigma_k)\}}) \\
&\qquad\qquad - \rho f I_{\{(r_k,\mu_k,\Sigma_k)\}}](w_s, \mu_i, \pi, c) p_s^i(d\mu) \Lambda_s(d\pi \times dc) ds
\end{aligned}$$

is an $\{\mathcal{F}_t^{w,r,\Lambda}\}$-martingale. As before, taking expectations, letting $t \to \infty$ and multiplying by -1 we have

$$\begin{aligned} \sum_{i=1}^{n} E\Bigg[\int_0^\infty \int_{\mathbb{R}^N\times\mathbb{R}^+} & e^{-\rho s}[L(fI_{\{(r_k,\mu_k,\Sigma_k)\}}) \\ & - \rho f I_{\{(r_k,\mu_k,\Sigma_k)\}}](w_s,\mu_i,\pi,c)p_s^i \,\Lambda_s(d\pi\times dc)\,ds\Bigg] \\ & = -\sum_{i=1}^{n} f(\tilde{w}_0)p_0^i \\ & = -f(\tilde{w}_0) \end{aligned} \tag{14}$$

for each $f \in \mathcal{D}$ and $I_{\{(r_k,\mu_k,\Sigma_k)\}} \in \mathcal{I}$. Now defining the expected discounted occupation measures $m^{(i)}$ on $\mathbb{R}^+ \times [0,1] \times \mathbb{R}^N \times \mathbb{R}^+$, for $i = 1,\ldots,n$, by

$$\begin{aligned} & m^{(i)}(G_1\times G_2\times G_3\times G_4) \\ & \quad = E\left[\int_0^\infty e^{-\rho s} I_{G_1}(w_s)I_{G_2}(p_s^i)\int_{\mathbb{R}^N\times\mathbb{R}^+} I_{G_3}(\pi)I_{G_4}(c)\Lambda_s(d\pi\times dc)\,ds\right] \end{aligned}$$

for all Borel sets $G_1 \subset \mathbb{R}^+$, $G_2 \subset [0,1]$, $G_3 \subset \mathbb{R}$ and $G_4 \subset \mathbb{R}^+$, (14) can be written as

$$\begin{aligned} & -f(\tilde{w}_0) \\ & \quad = \sum_{i=1}^{n} \int [L(fI_{\{(r_k,\mu_k,\Sigma_k)\}}) - \rho f I_{\{(r_k,\mu_k,\Sigma_k)\}}](w,(r_i,\mu_i,\Sigma_i),\pi,c) \\ & \qquad\qquad \cdot p\, m^{(i)}(dw\times dp\times d\pi\times dc). \end{aligned} \tag{15}$$

Again note that $\sum_i m^{(i)}(\mathbb{R}^+ \times [0,1] \times \mathbb{R}^N \times \mathbb{R}^+) = 1/\rho$. Also note that (4) can be written in terms of $\{m^{(i)}\}$ as

$$\sum_{i=1}^{n} \int U(cw)\, m^{(i)}(dw\times dp\times d\pi\times dc). \tag{16}$$

By letting m be the measure on $\mathbb{R}^+ \times \mathbb{M} \times [0,1] \times \mathbb{R}^N \times \mathbb{R}^+$ having

$$m(dw \times \{(r_i,\mu_i,\Sigma_i)\} \times dp \times d\pi \times dc) = m^{(i)}(dw\times dp\times d\pi\times dc),$$

the same existence result of Kurtz and Stockbridge ([6, Corollary 5.3]) therefore yields the following result.

THEOREM 4.1. *Let $m^{(i)}$, $i = 1,\ldots,n$, be measures on $\mathbb{R}^+ \times [0,1] \times \mathbb{R}^N \times \mathbb{R}^+$ satisfying (15) with total mass $1/\rho$. For $i = 1,\ldots,n$, let $m_0^{(i)}$ denote the state marginal on $\mathbb{R}^+ \times [0,1]$ and $\eta^{(i)}$ be the regular conditional distribution of (π,c) given (w,p) under $m^{(i)}$, and thus $m^{(i)}$, $m_0^{(i)}$ and $\eta^{(i)}$ are related by the relation*

$$m^{(i)}(dw\times dp\times d\pi\times dc) = \eta^{(i)}(w,p,d\pi\times dc)m_0^{(i)}(dw\times dp).$$

Then there exists a process $(w, p^1, \ldots, p^n)$ *with* $w_0 = \tilde{w}_0$ *such that for each* $f \in \mathcal{D}$ *and* $I_{\{(r_k,\mu_k,\Sigma_k)\}} \in \mathcal{I}$

$$\begin{aligned} e^{-\rho t} f(w_t) p_t^k \\ - \int_0^t \int_{\mathbb{R}^N \times \mathbb{R}^+} \sum_{i=1}^n e^{-\rho s} [L(f I_{\{(r_k,\mu_k,\Sigma_k)\}}) - \rho f I_{\{(r_k,\mu_k,\Sigma_k)\}}](w_s, \mu_i, \pi, c) \\ \cdot p_s^i(d\mu) \eta_s^{(i)}(d\pi \times dc) ds \end{aligned} \tag{17}$$

is an $\{\mathcal{F}_t^{w,p}\}$*-martingale and*

$$\begin{aligned} \sum_{i=1}^n E\left[\int_0^\infty e^{-\rho s} \int_{\mathbb{R}^N \times \mathbb{R}^+} U(cw_s) \eta^{(i)}(w_s, p_s^i, d\pi \times dc)\, ds\right] \\ = \sum_{i=1}^n \int_{\mathbb{R}^+ \times [0,1] \times \mathbb{R}^N \times \mathbb{R}^+} U(cw)\, m^{(i)}(dw \times dp \times d\pi \times dc). \end{aligned} \tag{18}$$

As a corollary, the partially observed portfolio optimization problem is equivalent to a linear program over the space of measures.

COROLLARY 4.2. *The problem of maximizing (4) over all solutions* (w, p, Λ) *satisfying (17) is a martingale for all* $f \in D$ *and* $I_{\{(r_k,\mu_k,\Sigma_k)\}} \in \mathcal{I}$ *with* $w_0 = \tilde{w}_0$ *is equivalent to the linear program of maximizing (16) over all measures* $\{m^{(i)} : i = 1, \ldots, n\}$ *satisfying (15) for each* $f \in \mathcal{D}$ *and* $I_{\{(r_k,\mu_k,\Sigma_k)\}} \in \mathcal{I}$ *and with combined total mass equal to* $1/\rho$*.*

REMARK 4.3. It should be noted that this linear programming approach to the partially observed control problem separates the control problem from the estimation of the current state of the market rates. The solution of the linear program determines the optimal decisions for each level of wealth and distribution $(p^1, \ldots, p^n)$ on the values of the market rates. It is still necessary to obtain accurate estimates of the conditional distributions based on the observed quantities.

REMARK 4.4. A careful analysis of the terms in (17) will remove the dependence on $I_{\{(r_k,\mu_k,\Sigma_k)\}} \in \mathcal{I}$ from the expression as will be seen when we now specialize to the simple bull and bear market.

4.1 Simple Bull and Bear Market

Consider the simple bull and bear market of a single stock in which the mean return rate μ_t takes two values μ_1 and μ_2, with $\mu_1 < r < \mu_2$, and has generator $Qf(\mu_i) = \sum_{j=1}^2 (f(\mu_j) - f(\mu_i)) q_{ij}$ corresponding to the transition rate matrix given by

$$\begin{pmatrix} -q_{12} & q_{12} \\ q_{21} & -q_{21} \end{pmatrix}, \tag{19}$$

and with the bond's interest rate r and the stock's volatility σ remaining constant. Setting $I_{\{(r_k,\mu_k,\Sigma_k)\}} = I_{\{\mu_1\}}$ the constraint (15) becomes

$$\begin{aligned} -f(\tilde{w}_0) &= \int [Af(w,\mu_1,\pi,c) - (q_{12}+\rho)f(w)]p\, m^{(1)}(dw \times dp \times d\pi \times dc) \\ &\qquad + \int f(w) q_{12}\, p\, m^{(2)}(dw \times dp \times d\pi \times dc) \\ &= \int \Bigg[(r + (\mu_1 - r)\pi - c) w f'(w) + (1/2)\sigma^2\pi^2\tilde{w}_0^2 f''(w) \\ &\qquad - (q_{12}+\rho)f(w) \Bigg] p\, m^{(1)}(dw \times dp \times d\pi \times dc) \\ &\qquad + \int f(w) q_{12}\, p\, m^{(2)}(dw \times dp \times d\pi \times dc) \end{aligned} \tag{20}$$

and similarly taking $I_{\{(r_k,\mu_k,\Sigma_k)\}} = I_{\{\mu_2\}}$ yields

$$\begin{aligned} -f(\tilde{w}_0) &= \int f(w) q_{21}\, p\, m^{(1)}(dw \times dp \times d\pi \times dc) \\ &\qquad + \int \Bigg[(r + (\mu_2 - r)\pi - c) w f'(w) + (1/2)\sigma^2\pi^2\tilde{w}_0^2 f''(w) \\ &\qquad - (q_{21}+\rho)f(w) \Bigg] p\, m^{(2)}(dw \times dp \times d\pi \times dc). \end{aligned} \tag{21}$$

Moreover, letting $m^{(1)*}$ and $m^{(2)*}$ denote a pair of optimal measures, the conditional distributions ζ_1^* and ζ_2^* satisfying, for $i = 1, 2$,

$$m^{(i)*}(dw \times dp \times d\pi \times dc) = \zeta_i^*(w, p, d\pi \times dc) m_0^{(i)*}(dw \times dp)$$

gives optimal portfolio allocations π and consumption rates c as a randomized function of the wealth w and the conditional probability that the mean return rate is μ_i

References

1. Buffington, J. and Elliott, R. J. American options with regime switching, preprint.
2. Di Masi, G. B., Kabanov, Y. M., and Runggaldier, W. J. (1994) Mean variance hedging of options on stocks with Markov volatility, Theory Probab. Appl. **39**, 173–181.
3. Guo, X. (1999) Insider information and stock fluctuations, Ph.D Thesis, Rutgers University, N.J.
4. Kurtz, T. G. (1987) Martingale problems for controlled processes, in Stochastic Modelling and Filtering (Rome, 1984), Lecture Notes in Control and Inform. Sci. **91**, 75–90, Springer, Berlin.
5. Kurtz, T. G. and Ocone, D. L. (1988) Unique characterization of conditional distributions in nonlinear filtering, Annals of Probab., **16**, 80–107.
6. Kurtz, T. G. and R. H. Stockbridge, R. H. (1998) Existence of Markov Controls and Characterization of Optimal Markov Controls, SIAM J. Control Optim. **36**, 609–653.
7. Kurtz, T. G. (1998) Martingale problems for conditional distributions of Markov processes, Electronic Journal of Probability, **3**, 1–29.
8. Merton, R. C. (1971) Optimum consumption and portfolio fules in a continuous-time model, J. Econom. Theory, **3**, 373–413.

Moment Problems Related to the Solutions of Stochastic Differential Equations

Dedicated to Professor Tyrone Duncan on the occasion of his 60th birthday

Jordan Stoyanov

School of Mathematics & Statistics, University of Newcastle, Newcastle upon Tyne, NE1 7RU, Great Britain. jordan.stoyanov@ncl.ac.uk

Abstract. Our goal is to analyze the moment uniqueness or non-uniqueness of the one-dimensional distributions of the solution processes of Itô type stochastic differential equations (SDE). We recall some criteria, classical and/or new, and apply them to derive results for the solutions of linear and nonlinear SDEs. Special attention is paid to the Brownian motion, stochastic integrals and geometric Brownian motion. Another possibility is to use the moment convergence theorem (Frećhet-Shohat) for finding explicitly the limit one-dimensional distributions of specific processes. Related moment problems are also outlined with the focus on functional transformations of processes and approximations of the solutions of perturbed SDEs.

1 Introduction

We consider stochastic processes X_t, $t \in [0,T]$ or $t \geq 0$ which are defined as the strong solutions of Itô type stochastic differential equations (SDE).

Let us explain briefly how the classical moment problem is involved when studying SDEs in general and perturbed SDEs in particular.

Suppose X is a random variable (r.v.) on a given probability space $(\Omega, \mathcal{F}, P)$ such that X with distribution function (d.f.) F have finite moments of all orders: $m_k := E[X^k] = \int x^k dF(x)$, $k = 1, 2, \ldots$ In the sequel we use $\mathcal{M}$ for the class of all d.f.s having finite moments of all orders and $\mathcal{L}(X)$ for the law (the d.f.) of the r.v. X. Here $F = \mathcal{L}(X)$.

We start with $F \in \mathcal{M}$, so the moment sequence $\{m_k\}$ is well-defined. The classical moment problem is to answer the following converse question: Is F the only d.f. with this moment sequence? If the answer is "yes", we say that the moment problem has a unique solution, or equivalently that F is M-det (moment-determinate). Otherwise, we say that the moment problem has a non-unique solution, and also that F is M-indet (moment-indeterminate) meaning that there is at least one d.f. G, such that

$$G \neq F \text{ but } \int x^k dG(x) = \int x^k dF(x) \text{ for all } k = 1, 2, \ldots$$

B. Pasik-Duncan: Stochastic Theory and Control, LNCIS 280, pp. 459–469, 2002.

Equivalently we can say that any $F \in \mathcal{M}$ generates only one moment sequence $\{m_k\}$. Thus the converse question is: Does $\{m_k\}$ generate only one d.f., F?

When studying the moment problem we distinct the following three cases depending on supp(F), the support of the d.f. F:

(a) Hamburger moment problem: $\text{supp}(F) = \mathbb{R}^1 = (-\infty, \infty)$.
(b) Stieltjes moment problem: $\text{supp}(F) = \mathbb{R}_+ = (0, \infty)$.
(c) Hausdorff moment problem: supp(F) is a bounded interval in $\mathbb{R}^1$.

For our further consideration we need the following classical example involving two most popular distributions one of which is M-det and the other one is M-indet.

Example. We say that the r.v. X has a lognormal distribution with parameters 0 and 1 and write $X \sim LogN(0,1)$, if X is positive and absolutely continuous with density

$$f(x) = \frac{1}{\sqrt{2\pi}}\frac{1}{x}\exp\left[-\frac{1}{2}(\ln x)^2\right] \text{ for } x > 0 \text{ and } f(x) = 0 \text{ for } x \le 0.$$

The moments of X are known explicitly: $m_k = E[X^k] = \int_0^\infty x^k f(x)dx = e^{k^2/2}$, $k = 1, 2, \ldots$ Thus $\mathcal{L}(X) \in \mathcal{M}$. It turns out that $\mathcal{L}(X)$, i.e. the lognormal distribution is M-indet. In fact, there are infinitely many absolutely continuous distributions on $(0, \infty)$ and also infinitely many purely discrete distributions with support a subset of $(0, \infty)$, they are all different of the lognormal distribution but all have the same moment sequence $m_k = e^{k^2/2}$, $k = 1, 2, \ldots$ Details can be seen in [4], [18] or [19].

Taking now the logarithmic transformation of X we see that the r.v. $Y := \ln X$ has a standard normal distribution: $Y \sim N(0,1)$. Hence $X = e^Y$ (another way to define $LogN(0,1)$). The normal distribution, however, is M-det: $N(0,1)$ is the only distribution with odd order moments $m_{2k+1} = 0$ and even order moments $m_{2k} = (2k-1)!!$, $k = 1, 2, \ldots$

Therefore even a frequently used functional transformation such as the logarithmic/expo- nential can change the moment determinacy of a distribution. Later we will exploit this fact when considering the geometric Brownian motion and the standard Brownian motion.

2 Criteria for Uniqueness and Non-Uniqueness

Below we have summarized basic and very useful criteria allowing us to characterize the moment determinacy of a distribution.

Criterion 1 (General). Suppose X is a r.v. with d.f. F. If $M(u) := E[e^{uX}] = \int e^{ux} dF(x)$ exists for all real $u \in (-u_0, u_0)$ for some $u_0 > 0$, i.e. in a proper neighborhood of zero, we say that X has a moment generating function (m.g.f.) M. In such a case the following two conclusions can be derived: (1) $F \in \mathcal{M}$, that is, the moments of X of all orders are finite and $m_k = M^{(k)}(0)$, $k = 1, 2, \ldots$; (2) F is the only d.f. with the moment sequence $\{m_k\}$. In other words, the existence of the m.g.f. M implies that the d.f. F is M-det.

Criterion 2 (Hausdorff Problem). We consider a r.v. X taking values in a bounded interval $(a, b) \subset \mathbb{R}^1$, $-\infty < a < b < \infty$. Since $\mathrm{supp}(F) = (a, b)$ is bounded, then $F \in \mathcal{M}$ and moreover, the m.g.f. $M(u) = \int_a^b e^{ux} dF(x)$ exists even for all real u. Hence $F = \mathcal{L}(X)$ is the only d.f. corresponding to the moment sequence $\{m_k\}$, where $m_k = E[X^k]$, $k = 1, 2, \ldots$

Criterion 3 (General). Assume that X is a r.v. such that $F = \mathcal{L}(X) \in \mathcal{M}$ and let $m_k := E[X^k] = \int x^k dF(x)$, $k = 1, 2, \ldots$ be the moment sequence. Suppose that the following *Carleman condition* is satisfied:

$$\sum_{k=1}^{\infty} \frac{1}{(m_{2k})^{1/2k}} = \infty \quad (\text{if } \mathrm{supp}(F) = \mathbb{R}^1)$$

or

$$\sum_{k=1}^{\infty} \frac{1}{(m_k)^{1/2k}} = \infty \quad (\text{if } \mathrm{supp}(F) = \mathbb{R}_+).$$

Then the d.f. F is M-det.

Clearly, if F is M-indet, then necessarily the Carleman quantity $\sum_{k=1}^{\infty} (m_{2k})^{-1/2k}$ or $\sum_{k=1}^{\infty} (m_{2k})^{-1/2k}$ is finite. Notice that non-uniqueness in the moment problem, or moment indeterminacy of the d.f. F, can eventually appear only if $\mathrm{supp}(F)$ is infinite.

Criterion 4 (Hamburger Problem). Assume that the r.v. X with values in $\mathbb{R}^1$ has density $f(x) = F'(x)$, $x \in \mathbb{R}^1$, which is positive and differentiable. We need the following logarithmic normalized integral called also *Krein quantity*:

$$K^{(H)} = \int_{-\infty}^{\infty} \frac{-\ln f(x)}{1 + x^2} dx.$$

(a) If $K^{(H)} < \infty$, then the d.f. F is M-indet.
(b) Let $K^{(H)} = \infty$. This together with the following *Lin condition*:

$$\frac{-x f'(x)}{f(x)} \nearrow \infty \text{ as } 0 < x_0 < x \nearrow \infty$$

implies that F is M-det.

Criterion 5 (Stieltjes Problem). Suppose the r.v. X takes values in $\mathbb{R}_+$ and has density $F'(x) = f(x), x \in \mathbb{R}^1$, which is positive. The Krein quantity is defined by

$$K_a^{(S)} := \int_a^\infty \frac{-\ln f(x^2)}{1+x^2} dx.$$

Here $a \geq 0$ with $a = 0$ only if $\lim_{x\to 0} f(x) > 0$.

(a) Let $K_0^{(S)}$ be finite: $K_0^{(S)} = \int_0^\infty [(-\ln f(x^2))/(1+x^2)]dx < \infty$. Then F is M-indet.
(b) Let $K_a^{(S)} = \infty$ for some $a > 0$. This and the above Lin condition for the density f implies that F is M-det.

Details about these criteria can be found in [15], [18], [10], [19], [14].

Let us recall here another result in which the moment uniqueness is essential.

Moment Convergence Theorem (Fréchet-Shohat). Suppose $F_1, F_2, \ldots$ is a sequence of d.f.s on $\mathbb{R}^1$. Let us assume that $m_k(n) := \int x^k dF_n(x) \to m_k$ as $n \to \infty$ for all $k = 1, 2, \ldots$ and that all these numbers are finite. Then:

(1) $\{m_k, k = 1, 2, \ldots\}$ is a moment sequence of some d.f., say F.
(2) If F is M-det, then the following weak convergence holds: $F_n \xrightarrow{w} F$ as $n \to \infty$.

3 Some Problems of Interest

In areas such as population theory and control theory we often meet stochastic phenomena where the problems of interest rely only on the one-dimensional distributions of the involved stochastic processes. Moreover, the solutions of important problems explicitly include the requirement for some distributions to be unique in terms of the moments. The recent paper [9] is a good reference to look at.

Problem A. Suppose X_t, $t \in [0, T]$ for a fixed $0 < T < \infty$ or $t \geq 0$ is a stochastic process such that for any t the r.v. X_t has finite moments of all orders, that is, $\mathcal{L}(X_t) \in \mathcal{M}$. This means that $m_k(t) = E[X_t^k]$, the kth order moment function, exists for all $k = 1, 2, \ldots$ The goal is to establish whether the one-dimensional distributions $\mathcal{L}(X_t)$ are M-det or M-indet. Then we want to answer the same question for functional transformations of the original process, that is, for the process $Y_t = H(X_t)$, $t \geq 0$, where H is a given function.

Problem B. Let the process X_t, $t \geq 0$, be such that $\mathcal{L}(X_t) \in \mathcal{M}$. Suppose that the kth order moment function $m_k(t) := E[X_t^k] \to m_k$ as $t \to \infty$ for all $k = 1, 2, \ldots$, with all $m_k(t)$ and m_k finite. Then $\{m_k, k = 1, 2, \ldots\}$

is the moment sequence of a d.f., say F. Now the problem is to identify the limit distribution $F := \lim_{t\to\infty} \mathcal{L}(X_t)$ by knowing its moment sequence $\{m_k\}$ and establish the moment determinacy of F. (If, for example, $X_t, t \geq 0$ is a Markov process, then F is associated with the invariant measure of the process and its ergodic distribution, see, for example [20].)

If F is M-det, then Fréchet-Shohat theorem implies that $\mathcal{L}(X_t) \xrightarrow{w} F$, $t \to \infty$. Additionally we can analyze the rate of convergence of $\mathcal{L}(X_t)$ to F in an appropriate metric.

Problem C. We consider a family of stochastic processes $X^\varepsilon = (X_t^\varepsilon, t \geq 0)$, where $\varepsilon \in (0, \varepsilon_0)$ for some fixed $\varepsilon_0 > 0$ with X_t^ε appropriately defined, for example, as the unique strong solution of a perturbed SDE. We assume that for each ε and each t the r.v. X_t^ε has finite moment function of any order k, meaning that the law $\mathcal{L}(X_t^\varepsilon) \in \mathcal{M}$. The goal is to study the limit behavior of X_t^ε as $\varepsilon \to 0$, or of a suitable time/space transformation $H_t^\varepsilon := H(X_t^\varepsilon, t, \varepsilon)$ as $\varepsilon \to 0$. It is of interest to find the limit kth order moment functions $\lim_{\varepsilon\to 0} m_k^\varepsilon(t) = m_k(t)$, $k = 1, 2, \ldots$, where $m_k^\varepsilon(t) = E[(H^\varepsilon(t))^k]$, $k = 1, 2, \ldots$, and derive properties, including about determinacy of the distribution corresponding to $\{m_k(t)\}$.

4 Some Results About Stochastic Integrals

Let $w_t, t \geq 0$ be a standard Brownian motion (BM) and $g(t, \omega)$, $t \geq 0$ an adapted and integrable process: we assume that $E\{\int_0^T [g(t, \omega)]^2\}dt < \infty$. Then the following Itô stochastic integral is well-defined:

$$I_t := \int_0^t g(s, \omega) dw_s, \qquad t \in [0, T].$$

It is of general interest to find conditions for moment determinacy or indeterminacy of the distributions $\mathcal{L}(I_t^n)$ and $\mathcal{L}(|I_t|^r)$ for positive integers n and real $r > 0$. We can ask similar questions for other functionals of the BM process, see the recent book [23].

Proposition 1. *Suppose the integrand g in $I_t = \int_0^t g(s)dw_s$, $t \in [0, T]$ is deterministic and integrable: $\int_0^T g^{2k}(t)dt < \infty$, $k = 1, 2, \ldots$ Then $\mathcal{L}(I_t^{2n+1}) \in \mathcal{M}$, $n = 1, 2, \ldots$, $\mathcal{L}(|I_t|^r) \in \mathcal{M}$, $r \geq 0$ for each $t \in [0, T]$ and the following properties hold:*

(1) *$\mathcal{L}(I_t)$ is M-det;*
(2) *$\mathcal{L}(I_t^2)$ is M-det;*
(3) *$\mathcal{L}(I_t^{2n+1})$ is M-indet for all $n = 1, 2, \ldots$;*
(4) *$\mathcal{L}(|I_t|^r)$ is M-det for $0 \leq r \leq 4$;*
(5) *$\mathcal{L}(|I_t|^r)$ is M-indet for $r > 4$.*

Proof. The definition of the stochastic integral implies that for deterministic integrand g, I_t, $t \in [0, T]$ is a Gaussian process. In particular, the r.v. $I_t \sim N(0, \sigma_t^2)$, where $\sigma_t^2 = \int_0^t g^2(s)ds$. Hence claim (1) is true. The r.v. $\xi = I_t/\sigma_t$ is already standard normal, $\xi \sim N(0,1)$, and its square ξ^2 is a r.v. which is χ^2-distributed with one degree of freedom. Since χ^2 distribution has a m.g.f., this implies claim (2).

Consider now the r.v. $Z_t := I_t^3$, the cube of the stochastic integral. Clearly, Z_t has finite moments of all orders, so the law $\mathcal{L}(Z_t) \in \mathcal{M}$. We want to show that $\mathcal{L}(Z_t)$ is M-indet. First, we use the fact that $\xi = I_t/\sigma_t \sim N(0,1)$ to derive explicitly the density h_t of the r.v. Z_t (the support is $\mathbb{R}^1$, so this is Hamburger case). Second, we use Criterion 4(a) by easily checking that the Krein quantity for h_t is finite; for details see [19]. This shows the moment indeterminacy of the law $\mathcal{L}(I_t^{2n+1})$ for $n = 1$. Similar arguments lead to the conclusion that the law $\mathcal{L}(I_t^{2n+1})$ is M-indet for any $n = 1, 2, \ldots$, so claim (3) is proved.

Finally, claims (4) and (5) are checked by finding the explicit expression for the density of the r.v. $|I_t|^r$ (the support is $\mathbb{R}_+$, so it is a Stieltjes case). For we need some details from [19]. It remains then to use Criterion 5, case (a) for $r > 4$ and case (b) for $0 \leq r \leq 4$. ∎

Remark 1. Another proof of claims (3), (4) and (5) in Proposition 1 can be given by combining the specific construction developed in [2] with a result from [21].

It is worth to mention the following curious, and perhaps not so well-known fact that I_t^3 and $|I_t|^3$ have different moment determinacy property: $\mathcal{L}(I_t^3)$ is M-indet while $\mathcal{L}(|I_t|^3)$ is M-det.

Corollary. In Proposition 1 it is interesting to take $g(t) = const = 1$ for all $t \in [0, T]$ in which case simply $I_t = w_t$, the BM. Hence the above claims from (1) to (5) can be easily reformulated already for the BM process. Clearly a modification of Proposition 1 is valid for arbitrary Gaussian processes with continuous or discrete time parameter.

Remark 2. It is more delicate to analyze the moment determinacy of the stochastic integral $I_t := \int_0^t g(s,\omega)dw_s$ for random integrand $g(s,\omega)$ assuming that $\int_0^T E\{[g(t,\omega)]\}^{2k}dt < \infty$ for all integers $k = 1, 2, \ldots$ We need some moment estimates for stochastic integrals and random time change technique, see for example [3], [11], [8], [6], [13].

5 Some Results About Stochastic Differential Equations

Suppose $X_t, t \geq 0$ is a stochastic process which is the unique strong solution of an Itô type SDE with drift and diffusion coefficients respectively $a(\cdot)$ and

$b^2(\cdot)$ and initial value x_0:

$$dX_t = a(t, X_t)dt + b(t, X_t)dw_t, \qquad X_0 = x_0, \ t \in [0, T].$$

Here $a(\cdot)$ and $b(\cdot)$ are "nice" functions: they are measurable functions satisfying the global Lipschitz condition and they grow in x not faster than linearly.

We assume that the initial value $X_0 = x_0$ is either a r.v. which is independent of the BM w and with finite moments of all orders, or simply we take x_0 to be a constant. Then for each t the moments of the r.v. X_t are all finite, hence $\mathcal{L}(X_t) \in \mathcal{M}$. The natural question to ask is whether or not the law $\mathcal{L}(X_t)$ is M-det or M-indet. We are going to show that there are cases when $\mathcal{L}(X_t)$ is M-det and other cases when $\mathcal{L}(X_t)$ is M-indet. Perhaps some of the results appear a little unexpected, especially when seeing that the solutions of a large class of SDEs are actually M-indet.

Let us start with one of the very popular processes, the geometric Brownian motion. Suppose that S_t, $t \geq 0$ is such a process, so it is described as follows:

$$dS_t = \mu S_t dt + \sigma S_t dw_t, \qquad S_0 = s_0, \ t \geq 0,$$

where μ and σ are real parameters. Let us assume that $s_0 = const > 0$, $\mu > 0$, and $\sigma > 0$. Then all values of the process S_t, $t \geq 0$ are positive and for each t the r.v. S_t has finite moments of all orders:

$$m_k(t) := E[S_t^k] = s_0^k \exp\left[atk + \frac{1}{2}\sigma^2 tk^2\right], \qquad k = 1, 2, \dots,$$

where $a = \mu - \frac{1}{2}\sigma^2$.

Proposition 2. *There exists a family of stochastic processes,*

$$S_t^{(\varepsilon)}, \qquad t \geq 0,$$

with $\varepsilon \in (-\varepsilon_0, \varepsilon_0)$, $0 < \varepsilon_0 \leq 1$, which obey the following properties:

(1) *For any t and ε, the r.v. $S_t^{(\varepsilon)}$ has finite moments of all orders.*
(2) *The moments of $S_t^{(\varepsilon)}$ are the same as those of S_t, that is,*

$$m_k^{(\varepsilon)} := E[(S_t^{(\varepsilon)})^k] = E[S_t^k], \qquad k = 1, 2, \dots$$

(3) *The process $S^{(\varepsilon)}$, eventually except for $\varepsilon = 0$, is not a geometric BM.*

Proof. The moment indeterminacy of the law $\mathcal{L}(S_t)$ can be shown in different ways. We need the probability density $f_t(x) = (P[S_t \leq x])'_x$ of the r.v. S_t:

$$f_t(x) = \begin{cases} 0, & \text{if } x \leq 0 \\ \dfrac{1}{\sigma\sqrt{t}\sqrt{2\pi}}\dfrac{1}{x}\exp\left[-\dfrac{1}{2\sigma^2 t}(\ln x - at)^2\right], & \text{if } x > 0. \end{cases}$$

The Krein quantity corresponding to f_t is finite and Criterion 5(a) tells us that the law $\mathcal{L}(S_t)$ is M-indet. The next step is to follow [4] and define a "new" function, $f_t^{(\varepsilon)}(x)$ for $x \in \mathbb{R}^1$ and $\varepsilon \in (-\varepsilon_0, \varepsilon_0)$, where $0 < \varepsilon_0 \leq 1$ as follows: $f_t^{(\varepsilon)}(x) = f_t(x)[1 + \varepsilon h(x)], x \in \mathbb{R}^1$. Here $h(x) = \sin((2\pi(\ln x - a))$, if $x > 0$ and $h(x) = 0$, if $x \leq 0$. It is not difficult to check that for each ε the function $f_t^{(\varepsilon)}$ is a probability density. Denote by F_t^ε the d.f. corresponding to $f_t^{(\varepsilon)}$. Now we define $S_t^{(\varepsilon)}$, $t \geq 0$, $\varepsilon \in (-\varepsilon_0, \varepsilon_0)$, where $0 < \varepsilon_0 \leq 1$, to be a stochastic process corresponding to the family $\mathbb{F}^\varepsilon$ with the property that for each ε and t the r.v. S_t^ε has a d.f. exactly F_t^ε. Finally we can show, for details see [18], that

$$m_k^{(\varepsilon)}(t) := E[(S_t^{(\varepsilon)})^k] = E[S_t^k] = \exp\left[atk + \frac{1}{2}\sigma^2 tk^2\right], \qquad k = 1, 2, \ldots$$

Clearly $f_t^{(\varepsilon)} \neq f_t$ for $\varepsilon \neq 0$. Hence the stochastic process $S^{(\varepsilon)}$ cannot be a geometric BM for any $\varepsilon \neq 0$. This completes the proof. ∎

Remark 3. Let us note that we can use a result from [21] and suggest another construction of a stochastic process, say X_t, which is not a geometric BM but obeys the property that for each t the r.v. X_t has the same moments as those of the geometric BM S_t.

Proposition 3. *Suppose that the process $X_t, t \geq 0$ is the unique strong solution of the SDE:*

$$dX_t = a(t, X_t)dt + b(t, X_t)dw_t, \qquad X_0 = x_0, \ t \in [0, T].$$

Here the drift and diffusion coefficients $a(t,x)$, $b^2(t,x)$, $t \in [0,T]$, $x \in \mathbb{R}^1$ satisfy the standard conditions: they are global Lipschitz in x and grow in x not faster than linearly (both conditions uniformly in t).

Suppose further that the diffusion coefficient is uniformly bounded:

$$b^2(t, x) \leq K \text{ for all } t \in [0, T], \ x \in \mathbb{R}^1, \text{ where } 0 < K < \infty.$$

Then for each $t \in [0, T]$:

(1) *The law $\mathcal{L}(X_t)$ is M-det (Hamburger problem).*
(2) *The law $\mathcal{L}(|X_t|)$ is M-det (Stieltjes problem).*

Proof. The easiest way is to show that for each $t \in [0, T]$ the r.v. X_t has a m.g.f., that is, that the function $E[e^{uX_t}]$ is well-defined for all $u \in (-u_0, u_0)$ for some fixed number $u_0 > 0$.

Indeed, if $|X_t| \leq 1$ a.s., then the quantity $E[e^{uX_t}]$ is finite for any $u > 0$ and any $u \leq 0$ and this implies the moment uniqueness of the distributions $\mathcal{L}(X_t)$ and $\mathcal{L}(|X_t|)$.

It remains to consider the case when $|X_t| > 1$. This together with the monotone property of the exponential function implies that for $u > 0$ we have

$E[e^{u|X_t|}] \leq E[e^{uX_t^2}]$. Now we see that under the above imposed conditions we can apply Theorem 4.7 from [11] and conclude that there is a $\delta = \delta(T) > 0$ for which

$$\sup_{t \in [0,T]} E[e^{\delta X_t^2}] < \infty.$$

Therefore the function $E[e^{u|X_t|}]$ is finite for $u \in (0, u_0)$, where $u_0 = \delta$. Hence $E[e^{u|X_t|}]$ and $E[e^{uX_t}]$ are well-defined for all $u \in (-u_0, u_0)$ and the existence of these two m.g.f.s for each $t \in [0, T]$ implies, according to Criterion 1, the moment uniqueness of $\mathcal{L}(X_t)$ and $\mathcal{L}(|X_t|)$.

Let us note that an alternative proof of the same statement can be given by using the moment estimates for the solutions of SDEs, see for example [8], and then check that the Carleman condition is satisfied and refer to Criterion 3. ∎

Remark 4. It turns out that the global Lipschitz condition $b^2(t, x) < K < \infty$ in Proposition 4 cannot be replaced by, say the usual linear growth condition $b^2(t, x) \leq K(1 + x^2)$ which is widely used for the existence and uniqueness of strong solutions of SDEs. Look at the geometric BM: its diffusion coefficient $b^2(t, x) = \sigma^2 x^2$ is not globally bounded (as required in Proposition 3). Hence, the moment indeterminacy property of the geometric BM does not appear as a surprise. Note that if we write the explicit expression for the geometric BM X_t as an exponential functional of the standard BM w_t we easily find that $E[e^{\delta X_t^2}] = \infty$ for any $\delta > 0$ and all $t \in [0, T]$. This shows that, for example, the r.v. X_t^2 does not have a m.g.f. This, however, still does not allow to make any conclusion about the determinacy of the d.f. $\mathcal{L}(X_t^2)$. In fact, $\mathcal{L}(X_t^2)$ is M-indet which follows from one of the results in [19] where it is shown that the distribution of any power of a lognormally distributed r.v. is M-indet. Alternatively, we can derive the same conclusion by using Criterion 5(a).

6 Comments on Related Topics

Let us mention that each of Problems A, B and C can be further extended and/or specified. For example, we can treat other classes of stochastic processes which are not connected with SDEs and try to answer questions, the same or similar to those discussed above.

The moment determinacy and indeterminacy of the one-dimensional distributions of the solutions of SDEs is related to the asymptotic behavior of these solutions as the time parameter $t \to \infty$. Hence we can expect an interesting involvement of specific asymptotic stability properties of the solutions of SDEs, see for example [1], [7], [13]. We expect the Ornstein-Uhlenbeck process to be essentially involved.

Several models of perturbed SDEs, including those previously considered in our papers [12], [5], [16], [17], can be studied by following the approach in this paper.

Another interesting question is about the moment determinacy or indeterminacy of the invariant measures of diffusion type processes, see for example [7], [8], [22].

The recent book [23] suggests a lot of interesting results when studying functionals of the Brownian motion, so of interest are questions about determinacy or indeterminacy properties of the distributions of such functionals.

Acknowledgments

I am using this opportunity to express my appreciation and thanks to Prof. Bozenna Pasik-Duncan for organizing the "Second Kansas Workshop in Stochastic Systems and Control" (University of Kansas, Lawrence, October 2001).

I am grateful to the anonymous referee and the Editor who made some useful comments and suggestions.

This study was partly supported by the EPSRC, Grant no. GR/R71719/01.

References

1. Arnold, L. (1974) Stochastic Differential Equations: Theory and Applications, John Wiley & Sons, New York.
2. Berg, C. (1988) The cube of the normal distribution is indeterminate, Annals of Probability, **16**, 910–913.
3. Gihman, I. I. and Skorohod, A. V. (1972) Stochastic Differential Equations, Springer-Verlag, Berlin. (Russian ed. 1968)
4. Heyde, C. C. (1963) On a property of the lognormal distribution, Journal of the Royal Statistical Society, Series B **29**, 392–393.
5. Kabanov, Yu., Pergamenshchikov, S., and Stoyanov, J. (1991) Asymptotic expansions for singularly perturbed stochastic differential equations, in New Trends in Probability and Statistics. In Honour of Yu. V. Prohorov, VSP, Utrecht (NL), 413–435.
6. Karatzas, I. and Shreve, S. E. (1991) Brownian Motion and Stochastic Calculus, 2nd ed., Spinger, New York.
7. Khasminskii, R. Z. (1980) Stochastic Stability of Differential Equations, Sijthoff & Noordhoff, Alphen aan den Rijn (NL). (Russian ed. 1969)
8. Krylov, N. V. (1980) Controlled Diffusion Processes, Springer, New York. (Russian ed. 1977)
9. Levanony, D. and Caines, P. E. (2002) On persistent excitation for linear systems with stochastic coefficients, SIAM Journal on Control & Optimization, to appear.
10. Lin, G. D. (1997) On the moment problems, Statistics & Probability Letters, **35**, 85–90.
11. Liptser, R. Sh. and Shiryaev, A. N. (2001) Statistics of Random Processes I: General theory, 2nd ed., Springer, New York. (Russian ed. 1974; first English ed. 1977)
12. Liptser, R. and Stoyanov, J. (1990) Stochastic version of the averaging principle for diffusion type processes, Stochastics & Stochastics Reports, **32**, 145–163.

13. Mao, X. (1997) Stochastic Differential Equations & Applications, Horwood, Chichester.
14. Pakes, A. G., Hung, W.-L., and Wu, J.-W. (2000) Criteria for the unique determination of probability distributions by moments, Australian & New Zealand Journal of Statistics, **43**, 101–111.
15. Slud, E. (1983) The moment problem for polynomial forms in normal random variables, Annals of Probability, **21**, 2200–2214.
16. Stoyanov, J. and Botev, D. (1996) Quantitative results for perturbed stochastic differential equations, Journal of Applied Mathematics and Stochastic Analysis, **9**, 255–261.
17. Stoyanov, J. (1997) Regularly perturbed stochastic differential equations with an internal random noise, Nonlinear Analysis: Theory, Methods & Applications, **30**, 4105–4111.
18. Stoyanov, J. (1997) Counterexamples in Probability, 2nd ed., John Wiley & Sons, Chichester.
19. Stoyanov, J. (2000) Krein condition in probabilistic moment problems, Bernoulli, **6**, 939–949.
20. Stoyanov, J. and Pirinsky, Ch. (2000) Random motions, classes of ergodic Markov chains and beta distributions, Statistics & Probability Letters, **50**, 293–304.
21. Stoyanov, J. (2001) Probability spaces and sets of random elements with prescribed independence/dependence structure, submitted.
22. Veretennikov, A. Yu. (2001) On polynomial mixing estimates for stochastic differential equations with a gradient drift, Theory of Probability and Its Applications, **45**, 160–163.
23. Yor, M. (2001) Exponential Functionals of Brownian Motion and Related Processes, Springer, Berlin.

$\mathcal{L}$-Transform, Normal Functionals, and Lévy Laplacian in Poisson Noise Analysis

Allanus H. Tsoi

Department of Mathematics, The University of Missouri, Columbia, MO 65211, USA

Abstract. In this paper we study Poisson noise with the help of generalized functions. We first discuss the $\mathcal{L}$-transform of (L^2)-Poisson functionals. We then consider an extension of the $\mathcal{L}$-transform with the help of Sobolev norms. Next we introduce the class of normal functionals and consider the Wick product. We consider the Hida derivatives and the Lévy Laplacian acting on these normal functionals . A relationship between the $\mathcal{L}$-transform and Y. Itô's $\mathcal{U}$-transform is given. Finally, we give a stochastic limit characterization of the Lévy Laplacian with the help of two-parameter Poisson processes.

1 Introduction

In this paper we study Poisson noise functionals. Poisson noise was studied by Hida [4]; Ito [16], [17]; Ito and Kubo [14], [15]; Us [23]; Kondratiev et al. [19], [20]; and by Saito and Tsoi [22]. Suppose $E = \mathcal{L}(\mathbb{R})$ denotes the Schwartz space of test functions on $\mathbb{R}$, endowed with its usual topology. Thus E is countably Hilbert and nuclear ([21]). Let $E' = \mathcal{L}'(\mathbb{R})$ be the dual of E, and $\mathcal{H} = L^2(\mathbb{R})$. Then we have the Gel′fand triple:

$$E \subset \mathcal{H} \subset E'$$

By the Bochner-Minlos theorem there exists a unique probability measure μ_p on E' corresponding to the characteristic functional:

$$L(\xi) = \exp\{\int_{-\infty}^{\infty} (e^{i\xi(t)} - 1)dt\}, \qquad \xi \in E. \tag{1.1}$$

The space $(E', \mathcal{B}(E'), \mu_p)$ is the Poisson space and μ_p is the Poisson measure.

A Poisson process is defined by:

$$N_t(x) = \lim_{n\to\infty} \langle x, \eta_n \rangle, \qquad x \in E'$$

where the limit is in $(L^2) = L^2(E', \mu_p)$ and the sequence $\{\eta_n\} \subset E$ converges in $L^1(\mathbb{R})$ and $L^2(\mathbb{R})$ to the indicator function $I_{(0,t]}$ if $t \geq 0$, and to $-I_{(t,0]}$ if $t < 0$. Note that if a sequence $\{\eta_n\}$ converges to $I_{(s,t]}$ in $L^1(\mathbb{R})$ and $L^2(\mathbb{R})$, then

$$\lim_{n\to\infty} \langle x, \eta_n \rangle = N_t(x) - N_s(x).$$

B. Pasik-Duncan: Stochastic Theory and Control, LNCIS 280, pp. 471–489, 2002.

Note that $N_t - N_s$ has Poisson distribution with mean $(t-s)$ and if $(s_1, t_1), \ldots,$ $(s_n, t_n]$ are disjoint, then $(N_{t_1} - N_{s_1}), (N_{t_2} - N_{s_2}), \ldots, (N_{t_n} - N_{s_n})$ are independent. It is well known that we may choose a right-continuous integer-valued version of the process. ([5]).

In Ito [16] (Theorem 2.9), it was shown that the space (L^2) has the Chaos decomposition (CD) property:

$$(L^2) = \sum_{n=0}^{\infty} \oplus (L^2)_n, \tag{1.2}$$

where

$$(L^2)_n = \left\{ \phi^n : \phi^n = \int_{\mathbb{R}^n_*} F^{(n)}(t_1, \ldots, t_n) dQ_{t_1} \ldots dQ_{t_n}, F^{(n)} \in \hat{L}^2(\mathbb{R}^n) \right\}, \tag{1.3}$$

where $Q_t = N_t - t$; $\hat{L}^2(\mathbb{R}^n)$ is the subspace of $L^2(\mathbb{R}^n)$ which consists of symmetric elements, and $\mathbb{R}^n_* = \{(t_1, \ldots, t_n) \in \mathbb{R}^n : t_i \neq t_j \text{ if } i \neq j$.

In Ito [16] and Ito and Kubo [14], the $\mathcal{U}$-transform on (L^2) was considered. If $\phi \in (L^2)$, the image $\mathcal{U}(\phi)$ is an E-functional defined by:

$$(\mathcal{U}\phi)(\xi) = L(\xi)^{-1} \int_{E'} \phi(x) e^{i\langle x, \xi \rangle} d\mu_p(x).$$

As a consequence (see [16], pg. 6–9), if $\phi \in (L^2)$ has the CD representation,

$$\phi = \sum_{n=0}^{\infty} \phi^n,$$

where

$$\phi^n = \int_{\mathbb{R}^n_*} F^{(n)}(t_1, \ldots, t_n) dQ_{t_1} \ldots, dQ_{t_n}, \qquad F^{(n)} \in \hat{L}^2(\mathbb{R}^n)$$

then

$$\mathcal{U}\phi = \sum_{n=0}^{\infty} \mathcal{U}\phi^n,$$

and

$$(\mathcal{U}\phi^n)(\xi) = \int_{\mathbb{R}^n} F^{(n)}(t_1, \ldots, t_n)(e^{i\xi(t_1)} - 1) \ldots (e^{i\xi(t_n)} - 1) dt_1 \ldots dt_n. \tag{1.5}$$

One important case which Ito considered in [16] is the Charlier polynomial functionals C_n based on E:

$$C_0(x) \equiv 1,$$

$$C_n(x; \eta_1, \ldots, \eta_n) = \frac{\partial^n}{\partial w_1 \ldots \partial w_n}$$

$$\times \exp\left[\left\langle x, \log\left(1+\sum_{j=1}^{n} w_j\eta_j\right)\right\rangle - \sum_{j=1}^{n} w_j\langle\eta_j\rangle\right]_{|w_1=\ldots=w_n=0} \tag{1.6}$$

where $\eta_1, \ldots, \eta_n \in E$ and $\langle\eta_j\rangle = \int_{-\infty}^{\infty} \eta_j(t)dt$.

The Charlier polynomial functionals have the multiple integral expressions:

$$C_n(x;\eta_1,\ldots,\eta_n) = \int \cdots \int \eta_1 \otimes \ldots \otimes \eta_n(t_1,\ldots,t_n)dQ_{t_1}\ldots dQ_{t_n} \tag{1.7}$$

and thus

$$(\mathcal{U}C_n(.,\eta_1,\ldots,\eta_n))(\xi) = \prod_{j=1}^{n} \int \eta_j(t_j)(e^{i\xi(t_j)}-1)dt_j. \tag{1.8}$$

Proposition 3.7 on page 11 in Ito [16] stated that if $\mathcal{M} \subset E$ is a complete orthonormal system in $L^2(\mathbb{R})$ and if S_n is the system of functions:

$$S_n = \{(n_1!\ldots n_k!)^{-\frac{1}{2}} C_n(x;\overbrace{\eta_1,\ldots,\eta_1}^{n_1},\ldots,\overbrace{\eta_k,\ldots,\eta_k}^{n_k})\}, \tag{1.9}$$

where $n_1 + \ldots + n_k = n$, $\eta_1, \ldots, \eta_k$ being different elements in $\mathcal{M}$, and if $S = \cup_n S_n$, then S_n and S are complete orthonormal systems in $(L^2)_n$ and (L^2) respectively.

In this paper we consider the $\mathcal{L}$-transform on (L^2). If $\phi \in (L^2)$ has the CD representation:

$$\phi = \sum_{n=0}^{\infty} \phi^n$$

$$\phi^n = \int_{\mathbb{R}_*^n} F^{(n)}(t_1,\ldots,t_n)dQ_{t_1}\ldots dQ_{t_n},$$

then $\mathcal{L}\phi$ is an E-functional defined by:

$$\mathcal{L}\phi = \sum_{n=0}^{\infty} \mathcal{L}\phi^n,$$

$$(\mathcal{L}\phi^n)(\xi) = \int_{\mathbb{R}^n} F^{(n)}(t_1,\ldots,t_n)\xi^{\otimes n}(t_1,\ldots,t_n)dt_1,\ldots,dt_n, \qquad \xi \in E.$$

An immediate consequence is that if $C_n(x;\eta_1,\ldots,\eta_n)$ is a Charlier polynomial functional, then by (1.7),

$$(\mathcal{L}C_n(.;\eta_1,\ldots,\eta_n))(\xi) = \int_{\mathbb{R}^n} \eta_1\otimes\cdots\otimes\eta_n(t_1,\ldots,t_n)\xi^{\otimes n}(t_1,\ldots,t_n)dt_1\ldots dt_n.$$

A natural consequence is that the Hida derivative on (L^2) with respect to the $\mathcal{L}$-transform coincides with the Hida derivative with respect to the $\mathcal{U}$-transform considered in Ito [16]. One advantage of studying the $\mathcal{L}$-transform is that via the $\mathcal{L}$-transform, we can introduce the Poisson normal functionals which serve as a natural and convenient tool to study the Lévy Laplacian. The $\mathcal{L}$-tranform method in this paper provides a general method to construct normal functionals and to study the Lévy-Laplacian for processes which possess the CD property. (see K. Ito [13], and Hida and Ikeda [9]). That is, if a process has the CD property, the we can introduce the $\mathcal{L}$-transform via equations (2.10) and (2.11), and hence we can introduce the normal functionals in a similar spirit as in equations (3.1)–(3.4). On the other hand, this paper about Poisson noise is unique in the sense that (i) The definition of the $\mathcal{U}$-transform is based on the Poisson measure, μ_p, and by Theorem 3.14 of this paper, the $\mathcal{U}$-transform of elements in (L^2) can be expressed in terms of the $\mathcal{L}$-transform of normal functionals, and (ii), The stochastic limit expression given in Theorem 4.33 in this paper uniquely reflects the Poisson property of the Lévy Laplacian.

This paper is divided into four main sections. In section 2 we first consider the $\mathcal{L}$-transform on (L^2), and then we extend the $\mathcal{L}$-transform with the help of Sobolev norms. In section 3 we introduce the class of Poisson normal functionals. We give a relation between the $\mathcal{U}$-transform and the $\mathcal{L}$-transform in Theorem 3.14. We then consider the Wick product of these functionals. In section 4 we consider the Hida derivatives of normal functionals and the Lévy Laplacian. Finally, we give a stochastic limit characterization of our Lévy Laplacian with the help of 2-parameter Poisson processes in Theorem 4.33.

2 The $\mathcal{L}$-Transform

For each integer $n \geq 1$, let $\{e_k^n\}_{k=1,2,\ldots}$ be an orthonormal basis of $\hat{L}^2(\mathbb{R}^n)$. If $F^{(n)} \in \hat{L}^2(\mathbb{R}^n)$, then

$$F^{(n)} = \sum_{k=1}^{\infty} \langle F^{(n)}, e_k^n \rangle_{\hat{L}^2(\mathbb{R}^n)} e_k^n. \tag{2.1}$$

Consider

$$E_k^n = \int_{\mathbb{R}_*^n} e_k^n(t_1, \ldots, t_n) dQ_{t_1} \ldots dQ_{t_n} \qquad \in (L^2)_n \tag{2.2}$$

If

$$\varphi^n = \int_{\mathbb{R}_*^n} F^{(n)}(t_1, \ldots, t_n) dQ_{t_1} \ldots dQ_{t_n} \qquad \in (L^2)_n, \tag{2.3}$$

then

$$\langle \varphi^n, E_k^n \rangle_{(L^2)_n} = E[\varphi^n E_k^n]$$

$$= n!\langle e_k^n, F^{(n)}\rangle_{\hat{L}^2(\mathbb{R}^n)}. \tag{2.4}$$

Note that

$$\varphi^n = \sum_{k=1}^{\infty}\int_{\mathbb{R}_*^n}\langle F^{(n)}, e_k^n\rangle_{\hat{L}^2(\mathbb{R}^n)} e_k^n dQ_{t_1}\dots dQ_{t_n}$$
$$= \sum_{k=1}^{\infty}\langle \varphi^n, E_k^n\rangle_{(L^2)_n} E_k^n. \tag{2.5}$$

If $k_1 \neq k_2$,

$$\langle E_{k_1}^n, E_{k_2}^n\rangle_{(L^2)_n} = 0.$$

Thus (E_k^n) forms an orthonornal basis of $(L^2)_n$. Since $(L^2)_n$ and $(L^2)_m$ are orthogonal subspaces of (L^2) for $n \neq m$, hence we can express $\varphi \in (L^2)$ as:

$$\varphi = \sum_{n=0}^{\infty}\sum_{k=1}^{\infty}\langle \varphi, E_k^n\rangle E_k^n$$

and $\{E_k^n, n = 0, 1, 2, \dots; k = 1, 2, \dots\}$ forms a complete orthonormal system for (L^2).

Definition 2.6: The $\mathcal{L}$-transform on (L^2) is defined as: for $\phi \in (L^2)$, $\mathcal{L}\phi$ is the E-functional given by:

$$(\mathcal{L}\phi)(\xi) = \sum_{n=0}^{\infty}\frac{1}{i^n}\frac{d^n}{dw^n}\int_{E^*}\sum_{j=1}^{\infty}\langle \varphi, E_j^n\rangle_{(L^2)} E_j^n e^{iw\langle x,\xi\rangle} d\mu_p(x)_{|w=0} \tag{2.7}$$

Remark 1. The above definition of the $\mathcal{L}$-transform is equivalent to:

$$(\mathcal{L}\phi)(\xi) = \sum_{n=0}^{\infty}\frac{1}{i^n}\frac{d^n}{dw^n}L(w\xi)\mathcal{U}(\sum_{j=1}^{\infty}\langle \varphi, E_j^n\rangle_{(L^2)} E_j^n)(w\xi)_{|_{w=0}}. \tag{2.8}$$

A direct computation from equation (2.7) gives the following:

Proposition 2.9. *If $\phi \in (L^2)$ has the CD representation:*

$$\varphi = \sum_{n=0}^{\infty}\varphi^n,$$
$$\varphi^n = \int_{\mathbb{R}_*^n} F^{(n)}(t_1, \dots, t_n) dQ_{t_1}\dots dQ_{t_n},$$
$$F^{(n)} \in \hat{L}^2(\mathbb{R}^n),$$

then

$$\mathcal{L}\varphi = \sum_{n=0}^{\infty}\mathcal{L}\varphi^n, \tag{2.10}$$

where

$$(\mathcal{L}\varphi^n)(\xi) = \int_{\mathbb{R}^n} F^{(n)}(t_1, \ldots, t_n)\xi^{\otimes n}(t_1, \ldots, t_n)dt_1 \ldots dt_n. \tag{2.11}$$

Remark 2. (i) Let $\{C_n\}$ be the Charlier polynomial functionals described in Section 1. An immediate consequence of Proposition 2.9 and equation (1.7) in Section 1 is:

$$\begin{aligned}&(\mathcal{L}C_n(., \eta_1, \ldots, \eta_n))(\xi)\\ &\quad= \int_{\mathbb{R}^n} \eta_1 \otimes \ldots \otimes \eta_n(t_1, \ldots, t_n)\xi^{\otimes n}(t_1, \ldots, t_n)dt_1, \ldots, dt_n.\end{aligned}$$

(ii) If we consider the subset $G \subset E$ given by:

$$G = \{\xi \in E : \sup_t \xi(t) < 1\},$$

then for $\xi \in G$,

$$(\mathcal{L}\phi)(\xi) = (\mathcal{U}\phi)(-i\log(1 + \xi(t))), \qquad \phi \in (L^2). \tag{2.12}$$

For each $n \geq 1$, let V_n be the space of E-functionals such that for each $\Psi_n \in V_n$, there exists an $F^{(n)} \in \hat{L}^2(\mathbb{R}^n)$ and $\Psi_n(\xi)$ has the representation:

$$\Psi_n(\xi) = \int_{\mathbb{R}^n} F^{(n)}(t_1, \ldots, t_n)\xi^{\otimes n}(t_1, \ldots, t_n)dt_1 \ldots dt_n. \tag{2.13}$$

From the above discussions we see that the $\mathcal{L}$-transform is the map:

$$\mathcal{L} : (L^2) \to \sum_{n=0}^{\infty} \oplus V_n$$

which is one-to-one and onto. For each n, $\mathcal{L} : (L^2)_n \to V_n$ is one-to-one and onto, and the inner product $(.,.)_{V_n}$ is induced by the inner product of $(L^2)_n$:

$$(\mathcal{L}\varphi^n, \mathcal{L}\psi^n)_{V_n} := (\varphi^n, \psi^n)_{(L^2)_n}$$

Consider an increasing sequence of positive real numbers $(\gamma_n)_{n=1,2,\ldots}$. Let $\hat{H}_n^{\frac{n+1+\gamma_n}{2}}$ be the symmetric Sobolev space with norm:

$$\|f\|_{\frac{n+1+\gamma_n}{2}} = \|(1 + x_1^2 + \ldots + x_n^2)^{\frac{n+1+\gamma_n}{4}} \tilde{f}\|_{L^2(\mathbb{R}^n)}, \tag{2.14}$$

where $\tilde{f}$ stands for the Fourier transform. Set

$$(L^2)^+_{n,\gamma_n} = \left\{\phi : \phi = \int_{\mathbb{R}^n} F^{(n)}(t_1, \ldots, t_n)dQ_{t_1} \ldots dQ_{t_n}, F^{(n)} \in \hat{H}_n^{\frac{n+1+\gamma_n}{2}}\right\} \tag{2.15}$$

and set

$$V_{n,\gamma_n}^{+} = \{\mathcal{L}\phi : \phi \in (L^2)_{n,\gamma_n}^{+}\}. \tag{2.16}$$

The spaces $(L^2)_{n,\gamma_n}^{+}$ and V_{n,γ_n}^{+} are endowed with topologies induced by $\hat{H}_n^{\frac{n+1+\gamma_n}{2}}$; i.e.,

$$\|\phi\|_{(L^2)_{n,\gamma_n}^{+}} = \|\mathcal{L}\phi\|_{V_{n,\gamma_n}^{+}} = \sqrt{n!}\|F^{(n)}\|_{\frac{n+1+\gamma_n}{2}}. \tag{2.17}$$

Let $\hat{H}_n^{-n+1+\gamma_n}$ be the dual space of $\hat{H}_n^{\frac{n+1+\gamma_n}{2}}$ with norm $\|.\|_{-\frac{n+1+\gamma_n}{2}}$.

Let $(L^2)_{n,\gamma_n}^{-}$ be the closure of $(L^2)_{n,\gamma_n}^{+}$ with respect to the norm $\sqrt{n!}\|.\|_{-\frac{n+1+\gamma_n}{2}}$.

If $\{\varphi_n\}$ is a Cauchy sequence with respect to the norm $\sqrt{n!}\|\cdot\|_{-\frac{n+1+\gamma_n}{2}}$, where

$$\varphi_n = \int_{\mathbb{R}^n} F_n(t_1,\dots,t_n)dQ_{t_1}\dots dQ_{t_n}\ ,\ F_n \in \hat{H}_n^{\frac{n+1+\gamma_n}{2}},$$

$$\varphi_m = \int_{\mathbb{R}^n} F_m(t_1,\dots,t_n)dQ_{t_1}\dots dQ_{t_n}\ ,\ F_m \in \hat{H}_n^{\frac{n+1+\gamma_n}{2}},$$

i.e., $\|\varphi_n - \varphi_m\|_{(L^2)_{n,\gamma_n}^{-}} = \|F_n - F_m\|_{-\frac{n+1+\gamma_n}{2}} \to 0$, as $n, m \to \infty$, then there exists $F_\infty \in \hat{H}_n^{-\frac{n+1+\gamma_n}{2}}$, such that

$$F_n \to F_\infty \text{ in } \hat{H}_n^{-\frac{n+1+\gamma_n}{2}}.$$

The limit of φ_n in $(L^2)_{n,\gamma_n}^{-}$ is denoted by φ_∞, and it has a symbolic representation:

$$\varphi_\infty = \int_{\mathbb{R}^n} F_\infty(t_1,\dots,t_n)dQ_{t_1}\dots dQ_{t_n}. \tag{2.18}$$

Now recall that

$$\begin{aligned} V_{n,\gamma_n}^{+} &= \{\mathcal{L}\phi : \phi \in (L^2)_{n,\gamma_n}^{+}\} \\ &= \Big\{\Phi(\xi) = \int_{\mathbb{R}^n} F^{(n)}(t_1,\dots,t_n)\xi^{\otimes n}(t_1,\dots,t_n)dt_1\dots dt_n, \\ &\qquad F^{(n)} \in \hat{H}_n^{\frac{n+1+\gamma_n}{2}}\Big\}. \end{aligned} \tag{2.19}$$

Define the norm $\|\ \|_{V_{n,\gamma_n}^{-}}$ on V_{n,γ_n}^{+} by:

$$\|\mathcal{L}\phi\|_{V_{n,\gamma_n}^{-}} := \sqrt{n!}\|F^{(n)}\|_{-\frac{n+1+\gamma_n}{2}}, \tag{2.20}$$

and let V_{n,γ_n}^{-} be the closure of V_{n,γ_n}^{+} with respect to this norm $\sqrt{n!}\|\cdot\|_{V_{n,\gamma_n}^{-}}$.

Suppose $\varphi_n \to \varphi_\infty \in (L^2)^-_{n,\gamma_n}$, where $\varphi_n \in (L^2)^+_{n,\gamma_n}$.
Then
$$\|\mathcal{L}\varphi_n - \mathcal{L}\varphi_m\|_{V^-_{n,\gamma_n}} = \|F_n - F_m\|_{-\frac{n+1+\gamma_n}{2}} \to 0.$$
where
$$\varphi_n = \int_{\mathbb{R}^n} F_n(t_1, \dots, t_n) dQ_{t_1} \dots dQ_{t_n}.$$
$$\varphi_m = \int_{\mathbb{R}_n} F_m(t_1, \dots, t_n) dQ_{t_1} \dots dQ_{t_n}.$$

Thus $\{\mathcal{L}\varphi_n\}$ is Cauchy in V^-_{n,γ_n}. Denote the limit by $\mathcal{L}\varphi_\infty$. I.e., if $\varphi_n \to \varphi_\infty \in (L^2)^-_{n,\gamma_n}$, define
$$\mathcal{L}\varphi_\infty := \lim_n \mathcal{L}\varphi_n \in V^-_{n,\gamma_n}. \tag{2.21}$$
Thus we have the Hida diagram:
$$\begin{array}{ccccc} (L^2)^+_{n,\gamma_n} & \cong V^+_{n,\gamma_n} & \cong \hat{H}_n^{\frac{n+1+\gamma_n}{2}} \\ \Downarrow & \Downarrow & \Downarrow \\ (L^2)_n & \cong V_n & \cong \hat{L}^2(\mathbb{R}^n) \\ \Downarrow & \Downarrow & \Downarrow \\ (L^2)^-_{n,\gamma_n} & \cong V^-_{n,\gamma_n} & \cong \hat{H}_n^{-\frac{n+1+\gamma_n}{2}} \end{array}$$

where the notation $\cong$ stands for isometries and $\Downarrow$ stands for continuous and dense injection.

Let
$$(L^2)^+_{(\gamma_n)} = \sum_{n=0}^{\infty} \oplus (L^2)^+_{n,\gamma_n}$$
$$(L^2)^-_{(\gamma_n)} = \sum_{n=0}^{\infty} \oplus (L^2)^-_{n,\gamma_n}$$

with norms:
$$\|\phi\|^2_{(L^2)^+_{\gamma_n}} = \sum_{n=0}^{\infty} \|\phi_n\|^2_{(L^2)^+_{n,\gamma_n}}$$
$$\|\phi\|^2_{(L^2)^-_{\gamma_n}} = \sum_{n=0}^{\infty} \|\phi_n\|^2_{(L^2)^-_{n,\gamma_n}}$$

Thus we have
$$(L^2)^+_{(\gamma_n)} \Rightarrow (L^2) \Rightarrow (L^2)^-_{(\gamma_n)}.$$

Sometimes, for simplicity, we omit writing the (γ_n) when the content is clear.

Let

$$V^+ = \{\mathcal{L}\phi : \phi \in (L^2)^+\}.$$
$$V^- = \{\mathcal{L}\phi : \phi \in (L^2)^-\}.$$

with the $\mathcal{L}$-transform naturally extended to $(L^2)^-$.

3 The Normal Functionals

The class of normal functionals plays an important role in the Gaussian white noise case. (see, e.g. Kuo [21]). In this section we utilize the extended $\mathcal{L}$-transform discussed in the last section to construct the Poisson normal functionals. Consider, for distinct real numbers $t_1, \dots, t_m$ and non-negative integers $n_1, \dots, n_m$, and for $\xi \in E$,

$$\prod_{j=1}^{m} \xi(t_j)^{n_j} = \int_{\mathbb{R}^n} \prod_{j=1}^{m} (\delta_{t_j}^{\otimes n_j}(u_1^j, \dots, u_{n_j}^j)\xi^{\otimes n_j}(u_1^j, \dots, u_{n_j}^j))du_1^1 \dots du_{n_m}^n \tag{3.1}$$

where $n_1 + \dots + n_m = n$.

Write

$$\sharp\dot{Q}(t_1)^{n_1} \dots \dot{Q}(t_m)^{n_m}\sharp = \mathcal{L}^{-1}\left(\prod_{j=1}^{m} \xi(t_j)^{n_j}\right). \tag{3.2}$$

Next, if $h \in L^2(\mathbb{R})$, define

$$\begin{aligned}
&\int_{\mathbb{R}} h(t)\sharp\dot{Q}(t)^n\sharp dt \\
&\quad = \mathcal{L}^{-1}\left(\int_{\mathbb{R}} h(t)\xi(t)^n dt\right) \\
&\quad = \mathcal{L}^{-1}\left(\int_{\mathbb{R}^n}\int_{\mathbb{R}} h(t)\delta_t^{\otimes n}(u_1, \dots, u_n)dt\xi^{\otimes n}(u_1, \dots, u_n)du_1, \dots, du_n\right)
\end{aligned} \tag{3.3}$$

More generally, if $h \in L^2(\mathbb{R}^m)$, define

$$\begin{aligned}
&\int_{\mathbb{R}^m} h(t_1, \dots, t_m)\sharp\dot{Q}(t_1)^{n_1} \dots \dot{Q}(t_m)^{n_m}\sharp dt_1 \dots dt_m \\
&\quad = \mathcal{L}^{-1}\left(\int_{\mathbb{R}^m} h(t_1, \dots, t_m)\xi(t_1)^{n_1} \dots \xi(t_m)^{n_m} dt_1 \dots dt_m\right) \\
&\quad = \mathcal{L}^{-1}\Bigg(\int_{\mathbb{R}^N}\int_{\mathbb{R}^m} h(t_1, \dots, t_m)\prod_{j=1}^{m} \delta_{t_j}^{n_j}(u_1^j, \dots, u_{n_j}^j)dt_1 \dots dt_m
\end{aligned}$$

$$\times \prod_{j=1}^{m} \xi^{\otimes n_j}(u_1^j, \dots, u_{n_j}^j) du_1^1 \dots du_{n_m}^m \Bigg), \tag{3.4}$$

where $n_1 + \ldots + n_m = N$.

Remark 3. (i) In (3.3), if we take $h(t) = \delta_{t_0}(t)$, then the RHS of (3.3) is just $\mathcal{L}^{-1}(\xi(t_0)^n)$ and according to (3.2), we have defined:

$$\int_{\mathbb{R}} \delta_{t_0}(t) \sharp \dot{Q}(t)^n \sharp dt = \sharp \dot{Q}(t_0)^n \sharp.$$

(ii) From the discussions in Section 2, we note that the norm of

$$\int_{\mathbb{R}^m} h(t_1, \dots, t_m) \sharp \dot{Q}(t_1)^{n_1} \dots \dot{Q}(t_m)^{n_m} \sharp dt_1 \dots dt_m$$

given in (3.4) is

$$\| \int_{\mathbb{R}^m} h(t_1, \dots, t_m) \prod_{j=1}^{m} \delta_{t_j}^{n_j}(u_1^j, \dots, u_{n_j}^j) dt_1 \dots dt_m \|_{\hat{H}_N^{-\frac{N+1+\gamma_N}{2}}}$$

Definition 3.5. A normal functional is a functional of the form:

$$\varphi(x) = \sum_{n_1=0}^{\infty} \cdots \sum_{n_m=0}^{\infty} \int_{\mathbb{R}^m} h_{n_1}, \dots, n_m(t_1, \dots, t_m) \sharp \dot{Q}(t_1)^{n_1} \dots \dot{Q}(t_m)^{n_m} \sharp \times dt_1 \dots dt_m \tag{3.6}$$

$\varphi(x)$ in (3.6) is said to have under m. The norm of $\varphi(x)$ is given by:

$$|||\varphi(x)|||_{\mathcal{N}}^2 = \sum_{n_1=0}^{\infty} \cdots \sum_{n_m=0}^{\infty} \Bigg\| \int h_{n_1 \dots n_m}(t_1, \dots, t_m) \times \prod_{j=1}^{m} \delta_{t_j}^{\otimes n_j}(u_1^j, \dots, u_{n_j}^j) \times dt_1 \dots dt_m \Bigg\|_{\hat{H}_{N_m}^{-\frac{N_m+1+\gamma_{N_m}}{2}}}, \tag{3.7}$$

where $N_m = n_1 + \ldots + n_m$, and $\mathcal{N}$ denotes the class of all normal functionals.

Definition 3.8. Let φ, ψ be two normal functionals. The wick product, denoted by $\varphi \diamond \psi$, of φ and ψ, is defined as:

$$\varphi \diamond \psi = \mathcal{L}^{-1}(\mathcal{L}\varphi \mathcal{L}\psi). \tag{3.9}$$

Remark 4. From Definition (3.8), if

$$\varphi(x) = \sum_{n_1=0}^{\infty} \cdots \sum_{n_m=0}^{\infty} \int_{\mathbb{R}^m} h_{n_1 \dots n_m}(t_1 \dots t_m) \sharp \dot{Q}(t_1)^{n_1} \dots \dot{Q}(t_m)^{n_m} \sharp \times dt_1 \dots dt_m,$$

$$\psi(x) = \sum_{k_1=0}^{\infty} \cdots \sum_{k_e=0}^{\infty} \int_{\mathbb{R}^e} g_{k_1 \dots k_e}(t_1 \dots t_e) \sharp \dot{Q}(t_1)^{k_1} \dots \dot{Q}(t_e)^{k_e} \sharp dt_1 \dots dt_e,$$

then

$$\begin{aligned}\varphi(x) \diamond \psi(x) = \sum_{k_1} \cdots \sum_{k_e} \sum_{n_1} \cdots \sum_{n_m} \int_{\mathbb{R}^{m+e}} & h_{n_1 \dots n_m}(t_1, \dots t_m) \\ & \times g_{k_1 \dots k_e}(s_1, \dots, s_e) \\ & \times \sharp \dot{Q}(t_1)^{n_1} \dots \dot{Q}(t_m)^{n_m} \\ & \times \dot{Q}(s_1)^{k_1} \dots \dot{Q}(s_e)^{k_e} \sharp \\ & \times dt_1 \dots dt_m ds_1 \dots ds_e \end{aligned} \tag{3.10}$$

From Section 1 we recall that if $\varphi \in (L^2)$ which has the CD representation:

$$\varphi = \sum \varphi^n \tag{3.11}$$

$$\varphi^n = \int_{\mathbb{R}_*^n}^n F^n(t_1 \dots t_n) dQ_{t_1} \dots dQ_{t_n}, \tag{3.12}$$

then the $\mathcal{U}$-transform of φ considered in Itô [16] is given by:

$$\mathcal{U}\varphi = \sum \mathcal{U}\varphi^n,$$

$$(\mathcal{U}\phi^n)(\xi) = \int_{\mathbb{R}^n} F^{(n)}(t_1 \dots t_n) \prod_{j=1}^{n} (e^{i\xi(t_j)} - 1) dt_1 \dots dt_n. \tag{3.13}$$

Next we present a theorem which gives a representation of $\mathcal{U}\varphi$ in terms of the $\mathcal{L}$-transform of normal functionals for $\varphi \in (L^2)$.

Theorem 3.14. *Suppose $\varphi \in (L^2)$ is given by (3.11) and (3.12), and for each n, each $F^{(n)}$ vanishes outside a bounded 'cube' T^n, where $T \subset \mathbb{R}$ is a bounded interval, then*

$$(\mathcal{U}\varphi^n)(\xi) = \sum_{j_1=1}^{\infty} \cdots \sum_{j_n=1}^{\infty} \frac{i^{j_1 + \dots + j_n}}{j_1! \dots j_n!} \times \mathcal{L}\biggl(\int_{\mathbb{R}_*^n} F^{(n)}(t_1, \dots, t_n) \sharp \dot{Q}(t_1)^{j_1} \dots \dot{Q}(t_n)^{j_n} \sharp$$

$$\times dt_1 \dots dt_n \Bigg). \tag{3.15}$$

Proof. Since $(e^{i\xi(t_1)} - 1) \dots (e^{i\xi(t_n)} - 1)$

$$= \sum_{j_1=1}^{\infty} \cdots \sum_{j_n=1}^{\infty} \frac{i^{j_1+\dots+j_n}}{j_1! \dots j_n!} \xi(t_1)^{j_1} \dots \xi(t_n)^{j_n}.$$

Noting that

$$\sum_{j_1=1}^{\infty} \cdots \sum_{j_n=1}^{\infty} \frac{1}{j_1! \dots j_n!} \int_{T^n} |F^n| |\xi(t_1)^{j_1} \dots \xi(t_n)^{j_n}| dt$$
$$\leq \sum_{j_1=1}^{\infty} \cdots \sum_{j_n=1}^{\infty} \frac{1}{j_1! \dots j_n!} \int_{T^n} |F^n(\mathbf{t})| d\mathbf{t} \ . \ \|\xi\|_\infty^{j_1} \dots \|\xi\|_\infty^{j_n}$$
$$\leq \|F^n\|_{L^1(T^n)} \times e^{n\|\xi\|_\infty}.$$

Thus applying Dominated Convergence theorem we obtain our theorem.

4 The Lévy Laplacian on Normal Functionals

For $\phi \in (L^2)$, define the operator ∂_t as:

$$\partial_t \phi = (\mathcal{L}^{-1} \frac{\partial}{\delta \xi(t)} \mathcal{L}) \phi \tag{4.1}$$

Then using the chaos decomposition expansion of $\phi(x) = C_n(x; \eta_1, \dots, \eta_n)$, it is easy to check that

$$\partial_t C_n(x; \eta_1, \dots, \eta_n) = C_n(x + \delta_t; \eta_1, \dots, \eta_n) - C_n(x; \eta_1, \dots, \eta_n)$$

for a Charlier polynomial functional $C_n(x; \eta_1, \dots, \eta_n)$. Since the Charlier polynomials span (L^2), hence for any $\phi \in (L^2)$,

$$\partial_t \phi(x) = \phi(x + \delta_t) - \phi(x). \tag{4.2}$$

Definition 4.3. Let $\mathcal{N}'$ be define as:

$$\mathcal{N}' = \left\{ \Phi : \Phi = \sum_{n=1}^{\infty} \int_{\mathbb{R}} h_n(t) \sharp \dot{Q}(t)^n \sharp dt : h_n \in L^2(\mathbb{R}) \text{ for each } n \geq 1 \right\}. \tag{4.4}$$

Let $\Phi = \sum_{n=1}^{\infty} \int_{\mathbb{R}} h_n(t) \sharp \dot{Q}(t)^n \sharp dt$ satisfy:

$$\sum_{n=1}^{\infty} \left\| n \int_{\mathbb{R}} \delta_t(s) h_n(s) \delta_s^{\otimes n-1}(u_1, \dots, u_{n-1}) ds \right\|^2_{\hat{H}_{n-1}^{-\frac{n+\gamma_n}{2}}} < \infty. \tag{4.5}$$

then

$$\partial_t \Phi = \sum_{n=1}^{\infty} n h_n(t) \sharp \dot{Q}(t)^{n-1} \sharp$$
$$= \sum_{n=1}^{\infty} \int_{\mathbb{R}}^{n} \delta_t(s) h_n(s) \sharp \dot{Q}(s)^{n-1} \sharp ds \tag{4.6}$$

i.e.

$$\mathcal{L}(\partial_t \Phi) = \sum_{n=1}^{\infty} \int_{\mathbb{R}} n\delta_t(s) h_n(s) \xi(s)^{n-1} ds$$
$$= \sum_{n=1}^{\infty} \int_{\mathbb{R}^{n-1}} n\delta_t(s) h_n(s) \delta_s^{\otimes n-1}(u_1, \dots, u_{n-1})$$
$$\times ds \xi^{\otimes n-1}(u_1, \dots, u_{n-1}) du_1 \dots du_{n-1}. \tag{4.7}$$

Thus if (4.5) is satisfied, then

$$\partial_t \Phi \in (L^2)^-.$$

If

$$\Phi = \sum_{n=1}^{\infty} \int_{\mathbb{R}} h_n(t) \sharp \dot{Q}(t)^n \sharp dt$$
$$\Psi = \sum_{m=1}^{\infty} \int_{\mathbb{R}} f_m(t) \sharp \dot{Q}(t)^m \sharp dt$$

Then

$$\Phi \diamond \Psi = \sum_{n=1}^{\infty} \sum_{m=1}^{\infty} \int_{\mathbb{R}^2} h_n(t) f_m(s) \sharp \dot{Q}(t)^n \dot{Q}(s)^m \sharp ds dt. \tag{4.8}$$

where the corresponding condition (3.7) is satisfied for $\Phi \diamond \Psi$. Using the definition (4.1), we can obtain the following:

$$\partial_t(\Phi \diamond \Psi) = (\partial_t \Phi) \diamond \Psi + \Phi \diamond (\partial_t \Psi) \tag{4.9}$$

provided the corresponding norms for $(\partial_t \Phi) \diamond \Psi$ and $\Phi \diamond (\partial_t \Psi)$ are finite. I.e., ∂_t is a derivation with respect to the Wick product $\diamond$. Next suppose $\Phi = \sum_{n=1}^{\infty} \int_{\mathbb{R}} h_n(t) \sharp \dot{Q}(t)^n \sharp dt \in \mathcal{N}_T^1 = \{\Phi \in \mathcal{N}' : h_n$ vanishes outside $T = [a, b] \subset \mathbb{R}$, for all $n \geq 1\}$

Write

$$H(\xi) = (\mathcal{L}\Phi)(\xi) = \sum_{n=1}^{\infty} \int_{\mathbb{R}} h_n(t) \xi(t)^n dt. \tag{4.10}$$

Then the functional derivative $H'(\xi)$ is:

$$H'(\xi)(t) = \sum_{n=1}^{\infty} n h_n(t) \xi(t)^{n-1} \tag{4.11}$$

Write

$$H'(\xi)(\eta) = \langle H'(\xi), \eta\rangle = \sum_n n \int_T h_n(t)\xi^{n-1}(t)\eta(t)dt. \tag{4.12}$$

Next,

$$H''(\xi)(t,\tau) = \delta_t(\tau)\sum_n n(n-1)h_n(\tau)\xi(\tau)^{n-2} \tag{4.13}$$

I.e.,

$$\begin{aligned} H''(\xi)(\eta\times\psi) &= \sum_n n(n-1)\int_T h_n(t)\xi(t)^{n-2}\eta(t)\psi(t)dt \\ &= \int_T H''_L(\xi,t)\eta(t)\psi(t)dt \end{aligned} \tag{4.14}$$

where

$$H''_L(\xi,t) = \sum_n n(n-1)h_n(t)\xi(t)^{n-2} \tag{4.15}$$

Definition 4.16. The Lévy Laplacian acting on $H(\xi)$ is defined as:

$$(\Delta_L^T H)(\xi) = \frac{1}{|T|}\int_T H''_L(\xi,t)dt \tag{4.17}$$

If $\Phi = \mathcal{L}^{-1}(H)$, then the Lévy Laplacian acting on Φ is defined by:

$$({}^{\mathcal{L}}\Delta_L^T)(\Phi) = \mathcal{L}^{-1}(\Delta_L^T H). \tag{4.18}$$

Remark 5. **(i)** We have to check $({}^{\mathcal{L}}\Delta_L^T)(\Phi) \in (L^2)^-$ by requiring the corresponding norm being finite.
(ii) If $\Phi = \mathcal{L}^{-1}(H(\xi)) = \int_T h(t)\sharp\dot{Q}(t)\sharp dt$, then $({}^{\mathcal{L}}\Delta_L^T)(\Phi) \equiv 0$.

Recall from Theorem 3.14 that if $\phi \in (L^2)$, then $\mathcal{U}\phi$ can be expressed as an infinite series of normal functionals. This allows us to define the Lévy Laplacian in terms of the $\mathcal{U}$-transform:

Definition 4.19. Given $\phi \in (L^2)$, the Lévy Laplacian ${}^{\mathcal{U}}\Delta_L^Z$ will respect to the $\mathcal{U}$-transform of ϕ is defined as:

$${}^{\mathcal{U}}\Delta_L^Z(\phi) = \mathcal{U}^{-1}(\Delta_L^Z(U\phi)(\xi)) \tag{4.20}$$

Theorem 4.21. *Suppose $\phi \in (L^2)$ is given by:*

$$\phi = \int_{\mathbb{R}} F(t)dQ_t,$$

where F vanishes outside the interval $Z = [A,B] \subset [0,\infty)$. Then ϕ satisfies the following:

$${}^{\mathcal{U}}\Delta_L^Z(\phi) + \frac{1}{|Z|}\phi = \frac{-1}{|Z|}\int_Z F(t)dt. \tag{4.22}$$

Proof. WLOG, take $Z = [0,1]$, write

$$V(\xi) = \mathcal{U}\phi(\xi) = \int_0^1 F(t)(e^{i\xi(t)} - 1)dt. \tag{4.23}$$

Note $F \in L^2([0,1]) \Rightarrow F \in L^1([0,1])$, and

$$\sum_{k=1}^{\infty} \frac{1}{k!} \int_o^1 |F(t)\xi(t)^k| dt \leq \sum_{k=1}^{\infty} \frac{1}{k!} \int_o^1 |F(t)|dt. \|\xi\|_\infty^k$$
$$\leq \|F\|_{L^1([0,1])} \times e^{\|\xi\|_\infty}$$

Thus

$$V(\xi) = \int_0^1 F(t) \sum_{k=1}^{\infty} \frac{i^k}{k!} \xi(t)^k dt = \sum_{k=1}^{\infty} \frac{i^k}{k!} \int_0^1 F(t)\xi(t)^k dt.$$

Thus

$$\begin{aligned} V(\xi + \lambda\eta) - V(\xi) &= \int_0^1 F(t)(e^{i(\xi(t)+\lambda\eta(t))} - 1 - e^{i\xi(t)} + 1)dt \\ &= \int_0^1 F(t)e^{i\xi(t)}(e^{i\lambda\eta(t)} - 1)dt \\ &= \sum_{\nu=1}^{\infty} \frac{i^\nu}{\nu!} \lambda^\nu \int_0^1 F(t)e^{i\xi(t)}\eta(t)^\nu dt \end{aligned}$$

Thus

$$\begin{aligned} V'(\xi)(\eta) &= \lim_{\lambda \to 0} \frac{V(\xi + \lambda\eta) - V(\xi)}{\lambda} \\ &= i \int_0^1 F(t)e^{i\xi(t)}\eta(t)dt. \end{aligned} \tag{4.24}$$

Similarly, we can show that

$$\begin{aligned} V''(\xi)(\eta, \zeta) &= i^2 \int_0^1 F(t)e^{i\xi(t)}\eta(t)\zeta(t)dt \\ &= i^2 \sum_{\nu=0}^{\infty} \frac{i^\nu}{\nu!} \int_0^1 F(t)\xi(t)^\nu \eta(t)\zeta(t)dt \end{aligned} \tag{4.25}$$

Thus $\tilde{\Delta}_L V(\xi) = -\int_0^1 F(t)e^{i\xi(t)}dt = -V(\xi) - \int_0^1 F(t)dt$.

I.e., ${}^{\mathcal{U}}\Delta_L^Z(\phi) = -\phi - \int_0^1 F(t)dt$.

It turns out that the Lévy Laplacian ${}^U\Delta_L^Z$ has numerous nice properties in addition to (4.22). (See also Saito and Tsoi [22]). Also if we treat $\phi(t) = \int_0^t F(t,u)dN_u$ as the interest rate in a financial market, then equation (4.22) gives us the interpretation of the Lévy Laplacian acting on the interest rate.

Finally, we give a stochastic limit characterization of the Lévy Laplacian. We first recall some facts from Kuo [21]:

Definition 4.26. Suppose Z is a finite interval in $\mathbb{R}$, $Z=[A,B]$, $A\neq B$. An orthonormal basis $\{e_k; k\geq 1\}$ for $L^2(Z)$ is called equally dense if

$$\frac{1}{m}\sum_{k=1}^{m} e_k^2(t) \ \rightarrow\ \frac{1}{B-A} \quad \text{in } L^2(Z). \tag{4.27}$$

The basis is called uniformly bounded if

$$\sup_{k\geq 1}\sup_{t\in Z}|e_k(t)|<\infty. \tag{4.28}$$

As in the proof of Theorem 12.15 in Kuo [21], we can show that for any equally dense and uniformly bounded orthonormal basis $\{e_k; k\geq 1\}$ for $L^2(Z)$, if

$$F(\xi)=\sum_{n=0}^{\infty}\int_{\mathbb{R}} h_n(t)\xi(t)^n dt$$

where each h_n vanishes outside Z, then

$$\Delta_L^Z F(\xi)=\lim_{m\to\infty}\frac{1}{m}\sum_{k=1}^{m} F''(\xi)(e_k,e_k), \tag{4.29}$$

Let $(\Omega,\mathcal{F},P)$ be a complete probability space and $T=\{(t_1,t_2): 0\leq t_i\leq \infty,\ i=1,2\}$. For each $k=1,2,\dots$, suppose $\{N_t^k : t\in T\}$ is a two-parameter standard Poisson process, and that the sequence $\{N^k\}_{k=1,2,\dots}$ is an independent sequence of Poisson processes. In particular, if

$$Q^k(t)=N^k(t)-E[N^k(t)], \qquad t=(t_1,t_2)\in T, \tag{4.30}$$

i.e.,

$$Q^k(t_1,t_2)=N^k(t_1,t_2)-t_1t_2, \tag{4.31}$$

then

$$\begin{aligned} E[Q^k(t_1,t_2)]&=0,\\ E[(Q^k(t_1,t_2))^2&=t_1t_2. \end{aligned} \tag{4.32}$$

and Q^j and Q^k are independent if $j\neq k$.

Theorem 4.33. *Let $Z=[A,B]\subset(0,\infty)$, $|Z|=B-A$. Suppose $\{e_k\}$ is an orthonormal basis for $L^2(Z)$ which is equally dense and uniformly bounded. Let*

$$F(\xi)=\sum_{n=0}^{\infty}\int_{\mathbb{R}} h_n(t)\xi(t)^n dt,$$

which each $h_n \in L^2(\mathbb{R})$ vanishes outside Z. Suppose F has the following expension:

$$F(\xi + \eta) = F(\xi) + F'(\xi)(\eta) + \frac{1}{2}F''(\xi)(\eta, \eta) + R(\eta), \qquad (4.34)$$

where $\mathbb{R}(\eta) = 0(|\eta|_0^2)$. Then

$$\Delta_L^Z F(\xi) = \lim_{m\to\infty} \lim_{t\to 0} \frac{E[F(\xi + M_t^{(m)})] - F(\xi)}{t}, \qquad (4.35)$$

where $M_t^{(m)}$ is defined as:

$$M_t^{(m)}(u) = 0 \qquad \text{if } u = 0,\ t \geq 0.$$

$$M_t^m(u) = \frac{1}{u^{1/2}m^{1/2}} \sum_{k=1}^m Q^k(t,u)e_k(u), \qquad u > 0,\ t \geq 0. \qquad (4.36)$$

where $\{Q^k\}$ is a sequence of independent, 2-parameter standard centered Poisson processes.

Proof.

$$F''(\xi)(M_t^{(m)},\ M_t^{(m)}) = \int_Z F_L''(\xi, u) \times \frac{1}{m}\sum_{k=1}^m\sum_{j=1}^m \frac{Q^k(t,u)Q^j(t,u)}{u} e_k(n)e_j(u)du$$

$$\begin{aligned} E[F''(\xi)(M_t^{(m)}, M_t^{(m)})] &= \frac{1}{m}\sum_{k=1}^m \int_Z F_L''(\xi,u)\frac{E[(Q^k(t,u))^2]}{u}e_k^2(u)du \\ &= \frac{t}{m}\sum_{k=1}^m \int_Z F_L''(\xi,u)e_k^2(u)du \\ &= \frac{t}{m}\sum_{k=1}^m F''(\xi)(e_k, e_k). \end{aligned} \qquad (4.37)$$

Thus

$$\lim_{m\to\infty}\lim_{t\to 0} \frac{E[F''(\xi)(M_t^{(m)}, M_t^{(m)})]}{t} = \Delta_L^Z F(\xi). \qquad (4.38)$$

It is easy to see that

$$E[F'(\xi)(M_t^{(m)})] = 0 \qquad \forall\, t > 0.$$

Finally,

$$E[|M_t^{(m)}|_0^2] = E[\int_Z \frac{1}{m}\sum_k\sum_j \frac{Q^j(t,u)Q^k(t,u)}{u} e_k(u)e_j(u)du]$$

$$= \frac{1}{m} \sum_{k=1}^{m} \int_Z e_k^2(u) du$$

$$= t \tag{4.39}$$

Thus $E[R(M_t^{(m)})] = o(t)$.

Thus the theorem is proved.

References

1. Albeverio, S., Hida, T., Potthoff, J., Röckner, M. and Streit, L. (1990), Dirichlet forms in terms of white noise analysis I-Construction and QFT examples. Rev. Math. Phys. 1, 291-312.
2. Gel'fand, I.M. and Vilenkin, N.Yȧ. (1968), Generalized Functions, Vol. IV, Academic Press, New York and London.
3. Hida, T. (1982), White noise analysis and its applications. In: Proc. Int. Mathematical Conf. ed. by L.H.Y. Chen et al. North-Holland, Amsterdam, 43-48.
4. Hida, T. (1975), Analysis of Brownian Functionals. Carleton Mathematical Lecture Notes No.13, Carleton University, Ottawa.
5. Hida, T. (1970), Stationary Stochastic Processes. Princeton University Press.
6. Hida, T. (1980), Brownian Motion. Springer Verlag, New York, Heidelberg, Berlin.
7. Hida, T. (1997), Random Fields and Quantum Dynamics. Preprint.
8. Hida, T. (1997), White Noise Analysis with Special Emphasis on Applications to Biology, World Scientific.
9. Hida, T. and Ikeda, N. (1967), Analysis on Hilbert Space with Reproducing Kernel Arising from Multiple Wiener Integral. In: Proc. 5th Berkeley Symp. on Math. Stat. and Prob. Vol. II, Part 1, 117-143.
10. Hida, T., Kuo, H.-H., Potthoff, J. and Streit, L. (eds.) (1990), White Noise-Mathematics and Applications. World Scientific, Singapore.
11. Hida, T., Potthoff, J. and Streit, L. (1989), White noise analysis and applications. In: Mathematics+ Physics, Vol. 3, L. Streit (ed.). World Scientific, Singapore.
12. Hida, T. Kuo, H.-H., Potthoff, J. and Streit, L.(1993), White Noise-An Infinite Dimensional Calculus. Kluwer Academic Publishers, Dordrecht-Boston-London.
13. Itô, K.(1956), Spectral Type of the Shift Transformation of Differential Processes with Stationary Increments. Trans. Amer. Math. Soc. 81, 253-263.
14. Itô, Y and Kubo, I. (1988), Calculus on Gaussian and Poisson White Noise. Nagoya Math. Journal, vol.111, 41-84.
15. Itô, Y. and Kubo, I. (1997). The Digest of "Calculus on Gaussian and Poisson White Noise". Preprint.
16. Itô, Y. (1988), Generalized Poisson Functionals. Probability Theory and Related Fields, 77, 1-28.
17. Itô, Y. (1988), Differential Operators Arising from Translation of Poisson Functionals. Austral. J. Statistics, 30A, 247-258.
18. Jeulin, T. and Yor, M. (1979), Inegalité de Hardy, semimartingales, et faux-amis. Springer LN Math 721, 332-359.

19. Kondratiev, Y.G., Streit, L., Westerkamp, W. and Yan, J.-A. (1995), Generalized Functions in Infinite Dimensional Analysis. International Institute for Advanced Studies, IIAS Report No.1995-002, Kyoto, Japan.
20. Kondratiev, Y.G., Da Silva,J.L., Streit,L., and Us,G.H. (1998), Analysis on Poisson and Gamma Spaces, Infinite Dim. Anal., Quantum Prob. Rel. Topics, Vol.1, 91-117.
21. Kuo, H.H. (1996), White Noise Distribution Theory. CRC Press, New York.
22. Saitô, K. and Tsoi, A (1999), The Lévy Laplacian Acting on Poisson Noise Functionals. Infinite Dim. Anal., Quantum Prob. Rel. Topics, Vol.2, No.4, 503-510.
23. Us, G.F. (1995), Dual Appell Systems in Poissonian Analysis. Preprint, Kyiv Unviersity, Kyiv, Ukraine.
24. Wu, L.(1987), Construction de l'opérateur de Malliavin sus l'espace de Poisson. Séminaire de Probabilités XXI, Springer LN Math 1247, 100-113.

Probabilistic Rate Compartment Cancer Model: Alternate versus Traditional Chemotherapy Scheduling

John J. Westman[1], Bruce R. Fabijonas[2], Daniel L. Kern[3], and Floyd B. Hanson[4]

[1] Department of Mathematics, University of California, Box 951555, Los Angeles, CA 90095-1555, USA
[2] Department of Mathematics, Southern Methodist University, P.O. Box 750156, Dallas, TX 75275-0156, USA
[3] Institute for Mathematics and Its Applications, University of Minnesota, 207 Church Street SE, Minneapolis, MN 55455, USA
[4] Laboratory for Advanced Computing, University of Illinois at Chicago, 851 Morgan St.; M/C 249, Chicago, IL 60607-7045, USA

Abstract. A four-compartment model for the evolution of cancer based on the characteristics of the cells is presented. The model is expanded to account for intrinsic and acquired drug resistance. This model can be explored to see the evolution of drug resistance starting from a single cell. Numerical studies are performed illustrating the palliative nature of chemotherapeutic treatments. Computational results are given for traditional treatment schedules. An alternate schedule for treatments is developed increasing the life expectancy and quality of life for the patient. A framework for the alternate scheduling is presented that addresses life expectancy, quality of life, and risk of metastasis. A key feature of the alternate schedule is that information for a particular patient can be used resulting in a personalized schedule of treatments. Alternate scheduling is compared to traditional treatment scheduling.

1 Introduction

Various treatment options may be open to a cancer patient such as surgery, chemotherapy, radiotherapy, and immunotherapy. These treatment modalities may be used in any combination and depend on the type and extent of the cancer in the patient. One of the most common modalities is chemotherapy which may be a primary treatment or a subsequent treatment following surgery as a suppressive therapy, see for example [10]. Chemotherapy is palliative in nature and most often cannot lead to a cure for the cancer due to drug resistance which may be either intrinsic, i.e. naturally occurring, or acquired in the presence of a cytotoxic or chemotherapeutic agent .

The four-compartment model used in this paper, described in detail in [29], allows for the exploration of treatment schedules so that a schedule may be determined that provides for the patient a greater life expectancy and

B. Pasik-Duncan: Stochastic Theory and Control, LNCIS 280, pp. 491–506, 2002.

higher quality of life such as the one proposed in [25]. A stochastic optimal scheduling or control problem can be used to determine such alternate schedules for treatments. In formulating a stochastic optimal control problem, the form for the dynamics of the system or for the cost functional can be structured in such a way as to obtain diverse results. The objectives that would be used for the cost functionals for traditional and alternate treatment scheduling are radically different. In traditional methods, as soon as the cancer is detected, the goal is to drive the cancer into remission as soon as possible. The goal of the alternate method is to increase the life expectancy of the patient and to maintain a minimum specified level for health and quality of life while ensuring that the risk of metastasis is kept to acceptable levels. Due to the nature of chemotherapeutic drugs and the development of resistance to them, both treatment strategies are subject to constraints based on the toxicity of the cytotoxic agent and the total number of treatments that can be administered. Additionally, traditional scheduling considers the time when a collection of treatments should be given whereas the alternate schedule uses individual treatments. The authors will do a more complete investigation of the optimal scheduling control problem in a future paper.

Two of the most crucial concerns with the use of chemotherapy are toxicity and drug resistance. The development of drug resistance occurs at the cellular level via several mechanisms [19,20]. Toxicity limits the dose and frequency by which treatments may be administered. Drug resistance, whether intrinsic or acquired, limits the effectiveness of the treatments. Therefore, if the roles of toxicity and resistance can be understood leading to a model which would relate their effects to the evolution of the cancer, then this information can be used to select appropriate treatments so as to minimize the spread of cancer and resistance while adhering to toxicity limits for a given patient with a given cancer. One way to reduce the development of drug resistance of a cancer to a cytotoxic agent is to supplement the treatment with additional treatment modes, such as radiotherapy or another cytotoxic agent which is not cross resistant with the first agent. An extension of the model used in this paper can be found in [28] which considers multiple cytotoxic agents to reduce the effects of drug resistance.

Typically, a reduction in cancer is seen with the initial administration of cytotoxic agents, but eventually the tumor begins to grow and expand in the presence of the agent. This implies that the available effectiveness of the agent is limited to a finite number of applications after which it can no longer control the growth of the cancer. The application of cytotoxic agents also destroys normal or good cells which negatively impacts the patient's health. Therefore the determination whether or not to administer the cytotoxic agent needs to consider effects on the cancer as well as that of normal tissue, see for example [1,2]. Traditional treatment schedules for chemotherapy consist of several applications of the cytotoxic agent relatively close together. After a treatment is administered, both the cancer cells and normal cells begin to grow again. The normal or good tissue used to measure the toxicity of the treatment is typically the white blood cell count, and indirectly the bone marrow cells responsible for it's production, see [3,11] for example. Clinically, it is easy to measure the white blood cell count before treatment; if the level of white blood cells is too small, the treatment may be delayed or given at a reduced dose.

Alternate optimal scheduling for cytotoxic agents has been the subject of many papers. However the dynamics employed for the evolution and treatment of the cancer have been lacking in the sense that, to the best of our knowledge, no source considers the heterogeneous nature of the cancer, the development of drug resistance, and appropriate Gompertzian growth dynamics. See [5,9,16–18,21–24,27] for various examples of optimal treatment scheduling for cytotoxic agents.

In Section 2 a summary of the compartment model for cancer subject to chemotherapy [29] is presented that accounts for the heterogenous nature of cancer and the evolution of drug resistance. A discussion of the alternate treatment scheduling is presented in Section 3 and a numerical example is given in Section 4 which is meant to provide a proof of concept for the optimal alternate scheduling.

2 Four-Compartment Model for Cancer Treatment

The following material is a summary of the four-compartment model presented in [29]. The compartments represent the heterogeneous nature of cancer subject to the development of drug resistance to a single cytotoxic agent. A key feature of this model is that the heterogeneous nature of the cancer as well as drug resistance are taken into account. To the best of the authors' knowledge, the only other treatment model to incorporate these factors is by Birkhead et al. [4] which is a deterministic system governed by exponential dynamics with limited interactions between the various compartments. In the model used here, more realistic Gompertzian dynamics (see [6,7,15]) are employed with all possible transitions between compartments allowed and the transition rates between compartments are both probabilistic in nature and dependent on the subpopulation sizes and time.

Cancer consists of three primary types of cells: the proliferating fraction, clonogenic fraction, and end cells (see for example [8]). End cells, denoted by E, primarily consist of somatic tissue, vascular and endothelial support cells, and necrotic tissue. These cells cannot further propagate the cancer directly, but may play a fundamental support role in the development of the cancer. The proliferating or growth fraction cells, denoted by P, are actively dividing. After the completion of mitosis by the parent cell, the daughter cells have a specified probability of becoming any of the primary types of cells which depends on the relative number of cells for each of the groups. The clonogenic cells, denoted by C, are quiescent or dormant cancer stem cells in the G_0 phase of the cell life cycle. With the proper stimulus, clonogenic cells can begin actively dividing, becoming proliferating cells, or can differentiate into support tissue.

The goal of chemotherapy is to move all of the cells from the proliferating and clonogenic fractions into the end cell compartment. Since end cells cannot directly propagate cancer, they are not considered in the model. The proliferating and clonogenic cells are further subdivided into susceptible and resistant subpopulations, denoted by the subscripts S and R, respectively. Define the indicator sets as

$$\mathcal{R} \equiv \{R,\ S\}\,, \qquad \mathcal{T} \equiv \{P,\ C\}\,, \qquad \text{and} \qquad \mathcal{I} \equiv \{PR,\ CR,\ PS,\ CS\}\,.$$

This results in a four-compartment model for the number of cells in the different subpopulations, $\{P_R(t),\ C_R(t),\ P_S(t),\ C_S(t)\}$, representing the *bulk* or *macroscopic* dynamics subject to treatment by a single cytotoxic agent. The effects of a given treatment is related to the quantity or dose of the cytotoxic agent given and is limited by potential toxic effects. In this formulation, the cytotoxic agent acts on the appropriate subpopulations and not on all cells. The maximum dose allowed for the cytotoxic agent will be given in order to kill as many cells as possible with effects assumed to be instantaneous. If the assumption of instantaneous effects is not realistic, for example intravenous infusion of the cytotoxic agent for 24 hours, then the mean time at which the majority of the agents act is used as the *effective* treatment time. The time at which treatment $i = 1, \dots, N$ is given is denoted by $\delta(t - t_i)$. The resulting treatment model is illustrated in Figure 1 and is given by the following system of equations:

$$\begin{aligned}
\frac{dP_S}{dt} &= \Big[(1 - \alpha_S - \mu_{PS,CR} - \mu_{PS,PR})P_S + \mu_{PR,PS}P_R\Big] f + \beta_S C_S - \delta_{PS} P_S \\
&\qquad + \sum_{i=1}^{N} \delta(t - t_i)(-\mu_{PS,PR,i} P_S + \beta_{S,i} C_S - \kappa_{P,i} P_S)\,, \\
\frac{dP_R}{dt} &= \Big[(1 - \alpha_R - \mu_{PR,CS} - \mu_{PR,PS})P_R + \mu_{PS,PR}P_S\Big] f + \beta_R C_R - \delta_{PR} P_R \\
&\qquad + \sum_{i=1}^{N} \delta(t - t_i)(\mu_{PS,PR,i} P_S + \beta_{R,i} C_R)\,, \qquad (1) \\
\frac{dC_S}{dt} &= \Big[\alpha_S P_S + \mu_{PR,CS} P_R\Big] f - \beta_S C_S - \delta_{CS} C_S \\
&\qquad - \sum_{i=1}^{N} \delta(t - t_i)(\beta_{S,i} + \kappa_{C,i}) C_S\,, \\
\frac{dC_R}{dt} &= \Big[\alpha_R P_R + \mu_{PS,CR} P_S\Big] f - \beta_R C_R - \delta_{CR} C_R - \sum_{i=1}^{N} \delta(t - t_i)\beta_{R,i} C_R\,,
\end{aligned}$$

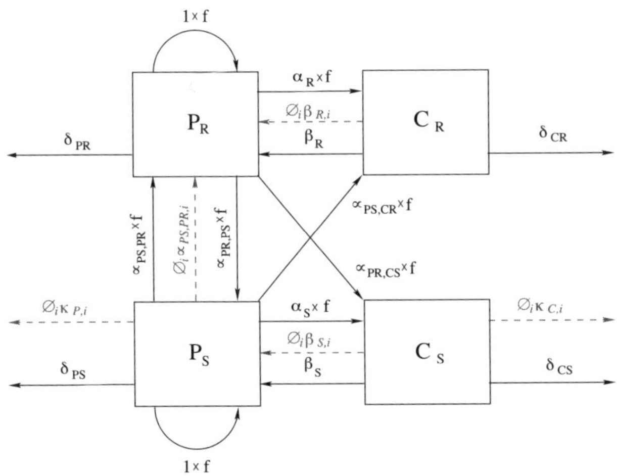

Fig. 1. Schematic representation of the four-compartment model subject to chemotherapy. The dashed lines refer to population migration due to chemotherapy. Compartments aligned in rows are either susceptible or resistant to the administered cytotoxic agent, and compartments aligned in columns are either proliferating or clonogenic sub-populations

where all coefficients are probabilistic rates and are functions related to the sizes of the subpopulations and time (P_R, P_S, C_R, C_S, t). The summation terms in (1) represent the effects of treatment. All daughter cells maintain their quality of drug resistance from the parent cell unless a mutation occurs in the parent cell which is inherited by the daughter cells after mitosis. The growth rate for the cancer is given by the Gompertzian form

$$f(P_R, P_S, C_R, C_S, t) = \lambda \log \left(\frac{K}{P_R(t) + P_S(t) + C_R(t) + C_S(t)} \right), \tag{2}$$

where λ is the growth rate and K is the carrying capacity for the proliferating and clonogenic cells. The allowed probabilities for the mutation rates are given by $\mu_{j,k}$ where $j, k \in \mathcal{I}$ are shown in Figure 1, if not shown then $\mu_{j,k} \equiv 0$. The probabilistic rate α_*, where $* \in \mathcal{R}$, represents the fraction of cells that become quiescent, that is clonogenic, after mitosis which implies that the $1 - \alpha_*$ represents the fraction of cells that remain proliferating. Under the appropriate stimulus, clonogenic cells begin to proliferate with a probability given by β_* for $* \in \mathcal{R}$. The loss rates from a given compartment are denoted by δ_J where $J \in \mathcal{I}$ and account for apoptosis, natural death of cells, and cells recruited to the end cell compartment to become vascular and endothelial cells.

The probabilistic *death* or *kill* rate of cells due to the ith treatment of the cytotoxic agent are given by $\kappa_{\star,i} \geq 0$ for $\star \in \mathcal{T}$. If $\kappa_{C,i} = 0$ then the cytotoxic agent is said to be *cycle-specific* otherwise the agent is *cycle-nonspecific*. A stimulus is created due to the death of a large number of cells by the cytotoxic agent which causes recruitment from the clonogenic to the proliferating fractions with a probabilistic rate for the ith treatment given by $\beta_{*,i}$ for $* \in \mathcal{R}$. In the presence of a cytotoxic agent, proliferating cells may acquire resistance at the completion of mitosis with probability $\mu_{PS,PR,i}$ for the ith treatment. In the treatment presented here, assuming the maximum dose for the cytotoxic agent, drug resistance is seen as semipermanent in the sense that once drug resistance is acquired a mutation must occur for the subsequent generation to lose or gain drug resistance. The concept of a drug resistance spectrum which depends on the dose of the cytotoxic agent is presented in Goldie and Coldman [14]. This implies that probabilities for both the kill rates and for acquiring resistance are dependent on the concentration of the drug at the site of the cancer. Note that all of the effects of treatment are indexed by the treatment number and therefore can change with the number of treatments given which allows for greater flexibility in modeling chemotherapy and can be used to generate the effects of the agents as presented in [14].

3 Treatment Schedulings

The ability for a cytotoxic agent to effectively treat cancer is limited by drug resistance, which can either be intrinsic or acquired. Drug resistance is inherited by daughter cells after mitosis and will be passed on to their progeny. This rapidly leads to a large subpopulation of the cancer cells that are *immune* to treatment. Cytotoxic agents destroy both cancerous and normal cells. Therefore, the benefit of a given treatment needs to consider the impact on the cancer as well as the overall health of the patient. This leads to a situation in which only a small number of treatments can be administered with overall positive impact such that additional treatments will have a nominal effect on the cancer and a negative impact on the patient.

Traditional treatment scheduling of cytotoxic agents is based on administering a *treatment cycle*. A treatment cycle consists of administering a fixed number of doses, denoted by n, of the cytotoxic agent at fixed time intervals of $\Delta t_i \equiv t_{i+1} - t_i$, see for example [10]. If t_i denotes the time for the ith treatment, then the next treatment would be given at time $t_{i+1} = t_i + \Delta t_i$. Only a few treatment cycles can be used with positive impact to the patient, the number of which is denoted by m. Thus, a total of $N = n \times m$ treatments may be given with positive impact to the patient. Once the decision is made to utilize chemotherapy, for example after initial detection or surgery, a treatment cycle is administered. If the treatment cycle does not lead to remission another treatment cycle is administered, provided one is available, until remission is achieved. Upon recurrence, clinical detection, of the cancer the process for administration of treatment cycles repeats. Treatment concludes when all of the m available treatments have been administered. The goal of traditional treatment schedules is to drive the cancer into remission. The use of a treatment cycle may have serious side effects in patients, leading to diminished quality of life due to cumulative toxic effects of the cytotoxic agents being administered relatively close together.

Traditional chemotherapy scheduling can waste the effectiveness of treatments by over-treating. That is, the benefit for the reduction of the cancer may be nominal or minimal relative to the detrimental side effects to the patient, reducing the patient's quality of life. Motivated by concepts of *managing* the cancer relative to the impact of treatment on the patient presented by Schipper et al. [25], an alternate treatment scheduling is considered. The goal is to determine the time at which the *next* treatment should be given. To do this, a *treatment level*, total number of cancer cells present, needs to be established. By doing this the time between treatments is expected to increase and should lead to an increased life expectancy and quality of life. The treatment level is then used to define a *next treatment time problem*. Additionally, a *stopping time problem* needs to be established to determine when treatments should be stopped because they are no longer beneficial in the overall sense to the patient.

A major concern for maintaining a large number of cancer cells is the risk of *metastasis*, the spread of the cancer to remote places in the body from the original site. This is a stochastic process that depends the size of the cancer, the level of angiogenesis (amount of vascularization and endothelial cells), as well as other factors. Metastasis is a process in which cancer cells detach, enter the blood stream, and are transported to remote locations where they attach and form new colonies of cancer. To counter the effects of metastasis, various investigators [12,13,26] suggest that angiogenesis inhibitors and antimetastatic drugs be given to the patient.

To determine when a treatment should be given, a stochastic optimal control problem needs to be solved for the maximum number of cancer cells present for treatment. This optimal control problem needs to based on the system (1) with extensions for stochastic effects such as metastasis which should be represented as a jump process as well as a background Gaussian random process to model small uncertainties and for a new state variables that represent the time of expected death, quality of life, and measure of health. The cost functional should be designed to determine the maximum level of cancer cells necessary for a treatment to be given to maximize the time of expected death while minimizing the risk of metastasis and while maintaining a specified minimum level for quality of life and health. Additionally, constraints for toxicity need to be imposed.

The practical implementation of the alternate treatment scheduling must initially drive the number of cancer cells below the treatment level using aggressive chemotherapy. Once this is achieved, the control problem should be implemented as a receding time horizon problem based on the discrete events of when treatments are given. In doing this, valuable information can be included about the patient in the decision making process for treatment scheduling. If the times between treatments is large enough, additional monitoring of the patient may be necessary to ensure the goals of treatment are met. This alternate scheduling should improve the life expectancy and quality of life for the patient by removing the cumulative negative side effects of chemotherapy. Hopefully, this new scheduling will lead to fewer patients who discontinue treatment.

The concepts presented here can be extended to consider the pharmacokinetic effects and properties of the treatments and allow for a new control variable in terms of the dose size. Another extension to this decision process would be the inclusion of multiple treatment modalities including for example, multiple cytotoxic agents, surgery, radiotherapy, and immunotherapy. The ultimate goal of cancer treatment should be to give the patient a near normal life expectancy and quality of life.

4 Numerical Example

From a macroscopic perspective, the movement of cells between the clonogenic and growth fractions can be simplified by considering the probabilistic bulk or net effects of the growth of cancer. In the case considered here, the proliferating compartments constitute 80% of the cancer. The values used in the numerical example are taken to be the constant limiting probabilities. Furthermore, the probabilistic rates of the system (1) are taken to be uniform since there is no conclusive evidence to suggest that the properties of resistant and susceptible cells behave differently in the way that they propagate.

The numerical example presented here considers the treatment of a cancer which began growing at time $t = 0$ with a single growth cell, so that the initial conditions for the system (1) are given by

$$P_S(0) = 1 \quad \text{and} \quad P_R(0) = C_R(0) = C_S(0) = 0 \, .$$

Clinical detection requires a cancer burden of 10^9 cells, death is anticipated at a cancer burden of 10^{12} cells. The patient is diagnosed with the cancer and treatment begins at time $t = 625$ days with a cancer burden of 2.58×10^{10} cells. In the absence of treatment, the probabilistic limiting distributions for the cancer cells are 20% for clonogenic cells and 80% for proliferating cells which represents an very aggressive cancer. These values are used as the uniform probabilistic rates of migration after mitosis from the proliferating to the clonogenic fractions so that $\alpha_R = \alpha_S = 0.20$. The uniform probabilistic rate of natural back migration from the clonogenic to the proliferating fraction is $\beta_S = \beta_R = 10^{-5}$. The probabilistic loss terms, accounting for natural death and recruitment to develop stromal tissues, for the proliferating and clonogenic cells are $\delta_{PS} = \delta_{PR} = 0.01925$ and $\delta_{CS} = \delta_{CR} = 0.017325$, respectively. The growth rate and the carrying capacity for the Gompertzian dynamics (2) are $\lambda = 0.00396$ and $K = 5 \times 10^{14}$. Mutations that lead to viable cells may either acquire or lose intrinsic resistance to all cytotoxic agents. The effects of the mutations occur after mitosis and are inherited by the daughter

Table 1. Summary of parameter values used in the numerical example

Parameter values used in the example	
$\alpha_R = \alpha_S = 0.20$	Probabilistic rate of migration from proliferating to clonogenic compartments
$\mu_{PR,PS} = \mu_{PS,PR} = 10^{-10}$	Probabilistic rate of intrinsic resistance
$\mu_{PR,CS} = \mu_{PS,CR} = 10^{-11}$	Probabilistic rate of cross compartment intrinsic resistance
$\beta_S = \beta_R = 10^{-5}$	Probabilistic rate of natural back migration from clonogenic to proliferating compartments
$\lambda = 0.00396,\ K = 5 \times 10^{14}$	Growth rate and overall carrying capacity used in f
$\delta_{PS} = \delta_{PR} = 0.01925$	Loss rates for proliferating compartments
$\delta_{CS} = \delta_{CR} = 0.017325$	Loss rates for clonogenic compartments
$P_S(0) = 1,$ $P_R(0) = C_R(0) = C_S(0) = 0$	Initial conditions
10^{12} cells	Expected number of cancer cells to cause death
10^9 cells	Population size at which cancer can be clinically detected
1.5×10^{10} cells	Number of cancer cells at which treatment is given for alternate scheduling
$\kappa_{P,i} = 98\%$	Cytotoxic agent's kill fraction for the proliferating compartment
$\kappa_{C,i} = 0$	Cytotoxic agent's kill fraction for the clonogenic compartment, i.e. cycle specific agent was used
$\beta_{R,i} = \beta_{S,i} = 90\%$	Probabilistic rate of cellular back migration from clonogenic to proliferating compartments due to chemotherapy
$\mu_{PS,PR,i} = 5 \times 10^{-9}$	Probabilistic rate of acquired drug resistance

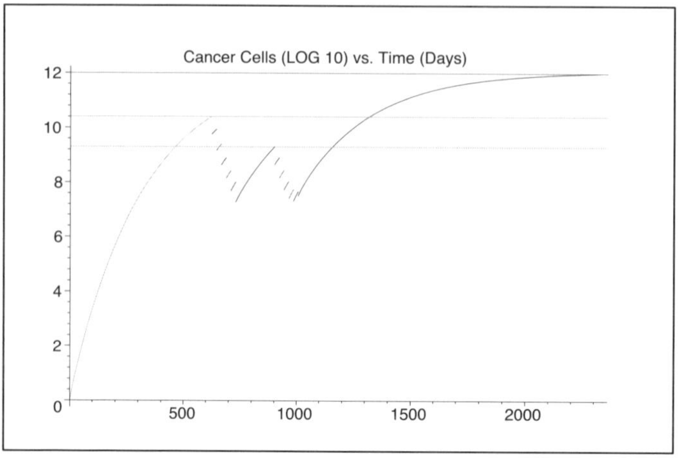

Fig. 2. Logarithm of the total population size of cancer subjected to chemotherapeutic regimen of two treatment cycles of a single cycle-specific cytotoxic agent. The horizontal lines represent, from bottom to top, the number of cells for clinical detection, beginning of treatment, and anticipated death, respectively

cells which may either remain in the proliferating compartment or go into the quiescent phase. The probabilistic rate for the development of intrinsic resistance after mitosis is $\mu_{PR,PS} = \mu_{PS,PR} = 10^{-10}$ and the probabilistic rate for mutations or repair mechanisms to occur so that drug resistance is lost after mitosis is $\mu_{PR,CS} = \mu_{PS,CR} = 10^{-11}$.

Treatment of the cancer uses a single cytotoxic agent with kill rates for each treatment i given by $\kappa_{P,i} = 98\%$, and $\kappa_{C,i} = 0\%$, i.e., the chemotherapeutic agent is proliferating cycle-specific, so the treatment does not kill any of the clonogenic cells. Treatments cause a stimulus from the large number of proliferating susceptible cells that are killed, which recruits quiescent cells from the clonogenic fraction to become proliferating cells with rates given by $\beta_{R,i} = \beta_{S,i} = 90\%$. The probability that a surviving susceptible cell after treatment i acquires drug resistance is $\mu_{PS,PR,i} = 5 \times 10^{-9}$. In the absence of treatment, the model predicts that death will occur at time $t = 1666$ days. The parameter values are summarized in Table 1.

4.1 Traditional Treatment Schedule

Treatment of the cancer uses a single cycle-specific cytotoxic agent such that a maximum of $m = 2$ clinically valuable treatment cycles of $n = 6$ treatments of the agent are given at intervals of 21 days for a total of $N = 12$ treatments. This two treatment cycle regimen is depicted in Figure 2. The first treatment begins at time $t = 625$ days and concludes at $t = 730$ days with a cancer burden of approximately 1.90×10^7 cells, which is considered remission since the cancer burden is not clinically detectable. Recurrence of

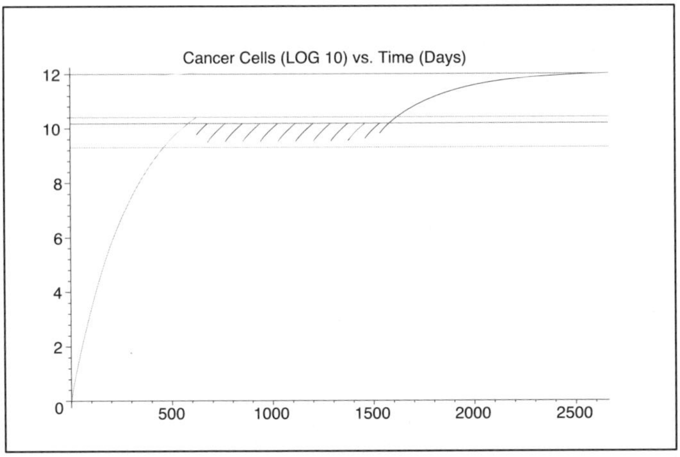

Fig. 3. Logarithm of the total population size of cancer subjected to 12 alternate scheduled chemotherapeutic treatments of a single cycle-specific cytotoxic agent. The horizontal lines represent, from bottom to top, the number of cells for clinical detection, level for alternate treatment, beginning of treatment, and anticipated death, respectively

the cancer occurs at time $t = 900$ days and another treatment cycle is administered. The cancer burden at the conclusion of the second treatment cycle is approximately 3.27×10^7 cells and the patient is once again in remission. After the 2 treatment cycles have been administered death is anticipated at time $t = 2359$ days.

4.2 Alternate Treatment Schedule

In the alternate treatment scheduling, 12 treatments are used with a treatment level of 1.5×10^{10} cells. Initially, treatments are given spaced at the traditional schedule, $\Delta t_i = 21$ days, until the number of cancer cells is below the treatment level, that is aggressive treatment is used. Subsequent treatments are given when the number of cancer cells reaches treatment level, and a toxicity constraint for scheduling treatments is imposed. Let the treatment i be given at time t_i, and the next time that the number of cancer cells reaches the treatment level be τ_{i+1}. Then, the time for treatment $i+1$ is determined by the toxicity constraint:

$$t_{i+1} = t_i + \max[\Delta t_i, \quad \tau_{i+1} - t_i] \,. \tag{3}$$

The alternate scheduling corresponding to the two treatment cycle traditional schedule is depicted in Figure 3. After treatment ends death is anticipated at $t = 2661$ days.

Table 2. Summary of traditional vs. alternate treatment scheduling, where 'percent increase' is relative to the anticipated death in the untreated case

Traditional Scheduling			Alternate Scheduling		
Number of Treatment Cycles	Anticipated Death (days)	Percent Increase	Equivalent Number of Treatments	Anticipated Death (days)	Percent Increase
1	2103	26.23%	6	2200	32.05%
2	2359	41.60%	12	2661	59.72%
3	2308	38.54%	18	2686	61.22%

4.3 Traditional vs. Alternate Treatment Scheduling

The goal of traditional treatment scheduling is to drive the cancer into remission as quickly as possible. In doing this, some treatments may be given without major benefit to the patient, that is the number of cancer cells killed is small relative to the remaining cancer cells. Alternate treatment scheduling seeks to have each treatment be as valuable as possible; however in doing this the patient is at risk for metastasis of the cancer. To minimize the risk of metastasis, angiogenic inhibitors and antimetastatic drugs should be used and the level of treatment should be selected so that the risk profile is acceptable to the patient.

A summary of the results for traditional and alternate is presented in Table 2. Note that 3 treatment cycles traditionally scheduled actually reduces the life expectancy of the patient which is attributed to drug resistance. In the alternate scheduling, even though 18 treatments leads to a longer life expectancy than 12 treatments the last 5 of the 18 treatments are scheduled at the toxicity threshold and above the treatment level which means that the quality of life for the patient would be diminished. Using 3 traditional treatment cycles or the equivalent of 18 alternate scheduled treatments is shown in Figure 4. In regions of traditional scheduling, the cumulative effects of the cytotoxic agents reduces the health of the patient since good or normal cells are killed as well weakening the patient, which may leave the patient susceptible to opportunistic infections. It is clear from Figure 4 that the third traditional treatment cycle and the last 5 alternate scheduled treatments provide no benefit to the patient and should have a significant negative impact on the patient's health, therefore they should not be administered. Table 3 lists the times for treatments for traditional and alternate schedules. The thirteenth alternate treatment is administered on day 1581,

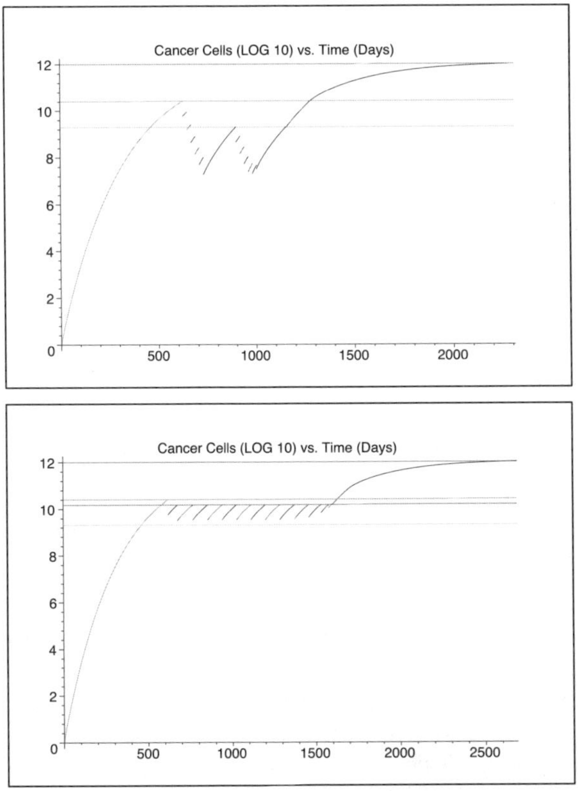

Fig. 4. Logarithm of the total population size of cancer subjected to (top figure) 3 traditional treatment cycles and (bottom figure) 18 alternate scheduled chemotherapeutic treatments of a single cycle-specific cytotoxic agent. The horizontal lines correspond to those in Figures 2 and 3. Note that the last traditional treatment cycle is of nominal value and that the last 5 alternate treatments are above the level of treatment

which accounts for the additional 25 extra days for anticipated death between the 12 and 18 treatments as listed in Table 2. Clearly, the alternate treatment schedule utilizing 13 treatments would be preferred mode of treatment scheduling. Note that the alternate treatment method not only increases the life expectancy and quality of life, but also allows for more treatments to be given with positive benefit to the patient.

5 Conclusions

A compartment model for the evolution of cancer subject to chemotherapy should include aspects for the heterogeneous nature of cancer and for the development drug resistance. In using such a model, alternate treatment

Table 3. Times of treatments for traditional and alternate scheduling

Treatment number	1	2	3	4	5	6	7	8	9	10	11	12
Traditional Schedule	624	646	667	688	709	730	900	921	942	963	984	1005
Alternate Schedule	625	680	770	857	944	1031	1118	1205	1292	1378	1462	1535

schedules can be tested against traditional schedules. Alternate treatment schedules allow for a better control of cancer management and can be tailored to the patient. The numerical example presented illustrates the benefit in terms of *life expectancy* as well as an increase in the *quality of life* since the times of treatments are spread over greater periods of time than that of traditional schedules.

Acknowledgements

The authors thank Professor Bozenna Pasik–Duncan for the invitation to this splendid Kansas Workshop on Stochastic Theory in honor of Professor Tyrone Duncan's 60th birthday, providing local support via the University of Kansas, the National Science Foundation and IEEE Control Society. The last author (FBH) acknowledges that this work was supported in part by the National Science Foundation Grant DMS–99–73231.

References

1. Afenya, E. K. (1996) Acute leukemia and chemotherapy: A modeling viewpoint, Mathematical Biosciences **138**, 79–100.
2. Afenya, E. K. and Bentil, D. E. (1998) Some perspectives on modeling leukemia, Mathematical Biosciences **150**, 113–130.
3. Barbolosi, D. and Iliadis, A. (2001) Optimizing drug regimens in cancer chemotherapy: A simulation study using a PK-PD model, Computers in Biology and Medicine **31**, 157–172.
4. Birkhead, B. G., Rakin, E. M., Gallivan, S., Dones, L., and Rubens, R. D. (1987) A mathematical model for the development of drug resistance to cancer chemotherapy, European Journal of Cancer Clinical Oncology **23**, 1421–1427.
5. Boldrini, J. L. and Costa, M. I. S.(2000) Therapy burden, drug resistance, and optimal treatment regimen for cancer chemotherapy, IMA Journal of Mathematics Applied in Medicine and Biology **17**, 33–51.
6. Calderón, C. P. and Kwembe, T. A. (1991) Modeling tumor growth, Mathematical Biosciences **103**, 97–114.
7. Calderón, C. P. and Kwembe, T. A. (1992) Diverse ideas in modeling tumor growth, Acta Cientifica Venezolana **43**, 64–75.

8. Calman, K. C., Smyth, J. F., and Tattersall, M. H. N. (1980) Basic Principles of Cancer Chemotherapy, Macmillan Press LTD, London.
9. Coldman, A. J. and Murray, J. M. (2000) Optimal control for a stochastic model of cancer chemotherapy, Mathematical Biosciences **168**, 187–200.
10. Fischer, D. S., Knobf, M. T., and Durivage, H. J. (1997) The Cancer Chemotherapy Handbook, Mosby, Chicago.
11. Fister, K. R. and Panetta, J. C. (2000) Optimal control applied to cell-cycle-specific cancer chemotherapy, SIAM Journal of Applied Mathematics **60**, 1059–1072.
12. Jain, R. K. and Carmeliet, P. (2001) Vessels of Death or Life. *URL: http://www.sciam.com/2001/1201issue/1201jain.html*
13. Goldfarb, R. H. and Brunson, K. W. (1992) Therapeutic agents for treatment of established metastases and inhibitors of metastatic spread: Preclinical and clinical progress, Current Opinion in Oncology **4**, 1130–1141.
14. Goldie, J. H. and Coldman, A. J. (1998) Drug Resistance in Cancer: Mechanisms and Models, Cambridge University Press, Cambridge.
15. Kendal, W. S. (1985) Gompertzian growth as a consequence of tumor heterogeneity, Mathematical Biosciences **73**, 103–107.
16. Martin, R. B., Fisher, M. E., Minchin, R. F., and Teo, K. L. (1990) A mathematical model of cancer chemotherapy with an optimal selection of parameters, Mathematical Biosciences **99**, 205–230.
17. Martin, R. B., Fisher, M. E., Minchin, R. F., and Teo, K. L. (1992) Optimal control of tumor size used to maximize survival time when cells are resistant to chemotherapy, Mathematical Biosciences **110**, 201–219.
18. Martin, R. B., Fisher, M. E., Minchin, R. F., and Teo, K. L. (1992) Low-intensity combination chemotherapy maximizes host survival time for tumors containing drug-resistant cells, Mathematical Biosciences **110**, 221–252.
19. Michelson, S. and Slate, D. (1992) A mthematical model of the P-glycoprotein pump as a mediator of multidrug resistance, Bulletin of Mathematical Biology **54**, 1023–1038.
20. Michelson, S. and Slate, D. (1994) A mathematical model for the inhibition of the multidrug resistance-associated P-glycoprotein pump, Bulletin of Mathematical Biology **56**, 207–223.
21. Murray, J. M. (1990) Optimal control for a cancer chemotherapy problem with general growth and loss functions, Mathematical Biosciences **98**, 273–287.
22. Murray, J. M. (1990) Some optimal control problems in cancer chemotherapy with a toxicity limit, Mathematical Biosciences **100**, 49–67.
23. Murray, J. M. (1995) An example of the effects of drug resistance on the optimal schedule for a single drug in cancer chemotherapy, IMA Journal of Mathematics Applied in Medicine and Biology **12**, 55–69.
24. Murray, J. M. (1997) The optimal scheduling of two drugs with simple resistance for a problem in cancer chemotherapy, IMA Journal of Mathematics Applied in Medicine and Biology **14**, 283–303.
25. Schipper, H., Turley, E. A., and Baum, M. (1996) A new biological framework for cancer research, Lancet **348**, 1148–1151.
26. Teicher, B. A. (1995) Angiogenesis and cancer metastases: Therapeutic approaches, Clinical Reviews in Oncology/Hematology **20**, 9–39.
27. Usher, J. R. and Henderson, D. (1996) Some Drug-resistant models for cancer chemotherapy. Part 1: Cycle-nonspecific drugs, IMA Journal of Mathematics Applied in Medicine and Biology **13**, 99–126.

28. Westman, J. J., Fabijonas, B. R., Kern, D. L., and Hanson, F. B. (2001) Cancer treatment using multiple chemotherapeutic agents subject to drug resistance. In: Proceedings of the 15th International Symposium on Mathematical Theory of Networks and Systems, accepted, 6 pages.
29. Westman, J. J., Fabijonas, B. R., Kern, D. L., and Hanson, F. B. (2001) Compartmental model for cancer evolution: Chemotherapy and drug resistance. Mathematical Biosciences, submitted, 20 pages.

Finite-Dimensional Filters with Nonlinear Drift. XII: Linear and Constant Structure of Wong-Matrix*

Dedicated to T.E. Duncan on the occasion of his 60th birthday

Xi Wu, Stephen S.-T. Yau, and Guo-Qing Hu

Department of Mathematics, Statistics, and Computer Science (M/C 249), University of Illinois at Chicago, 851 South Morgan Street, Chicago, IL 60607-7045, USA. wu@math.uic.edu, yau@uic.edu

Abstract. This is the first of final two papers in this series which will give complete classification of the finite dimensional estimation algebra of maximal rank (cf. Definition 2 in Sec. 2), a problem proposed by R. Brockett in his invited lecture at the International Congress of Mathematics in 1983. The concept of estimation algebra (Lie algebra) was first introduced by Brockett and Mitter independently. This concept plays a crucial role in the investigation of finite-dimensional nonlinear filters. Since 1990, Yau has launched a program to study Brockett's problem. He first considered Wong's anti-symmetric matrix $\Omega = (\omega_{ij}) = (\partial f_j/\partial x_i - \partial f_i/\partial x_j)$, where f denotes the drift term in equation (1). He solved the Brockett's problem when Ω matrix has only constant entries. Yau's program is to show that Ω matrix must have constant entries for finite dimensional estimation algebra. Recently Chen and Yau studied the structure of quadratic forms in a finite-dimensional estimation algebra. Let k be the quadratic rank of the estimation algebra and n be the dimension of the state space. They showed that the left upper corner (ω_{ij}), $1 \leq i,j \leq k$, of Ω matrix is a matrix with constant coefficients. In this paper, we shall show that the lower right corner (ω_{ij}), $k+1 \leq i,j \leq n$, of Ω matrix is also a constant matrix.

Keywords: finite-dimensional filter, estimation algebra of maximal rank, nonlinear drift.

1 Introduction

In the late 1970s, Brockett and Clark [2], Brockett [1], and Mitter [16] proposed the idea of using estimation algebras to construct finite dimensional nonlinear filters. The motivation came from the Wei-Norman approach [19] of using Lie algebraic method to solve time varying linear differential equations. In [20], the concept of Ω was introduced, which is defined as the matrix whose (i, j) element is $\omega_{ij} = \partial f_j/\partial x_i - \partial f_i/\partial x_j$, where f is the drift term of the state evolution equation. In [18] and [11], the estimation algebras for the nonlinear

* Research partially supported by NSF and U.S. Army Grants.

B. Pasik-Duncan: Stochastic Theory and Control, LNCIS 280, pp. 507–518, 2002.

filtering system when $\Omega = O$ are completely classified. In [24], Yau studied the case when Ω is a constant matrix. The necessary and sufficient conditions for an estimation algebra of such filtering systems to be finite dimensional are given. It turns out that such class of nonlinear filtering systems includes both Kalman-Bucy and Benes filters as special cases and the explicitly finite dimensional recursive filters are constructed via Wei-Norman approach. In [8], Chiou and Yau introduced the concept of an estimation algebra of maximal rank. They were able to classify all finite-dimensional estimation algebras of maximal rank state space dimension less than or equal to two. In [6] and [7], Chen, Yau, and Leung classified all finite-dimensional estimation algebras of maximal rank with state space dimension three and four.

Our approach of complete classification of finite-dimensional estimation algebras of maximal rank consists of two steps. The first step is to prove that for such an estimation algebra, all the entries in the Ω-matrix are degree one polynomials. The second step is to prove that in fact all the entries in Ω are constants. This finishes the classification because Yau [23][24] has given a complete classification of finite-dimensional estimation algebras of maximal rank when all the entries in Ω are constants. This paper is, in essence, a continuation of our previous papers [4], [5], [6], [7], [8], [13] and [24]. We strongly recommend that readers familiarize themselves with the results in [4]. However, we make every effort to have this paper as self-contained as possible without too much duplication of the previous paper.

Theorem 1 (Main Result). *Let E be a finite-dimensional estimation algebra of maximal rank, k be the maximal rank of quadratic forms in E and n be the dimension of the state space. Then ω_{ij} are constants for $1 \le i,j \le k$ or $k+1 \le i,j \le n$; ω_{ij} are degree one polynomials in $x_1, \dots, x_k$ for $1 \le i \le k, k+1 \le j \le n$ or $k+1 \le i \le n, 1 \le j \le k$.*

The fact that ω_{ij} are constants for $1 \le i,j \le k$, ω_{ij} are degree one polynomials in $x_1, \dots, x_k$ for $1 \le i \le k$ or $1 \le j \le k$, and ω_{ij} are degree one polynomials in $x_{k+1}, \dots, x_n$ for $k+1 \le i,j \le n$, were proved in [4]. The whole point of our main theorem states that ω_{ij} are constants for $k+1 \le i,j \le n$. It turns out that we prove the following general fact which is of independent interest other than nonlinear filtering theory.

Theorem 2. *Let $\eta_4(x_1, \dots, x_n)$ be a homogeneous polynomial of degree 4 in $x_1, \dots, x_n$. Let $H(\eta_4) = (\partial^2 \eta_4 / \partial x_i x_j)_{1 \le i,j \le n}$ be the Hessian matrix of η_4. Then $H(\eta_4)$ can not be decomposed as $\Delta \Delta^T$ where $\Delta = (\beta_{ij})_{1 \le i,j \le n}$ is an anti-symmetric matrix with entries of homogeneous polynomial of degree one satisfying the cyclic relation*

$$\frac{\partial \beta_{ij}}{\partial x_l} + \frac{\partial \beta_{li}}{\partial x_j} + \frac{\partial \beta_{jl}}{\partial x_i} = 0 \qquad \text{for all } 1 \le i,j,l \le n$$

unless η_4 and Δ are trivial, i.e., $H(\eta_4) = \Delta\Delta^T$ implies $\Delta = 0$.

This paper together with [25] completes our classification program described above. The important roles played by the estimation algebra in the existence and construction of finite dimensional filters can be found in [10], [12], [15] and [22]. In [14], Levine presented a unified approach to the existence and non-existence of universal finite dimensional filters dealing with transformations of the solution of the Duncan-Mortensen-Zakai (DMZ) equation (cf. equation (2.2)).

The second author was invited by Morningside Institute of Academia Sinica at Beijing to give eight hour lectures on nonlinear filtering theory in October 1997, where the result of this paper was presented. Yau's lecture series as well as other Morningside Institute lecture series in the same year by Professor Brockett, Professor Caines and Professor Kumar were attended by control theorists from all around the People's Republic of China. Yau particularly thanks Professor Lei Guo and Professor Peter Caines for their useful comments in October 1997 while he was giving a complete proof of classification of finite dimensional estimation algebras of maximal rank. Our results were announced in [22] which appeared in 1998.

2 Basic Concepts

In this section, we recall some basic concepts and results from previous papers [4] [8] [24]. The filtering problem considered here is based on the signal observation model

$$\begin{cases} dx(t) = f\big(x(t)\big)dt + g\big(x(t)\big)dv(t) & x(0) = x_0 \\ dy(t) = h\big(x(t)\big)dt + dw(t) & y(0) = 0 \end{cases} \tag{1}$$

in which x, v, y, and w are respectively R^n, R^p, R^m, and R^m valued processes and v and w have components that are independent, standard Brownian processes. We further assume that $n = p$; f, g and h are vector-valued, orthogonal matrix-valued and vector-valued C^∞ smooth functions. We shall refer to $y(t)$ as the observation at time t.

Let $\rho(t,x)$ denote the conditional probability density of the state given the observation $\{y(s) : 0 \le s \le t\}$. It is well known (see [10], for example) that $\rho(t,x)$ is given by normalizing a function $\sigma(t,x)$ that satisfies the following DMZ equation:

$$\begin{cases} d\sigma(t,x) = L_0\sigma(t,x)dt + \sum_{i=1}^{m} L_i\sigma(t,x)dy_i(t), \\ \sigma(0,x) = \sigma_0(x) \end{cases} \tag{2}$$

where

$$L_0 = \frac{1}{2}\sum_{i=1}^{n}\frac{\partial^2}{\partial x_i^2} - \sum_{i=1}^{n} f_i \frac{\partial}{\partial x_i} - \sum_{i=1}^{n}\frac{\partial f_i}{\partial x_i} - \frac{1}{2}\sum_{i=1}^{m} h_i^2$$

for $i = 1, \ldots, m$, L_i is the zero-degree differential operator of multiplication by h_i, and σ_0 is the probability density of the initial point x_0.

Equation (2) is a Stratonovich stochastic partial differential equation. In real applications, we are interested in constructing robust state estimators from observed sample paths with some property of robustness. Davis in [9] studied this problem and proposed some robust algorithms. In our case, his basic idea reduces to defining a new unnormalized density

$$u(t,x) = \exp\Big(-\sum_{i=1}^{m} h_i(x)y_i(t)\Big)\sigma(t,x).$$

Davis reduced (2) to the following time-varying partial differential equation, which is called the robust DMZ equation:

$$\begin{cases} \dfrac{\partial u}{\partial t}(t,x) = L_0 u(t,x) + \displaystyle\sum_{i=1}^{m} y_i(t)[L_0, L_i]u(t,x) \\ \qquad\qquad + \dfrac{1}{2}\displaystyle\sum_{i,j=1}^{m} y_i(t)y_j(t)[[L_0, L_i], L_j]u(t,x), \\ u(0,x) = \sigma_0(x) \end{cases} \tag{3}$$

where $[\,\cdot,\cdot]$ is the Lie bracket defined as follows.

Definition 1. If X and Y are differential operators, the Lie bracket of X and Y, $[X,Y]$, is

$$[X,Y]\psi = X(Y\psi) - Y(X\psi)$$

for any C^∞ function ψ.

Definition 2. The estimation algebra E of a filtering problem (1) is defined as the Lie algebra generated by $\{L_0, L_1, \ldots, L_m\}$. E is said to be an estimation algebra of maximal rank if, for any $1 \le i \le n$, there exists a constant c_i such that $x_i + c_i$ is in E.

Definition 3. We define matrix $\Omega = (\omega_{ij})$, where

$$\omega_{ij} = \frac{\partial f_j}{\partial x_i} - \frac{\partial f_i}{\partial x_j}, \qquad \forall 1 \le i,j \le n.$$

Clearly, Ω is skew symmetric and $\frac{\partial \omega_{jk}}{\partial x_i} + \frac{\partial \omega_{ki}}{\partial x_j} + \frac{\partial \omega_{ij}}{\partial x_k} = 0$ for every $1 \le i,j,k \le n$.

Define

$$\begin{aligned} D_i &= \frac{\partial}{\partial x_i} - f_i, \\ \eta &= \sum_{i=1}^{n} \frac{\partial f_i}{\partial x_i} + \sum_{i=1}^{n} f_i^2 + \sum_{i=1}^{m} h_i^2. \end{aligned}$$

Then

$$L_0 = \frac{1}{2}(\sum_{i=1}^{n} D_i^2 - \eta).$$

The following basic results play a fundamental role in the classification of finite-dimensional estimation algebras.

Theorem 3 ([17]). *Let E be a finite-dimensional estimation algebra. If a function ψ is in E, then ψ is a polynomial of degree ≤ 2.*

Theorem 4 ([24]). *Let E be a finite-dimensional estimation algebra of (1) such that ω_{ij} are constant functions. If E is of maximal rank, then E is a real vector space of dimension $2n+2$ with basis given by $1, x_1, \ldots, x_n, D_1, \ldots, D_n$ and L_0.*

Corollary 1 ([24]). *Let E be a finite-dimensional estimation algebra with maximal rank. Then E contains the real vector space spanned by $1, x_1, \ldots, x_n$, $D_1, \ldots, D_n$ and L_0.*

Definition 4. Let Q be the space of quadratic forms in n variables, namely, the real vector space spanned by $x_i x_j$, with $1 \leq i \leq j \leq n$. Let $X = (x_1, \ldots, x_n)^T$. For any quadratic form $p \in Q$, there exists a symmetric matrix A such that $p(x) = X^T A X$. The rank of the quadratic form p is denoted by $r(p)$ and is defined to be the rank of the matrix A. A fundamental quadratic form of the estimation algebra E is an element $p_0 \in E \cap Q$ with the greatest positive rank, that is, $r(p_0) \geq r(p)$ for any $p \in E \cap Q$. The maximal rank of quadratic forms in the estimation algebra E is defined to be $k = r(p_0)$ and is called the quadratic rank of E.

Lemma 1 ([4]). *If p is a quadratic form in estimation algebra E, then p is independent of x_j for $j > k$, that is, $\frac{\partial p}{\partial x_j} = 0$ for $k+1 \leq j \leq n$.*

For the convenience of the readers, we also list the following elementary lemmas without proof. The lemmas were proven in [18] and [8].

Lemma 2. (i) *$[XY, Z] = X[Y, Z] + [X, Z]Y$, where X, Y, and Z are differential operators;*

(ii) *$[gD_i, h] = g\frac{\partial h}{\partial x_i}$, where $D_i = \frac{\partial}{\partial x_i} - f_i$, g and h are functions defined on R^n;*

(iii) $[D_i^2, hD_j] = 2\frac{\partial h}{\partial x_i} D_i D_j - 2h\omega_{ij} D_i + \frac{\partial^2 h}{\partial x_i^2} D_j - h\frac{\partial \omega_{ij}}{\partial x_i}$.

3 Linear Structure of Ω

Let us first consider the elements of a finite-dimensional estimation algebra of E with maximal rank. By Corollary 1, we know that E contains elements: $1, x_1, \ldots, x_n, D_1, \ldots, D_n$ and L_0.

Lemma 3. *E contains the following elements:*

(i) $\omega_{ij} = [D_j, D_i] \in E, \qquad \forall 1 \leq i, j \leq n;$

(ii) $H_j := [L_0, D_j] = \sum_{i=1}^{n}(\omega_{ji}D_i + \frac{1}{2}\frac{\partial \omega_{ji}}{\partial x_i}) + \frac{1}{2}\frac{\partial \eta}{\partial x_j} \in E, \quad \forall 1 \leq j \leq n;$

(iii) $[H_j, D_l] = \sum_{i=1}^{n} \omega_{ji}\omega_{li} - \frac{1}{2}\frac{\partial^2 \eta}{\partial x_l \partial x_j} - \sum_{i=1}^{n} \frac{\partial \omega_{ji}}{\partial x_l} D_i - \frac{1}{2}\sum_{i=1}^{n} \frac{\partial^2 \omega_{ji}}{\partial x_l \partial x_i} \in E, \ \forall 1 \leq j, l \leq n.$

Lemma 4 ([4]). *Suppose that E is a finite dimensional estimation algebra. Then for any $1 \leq i,j,l \leq n$,*

$$\frac{\partial \omega_{ij}}{\partial x_l} + \frac{\partial \omega_{li}}{\partial x_j} + \frac{\partial \omega_{jl}}{\partial x_i} = 0.$$

Theorem 5 ([4]). *Suppose that E is a finite-dimensional estimation algebra of maximal rank. Then*

$$\Omega = (\omega_{ij}) = \begin{pmatrix} Constant & | \ P_1(x_1, \ldots, x_k) \\ ------ & | \ -------- \\ P_1(x_1, \ldots, x_k) & | \ P_1(x_{k+1}, \ldots, x_n) \end{pmatrix}.$$

(i) *ω_{ij}'s are constants for $1 \leq i, j \leq k$;*
(ii) *ω_{ij}'s are polynomials of degree one in $x_1, \ldots, x_k$ for $1 \leq i \leq k$ or $1 \leq j \leq k$;*
(iii) *ω_{ij}'s are polynomials of degree one in $x_{k+1}, \ldots, x_n$ for $k+1 \leq i, j \leq n$.*

The following lemma was proved in lemma 4.1 in [5] and Proposition 3.2 in [7]. The proof is elementary. It follows directly from Lemma 3 and Theorem 3 above.

Lemma 5 ([5], [7]). *Suppose that E is a finite dimensional estimation algebra of maximal rank. Then*

(i) $1, x_1, \ldots, x_n, D_1, \ldots, D_n, L_0 \in E$;
(ii) $\sum_{l=1}^{n} \omega_{jl}\omega_{il} - \frac{1}{2}\frac{\partial^2 \eta}{\partial x_j \partial x_i} \in E$ *for any $1 \leq i, j \leq n$;*

(iii) *η is a polynomial of degree 4.*

Lemma 6. *Let $\eta = \eta_4(x_{k+1}, \dots, x_n)+$ polynomial of degree 3 in $x_{k+1}, \dots, x_n$ variables with coefficients degree 4 polynomials in $x_1, \dots, x_k$ variables, where $\eta_4 = \eta_4(x_{k+1}, \dots, x_n)$ is a homogeneous polynomial of degree 4 in $x_{k+1}, \dots, x_n$ variables. Then for any $k+1 \le i, j \le n$,*

$$\sum_{l=k+1}^{n} \beta_{jl}\beta_{il} = \frac{1}{2}\frac{\partial^2 \eta_4}{\partial x_j \partial x_i},$$

where β_{ij} is the homogeneous polynomial of degree one part of ω_{ij}.

Proof. From Theorem 3 and Lemma 5, we know that for $k+1 \le i, j \le n$, $\sum_{l=k+1}^{n} \beta_{jl}\beta_{il} - \frac{1}{2}\frac{\partial^2 \eta_4}{\partial x_j \partial x_i}$ is the homogeneous polynomial of degree 2 part of $\sum_{l=1}^{n} \omega_{jl}\omega_{il} - \frac{1}{2}\frac{\partial^2 \eta}{\partial x_j \partial x_i}$ in $x_{k+1}, \dots, x_n$ variables. The result follows immediately from Lemma 1. *Q.E.D.*

The following notations and Lemma 7 were used in [7]. Define

$$\begin{aligned} \Delta :&= (\beta_{il}),\ k+1 \le i, l \le n,\ \text{an}(n-k) \times (n-k) \text{ anti-symmetric matrix} \\ &= \sum_{j=k+1}^{n} A_j x_j \end{aligned}$$

where $A_j = (A_j(p,q))$, $k+1 \le p, q \le n$, are $(n-k) \times (n-k)$ anti-symmetric matrix with constant coefficients. The anti-symmetry of Δ and A_j follows directly from that of Ω.

Lemma 7 Let $\Delta = \sum_{j=k+1}^{n} A_j x_j$ be an $(n-k) \times (n-k)$ anti-symmetric matrix. Then for $k+1 \le i, j, l \le n$,

(i) $\Delta\Delta^T = \frac{1}{2}H(\eta_4)$, where $H(\eta_4) = \frac{\partial^2 \eta_4}{\partial x_i \partial x_j}$, $k+1 \le i, j \le n$ is the Hessian matrix of $\eta_4 = \eta_4(x_{k+1}, \dots, x_n)$;
(ii) $\frac{\partial \beta_{il}}{\partial x_j} = A_j(i,l)$;
(iii) $A_i(j,l) + A_l(i,j) + A_j(l,i) = 0$;
(iv) $\sum_{l=k+1}^{n} [A_i(j,l)]^2 = \sum_{l=k+1}^{n} [A_j(i,l)]^2$
$= \frac{1}{2}\sum_{l=k+1}^{n} [A_i(i,l)A_j(j,l) + A_j(i,l)A_i(j,l)]$;
(v) $\sum_{l=k+1}^{n} A_j(j,l)A_l(j,i) = 0$.

Proof. It is clear that (i) follows from Lemma 6 while (ii) follows from the definition of A_j. (iii) is an immediate consequence of Lemma 4 and (ii).

For (iv), we observe that

$$\begin{aligned} &\frac{\partial^2}{\partial x_j^2}\left(\frac{\partial^2 \eta_4}{\partial x_i^2}\right) = \frac{\partial^2}{\partial x_i^2}\left(\frac{\partial^2 \eta_4}{\partial x_j^2}\right) = \frac{\partial^2}{\partial x_i x_j}\left(\frac{\partial^2 \eta_4}{\partial x_i x_j}\right) \\ &\Longrightarrow \frac{\partial^2}{\partial x_j^2}\left[\sum_{l=k+1}^{n} \beta_{il}^2\right] = \frac{\partial^2}{\partial x_i^2}\left[\sum_{l=k+1}^{n} \beta_{jl}^2\right] = \frac{\partial^2}{\partial x_i x_j}\left[\sum_{l=k+1}^{n} \beta_{il}\beta_{jl}\right] \end{aligned}$$

by (i)

$$\Longrightarrow 2\sum_{l=k+1}^{n}\left(\frac{\partial\beta_{il}}{\partial x_j}\right)^2 = 2\sum_{l=k+1}^{n}\left(\frac{\partial\beta_{jl}}{\partial x_i}\right)^2$$
$$= \sum_{l=k+1}^{n}\frac{\partial\beta_{il}}{\partial x_i}\frac{\partial\beta_{jl}}{\partial x_j} + \sum_{l=k+1}^{n}\frac{\partial\beta_{il}}{\partial x_j}\frac{\partial\beta_{jl}}{\partial x_i}$$
$$\Longrightarrow \sum_{l=k+1}^{n}[A_j(i,l)]^2 = \sum_{l=k+1}^{n}[A_i(j,l)]^2$$
$$= \frac{1}{2}\sum_{l=k+1}^{n}[A_i(i,l)A_j(j,l) + A_j(i,l)A_i(j,l)].$$

For (v), we observe that

$$\frac{\partial^2}{\partial x_i x_j}\left(\frac{\partial^2\eta}{\partial x_j^2}\right) = \frac{\partial^2}{\partial x_j^2}\left(\frac{\partial^2\eta}{\partial x_i x_j}\right)$$
$$\Longrightarrow \frac{\partial^2}{\partial x_i x_j}\left[\sum_{l=k+1}^{n}\beta_{jl}^2\right] = \frac{\partial^2}{\partial x_j^2}\left[\sum_{l=k+1}^{n}\beta_{il}\beta_{jl}\right] \qquad \text{by (i)}$$
$$\Longrightarrow 2\sum_{l=k+1}^{n}\frac{\partial\beta_{jl}}{\partial x_j}\frac{\partial\beta_{jl}}{\partial x_i} = 2\sum_{l=k+1}^{n}\frac{\partial\beta_{il}}{\partial x_j}\frac{\partial\beta_{jl}}{\partial x_j}$$
$$\Longrightarrow \sum_{l=k+1}^{n}A_j(j,l)A_i(j,l) = A_j(j,l)A_j(i,l)$$
$$\Longrightarrow \sum_{l=k+1}^{n}A_j(j,l)[A_i(j,l) + A_j(l,i)] = 0$$
$$\text{by anti-symmetry of } A_i$$
$$\Longrightarrow \sum_{l=k+1}^{n}A_j(j,l)A_l(j,i) = 0 \qquad \text{by (iii)}$$

Q.E.D.

4 Nonexistence of Nontrivial Solution of the Matrix Equation $\Delta\Delta^T = \frac{1}{2}H(\eta_4)$

In the following theorem, we shall prove that the matrix equation in Lemma 7 above has only trivial solution. The theorem is of independent interest besides nonlinear filtering theory. For $p \times p$ matrix with p less than or equal to 4, the theorem was proved in [7].

Theorem 6. *Let $\Delta = \sum_{j=k+1}^{n} A_j x_j$ be an $(n-k)\times(n-k)$ anti-symmetric matrix where $A_j = (A_j(p,q))$, $k+1 \le p, q \le n$, is an anti-symmetric matrix* *with constant coefficients. Suppose*

$$A_i(j,l) + A_l(i,j) + A_j(l,i) = 0 \qquad \text{for all } k+1 \le i,j,l \le n$$

Let $\eta_4 = \eta_4(x_{k+1}, \dots, x_n)$ *be a homogeneous polynomial of degree 4 in* $x_{k+1}, \dots, x_n$*. Let* $H(\eta_4) = (\frac{\partial^2 \eta_4}{\partial x_i \partial x_j})$*,* $k+1 \le i,j \le n$*, be the Hessian matrix of* η_4*. If* $\Delta\Delta^T = \frac{1}{2}H(\eta_4)$*, then* $\Delta \equiv 0$*, i.e.,* $A_j = O$ *for all* $k+1 \le j \le n$*.*

Proof. For any $k+1 \le j,p,q \le n$, we have

$$\begin{aligned}
&\frac{\partial^2}{\partial x_p \partial x_q}\left(\frac{\partial^2 \eta_4}{\partial x_j^2}\right) = \frac{\partial^2}{\partial x_j^2}\left(\frac{\partial^2 \eta_4}{\partial x_p \partial x_q}\right) \\
&\Longrightarrow \frac{\partial^2}{\partial x_p \partial x_q}\left[\sum_{l=k+1}^{n} (\beta_{jl})^2\right] = \frac{\partial^2}{\partial x_j^2}\left[\sum_{l=k+1}^{n} \beta_{pl}\beta_{ql}\right] \\
&\Longrightarrow 2\sum_{l=k+1}^{n} \frac{\partial \beta_{jl}}{\partial x_p}\frac{\partial \beta_{jl}}{\partial x_q} = 2\sum_{l=k+1}^{n} \frac{\partial \beta_{pl}}{\partial x_j}\frac{\partial \beta_{ql}}{\partial x_j} \\
&\Longrightarrow \sum_{l=k+1}^{n} A_p(j,l)A_q(j,l) = \sum_{l=k+1}^{n} A_j(p,l)A_j(q,l).
\end{aligned} \tag{1}$$

Observe that $(A_j^2)^T = (A_jA_j)^T = A_j^TA_j^T = (-A_j)(-A_j) = A_j^2$. Denote $A_j^2(j,l)$ the (j,l)-entry of A_j^2 matrix. Then for any $k+1 \le j \le n$,

$$\begin{aligned}
\sum_{l=k+1}^{n} [A_j^2(j,l)]^2 &= \sum_{l=k+1}^{n} A_j^2(j,l)A_j^2(l,j) \\
&= \sum_{l,p,q=k+1}^{n} A_j(j,p)A_j(p,l)A_j(l,q)A_j(q,j) \\
&= \sum_{p,q=k+1}^{n} A_j(j,p)A_j(q,j)\sum_{l=k+1}^{n} A_j(p,l)A_j(l,q) \\
&= -\sum_{p,q=k+1}^{n} A_j(j,p)A_j(q,j)\sum_{l=k+1}^{n} A_j(p,l)A_j(q,l) \\
&= -\sum_{p,q=k+1}^{n} A_j(j,p)A_j(q,j)\sum_{l=k+1}^{n} A_p(j,l)A_q(j,l) \\
&\qquad\qquad\qquad\qquad\qquad\qquad\qquad\qquad \text{by (4.1)} \\
&= \sum_{l=k+1}^{n}\{\sum_{q=k+1}^{n} A_j(j,q)A_q(j,l)\}\{\sum_{p=k+1}^{n} A_j(j,p)A_p(j,l)\} \\
&= 0 \qquad\qquad\qquad\qquad\qquad \text{by (v) of Lemma 7}
\end{aligned}$$

$$\Longrightarrow A_j^2(j,j) = 0 \qquad \text{for all } k+1 \le j \le n$$

$$\sum_{l=k+1}^{n}[A_j(j,l)]^2 = -\sum_{l=k+1}^{n} A_j(j,l)A_j(l,j) = -A_j^2(j,j) = 0$$
$$\Longrightarrow A_j(j,l) = 0 \qquad \text{for all } k+1 \leq j,l \leq n. \tag{2}$$

For any $k+1 \leq i,j \leq n$, we have by (4.2), (iv) of Lemma 7

$$\Longrightarrow \frac{3}{4}\sum_{l=k+1}^{n}[A_i(j,l)]^2 \leq \frac{1}{4}\sum_{l=k+1}^{n}[A_j(i,l)]^2 = \frac{1}{4}\sum_{l=k+1}^{n}[A_i(j,l)]^2$$

by (iv) of Lemma 7

$$\Longrightarrow \sum_{l=k+1}^{n}[A_i(j,l)]^2 = 0$$

Therefore, we have shown

$$A_i(j,l) = 0 \qquad \text{for all } k+1 \leq i,j,l \leq n$$

i.e., $\Delta = 0$. *Q.E.D.*

We finally remark that the Main Theorem in the introduction follows from Lemma 6, Lemma 7 and Theorem 4.

Added to the Proof. This paper was first completed in 1997. Due to various reasons it was not published, although it was circulated among many control theorists.

References

1. Brockett, R. W. (1981) Nonlinear systems and nonlinear estimation theory, in The Mathematics of Filtering and Identification and Applications, M. Hazewinkel and J. C. Willems, eds., Reidel, Dordrecht.
2. Brockett, R. W. and Clark, J. M. C. (1980) The geometry of the conditional density functions, in Analysis and Optimization of Stochastic Systems, O. L. R. Jacobs et. al., eds., Academic Press, New York, 399–309.
3. Chen, J. (1994) On ubiquity of Yau filters, Proceedings of the American Control Conference (Baltimore, Maryland), June 1994, 252–254.
4. Chen, J. and Yau, Stephen S.-T. (1996) Finite-dimensional filters with nonlinear drift VI: Linear structure of Ω, Mathematics of Control, Signals and Systems, **9**, 370–385.
5. Chen, J. and Yau, Stephen S.-T. (1997) Finite-dimensional filters with nonlinear drift VII: Mitter conjecture and structure of η, SIAM J. Control and Optimization **36**, 1116–1131.
6. Chen, J., Yau, S. S.-T., and Leung, C. W. (1996) Finite-dimensional filters with nonlinear drift IV: Classification of finite-dimensional estimation algebras of maximal rank with state space dimension 3, SIAM J. Control and Optimization **34**, 179–198.

7. J. Chen, S. S.-T. Yau, and C. W. Leung, Finite-dimensional filters with nonlinear drift VIII: Classification of finite-dimensional estimation algebras of maximal rank with state space dimension 4, SIAM J. Control and Optimization **35** (1997), 1132-1141.
8. Chiou, W. L. and Yau, S. S.-T. (1994) Finite-dimensional filters with nonlinear drift II: Brockett's problem on classification of finite-dimensional estimation algebras, SIAM J. Control and Optimization **32**, 297–310.
9. Davis, M. H. A. (1980) On a multiplicative functional transformation arising in nonlinear filtering theory, Z. Wahrsch Verw. Gebiete **54**, 125–139.
10. Davis, M. H. A. and Marcus, S. I. (1981) An introduction to nonlinear filtering, in The Mathematics of Filtering and Identification and Applications, M. Hazewinkel and J. S. Willems, eds., Reidel, Dordrecht.
11. Dong, R. T., Tam, L. F., Wong, W. S., and Yau, S. S.-T. (1991) Structure and classification theorems of finite-dimensional exact estimation algebras, SIAM J. Control and Optimization **29**, 866–877.
12. Hazewinkel, M. (1989) Lecture on linear and nonlinear filtering, in Analysis and Estimation of Stochastic Mechanical Systems, CISM Courses and Lectures 303, W. Shiehlen and W. Wedig, eds., Springer, Vienna 1988.
13. Hu, G. Q. and Yau, S. S.-T. Finite-dimensional filters with nonlinear drift XIII: Classification of finite-dimensional estimation algebras of maximal rank with state space dimension five, Asian J. Math., to appear.
14. Levine, J. (1991) Finite-dimensional realizations of stochastic PDE's and application to filtering, Stochastic and Stochastic Reports **37**, 75–103.
15. Marcus, S. (1984) Algebraic and geometric methods in nonlinear filtering, SIAM J. Control and Optimization **22**, 817–844.
16. Mitter, S. K. (1979) On the analogy between mathematical problems of nonlinear filtering and quantum physics, Ricerche Automat. **10**, 163–216.
17. Ocone, D. (1980) Topics in nonlinear filtering theory, Ph.D. thesis, Massachusetts Institute of Technology.
18. Tam, L.-F., Wong, W. S., and Yau, S. S.-T. (1990) On a necessary and sufficient condition for finite dimensionality of estimation algebras, SIAM J. Control and Optimization **28**, 173–185.
19. Wei, J. and Norman, E. (1964) On the global representation of the solutions of linear differential equations as a product of exponentials, Proc. Amer. Math. Sci. **15**, 327–334.
20. Wong, W. S. (1987) On a new class of finite-dimensional estimation algebras, Systems Control Lett. **9**, 79–83.
21. Wong, W. S. (1983) New classes of finite dimensional filters, Systems Control Lett. **3**, 155–164.
22. Wong, W. S. and Yau, S. S.-T. (1998) The estimation algebra of nonlinear filtering systems, in Mathematical Control Theory, J. Baillieul and J. C. Willems eds., Springer Verlag, 33–65.
23. Yau, S. S.-T. (1990) Recent results on nonlinear filtering: New class of finite dimensional filters, in Proceedings of the 29th Conference on Decision and Control at Honolulu, Hawaii, Dec. 1990, 231–233.
24. Yau, S. S.-T. (1994) Finite dimensional filters with nonlinear drift I: A class of filters including both Kalman-Bucy filters and Benes filters, Journal of Mathematical Systems, Estimation and Control **4**, 181–203.

25. Yau, S. S.-T. and Hu, G. Q. Finite-dimensional filters with nonlinear drift XIV: Classification of finite-dimensional estimation algebras of maximal rank with arbitrary state space dimension and Mitter conjecture, preprint.
26. Yau, S. S.-T. and Rasoulian, A. (1999) Classification of four-dimensional estimation algebras, IEEE Transaction on Automatic Control, **44**, 2312–2318.

The Stability Game

Kwan-Ho You[1] and E. Bruce Lee[2]

[1] Dept. of Electrical and Computer Engr., Sungkyunkwan University, Suwon, South Korea
[2] Dept. of Electrical and Computer Engr., University of Minnesota, Minneapolis MN 55455, USA

Abstract. Based on state space reachable sets we formulate a two-player differential game for stability. The role of one player (the bounded disturbance) is to remove as much of the system's stability as possible, while the second player (the control) tries to maintain as much of the system's stability as possible. To obtain explicit computable relationships we limit the control selection to setting basic parameters of the system in their stability range and the disturbance is a bounded scalar function. This stability game provides for an explicit quantification of uncertainty in control systems and the value in the game manifests itself as the L_∞-gain of the dynamic input/output disturbance system as a function of the control parameters. It has been discovered recently that the L_∞-gain can be expressed as an explicit parametric formula for linear second order games, and design charts here provide quantitative information in third order linear games. A model predictive scheme will be given for the higher order linear and nonlinear stability games.

1 Introduction

The stability game is a differential game based on the state model:

$$\dot{x} = f(x, w, \theta), \quad x(0) = x_0 \in \Re^n$$
$$y = c'x(t), \quad 0 \le t < \infty$$

where $y(t)$ is a scalar output variable given as a linear combination of the state variable components [1]. w is the disturbance (uncertainty) and θ is a vector of selectable parameters of the controller. $w = w(t)$ is a bounded but otherwise unstructured noise. Its worst case version will be realized in state feedback form for the linear second order games and can be found in model predictive feedback form for third and higher order games. θ takes its parameter values in the restraint set Θ and can be selected once the parametric value (L_∞-gain) of the game is determined for the worst case disturbance.

In the analysis of control systems one of the most important properties is associated with stable system responses. The system state model $\dot{x} = f(x, w, \theta)$ is assumed to have stable responses, $y(t) = c'x(t)$ when the disturbance is zero in each direction c. So it is assumed that $\dot{x} = f(x, 0, \theta)$ is stable asymptotically near the equilibrium state $x = 0$ for each fixed $\theta \in \Theta$. For explicit results the linear model with global stability properties will be

B. Pasik-Duncan: Stochastic Theory and Control, LNCIS 280, pp. 519–532, 2002.

our main concern here. The stable linear state models provide an explicit understanding of the stability game and point the direction for results in the nonlinear situation where computational approaches must be used [2], [3], [4].

Parameter optimization in systems subject to worst bounded disturbances has been initiated many years ago [5] [6]. This task was formulated as a min/max one of

$$\min_{\theta} \max_{w(t)} J(w, \theta)$$

for the state model $\dot{x} = f(x, w, \theta)$ with a number of payoff functions J.

In the stability game the payoff will be associated with state model reachable sets generated by the bounded disturbances for each fixed parameter value θ. To explain the reachable set usage in the stability game consider the linear state model $\dot{x} = Ax + Bw$ with A a stability matrix and with initial state $x(0) = 0$ and disturbance $w(t)$ having bound $|w(t)| \leq 1$. To counter stability of the equilibrium state $x = 0$ when starting at initial state $x(0) = 0$, one could select the bounded disturbance on a given time interval $[0, \hat{t}]$ to maximize the state response at the end time $\hat{t}$ in a given direction η.

This maximum point would be a boundary point of the disturbance reachable set $K(\hat{t}, 0)$. (Here $K(\hat{t}, 0)$ is the set of points reachable from $x = 0$ on the time interval $[0, \hat{t}]$ for the various disturbances $w(t)$ on $[0, \hat{t}]$ with $|w(t)| \leq 1$ - a closed, convex continuous in $\hat{t}$ set in $\Re^n$-see [9], p. 69) Thus the isochrone $T(x) = \hat{t}$, which is the boundary, provides solutions to all such maximization problems on $[0, \hat{t}]$ as η assumes all directions in $\Re^n$ with $\|\eta\| = 1$. This same task on the other time intervals involves the other isochrones $T(x) = t$, $0 \leq t < \infty$.

It seems natural for destabilization to use these isochrones to measure the distance from the stable equilibrium point $x = 0$, and to pick the bounded disturbance to maximize this distance. Thus our solution to the parametric stability game involves selecting the disturbance using time maximization by using the *maximum principle* [9] for each fixed parameter value θ.

This means that once the isochronal function $T(x) = t$ is known the disturbance is given in feedback form using gradient information about T. We will illustrate this in detail for a third order example using local gradient calculations. For second order linear systems there is an analytic representation for the isochrones in terms of the elementary functions [12] - and this leads to a feedback realization for the worst case disturbance as a bang-bang noise disturbance using a switch curve for each parameter vector $\theta \in \Theta$ [11], [12]. For the stable systems with a bounded disturbance the family of reachable sets from the origin reach to a maximum limit body which is the boundary of a bounded set. The isochrone $T(x) = \infty$ is this maximum limit body [7].

For initial points $x(0)$ outside the maximum limit body the disturbance can also be selected in a destabilization sense by maximizing the time it takes for the disturbed state response to reach an ε-nbh of the maximum limit body from the outside for each small $\varepsilon > 0$. To keep the stability stressing level high the disturbance is selected to maximize survival time outside the ε-nbh. There is also an isochronal distance outside the maximum limit body which can be similarly used, as was suggested above inside the maximum limit body; to get a feedback realization of the disturbance as a bang-bang signal with switching set by a switch curve or surface [12].

In the next section 2 we show details of the construction of reachable sets (isochrones) and their use in construction of the switch set for feedback disturbance selection for various values of the control stability vector θ. A third order example illustrates the model predictive approach and shows that value in the stability game is related to the size of the maximal limit body in the output direction—the L_∞-gain becomes the value of the game. Section 3 covers calculation of the L_∞-gain (L_1-norm) as a function of four controller parameters for the third order stable systems. Section 4 covers some applications of the L_∞-gain formula in simple feedback controller designs, and conclusions.

2 The Stability Game and Time Maximum Disturbances

Reachable set boundaries for the system state model provides the isochrones, which in turn determine the time maximum disturbance. The following Proposition 1 suggests a way to design a time maximum disturbance from the isochrones. Using standard state space model and associated optimization theory as in reference [9], page 146 we have the following general result concerning the gradient direction time maximizing disturbance, for isochrones interior to the maximal limit body.

Proposition 1. *Let $T(x)$ (is of class C^1) be the isochronal sets (boundary of reachable sets) of the dynamic linear state space system*

$$\dot{x}(t) = Ax(t) + Bw(t). \tag{1}$$

Then

$$\max_{w \in \Omega} \nabla T(x)[Ax + Bw] = 1. \tag{2}$$

If the restraint set Ω is a cube $|w^j| \leq 1$ in $\Re^m$, then at each point of x the time maximum disturbance $w^*(t)$

$$w^* = \mathrm{sgn}[\nabla T(x)\, B]. \tag{3}$$

Proof: Let x^* on $0 \leq t \leq t^*$ be the time maximum response steering the origin to a point x_1 by disturbance w^*. Then $T(x^*(t)) = t$ and so $\nabla T(x^*(t))\dot{x}^*(t) = \nabla(T(x^*(t))[Ax^*(t) + Bw^*(t)] = 1$ for $0 \leq t \leq t_1$. $\nabla T(x_1)$ means the outwards normal vector to the tangent hyperplane to the isochronal hyper-surface through x_1. Hence $\nabla T(x_1)$ is a positive scalar multiple of $\lambda(t^*)$ the adjoint response corresponding to the time maximum disturbance $w^*(t)$ steering $x_o = 0$ to x_1.

Then $Bw^*(t^*)$ has the maximal possible component along $\nabla T(x_1)$, or $\nabla T(x_1)Bw^*(t^*) = \max_{w\in\Omega}[\nabla T(x_1)Bw]$. Hence $\max_{w\in\Omega} \nabla T(x_1)[Ax_1 + Bw]$ $= \nabla T(x_1)[Ax+Bw^*] = 1$. Therefore at each point x we have $\max_{w\in\Omega} \nabla T(x)$ $[Ax + Bw] = 1$. Finally, let Ω be the cube $|w^j| \leq 1$ in $\Re^m$. Then the time maximum disturbance is $w^* = \mathrm{sgn}[\nabla T(x)B]$.

A similar result obtains for the isochrone family and gradient directions outside of the maximum limit cycle. We now implement this gradient solution to give the model predictive disturbance, by solving using Proposition 1 for the isochronal sets (wave front) until it passes to the given present state. The gradient information allows determination of the most stressful disturbance direction and value, which is then used for a small time duration. Then a new state will be obtained, and the procedure repeated - just as in model predictive control.

To illustrate the use of this proposition in the stability game a third order example with one essential control parameter is now studied. Consider a third order system with transfer function (from disturbance input to system output):

$$\frac{Y(s)}{W(s)} = G(s) = \frac{\omega_n^2}{(s+\zeta\omega_n)(s^2+2\zeta\omega_n s+\omega_n^2)}. \tag{4}$$

Assume the control parameter is the damping ratio $0 < \zeta < 1$ (it's stability range) and the natural frequency ω_n is taken to be 1 without loss of generality. Changing to state space (phase variable canonical form), this transfer function has state space realization from disturbance $w(t)$ to output $x_1(t)$ given as

$$\begin{aligned}
\dot{x_1}(t) &= x_2(t) \\
\dot{x_2}(t) &= x_3(t) \\
\dot{x_3}(t) &= -\zeta x_1(t) - (1+2\zeta^2)x_2(t) - 3\zeta x_3(t) + w(t) \\
x_1(0) &= x_1^o,\ x_2(0) = x_2^o,\ x_3(0) = x_3^o,\ |w(t)| \leq 1,\ 0 < \zeta < 1.
\end{aligned} \tag{5}$$

Our main task is to find the most stressful disturbance, that is, the one which maximizes time distance subject to the system dynamics and end point constraints. Using the maximum principle [9], the Hamiltonian H for time maximization is

$$H = 1 + \lambda_1 x_2 + \lambda_2 x_3 + \lambda_3[-\zeta x_1 - (1+2\zeta^2)x_2 - 3\zeta x_3 + w] \tag{6}$$

with the corresponding adjoint system:

$$\begin{aligned}
\dot{\lambda_1}(t) &= \zeta\lambda_3(t) \\
\dot{\lambda_2}(t) &= -\lambda_1(t) + (1+2\zeta^2)\lambda_3(t)
\end{aligned}$$

$$\dot{\lambda_3}(t) = -\lambda_2(t) + 3\zeta\lambda_3(t). \tag{7}$$

With the bounded disturbance magnitude, the most stressful disturbance to obtain maximum time is bang-bang type effort throughout the switching interval. From maximal principle [8], [9], it is obvious that the time maximum disturbance, $w^*(t)$ which maximizes H with respect to $|w(t)| \leq 1$ is of the form,

$$w^*(t) = \mathrm{sgn}[\lambda_3(t)] \tag{8}$$

The solution of the system (5) as a function of time for constant disturbance ($w(t) \equiv 1$) and initial conditions ($x_1^o,\ x_2^o,\ x_3^o$) at the origin for $0 \leq t \leq T_s$ is

$$\begin{aligned}
x_1(t) &= \frac{\Delta}{\zeta} - \frac{\Delta}{\zeta(1-\zeta^2)}e^{-\zeta t} + \frac{\zeta\Delta}{1-\zeta^2}e^{-\zeta t}\cos(\sqrt{1-\zeta^2}t) \\
&\quad -\frac{\Delta\sqrt{1-\zeta^2}}{1-\zeta^2}e^{-\zeta t}\sin(\sqrt{1-\zeta^2}t) \\
x_2(t) &= \frac{\Delta}{1-\zeta^2}e^{-\zeta t} - \frac{\Delta}{1-\zeta^2}e^{-\zeta t}\cos(\sqrt{1-\zeta^2}t) \\
x_3(t) &= -\frac{\Delta\zeta}{1-\zeta^2}e^{-\zeta t} + \frac{\Delta\zeta}{1-\zeta^2}e^{-\zeta t}\cos(\sqrt{1-\zeta^2}t) \\
&\quad +\frac{\Delta\sqrt{1-\zeta^2}}{1-\zeta^2}e^{-\zeta t}\sin(\sqrt{1-\zeta^2}t).
\end{aligned} \tag{9}$$

with $\Delta = 1$. When $w(t) \equiv -1$ on $t \in [0, T_s]$, use $\Delta = -1$. For the third order systems, the solutions of adjoint system variables $\lambda_1(t),\ \lambda_2(t),\ \lambda_3(t)$ are not in general periodic.

Even though it is hard to find the closed analytic form of the disturbance switch surface, we can construct the switch surface through computer calculations. We used the adjoint system equation, (7) with the modified backing out procedure [9]. However we could not use the switch surface to store the information concerning the time maximum disturbance directly. First it takes too much computational time and memory. Secondly to use a switch surface as a look-up table, there is some limitations to get the exact data needed to decide the location of current states. Moreover for nonlinear dynamic systems, it becomes even more difficult.

Now we consider the method of using the gradient of isochrones to construct the time maximum disturbance as done for second order systems in [13]. Let $T(x_1, x_2, x_3) = \mathit{constant}$ be the isochronal wave front from the origin; then the volume in R^n

$$T(x) = T(x_1, x_2, x_3) = t, \quad \text{for } t > 0$$

is the isochronal limit body with parameter t. The value in the stability game is the size of the maximal limit body in the output direction.

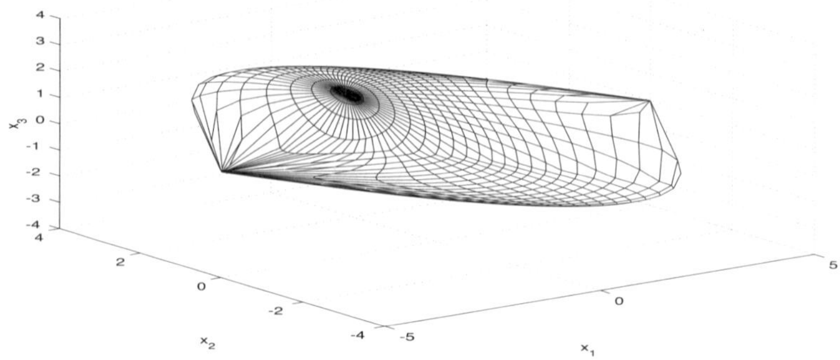

Fig. 1. Isochronal maximum limit body, $\zeta = 0.2$.

Figure 1 shows the isochronal maximum limit body for the system given by equation (5). All of the states approach to this limit body and oscillate along it by the time maximum disturbance. The game value is the maximum value of the output $x_1(t)$. This can be checked apparently from figure 1. For $\zeta = 0.2$, the maximum of $x_1(t)$ happens at $x_1 = 5$ as shown in figure 1. This maximum value is the L_∞-gain of the system and can be given in terms of the system's disturbance impulse response [10]. We can find the L_∞-gain from the reachable set by time maximum disturbance with the equation (10):

$$\begin{aligned} L_\infty - \text{gain} &= \int_{-\infty}^{\infty} |\text{unit impulse response}| dt \\ &= \frac{1}{1-\zeta^2} \int_0^{\infty} |e^{-\zeta t}(1 - \cos(\sqrt{1-\zeta^2} t))| dt \\ &= \frac{1}{\zeta} \end{aligned} \tag{10}$$

The game is now over since ζ can be selected to be 1 for minimization. (We will elaborate on the L_∞-gain calculation in the next section when $\omega_n \neq 1$.) The isochronal limit bodies at each step help us to decide the sign of the disturbance. The time maximum disturbance can be found following the result of Proposition 1 as follows:

$$w^*(x_1, x_2, x_3) = \text{sgn}[\nabla T(x_1, x_2, x_3)\ B]$$

where $B' = [0, 0, 1]$.

With $w^*(x_1, x_2, x_3)$ from the gradient of T, we did the simulation of state responses for the third order system. For the states inside of maximum limit body ($x_1^o = 0.1,\ x_2^o = 0.1,\ x_3^o = 0.1$), we did the simulation with a sufficiently small isochrone evaluation time (Δt). In figure 2, we evaluate the sign of w^* every 0.05 sec. The state trajectories come close and oscillate along the isochronal maximum limit body. This suggested disturbance design method

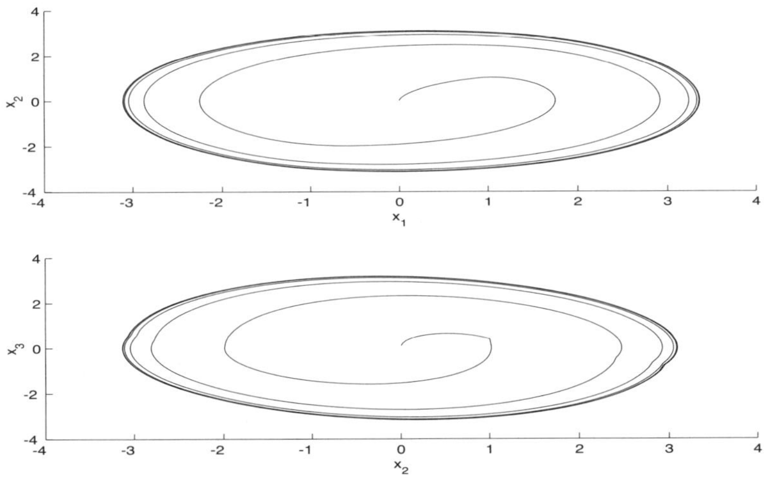

Fig. 2. State trajectory starting inside of maximum limit cycle with $\Delta t = 0.05$ sec, $\zeta = 0.2$, $(x_1^o = 0.1,\ x_2^o = 0.1,\ x_3^o = 0.1)$.

can be applied in the same way to initial points outside the maximum limit body. When we take the initial condition outside of isochronal maximum limit body $(x_1^o = 30,\ x_2^o = -10,\ x_3^o = -10)$, the disturbance w^* from the gradient of $T(x_1, x_2, x_3)$ makes all states move to the maximum limit body as shown in figure 3 and thereby maximize the time distance.

Even though we used a special third order model to illustrate how the theory works, the results are applicable to any third order linear system, or higher order linear system, in state space setting for which the maximum principle applies, and the reachable sets are convex. For nonlinear third order systems the procedure can still be applied, but special selection of feasible directions must be done (see [9] page 447 for a time optimal control example of this). We will now elaborate on the L_∞-gain which is associated with the time maximum disturbances as detailed above, and is the value in the stability game.

3 L_∞-Gain (L_1-Norm) Calculations for Third Order Systems

The value in the stability game as formulated above is the L_∞-gain of the disturbance input stable system. For the second order stable system with an added zero an explicit formula is given in reference [13] for the L_∞-gain as a function of parameters ζ, z and ω_n (damping ratio, zero location and undamped natural frequency).

In this section we will give detailed plots of the L_∞-gain for the third order system with transfer function (having an added zero and an added

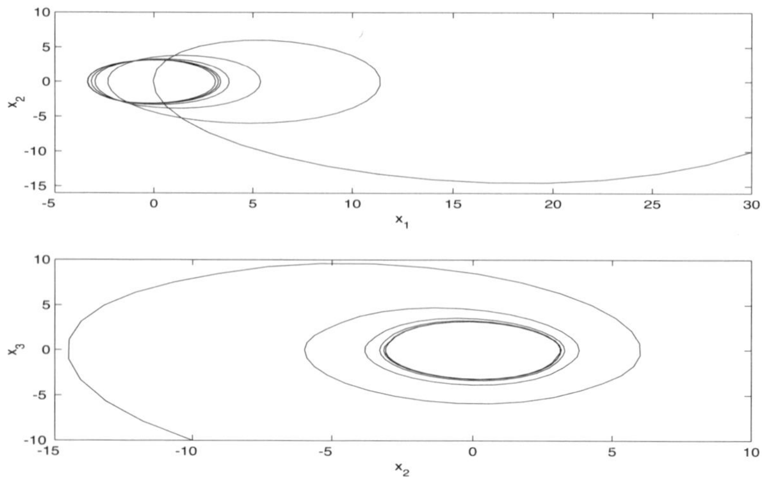

Fig. 3. State trajectory starting outside of maximum limit cycle with $\Delta t = 0.05$ sec, $\zeta = 0.2$, ($x_1^o = 30,\ x_2^o = -10,\ x_3^o = -10$).

pole to a complex pole second order system) given as

$$G(s) = \frac{K(p\omega_n^2/z)(s+z)}{(s+p)(s^2+2\zeta\omega_n s+\omega_n^2)} \tag{11}$$

Clearly the gain $K > 0$ can be used as an overall parameter to scale the size of any of the calculations and will therefore be set to a value of $K = 1$ in the subsequent developments, without loss of generality.

The L_1-norm (L_∞-gain) can be found easily for the third order model which is the second order complex pole system with an additional zero at $s = -z$ ($z > 0$) and with an additional pole at $s = -p$ ($p > 0$) having transfer function,

$$G(s) = \frac{(p\,\omega_n^2/z)(s+z)}{(s+p)(s^2+2\zeta\omega_n s+\omega_n^2)}, \tag{12}$$

$$\text{and } \Theta = \{0 < \zeta < 1,\ \omega_n > 0,\ p > 0,\ z > 0.\} \tag{13}$$

Θ is a parametric stability set. The unit impulse response is used to calculate the L_∞-gain ($z \neq p$) by the formula [13].

$$L_\infty - gain = \int_0^\infty \left| \frac{p\,\omega_n^2\,(z-p)}{z\,(\omega_n^2 - 2\zeta\omega_n p + p^2)} \left[e^{-pt} - e^{-\zeta\omega_n t}\cos(\sqrt{1-\zeta^2}\omega_n t) + \frac{z\,p - \zeta\omega_n p - \zeta\omega_n z + \omega_n^2}{(z-p)\,\sqrt{1-\zeta^2}\omega_n} e^{-\zeta\omega_n t}\sin(\sqrt{1-\zeta^2}\omega_n t) \right] \right| dt$$

Time scaling by letting $\tau = \omega_n t$ shows that only the ratios (p/ω_n) and (z/ω_n) enter the calculation and since the unit impulse response is nonnegative if p/ω_n is sufficiently small this integral gives L_∞-gain $=1$ for small p/ω_n.

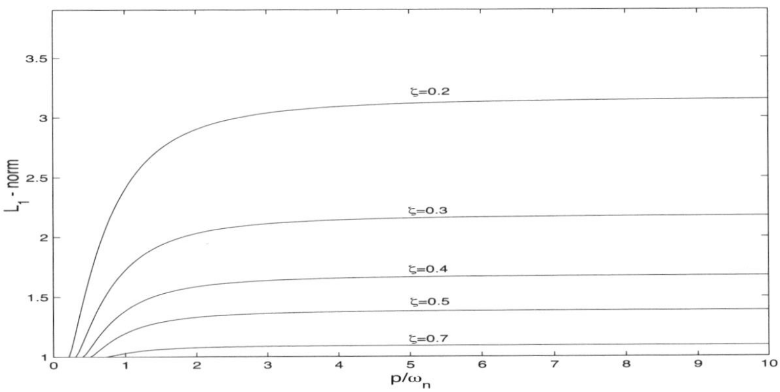

Fig. 4. L_∞-gain (L_1-norm) for the damped harmonic oscillator with an added pole. (here $z = \infty$).

We have plotted this as well as the other L_∞-gain numbers for a sequence of ζ values in figure 4. This L_1-norm is related to the size of the maximum limit body as discussed above. For the second order complex pole system with an additional zero at $s = -z$ $(z > 0)$ an explicit formula for the L_∞-gain is given in reference [13] and is one edge of the three dimensional plots given in this section. The other edge of the three dimensional plots happens when $z \to \infty$ and is shown as figure 4.

When the zero is taken to finite values the curves of figure 4 move upward as the finite zero moves inward. This is shown in figure 5. To cover the range from $z = 10$ to $z = 0.1$ we have plotted this further movement in figures 6, 7, 8 so that the L_1-norm values can be interpolated (extrapolated).

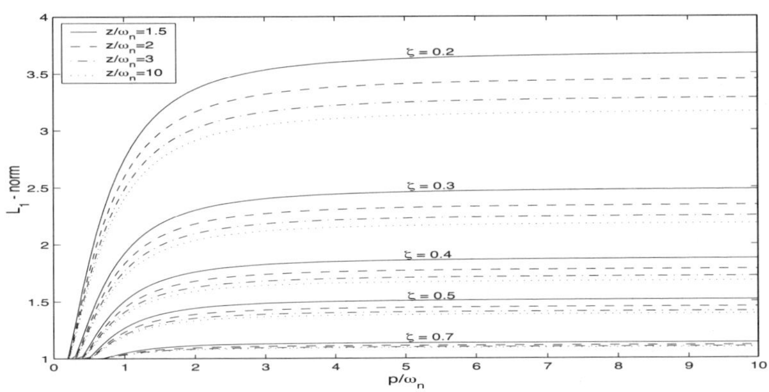

Fig. 5. L_∞-gain (L_1-norm) as a function of four parameters $\omega_n > 0$, $0 \leq p/\omega_n \leq 10$, ζ=0.2, 0.3, 0.4, 0.5 and 0.7 and z/ω_n=1.5, 2, 3, 10.

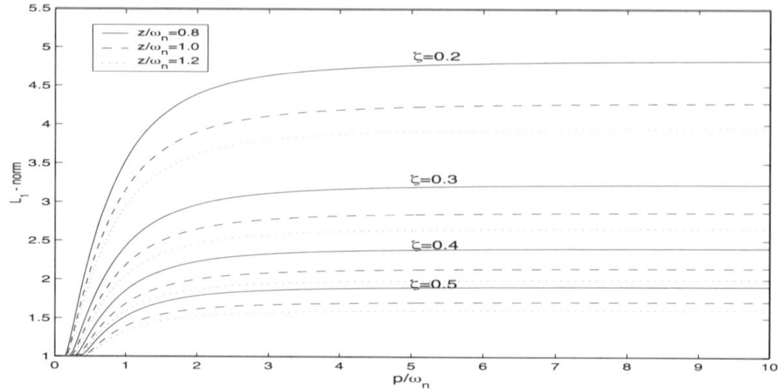

Fig. 6. L_∞-gain (L_1-norm) as a function of four parameters $\omega_n > 0$, $0 \le p/\omega_n \le 10$, ζ=0.2, 0.3, 0.4 and 0.5 and z/ω_n=0.8, 1.0, 1.2.

Three dimensional plots appear in figures 9, 10 for four different ζ values to show general features of the L_1-norm function. We used equation (11) as a general third order system. For example along the line $p/\omega_n = z/\omega_n$ the L_∞-gain is a constant value on each of the different plots (at $\zeta = 0.3$, or $\zeta = 0.5$ etc.). Also the rapid increase in L_1-norm as the zero moves toward the $j\omega$-axis is clearly evident. The general features of these three dimensional plots help in interpretation of the exact value of the L_∞-gain for a given set of parameters and enable these to be used as design charts as shown by examples in the next section.

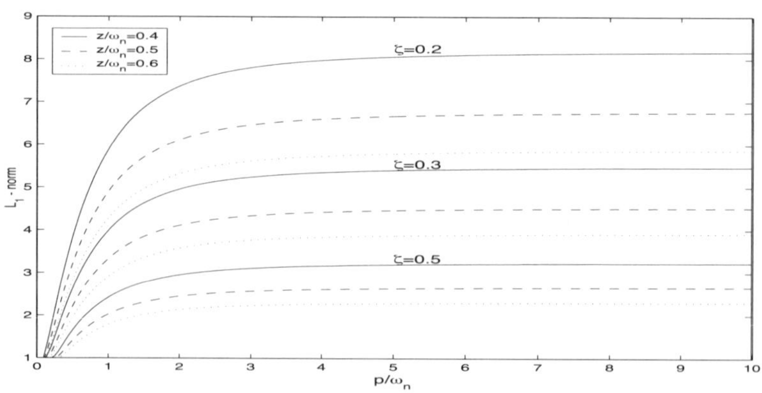

Fig. 7. L_∞-gain (L_1-norm) as a function of four parameters $\omega_n > 0$, $0 \le p/\omega_n \le 10$, ζ=0.2, 0.3 and 0.5 and z/ω_n=0.4, 0.5, 0.6.

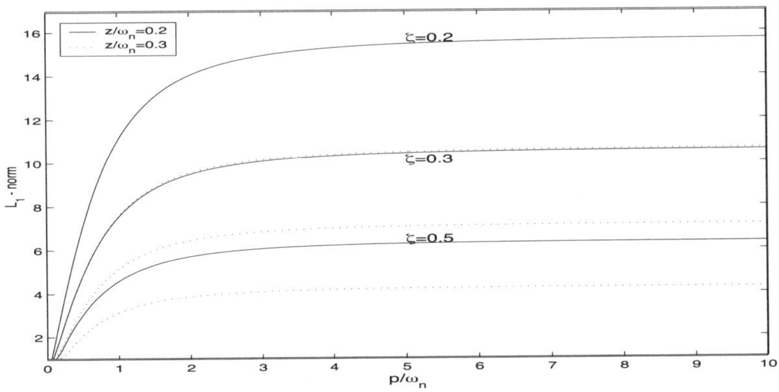

Fig. 8. L_∞-gain (L_1-norm) as a function of four parameters $\omega_n > 0$, $0 \le p/\omega_n \le 10$, ζ=0.2, 0.3 and 0.5 and z/ω_n=0.2, 0.3.

4 Conclusions with Example

The two player differential game with payoff function being system stability (the stability game) has been formulated and solved. The solution involves solving, for each controller parameter vector, for the isochronal function for the time maximizing disturbance. A model predictive scheme was given to find the disturbance in feedback form which provides the maximum destabi-

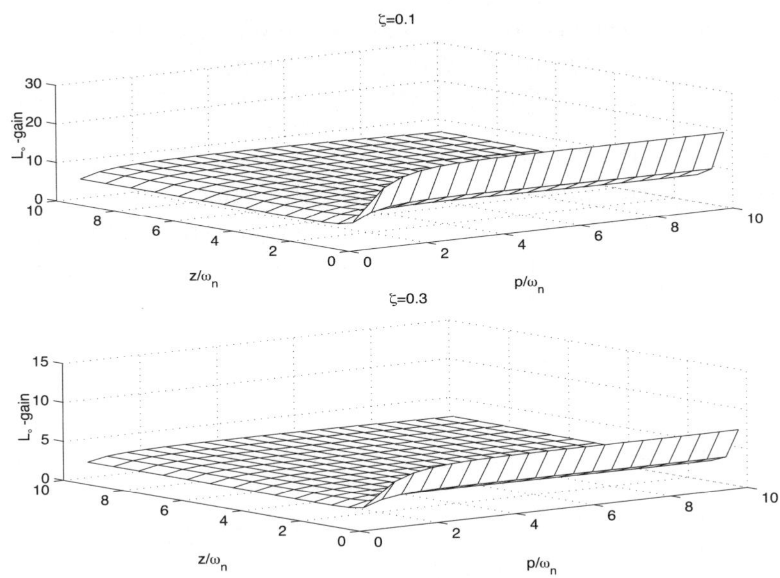

Fig. 9. L_∞-gain (L_1-norm) in three dimension as a function of four parameters $\omega_n > 0$, $0 \le p/\omega_n$, $z/\omega_n \le 10$, ζ=0.1 and 0.3.

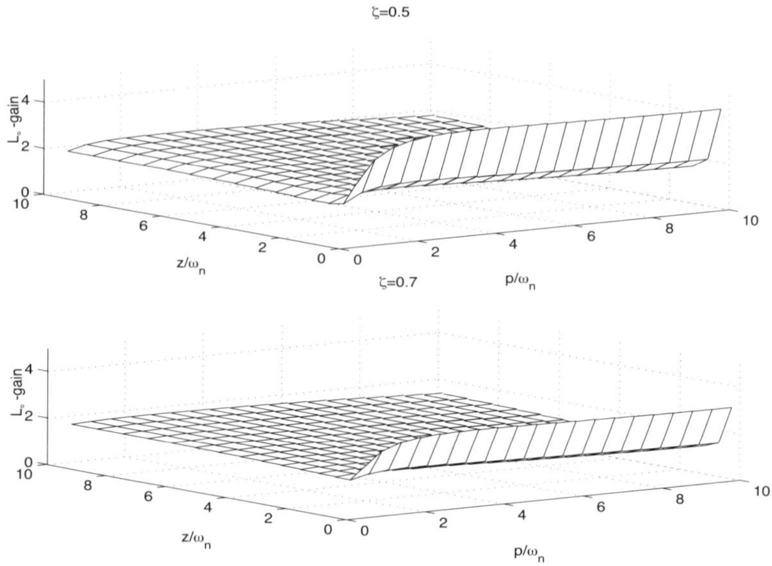

Fig. 10. L_∞-gain (L_1-norm) in three dimension as a function of four parameters $\omega_n > 0$, $0 \le p/\omega_n$, $z/\omega_n \le 10$, ζ=0.5 and 0.7.

lization for each controller parameter vector θ. The game value was shown to be the system's L_∞-gain for the disturbance inputs. This game value (L_∞-gain) can be computed as the L_1-norm.

For third order systems given in terms of natural controller parameter of a stable pole/zero transfer function of third order the value function was computed and displayed graphically as a function of four parameters $\{\zeta,\ \omega_n,\ p,\ z\}$ in their stable range. For each controller parameter vector θ in the stable range the feedback form of the disturbance can be calculated locally using the model predictive scheme for linear games.

We now give a second order example which shows all details of the stability game. Consider a second order system which has (from disturbance input $w(t)$ to system output $y(t)$) transfer function

$$\frac{Y(s)}{W(s)} = G(S) = \frac{(\omega_n^2/z)(s+z)}{s^2 + 2\zeta\omega_n s + \omega_n^2}. \tag{14}$$

The controller parameters $\{\zeta,\ \omega_n,\ z\}$ are in the stable range $0 < \zeta < 1$, $\omega_n > 0$, and $z > 0$. Suppose the restraint set $\Theta = \{0 \le \zeta \le 0.7,\ \zeta \le z \le 10,\ 5 \le \omega_n \le 100\}$. From reference [13] the L_∞-gain of this system is given in closed form as:

$$L_\infty - \text{gain} = \frac{1}{z}\int_0^\infty |e^{-\zeta t}\cos(\sqrt{1-\zeta^2}t) + \frac{z-\zeta}{\sqrt{1-\zeta^2}}e^{-\zeta t}\sin(\sqrt{1-\zeta^2}t)|dt$$

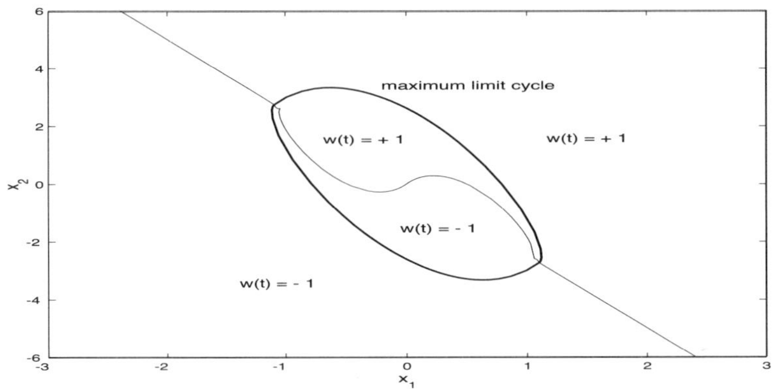

Fig. 11. Feedback realization for destabilizing disturbance $\zeta = 0.7$, $\omega_n = 5$, $z = 10$. ($\zeta = 0.7$, $z/\omega_n = 2$.)

$$= 1 + \frac{2\sqrt{z^2 - 2\zeta z + 1}}{z(1 - e^{\frac{-\zeta}{\sqrt{1-\zeta^2}}\pi})} e^{-\zeta\alpha}$$

$$\text{where } \alpha = \frac{1}{\sqrt{1-\zeta^2}} \left[\pi + \tan^{-1}\left(\frac{\sqrt{1-\zeta^2}}{\zeta - z}\right) \right]. \tag{15}$$

Here $z := z/\omega_n$ if ω_n is not equal to 1.

For the controller to minimize the L_∞-gain take $\zeta = 0.7$, $z = 10$, $\omega_n = 5$ with game value (L_∞-gain) of 1.1170. Only the feedback synthesis of the disturbance has to be found. This has also been found recently for the second order systems in closed form in reference [13] and is plotted in the state space (x_1-x_2 plane) when we used the state space realization of the transfer function

$$\dot{x}_1 = x_2(t) + \frac{\omega_n^2}{z} w(t)$$
$$\dot{x}_2 = -\omega_n^2 x_1(t) - 2\zeta\omega_n x_2(t) + \left(\omega_n^2 - 2\zeta\frac{\omega_n^3}{z}\right) w(t).$$

with output $y(t) = x_1(t)$. Figure 11 shows a disturbance switch curve and maximum limit cycle which is a result of using this disturbance switch curve.

The proposed disturbance selection method appears to be effective especially for third order systems. Even when we cannot predict the shape of switch surface precisely, the disturbance is decided by the gradient of isochronal surface with the current states and we calculate again the new isochronal wave front after Δt sec. (This is the basis of model predictive disturbance selection [14], [15], [16]). The essence of the idea is to sweep out the reachable set boundaries (isochrones) till the current state position is attained and assess the isochronal surface near that state position to determine what value of the disturbance will be most stressful.

References

1. Friedman A. (1971) Differential Games, Wiley, New York.
2. Ryan E. P. (1980) On the Sensitivity of a Time-Optimal Switching Function, IEEE Trans. on Automatic Control, **25**, 275–277.
3. Doyle J. C., Francis B. A. et al. (1992) Feedback Control Theory, Macmillan, New York.
4. Franklin G., Powell J. D. et al. (1994) Feedback Control of Dynamic Systems, 3rd edn, Addison-Wesley.
5. Koivuniemi A. J. (1966) Parameter Optimization in Systems Subject to Worst (Bounded) Disturbance, IEEE Trans. on Automatic Control, **11**, 427–433.
6. Polanski A. (2000) Destability Strategies for Uncertain Linear Systems, IEEE Trans. on Automatic Control, **45**, 2378–2382.
7. Wang H. H., Krstic M. (2000) Extreme Seeking for Limit Cycle Minimization, IEEE Trans. on Automatic Control, **45**, 2432–2437.
8. Kirk D. E. (1970) Optimal Control theory-An Introduction, Prentice-Hall, New Jersey.
9. Lee E. B., Markus L. (1967) Foundations of Optimal Control Theory, Wiley, New York.
10. Lee E. B., Luo J. C. (2000) On Evaluating the Bounded Input Bounded Output Stability Integral for Second Order Systems, IEEE Trans. on Automatic Control, **45**, 311–312.
11. Luo J. C., Lee E. B. (2000) A Closed Analytic Form for the Time Maximum Disturbance Isochrones of Second Order Linear Systems, System & Control Letters, **40**, 229–245.
12. You K. H., Lee E. B. (2000) Time Maximum Disturbance Switch Curve and Isochrones of Linear Second Order Systems with Numerator Dynamics, Journal of The Franklin Institute, **337**, 725–742.
13. You K. H., Lee E. B. (2001) Time Maximum Disturbance Design for Stable Linear Systems; A Model Predictive Scheme, IEEE Trans. on Automatic Control, **46**, 1327–1332.
14. Meadows E. S., Rawlings J. B. (1995) Topics in Model Predictive Control in Methods of Model Based Process Control, Kluwer, New York.
15. Rawlings J. B. (1999) Tutorial: Model Predictive Control Technology, Proc. of 1999 ACC., 662–676.
16. Martin G. (1999) Nonlinear Model Predictive Control, Proc. of 1999 ACC., 677–678.

Bayes Estimation via Filtering Equation for O-U Process with Discrete Noises: Application to the Micro-Movement of Stock Prices

Yong Zeng and Laurie C. Scott

Department of Mathematics and Statistics, University of Missouri at Kansas City, Kansas City, MO 64110-2499, USA

Abstract. A model of O-U process with discrete noises is proposed for the price micro-movement, which refers to the transactional price behavior . The model can be viewed as a multivariate point process and framed as a filtering problem with counting process observations. Under this framework, the whole sample paths are observable and are used for parameter estimation. Based on the filtering equation, we construct a consistent recursive algorithm to compute the approximate posterior and the Bayes estimates. Finally, Bayes estimates for a two-month transaction prices of Microsoft are obtained.

1 Introduction

Stock price models can be classified into two categories: macro- and micro-movement models. Macro-movement refers to daily, weekly, and monthly closing price behavior and micro-movement refers to transactional price behavior. There are both a strong connection and striking differences between the macro- and micro-movements. The macro-movement is an equally-spaced time series sample from the micro-movement and the overall shapes of both movements are the same. However, the micro-movement is irregularly-spaced in time and is high-frequency data. Furthermore, the impact of noises in micro-movement is large and noises must be modeled explicitly in micro-movement.

Stock price is distinguished from stock value and their distinction is noises. Black in [1] remarks that noise is contrasted with information, which influences stock value. Hasbrouck in [6] further points out that information has a long-term, or "permanent" impact on stock price while noise has no influence on stock value and has only short-term or "transitory" impact on stock price.

To build the micro-movement model, we follow this intuition: the micro-movement model is built from the macro-movement model by combining the noises in the high frequency manner, or economically, stock price is formed from stock value by incorporating the (financial) noises when a trade occurs. In this paper, the value process is assumed to be an O-U process and we focus on modeling three important types of noise in financial data: discrete, clustering and non-clustering noises.

B. Pasik-Duncan: Stochastic Theory and Control, LNCIS 280, pp. 533–548, 2002.

One natural way to model discrete noise is to round the continuous value to the nearest tick,[1] Even the estimation of the rounding model is difficult because of the complexity of its likelihood function due to the unobservability of the underlying value process.

Our model builds on the rounding model and has an important feature to be formulated as a filtering problem with counting process observations. Under this view, the whole sample paths are observable and complete information is used for estimation. We apply Bayesian estimation via filtering equation introduced in [8] to obtain the Bayes estimates.

The outline of the paper is as follows. We present the model in two ways in section 2. One is by construction, and the other is by formulating it as a filtering problem with counting process observations. In section 3, we first review the general theory of Bayesian estimation via filtering equation. Then, we construct the recursive algorithm for the model to compute Bayes estimates and show its consistency. In section 4, Bayes estimates for two-month Microsoft transaction prices are presented. We conclude in section 5.

2 The Model

The model is intended to fit four discrete-type sample characteristics of micro-movement that cannot be explained by the rounding model. First, the observed frequency for price changes that are more than a tick is larger than the frequency implied by the rounding model. Second, for highly traded stocks, several trades may take place within one second. However, their prices are not observed to be the same and the difference can even be two or more ticks. Third, there are outliers. The fourth one is price clustering. If a stock value process is assumed to be diffusion or jump-diffusion, and the discrete price is obtained by the rounding model, then we would expect approximately equal probability for each tick. But empirical findings are that integers are more common than halves; halves are more common than odd quarters; and odd quarters are more common than odd eighths (when the tick was one eighth dollar) (see [5]). This means prices are clustered on even eighths.

Evidence for all these four sample characteristics is presented in Section 4 when we summarize a two-month micro-movement data set of Microsoft.

Suppose that $X(t)$ is an unobserved value process for a stock, and it can be partially observed through the price process, $Y(t)$. $X(t)$ lives in a continuous state space while $Y(t)$ lives in a discrete state space given by the multiples of a tick, which is assumed to be $\frac{1}{M}$. Prices can only be observed at the irregularly spaced trading times, which are assumed to be driven by a Poisson process with a deterministic intensity, $a(t)$. The inhomogeneity assumption is more general than the naive assumption of homogeneity, it fits trade duration (waiting time) data better, and it can explain the observation that trading activity is higher near the opening and the closing than in the middle of the

[1] A tick is the minimum price variation in trading.

day. We restrict our attention to a deterministic intensity because it simplifies the filtering equation and even $a(t)$ itself disappears in the filtering equation and then also in the Bayesian estimation procedure.

The value process, $X(t)$, is assumed to be an O-U process, which in stochastic differential equation form is

$$dX(t) = (a + bX(t))dt + \sigma dW(t). \tag{1}$$

The deterministic trend is modeled by $(a + bX(t))dt$, and the information by $\sigma dW(t)$.

There are two methods to build our model from the value process. One by constructing Y from X via incorporating noises, and the other by formulating (X, Y) as a filtering problem with counting process observations. The former approach is intuitive and related to the *state space model*, while the latter approach is important in statistical analysis and is related to *hidden Markov models* (see [3]).

2.1 Construction of Y from X

Suppose trading times $t_1, t_2, \dots, t_i, \dots$ are generated by a Poisson process with the intensity $a(t)$. To simplify notation, set $x = X(t_i)$, the value at time t_i, and set $y = Y(t_i)$, the price at time t_i. We construct y from x in three steps.

Step 1: Incorporate Discrete Noise by rounding off x to its closest tick, $R[x, \frac{1}{M}]$. Without other noises, trades should occur at this tick, which is the closest tick to the stock value.

Step 2: Incorporate Non-clustering Noise by adding: $y' = R[x, \frac{1}{M}] + U$, where U is the non-clustering noise of ith trade at time t_i. We assume $\{U_i\}$, are independent of the value process, and they are *i.i.d.* with a doubly geometric distribution:

$$P\{U = u\} = \begin{cases} (1-\rho) & \text{if } u = 0 \\ \frac{1}{2}(1-\rho)\rho^{M|u|} & \text{if } u = \pm\frac{1}{M}, \pm\frac{2}{M}, \cdots \end{cases}.$$

We pick the doubly geometric distribution because it is uni-modal, symmetric and bell shaped with the implication that the trading price at a tick closer to the stock value is more likely to occur and trading prices with the same distance to the stock value have the same chances.

The non-clustering noise can explain three of the four discreteness-related sample characteristics. First, the non-clustering noise increases considerably the probability of the successive price changes that are more than a tick. Next, it allows the prices of trades occurring within the same second to differ and the difference can be two or more ticks. Finally, it produces outliers.

Step 3: Incorporate Clustering Noise by random biasing. After rounding the value process and adding the non-clustering noise, the fractional part of v' is still approximately uniformly distributed on all fractional parts. We bias y' through a random biasing function $b(\cdot)$ to produce price clustering. $\{b_i(\cdot)\}$ are assumed independent of $\{y_i'\}$ and serially independent given the sequence $\{y_i'\}$.

To be consistent with the data analyzed in Section 4, we construct a simple random biasing function only for the tick of $1/8$ dollar. We can generalize it to other ticks such as $1/100$ dollar easily.

The data to be analyzed has this clustering phenomenon: integers and halves are most likely and have about the same frequencies; odd quarters are the second most likely and have about the same frequencies; and odd eighths are least likely and have about the same frequencies. To generate such clustering, a random biasing function is constructed based on the following rule: if the fractional part of y' is even eighths, then y stays on y' with probability one; if the fractional part of y' is odd eighth, then y stays on y' with probability $1-\alpha-\beta$, y moves to the closest odd quarter with probability α, and moves to the closest half or integer with probability β. The detail of $b(\cdot)$ is presented in Appendix A.

In summary, the construction is

$$Y(t_i) = b_i(R[X(t_i), \frac{1}{M}] + U_i),$$

where the rounding function takes care of price discreteness, U_i takes care of non-clustering noise and the random biasing function b_i takes care of clustering noise. Note that information goes into the value process and has a permanent impact on price while noise goes into price only when trades occur and has only a transitory impact on price.

Through the construction, the transition probability from x to y, denoted by $g(y|x)$, can be computed through $g(y|x) = \sum_{y'} g(y|y')g(y'|x)$ where $g(y|y')$ (or $g(y'|x)$) is transition probability from y' (or x) to y (or y'). The detail of $g(y|x)$ is provided in Appendix A.

Lastly, the model can be framed as a filtering problem with counting process observations. This is important for statistical analysis, because under this framework, we are able to derive the filtering equation, which characterizes the posterior given the whole sample paths.

2.2 Counting Process Observations

In the construction, we view the prices in the order of trading occurrence over time. Alternatively, we can view them in the level of prices because of the price discreteness. We view the prices as a collection of counting processes as follows:

$$\boldsymbol{Y}(t) = \begin{pmatrix} N_1(\int_0^t \lambda_1(\boldsymbol{\theta}, X(s-), s-)ds) \\ N_2(\int_0^t \lambda_2(\boldsymbol{\theta}, X(s-), s-)ds) \\ \vdots \\ N_n(\int_0^t \lambda_n(\boldsymbol{\theta}, X(s-), s-)ds) \end{pmatrix}, \tag{2}$$

where $Y_k(t) = N_k(\int_0^t \lambda_k(\boldsymbol{\theta}, X(s-), s-)ds)$ is the observable counting process recording the cumulative number of trades that have occurred at the kth price level (denoted by y_k) up to time t, and $\boldsymbol{\theta} = (a, b, \sigma, \rho)$.

According to the theory of multivariate point processes, we make three mild assumptions to ensure the construction of the model is equivalent to the specification of counting process observations. The equivalence ensures that the statistical analysis based on the latter specification is also good for the model by the former construction. The first two assumptions are general since a large class of counting processes can be transformed into this setup by the technique of change of measures (See [2] page 165). The third one imposes a nice structure for the intensities of the model.

Assumption 1: *N_k's are unit Poisson processes under measure P.*

Then, the random time change implies that
$Y_k(t) = N_k(\int_0^t \lambda_k(\boldsymbol{\theta}, X(s-), s-)ds)$ is a counting process with intensity $\int_0^t \lambda_k(\boldsymbol{\theta}, X(s-), s-)ds$, and $Y_k(t) - \int_0^t \lambda_k(\boldsymbol{\theta}, X(s-), s-)ds$ is a martingale.

Assumption 2: *$X, N_1, N_2, \ldots, N_n$ are independent under measure P.*

Assumptions 1 and 2 imply that there exists a reference measure Q and that after a suitable change of measure to Q, $X, Y_1, \ldots, Y_n$ become independent, and $Y_1, Y_2, \ldots, Y_n$ become unit Poisson processes. The reference measure Q plays an important role in deriving the filtering equation and in proving the consistency of the recursive algorithm.

Assumption 3: *The structure of the intensity is $\lambda_k(\boldsymbol{\theta}, v, t) = a(t)g(y_k|x)$, where $a(t)$ is the total intensity and $g(y_k|x)$ is the transition probability from x to y_k, the kth price level at time t.*

Assumption 3 means that the total trading intensity $a(t)$ determines the overall rate of trade occurrence at time t and $g(y_k|x)$ determines the proportion of trading intensity at price y_k when the value is x. This intensity structure guarantees the equivalence of the two ways of modeling, because the conditional finite dimensional distributions of the price Y given the $\{X(s) : 0 < s < t\}$ are identical.

The final technical assumption ensures the uniqueness of the filtering equation.

Assumption 4: *The total intensity process, $a(t)$, is uniformly bounded below and above, namely, there exist positive constants, C_1 and C_2, such that $C_1 < a(t) \leq C_2$ for all $t > 0$.*

3 Bayes Estimation via Filtering Equation

There are six parameters, $(a, b, \sigma, \rho, \alpha, \beta)$. (a, b, σ) relate to the value process, ρ relates to the non-clustering noise, and (α, β) relate to the clustering noise. (α, β) can be estimated by the method of relative frequency and (a, b, σ, ρ) are estimated by Bayesian approach via filtering equation.

Bayes estimate, which is the posterior mean, is the least Mean Square Errors(MSE) estimate. Since the likelihood function of the model is unregular, the usual consistency, asymptotic normality and efficiency of maximum likelihood estimates (MLE) and Bayes estimates are unknown. Here, we choose Bayes estimation because the posterior, which contains more information than just an estimate like MLE, can be obtained. The core of the Bayesian estimation via filtering equation is to construct an algorithm to compute the conditional distribution, which becomes a posterior after a prior is assigned. The algorithm, which is based on the filtering equation, is naturally recursive with every trade. One basic requirement for the recursive algorithm is consistent, namely, the conditional distribution computed by the recursive algorithm converges to the true one determined by the filtering equation. This is guaranteed by a theorem on the convergence of conditional expectation.

We first review the filtering equation and the theorem on the convergence of conditional expectation, then construct the convergent recursive algorithm for the model in detail, and finally show its consistency.

3.1 Review of two Theorems

In [8], the filtering equation for a general model including this model is derived and the theorem on the convergence of conditional expectation for the general model is proven. Here, we state a simpler version of the two theorems applying to the model.

Suppose that $\boldsymbol{\theta}$ is a vector of parameters. One general assumption on $(\boldsymbol{\theta}, X)$ is made as follows.

Assumption 5: *$(\boldsymbol{\theta}, X)$ is the solution of a martingale problem for a generator $\mathbf{A}_{\boldsymbol{\theta}}$ such that*

$$M_f(t) = f(\boldsymbol{\theta}, X(t)) - \int_0^t \mathbf{A}_{\boldsymbol{\theta}} f(\boldsymbol{\theta}, X(s))ds$$

is a $\mathcal{F}_t^{\boldsymbol{\theta},X}$*-martingale, where* $\mathcal{F}_t^{\boldsymbol{\theta},X}$ *is the* σ*-field generated by* $(\boldsymbol{\theta}, X(s))_{0\le s\le t}$*.*

Set $\boldsymbol{\theta} = (a, b, \sigma, \rho)$, the generator for in the model is

$$\mathbf{A}_{\boldsymbol{\theta}} f(\boldsymbol{\theta}, x) = (a + bx)\frac{\partial f}{\partial x}(\boldsymbol{\theta}, x) + \frac{1}{2}\sigma^2 \frac{\partial^2 f}{\partial v^2}(\boldsymbol{\theta}, x). \tag{3}$$

Define $\mathcal{F}_t^{\boldsymbol{Y}} = \sigma\{\boldsymbol{Y}(s)\,|\,0 \le s \le t\}$, the σ-algebra generated by the observed sample path of the price $\boldsymbol{Y}$ up to time t. Define π_t as the conditional distribution of $(\boldsymbol{\theta}, X(t))$ given $\mathcal{F}_t^{\boldsymbol{Y}}$ and $\pi(f,t) = E^P[f(\boldsymbol{\theta}, X(t))|\mathcal{F}_t^{\boldsymbol{Y}}] = \int f(\boldsymbol{\theta}, x)\pi_t(d\boldsymbol{\theta}, dx)$. We denote $\mathbf{A}_{\boldsymbol{\theta}}$ simply as $\mathbf{A}$ in the rest.

Theorem1: *Suppose that* $\boldsymbol{Y}$ *is the counting process observations defined in Equation (2) with Assumptions 1 to 4 and* $(\boldsymbol{\theta}, X)$ *satisfies Assumption 5. Then,* π_t*, is the unique solution of the filtering equation : for every* $t > 0$ *and every* f *in the domain of* $\mathbf{A}$*, the generator,*

$$\begin{aligned}\pi(f,t) = \pi(f,0) &+ \int_0^t \pi(\mathbf{A}f, s)ds \\ &+ \sum_{k=1}^{n} \int_0^t \left[\frac{\pi(fg_k, s-)}{\pi(g_k, s-)} - \pi(f, s-)\right] dY_k(s),\end{aligned} \tag{4}$$

where $Y_k(s)$ *is the* k*th price level in* $\boldsymbol{Y}(s)$ *and* $g_k = g(y_k|x)$ *is the transition probability from* x *to* y_k*, the* k*-th price level.*

The filtering equation provides an effective way to characterize π_t and it is also the optimum filter in the sense of least MSE. Note that the total intensity $a(t)$ disappears in Equation (4).

For the second theorem, denote $(\boldsymbol{\theta}_{\boldsymbol{\epsilon}}, X_\epsilon) \Rightarrow (\boldsymbol{\theta}, X)$ as the weak convergence in the Skorohod topology as $\varepsilon = max(\epsilon_x, |\boldsymbol{\epsilon}|) \to 0$, where $|\boldsymbol{\epsilon}|$ is the norm of a vector. Then, $(\boldsymbol{\theta}_{\boldsymbol{\epsilon}}, X_\epsilon)$ is an approximate of $(\boldsymbol{\theta}, X)$. Define

$$\boldsymbol{Y}_\varepsilon(t) = \begin{pmatrix} N_1(\int_0^t \lambda_1(\boldsymbol{\theta}_{\boldsymbol{\epsilon}}, X_{\epsilon_x}(s-), s-)ds) \\ N_2(\int_0^t \lambda_2(\boldsymbol{\theta}_{\boldsymbol{\epsilon}}, X_{\epsilon_x}(s-), s-)ds) \\ \vdots \\ N_n(\int_0^t \lambda_n(\boldsymbol{\theta}_{\boldsymbol{\epsilon}}, X_{\epsilon_x}(s0), s-)ds) \end{pmatrix}, \tag{5}$$

and $\mathcal{F}_t^{\boldsymbol{Y}_\varepsilon} = \sigma\{\boldsymbol{Y}_\varepsilon(s)\,|\,0 \le s \le t\}$.

Theorem 2: *Suppose that* $(\boldsymbol{\theta}, X, \boldsymbol{Y})$ *is on the probability space* $(\Omega, \mathcal{F}, P)$ *and Assumptions 1 to 5 hold. Suppose that* $(\boldsymbol{\theta}_{\boldsymbol{\epsilon}}, X_{\epsilon_x}, \boldsymbol{Y}_\varepsilon)$ *is on* $(\Omega_\varepsilon, \mathcal{F}_\varepsilon, P_\varepsilon)$*, and*

Assumptions 1 to 5 also hold.
If $(\boldsymbol{\theta}_{\boldsymbol{\epsilon}}, X_{\epsilon_x}) \Rightarrow (\boldsymbol{\theta}, X)$ *as* $\varepsilon = \max\{\epsilon_x, |\boldsymbol{\epsilon}|\} \to 0$, *then*
(i) $\boldsymbol{Y}_{\varepsilon} \Rightarrow \boldsymbol{Y}$, *as* $\varepsilon \to 0$*; and*
(ii) $E^{P_\varepsilon}[F(\boldsymbol{\theta}_{\boldsymbol{\epsilon}}, X_{\epsilon_x}(t))|\mathcal{F}_t^{\boldsymbol{Y}_\varepsilon}] \Rightarrow E^P[F(\boldsymbol{\theta}, X(t))|\mathcal{F}_t^{\boldsymbol{Y}}]$ *as* $\varepsilon \to 0$, *for all* F *in the domain of* $\mathbf{A}$, *the generator.*

Theorem 2 implies that as long as $(\boldsymbol{\theta}_{\boldsymbol{\epsilon}}, X_{\epsilon_x})$ is an approximate of $(\boldsymbol{\theta}, X)$, $\boldsymbol{Y}_{\varepsilon}(t)$ is an approximate of $\boldsymbol{Y}(t)$ and that the conditional expectation of $(\boldsymbol{\theta}_{\boldsymbol{\epsilon}}, X_{\epsilon_x})$ given the observed sample path is close to that of $(\boldsymbol{\theta}, X)$. When we take "F" as an appropriate indicator function, $E^{P_\varepsilon}[F(\boldsymbol{\theta}_{\boldsymbol{\epsilon}}, X_{\epsilon_x}(t))|\mathcal{F}_t^{\boldsymbol{Y}_\varepsilon}]$ becomes the conditional probability mass function (pmf) for $(\boldsymbol{\theta}_{\boldsymbol{\epsilon}}, X_{\epsilon_x})$. Theorem 2, then, implies that the conditional pmf is close to the conditional distribution of $(\boldsymbol{\theta}, X(t))$. In the next subsection, the recursive algorithm is constructed to compute such a conditional pmf.

3.2 Recursive Algorithm

Theorem 1 gives the optimum filter and Theorem 2 provides the recipe to construct a consistent recursive algorithm as an approximate optimum filter. We apply the idea of the Markov chain approximation method (see [7] Chapter 12) in nonlinear filtering problem where the Markov chain is equally-spaced in time. Under the assumption that trades occur according to an inhomogeneous Poisson process, the simplest process to approximate $X(t)$ is the birth and death process. Therefore, we use the birth and death process approximation method in the construction. There are three major steps.

Step 1: construct $(\boldsymbol{\theta}_{\boldsymbol{\epsilon}}, X_{\epsilon_x})$, here $(\boldsymbol{\theta}_{\boldsymbol{\epsilon}} = (a_{\epsilon_a}, b_{\epsilon_b}, \sigma_{\epsilon_\sigma}, \rho_{\epsilon_\rho}))$.

First, we latticize the parameter spaces of a, b, σ, ρ and the state space of X. Suppose there are $n_a + 1$, $n_b + 1$, $n_\sigma + 1$, $n_\rho + 1$ and $n_x + 1$ lattices in the latticized spaces of a, b, σ, ρ and X respectively. For example, the discretization for b is, $b : [\alpha_b, \beta_b] \to \{\alpha_b, \alpha_b{+}\epsilon_b, \alpha_b{+}2\epsilon_b, \ldots, \alpha_b{+}j\epsilon_b, \ldots, \alpha_b{+} n_b\epsilon_b\}$ where $\alpha_b + n_b\epsilon_b = \beta_b$ and the number of lattices is $n_b + 1$. Define $b_j = \alpha_b + j\epsilon_b$, the jth lattice in the latticized parameter space of b. Similarly, define $a_i = \alpha_a + i\epsilon_a$, $\sigma_k = \alpha_\sigma + k\epsilon_\sigma$, $\rho_m = \alpha_\rho + m\epsilon_\rho$, and $x_l = x_l(t) = \alpha_x + l\epsilon_x$.

Then, we construct a birth and death process to approximate $X(t)$. This can be done by constructing a birth and death generator, $\mathbf{A}_\varepsilon$, such that $\mathbf{A}_\varepsilon \to \mathbf{A}$, the generator of the model defined in Equation (3). The generator involves first- and second-order differentiation and the finite difference approximation is applied. Then,

$$
\begin{aligned}
&\mathbf{A}_\varepsilon f(a_i, b_j, \sigma_k, \rho_m, x_l) \\
&\quad = (a_i + b_j x_l)\left(\frac{f(a_i, b_j, \sigma_k, \rho_m, x_l + \epsilon_x) - f(a_i, b_j, \sigma_k, \rho_m, x_l - \epsilon_x)}{2\epsilon_x}\right) \\
&\quad + \frac{1}{2}\sigma_k^2\left(\frac{f(a_i, b_j, \sigma_k, \rho_m, x_l + \epsilon_x) + f(a_i, b_j, \sigma_k, \rho_m, x_l - \epsilon_x)}{\epsilon_x^2}\right)
\end{aligned}
$$

$$
\begin{aligned}
&-\frac{1}{2}\sigma_k^2\left(\frac{2f(a_i,b_j,\sigma_k,\rho_m,x_l)}{\epsilon_x^2}\right)\\
&=a(a_i,b_j,\sigma_k,x_l)(f(a_i,b_j,\sigma_k,\rho_m,x_l+\epsilon_x)-f(a_i,b_j,\sigma_k,\rho_m,x_l))\\
&\quad+b(a_i,b_j,\sigma_k,x_l)(f(a_i,b_j,\sigma_k,\rho_m,x_l-\epsilon_x)-f(a_i,b_j,\sigma_k,\rho_m,x_l)),
\end{aligned}
\tag{6}
$$

where

$$
a(a_i,b_j,\sigma_k,x_l)=\frac{1}{2}\left(\frac{\sigma_k^2}{\epsilon_x^2}+\frac{a_i+b_jx_l}{\epsilon_x}\right),
$$

and

$$
b(a_i,b_j,\sigma_k,x_l)=\frac{1}{2}\left(\frac{\sigma_k^2}{\epsilon_x^2}-\frac{a_i+b_jx_l}{\epsilon_x}\right).
$$

Note that $a(a_i,b_j,\sigma_k,x_l)$ is the birth rate and $b(a_i,b_j,\sigma_k,x_l)$ is the death rate. These rates should be nonnegative all the time. If for some values of a, b, σ and X in their ranges, one of the rates becomes negative, then we can always make it positive by making ϵ_x smaller.

Clearly, $\mathbf{A}_\varepsilon \to \mathbf{A}$ and we have $(a_{\epsilon_a},b_{\epsilon_b},\sigma_{\epsilon_\sigma},\rho_{\epsilon_\rho},X_{\epsilon_x}) \Rightarrow (a,b,\sigma,\rho,X)$ as $\varepsilon \to 0$.

Now, we have the approximate model $(\boldsymbol{\theta}_{\boldsymbol{\epsilon}},X_{\epsilon_x}(t))$ of $(\boldsymbol{\theta},X(t))$. Then, we have the approximate $\boldsymbol{Y}_\varepsilon$ which is defined by Equation (5). Note that the counting process observations can be viewed as $\boldsymbol{Y}(t)$ defined by Equation (2) or $\boldsymbol{Y}_\varepsilon(t)$ depending on whether the driving process is $(\boldsymbol{\theta},X(t))$ or $(\boldsymbol{\theta}_\epsilon,X_{\epsilon_x}(t))$. When we model the parameters and the stock value as $(\boldsymbol{\theta},X(t))$, the counting process observations of stock price are regarded as $\boldsymbol{Y}(t)$. When we intend to compute the posteriors of the parameters and the stock value, we use $(\boldsymbol{\theta}_{\boldsymbol{\epsilon}},X_{\epsilon_x}(t))$ to approach $(\boldsymbol{\theta},X(t))$ and the counting process observations of stock price are regarded as $\boldsymbol{Y}_\varepsilon(t)$.

The recursive algorithm is to compute the posterior for the approximate model $(\boldsymbol{\theta}_{\boldsymbol{\epsilon}},X_{\epsilon_x},\boldsymbol{Y}_\varepsilon)$, which is close to the posterior of the model $(\boldsymbol{\theta},X,\boldsymbol{Y})$ by Theorem 2, when ε is small.

Step 2: Obtain the filtering equation for the approximate model. When (a,b,σ,ρ,X) is approximated by $(a_{\epsilon_a},b_{\epsilon_b},\sigma_{\epsilon_\sigma},\rho_{\epsilon_\rho},X_{\epsilon_x})$, $\mathbf{A}$ by $\mathbf{A}_\varepsilon$, and $\boldsymbol{Y}$ by $\boldsymbol{Y}_\varepsilon$, there also exists a probability measure P_ε to approximate P. It can be checked that Assumptions 1 to 5 hold for $(a_{\epsilon_a},b_{\epsilon_b},\sigma_{\epsilon_\sigma},\rho_{\epsilon_\rho},X_{\epsilon_x},\boldsymbol{Y}_\varepsilon)$. We define two terms which are the discretized approximations of π_t and $\pi(f,t)$. Define $\pi_{\varepsilon,t}$ as the conditional distribution of $(a_{\epsilon_a},b_{\epsilon_b},\sigma_{\epsilon_\sigma},\rho_{\epsilon_\rho},X_{\epsilon_x})$ given $\mathcal{F}_t^{\boldsymbol{Y}_\varepsilon}$ and define

$$
\begin{aligned}
\pi_\varepsilon(f,t)&=E^{P_\varepsilon}[f(a_{\epsilon_a},b_{\epsilon_b},\sigma_{\epsilon_\sigma},\rho_{\epsilon_\rho},X_{\epsilon_x}(t))|\mathcal{F}_t^{\boldsymbol{Y}_\varepsilon}]\\
&=\sum_{a,b,\sigma,\rho,X} f(a,b,\sigma,\rho,X)\pi_{\varepsilon,t}(a,b,\sigma,\rho,X),
\end{aligned}
\tag{7}
$$

where (a, b, σ, ρ, X) goes over all the lattices in the approximate state spaces at time t.

By Theorem 1, we obtain the filtering equation for the approximate model in a similar form:

$$\begin{aligned}\pi_\varepsilon(f,t) = \pi_\varepsilon(f,0) + \int_0^t \pi_\varepsilon(\mathbf{A}_\varepsilon f, s) ds \\ + \sum_{k=1}^n \int_0^t \left[\frac{\pi_\varepsilon(fg_k, s-)}{\pi_\varepsilon(g_k, s-)} - \pi_\varepsilon(f, s-)\right] dY_{\varepsilon,k}(s),\end{aligned} \tag{8}$$

where $Y_{\varepsilon,k}(s)$ is the k-th component of $\mathbf{Y}_\varepsilon(s)$.

The above filtering equation can be separated into the propagation equation: (assuming $\{t_i\}$ is the sequence of trading times,)

$$\pi_\varepsilon(f, t_{i+1}-) = \pi_\varepsilon(f, t_i) + \int_{t_i}^{t_{i+1}-} \pi_\varepsilon(\mathbf{A}_\varepsilon f, s) ds, \tag{9}$$

and the updating equation:

$$\pi_\varepsilon(f, t_{i+1}) = \frac{\pi_\varepsilon(fg_k, t_{i+1}-)}{\pi_\varepsilon(g_k, t_{i+1}-)}. \tag{10}$$

The above two Equations (9) and (10) are the key equations to derive the recursive algorithm.

Step 3: Convert the filtering equation of the approximate model to the recursive algorithm. Assume $a_\epsilon = a_{\epsilon_a}, b_\epsilon = b_{\epsilon_b}, \sigma_\epsilon = \sigma_{\epsilon_\sigma}, \rho_\epsilon = \rho_{\epsilon_\rho}$, and $X_\epsilon = X_{\epsilon_x}$. Define the indicator function:

$$\begin{aligned}&\mathbf{I}(a_i, b_j, \sigma_k, \rho_m, x_l) \\ &\quad = \mathbf{I}_{\{a_\epsilon = a_i, b_\epsilon = b_j, \sigma_\epsilon = \sigma_k, \rho_\epsilon = \rho_m, X_\epsilon(t) = x_l\}}(a_\epsilon, b_\epsilon, \sigma_\epsilon, \rho_\epsilon, X_\epsilon(t)).\end{aligned} \tag{11}$$

Define the approximate posterior at $(a_i, b_j, \sigma_k, \rho_m, x_l)$ at time t as

$$p(a_i, b_j, \sigma_k, \rho_m, x_l;\, t) = E^P\left[\mathbf{I}(a_i, b_j, \sigma_k, \rho_m, x_l) | \mathcal{F}_t^{\mathbf{Y}_\varepsilon}\right]. \tag{12}$$

The core of the conversion is to take f as the above indicator in Equations (9) and (10). Observe that

$$\begin{aligned}&E^P\left[a(a_\epsilon, b_\epsilon, \sigma_\epsilon, X_\epsilon(t))\mathbf{I}(a_i, b_j, \sigma_k, \rho_m, x_l + \epsilon_x) | \mathcal{F}_t^{\mathbf{Y}_\varepsilon}\right] \\ &\quad = a(a_i, b_j, \sigma_k, x_{l-1}) p(a_i, b_j, \sigma_k, \rho_m, x_{l-1}; t),\end{aligned}$$

and

$$\begin{aligned}&E^P\left[b(a_\epsilon, b_\epsilon, \sigma_\epsilon, X_\epsilon(t))\mathbf{I}(a_i, b_j, \sigma_k, \rho_m, x_l - \epsilon_x) | \mathcal{F}_t^{\mathbf{Y}_\varepsilon}\right] \\ &\quad = b(a_i, b_j, \sigma_k, x_{l+1}) p(a_i, b_j, \sigma_k, \rho_m, x_{l+1}; t).\end{aligned}$$

Then, along with two similar results, $\pi_\varepsilon(\mathbf{A}_\varepsilon \mathbf{I}, t)$ in Equation (9) becomes explicit as

$$\begin{aligned}\pi_\varepsilon(\mathbf{A}_\varepsilon \mathbf{I}, t) = & \, a(a_i, b_j, \sigma_k, x_{l-1}) p(a_i, b_j, \sigma_k, \rho_m, x_{l-1}; t) \\ & + b(a_i, b_j, \sigma_k, x_{l+1}) p(a_i, b_j, \sigma_k, \rho_m, x_{l+1}; t)) \\ & - (a(a_i, b_j, \sigma_k, x_l) + b(a_i, b_j, \sigma_k, x_l)) p(a_i, b_j, \sigma_k, \rho_m, x_l; t).\end{aligned}$$

Therefore, the propagation equation in Equation (9) becomes

$$\begin{aligned}& p(a_i, b_j, \sigma_k, \rho_m, x_l; t_{i+1}-) \\ & \quad = p(a_i, b_j, \sigma_k, \rho_m, x_l; t_i) \\ & \qquad + \int_{t_i}^{t_{i+1}-} \Big(a(a_i, b_j, \sigma_k, x_{l-1}) p(a_i, b_j, \sigma_k, \rho_m, x_{l-1}; s) \\ & \qquad - \big(a(a_i, b_j, \sigma_k, x_l) + b(a_i, b_j, \sigma_k, x_l)\big) p(a_i, b_j, \sigma_k, \rho_m, x_l; s) \\ & \qquad + b(a_i, b_j, \sigma_k, x_{l+1}) p(a_i, b_j, \sigma_k, \rho_m, x_{l+1}; s) \Big) ds.\end{aligned} \tag{13}$$

When a trade occurs at time t_{i+1} at the k_0th price level, p_{k_0}, the updating equation in Equation (10) becomes

$$\begin{aligned}& p(a_i, b_j, \sigma_k, \rho_m, x_l; t_{i+1}) \\ & \quad = \frac{p(a_i, b_j, \sigma_k, \rho_m, x_l; t_{i+1}-) g(y_{k_0} | x_l, \rho_m)}{\sum_{i',j',k',m',l'} p(a_{i'}, b_{j'}, \sigma_{k'}, \rho_{m'}, x_{l'}; t_{i+1}-) g(y_{k_0} | x_{l'}, \rho_{m'})},\end{aligned} \tag{14}$$

where the summation is over the latticized spaces of a, b, σ, ρ and $X(t_{i+1}-)$. $g(y_{k_0}|x_l, \rho_m)$ is the transition probability from x_l to y_{k_0}, which also depends on ρ_m. Note that $g(y|x)$ is specified in Equation (17) in Appendix A.

Next, we make Equation (13) a recursive algorithm. Note Equation (13) is deterministic and we employ Euler scheme for approximation. After excluding the probability-zero event that two or more jumps occur at the same time, there are two possible cases for the inter-trading time. Case 1, if $t_{i+1} - t_i \leq LL$, the length controller, then we can approximate $p(a_i, b_j, \sigma_k, \rho_m, x_l; t_{i+1}-)$ as

$$\begin{aligned}& p(a_i, b_j, \sigma_k, \rho_m, x_l; t_{i+1}-) \\ & \quad \approx p(a_i, b_j, \sigma_k, \rho_m, x_l; t_i) \\ & \qquad + \Big[a(a_i, b_j, \sigma_k, x_{l-1}) p(a_i, b_j, \sigma_k, \rho_m, x_{l-1}; t_i) \\ & \qquad\quad - (a(a_i, b_j, \sigma_k, x_l) + b(a_i, b_j, \sigma_k, x_l)) p(a_i, b_j, \sigma_k, \rho_m, x_l; t_i) \\ & \qquad\quad + b(a_i, b_j, \sigma_k, x_{l+1}) p(a_i, b_j, \sigma_k, \rho_m, x_{l+1}; t_i) \Big] (t_{i+1} - t_i).\end{aligned} \tag{15}$$

Case 2, if $t_{i+1} - t_i > LL$, then we can choose a fine partition $\{t_{i,0} = t_i, t_{i,1}, \ldots, t_{i,n} = t_{i+1}\}$ of $[t_i, t_{i+1}]$ such that $\max_j |t_{i,j+1} - t_{i,j}| < LL$ and then approximate $p(a_i, b_j, \sigma_k, \rho_m, x_l; t_{i+1}-)$ by applying repeatedly the recursive algorithm given by Equation (15) from $t_{i,0}$ to $t_{i,1}$, then $t_{i,2}, \ldots$, until $t_{i,n} = t_{i+1}$.

Equations (14) and (15) consist of the recursive algorithm we employ to calculate the posterior at time t_{i+1} for $(a, b, \sigma, \rho, X(t_{i+1}))$ based on the posterior at time t_i.

Finally, we choose a reasonable prior. We assume the independence between $X(0)$ and (a, b, σ, ρ). Set $P\{X(0) = Y(t_1)\} = 1$ where $Y(t_1)$ is the first trade price of a data set because they are very close. If there is no special information of (a, b, σ, ρ) available, we may simply assign uniform distributions to the latticized state space of (a, b, σ, ρ) and obtain the prior at $t = 0$ as,

$$p(a_i, b_j, \sigma_k, \rho_m, x_l; 0) = \begin{cases} \frac{1}{(1+n_a)(1+n_b)(1+n_\sigma)(1+n_\rho)} & \text{if } x_l = Y(t_1) \\ 0 & \text{otherwise} \end{cases}.$$

3.3 Consistency of the Recursive Algorithm

Reviewing the construction of the recursive algorithm, we notice that there are two approximations. One is to approach the integral in the propagation equation (13) by Euler scheme, whose convergence is well-known. The other one, which is more important, is the approximation of the filtering Equation (4) (the optimum filter) by the filtering Equation (8) of the approximate model (the approximate optimum filter). Since $(\boldsymbol{\theta}_{\boldsymbol{\epsilon}}, X_{\epsilon_x}) \Rightarrow (\boldsymbol{\theta}, X)$ by construction, Theorem 2 guarantees the convergence of the filtering Equation (8) to filtering Equation (4) in the sense of weak convergence in the Skorohod topology, that is, the consistency of the approximate filter.

In this subsection, denote $p(a_i, b_j, \sigma_k, \rho_m, x_l; t)$ defined in Equation (12) as $p_\varepsilon(a_i, b_j, \sigma_k, \rho_m, x_l; t)$. Define the neighborhood of b_j as $N_{b_j} = \{b_j - 0.5\epsilon_b \leq b < b_j + 0.5\epsilon_b\}$ and similarly for the neighborhoods of a_i, σ_k, ρ_m and x_l. Define the true posterior in the neighborhood of $(a_i, b_j, \sigma_k, \rho_m, x_l)$ as

$$\begin{aligned} &p(a_i, b_j, \sigma_k, \rho_m, x_l; t) \\ &\quad = E^P\left[\mathbf{I}_{\{a \in N_{a_j}, b \in N_{b_j}, \sigma \in N_{\sigma_k}, \rho \in N_{\rho_m}, X(t) \in N_{X_l(t)}\}}(a, b, \sigma, \rho, X(t)) | \mathcal{F}_t^{\mathbf{Y}}\right]. \end{aligned}$$

Note that the indicator function defined above becomes the indicator function defined in Equation (11) in the approximate model. Therefore, Theorem 2 implies $p_\varepsilon(a_i, b_j, \sigma_k, \rho_m, x_l; t) \to p(a_i, b_j, \sigma_k, \rho_m, x_l; t)$ as $\varepsilon \to 0$.

4 A Real Data Example

In this section, we apply the recursive algorithm described in Section 3 to a two-month Microsoft transaction data.

Table 1. Fractional Parts of Prices

Frac part	0	1/8	1/4	3/8	1/2	5/8	3/4	7/8	Total
Frequency	11218	2791	9136	2199	10009	2826	9376	2382	49937
Rel. Freq.	.2246	.0559	.1830	.0440	.2004	.0566	.1878	.0477	1.0000

4.1 Data Description

The data are extracted from the Trade and Quote (TAQ) database distributed by NYSE. We apply standard procedures to filter the data but with one important exception. Previous studies cannot handle multiple trades at a given point in time and they exclude those trades with zero time duration. Autoregressive conditional duration model proposed by Engle and Russell in [4] is such an example. The present method can handle such cases and we keep all zero durations. The final sample has 49,937 observations. As shown in [9], the data has the four discrete-type characteristics modeled, and the distribution of trade duration (or, trade waiting time) although does not fit well by an exponential distribution, it fits well by a mixture of exponential distributions, which is allowed by the model.

Table 1 presents the frequencies and relative frequencies of the fractional parts of price. Price clustering is obvious. Since the frequencies of integer and half are close, we may assume the chances to move from odd eighths to integers and halves are the same and fit the model. From the Table 1, we can estimate α and β by the method of relative frequency. Let f_i be the observed frequency of the fractional parts such that $r(y) = i$, where $r(y)$ is defined by Equation (16) in Appendix A. Assuming that the fractional part of $X(t)$ is uniformly distributed and matching the observed frequencies with the theoretical frequencies, we have $f_1 = \frac{1}{4}+\frac{1}{2}\beta,\ \ f_2 = \frac{1}{4}+\frac{1}{2}\alpha,\ \ f_3 = \frac{1}{2}(1-\alpha-\beta)$. with the unique solution: $\hat{\beta} = 2(f_1 - (1/4))$, $\hat{\alpha} = 2(f_2 - (1/4))$. Then, the estimates are $\hat{\alpha} = 2(.3707 - .25) = .2414$, $\hat{\beta} = 2(.4251 - .25) = .3502$.

4.2 Bayes Estimates for Microsoft Data

A Fortran program is constructed to compute the approximate posterior and the Bayes estimates, which are the posterior means. Standard simulation study is done to test its correctness.

It took about one day to run through the two-month data set on a Compaq XP1000 computer. It is faster than the real-time estimates require. The Bayes estimates are computed for second. Assuming 260 business days per year, the annualized estimates then are obtained. Both estimates for the Microsoft data are summarized in Table 2. Note that a and b are not significantly different

Table 2. Bayes Estimates of MSFT, Jan. and Feb. 1994. Numbers in "()" are standard errors and the annualized factor is 260.

Time Unit	a	b	σ	ρ
Second	-2.901e-10	6.970e-8	8.836e-3	0.2262
	(1.014e-6)	(1.003e-6)	(1.085e-4)	(0.0023)
Year	-1.765e-3	0.4240	21.79	0.2262
	(6.169)	(6.102)	(0.26)	(0.0023)

from zero, and the standard errors of a and b is much larger than those of σ and ρ (comparing to the Bayes estimates). This makes sense because a and b are trend parameters, and their accuracy of estimation depends on the length covered by the data (40 days); but the accuracy of estimating σ and ρ mainly depend on the number of observations (49,937 data).

5 Conclusions

In this paper, we propose a micro-movement model with three types of discrete noises based on an O-U process. The model has an important feature in that it can be formulated as a filtering problem with counting process observations. Under this formulation, the whole samples are observable, and complete information is used in Bayes estimation via filtering equation for the parameters of the model. A consistent recursive algorithm is developed to compute the approximated posterior and then the Bayes estimates. The recursive algorithm is fast and feasible for large data sets and it has the recursive feature allowing quick and easy update. The recursive algorithm is applied to Microsoft's transaction data and we obtain Bayes estimates and provide strong affirmative evidence that the price clustering noise matters for parameter estimation.

The model and its Bayes estimation via filtering equation can be extended to diffusion or jump-diffusion process for the value process, and other kinds of noise according to the sample characteristics of data. The model and the Bayes estimation can be applied to other asset markets such as exchange rates and commodity prices. It can also apply to assess the quality of security market, and to compare information flows and noises in different periods and different markets.

References

1. Black, F. (1986) Noise, J. of Finance **41**, 529–543.
2. Bremaud, P. (1981) Point Processes and Queues: Martingale Dynamics. Springer-Verlag, New York.
3. Elliott R. J., Lakhdar A., et al. (1995) Hidden Markov Models: Estimation and Control. Springer-Verlag, New York.
4. Engle, R. and Russell, J. (1998) Autoregressive conditional duration: A new model for irregularly spaced transaction data, Econometrica **66**, 1127–1162.
5. Harris, L. (1991) Stock price clustering and discreteness, Rev. Fin. Studies **4**, 389–415.
6. Hasbrouck J. (1996) Modeling market microstructure time series, in Handbook of Statistics, editted by G. S. Maddala and C. R. Rao. **14**, 647–692.
7. Kushner H. J. and Dupuis P. G. (1994) Numerical Methods for Stochastic Control Problems in Continuous Time. Springer-Verlag, New York.
8. Zeng, Y. (2001) A partially-observed model for micro-movement of stock price with Bayes estimation via filtering equation, Working Paper. Department of Mathematics and Statistics, University of Missouri at Kansas City.
9. Zeng, Y. (2001) Bayesian estimation for a simple micro-movement stock price model with discrete noises, Working Paper. Department of Mathematics and Statistics, University of Missouri at Kansas City.

Appendix

To formulate the biasing rule, we first define a classifying function $r(\cdot)$,

$$r(y) = \begin{cases} 3 \text{ if the fractional part of } y \text{ is odd eighth} \\ 2 \text{ if the fractional part of } y \text{ is odd quarter} \\ 1 \text{ if the fractional part of } y \text{ is a half or zero} \end{cases} . \tag{16}$$

The biasing rules specify the transition probabilities from x' to p, $g(y|x')$as below: $g(x'|x') = 1$, if $r(x') = 1$ or 2; $g(x'|x') = 1 - \alpha - \beta$, if $r(x') = 3$; $g(x'+\frac{1}{8}|x') = \alpha$, if $x' - \text{floor}(x') = \frac{1}{8}$ or $\frac{5}{8}$; $g(x'-\frac{1}{8}|x') = \alpha$, if $x' - \text{floor}(x') = \frac{3}{8}$ or $\frac{7}{8}$; $g(x' - \frac{1}{8}|x') = \beta$, if $x' - \text{floor}(x') = \frac{1}{8}$ or $\frac{5}{8}$; $g(x' + \frac{1}{8}|x') = \beta$, if $x' - \text{floor}(x') = \frac{3}{8}$ or $\frac{7}{8}$. Note that the floor(x) function gives the integer part of x.

Then, $g(y|x)$, the transition probability can be computed through $g(y|x) = \sum_{x'} g(y|x')g(x'|x)$ where $g(y|x')$s are just defined above and $g(x'|x) = P\{U = 8(x' - R[x, \frac{1}{8}])\}$. Suppose $D = 8|y - R[x, \frac{1}{8}]|$. Then, $g(y|x)$ is

$$g(y|x) = \begin{cases} (1-\alpha-\beta)(1-\rho) & \text{if } r(y) = 3 \text{ and } D = 0 \\ \frac{1}{2}(1-\alpha-\beta)(1-\rho)\rho^D & \text{if } r(y) = 3 \text{ and } D \geq 0 \\ (1-\rho)(1+\alpha\rho) & \text{if } r(y) = 2 \text{ and } D = 0 \\ \frac{1}{2}(1-\rho)[\rho+\alpha(2+\rho^2)] & \text{if } r(y) = 2 \text{ and } D = 1 \\ \frac{1}{2}(1-\rho)\rho^{D-1}[\rho+\alpha(1+\rho^2)] & \text{if } r(y) = 2 \text{ and } D \geq 2 \\ (1-\rho)(1+\beta\rho) & \text{if } r(y) = 1 \text{ and } D = 0 \\ \frac{1}{2}(1-\rho)[\rho+\beta(2+\rho^2)] & \text{if } r(y) = 1 \text{ and } D = 1 \\ \frac{1}{2}(1-\rho)\rho^{D-1}[\rho+\beta(1+\rho^2)] & \text{if } r(y) = 1 \text{ and } D \geq 2 \end{cases} . \tag{17}$$

Hybrid Filtering

Q. Zhang

University of Georgia, Athens, GA 30602, USA

Abstract. This paper is concerned with filtering of a hybrid model with a number of linear systems coupled by a hidden switching process. The most probable trajectory approach is used to derive a finite-dimensional recursive filter. Such scheme is applied to nonlinear systems using a piecewise-linear approximation method. Numerical examples are provided and computational experiments are reported.

1 Introduction

There are many real-world applications of hybrid models including target tracking, telecommunication, and manufacturing. Owing to various disturbances, the state variables of these systems are usually not completely observable. Hybrid filtering is crucial to control of these systems. Let $x(t)$, $t \geq 0$, denote a signal process which is not directly observable and let $\theta(t) \in \{1, 2, \dots, m\}$, $t \geq 0$, be a hidden switching process which represents the mode of the hybrid system. Consider the case that a function of $\theta(t)$ and $x(t)$ with an additive noise is observable. Let $y(t)$ denote such observation process. The signal $x(t) \in \mathbb{R}^{n_1}$ and observation $y(t) \in \mathbb{R}^{n_2}$ are governed by equations of the following form:

$$\begin{cases} \dot{x}(t) = A(t,\theta(t))x(t) + A_1(t,\theta(t)) + \text{system noise}, \\ y(t) = H(t,\theta(t))x(t) + H_1(t,\theta(t)) + \text{observation noise}, \end{cases} \tag{1}$$

where $0 \leq t \leq T < \infty$ and $A(\cdot, i)$, $A_1(\cdot, i)$, $H(\cdot, i)$, $H_1(\cdot, i)$, $i = 1, \dots, m$, are matrices of appropriate dimension. The objective of the filtering problem is to find an estimate of $x(t)$ based on the observation up to time t.

Traditionally, both the system noise and the observation noise are modeled as Brownian motions. The corresponding filtering problem is to find the conditional mean of $x(t)$ given the observation up to time t. A number of related models along this line are considered in the literature. For example, Hijab [9] considered the optimal filtering when $\theta(t)$ is an unknown constant. Zhang [14] considered the nonlinear filtering problem using statistical methods and obtained near-optimal filters for models with small observation noise; see also Fleming and Pardoux [6] and Haussmann and Zhang [8] for related problems with small observation noise. Recently, Miller and Runggaldier [11] studied the optimal filtering problem. Assuming, in addition, the jump times of the Markov chain to be observable, they obtained exact optimal scheme.

B. Pasik-Duncan: Stochastic Theory and Control, LNCIS 280, pp. 549–564, 2002.

However, the algorithm is not practical from a computational point of view because of its "branching" type of nature in the sense that the filtering dimension increases after each jump of the underlying Markov chain.

Related discrete-time models were considered in Blom and Bar-Shalom [2] and Costa [3]; see also Dufour and Elliott [4] for a slightly different model with additional observation equations. In [3], a linear minimum mean square error (LMMSE) estimator was obtained. The idea is to consider an augmented system so as to estimate the pair of the switching and signal processes simultaneously. In [2], the authors proposed the so-called Interactive Multiple Models (IMM) algorithm to compute the conditional mean. The IMM algorithm is a popular scheme in applications including target tracking. It performs better than most existing algorithms in the literature. However, it lacks theoretical justification for the optimality of these filters; see Bar-Shalom and Li [1] for further discussions.

In this paper, we consider the hybrid model with non-Gaussian disturbances. This model is more realistic because the Gaussian assumption is often hard to justify in practice. Moreover, since the model allows virtually arbitrary disturbances, it is ready for systems with mixed-times such as continuous-time signal and discrete-time observation; see Zhang [16] for details.

An interesting approach in filtering is that of the most probable trajectory (MPT) estimate; see, for example, Mortensen [12]. See also Fleming and McEneaney [5] in connection with robust filtering. An advantage of the MPT approach is that it does not require any specific conditions on the noise types. The disadvantage is, however, that such approach seems only suitable for deterministic systems. In this paper, we present a modification of the MPT approach to account for random switching. This is accomplished by a shift transformation introduced in [16]. This transformation allows us to convert the hybrid system into an averaged system and to derive the recursive (exact) optimal filter in the sense of the MPT estimate. It is interesting to note that such MPT approach leads to a *finite-dimensional* hybrid filter in contrast to the infinite-dimensional filter with the usual conditional mean approach. Another useful feature of the model considered in this paper is the uncertainty of the switching probability distribution. The MPT approach allows us to identify the most probable distribution as well among a finite number of candidates. This paper summarizes results in [16,10] for continuous-time models. Related discrete-time filtering is treated in Zhang [15].

The rest of the paper is organized as follows. In the next section, we present the MPT filtering. In Section 3, we give three numerical examples. First, we demonstrate the performance of the MPT scheme and compare it with the IMM, LMMSE and quadratic variation test (QVT) algorithms. Then we consider two nonlinear models and provide procedures for hybrid approximations. Numerical experiments are also reported in these cases.

2 The MPT Filtering

Let us consider the following linear hybrid model:

$$\begin{cases} \dot{x}(t) = A(t,\theta(t))x(t) + A_1(t,\theta(t)) + u(t,\theta(t)), \\ y(t) = H(t,\theta(t))x(t) + H_1(t,\theta(t)) + v(t,\theta(t)). \end{cases} \tag{2}$$

The disturbances $u(t,\theta(t))$ and $v(t,\theta(t))$ are defined as follows: For each $i \in \mathcal{M}$,

$$u(t,i) = \dot{x}(t) - A(t,i)x(t) - A_1(t,i),$$

$$v(t,i) = y(t) - H(t,i)x(t) - H_1(t,i).$$

In (2), $\theta(t)$ is a stochastic process with unknown probability distribution. We assume that its distribution is among a finite number of candidate distributions. Let $\mathcal{N}_0 = \{1,2,\ldots,n_0\}$, for some n_0, and let $\mathcal{P} = \{\phi^{(1)}\cdot,\ldots,\phi^{(n_0)}\cdot\}$ denote the set of such candidate distributions on $\mathcal{M}$, i.e., for $r \in \mathcal{N}_0$, $\phi^{(r)}(t) = (\phi_1^{(r)}(t),\ldots,\phi_m^{(r)}(t))$ with $\phi_i^{(r)}(t) \geq 0$ and $\sum_{i=1}^m \phi_i^{(r)}(t) = 1$. The choice of $\mathcal{P}$ should be based on prior information regarding the distribution of $\theta(t)$.

The objective is to find the MPT estimate of $x(\tau)$ based on the observation up to time τ. In view of the procedure described in Mortensen [12], the problem is to choose $u(\cdot,i)$, $v(\cdot,i)$ and $r \in \mathcal{N}_0$ to minimize

$$\begin{aligned} &\int_0^\tau \sum_{i=1}^m \phi_i^{(r)}(t)\Big(u'(t,i)M(t,i)u(t,i) + v'(t,i)N(t,i)v(t,i)\Big)dt \\ &\qquad + (x(0)-\hat{x}_0)'D(x(0)-\hat{x}_0), \end{aligned} \tag{3}$$

where $\hat{x}_0$ is an initial estimate of x_0 and $M(t,i)$, $N(t,i)$ and D are symmetric nonnegative definite matrices. Throughout, we assume $M(t,i)$ and D to be positive definite.

For each $r \in \mathcal{N}_0$, let $\hat{x}^{(r)}\cdot$ denote the optimal trajectory. Then the MPT estimate (for a fixed $r \in \mathcal{N}_0$) at time τ is given by $\hat{x}^{(r)}(\tau)$. In order to obtain the optimal filter, one needs to identify the most probable distribution in $\mathcal{P}$. Let $\hat{r}(\tau) \in \mathcal{N}_0$ denote the index of such distribution. Then the optimal filter is given by $\hat{x}(\tau) = \hat{x}^{(\hat{r}(\tau))}(\tau)$.

The basic idea of the MPT approach is to "average out" the effect of the switching $\theta(t)$, then derive an estimate for $x(t)$ based on the averaged model. This approach begins with the so-called shift transformation, (see [16]), which changes the drift term in (2) and the corresponding cost function (3).

For notational simplicity, we define

$$\overline{F}^{(r)}(t) = \sum_{i=1}^{m} \phi_i^{(r)}(t) F(t,i)$$

for any matrix function $F(t,i)$ and $r \in \mathcal{N}_0$. Similarly,

$$\overline{F_1 F_2}^{(r)}(t) = \sum_{i=1}^{m} \phi_i^{(r)}(t) F_1(t,i) F_2(t,i)$$

for matrix functions $F_1(t,i)$ and $F_2(t,i)$. Since there is no requirement on the system noise, we can shift the terms A and A_1 in (2) and center them at $\overline{A}^{(r)}(t)$ and $\overline{A}_1^{(r)}(t)$, respectively as follows:

$$\dot{x}(t) = \overline{A}^{(r)}(t)x(t) + \overline{A}_1^{(r)}(t) + u(t),\ 0 \le t \le \tau, \tag{4}$$

where

$$\begin{aligned} u(t) &= u^{(r)}(t) \\ &= (A(t,\theta(t)) - \overline{A}^{(r)}(t))x(t) + (A_1(t,\theta(t)) - \overline{A}_1^{(r)}(t)) + u(t,\theta(t)). \end{aligned}$$

Accordingly we replace the system noise $u(t,i)$ by

$$(\overline{A}^{(r)}(t) - A(t,i))x(t) + (\overline{A}_1^{(r)}(t) - A_1(t,i)) + u(t)$$

and the observation noise $v(t,i)$ by $y(t) - H(t,i)x(t) - H_1(t,i)$, in the cost function defined in (3), respectively. Let

$$\begin{aligned} &L^{(r)}(t,x,u,y) \\ &\quad = \sum_{i=1}^{m} \phi_i^{(r)}(t) \Big\{ [(\overline{A}^{(r)}(t) - A(t,i))x + (\overline{A}_1^{(r)}(t) - A_1(t,i)) + u]' \\ &\qquad\qquad \times M(t,i)[(\overline{A}^{(r)}(t) - A(t,i))x \\ &\qquad\qquad\qquad + (\overline{A}_1^{(r)}(t) - A_1(t,i)) + u] \\ &\qquad\qquad + (y - H(t,i)x - H_1(t,i))' N(t,i) \\ &\qquad\qquad \times (y - H(t,i)x - H_1(t,i)) \Big\}, \\ &\Phi(x) = (x - \hat{x}_0)' D (x - \hat{x}_0). \end{aligned}$$

Then the cost function in (3) can be written as

$$J^{(r)}(\tau, x, u\cdot) = \int_0^\tau L^{(r)}(t, x(t), u(t), y(t))dt + \Phi(x(0)).$$

For each fixed r, the objective of the problem is to choose an admissible $u\cdot$ to minimize $J^{(r)}(\tau, x, u\cdot)$ subject to (4) with $x(\tau) = x$. For each $r \in \mathcal{N}_0$, let $V^{(r)}(\tau, x)$ denote the value function of the control problem, i.e.,

$$V^{(r)}(\tau, x) = \inf_{u\cdot} J^{(r)}(\tau, x, u\cdot).$$

It is easy to obtain the associated Hamilton-Jacobi-Bellman (HJB) equations as follows:

$$\begin{cases} \dfrac{\partial V^{(r)}(\tau, x)}{\partial \tau} = \min_u \Big\{ L^{(r)}(\tau, x, u, y(\tau)) \\ \qquad\qquad - (\overline{A}^{(r)}(\tau)x + \overline{A}_1^{(r)}(\tau) + u)' \dfrac{\partial V^{(r)}(\tau, x)}{\partial x} \Big\}, \\ V^{(r)}(0, x) = \Phi(x), \ r \in \mathcal{N}_0. \end{cases} \tag{5}$$

Now, let

$$\hat{x}^{(r)}(\tau) = \text{argmin}\{V^{(r)}(\tau, x) : \ x \in \mathbb{R}^{n_1}\}.$$

We define $V^{(r)}(\tau) = V^{(r)}(\tau, \hat{x}^{(r)}(\tau))$. Then

$$V^{(r)}(\tau) = \min\{V^{(r)}(\tau, x) : \ x \in \mathbb{R}^{n_1}\}.$$

Let

$$\hat{r}(\tau) = \text{argmin}\{V^{(r)}(\tau) : \ r \in \mathcal{N}_0\}.$$

Then it is easy to show that

$$(\hat{r}(\tau), \hat{x}(\tau)) = \text{argmin}\left\{V^{(r)}(\tau, x) : \ r \in \mathcal{N}_0, x \in \mathbb{R}^{n_1}\right\}.$$

Namely, $(\hat{r}(\tau), \hat{x}(\tau))$ is an optimal MPT filter. Next, we give recursive filtering equations.

Define $\hat{x}^{(r)}$, $R^{(r)}$ and $q^{(r)}$ as follows:

$$\begin{cases} \dot{R}^{(r)} = -R^{(r)} \left[\overline{A'MA}^{(r)} - \overline{A'M}^{(r)} (\overline{M}^{(r)})^{-1} \overline{MA}^{(r)} + \overline{H'NH}^{(r)}\right] R^{(r)} \\ \qquad +(\overline{M}^{(r)})^{-1} + R^{(r)} \overline{A'M}^{(r)} (\overline{M}^{(r)})^{-1} \\ \qquad +(\overline{M}^{(r)})^{-1} \overline{MA}^{(r)} R^{(r)}, \\ R^{(r)}(0) = D^{-1}, \end{cases}$$

(6)

$$
\begin{cases}
\dot{\hat{x}}^{(r)} = -R^{(r)}\Big[\overline{A'MA}^{(r)} - \overline{A'M}^{(r)}(\overline{M}^{(r)})^{-1}\overline{MA}^{(r)} + \overline{H'NH}^{(r)}\Big]\hat{x}^{(r)} \\
\qquad +(\overline{M}^{(r)})^{-1}\overline{MA}^{(r)}\hat{x}^{(r)} \\
\qquad +(\overline{M}^{(r)})^{-1}\overline{MA_1}^{(r)} + R^{(r)} \times \Big(\overline{H'N}^{(r)}y - \overline{A'MA_1}^{(r)} \\
\qquad -\overline{H'NH_1}^{(r)} + \overline{A'M}^{(r)}(\overline{M}^{(r)})^{-1}\overline{MA_1}^{(r)}\Big), \\
\hat{x}^{(r)}(0) = \hat{x}_0,
\end{cases}
\tag{7}
$$

and

$$
\begin{cases}
\dot{q}^{(r)} = -(\hat{x}^{(r)})'(R^{(r)})^{-1}(\overline{M}^{(r)})^{-1}(R^{(r)})^{-1}\hat{x}^{(r)} + y'\overline{N}^{(r)}y \\
\qquad +\overline{A_1'M}^{(r)}(\overline{M}^{(r)})^{-1}(R^{(r)})^{-1}\hat{x}^{(r)} + \overline{A_1'MA_1}^{(r)} \\
\qquad +(\hat{x}^{(r)})'(R^{(r)})^{-1}(\overline{M}^{(r)})^{-1}\overline{MA_1}^{(r)} + \overline{H_1'NH_1}^{(r)} \\
\qquad -\overline{A_1'M}^{(r)}(\overline{M}^{(r)})^{-1}\overline{MA_1}^{(r)} - y'\overline{NH_1}^{(r)} - \overline{H_1'N}^{(r)}y, \\
q^{(r)}(0) = \hat{x}_0'D\hat{x}_0.
\end{cases}
\tag{8}
$$

It is shown in [10] that

$$V^{(r)}(\tau) = -(\hat{x}^{(r)}(\tau))'(R^{(r)}(\tau))^{-1}\hat{x}^{(r)}(\tau) + q^{(r)}(\tau).$$

The optimal filter can be obtained following these two steps:
Step 1: Compute $R^{(r)}(\tau)$ and $\hat{x}^{(r)}(\tau)$ by solving the equations (6) and (7);
Step 2: Compute $q^{(r)}(\tau)$ by solving the equation (8) and find $\hat{r}(\tau)$ that minimizes $V^{(r)}(\tau)$.

The most probable distribution is $\phi^{(\hat{r}(\tau))}$. and the optimal filter is given by

$$(\hat{r}(\tau), \hat{x}(\tau)) = \Big(\hat{r}(\tau), \hat{x}^{(\hat{r}(\tau))}(\tau)\Big).$$

Remark 1. To find $\hat{r}(\tau)$, one only has to solve the differential equation (8) in addition to carry out the procedure in Step 1. Therefore, the computation effort for $\hat{r}$ increases linearly as the candidate distributions n_0 increases.

Remark 2. Note that, for any fixed r, $\hat{x}^{(r)}(\tau)$ is *not* a Kalman filter. If, in particular, the matrices A, A_1, H, H_1, M and N are independent of i, then the optimal filter reduces to the usual Kalman filter. In this case, if we compare our filter with the standard Kalman filter in the corresponding diffusion model, it is easy to see that the matrix M corresponds to the covariance of the signal noise, N to that of the observation noise, and D to that of the initial error, respectively. In fact, the corresponding diffusion model is given by

$$\begin{cases} dx(t) = Ax(t)dt + A_1 + \sigma du(t), \\ dy(t) = Hx(t) + H_1 + \sigma_1 dv(t), \end{cases}$$

here $u\cdot$ and $v\cdot$ are independent standard Brownian motions. Then

$$M = (\sigma\sigma')^{-1},\ N = (\sigma_1\sigma_1')^{-1},\ \text{ and } D = (E(x(0) - \hat{x}_0)(x(0) - \hat{x}_0)')^{-1}.$$

Therefore, if more information on the distribution of these disturbances is available, one may make use of it and come up with better filtering estimates.

3 Numerical Examples

In this section we present numerical examples and simulation results. First, we consider the one-dimensional model of (2) with $\theta(t) \in \mathcal{M} = \{1, 2\}$ and the following specifications:

$$A(t,1) = -0.3,\ A(t,2) = 0.3,\ H(t,1) = 1,\ H(t,2) = 2,$$

$$A_1(t,1) = A_1(t,2) = 0,\ H_1(t,1) = H_1(t,2) = 0,$$

$$M(t,1) = M(t,2) = 1,\ N(t,1) = N(t,2) = 1, D = 1,\ \hat{x}_0 = 0.$$

Moreover, the initial distribution of $\theta\cdot$ is $(1/2, 1/2)$. We compute the optimal filter by solving the equations (6), (7) and (8) using the well-known Runge-Kutta method. We take the time horizon $T = 10$ and discretize these differential equations with step size $h = 0.01$. The total number of iterations for each sample path is $T_h = T/h = 1000$. All of our results in this paper are based on computations with 100 sample paths.

We compare our filtering schemes with the IMM algorithm in [2], the LMMSE in [3] and the QVT algorithm in [8].

Let $|\hat{x}_{\text{IMM}} - x|$, $|\hat{x}_{\text{LMMSE}} - x|$, $|\hat{x}_{\text{QVT}} - x|$, and $|\hat{x}_{\text{OPT}} - x|$ denote the errors when using the IMM, the LMMSE, the QVT, and the optimal filter, respectively.

We consider the case with $\mathcal{N}_0 = \{1\}$ and $Q^{(1)} = Q = \begin{pmatrix} -5 & 5 \\ 2 & -2 \end{pmatrix}$. Our numerical results are illustrated by a trajectory of $x\cdot$ and the corresponding

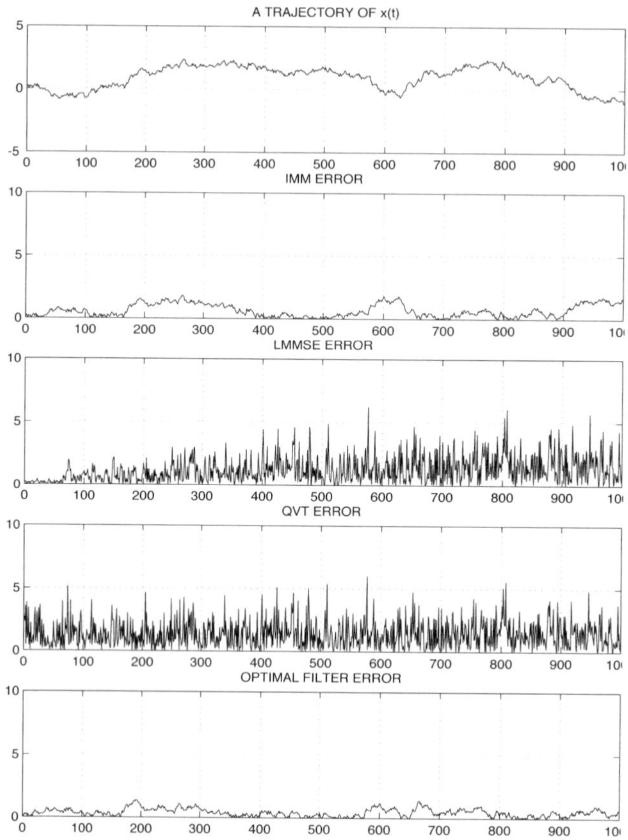

Fig. 1. Errors of the IMM, LMMSE, QVT and Optimal Schemes

errors. Their graphs are given in Fig. 1 and the numerical results are summarized in Table 1. Clearly, our optimal filtering scheme outperforms other three schemes in this case.

Next, we consider applications of the MPT filter to some nonlinear systems. The basic idea here is to construct an appropriate hybrid linear model of form (2), based on the structure of the nonlinear system under consideration, to approximate the nonlinear system. In the construction of such linear

Table 1. Average Errors of Various Schemes

$\|\hat{x}_{\mathrm{IMM}} - x\|$	$\|\hat{x}_{\mathrm{LMMSE}} - x\|$	$\|\hat{x}_{\mathrm{QVT}} - x\|$	$\|\hat{x}_{\mathrm{OPT}} - x\|$
0.665	1.213	1.382	0.435

hybrid model, the main step is the selection of the switching process $\theta\cdot$. In particular, we need to determine the following specifications.

(1) The number of states of $\theta(t)$.
(2) The matrices $A(t,i)$, $A_1(t,i)$, $H(t,i)$, and $H_1(t,i)$.
(3) The probability distribution of θ. As mentioned before, the exact probability distribution of θ is not necessary given. Instead, only a group of candidate distributions that characterize θ need to be specified.

A Piecewise-linear System. Let $x\cdot$ and $y\cdot$ denote the signal and observation processes given by the following piecewise-linear system:

$$\begin{cases} \dot{x}(t) = \begin{cases} A(t,1)x(t) + A_1(t,1) + u(t,1), & \text{if } x(t) \geq 0 \\ A(t,2)x(t) + A_1(t,2) + u(t,2), & \text{if } x(t) < 0, \end{cases} \\ y(t) = \begin{cases} H(t,1)x(t) + H_1(t,1) + v(t,1), & \text{if } x(t) \geq 0 \\ H(t,2)x(t) + H_1(t,2) + v(t,2), & \text{if } x(t) < 0. \end{cases} \end{cases} \tag{9}$$

where

$$u(t,1) = (\dot{x}(t) - A(t,1)x(t) - A_1(t,1))I_{\{x(t)\geq 0\}},$$

$$u(t,2) = (\dot{x}(t) - A(t,2)x(t) - A_1(t,2))I_{\{x(t)<0\}},$$

$$v(t,1) = (y(t) - H(t,1)x(t) - H_1(t,1))I_{\{x(t)\geq 0\}},$$

$$v(t,2) = (y(t) - H(t,2)x(t) - H_1(t,2))I_{\{x(t)<0\}}.$$

A natural way to define the switching process θ is to choose $\mathcal{M} = \{1,2\}$ and define $\theta(t) = \begin{cases} 1 & \text{if } x(t) \geq 0 \\ 2 & \text{if } x(t) < 0. \end{cases}$

The next step is to determine the set of distributions $\mathcal{P}$. If the distribution of the events $\{x(t) \geq 0\}$ and $\{x(t) < 0\}$ is available for each t, then one may use such information to come up with $\phi(t)$. Otherwise, one may consider the "canonical form" $\mathcal{P} = \{\phi^{(1)}\cdot, \phi^{(2)}\cdot, \phi^{(3)}\cdot\}$ where $\phi^{(1)}(t) = (1,0)$, $\phi^{(2)}(t) = (0,1)$ and $\phi^{(3)}(t) = (1/2,1/2)$. Here $\phi^{(1)}$ corresponds to the case when $\theta(t) = 1$, $\phi^{(2)}$ is associated with the case when $\theta(t) = 2$, and $\phi^{(3)}$ corresponds to the situation when $\theta(t)$ switch between 1 and 2 frequently over a period of time.

In connection with piecewise linear models, Fleming and Poudoux [6] considered the filtering of similar system with small observation noise. They first use the QVT to detect the intervals of linearity of the observation function, then with such estimate, choose among a set of extended Kalman filters to estimate the conditional mean $\hat{x}_k$. They showed that if the observation noise is small enough, such filtering is asymptotically optimal. In Fleming et al. [7], they studied the following example:

$$\begin{cases} \dot{x}(t) = ax(t) + u(t), \\ y(t) = \begin{cases} \alpha x(t) + \varepsilon v(t), & \text{if } x(t) \geq 0 \\ \beta x(t) + \varepsilon v(t), & \text{if } x(t) < 0, \end{cases} \end{cases} \tag{10}$$

where a, α, β are constants and ε is a small positive number measuring the magnitude of the observation noise.

Using the hybrid setting, it is easy to get the equivalent hybrid model of (10) as follows:

$$\begin{cases} \dot{x}(t) = A(\theta(t))x(t) + u(t, \theta(t)), \\ y(t) = H(\theta(t))x(t) + v(t, \theta(t)), \end{cases}$$

where $A(1) = A(2) = a$, $u(t, 1) = u(t, 2) = u(t)$, $H(1) = \alpha$, $H(2) = \beta$ and

$$u(t) = \dot{x}(t) - ax(t),$$

$$v(t, 1) = (y(t) - \alpha x(t))I_{\{x(t) \geq 0\}},$$

$$v(t, 2) = (y(t) - \beta x(t))I_{\{x(t) < 0\}}.$$

To compare the performance of the MPT scheme with the QVT, we need to solve the filtering equations (6), (7) and (8). Again, we take the time horizon $T = 10$ and discretize these differential equations with step size $h = 0.01$. The total number of iterations for each sample path is $T_h = T/h = 1000$.

We consider the piecewise linear model (10) with the following specifications:

$$a = -0.25, \ \alpha = 1, \ \beta = -2,$$

$$M(t, 1) = M(t, 2) = 1, \ N(t, 1) = N(t, 2) = 1, \ D = 1, \ \hat{x}_0 = 0.$$

In this example, we choose u, v to be independent Gaussian processes such that $u(t) \sim N(0, 1)$, $v(t, 1) \sim N(0, 1)$, and $v(t, 2) \sim N(0, 1)$.

Our numerical results are illustrated by a trajectory of $x\cdot$ and the corresponding estimation errors of MPT and QVT approaches. Their graphs are given in Fig. 2. In Table 2, the comparisons of the averaged errors of MPT and QVT for different values of ε are provided. We can see that for small ε,

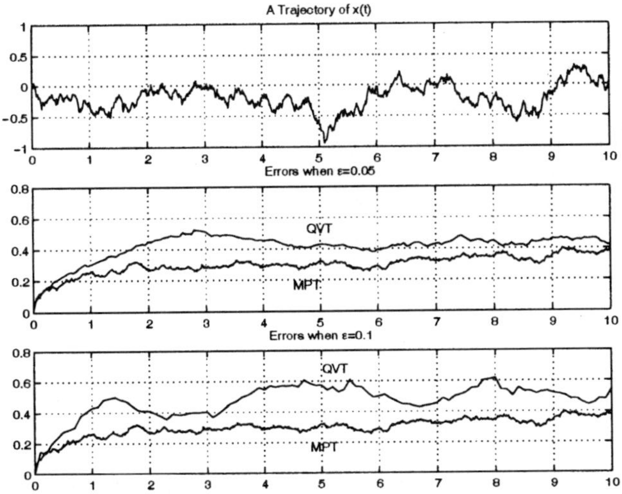

Fig. 2. Comparisons of MPT and QVT

both schemes yield small filtering errors. However, as ε gets larger, the error of QVT grows much faster than that of MPT does. That indicates that for large observation noise, MPT out-performs the QVT approach.

A Nonlinear System. Consider the following one-dimensional nonlinear system:

$$\begin{cases} \dot{x}(t) = x(t)(x^2(t) - 1)(4 - x^2(t)) + \cos(\frac{\pi t}{2}) + u(t), \\ y(t) = x(t) + v(t). \end{cases} \tag{11}$$

Note that here the dominating term $x(t)(x^2(t) - 1)(4 - x^2(t))$ is a highly nonlinear function. The disturbances u and v are defined as

$$u(t) = \dot{x}(t) - x(t)(x^2(t) - 1)(4 - x^2(t)) - \cos(\frac{\pi t}{2}),$$

$$v(t) = y(t) - x(t).$$

Table 2. Averaged Errors of QVT and MPT for Various ε

ε	0.01	0.02	0.05	0.1
QVT	0.479	0.492	0.534	0.597
MPT	0.454	0.454	0.455	0.457

Note that $f(x) = x(x^2-1)(4-x^2)$ has five zero points $x = -2, -1, 0, 1, 2$ and four extrema (See Fig. 3). A simple calculation finds the x-coordinates of these extrema $x_1 \approx -1.644$, $x_2 \approx -0.544$, $x_3 \approx 0.544$, and $x_4 \approx 1.644$. Let $I_1 = (-\infty, x_1]$, $I_2 = (x_1, x_2]$, $I_3 = (x_2, x_3]$, $I_4 = (x_3, x_4]$, and $I_5 = (x_4, \infty)$. Then $I_1 \cup I_2 \cup I_3 \cup I_4 \cup I_5 = \mathbb{R}^1$. That is, the real line is split into five intervals. Each interval I_i contains one zero point of $f(x)$. In view of Fig. 3, it is reasonable to approximate $f(x)$ over I_i by a linear function. Accordingly, we use the Taylor expansion at these zero points to get the piecewise-linear approximation of $f(x)$. We have:

$$f(x) \approx \begin{cases} -24(x+2), & x \in I_1, \\ 6(x+1), & x \in I_2, \\ -4x, & x \in I_3, \\ 6(x-1), & x \in I_4, \\ -24(x-2), & x \in I_5. \end{cases}$$

Define $\theta(t) = i$ if $x(t) \in I_i$, for $i = 1, 2, 3, 4, 5$. Using such $\theta\cdot$ we can write the equations in (11) equivalently as

$$\begin{cases} \dot{x}(t) = A(\theta(t))x(t) + A_1(t, \theta(t)) + u(t, \theta(t)), \\ y(t) = H(\theta(t))x(t) + v(t, \theta(t)), \end{cases} \tag{12}$$

where

$$\begin{aligned} &A(1) = -24, \; A_1(t,1) = -48 + \cos(\frac{\pi t}{2}), \; H(1) = 1, \\ &A(2) = 6, \quad A_1(t,2) = 6 + \cos(\frac{\pi t}{2}), \quad H(2) = 1, \\ &A(3) = -4, \; A_1(t,3) = \cos(\frac{\pi t}{2}), \quad H(3) = 1, \\ &A(4) = 6, \quad A_1(t,4) = -6 + \cos(\frac{\pi t}{2}), \; H(4) = 1, \\ &A(5) = -24, \; A_1(t,5) = 48 + \cos(\frac{\pi t}{2}), \; H(5) = 1, \end{aligned}$$

and

$$\begin{aligned} u(t,1) &= f(x(t)) + 24(x(t)+2) + u(t), \\ u(t,2) &= f(x(t)) - 6(x(t)+1) + u(t), \\ u(t,3) &= f(x(t)) + 4x(t) + u(t), \end{aligned}$$

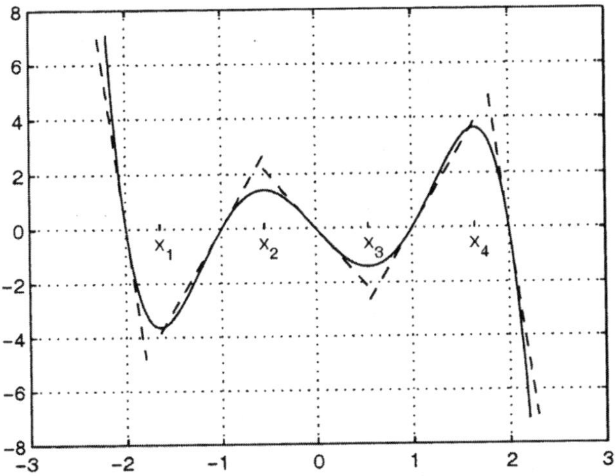

Fig. 3. $f(x)$ and Its Piecewise Linear Approximation

$$\begin{aligned} u(t,4) &= f(x(t)) - 6(x(t)-1) + u(t), \\ u(t,5) &= f(x(t)) + 24(x(t)-2) + u(t), \\ v(t,1) &= v(t,2) = v(t,3) = v(t,4) = v(t,5) = v(t). \end{aligned} \tag{13}$$

In general it is difficult to obtain the distribution of $x(t)$ (and so is the distribution of $\theta(t)$). We use the canonical form

$$\mathcal{P} = \{\phi^{(1)}\cdot, \phi^{(2)}\cdot, \phi^{(3)}\cdot, \phi^{(4)}\cdot, \phi^{(5)}\cdot, \phi^{(6)}\cdot\}$$

given following as the group of candidate distributions:

$$\begin{aligned} &\phi^{(1)}\cdot = \{1,0,0,0,0\},\ \phi^{(2)}\cdot = \{0,1,0,0,0\}, \\ &\phi^{(3)}\cdot = \{0,0,1,0,0\},\ \phi^{(4)}\cdot = \{0,0,0,1,0\}, \\ &\phi^{(5)}\cdot = \{0,0,0,0,1\},\ \phi^{(6)}\cdot = \{\tfrac{1}{5},\tfrac{1}{5},\tfrac{1}{5},\tfrac{1}{5},\tfrac{1}{5}\}. \end{aligned}$$

Here $\phi^{(i)}$ corresponds to $r(t) = i$, for $i = 1, \ldots, 5$. That is, $x(t)$ in most time falls in I_i. $\phi^{(6)}$ corresponds to the case when $\theta(t)$ jumps equally likely among the set $\{1, 2, 3, 4, 5\}$.

Remark 3. Using the zero points of $f(x)$ to expand the function is not the only way to do the linear approximation. One can choose another set of points and expand $f(x)$ accordingly. In addition, the choice of the intervals I_i is not unique either. There are other ways to choose the intervals and the distribution $\mathcal{P}$. Intuitively, a better approximation to a nonlinear function can be achieved by increasing the number of states of $\theta(t)$ because a finer approximation is usually associated with smaller system noise in (**??**). The disadvantage is, however, the increase in the computational budget.

We next consider the nonlinear filtering of (11) by using both the MPT and the EKF approaches. One interesting aspect to present here is how the choice of the initial estimate $\hat{x}_0$ affects the filtering outcome.

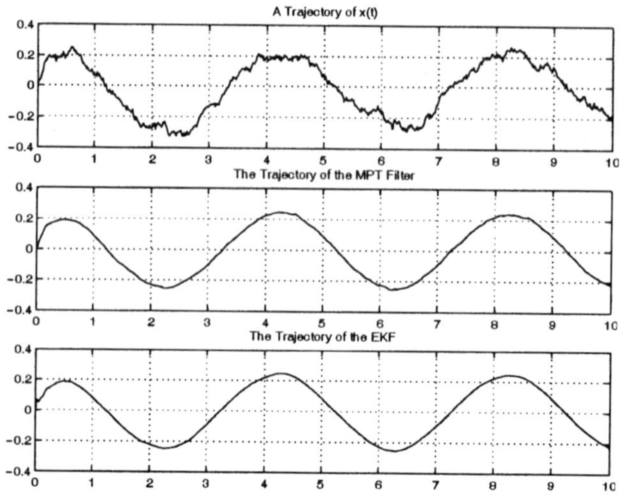

Fig. 4. MPT and EKF Estimation for the Nonlinear Model

Note that the solution of the differential equation in (11) depends heavily on the initial value $x(0)$. For different $x(0)$, the solution would possibly fall in a different "attraction" zone. One task to implement the filtering algorithms is to make an initial guess for $\hat{x}_0$. If *a priori* information about $x(0)$ is available, then we can make use of it. Otherwise, we just set, for example, $\hat{x}_0 = 0$ and let the algorithm to take care of the rest. First we assume that the actual initial value is $x(0) = 0$. In this case, $\hat{x}_0 = x(0)$.

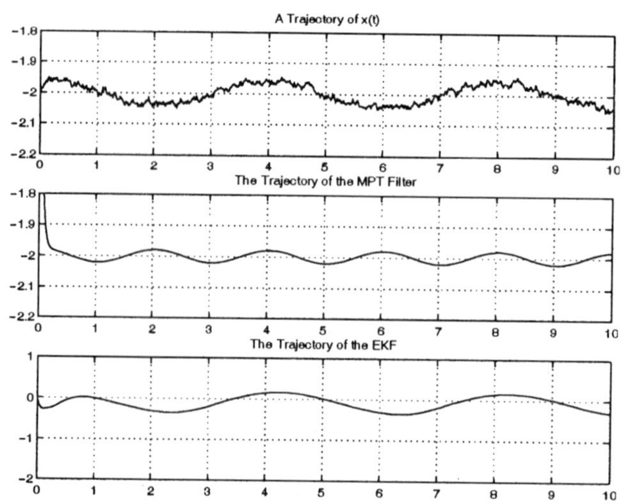

Fig. 5. MPT and EKF Estimation

The other parameters of the MPT filtering are specified as follows:

$$M(t,i) = 1, \quad N(t,i) = 1, \qquad i = 1,2,3,4,5, \ D = 1.$$

Our numerical results are illustrated by a trajectory of $x\cdot$ and the corresponding trajectory of the MPT estimator $\hat{x}\cdot$. For the purpose of comparison, the corresponding trajectory of using EKF is also shown. These graphs are given in Fig. 4 in which both the MPT approach and the EKF perform almost identically.

Next, we consider the case when $x(0) = -2$, which is quite different from our initial guess $\hat{x}_0 = 0$. In this case, the state trajectory $x(t)$ falls in a different zone around -2. Fig. 5 shows a trajectory of $x(t)$ in this case and the corresponding filtering trajectories of MPT and EKF, respectively. In Fig. 5, the EKF got totally lost and led to a completely different trajectory estimation. However, our MPT approach works well and gives a decent estimation of $x(t)$. That demonstrates that the EKF approach is sensitive to the initial choice of $\hat{x}_0$, while on the other hand our MPT scheme is adaptive and robust to the initial choice $\hat{x}_0$, which is desirable in practice.

4 Conclusions

In this paper we considered hybrid filtering models and presented the finite-dimensional recursive optimal filter. A major advantage of these filters are that they do not require the Gaussian assumptions. As a result, these filters can be easily modified to deal with some highly nonlinear systems. The filtering schemes obtained are simple in structure and yet very powerful for dealing with a variety of models with superb numerical performance.

References

1. Bar-Shalom, Y. and Li, X. R. (1996) Estimation and Tracking: Principles, Techniques, and Software, Artech House Publishers, Norwood, MA.
2. Blom, H. A. P. and Bar-Shalom, Y. (1988) The interacting multiple model algorithm for systems with Markovian switching coefficients, IEEE Trans. Automat. Contr. **33**, 780–783.
3. Costa, O. L. V. Linear minimum mean square error estimation for discrete-time Markovian jump linear systems, IEEE Trans. Automat. Contr. **39**, 1685–1689, 1994.
4. Dufour, F. and Elliott, R. J. (1997) Adaptive control of linear systems with Markov perturbations, IEEE Trans. Automat. Contr. **43**, 351–372.
5. Fleming, W. H. and McEneaney, W. M. (1997) Risk sensitive and robust nonlinear filtering, Proc. 36th IEEE CDC, San Diego, CA.
6. Fleming, W. H. and Pardoux, E. (1989) Piecewise monotone filtering with small observation noise, SIAM J. Contr. Optim. **27**, 1156–1181.

7. Fleming, W. H., Ji, D., Salame, P., and Zhang, Q. (1991) Piecewise monotone filtering in discrete time with small observation noise, IEEE Trans. Auto. Contr. **36**, 1181–1186.
8. Haussmann, U. G. and Zhang, Q. (1990) Stochastic adaptive control with small observation noise, Stochastics Stochastics Rep. **32**, 109–144.
9. Hijab, O. (1983) The adaptive LQG problem—Part I, IEEE, Trans. Automat. Contr. **28**, 171–178.
10. Liu, R. H. and Zhang, Q. (2001) Nonlinear filtering: A hybrid approximation scheme, IEEE Trans. Aerospace Electr. Sys. **37**, 470–480.
11. Miller, B. M. and Runggaldier, W. J. (1997) Kalman filtering for linear systems with coefficients driven by a hidden Markov jump process, Sys. Contr. Lett. **31**, 93–102.
12. Mortensen, R. E. (1968) Maximum-likelihood recursive nonlinear filtering, J. Optim. Theory Appl. **2**, 386–394.
13. Yin, G. and Zhang, Q. (1998) Continuous-Time Markov Chains and Applications: A Singular Perturbation Approach, Springer-Verlag, New York.
14. Zhang, Q. (1998) Nonlinear filtering and control of a switching diffusion with small observation noise, SIAM J. Contr. Optim. **36**, pp. 1738–1768.
15. Zhang, Q. (1999) Optimal filtering of discrete-time hybrid systems, J. Optim. Theory Appl. **100**, 123–144.
16. Zhang, Q. (2000) Hybrid filtering for linear systems with non-Gaussian disturbances, IEEE Trans. Automat. Contr. **45**, 50–61.

Lecture Notes in Control and Information Sciences

Edited by M. Thoma and M. Morari
1998–2002 Published Titles:

Vol. 231: Emel'yanov, S.V.; Burovoi, I.A.; Levada, F.Yu.
Control of Indefinite Nonlinear Dynamit Systems
196 pp. 1998 [3-540-76245-0]

Vol. 232: Casals, A.; de Almeida, A.T. (Eds)
Experimental Robotics V: The Fifth International Symposium Barcelona, Catalonia, June 15–18, 1997
190 pp. 1998 [3-540-76218-3]

Vol. 233: Chiacchio, P.; Chiaverini, S. (Eds)
Complex Robotic Systems
189 pp. 1998 [3-540-76265-5]

Vol. 234: Arena, P.; Fortuna, L.; Muscato, G.; Xibilia, M.G.
Neurat Networks in Multidimensional Domains: Fundamentals and New Trends in Modelling and Control
179 pp. 1998 [1-85233-006-6]

Vol. 235: Chen, B.M.
H∞ Control and Its Applications
361 pp. 1998 [1-85233-026-0]

Vol. 236: de Almeida, A.T.; Khatib, O. (Eds)
Autonomous Robotic Systems
283 pp. 1998 [1-85233-036-8]

Vol. 237: Kreigman, D.J.; Hagar, G.D.; Morse, A.S. (Eds)
The Confluence of Vision and Control
304 pp. 1998 [1-85233-025-2]

Vol. 238: Elia, N.; Dahleh, M.A.
Computational Methods for Controller Design
200 pp. 1998 [1-85233-075-9]

Vol. 239: Wang, Q.G.; Lee, T.H.; Tan, K.K.
Finite Spectrum Assignment for Time-Delay Systems
200 pp, 1998 [1-85233-065-1]

Vol. 240: Lin, Z.
Low Gain Feedback
376 pp. 1999 [1-85233-081-3]

Vol. 241: Yamamoto, Y.; Hara S.
Learning, Control and Hybrid Systems
472 pp. 1999 [1-85233-076-7]

Vol. 242: Conte, G.; Moog, C.H.; Perdon A.M.
Nonlinear Control Systems
192 pp. 1999 [1-85233-151-8]

Vol. 243: Tzafestas, S.G.; Schmidt, G. (Eds)
Progress in Systems and Robot Analysis and Control Design
624 pp. 1999 [1-85233-123-2]

Vol. 244: Nijmeijer, H.; Fossen, T.I. (Eds)
New Directions in Nonlinear Observer Design
552 pp. 1999 [1-85233-134-8]

Vol. 245: Garulli, A.; Tesi, A.; Vicino, A. (Eds)
Robustness in Identification and Control
448 pp. 1999 [1-85233-179-8]

Vol. 246: Aeyels, D.; Lamnabhi-Lagarrigue, F.; van der Schaft, A. (Eds)
Stability and Stabilization of Nonlinear Systems
408 pp. 1999 [1-85233-638-2]

Vol. 247: Young, K-D.; Özgüner, Ü. (Eds)
Variable Structure Systems, Sliding Mode and Nonlinear Control
400 pp. 1999 [1-85233-197-6]

Vol. 248: Chen, Y.; Wen C.
Iterative Learning Control
216 pp. 1999 [1-85233-190-9]

Vol. 249: Cooperman, G.; Jessen, E.; Michler, G. (Eds)
Workshop on Wide Area Networks and High Performance Computing
352 pp. 1999 [1-85233-642-0]

Vol. 250: Corke, P.; Trevelyan, J. (Eds)
Experimental Robotics VI
552 pp. 2000 [1-85233-210-7]

Vol. 251: van der Schaft, A.; Schumacher, J.
An Introduction to Hybrid Dynamical Systems
192 pp. 2000 [1-85233-233-6]

Vol. 252: Salapaka, M.V.; Dahleh, M.
Multiple Objective Control Synthesis
192 pp. 2000 [1-85233-256-5]

Vol. 253: Elzer, P.F.; Kluwe, R.H.; Boussoffara, B.
Human Error and System Design and Management
240 pp. 2000 [1-85233-234-4]

Vol. 254: Hammer, B.
Learning with Recurrent Neural Networks
160 pp. 2000 [1-85233-343-X]

Vol. 255: Leonessa, A.; Haddad, W.H.; Chellaboina V.
Hierarchical Nonlinear Switching Control Design with Applications to Propulsion Systems
152 pp. 2000 [1-85233-335-9]

Vol. 256: Zerz, E.
Topics in Multidimensional Linear Systems Theory
176 pp. 2000 [1-85233-336-7]

Vol. 257: Moallem, M.; Patel, R.V.; Khorasani, K.
Flexible-link Robot Manipulators
176 pp. 2001 [1-85233-333-2]

Vol. 258: Isidori, A.; Lamnabhi-Lagarrigue, F.; Respondek, W. (Eds)
Nonlinear Control in the Year 2000 Volume 1
616 pp. 2001 [1-85233-363-4]

Vol. 259: Isidori, A.; Lamnabhi-Lagarrigue, F.; Respondek, W. (Eds)
Nonlinear Control in the Year 2000 Volume 2
640 pp. 2001 [1-85233-364-2]

Vol. 260: Kugi, A.
Non-linear Control Based on Physical Models
192 pp. 2001 [1-85233-329-4]

Vol. 261: Talebi, H.A.; Patel, R.V.; Khorasani, K.
Control of Flexible-link Manipulators Using Neural Networks
168 pp. 2001 [1-85233-409-6]

Vol. 262: Dixon, W.; Dawson, D.M.; Zergeroglu, E.; Behal, A.
Nonlinear Control of Wheeled Mobile Robots
216 pp. 2001 [1-85233-414-2]

Vol. 263: Galkowski, K.
State-space Realization of Linear 2-D Systems with Extensions to the General nD ($n>2$) Case
248 pp. 2001 [1-85233-410-X]

Vol. 264: Baños, A.; Lamnabhi-Lagarrigue, F.; Montoya, F.J
Advances in the Control of Nonlinear Systems
344 pp. 2001 [1-85233-378-2]

Vol. 265: Ichikawa, A.; Katayama, H.
Linear Time Varying Systems and Sampled-data Systems
376 pp. 2001 [1-85233-439-8]

Vol. 266: Stramigioli, S.
Modeling and IPC Control of Interactive Mechanical Systems – A Coordinate-free Approach
296 pp. 2001 [1-85233-395-2]

Vol. 267: Bacciotti, A.; Rosier, L.
Liapunov Functions and Stability in Control Theory
224 pp. 2001 [1-85233-419-3]

Vol. 268: Moheimani, S.O.R. (Ed)
Perspectives in Robust Control
390 pp. 2001 [1-85233-452-5]

Vol. 269: Niculescu, S.-I.
Delay Effects on Stability
400 pp. 2001 [1-85233-291-316]

Vol. 270: Nicosia, S. et al.
RAMSETE
294 pp. 2001 [3-540-42090-8]

Vol. 271: Rus, D.; Singh, S.
Experimental Robotics VII
585 pp. 2001 [3-540-42104-1]

Vol. 272: Yang, T.
Impulsive Control Theory
363 pp. 2001 [3-540-42296-X]

Vol. 273: Colonius, F.; Grüne, L. (Eds)
Dynamics, Bifurcations, and Control
312 pp. 2002 [3-540-42560-9]

Vol. 274: Yu, X.; Xu, J.-X. (Eds)
Variable Structure Systems: Towards the 21st Century
420 pp. 2002 [3-540-42965-4]

Vol. 275: Ishii, H.; Francis, B.A.
Limited Data Rate in Control Systems with Networks
171 pp. 2002 [3-540-43237-X]

Vol. 276: Bubnicki, Z.
Uncertain Logics, Variables and Systems
142 pp. 2002 [3-540-43235-3]

Vol. 277: Sasane, A.
Hankel Norm Approximation for Infinite-Dimensional Systems
150 pp. 2002 [3-540-43327-9]

Vol. 278: Chunling D. and Lihua X. (Eds)
H_∞ Control and Filtering of Two-dimensional Systems
161 pp. 2002 [3-540-43329-5]

Vol. 279: Engell, S.; Frehse, G.; Schnieder, E. (Eds)
Modelling, Analysis, and Design of Hybrid Systems
516 pp. 2002 [3-540-43812-2]

Printing (Computer to Film): Saladruck Berlin
Binding: Stürtz AG, Würzburg